1001 NATURAL WONDERS YOU MUST SEE BEFORE YOU DIE

GENERAL EDITOR MICHAEL BRIGHT

PREFACE BY KOÏCHIRO MATSUURA
DIRECTOR-GENERAL OF UNESCO

A Quint**essence** Book

This edition for Great Britain published in 2010 by Cassell Illustrated,
a division of Octopus Publishing Group Limited.
2-4 Heron Quays,
London E14 4JP

A CIP catalogue record for this book is available from the British Library

ISBN: 978-1-84403-674-5

QSS.KNAT2

The moral right of the contributors of this work has been asserted
in accordance with the Copyright, Designs and Patents Act of 1988.

This book was designed and produced by
Quint**essence**
226 City Road
London EC1V 2TT

2009 Edition

Editor:	Philip Contos
Designer:	Rod Teasdale

Original Edition

Senior Editor:	Catherine Osborne
Project Editor:	Jenny Doubt
Editors:	Ruth Patrick, Marianne Canty
Art Director:	Roland Codd
Designers:	Ian Hunt, James Lawrence

Editorial Director:	Jane Laing
Publisher:	Tristan de Lancey

Colour reproduction by Pica Digital Pte Ltd, Singapore
Printed in China by Toppan Leefung Printing Ltd

CONTENTS

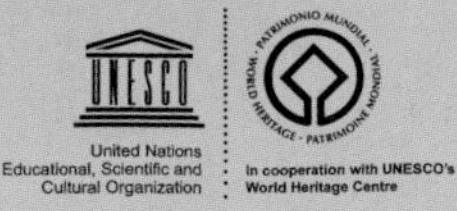

PREFACE

BY KOÏCHIRO MATSUURA, DIRECTOR-GENERAL OF UNESCO

The Earth is a complex, fascinating place. From its vast oceans to its towering mountains, arid deserts, and lush forests, the beauty and wonder of our planet is truly staggering. This book excellently captures spectacular phenomena and little-known marvels in every part of the globe.

At UNESCO, the goal of the 1972 World Heritage Convention is the identification and preservation of the most extraordinary cultural and natural sites in the world. In these pages you will discover many of them, but also many other spectacular natural wonders from around the planet. When a country adopts the Convention, it commits to protecting sites both on its own territory and in other countries. To date, 185 countries around the world have undertaken this commitment. UNESCO provides the Secretariat of the Convention, assisting countries in protecting sites on the World Heritage List and helping to propose new sites for inscription.

The World Heritage List currently includes 878 sites, 174 of them natural sites and 25 mixed, meaning they are both cultural and natural sites. Some of these are cultural landscapes, areas where humans have interacted harmoniously with the environment so that both people and nature flourish, such as Mount Huangshan in China and the mystic beauty of Machu Picchu in Peru.

Among other projects, the World Heritage Center has specific programs dedicated to preserving certain irreplaceable aspects of our natural heritage, such as the World Heritage Forest Program. With the inscription of Mexico's Monarch Butterfly Biosphere Reserve on the World Heritage List in 2008, the number of World Heritage Forest sites increased to 97. These range in size from 18 hectares (Vallée de Mai, Seychelles), to 8.8 million hectares (Lake Baikal, Russia), and now have a total surface area of over 76 million hectares (1.5 times the surface area of France) and represent over 13 percent of government protected forests worldwide.

The World Heritage Center's Marine Program is dedicated to protecting the marine environment, and now supports over 30 sites on the World Heritage List, although many more sites are under preparation for inscription. From Australia's Great Barrier Reef to Argentina's Península Valdés, the world's oceans, seas, and coral reefs are under increasing threat from a variety of sources, including over-fishing, inappropriate fishing practices, coastal development, and pollution. Relatively intact marine ecosystems are becoming scarcer, and with less than 0.5 percent of marine areas worldwide under any form of protection, urgent action is more important than ever.

The Tourism Program at the World Heritage Center works on developing sustainable tourism projects so that visitors can enjoy experiencing World Heritage sites without adversely affecting them. Partnerships with tour operators, where tour guides are trained to inform visitors of the values of sites and how to appreciate them, as well as other initiatives aimed at informing the public and showing everyone how they can contribute positively to the environment, are underway.

There are also programs at UNESCO targeted at studying climate change and its effects, uniting experts from around the world to propose solutions for World Heritage sites that can serve as examples for protecting other sites as well. Regular monitoring seeks to ensure the ongoing conservation of the outstanding values for which sites were inscribed, and the List of World Heritage in Danger calls attention to sites that are imperiled by all kinds of threats, both natural and human-made—deforestation, erosion, earthquakes and other natural disasters, poaching, uncontrolled urbanization, and armed conflict, among others. At present, 30 World Heritage sites, including 13 natural sites, are on this List.

On an individual level, each of us can take action to mitigate the threats to our natural environment. We can be responsible tourists: leaving no trace of our visits, and investing in eco-friendly programs that work toward conserving areas and supporting local communities. Our attitude is crucial—if we each make an effort to have a positive impact on our environment, and teach our children to do the same, it will have huge long-term ramifications.

This marvelous book is a step in that direction. Whether you are an experienced traveler or someone who loves to dream about far-flung corners of the world, it will allow you to learn about the most incredible places on Earth and, hopefully, inspire you to help conserve them.

I wish you an adventurous read.

K. Matsuura

INTRODUCTION

BY MICHAEL BRIGHT, GENERAL EDITOR

Imagine being able to fully explore the length and breadth of the world: *1001 Natural Wonders* is a gateway to this adventure. With a flip of the page, you can plummet from lofty mountain peaks to the dark depths of subterranean worlds, trek across red hot deserts to the wilds of tropical rainforests, swim emerald green waters of sheltered lagoons, explore fish-rich coral reefs, and witness icebergs calving from giant glaciers and the lava flows of mighty volcanoes. With this book, your journey can come alive.

Many of us are content to be armchair travelers—we enjoy the sights and sounds of foreign places on film and television, flick through exotic holiday sections in the weekend papers, or share the adventures of an intrepid explorer in a favorite, well-thumbed book. But it doesn't have to end there—this edition of *1001 Natural Wonders*, with new entries and eye-catching photographs, not only introduces you to 1001 of the world's most spectacular locations, it also provides information on protected areas and some of the endangered species that live there. It contains facts on geological history, unique plants and animals, local customs and folklore, and danger zones.

Within these spectacular landscapes hides a treasure trove of flora and fauna. In Canada, migrating polar bears play-fight alongside the frigid waters of Hudson Bay, and in Mexico entire trees are covered with hibernating butterflies. In the watery wilds of the Amazon, pink river dolphins and toothy piranhas swim through the treetops in the flooded forest. While crossing the Mara River, vast herds of wildebeest struggle to survive in waters overflowing with hungry crocodiles. In Oman, swifts, doves, and birds of prey swoop in and out of the Well of Birds. The mysterious waters of Scotland's Loch Ness are home to the legend of the elusive Loch Ness monster. Moving east to Asia, cranes perform elaborate courtship dances in the wetlands of Japan, and Komodo dragons, the largest lizards in the world, patrol the shores of Komodo Island.

With many species under threat from climate change, poaching, and pollution, national parks and nature reserves have been established worldwide to protect the fragile ecosystems that they inhabit. Many key sites are included on the UNESCO World Heritage List: the Pantanal, a vast wetland area in South America, important for bird migrations the length and breadth of the Americas; the

UNESCO World Heritage sites in this book are indicated by this symbol:

Galapagos Islands, where Charles Darwin collected the specimens that would help him formulate his theory of evolution; the Great Barrier Reef, the largest coral reef system in the world; the Jurassic Coast of southern Britain, where fossils of giant ammonites and ichthyosaurs are to be found; Hawaii's Mount Kilauea, the world's most active volcano; and the Serengeti, where one of the greatest migrations in the world occurs at roughly the same time every year. *1001 Natural Wonders* not only highlights the vulnerability of these natural habitats, but also our own vulnerability in the face of natural forces at work on our ever-changing planet.

We think of the ground beneath our feet as solid and permanent, but as anyone who has experienced an earthquake or volcanic eruption will tell you, our planet is far from stable. Earth is constantly on the move, both in space and deep within its crust. On a grand scale, the continents and the ocean floor are constantly shifting, marking the planet with geological wonders. Mountain chains are thrust high into the sky so that deposits that were once at the bottom of the sea are now tens of thousands of feet above sea level. Volcanoes release their fury in explosive bursts of molten lava and incandescent clouds of searing gases, and there are mudpots and hot springs that bubble and simmer, and geysers that spew out boiling hot fountains of water. Wind, water, and ice sculpt the rocks into all manner of extraordinary shapes and sizes, creating towering sea stacks, jagged pinnacles, rounded monoliths, and striated gorges.

News headlines from around the world remind us in graphic and often horrific detail of this continual movement, creation, and destruction: the Indian Ocean tsunami of December 26, 2004, occurred when an earthquake below the Indian Ocean ruptured the seafloor and created a wave powerful enough to ravage 13 countries and rob hundreds of thousands of people of their lives and still more of their homes and livelihoods; the Pakistan earthquake of October 2005 turned the awesome mountains around the troubled region of Kashmir into a place of tragedy; and the procession of hurricanes in the Caribbean and Gulf of Mexico in September 2008 resulted in death and destruction for tens of thousands of people. These devastating and heart-wrenching natural disasters demonstrate the awesome power of nature—its potential to reshape the world in mere moments and not just over vast stretches of time.

Many of the natural wonders in this book are the result of such historic cataclysms, and all too often we become aware that the processes that created them continue today.

From time immemorial, however, humankind has attempted to bring order to the natural world where no such order was intended. Commonly, people have divided the world, rearranged it, and divided it again. Country boundaries come and go as wars and other civil strife have an impact on territory and possessions. Entire countries are here one day and gone the next. Not surprisingly, no political, cultural, or scientific attempt to classify the breadth of the world has ever proved entirely satisfactory.

These 960 pages delineate the world according to Mother Nature. The natural wonders at which we marvel have been shaped more by tectonic plate movements, volcanic eruptions, and the erosive power of water and ice than by the international boundaries and designated parks that claim and protect them. In order to preserve this perspective, and not let the politics of disputed boundaries overshadow the wonders of the natural world, international politics has informed few of the decisions made while compiling this book. However, where possible, each entry is organized by continent and country and then arranged from the most northerly latitudinal point south. In doing so, we intend that Earth's natural geography be the guiding principle of *1001 Natural Wonders.*

This structure has created its own complications; as slippery as water, locations have sometimes fallen away from the organization we have used to unify the entries—in these circumstances, we have let common sense guide us. Although a part of the United States, for example, there is little logic in isolating Hawaii's entries from those of the surrounding Pacific islands of Oceania. However, where they help clarify location, or elucidate some aspect of the wonder contained in these pages, the broadest categorization of state or administrative region has been indicated. Let these be your compass when navigating your way through the book.

The language of place names raises similar questions. Although many place names can be rendered in the English language, it has sometimes been appropriate to use alternative, transliterated, or local tribal names according to the dictates of indigenous and colonial histories, or indeed, of different alphabets.

Separating the natural world from the artifices of man has revealed the extraordinary relationship between the two. By virtue of their height, shape, location, or exposure, many of these places have become centers of religion, culture, or trade. Equally, tourism and the human histories that are attached to these places have helped to play a part in mythologizing them. The man-made Inca ruins at Machu Picchu, for example, can hardly be separated from the natural history that first carved its plateau perched atop a mountain in the Peruvian highlands.

Human influence is not confined to ancient buildings. What might appear, at first sight, to be a natural landscape may, in reality, be man-made: the heaths of southern England, the Low Countries, and western France are carefully maintained to exclude trees. Scottish moors are burned regularly to ensure fresh heather for grouse, and in Yosemite National Park burning the undergrowth contributes to the park's unique beauty. Even the wide-open moorlands of southwest England are the result of Bronze Age farmers having removed the forests. In some places, nature has reclaimed what was originally hers. Who would have believed that wilderness areas such as the mighty Amazon were once home to great civilizations with garden cities, villages, roads, and farms that have since been enveloped by trees and the tangle of undergrowth?

The wilderness areas of the world undoubtedly hide some of its most spectacular gems, and the journey to and from these often remote places can be as exhilarating and rewarding as the destination itself. A ride upriver in a dugout canoe is the only way to get to the base of the awesome Angel Falls, whereas you need a helicopter to take you to the summit of the flat-topped, rocky monolith over which the falls cascade so spectacularly. A good, sturdy pair of walking boots and an entire camp is required to reach the top of Mount Kinabalu in Borneo, but a comfortable, if not luxurious stay, in a modern resort is now the norm when visiting the island's bat-filled, cathedral-like Mulu Caves.

And if you can't get out there to see the real thing, there's always *1001 Natural Wonders.* Tapping into its list is like planning the imaginary journey of a lifetime. Many parts of our natural world are breathtakingly beautiful, and are here for all to see. Let *1001 Natural Wonders* be the first step in releasing the adventurer in you.

Index of Locations

I

NORTH AMERICA

North America is a continent of contrasts—from the lofty mountain peaks of the Cascade Mountain range to the watery paradise of Florida's Everglades. This chapter takes you on a tour of all the greats—get up close and personal with giant glaciers, polar bears, and wooly bison; hike through the wilds of Yellowstone Park; journey deep into caves rich with stalagmites and stalactites; and experience the beauty of sun-kissed deserts.

LEFT *Water rushing over the spectacular Horseshoe Falls—part of Niagara Falls—into Lake Ontario.*

ELLESMERE ISLAND

NUNAVUT, CANADA

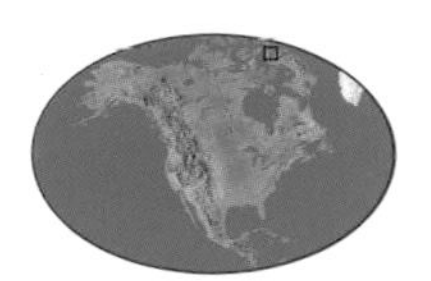

Area of Ellesmere Island: 75,767 sq mi (196,235 sq km)

Highest peak (Mt. Barbeau): 8,583 ft (2,616 m)

Length of Lake Hazen: 44 mi (70 km)

Ellesmere is a vast desert island—the 10th largest island in the world, not a tropical paradise but an icy wilderness at the top of the world, a place of undulating ice fields, rugged gray-black mountains, and boulder-strewn glaciers. For nearly five months a year the sun is not seen here at all, but in midsummer it shines for 24 hours a day, not even dipping below the northern horizon. The island's northern extremity is Cape Columbia, just a stone's throw—497 miles (800 kilometers)—from the North Pole. Its highest point is the peak of Mount Barbeau, 8,583 feet (2,616 meters) above sea level. Deep fjords, such as Archer Fjord, indent its coastline. The cliffs here plunge 2,300 feet (700 meters) to the tumultuous sea below. Winter temperatures can drop to -49°F (-45°C), and the sea freezes solid. The land, however, remains dry for most of the year. Surprisingly, there is little precipitation; no more than 2.5 inches (60 millimeters) a year. Summer temperatures—from late June to late August—rarely rise above 45°F (7°C), although it can be warmer on days without clouds. It is a true wilderness, with just three settlements—Eureka, Alert, and Grise Fjord. MB

MACKENZIE DELTA

NORTHWEST TERRITORIES, CANADA

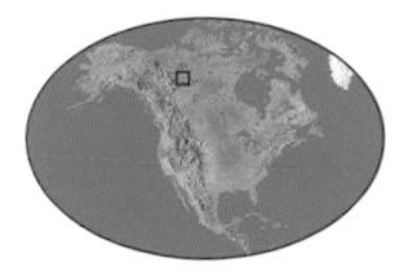

Length of Mackenzie River: 1,100 mi (1,800 km)

Length of Mackenzie Delta: 131 mi (210 km)

Depth of Great Slave Lake: 2,015 ft (614 m)

Mackenzie River flows into the Beaufort Sea across a delta front which is about 50 miles (80 kilometers) wide. During the dark, cold days of winter it is hard to see that a delta exists at all. The river is frozen and blends in with the flat coastal plain. But by springtime the ice has melted, revealing a fan-shaped network of rivers, streams, lakes, and islands. The layout is never the same, the sand and mud changing the course of channels and building or eroding islands. The most recognizable features are the conical mounds, known as pingos. There are over a thousand of these dotted around the delta, the largest concentration in the world. At the center of each is a block of solid ice that pushes up the soil into a hillock. The mounds grow each year, but a number of them collapse in the spring when the heat melts their ice core—the center caves in, resulting in a high-sided pond. The oldest mound on record was 1,300 years old and reached a height of 160 feet (50 meters). The Mackenzie runs from the Great Slave Lake (North America's deepest) to the sea, but its catchment area is the size of Europe. MB

GROS MORNE NATIONAL PARK

NEWFOUNDLAND, CANADA

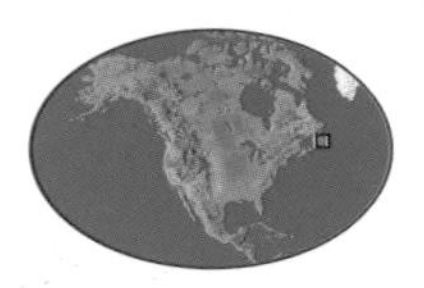

Area of Gros Morne National Park:
700 sq mi (1,813 sq km)

Average temperature (summer):
68°F (20°C)

Average temperature (winter):
17°F (-8°C)

Sometimes referred to as the "Galapagos of Geology," this national park in Newfoundland's western highlands contains some of the oldest rocks in the world, which provide an illuminating insight into the geological evolution of Earth. The bedrock here tells the 1.2 billion-year-old story of shifting and colliding continental plates that once united North America with Europe and Asia. Scientists have discovered that Gros Morne's ancient Long Range Mountains (20 times older than the Rocky Mountains) are part of the same mountain range that runs through Scotland on the other side of the Atlantic Ocean. The ancient rocks of Gros Morne have been worn down by successive waves of advancing and retreating ice over the past two million years, leaving the rounded summits and natural beauty of Gros Morne, Big Hill, and Kildevil Mountains. The result is a remarkable landscape of ancient mountains, fjord valleys, deep glacial lakes, coastal bogs, and, along the coast, wave-carved cliffs.

By traveling from the warmer coastal lowlands up to the alpine barrens of the Long Range Mountains, one can encounter a unique mixture of temperate, boreal, and arctic plant and animal species. The lower elevations are home to black bear and moose; the uplands to animals, such as arctic hare and woodland caribou, that have adapted to the colder weather. All of these animal species found their way to the island in the last 15,000 years, after the ice sheets retreated following the end of the last ice age. Nine of the island's 14 land mammals are subspecies, subtly different from their relatives on the mainland.

One of the most awesome features of Gros Morne is Western Brook Pond, a deep fjord-like canyon containing a freshwater lake. The canyon was shaped by the great ice sheet that once covered all of Newfoundland. Meltwater from the ice sheet flowed down the canyon to the sea. But once the ice retreated, the land, relieved of the weight of the ice, lifted up and raised the fjord shoreline above sea level. The "pond" has since filled with runoff that still cascades into it from the plateau above in the form of spectacular waterfalls. The park has numerous excellent hiking trails through the wild, uninhabited mountains, as well as several campgrounds near the sea. JK

Sometimes referred to as the "Galapagos of Geology," this national park in Newfoundland's western highlands contains some of the oldest rocks in the world, which provide an illuminating insight into the geological evolution of Earth.

RIGHT *Gros Morne National Park has a number of deep fjord-like lakes, with Western Brook Pond being the largest.*

GULF OF ST. LAWRENCE

QUEBEC, CANADA

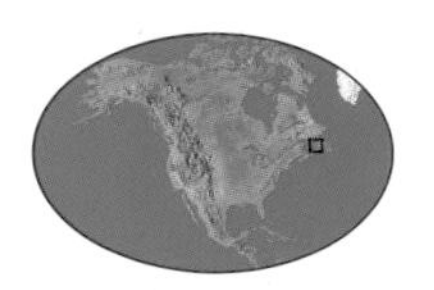

Magdalen Islands: 9 main islands—Alright, Amherst, Brion, Coffin, East, Entry, Grindstone, Grosse, Wolf

Area of Gulf of St. Lawrence: 59,846 sq mi (155,000 sq km)

Each spring, starting in late February or early March, female harp seals move out onto the sea ice, each one giving birth to a single, snowy-white pup. The pupping grounds of the "Gulf Herd" are near the Magdalen Islands (the "Front Herd" are found off Labrador), and here there can be up to 2,000 female seals per square kilometer. Their pups are known as "whitecoats," and they are fed a rich milk containing 45 percent fat (compared to 4 percent in cow's milk). They put on weight rapidly and are weaned in as little as 12 days, and then abandoned by their mothers. Why the nursing time is so short is not clear, although it is an effective way to get the youngsters ready to swim before the ice breaks up in mid-March. This way, they spend the minimum time on the ice, where they are vulnerable to polar bears on the lookout for food. The seals are also the target of a controversial cull. A Seal Interpretive Center on the Magdalen Islands explains the environmental and social aspects of seals in the area, and helicopter tours take visitors onto the ice for close encounters with baby seals. MB

WESTERN BROOK POND

NEWFOUNDLAND, CANADA

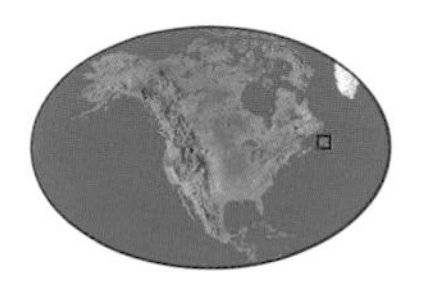

Length of Western Brook Pond: 10 mi (16 km)

Depth of pond: 545 ft (166 m)

Age of pond: 11,000 years

Cut deep into Newfoundland's Long Range Mountains is a canyon over 2,000 feet (600 meters) deep. An ice age glacier gouged out the existing river valley, deepening and widening it. When the ice melted about 11,000 years ago, the bottom of the canyon filled with water and the Western Brook Pond was created. Many of Newfoundland's surface waters were formed in the same way and are labeled "ponds" or "brooks," such as Parson's Pond or Main Brook, but "pond" is certainly an understatement, for Western Brook Pond is really a 10-mile- (16-kilometer-) long lake that winds through steep-sided mountains.

Today, it is 545 feet (166 meters) deep, and in spring and summer waterfalls cascade down the canyon walls raising the level of its waters. In winter the place is frozen solid, with temperatures dropping to 14°F (-10°C). Tourist boats ply the lake in summer, where you can observe the glacier-carved fjord in all its glory. Nearby, a marshy area is home to Newfoundland's floral emblem, an insect-eating pitcher plant. There is also plenty of wildlife—salmon, Brook trout, and arctic char swim the lake's waters and an unusual colony of gulls nest on the cliffs. **MB**

HELL'S GATE

BRITISH COLUMBIA, CANADA

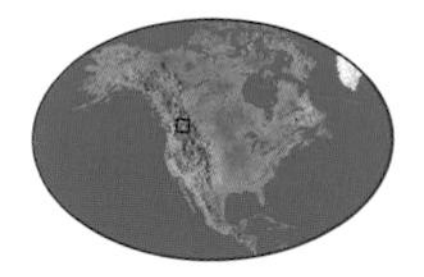

Width of Hell's Gate: 110 ft (35 m)

Depth of Hell's Gate: 500 ft (150 m)

Flow rate of Fraser River: 3.9 million gal/sec (15,000 cubic m/sec)

After his death-defying boat ride down some of the most fearsome rapids in North America, the explorer Simon Fraser wrote in 1808: "We had to travel where no human being should venture—for surely we have encountered the gates of hell." Hell's Gate is a narrow passageway in British Columbia's Cascade Mountain Range, cut by the mighty Fraser River on its way to the Pacific Ocean. Here the river is a narrow 110 feet (35 meters) wide and is constrained by enormous granite walls 500 feet (150 meters) high. The result is a spectacle of violence, of white-water rushing at breakneck speed and tearing through the canyon with a deafening roar. The flow of water is twice the volume of Niagara Falls. This is nature at its most magnificent. The Trans-Canada highway runs right beside Hell's Gate, so there is easy access to this natural wonder. An aerial tramway has also been constructed to take thrill seekers on a gondola ride over the top of the rapids, dropping 500 feet (150 meters) from the road to the water's edge. Even braver visitors should try the suspension bridge that hovers over the raging torrent. They soon discover why this is one of western Canada's most popular attractions. **JK**

BURGESS SHALES

BRITISH COLUMBIA, CANADA

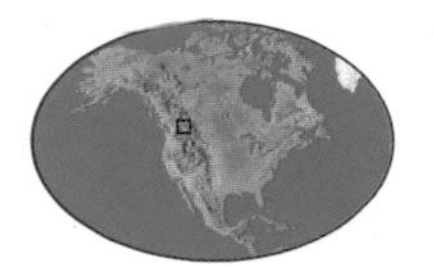

Age of Burgess Shales: 540 million years

Rock type: shale

Habitat: fossil tropical reef with exquisite invertebrate fossils entombed in mudslides

On the perimeter of a great carbonate reef at the tip of the North American continent, in a warm shallow sea, there once lived a collection of the most extraordinary creatures Earth has ever known. After nearly two billion years of simple life forms, a huge spectrum of complex body shapes had evolved in just 10 to 20 million years. Turbulent mudslides buried these animals, cutting them off from decomposing bacteria and preserving them in perfect conditions.

Once buried under 6 miles (10 kilometers) of rock, the re-exposure of the Burgess Shale fossils began 175 million years ago. In 1909, a paleontologist excavated the dark fossil-bearing strata in a 320-foot (100-meter) high limestone cliff. Consisting of over 120 different animal types, the fossils revealed the past diversity of life and gave us the present notion of an early "experimental" phase of evolution.

The site forms part of Yoho National Park, itself part of the Canadian Rocky Mountain Parks World Heritage site. Hikes can be organized at the Walcott and Raymond quarries, but private collection of fossils or shale is strictly prohibited in case they contain an exciting new find. **AB**

CATHEDRAL GROVE

BRITISH COLUMBIA, CANADA

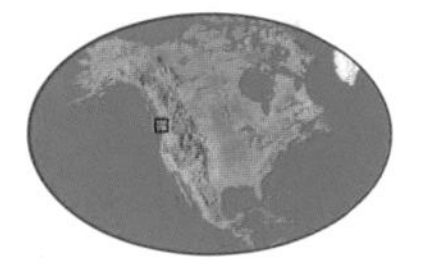

Area of Cathedral Grove: 388 acres (157 ha)

Annual rainfall: 120 in (300 cm)

Tallest trees: 250 ft (76 m)

Cathedral Grove is aptly named, for a walk through this old-growth, temperate rainforest on Vancouver Island is bound to stir the soul. The forest is composed mostly of mature Douglas fir trees mixed with ancient western red cedar, western hemlock, and balsam fir. The trees are 300 to 400 years old, but some go back as far as 800 years. These older trees stand like giant sentinels in the forest, reaching 250 feet (76 meters) high and with trunks measuring over 30 feet (9 meters) in circumference. The rainforest contains trees of various sizes, species, and ages, with a large number of dead standing and fallen trees. It's hard not to be overwhelmed by the sheer beauty of the place—the tops of the trees form a cathedral-like ceiling high above the ground and beams of sunlight break through the thick canopy, lighting up the mists that swirl above the soft green ferns that cover the forest floor. Cathedral Grove is easily accessible by road and a popular destination for those wanting to experience the grandeur of an old-growth forest. You can also enjoy a swim in the fresh waters of Cameron Lake, or picnic along its shores. The lake also has an abundance of fish for those keen on fishing. **JK**

BANFF NATIONAL PARK

ALBERTA, CANADA

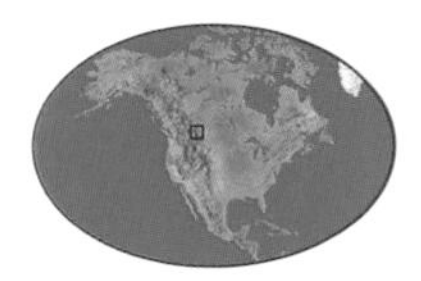

Area of Banff National Park:
2,580 sq mi (6,680 sq km)

Area of Columbia Icefield:
125 sq mi (325 sq km)

Banff is Canada's oldest national park, first known as Hotsprings Reserve, and it stretches along the eastern edge of the Rocky Mountains in Alberta. It is a place of lakes, mountains, and glaciers. The mountains are young—from 45 to 120 million years—and include magnificent peaks such as Mount Amery in the north of the park. Farther north is the Columbia Icefield, the largest ice field on the North American mainland. Its glaciers feed rivers that flow to three oceans—the Arctic, Atlantic, and Pacific. Some glaciers push down to lakes, such as Lake Louise. The runoff fills the lake with suspended sediments that bend the sunlight and make the waters appear a bright blue-green. Meltwater also seeps through the rocks and is heated, pressurized, and pushed up to the surface to form the hot springs that first attracted visitors here over a hundred years ago. The slopes are clothed in thick forests of conifers, while higher elevations have gnarled survivors that finally give way to barren, rocky mountain tops. Wildlife is varied—from hummingbirds to grizzly bears, eagles to moose, and it is all best seen from the over 930 miles (1,500 kilometers) of trails that crisscross the park. **MB**

THE DRUMHELLER BADLANDS

ALBERTA, CANADA

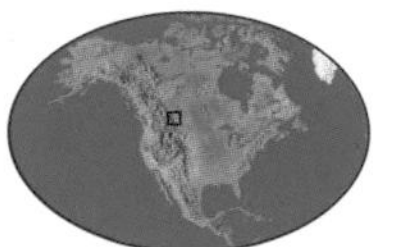

Dinosaur species discovered: 150

Age of Red Deer River Valley: 13,000 years

Notable features: badlands, hoodoos, canyons, coulees

The torn and twisted landscape of the Drumheller Badlands stretches like a giant scar through the rolling farmlands of southern Alberta. It is a conglomeration of gulches, buttes, gulleys, and canyons, all eroded from multicolored layers of sandstone, mudstone, coal, and shale that date back 70 million years. The impact of visiting this place can be overwhelming, as if one has been transported back in time to another era, or even another planet.

This ancient world is referred to as "Badlands" because it has no agricultural use, but it is a treasure trove for dinosaur hunters. Throughout these hills, fossil hunters have discovered some of the greatest dinosaur fossils ever recorded, including complete skeletons of the king of the dinosaurs, *Tyrannosaurus rex*. Not surprisingly, they call this the "dinosaur capital of the world." Dozens of fossilized dinosaur skeletons are displayed in the world-class Royal Tyrrell Museum of Paleontology in the heart of the Badlands.

Though its landscape may be harsh and barren today, Drumheller was a very different place 70 million years ago, when a warm, almost tropical climate supported a wide variety of wildlife, including many dinosaurs. Following the mass extinction at the end of the Cretaceous Period, which saw the dinosaurs wiped out, temperatures turned much colder. Successive ice ages during the past two million years exposed the land to relentless erosion by wind, water, and ice. Conveniently for scientists, this action exposed the area's ancient sediments and the fossil bounty that lay within.

In more recent times, the Badlands became part of the folklore of Alberta, providing shelter from the elements for the Cree and Blackfoot peoples and a convenient hideaway for outlaws trying to avoid the authorities. Today they are a major attraction to travelers hungry for a different experience to the relative monotony of the surrounding prairie wheat fields.

To experience the many moods of the Badlands it is best to visit more than once, and at different times of the day. At sunrise they glow pink, while at noon they are bleached white by the high sun, only to turn golden in late afternoon, and finally to fiery orange and deep purple at sunset. JK

The torn and twisted landscape of the Drumheller Badlands stretches like a giant scar through the rolling farmlands of southern Alberta. The impact of visiting this place can be overwhelming, as if one has been transported back in time.

RIGHT *Named the dinosaur capital of the world, the Drumheller Badlands has some of the greatest dinosaur fossils ever recorded.*

MORAINE LAKE

ALBERTA, CANADA

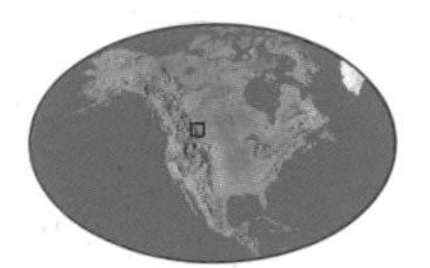

Elevation of Moraine Lake: 6,300 ft (1,920 m)

Type of lake: glacial

Age of surrounding mountains: 120 million years

Walter Wilcox wrote that "no scene had ever given me an equal impression of inspiring solitude and rugged grandeur." He discovered and named Moraine Lake in 1899; so impressed with the scene before him, he claimed it was the most beautiful lake he had ever seen and that the time spent contemplating the view was the happiest half hour of his life. It is easy to see why, for this amazing crystalline lake is a wonder to behold. Above it tower the ice-capped peaks of Mount Wenkchemna, whose 3,000-foot (900-meter) sheer walls enclose the eastern side of the lake. For a time this spectacular view was even featured on the back of the Canadian $20 bill.

The lake was created not by the moraine (or debris) of a glacier, as its name implies, but by a large rockslide from neighboring Mount Babel. Its wondrous, iridescent blue color comes from fine particles of glacial till, also known as "rock flour," that flow into the lake during the summer melt from glaciers high in the mountain. The particles absorb all colors

of the visible spectrum except blue, which is reflected back. Not surprisingly, this lake is called the "Jewel of the Rockies."

The area is part of Banff National Park, which was established in 1885 and was the first of Canada's national parks. This is an excellent place to see a wide variety of wildlife including black and grizzly bear, bighorned sheep, mountain goat, elk, and moose. Moraine Lake is also the starting point for a number of excellent hiking trails into the surrounding mountains. One of them ascends over 2,300 feet (700 meters) above the surface of the lake—one of the highest points reached by a major trail in the Canadian Rockies. Moraine Lake is only 9 miles (15 kilometers) from its more famous neighbor, Lake Louise, but much less visited, making this destination a decidedly better alternative during the busy tourist season. It has a much-admired lodge on its shores, constructed in a post-and-beam design and graced with grand windows that provide generous views of the lake and the surrounding mountains. Guests are beguiled by a natural setting that offers canoeing, hiking, nature watching, and climbing. **JK**

BELOW *The beautiful blue waters of Moraine Lake.*

NAHANNI RIVER

NORTHWEST TERRITORIES, CANADA

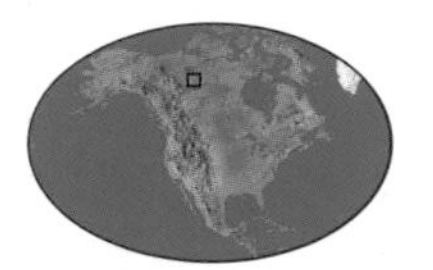

Nahanni River flow rate: 17.25 mph (28 kph)

Height of Virginia Falls: 316 ft (96 m)

The accessible part of the Nahanni River is a 130-mile (210-kilometer) stretch between Nahanni Butte and Virginia Falls. The opening of First Canyon unveils the Kraus Hotsprings, and an unexpected world of lush meadows and spring flowers. The canyon itself has steep limestone walls that rise 4,000 feet (1,200 meters) on either side. Caves scar the cliffs, including Valerie Grotte, in which the skeletons of 100 Dall's sheep were found. Next comes Deadmen Valley, named for the headless skeletons of gold prospectors that were discovered here in 1906. Second Canyon slices through Headless Range, where black bears and Dall's sheep are common. The narrow Third Canyon cuts through the Funeral Range and into Hell's Gate, a white-knuckle boat ride through rapids and whirlpools. At the end of the journey are the thunderous twin cataracts of Virginia Falls, with a drop of 316 feet (96 meters). Farther upstream, and only reached by light aircraft, is Rabbitkettle Hotsprings. These deposit tufa in terraces, each ledge with a mirror-like pool and fringed with mosses and low-growing flowers. Nahanni is a World Heritage site and to keep it remote can only be accessed by boat or plane. MB

CHURCHILL

MANITOBA, CANADA

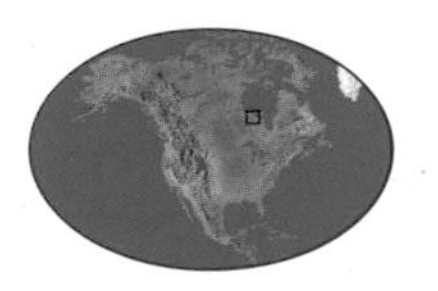

Western Hudson Bay polar bear population: 1,200

Churchill human population: 800 to 1,200

Nearest city: Winnipeg—604 mi (966 km) by air

Churchill in Manitoba is the "polar bear capital" of the world. It stands beside Hudson Bay and is the migration route of polar bears returning from their summer sojourn to their winter quarters on the sea ice. If the ice is late in forming, the citizens of Churchill have a bunch of delinquent bears on their hands, and the biggest draw is the local garbage dump. Churchill, however, has turned its problem bears into an asset—over 15,000 visitors a year head here to watch wild polar bears. Tundra Buggies, set high above the ground, carry tourists to the "hot spots" where they come face-to-face with the world's biggest carnivore. The best time to visit is in late October to mid-November, and you should book a year in advance. Polar bears are strictly solitary but here they tolerate each other until they can go their separate ways. Some young males indulge in "play fights," practicing for bloodier contests out in the bay. They rise up on their hind legs, towering an impressive 11 feet (3.3 meters) above the ground. When the ice forms in late November, the polar bears and tourists disappear. MB

RIGHT *A lone polar bear crosses the ice in Churchill.*

THE BAY OF FUNDY

NEW BRUNSWICK / NOVA SCOTIA, CANADA

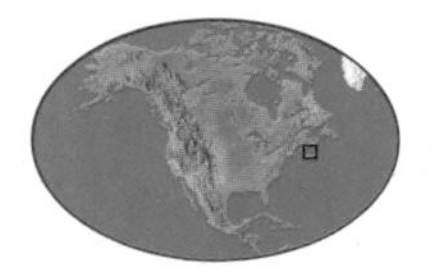

Length of Bay of Fundy: 168 mi (270 km)

Average depth of bay: 246 ft (75 m)

Size of average tide: 100 billion tons of water

Twice a day the Bay of Fundy puts on a magnificent show unrivaled anywhere else in the world. The largest tides carry a quantity of water equivalent to all the rivers of the world into the bay. At the head of the bay the tides rise 53 feet (16 meters). It is an extraordinary spectacle, complete with rip currents, seething upwellings, and violent whirlpools. At low tide the water can retreat as far back as three miles (five kilometers). A few hours later, over 50 feet (15 meters) of water inundates the area.

The reason for the high tides lies in the unique funnel shape and depth of the bay. The water moves in sync with the ocean tides outside, and the natural oscillation of water sloshing back and forth inside this basin corresponds exactly with the rhythm of the Atlantic tides. The result is "resonance," where the height of the incoming tide is reinforced by the oscillation of existing water in the bay. The phenomenal tides have left their mark on the bay. At Hopewell Rocks, the tides have sculpted towering statues of red sandstone, and at St. Martin's, tidal action has carved enormous sea caves. The bay is also rich in nutrients, providing a wealth of food to eight species of whale, as well as thousands of shorebirds. JK

NIAGARA FALLS

ONTARIO, CANADA / NEW YORK, UNITED STATES

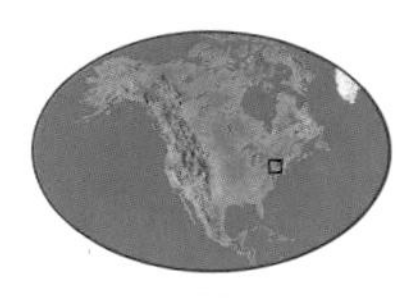

Height of Niagara Falls: 180 ft (55 m)

Age of Niagara Falls: 10,000 years

Probably the most famous waterfall in the world, Niagara is 180 feet (55 meters) high and 2,200 feet (671 meters) wide, and consists of two cascades separated by Goat Island—the American Falls on the eastern side and the Canadian Horseshoe Falls to the west. The water comes from Lake Erie, one of the Great Lakes, and flows casually for 22 miles (35 kilometers) before pitching into a series of rapids that end in the famous Niagara Falls that feed Lake Ontario. Just after the last ice age, about 10,000 years ago, the falls were 7 miles (11 kilometers) downstream, but the constant erosion caused by 7,000 tons of water per second passing over the bedrock has caused them to retreat at a rate of about 4 feet (1.2 meters) per year. Many performers have tried to "ride" the falls; the first was Sam Patch, who jumped from Goat Island in 1829 and amazingly survived the ordeal. Annie Edson Taylor was the first to go over in a barrel in 1901. She tried to make her fortune by lecturing as the "Queen of the Mist" but never really succeeded. The *Maid of the Mists*, however, is the tour boat that takes visitors from both the Canadian and U.S. sides into the maelstrom of Niagara Falls. MB

THE GREAT LAKES

CANADA / UNITED STATES

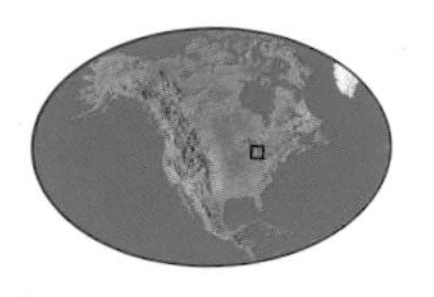

Area of Great Lakes: 94,000 sq mi (243,460 sq km)

Length of coastline: 10,000 mi (16,093 km)

Deepest lake (Lake Superior): 1,332 ft (406 m)

The five Great Lakes—Superior, Michigan, Huron, Erie, and Ontario—encompass the largest fresh water surface on Earth and hold a staggering 6 quadrillion gallons (23 quadrillion liters) of water, about one-fifth of the world supply. If spread across the lower 48 states of America, the water would still be 9.5 feet (3 meters) deep. Lake Superior is the largest of the lakes, so big in fact that it could contain all the other Great Lakes, plus three additional lakes the size of Lake Erie.

The lakes are a testament to the power of the great ice sheets that shaped much of North America during the ice ages. The lake basins are composed mainly of soft sandstone and shale, which were easily gouged out by the mile high glaciers that once covered this region. When the ice retreated, these huge lakes were left behind.

Beginning with Lake Superior, the largest of the Great Lakes, water flows through channels between the lakes and onward down the St. Lawrence River to the Atlantic Ocean,

which is over 1,000 miles (1,600 kilometers) away. Surrounding the lakes are a wonderful variety of habitats, including coastal marshes, rocky shorelines, lake-plain prairies, savannas, forests, fens, and innumerable wetlands. The world's largest freshwater dunes line the shores of Lake Michigan, and Lake Huron has over 30,000 small islands. The Great Lakes are also home to world-class fishing with about 180 native species of fish, including bass, northern pike, lake herring, whitefish, walleye, and lake trout. Lake Erie is the warmest and most biologically productive. Its walleye fishing is considered the best in the world. The green, leafy forests around the Great Lakes are also home to a rich mix of wildlife, including white-tailed deer, beaver, muskrat, weasel, fox, black bear, bobcat, wolves, and moose. With more than 30 million people living on its shores, in eight U.S. states (Minnesota, Wisconsin, Illinois, Indiana, Michigan, Ohio, New York, and Pennsylvania) and the Canadian province of Ontario, the lakes are a much-appreciated year-round recreational paradise. **JK**

BELOW *The sheer size of the Great Lakes is clear in this photo of Lake Superior's shoreline.*

BROOKS RANGE

ALASKA, UNITED STATES

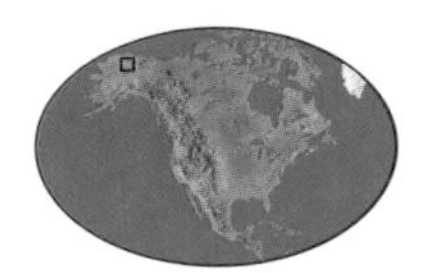

Length of Brooks Range: 600 mi (1,000 km)

Highest peak: 8,500 ft (2,600 m)

Area of the Arctic National Park: 8,472,506 acres (3,428,823 ha)

At the northern tip of the Rocky Mountains is the Brooks Range of Alaska. It is a true wilderness area, the preserve of grizzly and black bears, Dall's sheep, wolves, moose, and caribou. The southern slopes are forested with skeleton-like boreal forest, while the north-facing side forms the Alaskan North Slope of frozen tundra. All the plants here grow close to the ground, where the effects of the icy-cold, drying winds are reduced and heat is trapped in the clumps they form. In winter, temperatures drop to -49°F (-45°C). It is here that an estimated 160,000 caribou are on the move each year in a natural wildlife spectacle reminiscent of the great migration in Africa's Serengeti. Known as the Porcupine herd, because they winter in valleys that feed the Porcupine River, they migrate north each year toward the coastal plains where they drop their calves and graze on lichens known as "reindeer moss." Protection in the western part of the Brooks Range is afforded by the Gates of the Arctic National Park and Preserve, which is named after the natural pass between Boreal Mountain and Frigid Crags. Settlements are few and far between, accessible only by air, snowmobile, or dogsled. MB

McNEIL RIVER FALLS

ALASKA, UNITED STATES

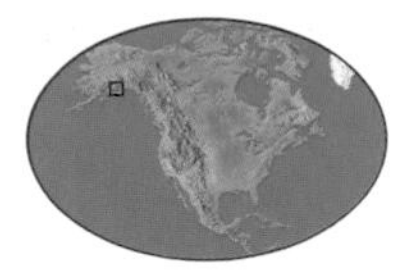

Protected area: McNeil River State Game Sanctuary

Area of sanctuary: 114,400 acres (46,298 ha)

In late summer (July and August), about 150 grizzly bears head for the McNeil River, 250 miles (402 kilometers) southwest of Anchorage, Alaska. They are going fishing, fattening up for winter on the run of chum salmon that heads upstream to spawn at this time of year. The salmon are slowed down by the McNeil Falls, and it is here that 30 to 40 bears a day can be seen scooping salmon out of the river and eating them. One year 70 bears were spotted together at one time, with as many as 144 known individuals entering the area during the season. They gather in and around the cataract, some perching on the water's edge, others sitting or standing in the river itself, some diving under the water for salmon. Youngsters run and belly-flop but rarely catch anything, whereas adult bears catch plenty of fish—a female can average 75 pounds (34 kilograms) of salmon a day. One bear was seen to catch 90 fish in a single day. Access is by air from Homer 100 miles (161 kilometers) away, and only 250 people are allowed into the area each season. Selection is by a lottery. MB

MOUNT KATMAI

ALASKA, UNITED STATES

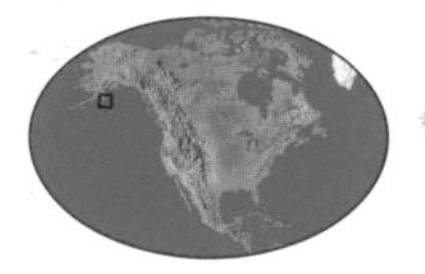

Height of Mt. Katmai: 6,715 ft (2,047 m)

Height of Novarupta: 2,758 ft (841 m)

Length of Valley of Ten Thousand Smokes: 12 mi (20 km)

On June 7, 1912, a huge volcanic explosion lit up the sky with a yellow day-like light over the mainland opposite Kodiak Island. An estimated 33 million tons of debris was propelled into the air. Dust and ash rose into the stratosphere and were carried around the world. It was not until 1915 and 1916 that expeditions arrived to see what had happened. They found nothing alive, just mud and ash. The summit of Mount Katmai had vanished. Today, all that remains is a huge caldera 8 miles (13 kilometers) across and 3,700 feet (1,128 meters) deep, containing a lake of blue-green water. Nearby at the time, an expedition discovered a valley with fissures in its floor from which sulfurous steam poured. It was named Valley of Ten Thousand Smokes, and it housed a small volcano that had drained molten rock from Katmai and caused its summit to collapse. All the violent activity, in fact, had been the work of this new volcano, named Novarupta. It smothered the valley in a 700-foot (215-meter) thick layer of ash, and steam from the buried river gave rise to the "Ten Thousand Smokes." Today, the valley has few smokes and remains a desert. It became a national monument in 1918. MB

BEAR GLACIER

ALASKA, UNITED STATES

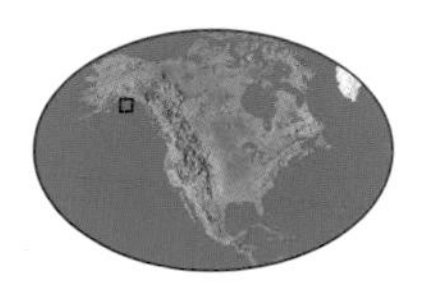

Width of Bear Glacier: 2 mi (3.2 km)

Depth of glacier: 4,000 ft (1,220 m)

Annual snowfall: 80 in (203 cm)

Bear Glacier is one of the 30 magnificent glaciers that flow from the Harding Icefield in Alaska's Kenai Fjords National Park. It is the only Harding glacier that doesn't reach the sea. Instead, this enormous glacier ends in a freshwater lagoon that has formed behind its terminal moraine. The lagoon is full of fantastically-shaped icebergs that the glacier calves daily from its leading edge. These are some of Alaska's biggest icebergs.

The best way to see and experience the glacier and its icebergs is by boat, only a 10-mile (16-kilometer) ride away from the town of Seward. Some tour companies even provide kayaks so that visitors can get close to the icebergs, and hear the dripping of melting ice and the cracking of air bubbles. Water from the lagoon flows into Resurrection Bay, creating a milky-white path in the seawater. There is a distinct line where the cold white glacier-meltwater meets the blue ocean water. The bay is a wildlife paradise, home to killer whales, humpbacks, sea lions, and sea otters. Thousands of seabirds such as puffins, murres, and eagles nest in the steep cliffs above. JK

MENDENHALL GLACIER

ALASKA, UNITED STATES

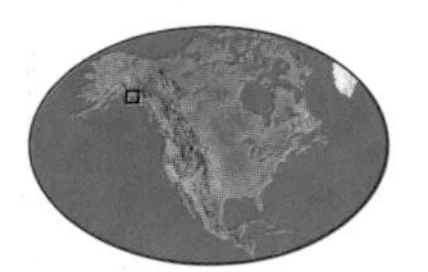

Highest elevation of Mendenhall Glacier: 5,500 ft (1,676 m)

Lowest elevation: 100 ft (30 m)

Average annual snowfall: 100 ft (30 m)

The Mendenhall Glacier is just one of 38 large glaciers that form the Juneau Icefield, in southeast Alaska. This enormous expanse of ice covers over 1,500 square miles (3,885 square kilometers). The naturalist John Muir described Mendenhall as one of the most beautiful of all Alaska's glaciers. It is also the most accessible, located only 13 miles (21 kilometers) away from the city of Juneau by road. The maritime climate of this region ensures that an abundant snowfall exceeding 100 feet (30 meters) falls on the glacier each year. Over time the compacted snow turns to ice, and so renews the glacier.

Like all glaciers, Mendenhall is always on the move, a frozen river sliding 5,400 feet (1,646 meters) down the Coast Mountains at a rate of about 2 feet (0.6 meters) per day, scouring the bedrock. Ice in this impressive glacier takes 250 years to travel from the glacier's summit to its terminus, which is 13 miles (21 kilometers) away at the 200-foot- (60-meter-) deep Mendenhall Lake. Here, great chunks of blue ice the size of buildings break off suddenly from the glacier's leading edge and fall into the water, where they form large, floating icebergs. The glacier is over 200 feet (60 meters) thick at its terminus, with more than 100 feet (30 meters) poking above the water and another 100 feet (30 meters) below the water's surface.

The Mendenhall Glacier, along with the rest of the Juneau Icefield, began to form over 3,000 years ago during a "mini–ice age" and grew continually until the late 18th century. Since then, warmer temperatures have caused it to slowly retreat, because the rate of renewal of glacier ice is slower than the rate of melting at its lower levels. Since 1767, it has receded 2.5 miles (4 kilometers). At its present rate of retreat (approximately 25 feet [7.5 meters] per year), the glacier will take several centuries to completely disappear. The land revealed beneath the departing glacier, barren for hundreds of years, soon begins to support vegetation and, in due course, wildlife.

The glacier is named after Thomas Mendenhall, a noted scientist who was responsible for surveying the international boundary between Canada and Alaska. More adventurous visitors can charter a helicopter for the most dramatic views of the landscape, or even take a dogsled ride on the glacier. JK

The naturalist John Muir described Mendenhall as one of the most beautiful of all Alaska's glaciers. Here, great chunks of blue ice the size of buildings break off suddenly from the glacier's leading edge and fall into the water.

RIGHT *The Mendenhall Glacier near Juneau, Alaska.*

PORTAGE GLACIER

ALASKA, UNITED STATES

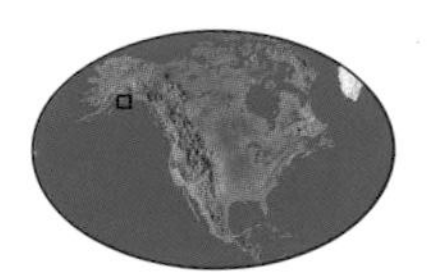

Area of Portage Glacier: 11.5 sq mi (30 sq km)

Type of glacier: valley glacier

Depth of Portage Lake: 800 ft (244 m)

They say that Alaska must be discovered slowly, one adventure at a time, and there is no doubt that Portage Glacier, located 50 miles (80 kilometers) south of Anchorage, is one example of the many beauties to be explored.

This beautiful glacier flows out of the Chugach Mountains at the western end of Prince William Sound. It is just one of 100,000 glaciers in Alaska. It is easy to reach by road, making it one of Alaska's most popular tourist attractions. The trip there along the Turnagain Arm of Cook Inlet is a wonder in itself. The inlet has one of the fastest changing tides in the world and plenty of wildlife in the surrounding countryside, including Dall's sheep, moose, bald eagles, and black bears. Beluga whales are a frequent sight in the inlet's cold waters.

Portage Glacier ends in a wondrous iceberg-filled lake, Portage Lake, which was created by the glacier as it retreated. Ice regularly breaks off the face of the glacier, renewing the iceberg population of the lake. A boat tour runs from the Begich Boggs Visitors Center between May and September, taking tourists through the magnificently-shaped icebergs up to the face of the glacier. It is a wonderful example of nature's vast beauty. JK

ALEXANDER ARCHIPELAGO

ALASKA, UNITED STATES

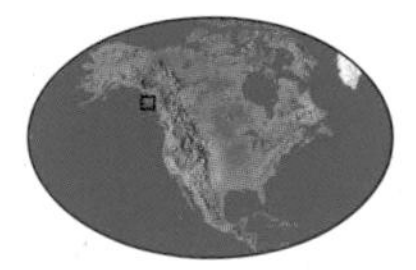

Area of Alexander Archipelago: 13,054 sq mi (33,811 sq km)

Number of islands: 1,100

Highest point (Baranof Island High Point): 5,390 ft (1,643 m)

The Alexander Archipelago is a dense area of peninsulas and islands off southeast Alaska where the largest aggregation of feeding humpback whales in the North Pacific can be found from May to September. The feeding behavior of a group of these humpback whales is remarkable, the huge whales operating swiftly and succinctly as a deadly team to hunt down their prey.

The whales feed in small groups of around seven, with each whale occupying its own place in the formation. They round up prey by surrounding them with a cylindrical curtain of bubbles. They submerge below the shoal of fish for about two-and-a-half minutes while they blow the bubbles. The fish are startled by the light reflections in the bubbles and concentrate in the center of what is known as a "bubble net." Then, one whale emits a deafening scream that panics the fish, and the rest of the giants swim up the center of the cylinder with their mouths agape. They gulp in a bouillabaisse of fish and seawater, and as they break the surface, fish fly everywhere in a desperate attempt to escape the cavernous maws. This spectacular sight can be observed from the many whale-watching boats that ply these waters every summer. MB

GLACIER BAY

ALASKA, UNITED STATES

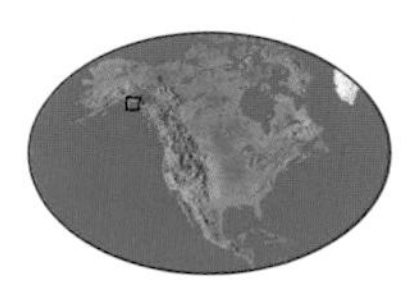

Area of Glacier Bay: 350 sq mi (905 sq km)

Height of Mt. Fairweather: 15,320 ft (4,670 m)

In 1794, when Captain George Vancouver of HMS *Discovery* visited this area, he could not see a bay at all. Instead, he saw the end wall of a massive glacier, 10 miles (16 kilometers) wide and 300 feet (90 meters) high. By 1879, when naturalist John Muir arrived, the glacier had retreated 48 miles (77 kilometers), leaving what is now called Glacier Bay.

Today the bay is part of Glacier Bay National Park, which consists of fjords, forests, and 16 giant glaciers that stand astride the U.S.–Canada border. The glaciers are retreating rapidly—1,320 feet (400 meters) every year. Each summer, they calve huge icebergs, which crash into the Pacific.

In the background are the mountains, including Mount Fairweather, the highest peak in the area. Marine mammals, such as harbor seals and killer whales, are in abundance, but the true stars of the show are the humpback whales. They arrive here from their California breeding grounds each summer to feed in the fish-rich waters. Whales leap clear of the water and then plunge back down in a mountain of spray. Several whales may feed together, pushing up through the surface in an explosion of gaping mouths and flailing flippers. MB

MOUNT McKINLEY

ALASKA, UNITED STATES

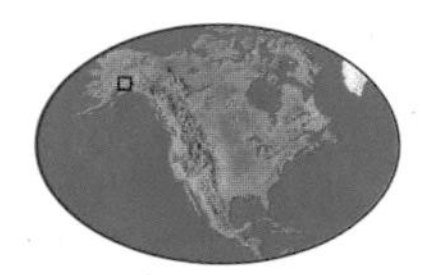

Alternative name: Denali ("The High One")

Height of Mt. McKinley: 20,320 ft (6,194 m)

First climbed: 1913

This huge, snowcapped mountain dominates Alaska and is a beauty to behold, with five giant glaciers on its flanks and dozens of permanent snowfields hundreds of feet thick. More than half the mountain is buried under snow and ice. If approached from the south, this colossus rises 18,000 feet (5,486 meters) in just 12 miles (19 kilometers). This is a greater vertical relief than that of Mount Everest. You could say that in terms of elevation change, McKinley is the greatest climb in the world.

The mountain is part of the 600-mile- (966-kilometer-) long Alaska Mountain Range, which began forming 65 million years ago. Most of the other mountains here are sedimentary rock, but McKinley is an uplifted mass of granite and shale. It is popular with climbers. Technically it is not too difficult a climb, but it has some of the worst weather of any mountain on Earth due to its high latitude. Extreme care must be taken for those who attempt the climb. Temperatures can plummet to -95°F (-71°C). The mountain is located in Denali National Park, one of America's greatest wildernesses. JK

BELOW *The snowcapped peaks of Mount McKinley.*

BERING STRAIT

ALASKA, UNITED STATES / SIBERIA, RUSSIA

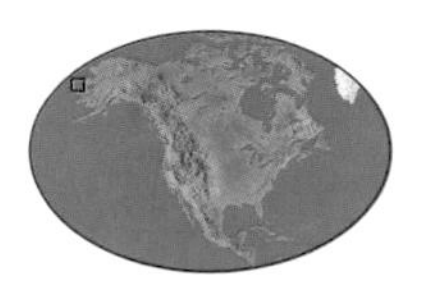

Area of Chirikov Basin: 8,494 sq mi (22,000 sq km)

Area of gray whale feeding: 463 sq mi (1,200 sq km)

South of the Bering Strait, the relatively narrow stretch of water that separates Russia from North America in the Arctic, is the Chirikov Basin. This is the permanent home of walruses, belugas, narwhals, and seals, but the most impressive summer visitors are the gray whales—all 22,000 of them. They arrive here from their winter breeding sites farther south and come for five months to fatten up on the rich stock of seafood that fills northern seas at this time of the year. Gray whales are unusual among whales because they feed on the bottom of the ocean, scooping up great mouthfuls of silt that contains burrowing shrimp-like creatures known as amphipods. From the air, it is possible to see the great grooves in the shallow seafloor where the whales have plowed their deep furrows. These grooves are 11 to 55 square feet (1 to 5 square meters) in area and 4 inches (10 centimeters) in depth. The muddy bottom also contains clams that are dug out by the 200,000 walruses that live in the area. They make long furrows in which they detect the shellfish with their sensitive whiskers and then suck them up into their mouths. Cruises enter the area each summer so visitors can witness this feeding spectacle. MB

MOUNT RAINIER

WASHINGTON, UNITED STATES

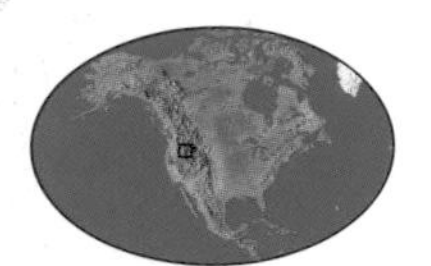

Height of Mt. Rainier: 14,410 ft (4,392 m)

Type of volcano: composite (highly explosive)

Mount Rainier is the most prominent peak in the Cascade Mountain Range, towering 3 miles (5 kilometers) above the lowlands to the west. It is also twice as high as the neighboring mountains, making it the unmistakeable landmark of Washington State.

Mount Rainier is an active volcano that last erupted 150 years ago—at about one million years old, it is also quite young. Its slopes, blanketed in snow and ice, cover an impressive 35 square miles (91 square kilometers), and tower over the green fir and hemlock forests below. The habitat on its lower slopes is diverse, with alpine meadows and subalpine heather communities that date back to the end of the last ice age 10,000 years ago. It also has superb stands of old-growth forest with trees that are 1,000 years old. Mount Rainier is located in the center of Mount Rainier National Park, and is a recreational paradise for camping and hiking in the summer and snowshoeing and skiing in the winter. The area is a temperate rainforest, with enough rainfall to support 382 lakes and 470 rivers and streams. JK

RIGHT *Mount Rainier behind the aptly named Reflection Lake.*

GRAND COULEE

WASHINGTON, UNITED STATES

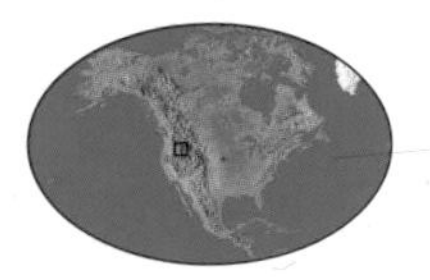

Length of the Grand Coulee canyon: 50 mi (80 km)

Width of canyon: 6 mi (10 km)

Depth of canyon: 900 ft (270 m)

The Grand Coulee is one of the youngest and most unusual natural wonders in North America. This amazing canyon is the largest of several coulees that cut through the Columbia Plateau of eastern Washington State.

For many years this peculiar canyon befuddled scientists who could not explain its formation. Its steep vertical walls and uneven topography mean that it is very new and could not have been formed by millions of years of river erosion, as first thought. In fact, not only is Grand Coulee bone dry and lacking a river, but the floor of the coulee slopes upward, in opposition to the prevailing downward slope of the Columbia Plateau.

Only after years of dedicated fieldwork by scientist J. Harlen Bretz was the explanation finally discovered: The Grand Coulee was created during the last ice age by the largest recorded floods known to science. At intervals of about 50 years, a towering wall of water over 2,000 feet (600 meters) high broke through an ice dam in the Rocky Mountains and cascaded through Washington State and out to the Pacific Ocean. The water exerted so much force that it ripped up the bedrock, creating the coulees that we see today. JK

DRY FALLS

WASHINGTON, UNITED STATES

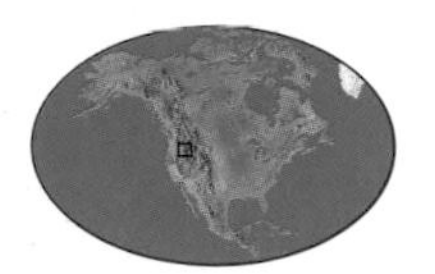

Height of Dry Falls: 400 ft (120 m)

Width of falls: 4 mi (6 km)

Rock type: volcanic basalt

Dry Falls is one of the most unusual and extraordinary geological wonders in North America. Situated in the middle of the Grand Coulee canyon, this four-mile (six-kilometer) wide, 400-foot (120-meter) high cliff was once the world's largest waterfall. Not a drop of water flows over Dry Falls today, but 15,000 years ago enormous floods escaping from a gigantic ice dam to the northeast thundered across the landscape. The floodwaters ripped up bedrock located approximately 11 miles (18 kilometers) farther south, eroding the cliff back to this point in the process.

This is the same erosion process that we see at Niagara Falls today, only the events at Dry Falls would have happened in just one day for each flood. It is not certain exactly how many separate floods occurred, but there were probably dozens. At their peak, the floodwaters flowing over this cataract were 1,000 feet (300 meters) high, with a volume 10 times that of Niagara. Today, Dry Falls overlooks a desert landscape with a few tame lakes at its base. At first glance it seems impossible that this amazing place was formed from such titanic violence. JK

UPPER SKAGIT RIVER

WASHINGTON, UNITED STATES

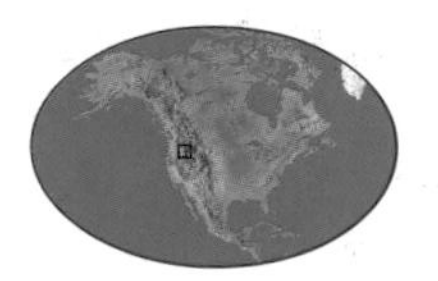

Upper Skagit River Bald Eagle Festival: early February

Number of bald eagles: up to 400

The Upper Skagit River is an unforgiving place. Ice forms at the edge of the river, the sky is an ominous gray, and large snowflakes drift and accumulate on the ground below. But high above, sitting in the trees lining the riverbank, is a myriad of brown and white shapes. They are bald eagles, and they flock here in their hundreds—on some days there can be as many as 400. They sit in the bare branches of cottonwood trees, conserving energy and waiting for the thousands of spawned out, dead, and dying salmon to be swept down from the breeding redds in the headwaters of the river.

The eagles arrive two weeks after the salmon have spawned, and come from as far away as Yukon and Alaska. They gather here for the "easy" food, but will fight for the right to a carcass. As they clash, the birds perform extraordinary maneuvers in the air—casualties end up flat on their backs. Peak activity is late December and early January, and the key place to visit is the Skagit River Bald Eagle Interpretive Center in Rockport. If visiting, wear warm, waterproof, outdoor clothing, but do not approach the birds too closely or startle them. MB

MOUNT ST. HELENS

WASHINGTON, UNITED STATES

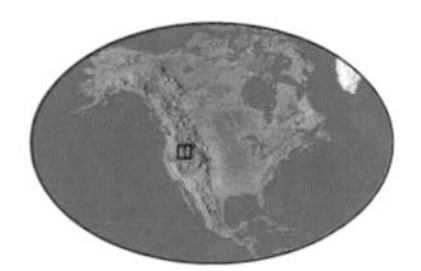

Height of Mt. St. Helens: 8,364 ft (2,549 m)

Width at base: 6 mi (10 km)

Type of volcano: composite (highly explosive)

Mount St. Helens is the youngest and most active of several snowcapped volcanoes that dominate the Pacific Northwest. It was once renowned for its beauty as a volcano, until it erupted suddenly on May 18, 1980, losing the top 1,320 feet (400 meters) of its summit, and becoming the worst volcanic disaster in recorded U.S. history. The world's largest landslide filled valleys and river channels, and pyroclastic flows scorched 230 square miles (596 square kilometers) of forest.

Yet the story of Mount St. Helens is one of life rising, literally, from the ashes. The volcano and the surrounding area have been conserved as a national monument. People can visit the volcano and see firsthand not only the awesome evidence of its destructive power, but also the remarkable recovery of the land as life returns. For spectacular views, visitors can drive to Windy Ridge, or they can look into the crater from the Johnston Ridge Observatory. For the more adventurous, mountain climbing to the summit and exploring the crater is possible with a permit. The largest eruption since 1980 occurred on October 1, 2004—a pale gray column of steam and ash spewed into the sky for 24 minutes. JK

MASSACRE ROCKS

IDAHO, UNITED STATES

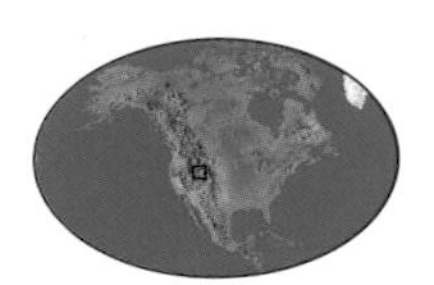

Massacre Rocks State Park: 990 acres (400 ha)

Elevation of Massacre Rocks: 4,400 ft (1,340 m)

Established as a state park: 1967

When wagon trains heading west in the mid-19th century approached the break in the rocks ahead of their trail, they were ready for attacks by the local Shoshone Indians. On August 9–10, 1862, 10 emigrants were killed, and so the natural thoroughfare became known as the Gate of Death or Devil's Gate, and the rocky hills became Massacre Rocks. The area is in the plain of the Snake River, and was once part of the Oregon Trail. Nearby Register Rock was a natural "rest stop" for travelers, and the names of these pioneers are carved into the stone.

In geological terms, the Devil's Gate Pass is all that remains of a basaltic volcano. The pass itself was carved out at the time of the Bonneville Flood about 15,000 years ago, when water from Lake Bonneville, which once covered most of Utah, burst through the pass and along what is now the channel of the Snake River. The flow of water was thought to be four times that of today's Amazon, making it one of the largest and most violent floods in history. The basalt boulders it broke away and carried are now scattered around the Idaho landscape. Massacre Rocks is located 10 miles (16 kilometers) west of American Falls. MB

ST. MARY LAKE

MONTANA, UNITED STATES

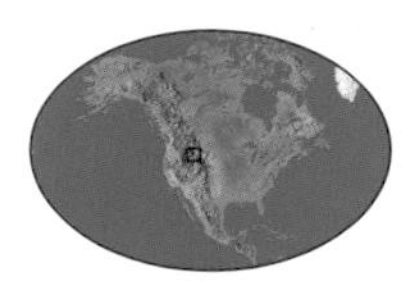

Length of St. Mary Lake: 9 mi (14.5 km)

Width of lake (widest point): 1 mi (1.6 km)

St. Mary Lake is a beautiful, glacial blue lake in one of the most perfect settings imaginable. On three sides it is surrounded by the steep Rocky Mountains, and on its eastern shore it gives way to rolling prairie and forested hills. The lake is part of Glacier National Park in northern Montana. Fed by snowmelt from the surrounding mountains, St. Mary Lake is exceptionally clear, not to mention quite cool throughout the summer. This lake lies on the eastern side of the Continental Divide that runs through Glacier National Park. Most of the rain from prevailing westerlies falls on the western side of the Divide, so the eastern part of the park is in the mountains' rain shadow and is therefore more arid.

Winds blowing down from the mountains are almost a permanent feature here, making boating on the lake a turbulent experience. St. Mary Lake is a popular camping and recreation area. Hiking trails beginning at the lake lead up into some of the most scenic, uplifting landscape in the park. The lake itself has great fishing with lake trout, rainbow trout, cut-throat trout, and whitefish. JK

BELOW *The early morning sun illuminates St. Mary Lake.*

NATIONAL BISON RANGE

MONTANA, UNITED STATES

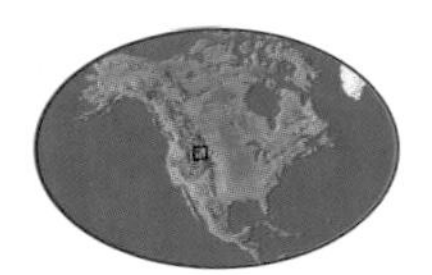

Area of National Bison Range: 29 sq mi (75 sq km)

Highest point: 4,600 ft (1,400 m)

Number of bison: 350 to 500

The National Bison Range in Montana's Rocky Mountains is one of the oldest wildlife refuges in the United States. It is home to some of the very last remaining plains bison, whose population plummeted from 50 million to less than 1,000 because of overhunting. The refuge encompasses a group of magnificent hills in the beautiful Flathead Valley and includes a rich variety of habitats such as prairie grasslands, mountain forest, wetlands, and river bottom woodland. Take the Red Sleep Drive for a 2,000-foot (600-meter) climb to the highest spot in the range and spectacular views of the endless peaks of mountains that surround the range. The bison are, of course, the highlight here. The breeding season from mid-July through August is one of the best times to visit, when the large bulls roar and do battle with one another. Calves are born from mid-April through May. The bison are hardy animals, with thick, heavy coats that are so well insulated that snow can lay on their backs without melting. The range is also home to 50 species of other mammals, including mountain lions, elk, black bear, and coyote. JK

LAKE McDONALD

MONTANA, UNITED STATES

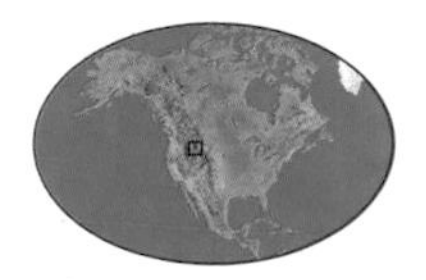

Length of Lake McDonald: 10 mi (16 km)

Width of lake: 1 mi (1.6 km)

Depth of lake: 472 ft (144 m)

Lake McDonald in northern Montana is the largest lake in Glacier National Park. Surrounded on three sides by towering Rocky Mountains that rise 6,000 feet (1,800 meters) into the sky, the view is sensational, with white alpine glaciers hugging the jagged upper slopes of the mountains and lush green forests covering the lower slopes. It acquired the name "McDonald" after a trader called Duncan McDonald carved his name into a birch tree beside the lake in 1878.

The lake is a sparkling reminder of the last ice age. Its deep basin was carved out by a giant glacier that once filled the entire valley, and its shimmering waters reflect images of the surrounding Rockies. The Lewis Mountain Range to the east of the lake is the Continental Divide, which acts as a rain block to the clouds whose moisture supports a rich, dense forest of western red cedar and hemlock trees. The lake is a perfect spot to explore Glacier National Park and its 35 large alpine glaciers. Especially stunning are the park's 1,000 waterfalls and the beautiful alpine meadows that burst with wildflowers. Around the lake, Rocky Mountain sheep, mountain goats, bald eagles, and the occasional grizzly bear can also be spotted. **JK**

MAHAR POINT

MAINE, UNITED STATES

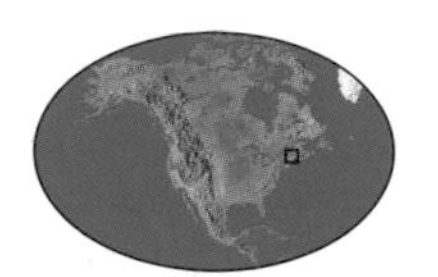

Length of channel: 0.5 mi (0.8 km)
Width of channel: 899 ft (274 m)
Speed of current: 16 mph (25 kph)

Mahar Point, on the coast of Maine, provides a fine vantage from which to view one of the most fascinating natural spectacles in America: the "reversing falls" of Cobscook Bay. The name "Cobscook" comes from a local Native American word meaning "boiling tides." Twice a day the tide sweeps through a narrow 899-foot (274-meter) channel that connects Cobscook Bay to two smaller bays, Whiting and Dennys Bay. After the tide turns, it surges right back out again.

The tidal change here is huge, about 20 feet (6 meters), and flows at 16 miles per hour (25 kilometers per hour). The water flowing through the channel runs a 0.5-mile (0.8-kilometer) gauntlet of jutting rocks that cause this reversing of the falls. For six hours the water roars through the channel until slack tide, when suddenly the channel is as calm as a mirror. Then, when the tide reverses direction, the current grows, the water level falls, and within 10 minutes the ripples and white water appear again around the jutting rocks. Mahar Point is on the western side of the Bay of Fundy, which has the largest tides in the world. To see the falls reverse, plan to be there an hour before high tide. JK

OLD SOW WHIRLPOOL

MAINE, UNITED STATES / NEW BRUNSWICK, CANADA

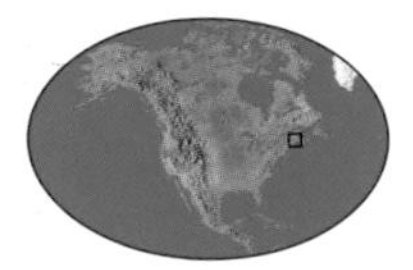

Diameter of Old Sow Whirlpool: 250 ft (76 m)
Volume of tide: 40 billion cu ft (1.13 billion cu m)
Speed of current: 7 mph (11 kph)

The Old Sow Whirlpool is the second largest in the world. It occurs in a narrow strait, called Western Passage, in Passamaquoddy Bay. This whirlpool reaches 250 feet (76 meters) in diameter and is an awesome display of power. It is part of a larger 7-mile- (11-kilometer-) wide swathe of turbulent ocean water that roils with fast currents, boils, and gyres that are collectively known as "the Piglets." Old Sow is caused by the unusual topography of the bay, and appears during the incoming flood tide when it is running high. The position of Deer Island and the Maine coastline forces the tide to make a 90° right-hand turn. The tidal waters then run smack into an undersea mountain and are forced to go around it. A counter current from the St. Croix River to the north adds to the mayhem. The Old Sow area starts to roil about three hours before high tide and continues for a couple of hours. How Old Sow got its name is a subject of much debate, but the most likely explanation is that "sow" is a corruption of the word "sough" (pronounced "suff"), which has two meanings: a sucking noise or a drain. The best place on land to observe Old Sow is at the southern end of Deer Island. JK

CRATER LAKE

OREGON, UNITED STATES

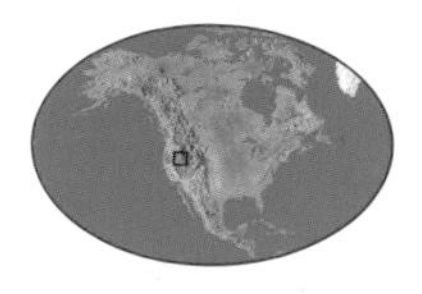

Diameter of Crater Lake: 5 mi (8 km)
Elevation of lake: 6,176 ft (1,882 m)
Height of sides: 2,000 ft (600 m)

In 1902 Crater Lake became the fifth national park in the United States. It consists of a large hole, created by the fierce volcanic collapse of a multi-cratered summit called Mount Mazama over 7,000 years ago. Today the crater is filled with a tranquil lake, but pushing up from its waters is a lava and ash cone—Wizard Island—a reminder of what could lie beneath, and another island—Phantom Ship—with rocky pinnacles and skeletal trees.

Local legend holds that Mount Mazama was the battleground on which the Chief of the World Above took on his equal from the World Below. Rocks flew, forests were torched, and earthquakes rocked the land in a scene reminiscent of the huge geological activity that actually occurred. Today the lake is 1,932 feet (589 meters) deep, but is not fed and therefore does not lose water by way of streams or rivers. Evaporation in summer is balanced by the winter fall of snow and rain. Crater Lake sits in the Cascade Range at an elevation of 6,176 feet (1,882 meters), where 50 feet (15 meters) of snow falls each winter—a long winter that lasts from September to July. MB

MULTNOMAH FALLS

OREGON, UNITED STATES

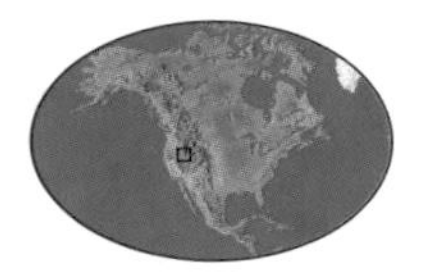

Height of Multnomah Falls: 620 ft (189 m)

Width of falls: 30 ft (9 m)

Form: tiered falls

Multnomah Falls plummets 620 feet (189 meters) from a breathtaking precipice on Larch Mountain, in the Columbia River Gorge, making it the second-highest year-round waterfall in the United States after Yosemite Falls. It is composed of two cataracts, with a small pool between them. The much taller upper falls is thinner and longer, while the lower section is wider and stronger. Fed by crystal clear underground springs, these flow down the mountain in Multnomah Creek and cascade over this point in spectacular fashion. The flow over the falls is at its highest during winter and spring, but unusually cold weather can turn this water spectacle into a sensational frozen icicle.

The cliff face has excellent exposures of five separate basalt lava flows that show how this area was created by great upwellings of magma around 12 to 16 million years ago. Native legend tells the story of a young maiden who threw herself off the cliff here as a sacrifice to the Great Spirit, to save her people from succumbing to a deadly illness. The next day, when he discovered her body, her father prayed to the Great Spirit to give him a sign that his daughter had been welcomed into the land of the spirits. Almost at once, a silvery-white stream of water flowed out of the forest and over the cliff as a high and beautiful waterfall.

Multnomah Falls is famed for its beauty and is Oregon's number one tourist attraction. The elegant Benson Bridge crosses the falls between its lower and upper cataracts, providing visitors with great views. It was crafted by Italian stonemasons in 1914 and is named after Simon Benson, the original owner of the falls.

Multnomah Falls is famed for its beauty and is Oregon's number one tourist attraction. The flow over the falls is at its highest during winter and spring, but unusually cold weather can turn this water spectacle into a sensational frozen icicle.

More adventurous visitors can hike up a narrow winding footpath to the top of the falls, which are straddled by a wooden observation platform. From here, you can look down over the falls as the water flows over the edge to begin its long descent. You can also enjoy views of the small rapids and mini-falls where the creek first emerges from the woods, heading toward the edge of the cliff. The climb to the top is well worth the effort, allowing you to take in views of the Columbia River Gorge. **HL**

RIGHT *Benson Bridge, from which visitors are treated to an excellent view of both the upper and lower cataracts of Multnomah Falls.*

MOUNT HOOD

OREGON, UNITED STATES

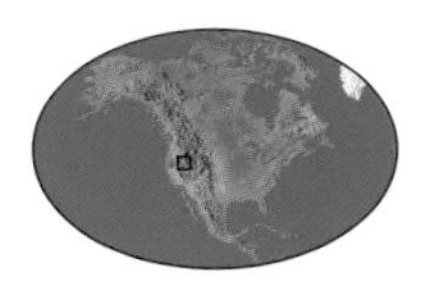

Height of Mt. Hood: 11,239 ft (3,426 m)

Age of Mt. Hood: 500,000 years

Type of volcano: composite (highly explosive)

Rising 11,239 feet (3,426 meters), Mount Hood is the tallest mountain in Oregon and one of the most-climbed peaks in the Pacific Northwest. Twelve glaciers ensure that it remains permanently white throughout the year. Only 45 miles (72 kilometers) east of Portland, this volcano draws backpackers in the summer and skiers during the winter. A member of Captain George Vancouver's 1792 naval expedition named it after British admiral Samuel Hood, but local Northwest Native Americans traditionally called it "Wy'East."

The main cone of this impressive volcano formed about 500,000 years ago. Scientists agree that, like all Cascade Range volcanoes, Mount Hood is only "resting." Its last major eruption was between 1754 and 1824, when mudslides and pyroclastic flows surged down its southern slopes. Nestled in its crater is a steaming lava dome called Crater Rock that caps the molten rock bubbling below. Crater Rock stands 1,320 feet (400 meters) across and 558 feet (170 meters) high. Warm fumaroles along its base emit sulfurous gases and steam. The area around Mount Hood is a designated wilderness area and is rich in wildlife. Visitors can get a great view from the famous Timberline Lodge located on the volcano's southern slope. JK

THE COLUMBIA RIVER GORGE

OREGON, UNITED STATES

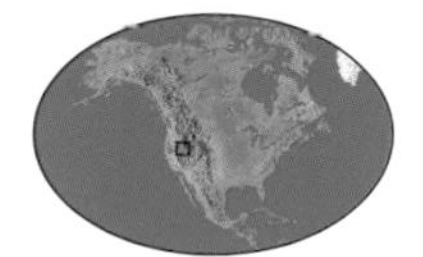

Length of Columbia River Gorge: 80 mi (128 km)

Depth of gorge: 4,000 ft (1,200 m)

Age of gorge: 10 million years

Once known as the most treacherous leg of the Oregon Trail, the Columbia River Gorge is a spectacular river canyon that cuts through the Cascade Mountain Range. It has a wealth of scenic wonders with river views, waterfalls, steep basalt cliffs, snowcapped mountains, and verdant forests. Over the past 10 million years the Columbia River has left its mark on this glorious gorge. The river was strong enough to erode the hard basalt rock that covers the entire area. About 15,000 years ago the gorge was widened and scoured by enormous floods (the world's largest) that swept out from melting continental ice sheets to the northeast at speeds of 85 miles per hour (136 kilometers per hour). An archeological dig in the middle of the gorge, at Five Mile Rapids, has found evidence of human occupation more than 10,000 years old.

The Columbia River Gorge hosts a rich salmon habitat, and for the more adventurous the 30-knot winds that blow through the canyon are ideal for windsurfers. Visitors with an eye for scenery can travel the Historic Columbia River Highway that meanders the length of the gorge. Along the route, especially in the western end of the Columbia Gorge, are numerous scenic waterfalls. The southern stretch alone has 70. JK

BADLANDS

SOUTH DAKOTA, UNITED STATES

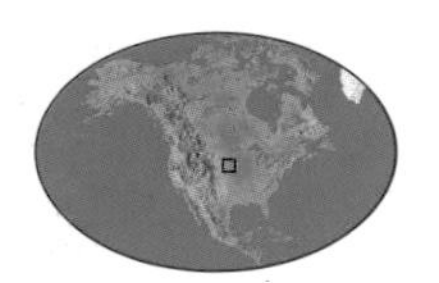

Area of Badlands: 381 sq mi (989 sq km)

Age of Badlands: 5 million years

The Badlands of South Dakota are a natural masterpiece of wind and water sculpture. This extraordinarily eroded landscape contains a profusion of buttes, pinnacles, and spires carved out of an underlying plateau of soft sediments and volcanic ash. Early settlers gave the Badlands their fitting name, for it would be almost impossible to grow crops in these battered hills. Fortunately, the Badlands' scientific and pictorial merits have long been recognized, and the area is now part of the Badlands National Park, which also protects America's largest area of mixed grass prairie.

The Badlands' sediments were deposited in layers beginning 75 million years ago when shifting continents raised the Black Hills to the west. Sand, silt, and clay measuring thousands of feet deep were then deposited on the plains, along with several layers of volcanic ash, until five million years ago, when the White River began eroding to gradually reveal the stark landscape we see today. The Badlands are also a showcase for the best deposits of fossilized mammals in the world, dating back 35 million years. JK

BELOW *The barren landscape of South Dakota's Badlands.*

SPEARFISH CANYON

SOUTH DAKOTA, UNITED STATES

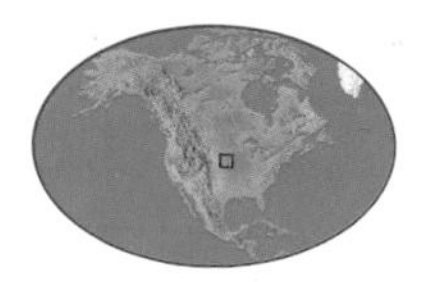

Length of Spearfish Canyon: 20 mi (32 km)

Width of canyon: 1 mi (1.6 km)

Spearfish Canyon was described as "the most magnificent canyon in the west" by one of America's most renowned architects, Frank Lloyd Wright. He noted correctly that much of the canyon's magnificence is due to the convergence of four North American plant biomes: Rocky Mountain pine forest, northern spruce forest, eastern aspen and birch forest, and plains oak and cottonwood forest. The origins of the canyon go back 62 million years, but most of what we see today was created in the past five million years. The canyon cuts through the Black Hills of South Dakota and has 17 side gulches that preserve much of the pristine beauty of the landscape. It is narrow—only 1 mile (1.6 kilometers) wide, and towers above the visitor on its floor.

As Spearfish Creek twists and turns through the canyon it reveals one beautiful vista after another. Tributaries flow into the creek, but some do not erode down through the sediment as quickly as the main stream and so become hanging valleys, their water plummeting as a cascade, like the lovely Bridal Veil Falls. The canyon and the Black Hills were among the last places in the American West to be colonized by settlers. JK

YELLOWSTONE NATIONAL PARK

WYOMING, UNITED STATES

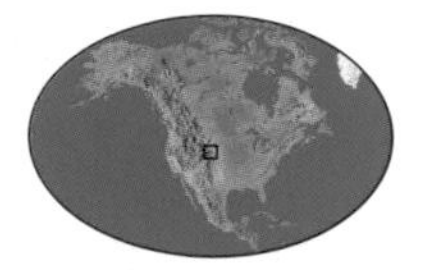

Area of Yellowstone National Park: 3,500 sq mi (9,000 sq km)

Height of Old Faithful waterspout: up to 200 ft (60 m)

Age of geysers: 600,000 years

When President Grant made the declaration in 1871, Yellowstone became the first national park to be established not only in the United States, but also in the world. It is an extraordinary place, with canyons, lakes, and its famous geysers, hot springs, and boiling mudpots. The natural assets are endless—a black glass mountain known as Obsidian Cliff from which Shoshone warriors made their arrowheads; boiling mud pools at Fountain Paint Pot; calcite terraces like layers of sugar-icing at the Minerva Terraces; the Grand Canyon of Yellowstone in which the Yellowstone River drops in a succession of falls from Yellowstone Lake; Specimen Ridge with its petrified trees that were covered by volcanic ash; and the Grand Prismatic, the largest of Yellowstone's hot springs, its waters a rainbow of vibrant colors caused by heat-resistant algae and bacteria.

But the most celebrated features are the spectacular geysers, the most famous being Old Faithful, which forces a fountain of hot water and steam 200 feet (60 meters) into the air every 90 minutes. An even higher jet is produced by Steamboat, the tallest geyser in the world, but it is unreliable, bursting forth anywhere between five days and five years apart. Riverside Geyser is known to throw up a curl of boiling hot spray right over the Firehole River. The engine for all this activity is a dome of molten rock that is no more than three miles (five kilometers) below the surface. It provides the heat that turns percolating water to steam, and drives out the dramatic columns of water. About 600,000 years ago, this time bomb below erupted in a cataclysmic explosion that covered much of North America in ash. But the volcanic chamber collapsed, leaving a huge caldera. Minor eruptions burst through its floor, filling it with lava and ash. Today, the whole thing is ready to go off again. But in the meantime, visitors can get into the action. Boardwalks and clearly signed paths guide you safely through the maze of geological highlights. There is wildlife to see too—you can stand cheek by jowl with formidable bison, spot packs of wolves in the distance, and chance upon grizzlies, coyotes, and moose. MB

Yellowstone was the first national park to be established in the United States, and it is an extraordinary place, with canyons, lakes, geysers, hot springs, and boiling mudpots. The most celebrated features are the spectacular geysers.

RIGHT *Old Faithful spurts water 200 feet (60 meters).*

MAMMOTH HOT SPRINGS

WYOMING, UNITED STATES

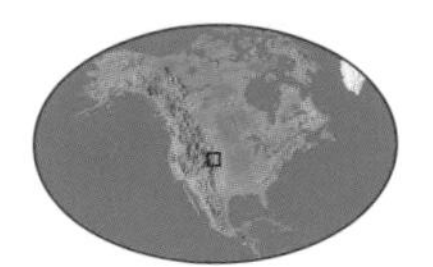

Daily calcium carbonate deposit: 2 tons

Average water flow: 71 cu ft per min (2 cu m per min)

Mammoth Hot Springs consists of about 50 colorful and fantastic looking hot springs bubbling up through the chalky bedrock of Yellowstone National Park. Located in the park's northwest corner, this complex of springs is a natural work of art, a living sculpture. They have fantastically shaped terraces, with a spectacular kaleidoscope of colors. The steaming hot water dissolves the soft limestone of the bedrock deep underground and then deposits it on the surface as a white mineral called travertine when the water cools. This mineral builds up at the phenomenal rate of about 1 inch (2.54 centimeters) per year and today some terraces are 300 feet (90 meters) high. The mineral deposits give each hot spring its own unique shape and look, and as long as the water keeps flowing, each will continue changing shape and appearance. Travertine is white when it is deposited, but these springs are home to heat-loving bacteria and algae that give the terraces brilliant yellow, brown, and green colors. Mammoth Hot Springs is fed by rainwater and snow falling high on the slopes in and around Yellowstone. This cold ground water seeps deep into the earth where it is warmed by heat radiating from the magma chamber before rising back to the surface. **JK**

GRAND TETON NATIONAL PARK

WYOMING, UNITED STATES

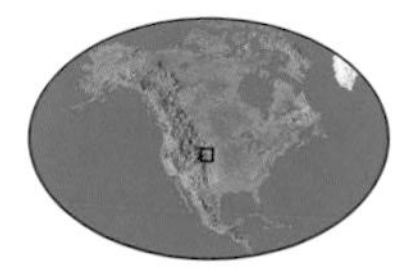

Established as a national park: 1929 (enlarged 1950)

Highest point (Grand Teton): 13,770 ft (4,197 m)

The Grand Teton National Park protects some of the most stunning mountain scenery in the United States. The Teton Range, the highest mountains in Wyoming, rise abruptly from the Jackson Hole valley floor, their jagged peaks reflected in the valley's lakes. The mountains are much younger than the rest of the Rockies—there were no mountains here about nine million years ago. Glaciers are found among the high peaks, and Grand Teton is the highest at 13,770 feet (4,197 meters).

There is a profusion of wildlife. Among the large mammals regularly seen are bison, moose, elk, pronghorn, beaver, and black bear. Grizzly bears are present in the northern part of the park and bighorn sheep are found on the high slopes. Wolves from Yellowstone have also been seen here in recent winters. Bird species include bald eagle, osprey, white pelican, and trumpeter swan. The Snake River cut-throat is a type of trout found only in the river that flows through Jackson Hole. The Grand Teton National Park was first established in 1929 and was enlarged in 1950. The best time to visit is from June to September—due to heavy snowfalls, most visitor accommodation is closed for the rest of the year. RC

GRAND PRISMATIC SPRING & FIREHOLE RIVER

WYOMING, UNITED STATES

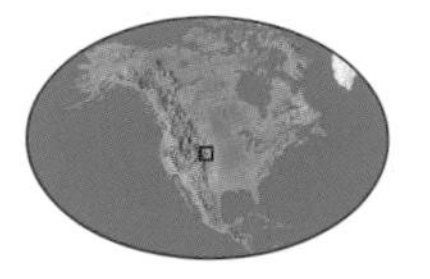

Depth of Grand Prismatic Spring: 160 ft (49 m)

Average temperature: 167°F (75°C)

Discharge rate: 560 gal per min (2,120 l per min)

Firehole River, in Yellowstone National Park, flows through one of the most unusual and remarkable landscapes in the world, which Rudyard Kipling described as "Hell's Half Acre." It begins as a series of small cold-water springs just south of Old Faithful geyser, and passes through a steaming land of geysers and hot springs which dramatically change its temperature and mineral composition.

Yellowstone contains 3,000 hot springs and geysers—the highest concentration in the world—and none is more beautiful than the Grand Prismatic Spring, which Firehole River runs past. This gigantic, steaming hot spring is the largest and most fascinating in Yellowstone. It is nearly 380 feet (116 meters) in diameter, and sits upon a large limestone mound, surrounded by a series of step-like terraces. The spring is a rainbow of colors—at its center, where the water is hottest, it is deep blue, turning paler blue farther out, and then green at its shallower, cooler edge where algae grows. Each type of bacteria adapts to a narrow temperature range, and heat-loving cyanobacteria (blue-green algae) thrive on the spring's heated terraces producing succeeding bands of yellow, orange, and red. On a sunny day, the steam rising above Grand Prismatic Spring reflects a rainbow that can be seen from half a mile away.

Nearby is the now dormant Excelsior Geyser that was once the largest geyser in the world, shooting water 300 feet (90 meters) into the air. This is now a very productive thermal spring, which discharges over 4,000 gallons (15,000 liters) of hot water per minute into the river.

Firehole River, in Yellowstone National Park, flows through one of the most unusual and remarkable landscapes in the world. Rudyard Kipling described it as "Hell's Half Acre." The river passes through a steaming land of geysers and hot springs.

Firehole River was named by early trappers who associated the billowing steam of its hot springs and geysers with underground fires; and their term for a mountain valley was "hole," hence "Firehole." The river is famed as a world-class fly-fishing habitat and has healthy populations of brown, brook, and rainbow trout. The grassy banks of the river are a favorite feeding ground for herds of bison. The early evening is an especially good time to see these majestic animals silhouetted against the white plumes of the hot springs. JK

RIGHT *Green and blue bacteria give the spring its bright color.*

DEVIL'S TOWER

WYOMING, UNITED STATES

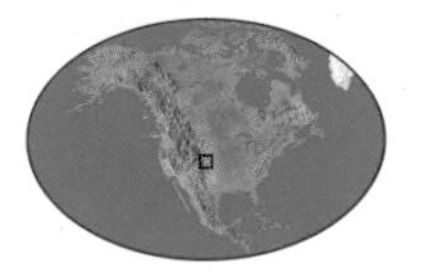

Age of Devil's Tower: 50 million years	
Height of tower: 865 ft (265 m)	
Width of tower: 300 ft (90 m)	

Stephen Spielberg chose it for a starring role in *Close Encounters of the Third Kind* in 1977; local rancher William Rogers conquered it using a ladder in 1893; and in the 1940s a parachutist sat there for six days when he lost the rope that should have allowed him to climb down. "It" is a gigantic monolith in northeast Wyoming known as Devil's Tower. Formed from volcanic lava thrust up through the underlying bedrock, it can be seen from over 100 miles (160 kilometers) away, the color of its rocks changing with the time of day and with the seasons.

It was formed about 50 million years ago when hot, molten lava cooled, contracted, and formed vertical rows of hexagonal columns. Softer rocks that surrounded the intrusion were worn away, leaving the tower exposed above the surrounding land. The local Kiowa people believe that Devil's Tower was formed when a bear chased a group of girls onto a low rock. The rock rose up, carrying the girls out of the bear's reach, but the bear scratched at the mound with its powerful claws, leaving furrows down the tower's side. Eventually, the bear died, and the girls lived forever as the seven stars of the Pleiades. **MB**

AGATE FOSSIL BEDS

NEBRASKA, UNITED STATES

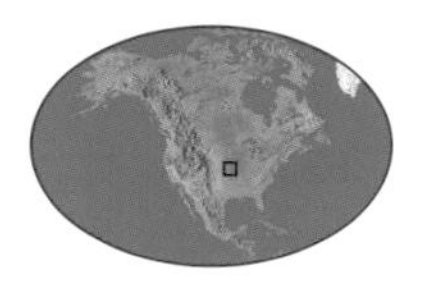

Area of fossil beds: 3,055 acres (1,236 ha)

Habitat type: fossil Miocene-age bluffs and river exposures

Rock type: sedimentary with occasional thin agate inclusions

Twenty to thirty million years ago, the "badlands" of Nebraska resembled a moist African savanna. In addtion to a variety of fossilized bones and remnants of various ancient grasses, the buttes of the region preserved trace fossils such as footprints, along with strange formations called either "Devil's Corkscrews" or "Devil's Augers."

Up to 10 feet (3 meters) deep, these formations are the helical burrows of a terrestrial beaver. Flash floods subsequently filled in these burrows with fine sediment. Drying very hard, this resulted in casts that weathered above the surrounding rock.

At first these trace fossils were a real puzzle to geologists—until one was found with a still-entombed beaver at the bottom. Later, fossil hunters managed to find some containing the skeletons of an ancient ferret-like animal, the predator against which the helices were made.

The beaver burrows, bones, and other paleontological wonders of this national park are preserved because the region managed to escape glaciation. Had it not done so, we would never have known about these spiraling, land-living beavers. **AB**

MONTEREY CANYON

CALIFORNIA, UNITED STATES

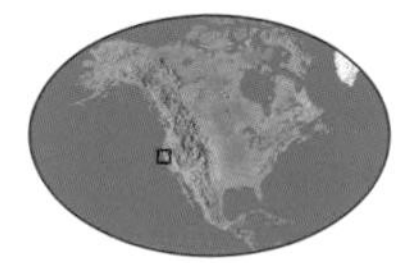

Length of Monterey Canyon: 292 mi (470 km)

Widest point of canyon: 7.5 mi (12 km)

Age of canyon: 15,000 to 20,000 years (most recent version)

Two hours south of San Francisco is Monterey Canyon, the largest and deepest sea canyon on North America's Pacific coast. Situated offshore near Monterey Bay's Moss Landing, this massive canyon quickly reaches depths of up to 12,000 feet (3,600 meters). The canyon was formed by ice age river erosion down exposed sea cliffs along an old earthquake fault and maintained by erosion catalyzed by subsurface freshwater springs.

The canyon has three distinctive habitats: sheer, vertical canyon walls rich in corals and sponges, which provide shelter for other animals; the mid-water, which is rich in jellyfish (some of which are the size of watermelons), specialist grazers such as owlfish, and predatory fangtooth and gulper eels; and the scavenger-dominated, sediment-buried seafloor. The canyon acts as a conduit for eroded soil to the sea and deflects nutrient-rich waters up from the deep. Consequently, the waters are a rich feeding ground for whales and other marine life. It is impossible to view this area without special equipment, but Monterey Bay Aquarium has several exhibits dedicated to this natural phenomenon. **AB**

MONTEREY BAY MARINE SANCTUARY

CALIFORNIA, UNITED STATES

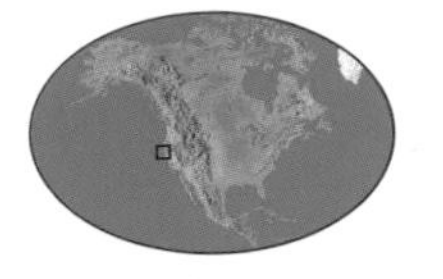

Area of Monterey Bay Marine Sanctuary: 5,360 sq mi (13,730 sq km)

Habitats: rocky coves, kelp forests, offshore islands, deep-sea canyons

Established as marine sanctuary: 1992

The largest of the United States's 14 National Marine Sanctuaries, Monterey Bay Marine Sanctuary stretches along central California's Pacific coast and extends an average of 30 miles (50 kilometers) into the ocean. The area's waters are enriched by deep-water upwellings brought in by the great submarine conduit of Monterey Canyon. Inshore, the sanctuary is famous for its giant kelp forests, individual plants of which can live up to 10 years. In the bay's cool, calm, nutrient-rich sunlit waters they can grow up to 2 feet (0.6 meters) per day. Spatially complex and biologically very productive, the kelp forms the basis for a unique ecosystem housing many unusual species of invertebrates and fish. It is also home to the endangered sea otter. Locally common and curious (and quite easily seen), sea otters use stone tools to feed on the kelp's abundant crabs, sea urchins, and sea snails. Other notable inhabitants include great white sharks, California sea lions, elephant seals, and a host of resident and migratory seabirds. **AB**

SAN ANDREAS FAULT

CALIFORNIA, UNITED STATES

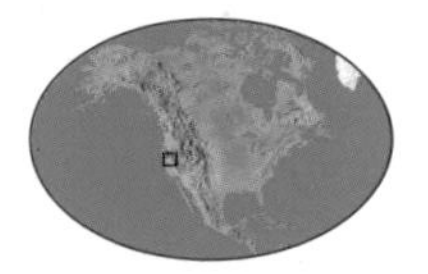

Length of San Andreas Fault: 800 mi (1,300 km)

Depth of fault: at least 10 mi (16 km)

Type of fault: strike-slip geological fault

Famously running right through San Francisco, the San Andreas Fault is one of the longest and most active geological faults in the world. Lying to the west of California's coastal mountain range, it is also one of the most visible, revealing its presence with many long linear lakes, strangely twisted formations of sedimentary rocks, streams that turn sharply, and road lines and fences warped from lateral strike-slip. Some 20 million years old, the fault extends 10 miles (16 kilometers) into the planet's crust and forms the "master" fault of an intricate fault network that cuts through rocks of the California coastal region.

Caused by tension generated as the Pacific plate moves northwest relative to the North American plate, the fault is a zone of crushed and broken rock a few hundred feet to 1 mile (1.6 kilometers) wide. Minor shocks are very frequent. Rocks along the fault have been displaced up to 350 miles (563 kilometers) in the last 20 million years. Average progress north is 2 inches (5 centimeters) a year. The 1906 San Francisco earthquake involved a 21-foot (6.4-meter) displacement. **AB**

RIGHT *The San Andreas Fault vanishes into the distance.*

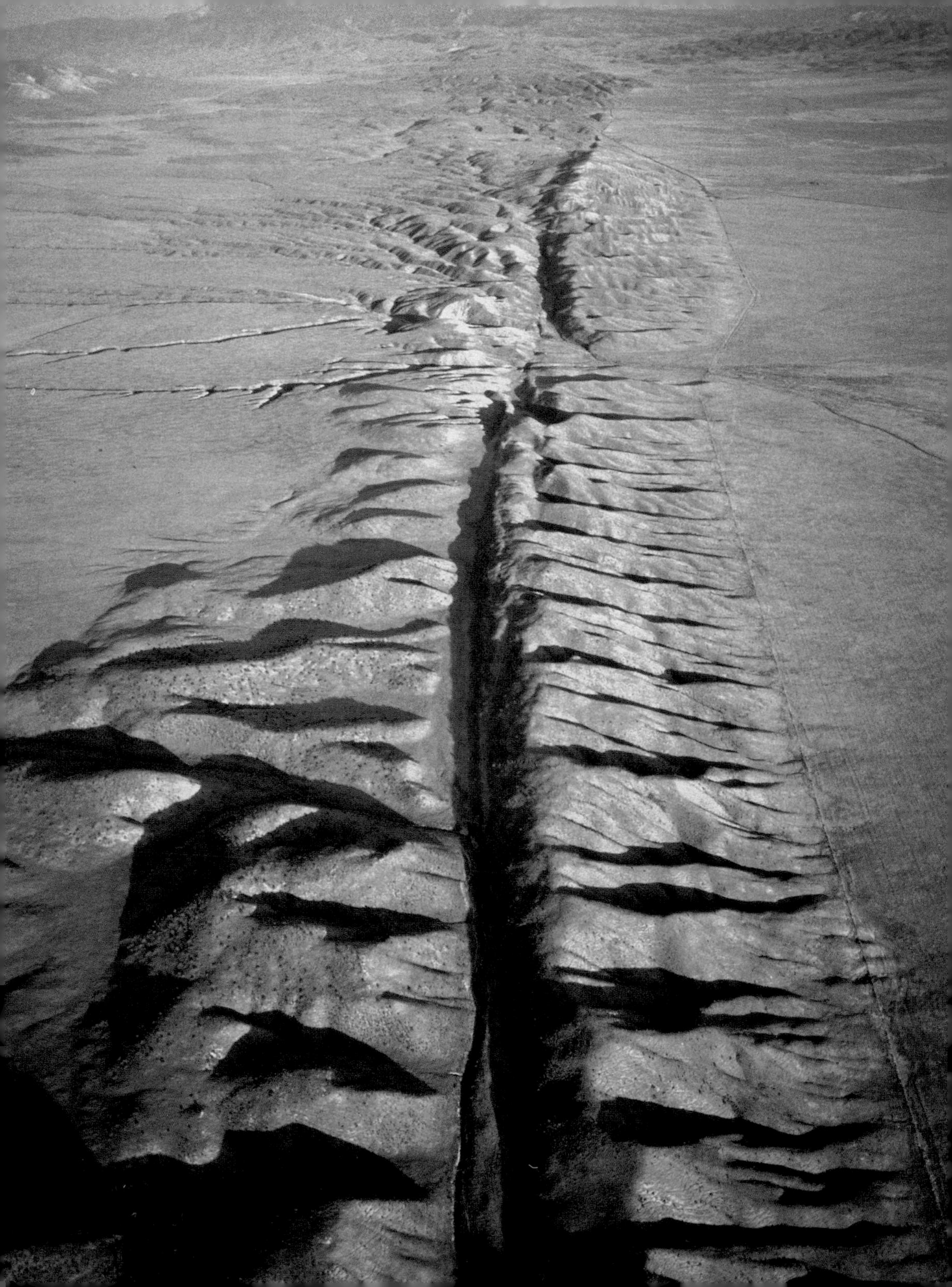

YOSEMITE NATIONAL PARK

CALIFORNIA, UNITED STATES

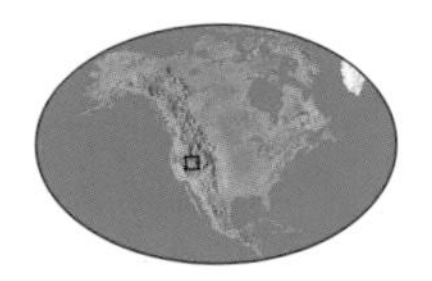

Area of Yosemite National Park: 1,200 sq mi (3,079 sq km)

Height of El Capitan: 3,000 ft (900 m)

Height of Yosemite Falls: 2,425 ft (739 m)

The name "Yosemite" traditionally refers to the grizzly bears that used to roam here. The largest predator today is the black bear, although one kind of "grizzly" does still exist—the Grizzly Giant, a 2,700-year-old sequoia that still grows in a stand of giant trees known as Mariposa Grove. The trees make Yosemite glow in fall, with black oak, incense cedar, and ponderosa pine all represented in the forested slopes and valleys of some of the most spectacular scenery in the world. Since 1901, tourists have been flocking to the park to walk through the magnificent Yosemite Valley beside the Merced River. Here you can not only see the world's tallest unbroken cliff—El Capitan—rise about 3,000 feet (900 meters), but also the spectacular Yosemite Falls drop 2,425 feet (739 meters) over three cliffs, making it the sixth highest in the world. The region was formed about 10 million years ago when major earth movements pushed the land upward while rivers cut downward. About three million years ago the glaciers of the ice age carved the valleys deeper and wider. When they melted, they left Yosemite Valley and the Grand Canyon of the Tuolumne River looming to its north. MB

GLACIER POINT

CALIFORNIA, UNITED STATES

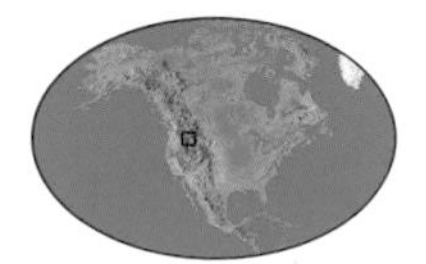

Elevation of Glacier Point: 7,214 ft (2,199 m)

Height above Yosemite Valley: 3,200 ft (975 m)

Glacier Point is located vertically above the Yosemite Valley and affords a stunning panoramic vista of the park below it, which spreads out to the north and east. Photographed by literally hundreds of thousands of visitors, the magnificent view has become known as the quintessential Yosemite scene. Far below this sky-scraping vantage point are meadows and forests flanked by precipitous cliffs surrounding the Merced River, and opposite are the Upper and Lower Yosemite Falls. The Nevada and Vernal Falls are clearly visible in the Little Yosemite Valley, as is the steep watercourse of the Tenaya Creek.

Separating these two deep, glacially-carved, U-shaped valleys is another Yosemite icon, the massive ice-polished granite sentinel of Half Dome. Glacier Point is one of the highest elevation viewpoints that can be easily reached by car in the Sierra Nevada Mountain range. The paved 15-mile (24-kilometer) long road forks west at Chinquapin Junction, south of the park entrance known as Wawona Tunnel. Visitors should note that the road can become impassable due to snow and is often closed to traffic during the winter months. For the more intrepid, a walking path (the Four Mile Trail) connects Glacier Point with Southside Drive on the valley floor. DL

RANCHO LA BREA TAR PITS

CALIFORNIA, UNITED STATES

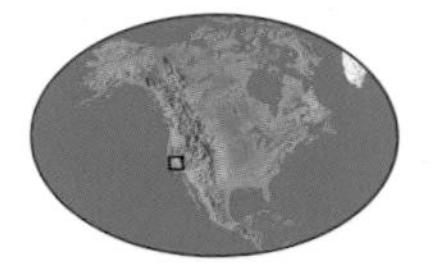

Age of Rancho La Brea tar pits: 40,000 years

Rock type: crude oil seep, with entombed fossils

Number of pools: 100

At Rancho La Brea, semi-solid asphalt oozes from the ground to create tar pits. Formed beneath the sea over millions of years, the crude oil began seeping through rock fissures after earthquakes raised California's seabed 40,000 years ago. Incredibly sticky but with a deceiving layer of water or leaves on top, the pools acted like giant flypapers. La Brea's pools have been fooling herbivores, carnivores, and scavengers for millennia—those that entered the pools became trapped and suffocated in the deep glutinous deposits, entombing extinct species such as the giant ground sloth, Yesterday's camel, tapir, mammoths, saber-toothed cat, mastodons, American lion, and dire wolves. La Brea's special conditions have also preserved the fossils of over 100,000 birds, including huge condors and eagles, and the giant carnivorous teratorns. In addition, fossils of plants, snails, mice, frogs, and insects abound. With literally millions of fossils excavated, La Brea is one of the best-known fossil communities in the world. AB

SENTINEL DOME

CALIFORNIA, UNITED STATES

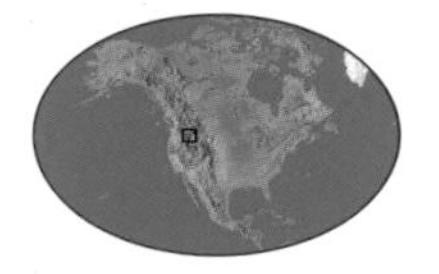

Height of Sentinel Dome: 8,122 ft (2,476 m)

Elevation from surrounding land: 460 ft (140 m)

Age: 150 to 210 million years (Triassic)

Sentinel Dome in Yosemite National Park is not a specific natural wonder in its own right, but climb to the top of this high granite dome and you will be given an extraordinary 360° view of the High Sierra mountains, giant sequoia groves, Yosemite Valley, and the waterfalls that makes America's first national park a true wonder to behold. The 460-foot (140-meter) climb up Sentinel Dome is relatively easy, especially on its northeast side, and will lead you to the second highest viewpoint in the park. From here it is not difficult to understand why Yosemite has inspired, and continues to inspire, artists, conservationists, and millions of casual tourists.

One of the most rewarding sights is the view of Yosemite Falls, the highest waterfall in North America, with a drop of 2,425 feet (739 meters). May is the best time to see the full-sized waterfall at its peak. Sentinel Dome is the exposed part of an enormous intrusion of igneous (granite) rock from deep within the Earth's mantle. Over time, erosion by glaciation and other natural forces has removed the overlying rock and peeled away successive layers of the granite dome as if it were an onion. JK

HALF DOME

CALIFORNIA, UNITED STATES

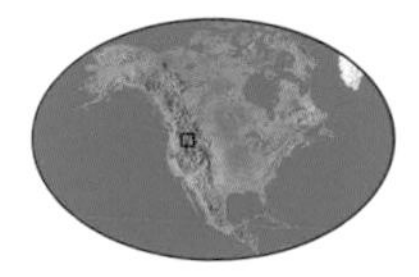

Height of Half Dome: 8,852 ft (2,698 m)

Length of Half Dome Trail: 17 mi (27 km), round trip

Trail trees: ponderosa pine, cedar, fir

With its vast, flat face cut by glaciers during the ice age, the impressive Half Dome in Yosemite National Park stands out among the other peaks in the area. Situated at the other end of Yosemite Valley from the famous falls, Half Dome is an enormous split monolith of granite named for its rounded back. On one side, a sheer rock wall, about 2,200 feet (670 meters) high, rises from the valley floor. A tourist trail to the summit spirals around the back of the mountain, although serious climbers go straight up the rock face and often get there quicker than the trekkers.

Day-hikers should leave at sunrise to be at Nevada Falls by 9 am, and the summit by 1 pm at the latest. The rounded summit is marked with a safety cable trail that is erected only between May and October. The trek down is difficult, although the John Muir Trail is suitable for those with wobbly legs. Those hikers in good shape take about 10 hours to make the round-trip, while the less physically fit can do it in 12 hours. Do not pass the lightning warning sign if there are thunderstorms—lightning strikes Half Dome at least once a month. The Yosemite Visitors' Center has the latest information on trail status. MB

BRIDALVEIL FALL

CALIFORNIA, UNITED STATES

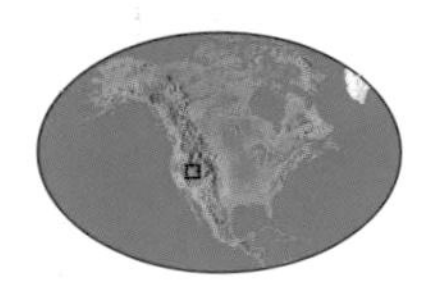

Height of Bridalveil Fall: 745 ft (227 m)

Annual rainfall: 36 to 47 in (900 to 1,200 mm)

Plunging down into the massive glacial ravine that is the Yosemite Valley, the Bridalveil Fall often appear to be falling sideways. This is due to the brisk winds that frequently blow across the sheer granite cliffs. Because of this, the Ahwahneechee Native Americans called this waterfall *Pohono*, which means "Spirit of the Puffing Wind."

The first Europeans to see the falls were probably The Mariposa Battalion. They were part of an Indian-hunting military expedition formed in 1851 that was formed to protect the rights of Sierra Nevada gold miners who had been attacked by the local Miwok natives defending their homeland.

The Bridalveil Fall are the first of several magnificent falls seen by the millions of visitors that visit the park every year. With heavy rain during most months, the falls flow all year round. Although they can be seen from the road at Bridalveil meadow, the best viewing spots are at the base of the falls from a vantage point just a few minutes walk from the road. On sunny afternoons, light catching the spray of the falls creates double rainbows.

The most famous viewpoint is from the overlook on the Wawona Road. From here, most of the park's iconic features—El Capitan, Half Dome—as well as Bridalveil Fall can be seen. DL

KINGS CANYON

CALIFORNIA, UNITED STATES

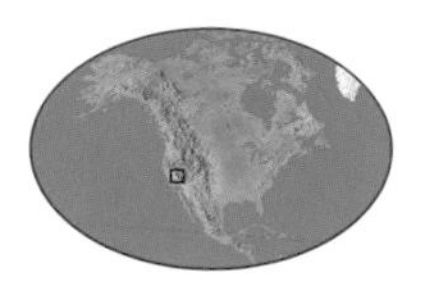

Area of Kings Canyon: 456,552 acres (184,748 ha)

Depth of canyon: 1,500 to 14,494 ft (457 to 4,418 m)

Rock type: metamorphosed ophiolite with granitic intrusions

Almost 8,200 feet (2,500 meters) deep, Kings Canyon is North America's deepest river canyon. Partly the result of erosion by the Kings River, and the action of glaciers over several ice ages scouring their way down the valley, the river drops 13,290 feet (4,051 meters) along its course, the greatest vertical drop for any river in the United States. Thunderously powerful after the spring melt, it continues to grind down the region's rock. Located in California's southern Sierra Nevada, Kings Canyon is mostly granite, with black pillow lavas and delicate green serpentinite offsetting beautiful, light blue-gray marble bands—the remnants of a seafloor uplifted 200 million years ago. Plant life includes alpine flower meadows, especially at Zumwalt Meadow on the Kings River shore, and groves of giant sequoia trees, including the General Grant tree, which since 1925 is the official U.S. national Christmas tree. Lying south of Yosemite, Kings Canyon forms the core of Kings Canyon National Park, a 1940 enlargement of the 1890 General Grant National Park. Visually stunning and with an enormous diversity of wildlife habitats, as well as over 800 miles (1,287 kilometers) of hiking trails, the park is contiguous with Sequoia National Park and the two are managed as a single unit. **AB**

LAKE TAHOE

CALIFORNIA / NEVADA, UNITED STATES

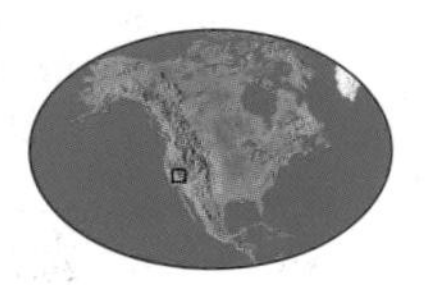

Length of Lake Tahoe: 22 mi (35 km)

Width of lake: 12 mi (19 km)

Volume of lake: 39 trillion gallons (148 trillion liters)

Lake Tahoe is unquestionably one of the most beautiful lakes in the world. Situated high up in the Sierra Nevada Mountains, the clear blue waters of this pristine alpine lake are ringed by stunning snowcapped peaks. The effect is a masterpiece of nature. The clarity of Lake Tahoe is extraordinary—it is possible to see to depths of 75 feet (23 meters). Most of the water in the lake comes from snowmelt and rainwater flowing in from 63 streams, and because of its 6,220-foot (1,896-meter) elevation, no sediment-filled rivers disturb its crystalline appearance. The water drains through lakeside marshes and meadows that act as water filtering systems, which helps to preserve its purity. Lake Tahoe is also very deep, reaching 1,640 feet (500 meters) at its greatest depth, making it the second deepest lake in the United States, after Oregon's Crater Lake. If drained, it would take 700 years for the lake to be naturally refilled.

"Tahoe" is derived from a Native American word meaning "big water." The lake was created between 5 and 10 million years ago when two parallel faults on either side of the valley where the lake now sits shifted and uplifted, causing the valley floor to drop by several thousand feet. A river flowed north through this sunken valley until an enormous lava flow from a volcanic eruption blocked its northern exit. With nowhere to go, the water filled the deep valley over thousands of years. Ancient waterlines on rocks high above the lake basin show that the lake level was once 800 feet (244 meters) higher than today. A small river called the Truckee River eventually forced its way around the volcanic debris and remains today the only outlet for the lake.

Situated high up in the Sierra Nevada Mountains, the clear blue waters of Lake Tahoe are ringed by stunning snowcapped peaks. The effect is a masterpiece of nature, and the lake's cool waters and surrounding forests have become a recreational paradise.

The cool waters of Lake Tahoe and its surrounding mixture of conifer and hardwood forests have become a recreational paradise for thousands of visitors from Nevada and California—the border between the two states runs right through the center of the lake. The Sierras are filled with long hiking trails and old logging roads for mountain biking. Numerous boating facilities along the lake shoreline offer excursions, fishing, and tours of the lake. Its beaches are also considered to be world-class, and in winter the surrounding mountain resorts play host to some of the best skiing in America. JK

RIGHT *The clear blue shallows of Lake Tahoe.*

JOSHUA TREE NATIONAL PARK

CALIFORNIA, UNITED STATES

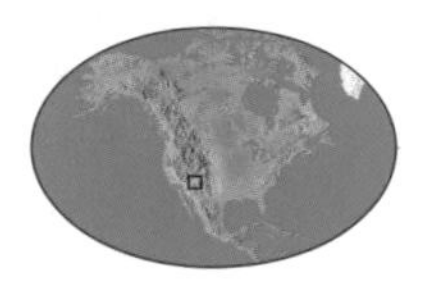

Area of national park: 789,745 acres (319,600 ha)

Maximum elevation (Key's View): 5,185 ft (1,580 m)

Habitat: low cactus desert, high cool desert, juniper-pinyon scrub, oases

Joshua Tree National Park embraces two desert types: the Colorado (the park's low, hot, dry eastern portion) and the Mojave Desert (the higher, cooler, moister, western part). The Colorado has creosote bush, ocotillo, and cholla cactus. The Mojave has Joshua Tree forests. At an altitude of 4,000 feet (1,200 meters), cool juniper-clad canyons occur. The park's five fan palm oases host abundant wildlife, especially migrant warblers, including Nashville, MacGillivray's, and orange-crowned warblers. An important part of the Pacific Flyway, Joshua Tree has distinct sets of summer and winter bird visitors, while residents include roadrunners, phainopeplas, verdins, cactus and rock wrens, burrowing owl, Le Conte's thrashers, Gambel's quail, and prairie falcon. Joshua Tree also has bighorn sheep, rattlesnakes, bobcats, jackrabbits, kangaroo rats, tarantulas, scorpions, and clouds of migrating butterflies. The park protects 501 archeological sites including rock paintings, covering the 5,000-year history of human occupation, including Pinto, Chemehuevi, and Cahuilla peoples. Renowned for fascinating geology, including strange granite spheres and beautiful erosion fans, it is a spectacular place to view meteor showers. For magnificent displays of ephemeral desert wildflowers, visit in March or early April. It was established as a national park in 1994. The nearest town is Twentynine Palms, which is three hours from Los Angeles. AB

GIANT REDWOODS

CALIFORNIA, UNITED STATES

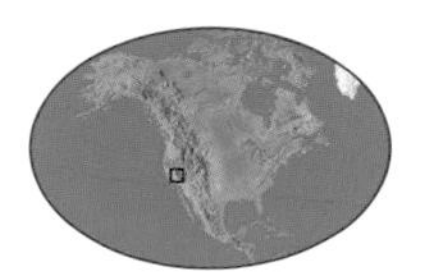

Height of Redwood trees: up to 370 ft (112 m)

Weight of trees: up to 3,300 tons

Locality: northern-Pacific coast of North America

Redwood trees are members of the yew family, and are probably the largest living things on Earth. There are three kinds: the coast redwood, giant redwood (also called Sequoias), and dawn redwood. The first two occur in California and include, respectively, the world's tallest, and most massive trees. Native to China, dawn redwoods rarely exceed 200 feet (60 meters). Coastal redwoods occur in the fog belt of coastal California and Oregon. Giant Redwoods only occur on California's Sierra Nevada mountains.

California's unique coastal climate of fog and rain helps these enormous trees grow to great heights. In 2002, the tallest living Giant Redwood measured 369.5 feet (112.6 meters) —an amazing 54 feet (16 meters) higher than the Statue of Liberty. An estimated 800 to 1,000 years old, it grows near Ukiah, California. It is surrounded by several others exceeding 350 feet (107 meters). The widest is the Del Norte Titan which has a basal diameter of 23.6 feet (7.2 meters). However, because its trunk is broader overall, Sequoia National Park's tall 275-foot (84-meter) General Sherman Giant Redwood is largest by volume. It weighs an estimated 2,000 tons. The biggest redwood ever recorded weighed some 3,300 tons and blew over in a storm in 1905. **AB**

MONO LAKE AND CRATERS

CALIFORNIA, UNITED STATES

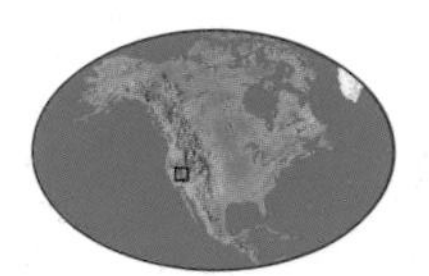

Area of Mono Lake: 70.5 sq mi (183 sq km)

Elevation of lake: 6,390 ft (1,948 m)

Vegetation: sagebrush

Located in the Great Basin drylands, Mono Lake is the last remnant of Lake Lahontan, one of two large lakes that once flooded the area during the ice age. Nearby are the Mono Craters, a long line of over 20 extinct 1,000-year-old volcanoes, each with a tiny central lake. Today, Mono Lake collects salts eroded from several hundred miles around, making it three times saltier than seawater.

Mono's best features are the tufa towers. Reaching up to 33 feet (10 meters) these are formed when fresh water, acidified from passage through the Mono Craters' volcanic deposits, wells up in the alkaline lake. This causes calcium carbonate dissolved in the lake's waters to precipitate and accrete in alien-looking white and gray limestone towers. Though it appears barren, Mono Lake is one of North America's most productive ecosystems. An annual algae bloom feeds brine shrimp and brine flies, attracting over 80 bird species, from both the Equator and the Arctic Circle, including eared grebes and 80 percent of the world's Wilson's phalarope. At 760,000 years, Mono is North America's oldest lake. **AB**

RIGHT *The unusual tufa towers of Mono Lake.*

CHANNEL ISLANDS

CALIFORNIA, UNITED STATES

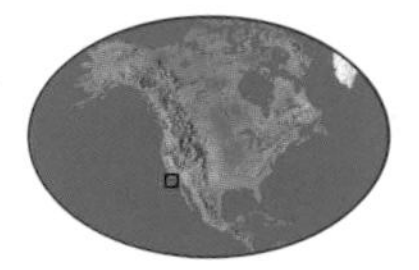

Area of Channel Islands National Park: 249,561 acres (598,946 ha), half of which is ocean

Rock type: volcanic origin islands, with sea caves, lava tubes, rock pools

A chain of eight islands off the coast of southern California, the Channel Islands have been called "America's Galapagos." Over 2,000 species of plants and animals have so far been recorded there. One hundred and forty-five of these are unique to the islands, including four species of mammals. The islands of Anacapa, Santa Barbara, Santa Cruz, San Miguel, and Santa Rosa make up the Channel Islands National Park. Providing fine hiking trails, the park protects the rich fauna, flora, and archeological sites. October to March is the best time to visit, with migrating grey whales and spectacular wildflower displays. Each island has its own specialities: Anacapa and Santa Barbara have whale and bird watching, scuba and snorkeling; San Miguel features bird and seal watching, wildflowers, and fossil forests; Santa Rosa offers beautiful sea kayaking; and Santa Cruz has fossils and good general wildlife observation. The seas are cold—offshore, the marine communities are rich in fish and marine invertebrates; they host resident sea lions and seals, as well as visiting whales and dolphins. The nearest mainland town is Ventura, which is 70 miles (112 kilometers) north of Los Angeles. **AB**

DEATH VALLEY NATIONAL PARK

CALIFORNIA, UNITED STATES

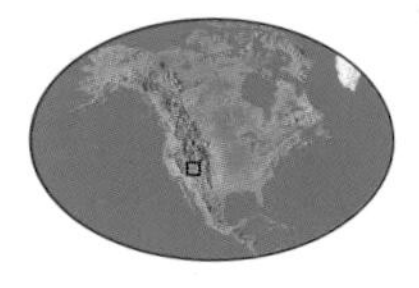

Hottest recorded temperature: 134°F (57°C)

Lowest point: 282 ft (86 m) below sea level

Highest point: 11,049 ft (3,368 m) above sea level

Death Valley in southeastern California is one of the hottest places on Earth. It is also the driest place in North America and contains the lowest point in the western hemisphere. The valley is a large 156-mile (250-kilometer) trough between the Amargosa and Paramint mountain ranges. Despite its forbidding name, the landscape of saltpans, sand dunes, canyons, and mountains has a rugged beauty and is home to a surprising variety of plants and animals that have adapted to its harsh environment.

The terrain of this huge national park rises dramatically from its lowest point in the Badwater Basin saltpan to the highest peak in the park—Telescope Peak, only 15 miles (24 kilometers) away. Although the vegetation is sparse on the valley floor and lower slopes, there is more abundant vegetation where water is available at higher altitudes. The larger mammals found in the park include desert bighorn, mountain lion, and bobcat, along with more numerous smaller (largely nocturnal) species. There are also many species of desert reptiles—and even pupfish that can survive in hot water pools. The best time to visit the park is from October to April because the summers can be unbearably hot, averaging over 100°F (38°C). RC

VERNAL POOLS

CALIFORNIA, UNITED STATES

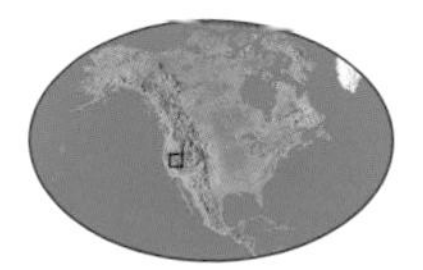

Habitat: seasonal plant-rich freshwater pools

Oldest pool: 100,000 years

Vernal Pools are a special seasonal habitat with many unique species of flowers and insects. They require a short, very wet winter followed by 8 to 10 hot dry months, flood prone grasslands, and areas of impervious soil to promote pool formation. Such conditions occur only in the western United States, parts of Chile, Australia, South Africa, and southern Europe. Formerly extensive, California's vernal pool communities are now one of the world's rarest and most threatened ecosystems. The pools are stable once formed, and some are 100,000 years old. They lie over hardpan formed from million-year old volcanic eruptions. Individual pools have endemic species of plant and freshwater shrimp.

California's pools support 200 plant species, half of which occur nowhere else. Specialized beetles and solitary bees are the most common pollinators. These too are unique to the vernal pool system. Endemic plants include frayed downingias, Fremont goldfields, meadowfoam, and delta wooly-marbles. These and other plants adapt to the moisture conditions of each vernal pool, and the best time to visit is February to May, when each pool blooms in concert providing a spectacular display of concentric rings of color. The best places to see the pools are Mather Field near Sacramento and Jepson Prairie Reserve. Guided tours are given at both sites. AB

BRISTLECONE PINE—WHITE MOUNTAINS

CALIFORNIA, UNITED STATES

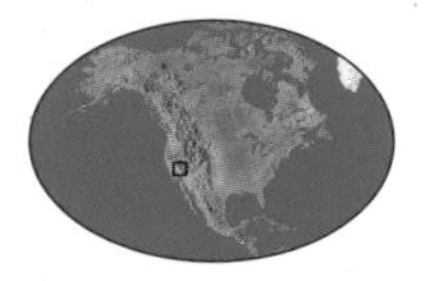

Oldest tree: the Methuselah tree, over 4,700 years old

Largest bristlecone: the Patriarch tree, 36 ft (11 m) in girth, but possibly two trees that have grown together

Resembling a growing rock rather than a plant, the bristlecone pine is the longest-living tree on the planet. These trees grow in the White Mountains, just east of the Sierra Nevada. The oldest specimens alive today have managed to survive in this inhospitable environment for almost five millennia.

Among the pristine, ultraviolet-drenched lunar landscape, young trees flourish, their branches heavily clothed in needles and resin-rich bristled cones exuding a pine-fresh scent. As they age, time and the elements take their toll, battering, sand-blasting, and polishing the hardiest specimens for century upon century. It is here that the secret of the pine's longevity lies, for those trees growing in the toughest environments do so only very slowly, forming extraordinarily dense wood in the process. When these bristlecones finally expire, their weathered remains stand for another thousand years or more, until at last they are eroded away by the wind and the ice.

Bristlecones have played a key role in helping to understand past climates. Tree ring chronologies based on living and dead bristlecones have enabled the construction of a record of change stretching back over 9,000 years.

The work was pioneered by Dr. Edmund Schulman, a scientist at the University of Arizona, who first discovered the bristlecones' staggering lifespan in the 1950s. Because climatic differences from one year to the next cause a distinct pattern of tree growth, which can be observed in cross-section through the trunk, it is possible to infer relative past growing conditions. By matching recent sections of dead trees with early sections from living ones, the chronology can be extended back much further.

In recognition of Schulman's pioneering work, a grove of these unusual trees has been named in his honor. Schulman's Grove contains the first tree dated at over 4,000 years of age. Nearby, a visitor center, picnic area, and amenities allow the curious traveler to explore this strange, beautiful landscape. A series of self-guided walks through the heart of the Ancient Bristlecone Pine Forest—a designated botanical area—can be found in and around Schulman Grove. NA

Resembling a growing rock rather than a plant, the bristlecone pine is the longest-living tree on the planet. The oldest specimens alive today have survived almost five millennia.

RIGHT *The gnarled trunk of a bristlecone pine.*

MOUNT LASSEN

CALIFORNIA, UNITED STATES

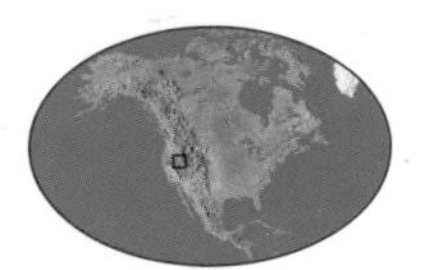

Area of Mt. Lassen: 106,372 acres (43,049 ha)

Height of Mt. Lassen: 10,453 ft (3,186 m)

Rock type: volcanic

Mount Lassen is almost entirely encircled by the remains of Tehama, a huge volcano that erupted about 350,000 years ago. Lassen arose to relieve geological pressures that are still present and last erupted in 1915. Both peaks stand within the collapsed caldera of Maidu, an extinct larger volcano. Lassen is covered in volcanic glass, or obsidian, and was once sacred to the region's Yahi people, who annually migrated to its slopes to escape the lowland summer heat.

Lassen is part of the High Cascade system, which also includes California's Shasta, Washington's Rainier, and Oregon's Hood, all of which began erupting several million years ago above an ocean floor slab that is still sinking into a trench off California's northern coast. The volcano is the kingpin of Lassen Volcanic National Park, which contains painted dunes, mudpots, gas vents, lava flows of various types, and a variety of volcanic cones. The most beautiful lava flows were exposed by the same Pleistocene glacial action that carved Emerald Lake, at the foot of Lassen peak. There is an extensive trail system, along with over 700 plant species and 250 vertebrate animals. **AB**

BEACHES OF SOUTHERN CALIFORNIA

UNITED STATES / MEXICO

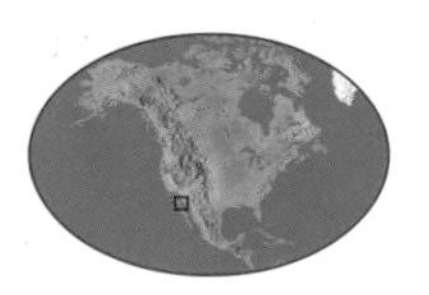

Main runs of the grunion fish: between Point Conception, California, and Point Abreojos, Baja California

Number of eggs deposited: 1,600 to 3,600 eggs during one spawning run

In the spring and summer months, from March through to August, millions of small silvery fish appear along the high tide line of most of the beaches of Southern California. The fish are the 6-inch (15-centimeter) long grunion, and appear two to six nights after the full or new moon in order to reproduce.

The grunion is the only fish that completely leaves the water to deposit its eggs. The spawning run takes place during the highest tides when the fish use the waves to travel as far up the beach as they can. The female first digs into the sand and waits with just her head protruding until a male curls his body around her to simultaneously produce eggs and milt. The male is then swept away, but the female may wait and have her eggs fertilized by more than one male. She finally emerges and is carried back to the sea on the next wave. The eggs are deposited in the nest on the highest tides, and incubate in the sand during the lower tides. They are washed out at the next very high tide. **MB**

MONUMENT VALLEY

UTAH / ARIZONA, UNITED STATES

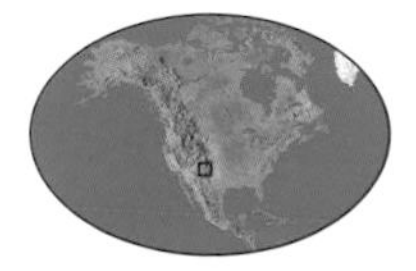

Elevation of Monument Valley: 5,200 ft (1,585 m)

Area of valley: 30,000 acres (12,141 ha)

Average rainfall: 8 in (200 mm)

The spellbinding scenery of Monument Valley has captivated the imagination of almost everyone who has ever seen a classic western movie. Here, amid the magnificent red sandstone buttes and mesas, Hollywood has shot some of its greatest films, such as *Stagecoach*, making this place one of the most famous landscapes in the world. Monument Valley is part of the Navajo Nation Tribal Park, which straddles the border between Arizona and Utah.

Fifty million years ago this was one solid plateau of hard sandstone, interspersed with several volcanoes. Over time, erosion by wind and water cut into and peeled away the surface of the plateau. The softer rock eroded, leaving behind the harder buttes and mesas. The volcanoes were leveled, and today only their hardened igneous cores remain as fantastic monoliths, up to 1,500 feet (450 meters) high.

Approach Monument Valley from the north for the most famous view. A Navajo guide will take you along a straight, empty road leading toward the 1,000-foot- (300-meter-) high, stark, red cliffs on the horizon. The highlight is the Totem Pole, a spire of rock 300 feet (90 meters) high, but only 6 feet (1.8 meters) wide. JK

ARCHES NATIONAL PARK

UTAH, UNITED STATES

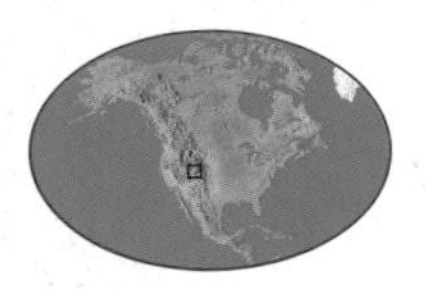

High desert elevation: between 6,430 ft (1,960 m) and 8,858 ft (2,700 m)

Designated as a national monument: 1929

The world's greatest diversity of natural arches and other geological formations has made Arches National Park in Utah the jewel of a unique region. Located in a high desert, extremes of temperature, water, and wind erosion have finely sculpted the region's multi-hued sandstone into no less than 2,400 named arches, as well as a variety of unusual balanced rocks, monoliths, and pinnacles. At Arches, a hole in the rock must be at least three feet (0.9 meters) across for it to be officially listed and mapped as such.

Delicate Arch is the symbol of the park. It is, perhaps, one of the most iconic American landmarks and has appeared in an endless array of books, films, postcards, and calendars. It has a span of approximately 33 feet (10 meters) and is 50 feet (15 meters) high. The twin arches of Double Arch, seen in many western films, crisscross a gap between two rocky outcrops at a height of 150 feet (45 meters). Although not as well known, the nearby Landscape Arch spans an incredible 290 feet (88 meters). It was here in 1991 that a more dramatic development in the park's geological evolution occurred. A massive slab of rock 66 feet (20 meters) long, 10 feet (3 meters) wide, and three feet (0.9 meters) thick plummeted from the underside of the arch, leaving a relatively thin ribbon of rock to support the arch. In terms of geological chronology, Landscape Arch's days are numbered.

Tourism has only recently been introduced to the area, but there is evidence that people lived here thousands of years ago. Paleolithic hunter-gatherers migrated to the area about 10,000 years ago and discovered the rock-type perfect for making stone tools. No remains of ancient habitations have been found in Arches, although inscriptions and petroglyphs have been discovered carved into the region's rocks. The first Europeans in the area were Spaniards who arrived in the 18th century and began marking trails between their various regional missions. Permanent settlements, however, were not established until the late 19th century.

Today, numerous four-wheel-drive tracks and walking trails weave in and out of the arches and canyons. Note that travels in this harsh backcountry are not to be undertaken lightly, as water is scarce and the temperature often reaches a scorching 104°F (40°C) during the summer months. **DL**

The world's greatest diversity of natural arches and other geological formations has made Arches National Park in Utah the jewel of a unique region. Extremes of temperature, water, and wind erosion have finely sculpted no less than 2,400 named arches.

RIGHT *The famous Delicate Arch in Arches National Park.*

GREAT SALT LAKE

UTAH, UNITED STATES

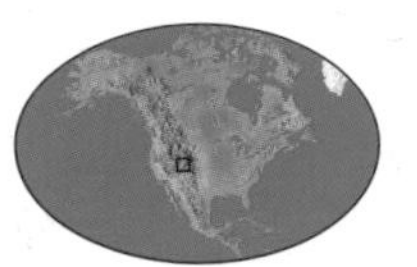

Size of Great Salt Lake: 1,616 sq mi (4,184 sq km)

Length of lake: 75 mi (121 km)

Width of lake: 35 mi (56 km)

Great Salt Lake is the largest lake west of the Mississippi, but this briny body of water is actually the remnant of a much larger prehistoric ice age lake called Lake Bonneville. As Lake Bonneville evaporated, the dissolved salts in its water became increasingly concentrated. Today water fails to flow out of Great Salt Lake because it sits in a depression in Utah's Great Basin. Consequently, the lake resembles the ocean in composition rather than that of any freshwater lake. The amount of dissolved salt in Great Salt Lake is nearly five billion tons, so much in fact that in places, especially in the northern arm of the lake, one can float with ease on the water surface.

Great Salt Lake is a wildfowl paradise. Ducks, geese, gulls, pelicans, and dozens more species live in the marshes and wetlands surrounding the lake. Their numbers reach into the millions. The lake is loaded with tiny brine shrimp—providing the main source of food for the birds. To the west of the lake are the Bonneville Salt Flats, a wide salt-covered lakebed, and one of the flattest places on Earth. It was here that Gary Gabolich's rocket car, "Blue Flame," attained a spectacular 622.4 miles per hour (1,001 kilometers per hour). JK

CANYONLANDS NATIONAL PARK

UTAH, UNITED STATES

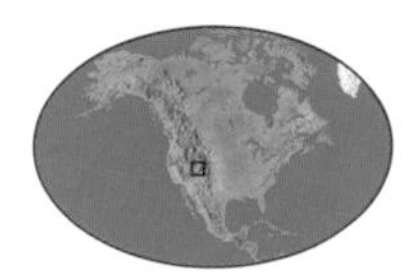

Depth of Canyonlands National Park: over 2,133 ft (650 m) deep

Highest point (Cathedral Point): (Needles district) 7,120 ft (2,170 m)

Canyonlands National Park preserves a rugged landscape of colorful sandstones eroded into a showcase of geological wonders. Rivers divide the park into four distinct areas: The Island in the Sky, the Needles, the Maze, and the rivers themselves. The park's artifacts suggest the presence of inhabitants 10,000 years ago, while official exploration of the Colorado and Green rivers occurred in 1869.

The area formed when material was deposited from a variety of sources over hundreds of millennia. Movements in Earth's crust altered surface features, and the North American continent slowly migrated north from the equator, changing the environment. Current-day Utah was flooded by shallow inland seas, covered by mudflats, and buried by sand dunes, forming layers of sedimentary rocks. Movements in the planet's crust subsequently caused the area to rise. The Colorado and Green rivers then began to cut deep canyons which were filled with sediment from storms, eroding the landscape into a labyrinth of tributary canyons and washes that mark the landscape today.

When traveling to the remote and unforgiving Canyonlands make sure you are suitably equipped and plan well ahead. The Maze area is only navigable by four-wheel-drive vehicles from rough roads to the west of the park. DL

DEAD HORSE POINT STATE PARK

UTAH, UNITED STATES

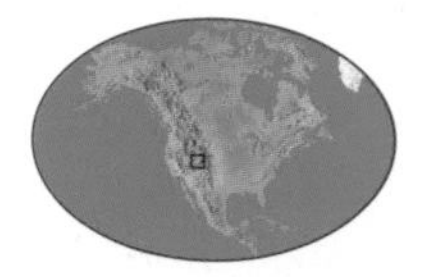

Elevation of Dead Horse Point State Park: 6,000 ft (1,800 m)

Height of Dead Horse Point: 2,000 ft (600 m)

Area of park: 5,362 acres (2,170 ha)

Herds of wild mustang horses once lived on the flat-topped mountain of Dead Horse Point. Its topography provided a natural corral into which cowboys could drive the horses. Once the horses were roped and broken, the best were sold and the remainder ("broomtails") were let go. Legend has it that one group of broomtails was accidentally left corraled here, and all died of thirst, within sight of the Colorado River below. Their skeletons bequeathed a name to this scenic wonder. Dead Horse Point is 23 miles (37 kilometers) south of Moab, Utah, and was designated a state park in 1959. It has the most spectacular views of all of Utah's state parks. The lookout at the Point towers 2,000 feet (600 meters) above the surrounding plateau and provides a magnificent panorama of nearby Canyonlands National Park, where canyon erosion has occurred on a grand scale. The spires and bluffs in the distant landscape are all the product of 150 million years of slow erosion by the Colorado River, which meanders in a big gooseneck directly below Dead Horse Point. JK

ZION CANYON

UTAH, UNITED STATES

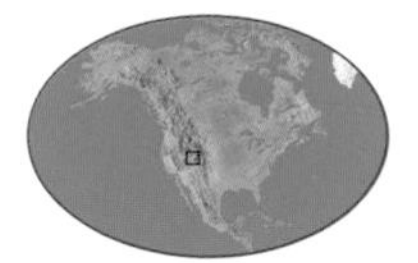

Length of Zion Canyon:	15 mi (24 km)
Width of canyon:	1,319 ft (402 m)
Annual rainfall:	15 in (38 cm)

Zion Canyon is a magnificent chasm with steep vertical walls that cut through the soft red sedimentary rocks of southwestern Utah. The canyon's depth is such that sunlight rarely reaches the bottom. It has been carved out by the North Fork of the Virgin River over the past four million years. Geologists estimate that the Virgin River has the power to cut down another thousand feet into the bedrock.

The canyon is sanctuary-like, with lush hanging gardens, waterfalls, and impressive sandstone columns and rock pyramids such as East Temple. Moreover, the Great White Throne is a towering rock rising 2,460 feet (750 meters) from the canyon floor.

As Zion National Park's largest canyon, Zion Canyon has numerous superb hiking trails that range from easy walks to extremely difficult technical climbs. When the water level is low, visitors can hike through the Narrows at the head of the canyon. But care must be taken—the canyon walls are so close to each other that the water level may rise 26 feet (8 meters) during a flash flood. A climb to Angels Landing will provide breathtaking views of the cliff and canyon landscape that make this place the wonder that it is. JK

NATURAL BRIDGES

UTAH, UNITED STATES

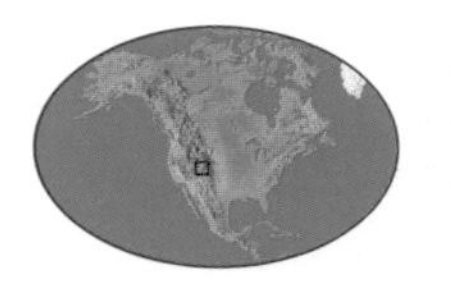

Span of Sipapu: 268 ft (82 m)

Span of Kachina: 204 ft (62 m)

Span of Owachomo: 180 ft (55 m)

There are three natural bridges in this area: Sipapu, Kachina, and Owachomo. The largest and most spectacular is Sipapu, meaning "the opening between worlds" in the Hopi language. Its smooth sides are the result of erosion by endless floods that carried rocks and sand downstream. Four thousand tons of sandstone fell from Kachina ("middle bridge") in June 1992, showing just how fragile these bridges are. Ancient rock paintings have been discovered at its base. The smallest and thinnest bridge is Owachomo ("rock mound," which refers to the rocks at one end of the span). It is the most elegant but fragile of the three natural arches. Each represents a stage in the life of a natural bridge, Sipapu being the youngest and Owachomo the oldest.

Natural Bridges was designated a national monument by President Theodore Roosevelt in 1908, making the area Utah's first national park. If you are lucky, you may see bobcats, coyote, bears, or mountain lions during your visit. A 9-mile- (14.5-kilometer-) long road overlooks all three bridges, offering fantastic views of the area, and the monument is located about 42 miles (68 kilometers) west of Blanding. MB

HOODOOS

COLORADO / UTAH, UNITED STATES

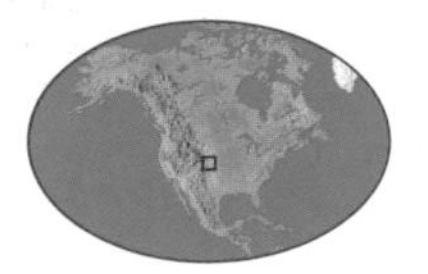

Bryce Canyon National Park hoodoo highlights: Wall of Windows, the Chessmen, Tower Bridge, the Poodle

Zion Canyon National Park hoodoo highlights: Checkerboard Mesa, East Temple Mesa, Weeping Rock, Kolob Arch

A hoodoo is a slender pinnacle of soft shale or mudstone that is capped by a harder, more resistant rock layer, usually sandstone or limestone. They generally occur in groups during the dissection of a plateau. As the surrounding softer layers are eroded by water and wind, the cap protects the material immediately beneath, creating a vertical spire. Exceptionally tall hoodoos, many over 100 feet (30 meters) high are a conspicuous feature of Utah's Bryce Canyon and Zion National Parks, although they are also fairly common in badlands topography all across Utah and Colorado to the west of the Rocky Mountains.

Native Americans likened the hoodoos to human figures. An old Apache fable recounts how the Creator let loose a great deluge of rain because He was upset with the world and its people, and decided to start anew. He favored the Apache, and was willing to grant them a reprieve. However, a group of selfish men took advantage, and rushed up to the mountains without a thought to help the children, elders, and women escape from the approaching flood. The Creator was so enraged that He punished those that abandoned their tribe by turning them into stone. DL

BRYCE CANYON

UTAH, UNITED STATES

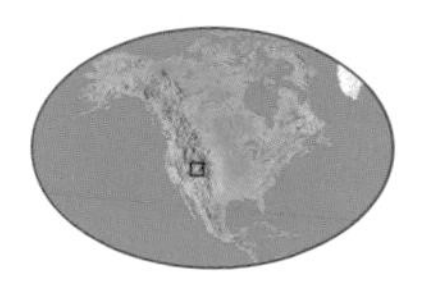

Alternative name: Unka-timpe-wa-wince-pock-ich

Area of canyon: 56 sq mi (144 sq km)

Highest point (Rainbow Point): 9,105 ft (2,775 m)

Bryce Canyon's landscape of brightly colored rock pinnacles, canyons, and ravines is a breathtaking spectacle. Here is a geological fantasyland of bizarre rock formations known as fins, windows, slot canyons, and tall totem-pole-shaped hoodoos. The canyon consists of a series of horseshoe-shaped amphitheaters carved from the eastern edge of the Paunsaugunt Plateau in southern Utah. The extraordinary rock formations were eroded by water, ice, and snow. The canyon rim is still receding by about 12 inches (300 mm) every 50 years. Weathering of minerals in the rocks has produced a wide range of colors that change throughout the day: red and yellow come from oxidized iron, while blue and purple are caused by oxides of manganese.

The best time to visit is early or late in the day when the shadows are long and the colors appear to glow. The park is open throughout the year but exploring the canyon on foot is best from May to October. The local Paiute name for it translates as "red rocks standing up like men in a bowl-shaped canyon." The area was named after Scottish pioneer Ebenezer Bryce who built a ranch in the canyon in the 1870s and once memorably described it simply as "a hell of a place to lose a cow." **RC**

THOR'S HAMMER

UTAH, UNITED STATES

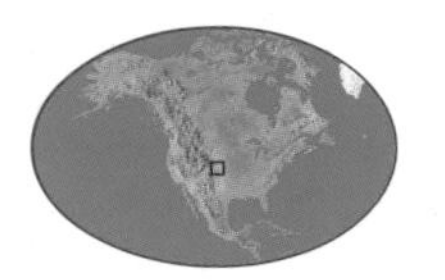

Height of Thor's Hammer: 150 ft (45 m)

Age of rock: 64 million years

According to ancient Viking mythology, Thor, the Norse God of thunder, would create earthquakes and volcanic eruptions by striking Earth with his mighty hammer. So it is rather fitting that one of the tallest stone columns, or hoodoos, in Bryce Canyon is named Thor's Hammer. This amazing pinnacle of rock is just one of hundreds of hoodoos here. It stands 150 feet (45 meters) high. Its top is capped with a large piece of hard mudstone—the hammerhead. The column below—the handle—is a narrow pillar of softer limestone. The harder hammerhead protects the soft handle from the erosional impact of stone-splitting frosts and falling rain.

Thor's Hammer is a favorite among visitors. It stands alone on a ridge, not far from Sunset Point. Most of the other hoodoos here are packed tightly together in a city-like landscape of colorful spires, but Thor's Hammer draws attention to itself, its hammerhead sitting precariously atop a high, thin pillar. In time, however, the power of erosion will prevail, the softer handle will crumble, and the hammerhead will come crashing to the ground. Thor's Hammer will have struck Earth for the last time. JK

THE MITTENS

UTAH / ARIZONA, UNITED STATES

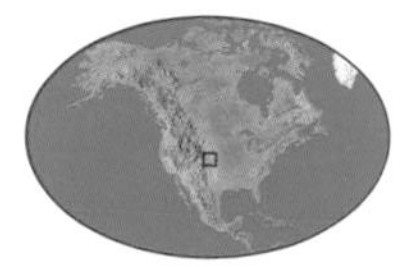

Composition of base: organ rock shale

Composition of middle: de Chelly sandstone

Composition of cap: shinarump siltstone

The Mittens are the iconic image of the world-famous Monument Valley. These two red buttes, named East and West Mitten, stand 1,000 feet (300 meters) above the flat, barren Colorado Plateau. They were named, appropriately, because they look like a pair of giant stone mittens, complete with a "thumb" of rock separate from the main "hand."

The buttes are composed of three main layers of sedimentary rock. The lowest layer is a form of shale called organ rock that erodes out in tiered horizontal terraces, forming a rounded scree slope around the foundation of the butte. The middle layer is a great vertical section of soft sandstone that would easily erode away if it were not protected by the harder layer of Shinarump siltstone that caps the monument. The best time to see the Mittens is at sunrise and sunset, when their stone surfaces glow a beautiful red color with the crimson light of the sun. Next to the Mittens is a third butte, Merrick Butte, named after a prospector who came to the valley in search of a lost silver mine. Legend says that he found the silver but lost his life at the hands of Navajo warriors who had previously warned him to stay off their land. JK

DINOSAUR NATIONAL MONUMENT

UTAH / COLORADO, UNITED STATES

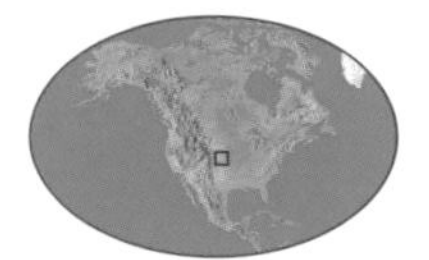

Area of Dinosaur National Monument: 325 sq mi (842 sq km)

Elevation of monument: 4,500 to 7,000 ft (1,372 to 2,134 m)

Number of fossils on view: 1,600

In 1909 Earl Douglas traveled to the sedimentary soils of Utah's northeastern plateau in hopes of discovering dinosaur bones. Here Douglas found so many dinosaur fossils that President Woodrow Wilson declared the area a national monument. Containing thousands of dinosaur bones from 150 million years ago, more than half of all dinosaur species known to have lived in North America at that time are now found here.

During the Jurassic era, the area was a 200-foot- (60-meter-) long sandbar in a river, and the surrounding area was home to many dinosaurs. When they died, their remains accumulated on the sandbar where mud and silt slowly covered them. Over time silica minerals seeped into the dinosaur bones, turning them to rock and preserving their features. There they lay buried until 70 million years ago, when an upheaval of the Rocky Mountains tilted the land up to reveal their fossilized bones.

A museum, the Quarry Visitors Center, has been built around the dig where Douglas excavated his fossils. Visitors can watch paleontologists working on the fossilized bones and can visit the laboratory where fossils are treated. This is the most productive Jurassic dinosaur quarry in the world. JK

BLACK CANYON OF THE GUNNISON

COLORADO, UNITED STATES

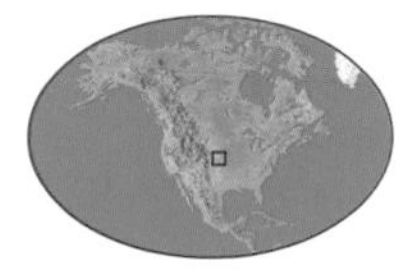

Depth of Black Canyon of the Gunnison: 1,970 ft (600 m)

Width of canyon: 1,476 ft (450 m)

Length of canyon: 12 mi (20 km) unspoiled by dams

The Black Canyon of the Gunnison's rugged landscape formed slowly when water and tumbling rock scoured down through hard crystalline rock. No other canyon in North America is as narrow or can equal its sheer vertical walls. It is named Black Canyon because very little direct sunlight penetrates its depths.

In 1901 Abraham Lincoln Fellows and William W. Torrence made the first successful expedition through the Black Canyon. Today the gorge is the domain of expert kayakers and rafters. Where it cuts through the gorge, the Gunnison River is categorized as a severe Class V rapid and has claimed the lives of many unwary and inexperienced adventurers.

The canyon was once an impressive 50 miles (80 kilometers) long but three dams have been built upstream, which has left just 12 miles (20 kilometers) unspoiled. The roads around the canyon rim feature several overlooks down to the strangely frightening dark, jagged, and menacing rocks below. There are three difficult trails leading down the cliffs to the river itself. Approach the river from the north or the south, but the easiest access is along the south rim. DL

GREAT SAND DUNES NATIONAL PARK

COLORADO, UNITED STATES

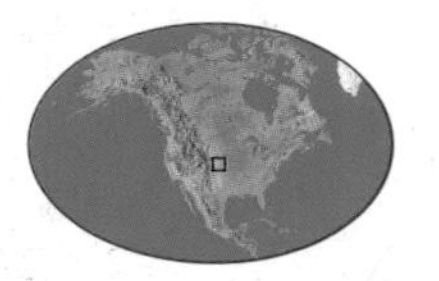

Area of Great Sand Dunes National Park: 35 sq mi (91 sq km)

Age of sand dunes: 12,000 years

Highest dune: 750 ft (229 m)

To encounter the giant, golden sand dunes of southern Colorado is as surprising as it is uplifting. These wind-sculpted hills rise up suddenly, over 700 feet (213 meters) from the floor of the San Luis Valley, between the Rio Grande and the Sangre de Cristo mountains. They are the tallest dunes in the United States, covering an area of over 35 square miles (90 square kilometers). They are particularly breathtaking to see in the early morning and early evening when sunlight highlights their sinewy outlines and enhances their rich golden color.

To encounter the giant, golden sand dunes of southern Colorado is as surprising as it is uplifting. They are particularly breathtaking to see in the early morning and early evening, when the sunlight highlights their sinewy outline and rich golden color.

Scientists believe that dune formation began in the Pleistocene era when glaciers formed in mountain valleys, carrying ice and rock far into the San Luis Valley. Twelve thousand years ago, a warming climate melted these glaciers, creating rivers and streams which carried more large quantities of silt, gravel, and sand into the San Luis Valley. Prevailing winds from the southwest mountain passes of Music, Medano, and Mosca carried the sand to the eastern edge of the valley where the Sangre de Cristo Mountains acted as a barrier, forcing the winds to lose steam and release their clutches of sand. Winds continue to carry new sand into the valley today, which means the dunes are gradually being enlarged. However, unlike most sand dunes, they keep their shape because, amazingly, they are quite moist and dense under their surface. That is because they act like a sponge, sucking water from the high water table and nearby creeks.

The best way to enjoy the dunes is to climb them, but care must be taken since the surface of the sand can be extremely hot; it can easily reach temperatures of over 100°F (38°C) in the summer months. The tallest dune is easy to reach, only half a mile from the edge. There are no marked trails, so visitors are free to choose their own paths through the dunes. One of the dunes' extra attractions is Medano Creek, a small snow-fed stream from the Sangre de Cristo mountains that flows along the eastern border of the dunes during the spring. For several hundred feet the creek babbles over a stretch of sand in an unpredictable manner. In one place it will flow freely, over a foot deep, and then it will suddenly disappear into the sand, only to pop up again several feet away. JK

RIGHT *Despite their barren appearance, the sand dunes support a varied ecosystem of plants and wildlife.*

FLORISSANT FOSSIL BEDS

COLORADO, UNITED STATES

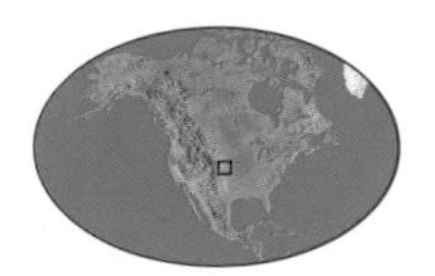

Age of Florissant Fossil Beds: 34 million years

Sedimentary type: shale

Number of insect species: 1,100

The Florissant Fossil Beds lie in a high mountain valley near Colorado Springs. They contain preserved fossils of plants and animals that lived here 34 million years ago. From huge petrified redwoods to perfectly preserved butterflies, the Fossil Beds are an illuminating treasure trove of creatures from another world. The trees were petrified following huge volcanic eruptions that buried them in ash. The largest, named *Rex arborae* (King of the Forest), is an enormous stump standing over 14 feet (4 meters) high with a circumference of 74 feet (23 meters).

A lake later formed in the valley, and its fine silt bottom collected plants and animals—especially insects—that died in the lake. In time the silt hardened into layers of shale, preserving in detail the delicate features of these organisms. Paleontologists have collected more than 60,000 fossil specimens, some so well-preserved that they clearly show the minute features of an insect's antenna, leg, hairs, and even the pattern on the wings of a butterfly. Hundreds of species of plant and over 1,100 species of insect have been identified so far. The park has over 14 miles (27 kilometers) of trails. **JK**

CAVE OF THE WINDS

COLORADO, UNITED STATES

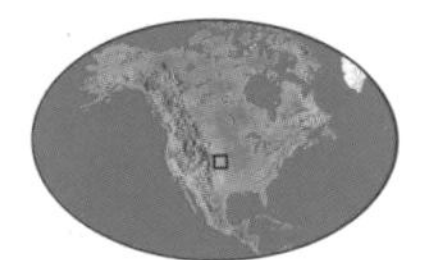

Height of Cave of the Winds: 50 ft (15 m)

Other caverns: Bridal Chamber, Temple of Silence, Valley of Dreams, Oriental Gardens

Colorado's Cave of the Winds is an unlikely stage for a Wild West showdown with a six-shooter, but it became just that in 1882 when George Washington Snider confronted others in the area who had designs on his show cave.

The story starts in the 1870s when another cave was found in Williams Canyon, and a local quarryman guarded the entrance and charged 50 cents to anyone who wanted to enter. Two local boys who could not afford the fee set about finding their own cave and on an old trail they stumbled on what was to be later named Cave of the Winds. The entrepreneurial Mr. Snider, eager to get into the early tourism industry, bought the land and took a shovel to enlarge the site. After several days of digging underground he uncovered a large cavern with glistening stalactites and flowstone formations.

In 1881 the cave was opened to the public with tours available twice daily. Snider explored much of the caverns seen on tours today but modern cavers have extended the cave to over 2 miles (3.2 kilometers) of passages, and discovered more of the cave's secrets including a number of incredible helictites—calcite crystals growing in unusual shapes. **DL**

DELAWARE BAY

DELAWARE / NEW JERSEY, UNITED STATES

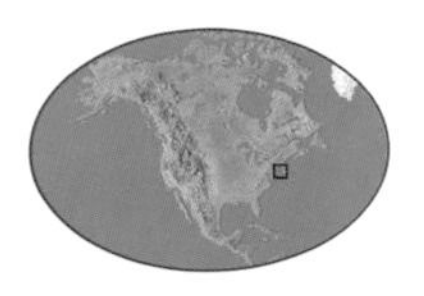

Migratory birds: red knot, semipalmated sandpiper, ruddy turnstone, sanderling

Area of bay: 782 sq mi (2,025 sq km)

Vegetation: salt marsh, mudflats, sandy and shingle beaches

In late spring and early summer, Delaware Bay is the scene of an incredible emergence of "living fossils." On the high lunar tides that coincide with the full and new moon during May and June, hundreds of thousands of horseshoe crabs clamber on to the beaches to bury their eggs on the shore. The crabs are not true crabs at all, but more closely related to spiders and resemble creatures called trilobites that lived in the prehistoric seas over 250 million years ago.

On average, each female deposits 3,650 eggs in each nest, but will return to the beach several times to bury more clutches. The density of crabs is so high that crabs dig up the eggs of previous nesters. But those eggs do not go to waste. Up to a million birds stop here on their journey from South America (where they spend the winter) to the Arctic (where they nest), making Delaware Bay the second biggest stopover location in the western hemisphere (only Copper River in Alaska hosts more birds). The eggs give the birds the essential fat they need to be able to continue their journey. Unfortunately, overzealous harvesting of the crabs has resulted in a declining population of migratory birds. MB

NATURAL BRIDGE

VIRGINIA, UNITED STATES

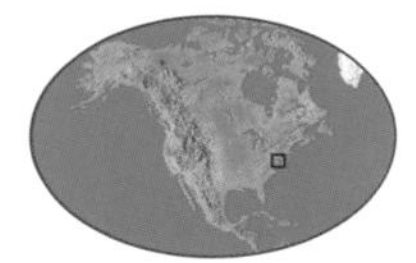

Width of Natural Bridge: 100 ft (30 m)

Thickness of Natural Bridge: 50 ft (15 m)

Thomas Jefferson described the Natural Bridge of Virginia's Shenandoah Valley as "the most sublime of Nature's works." This massive limestone arch is quite simply enormous. It towers 215 feet (66 meters) high, with a span of 150 feet (45 meters). And it is so wide that people have even constructed a convenient road across it to span the chasm below. Jefferson was so enamored with this limestone monument that he purchased it from King George III in 1774 for 20 shillings. He wanted to preserve the stone monument for all to see—the first major initiative to preserve nature in the United States.

The Natural Bridge is part of a wide network of limestone caverns in Virginia that have been hollowed out by millions of years of erosion. Geologists believe it is all that remains of the roof of a former underground cavern that collapsed in on itself. According to legend, the Monacan tribe native to the area were being pursued through the Virginia forests by the powerful Shawnee and Powhattan tribes. Finding their way blocked by a deep canyon, the Monacans knelt and prayed to the Great Spirit. When they arose, they found a great stone bridge spanning the gorge. JK

MAMMOTH CAVES

KENTUCKY, UNITED STATES

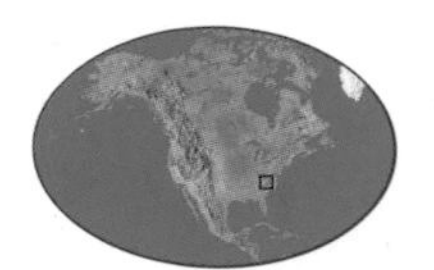

Length of explored caves: 350 mi (560 km)

Rock type: limestone

Number of animal species: 200

Mammoth Caves holds the distinction of being three times longer than any other known cave system in the world, and geologists estimate there could still be 620 miles (1,000 kilometers) of unexplored passageways. This is an irregular limestone or karst cave system, consisting of a thick layer of limestone approximately 700 feet (213 meters) deep that is easily eroded by percolating groundwater.

The dark channels of Mammoth Caves were once the pathways of underground waterways that fed into the nearby Green River. As the river eroded through the bedrock, so too did the underground waterways, forming the mind-boggling complex of caves we see today. The limestone was deposited over a period of 70 million years on an ancient seabed that covered the region 350 million years ago. Billions of animal shells were deposited and later compressed into limestone bedrock.

The size of the Mammoth Caves mirrors its extraordinarily diverse range of life. Over 200 species of animal have been found living in the caves—such as Mudpuppy salamanders, catbirds, eyeless fish, and Stinkpot turtles. Forty-two are exclusively cave dwellers seen nowhere else in the world. JK

HUACHUCA MOUNTAINS

ARIZONA, UNITED STATES

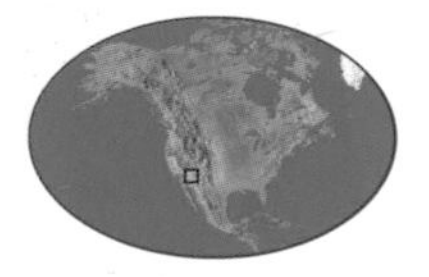

Highest mountain (Miller's Peak): 9,455 ft (2,882 m)

Annual rainfall: 20 in (51 cm)

Number of plant species: 400

The Huachuca Mountains are one of 40 "sky islands" that rise from Arizona's desert grassland. They are a welcome haven for bird life, as well as more than 60 species of reptile, 78 species of mammal, and the endangered Ramsay Canyon leopard frog. A combination of mixed scrub and grassland at lower elevations and oak and pine forests at higher levels produces this biological diversity.

The tallest mountain in the Huachucas is Miller's Peak. Well-maintained trails lead past sheer cliffs to some of the most exceptional 360° panoramas in southern Arizona. Another highlight of the Huachucas is Ramsay Canyon Preserve, a cool sanctuary with a year-round spring-fed stream. Fourteen species of hummingbird visit here from spring to fall, as well as white-tailed deer, coati, javelina, and black bear.

Once visited by the famous warrior Geronimo, the Huachuca Mountains are part of American folklore. This was also the site of Spanish conquistador Francisco Vásquez de Coronado's first visit to the southwest. He was searching for the mythical seven golden cities of Cibola. Although he never found them, he did discover this biological treasure. JK

OAK CREEK CANYON

ARIZONA, UNITED STATES

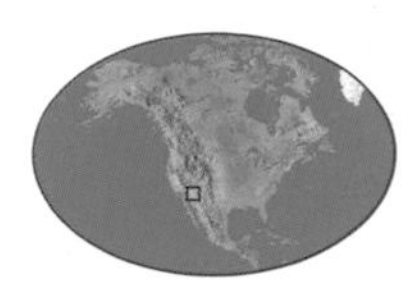

Depth of Oak Creek Canyon: 2,500 ft (762 m)

Rock type: sandstone and limestone

Length of canyon: 14 mi (22.5 km)

Oak Creek Canyon, near Flagstaff Arizona, is blessed with a high elevation that ensures cooler temperatures and more rainfall than the hotter, mainly desert habitat common to Arizona. Here lush woodlands cover the canyon floor, making it a favorite escape during the summer months. The canyon is deep and narrow. A stream flows through it year round, reflecting the red sandstone of the lower walls and the white limestone higher up. Like all sandstone canyons, the walls of Oak Creek Canyon have been eroded into a myriad of rounded arched shapes where water has carved through softer sections.

The West Fork of Oak Creek Canyon is renowned as one of the most beautiful hikes in Arizona. The combination of the flowing stream, narrow cliffs, and the greenery of the woodland make for astounding scenery. In the fall, the tree foliage turns beautiful hues of orange and red. The first three miles (five kilometers) of the canyon hike are an easy climb, but for the next 11 miles (18 kilometers) it's more challenging—albeit an exhilarating hike requiring boulder hopping, scrambling up canyon walls, wading, and even swimming through several cold, deep pools. JK

GRAND CANYON

ARIZONA, UNITED STATES

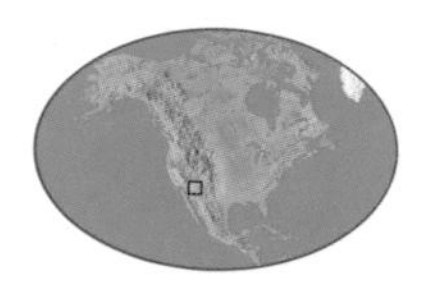

Area of Grand Canyon: 1,217,403 acres (492,683 ha)

Elevation of Point Imperial: 8,803 ft (2,683 m)

The Grand Canyon is considered one of the world's finest examples of arid-land erosion. The Colorado River and strong winds have carved and sculpted these plateaus into a labyrinth of twisting ravines, stripping back successive layers of rock and revealing a window on two billion years of the planet's geological history. Other forces that have contributed to the formation of the Grand Canyon are volcanism (volcanic activity), continental drift, and ice. Around 17 million years ago, pressures deep within Earth uplifted the landmass known today as the Colorado Plateau. This rising of the plateau, combined with five million years of erosion, created one of the world's deepest gorges, producing the breathtaking natural spectacle we see today.

The canyon is about 1 mile (1.6 kilometers) deep and 9 miles (15 kilometers) wide, and it runs for a staggering 280 miles (450 kilometers) across two states. Its striped walls change color with the passing day—silver-gold in the morning, muted browns at midday, deep crimson at sunset, and a cool

indigo in moonlight; no matter what time of day you visit the canyon, you are in for a visual treat.

Far above the canyon floor, the Colorado River appears like a shimmering thread, winding its way through a maze of tiered rock formations. This bird's-eye view belies the Colorado River's true size and power—it can swell to a raging torrent, especially during flash floods. However, its erosive power has been severely constrained by the creation of the Glen Canyon Dam.

With its diverse range of habitats and climates, the Grand Canyon is a valuable wildlife reserve. There are over 355 bird, 89 mammal, 47 reptile, 9 amphibian, and 17 fish species in the park. These range from mountain lion to pink rattlesnake.

Most visitors go to the South Rim, which is open all year round and is where Desert View Drive follows a 26-mile (42-kilometer) stretch of the rim. North Rim is open from mid-May to October, its highest point being Point Imperial, which overlooks the Painted Desert. Cape Royal faces east and west and is spectacular at sunrise and sunset. From here you can see the Colorado River framed by the natural arch known as Angel's Window. **MB**

BELOW *The majestic vista of the Grand Canyon.*

PETRIFIED FOREST NATIONAL PARK

ARIZONA, UNITED STATES

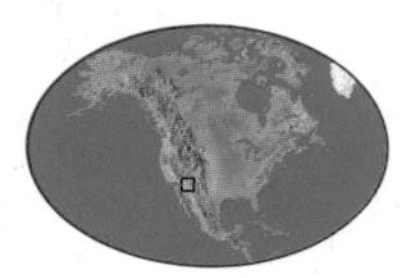

Area of Petrified Forest National Park: 146 sq mi (378 sq km)

Elevation of park: 5,500 ft (1,676 m)

Average annual rainfall: 10 in (25 cm)

The Petrified Forest holds the largest and most beautifully preserved collection of petrified wood ever found. Here, the slow process of fossilization has turned large trees into solid stone. Some 225 million years ago, these trees were part of an ancient forest, which was home to giant fish-eating amphibians, large reptiles, and early dinosaurs. After falling, the trees were washed downstream onto a flood plain at this location in northeast Arizona, and subsequently buried by silt and volcanic ash. Many of these trees rotted away, but the ones that survived were transformed into the beautiful fossilized logs we see today. Dissolved silica from the volcanic ash slowly filled or replaced the cell walls, crystallizing the trees into mineral quartz.

The process was often so precise that it preserved every detail of the log surface and, occasionally, the internal cell structures. Iron-rich minerals combined with quartz during the petrification process to give the trees a brilliant rainbow of colors. Today fossilized logs lie strewn across the clay hills and exposed in cliff faces. The petrified trees are hard and brittle and break easily when they are subjected to stress.

The park is a window to the past. Besides the trees, it has also preserved a wonderful collection of dinosaurs from the Triassic period, when the "Age of Dinosaurs" was just beginning. Visitors can see these fossils in the Rainbow Forest Museum, alongside the giant reptiles and giant amphibians that once called this place home. The Petrified Forest also contains many fine examples of rock art which early people carved onto the surfaces of boulders, canyon walls, and rock shelters. The range of images is staggering: human forms, feet and handprints, cougars, birds, lizards, snakes, bats, coyotes, bear paws, bird tracks, cloven hooves, and numerous geometric shapes. These petroglyphs may commemorate important events, mark clan boundaries, document natural events such as the summer solstice, and some may even be doodles.

The park also has a climate of extremes—half of the 10 inches (250 millimeters) of annual rain arrives via violent thunderstorms in July, August, and September. JK

The Petrified Forest holds the largest and most beautifully preserved collection of petrified wood ever found. Here, the slow process of fossilization has turned large trees into solid stone.

RIGHT *The stunning colors of the fossilized trees in the Petrified Forest National Park.*

THE PAINTED DESERT

ARIZONA, UNITED STATES

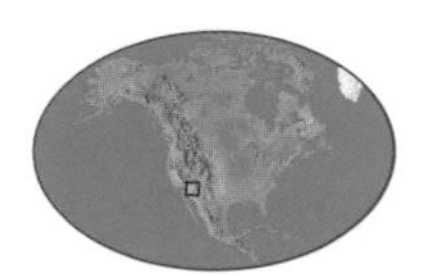

Length of the Painted Desert: 160 mi (257 km)

Age of desert: 225 million years

Rate of surface erosion: 0.25 in (6 mm) per year

The heavily eroded badland hills of the Painted Desert have a multi-layered effect because their soil contains a kaleidoscope of red, orange, pink, blue, white, lavender, and gray minerals. It is a feast for the eyes, especially toward sunset when the colors shine at their most spectacular. The Painted Desert is part of the Chinle Formation, which comprises soft sandstone sediments from the floor of an extinct water body dating back 225 million years. The rate at which the sediments were deposited determined the concentrations of iron and aluminum minerals in each layer—hence the colors. Slowly deposited soils turned red, orange, and pink, while rapidly deposited soil containing less oxygen created blue, gray, and lavender hues. Arizona's torrential summer monsoons ensure continual erosion and new exposures of color.

This is an arid land, sparsely vegetated with flat-topped mesas and buttes standing out from the hills. Eight vantage points along the rim give sweeping views of the landscape. One of the most striking areas in the park is the fabulous Blue Mesa in the east-central section. This area resembles a fantastic lunar landscape with wildly sculpted hills and striated rocks. The Painted Desert is the northern part of the Petrified Forest National Park. JK

SAGUARO NATIONAL PARK

ARIZONA, UNITED STATES

Area of Saguaro National Park: 143 sq mi (370 sq km)

Habitat: desert

Dominant plant: Saguaro cactus

The Saguaro National Park is part of the Sonora Desert that covers much of southwest America and northwest Mexico. The iconic Saguaro cactus—which can grow to a height of 50 feet (15 meters), weigh over 10 tons, and live up to 200 years—dominates the scenery of this national park. The cacti cover the valley floor but can also be found on the slopes of the Rincon and West Tucson mountains surrounding the park.

Without doubt, the long arms of the Saguaro cactus are the plant's most distinctive feature. Beginning in mid-April, large white flowers open in the middle of the night and last one day before they wilt. In that short time they attract a host of nectar-feeding bats, birds, and insects which pollinate the plant. The plant then produces bright red fruits packed with seeds. Both the flesh and seeds of the fruit are consumed by these desert animals. For the best chance of survival, Saguaro seedlings need the protective cover of a "nurse plant" to give them shade and additional moisture. They begin to grow arms at around 75 years of age.

Saguaro National Park has over 150 miles (241 kilometers) of different hiking trails. Wherever it is that these trails lead, the majestic, multi-armed Saguaro cactus will be there for company. JK

ANTELOPE CANYON

ARIZONA, UNITED STATES

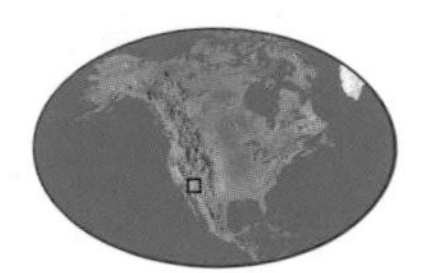

Length of Antelope Canyon: 5 mi (8 km)

Geological type: slot canyon

Rock type: Navajo sandstone

The spectacular Antelope Canyon has been described by landscape photographers as "a place of celebration for the eye, mind, and spirit." This little-known sandstone canyon is a natural work of art where light, color, and shape intermingle in an awesome display of exquisite beauty that changes throughout the day. The canyon is a narrow crevice divided into two sections (Upper and Lower Canyon) that have been scoured out of a mesa. It is possible to walk into the Upper Antelope Canyon, but the Lower Canyon is only accessible by climbing down ladders from a narrow slot in the ground. The effect of the light playing on the canyon walls is staggering. Strong orange and yellow colors brighten the upper reaches, but as the light diminishes, the lower walls turn to shades of purple and blue. The contrast of light and shade emphasize the canyon's rounded contours in a harmonious feast for the eyes.

The best time to visit is at midday, when the sun is directly overhead and single beams of light shine right down to the canyon floor. Indeed, certain points in the canyon are famous for these beautiful shafts of light that occur for only a few inspiring minutes of the day.

Antelope Canyon is known as a "slot canyon"; these canyons normally start as narrow fissures on the surface of a sandstone plateau. If the fissure is on a slope, the powerful erosive force of flowing water can turn the fissure into a drainage channel, carving down into the sandstone. The result at Antelope Canyon is a narrow, deep channel with wild, undulating contours and hollows that range from two to five feet (0.6 to 1.5 meters) wide and up to 165 feet (50 meters) deep.

Antelope Canyon is a natural work of art where light, color, and shape intermingle in an awesome display of exquisite beauty that changes throughout the day. Certain points are famous for beautiful shafts of light that occur for only a few minutes each day.

Now a tribal park within the Navajo Nation, the canyon is known to the Navajo as *Tse Bighanilia* ("the place where water runs through rocks"). The English name refers to the herds of pronghorn antelope that used to roam here.

To really appreciate the wonder of this place, try sitting in one of the canyon's darkened chambers in the morning and watching it come to life in a blaze of color, light, and shadow as the sun passes overhead. Access is only permitted with a guide since flash floods can strike suddenly, even in apparently good weather. JK

RIGHT *A shaft of light illuminates the beauty of the canyon.*

CANYON DE CHELLY NATIONAL MONUMENT

ARIZONA, UNITED STATES

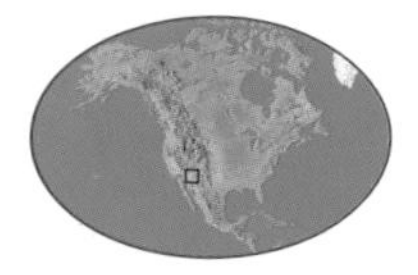

Area of Canyon de Chelly:
130 sq mi (338 sq km)

Depth of Canyon de Chelly:
800 ft (240 m)

Canyon de Chelly in northern Arizona was designated a national monument in 1931 to protect the spectacular branched sandstone canyon and its ancient Native American cliff-dwelling ruins. The site is known to have been inhabited for more than 1,500 years, and the remains of over 700 ruins (including pit houses and pueblos) have since been found. The first cliff dwellings were built around 1060 C.E. by the Anasazi and were abandoned toward the end of the 13th century. The site was later occupied by the Hopi and the Navajo and is situated on the Navajo Indian Reservation.

In places, the vertical red sandstone walls plunge 800 feet (244 meters) from the Defiance Plateau to the canyon bottom. De Chelly sandstone was not laid down horizontally, but formed in deserts of the Permian Period and has sloping lines in the rock typical of windblown sand deposits. The national monument is jointly managed by the Navajo Tribal Authority and the National Park Service. Navajo guides provide tours on foot, by horse, or four-wheel-drive vehicles. **RC**

BELOW *Pictographs at Standing Cow in Canyon de Chelly.*

CHIRICAHUA NATIONAL MONUMENT

ARIZONA, UNITED STATES

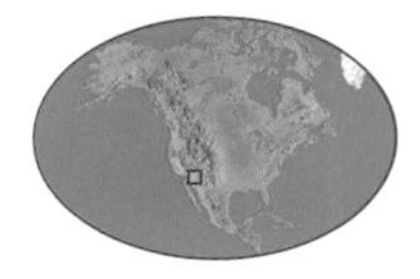

Alternative name: Land of Standing-Up Rocks

Age: 27 million years

Geological type: rhyolitic tuff

The landscape of Chiricahua National Monument is a myriad of dramatic pinnacles and spires (hoodoos). The rocks were laid down 27 million years ago when an enormous volcanic eruption deposited 2,000 feet (600 meters) of ash and pumice. Over time this mixture fused into rhyolitic tuff, which eroded to produce the hoodoos we see today. Trails have been forged through the monument to take visitors to hoodoos such as Totem Pole and Big Balanced Rock.

The monument is part of the Chiricahua Mountain Range, a chain of extinct volcanoes 120 miles (193 kilometers) east of Tucson that rise 7,800 feet (2,377 meters) from desert grassland. The cooler elevations of the mountains or "sky islands" support life that cannot survive on the hotter desert floor.

Situated at the meeting place of four major North American biomes—the Sonoran and Chihuahuan deserts, and Rocky and Sierra Madre mountain ranges—Chiricahua juxtaposes pine and spruce trees with yucca and prickly pear, as well as Douglas fir and Arizona cypress. Wildlife includes over 300 species of bird, as well as javelina, coatimundi, hog-nosed skunk, bear, and mountain lion. JK

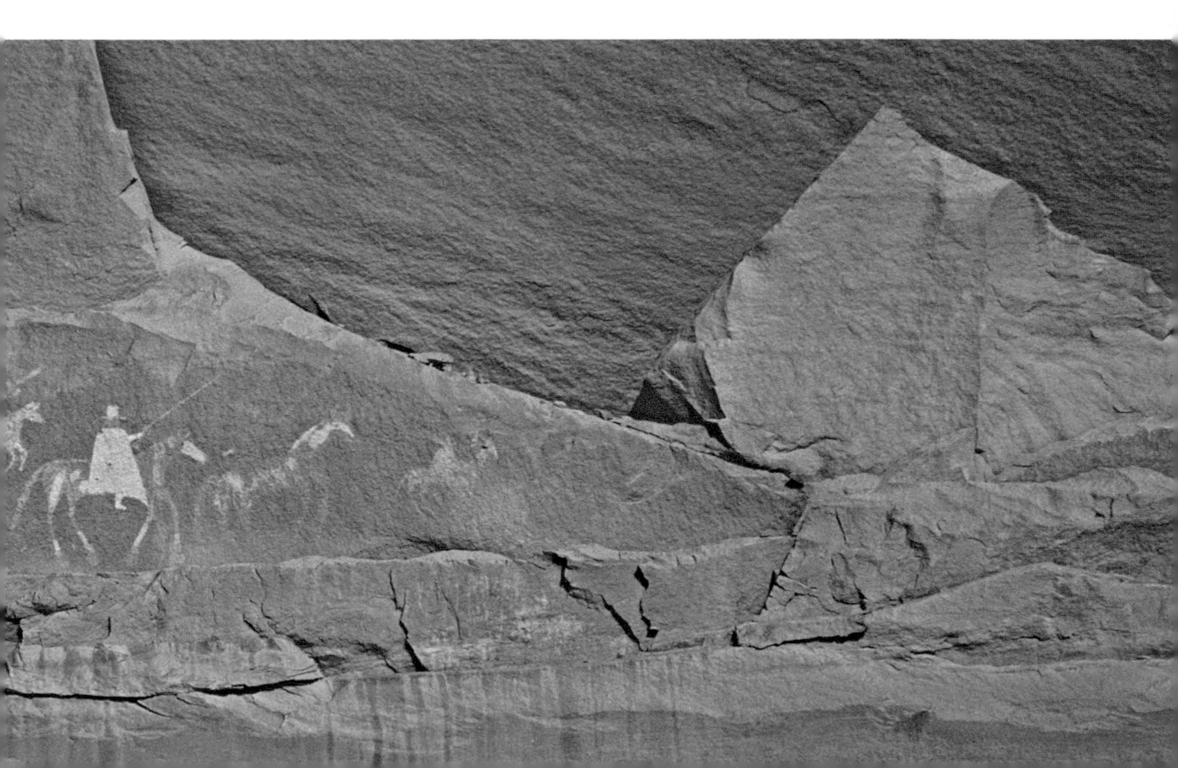

KARTCHNER CAVERNS

ARIZONA, UNITED STATES

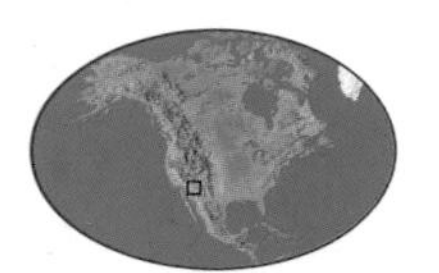

Total cavern length: 2.5 mi (4 km)
Bat numbers: 1,000
Temperature: 68°F (20°C) year round

Below the Sonoran Desert in southeast Arizona is a cave system with a delicate microclimate that has remained unchanged for 200,000 years. Until recently, adventurers passing among the giant Saguaro cactus and creosote were unaware of the enormous chambers and cave passages beneath them, yet this subterranean labyrinth boasts miles of passages, caverns the length of soccer fields, and a myriad of incredible mineral formations. The Kartchner Caverns were discovered by two amateur cavers who entered a small sinkhole in 1974. Inside they found calcite formations, huge stalactites, stalagmites, and columns, as well as the world's second-longest soda straw—a delicate crystal tube 21 feet (6.5 meters) long, growing precariously from the roof of the cave.

In order to keep the cave safe, the cavers kept their discovery secret for 14 years. While much of the system is wild and undeveloped, guided tours now take visitors into the heart of the cave. In the winter months, The Big Room—the largest of the chambers—is closed to the public while thousands of bats roost and give birth. DL

METEOR CRATER

ARIZONA, UNITED STATES

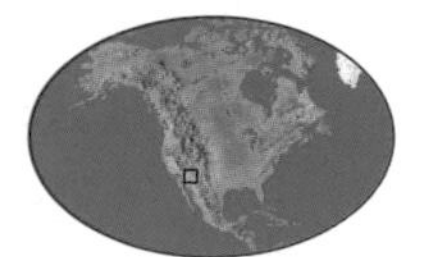

Diameter of meteor crater: 4,150 ft (1,265 m)
Depth of crater: 575 ft (175 m)
Age: approx. 22,000 to 50,000 years

When a huge saucer-shaped crater in the high desert plateau of Arizona was first reported in 1871, it was thought to be an extinct volcano. However, the discovery of iron fragments in the crater in the 1890s led geologists to conclude that it was not of volcanic origin. The Philadelphia mining engineer Daniel Barringer explored the site in 1903, and was so convinced that it was a meteor crater, he dedicated 26 years of his life to searching (unsuccessfully) for a buried meteorite.

In 1960 two rare forms of silica mineral which form only under intense heat and pressure, coesite and stichovite, were discovered. Their discovery confirmed the meteor impact theory and led scientists to believe that most of the material that made up the meteor vaporized on impact. Although not the largest terrestrial impact site, Meteor Crater is the best-preserved meteor impact site on Earth. The crater was designated a Natural Landmark in 1968. Estimates of the size of the meteorite and date of collision vary, but the impact must have been devastating to produce such a large crater. RC

RIGHT *Scientists believe that the meteor that caused this impressive crater weighed 70,000 tons.*

THE SONORAN DESERT

ARIZONA, UNITED STATES / MEXICO

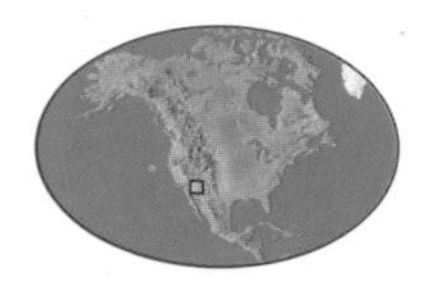

Area of Sonoran Desert: 100,000 sq mi (260,000 sq km)

Average annual rainfall: 10 in (25 cm)

Local people say that no one can walk the Sonoran Desert and not be changed; it has a natural power that both uplifts and humbles the spirit. Ringed by mountains and stretching across not just two states (Arizona and California), but two countries (the United States and Mexico), this wonderful desert has more plants and animals than any other North American desert. Two distinct rainy seasons are the reason for its rich biodiversity. In winter, when cold weather fronts from the Pacific Ocean bring widespread gentle rains, the desert turns into a wildflower paradise, especially in the west, where annuals like poppies and lupins burst into life. In summer, from July to September, wet tropical air blows in from the Gulf of Mexico causing violent thunderstorms and localized flooding. The surrounding mountains trap the rain clouds, thus providing the area with enough moisture to support a rich diversity of plants and animals.

This desert is home to over 2,500 species of plant and 550 species of vertebrate. About half

of the Sonoran Desert's plant life is tropical in origin, with life cycles that are closely tied in with the summer monsoon. Despite all the rain, this is still the hottest of North America's four major deserts.

A number of remarkable plants and animals distinguish it further. The Saguaro cactus, with its distinctive crooked "arms," is one of its best-known plants, and found nowhere else in the world. It can live for more than 200 years and grows so slowly that the first arm only appears after 75 years. Ironwood trees, bursage, palo verde, creosote, and mesquite inhabit the hottest parts of the desert.

Despite the extreme heat of the desert, there are a number of animals that manage to survive in this climate—the Mexican gray wolf (*el lobo*), mountain lion, great-horned owl, golden eagle, roadrunner, and the rattlesnake.

The Sonoran Desert is vast, but one way to appreciate it is by visiting the Arizona-Sonoran Desert Museum, near Tucson. This 20-acre (8-hectare) outdoor site is at once a zoo, a natural history museum, and a botanical garden, showcasing the most interesting plants and animals of the Sonoran Desert. JK

BELOW *Cacti and palo verde dot the arid desert habitat.*

BISTI BADLANDS & DE-NA-ZIN WILDERNESS

NEW MEXICO, UNITED STATES

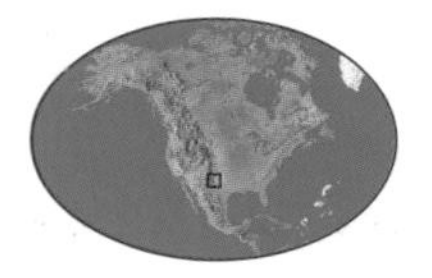

Area of Bisti Badlands: 38,381 acres (15,533 ha)

Designated as a national wilderness area: 1984

Average elevation: 6,300 ft (1,920 m)

The strange rock formations and eroded landscape of the Bisti Badlands are hidden in the desert of northwestern New Mexico. The hills are made up of layers of sandstone, mudstone, coal, and shale that have been eroded into a maze of mounds, ravines, caves, and hoodoos. In the past the climate was different, as is evident by the area's many well-preserved fossils. Petrified wood and the fossilized bones and teeth of a wide variety of animals can still be found here. You can view these fossils, but it is illegal to remove them from the site. Nowadays the desert environment is home to a variety of reptiles, small mammals, and birds of prey.

Ten miles (16 kilometers) farther is the more remote De-Na-Zin Wilderness. There are no established trails, and you should be careful when exploring the area—many of the rock features are easily damaged and are now known to be unstable. The abandoned buildings of the Bisti trading post heighten the desolate character of this place. **RC**

BANDERA ICE CAVES

NEW MEXICO, UNITED STATES

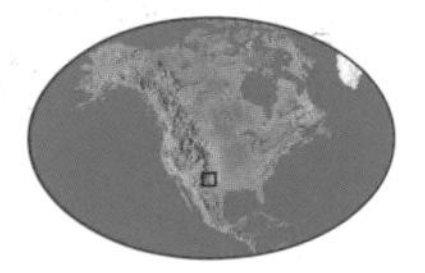

Bandera Ice Caves (ice thickness): 20 ft (6 m)

Bandera volcanic cone depth: 820 ft (250 m)

Age: 10,000 years

The Bandera Volcano is New Mexico's "Land of Fire and Ice." The deep cone is the remnant of a violent eruption which occured around 10,000 years ago, and the volcano's lava tube cave system was once a labyrinth of natural pipes over 19 miles (30 kilometers) long.

Today much of the complex has collapsed, although many short spectacular sections remain, including the Bandera Ice Cave. A walk up the lava trail through the gnarled and twisted juniper, fir, and Ponderosa pine trees leads visitors to a partially collapsed lava tube and a natural ice box.

Inside the ice box, the temperature never rises above freezing, and the floor glistens blue-green with layers of natural ice that reflect light from the entrance. A form of arctic algae causes the green hue, whereas the ice is created when snow meltwater and rain seep underground and freeze in the cold, dense winter air that sinks into the cave. In the summer, the cave is insulated from the harsh desert sun by overhead rocks. The oldest layers of ice have been dated to 170 C.E. Ancient Pueblo Indians used to take ice from the cavern, as did early western settlers. **DL**

WHITE SANDS NATIONAL MONUMENT

NEW MEXICO, UNITED STATES

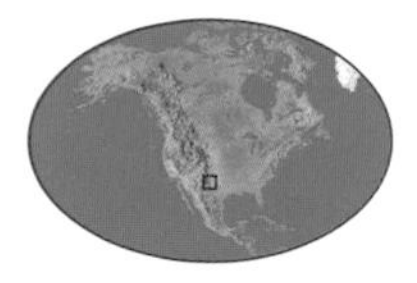

Dune height: up to 60 ft (18 m)

Dune movement: up to 30 ft (9 m) per year

Designated as a national monument: 1933

In the heart of the Tularosa Basin in the southwest of New Mexico lies a desert of glistening white dunes. Unlike sand dunes, which are composed of silica, the White Sands are dunes of gypsum—this is the largest pure gypsum dune system in the world. The origin of the gypsum (calcium sulfate) is Lake Lucero, an ephemeral lake to the west of the dunes. Evaporation from the lake leaves behind gypsum deposits that are transported by wind to form the dunes. Theoretically, about 80 inches (205 centimeters) of water a year evaporates from the lake. The most active dunes can move up to 30 feet (9 meters) per year. A few species of plant such as yucca and cottonwood manage to survive on the edges of the shifting dunes. There are a few resident animal species, some of which (including the bleached earless lizard and the Apache pocket mouse) have evolved white coloration that acts as an effective camouflage against the sand.

The gypsum dunes cover 275 square miles (712 square kilometers), 40 percent of which are contained within the national monument site. The remaining are located on the adjacent White Sands Missile Range, which belongs to the military and is not open to the public. RC

CITY OF ROCKS

NEW MEXICO, UNITED STATES

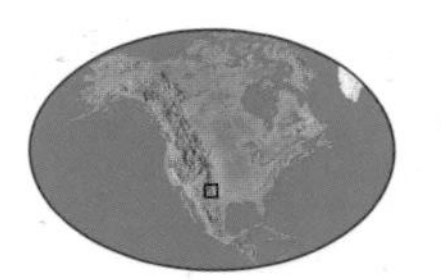

Annual rainfall: 16 in (40 cm)

Rock type: volcanic tuff

Rock height: 40 ft (12 m)

City of Rocks State Park is a small—only 1 square mile (2.7 square kilometers)—but utterly fascinating geological wonder composed of large, naturally sculptured rock columns and pinnacles up to 40 feet (12 meters) high. When approached from a distance, the giant rocks appear like a city on the horizon—a dense conglomeration rising out of a mesa. Most of the rocks are sufficiently spaced to allow visitors to walk among them, or alternatively to find shelter from the burning summer sun.

These rocks are 35 million years old and volcanic in origin. They were originally part of a very large flow of hot volcanic ash, or tuff, that surged over the area and gradually became welded into a solid stratum of rock. Over time, rain and wind have eroded the softer substrate, leaving rows of these harder rocks behind. Some people say they resemble huge molars, sticking out of the desert floor, some crooked, some straight. Many stand alone. Others are grouped together, and some appear to be leaning on each other for support. These rock shapes are so rare that only six other places in the world have similar formations.

The park is located 28 miles (45 kilometers) northeast of the town of Deming, in the scenic Mimbres Valley of the Chihuahuan Desert in southeastern New Mexico. The "City" has 35 species of bird living among its rocks, nesting in cavities and crevasses. These include bald and golden eagles, hawks, horned owls, cactus wrens, roadrunners, and finches. Numerous reptiles, including rattlesnakes, collared lizards, and hognosed snakes, also live here, as do many desert mammals, including ground squirrels, jackrabbits, kangaroo mice, pack rats, and coyotes.

The City of Rocks has also attracted human inhabitants, beginning in 750 C.E. with the Mimbres Indians who lived here until 1250 C.E. The Mimbres found the rocks useful for grinding grain, and many of the rocks still bear the scars of this activity. Some rocks still have the crosses that were carved into them in later centuries by passing Spanish conquistadors, while the desert floor bears wagon ruts left behind by emigrants heading west to California. Today the City is extremely popular with rock climbers and boasts over 500 formal climbing routes. JK

When approached from a distance, the giant rocks appear suddenly as a dense conglomeration rising out of a relatively flat mesa, like a city on the horizon. Some people say they resemble huge molars sticking out of the desert floor.

RIGHT *The unusual rock formations of City of Rocks.*

SHIPROCK PEAK

NEW MEXICO, UNITED STATES

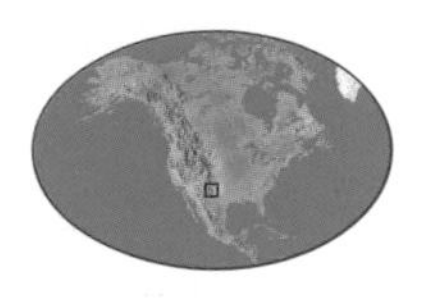

Alternative name: Tse Bitai

Height of Shiprock Peak: 1,583 ft (482 m)

Rock type: volcanic breccia

The Navajo call this amazing formation *Tse Bitai* ("the winged rock") and tell a legend of how their people were once flown to safety from enemies when the rock rose up and transported them to this location. Newcomers later called it Shiprock Peak because of its resemblance to a 19th-century clipper ship.

This stone monument is actually the solidified basalt lava core of a 30-million-year-old volcano. Its main peak rises an impressive 1,583 feet (482 meters) above the New Mexican plain. The smaller surrounding pinnacles were once the auxiliary vents of the volcano. They are composed of jagged fragments of rock and ash that have been welded together by extreme heat into a rock type known as "breccia." Shiprock Peak must have had a violent birth, as jagged rocks are a sure sign of the explosive nature of the eruption. Located 13 miles (21 kilometers) southwest of the town of Shiprock, Shiprock Peak is part of both the Navajo and Chuska volcanic fields that cover northeastern Arizona and northwestern New Mexico. **JK**

BELOW *The jagged outline of Shiprock Peak.*

THE KNEELING NUN

NEW MEXICO, UNITED STATES

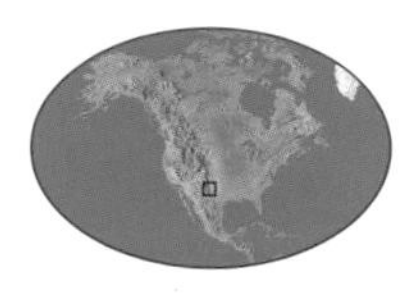

Height of the Kneeling Nun: 90 ft (27 m)

Rock type: volcanic rhyolite tuff

The legend of the Kneeling Nun tells how a local nun nursed an injured Spanish soldier back to health and then, despite her vows, fell in love with him. As a result, she was cast out from her convent and then turned to stone, to spend eternity kneeling atop a mountain in prayer. A fanciful story, perhaps, but the giant rock formation on Kneeling Nun Mountain really does look like a nun with her head bowed before an altar.

The real source of this geological formation is not as passionate—but it is dramatic. Thirty-five million years ago, a volcanic eruption a thousand times greater than the 1980 eruption of Mount St. Helens sent a burning hot flow of pumice, ash, and gas surging across this landscape. The volcanic debris hardened into solid rock and then slowly eroded after being uplifted by the formation of the Santa Rita Range. Wind, rain, and winter frosts then eroded the volcanic deposit to reveal this unusual stone monument. The Kneeling Nun overlooks Chino Mine and is located in western New Mexico, which is about 20 miles (32 kilometers) east of Silver City. JK

BLUE HOLE

NEW MEXICO, UNITED STATES

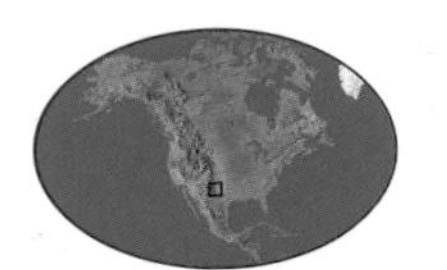

Flow rate of water through the Blue Hole (per minute): 3,000 gal (11,350 l)

Elevation of spring: 4,600 ft (1,400 m)

Temperature: 64°F (18°C)

One of the most remarkable scuba-diving experiences in the American southwest is contained in this land-locked geological phenomenon in New Mexico. The Blue Hole is a huge 80-foot- (25-meter-) deep natural artesian spring formed within a limestone chasm. It is also known as "Nature's Jewel." Water gushes into the Blue Hole at a prodigious rate—3,000 gallons (11,350 liters) per minute—so the water is remarkably clear; so clear in fact that it is possible to see to the bottom when the water is not being disturbed by divers.

Blue Hole's diameter is 80 feet (25 meters) at its surface, which increases gradually to 130 feet (40 meters) at its base. So there is ample room for many people to take the plunge in its cool waters. The temperature remains a constant 64°F (18°C) and the water recycles every six hours. Blue Hole is near the town of Santa Rosa, and is just one of several spring-fed lakes and clear mineral springs that abound here. The surrounding countryside is semi-desert, making this place a welcome oasis. A privately operated dive center near Blue Hole provides divers with tank refills, equipment rentals, and permits. **JK**

LECHUGUILLA CAVE

NEW MEXICO, UNITED STATES

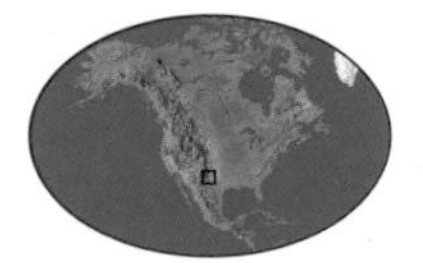

Depth of Lechuguilla Cave: 1,604 ft (489 m)

Length of cave: 122 mi (196 km)

Age: 6 million years

Discovered when humid gusts of air leaked out of an abandoned, rubble-laden mining pit called Misery Hole, Lechuguilla Cave in southern New Mexico is today recognized as one of the most important caves ever uncovered. From delicate crystals and loops of corroded stone that drip down walls to giant "chandeliers" of fragile gypsum that hang precariously overhead, Lechuguilla's gallery of formations is entirely unique. A true journey into the otherworldly, it also houses rare microbes, which NASA scientists and medical researchers are probing both for clues into life on Mars and cures for cancer.

Unlike most caves, which are created as rainwater dissolves the limestone below, Lechuguilla was carved out over millions of years by chemical reactions and rock-eating microbes and is the fifth longest cave complex in the world. Sulfur-burning bacteria converted gaseous emissions from deep oil reservoirs into sulfuric acid which, along with bacteria that tap into iron and manganese, gutted what is today one of the most beautiful cave complexes in the world. To preserve its rare formations and unique microbial life, Lechuguilla is closed to the public. **AH**

CIMARRON CANYON

NEW MEXICO, UNITED STATES

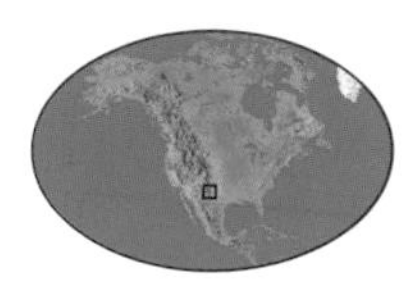

Area of Cimarron Canyon: 52 sq mi (134 sq km)

Elevation of canyon: 8,000 ft (2,400 m)

Annual rainfall: 12.5 in (32 cm)

The imposing granite cliffs of Cimarron Canyon are a dominant scenic feature of this beautiful park. Located high in the mountains of New Mexico, this is a cool, enjoyable place to take a break from the blasting heat that occurs at lower elevations. A sparkling river runs through the canyon for 12 miles (19 kilometers), adding an additional soothing element. The high cliffs extend horizontally to form a palisade above the canyon, something akin to the battlements of an old castle. At 400 feet (120 meters) high, this is a climber's paradise, although it is recommended that only experienced climbers attempt the climb because the rock is not stable. A special permit from park officials is also required. The canyon has several hiking trails, as well as cross-country skiing in winter. Cimarron Canyon is located in the northeast region of the state and is part of the Colin Neblett Wildlife Area—the largest wildlife refuge in the state, with elk, deer, bear, wild turkey, and grouse, plus many unusual birds like the rufous hummingbird and pygmy nuthatch. This river is also renowned as a fly-fishing paradise with plenty of brown and rainbow trout. JK

THE VALLEY OF FIRES

NEW MEXICO, UNITED STATES

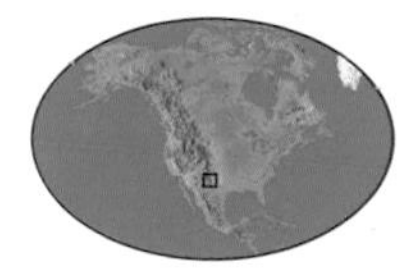

Alternative name: Malpais ("Badland")

Area of the Valley of Fires: 125 sq mi (324 sq km)

Rock type: olivine basalt

Between 1,000 and 1,500 years ago in New Mexico, a series of fractures opened on the floor of the Tularosa Basin and released thick flows of lava, thus forming a huge, black, primordial-looking terrain. In the Valley of Fires it reached a thickness of 165 feet (50 meters) and buried everything in its path, apart from a few sandstone hills that jut above the lava surface like misplaced islands.

The Valley of Fires is a good place to see the different rock formations that lava can make as it flows and cools. In some places the rock is rough and sharp, in others the surface has a smoother, ropy texture created by lava with more dissolved gas in it. The valley also has eight lava tubes, where molten rock once flowed through subsurface channels.

Though not fit for human use, many plants and animals have successfully colonized the Valley of Fires' pitted and creviced surfaces. Predators are plentiful too—look out for the great horned owl that not only hunts here, but also uses the flow as a nesting site.

Many rodents and reptiles have developed abnormally dark coloration to blend in with the dark lava. Better camouflage means less chance of being caught by predators. JK

CARLSBAD CAVERNS

NEW MEXICO, UNITED STATES

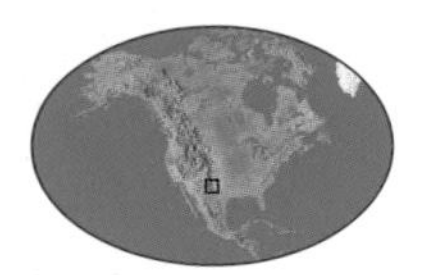

Number of caves: 100

Temperature: 56°F (13°C)

Rock type: limestone

Carlsbad Caverns is a vast network of caves within the Guadalupe Mountains of New Mexico. These large and deep caves are decorated with spectacular limestone columns, stalactites, and stalagmites.

The caverns are the hollowed-out remains of a fossilized reef from the Permian Age 250 million years ago. Once under a shallow inland sea, when the area was uplifted a few million years ago, rainwater seeped into cracks in the reef while hydrogen sulfide gas seeped up from vast underground oil and gas deposits; this combination was extremely corrosive and created the giant caverns we see today.

The caverns were discovered by local cowboy Jim White in 1898. He explored and led tour groups through the caverns, but his trips were not for the fainthearted—they began by lowering the visitors down 170 feet (52 meters) in a bucket. Today the caverns are more accessible, with trails and tours offered year round. The one million migratory Mexican bats are another highlight of the caverns. There can be as many as 300 crammed tightly into just one square foot (0.09 square meters). JK

RIGHT *Large stalagmites at Carlsbad Caverns.*

LAS HUERTAS CANYON

NEW MEXICO, UNITED STATES

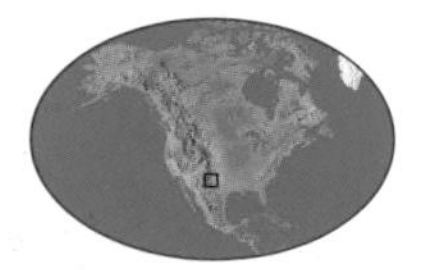

Alternative name: Land of Enchantment

Average rainfall: 14 in (356 mm)

Age of Sandia Mountains: 2 to 25 million years

Las Huertas Canyon was carved from the western slopes of the Sandia Mountains by a bubbling spring-fed creek. Las Huertas is Spanish for "the gardens," and was named by early Spanish settlers to New Mexico in 1765, who saw this lovely canyon as exactly that. Its steep slopes have a wonderful range of plants, from elders, cottonwoods, and willows at higher elevations, to piñon pine, juniper, and numerous heat-tolerant grasses and shrubs at lower levels.

Combining a mild climate, abundant sunshine, rich sunsets, and wonderful scenery, the canyon and its surrounding mountains are a visual delight throughout the day, their contours and features changing with the angle of the sun. The color of the mountains is especially pink during the early morning and early evening, something noted by early explorers who named them *Sandia*, which means "watermelon" in Spanish. Las Huertas Canyon is also subject to sudden, violent thunderstorms in the summer, so care must be taken on the trails along the canyon's exposed ridges—especially because New Mexico leads the United States in lightning deaths per capita. JK

SLAUGHTER CANYON CAVE

NEW MEXICO, UNITED STATES

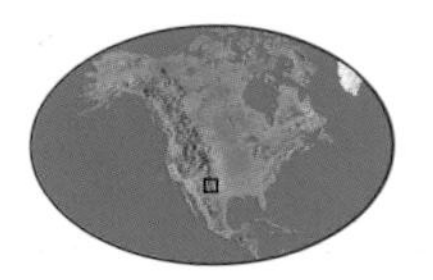

Temperature of Slaughter Canyon Cave: 56°F (13°C)
Rock type: limestone
Rock age: 250 million years

Slaughter Canyon Cave is an ideal place for strenuous adventure and wild caving. Located inside Carlsbad Caverns National Park, this large underground cavern has only primitive trails and no artificial lighting. The entrance to the cave is at the end of a steep 500-foot (150-meter) hike through the hot desert. But it is worth the effort because this cave has some magnificent features. Highlights include the Monarch, one of the world's tallest limestone columns, which stands 90 feet (27 meters) high. Another column, called the Christmas Tree, has the distinctive triangular shape of a fir tree and is covered with a white limestone frosting and decorated with sparkling calcite crystals.

An unusual delicate rimstone dam, only ankle-high, is another favorite attraction. It looks like a miniature Great Wall of China, and was formed by calcium carbonate deposits which consolidated around a pool of water. The cave was discovered in 1937 by a farmer whose goats used it to take shelter from a storm, and can now be explored with the help of a park ranger. The only light sources on this atmospheric 1.25-mile (2-kilometer) trek are from the headlamps of visitors' hardhats. **JK**

SODA DAM

NEW MEXICO, UNITED STATES

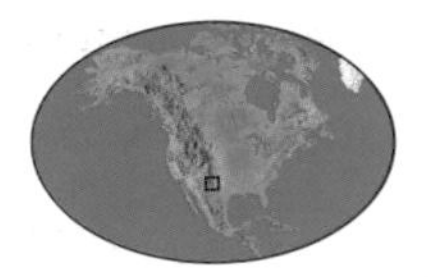

Length of Soda Dam: 300 ft (90 m)
Height of dam: 50 ft (15 m)
Oldest deposits: 1 million years

The Soda Dam looks like a giant rock that tumbled into the Jemez River. But the story of this geological formation in north-central New Mexico is even more bizarre. It started forming one million years ago when water from hot underground springs reached the surface and cooled, leaving a thick mineral deposit of calcium carbonate, or travertine. This deposit has since grown, and now forms a 330-foot- (100-meter-) long limestone structure that has overgrown the river.

The Soda Dam lies along a deep fault in the Jemez volcanic field, which last erupted 130,000 years ago. A body of molten hot igneous rock deep below the surface heats the groundwater sufficiently to dissolve minerals from the limestone bedrock.

Inside the dam is a warm, humid cave that is accessible through a small entrance. Here the spring waters bubble up with their cargo of travertine. Over the years, the river has managed to cut a path through the dam in order to continue its course. Yet while the river water erodes the dam, the spring water contributes to its unique structure—nature therefore simultaneously building and destroying this fascinating composition. **JK**

THE LOST SEA

TENNESSEE, UNITED STATES

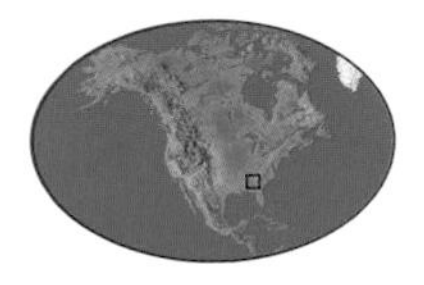

Length of the Lost Sea: 800 ft (244 m)
Width of lake: 220 ft (67 m)
Temperature: 58°F (14°C)

The Lost Sea is the largest underground lake in the United States. It lies deep within a cave system called Craighead Caverns, in the mountains of east Tennessee. The actual caverns have provided shelter and secrecy for a number of people: the Cherokee used to inhabit a cave called the Council Room and left behind a wide range of artifacts; soldiers stored ammunition here in the Civil War; and during Prohibition one of the caverns was used as a speakeasy.

Today the prime attraction is the magnificent underground lake, which visitors can explore on electric-power, glass-bottom boats. The exact size of the lake is unknown because the cavern connects to larger flooded caverns directly below. Exploring these deep underwater caverns can be dangerous, and as a result divers have only been able to map 13 acres (5 hectares) of the lake. Another fascinating attraction is the rare crystalline formations called anthodites. Composed of a form of calcium carbonate called aragonite, these delicate, hair-like objects are also known as "cave flowers." They are found in only a few of the world's caves, and have earned this cave's national landmark designation. **JK**

NATURAL BRIDGE CAVERNS

TEXAS, UNITED STATES

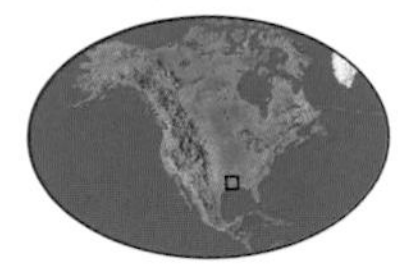

Age of Natural Bridge Caverns: 12 million years
Constant temperature: 70°F (21°C)
Cave type: karst

The entrance to the Natural Bridge Caverns is a giant sinkhole which is spanned by the caverns' namesake—a 60-foot (18-meter) limestone bridge. A crawlway descends into a wonderland of huge chambers, massive limestone columns, and delicate crystalline formations. The largest chamber, the Hall of the Mountain Kings, is 350 feet (107 meters) long, 100 feet (30 meters) high, and 100 feet (30 meters) wide. The caverns' first explorers found 5,000-year-old artifacts. They also stumbled across an 8,000-year-old skeleton of an extinct grizzly bear.

The caverns formed 12 million years ago when rainwater dissolved the limestone substrate. They are full of strange formations created by calcite precipitating out of the dripping water and accumulating into a myriad of shapes. It takes approximately 100 years for one of these formations to grow 1 cubic inch (16 cubic centimeters).

The developed portion of the cave, with a half-mile of trails and 35,000 watts of indirect lighting, takes the visitor 260 feet (79 meters) below ground level. The more adventurous can crawl among cave formations buried 160 feet (49 meters) into the subterranean world. **JK**

THE BASIN OF THE CHISOS MOUNTAINS

TEXAS, UNITED STATES

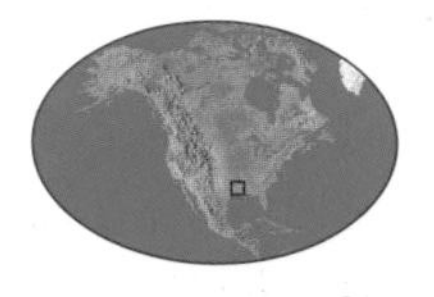

Height of tallest mountain: 7,835 ft (2,388 m)

Number of bird species: 434

Number of mammal species: 78

The Basin of the Chisos Mountains is part of a great rift valley that stretches from Colorado to Mexico. The basin and its mountains sit in a sunken block in the rift, surrounded on either side by more mountains. The result of this geological convolution is a land of contrasts—of desert lowlands and moist mountain woodlands. But most of all it is a land of panoramic vistas. The Chisos Mountains rise 2,000 feet (610 meters) from the basin floor to an elevation above sea level of nearly 8,000 feet (2,440 meters), and are covered in oak forests, ponderosa pine, juniper trees, and aspen.

These mountains represent the only moisture and moderate temperatures in a desert landscape and are home to black bears, cougars, rare birds, and many plants and animals that live nowhere else in the world. This entire region lies within the Chihuahuan Desert, a vast area covering most of northern Mexico, western Texas, and parts of New Mexico. Temperatures in the desert above the

basin are often over 100°F (38°C) during the summer, but the basin is often 50°F (10°C) cooler thanks to its sunken position below the floor of the desert.

This is one of the most remote points in Texas, but it is worth the journey to experience its wild beauty, mountain-desert vistas, and magnificent starlit skies. One of the highlights is the sunset at "Window of the Chisos Basin," which has a stunning view through high buttes toward Casa Grande Peak, a mountain on the east side of the basin. Here the sunset can be so fierce it turns the mountain and the desert into a red, Martian-like landscape.

The basin and its mountains are part of Big Bend National Park, where the Rio Grande River takes a wide curve across the great Chihuahuan Desert. The park encompasses 1,200 square miles (3,108 square kilometers) of some of the most rugged country in North America.

The best time to visit is between November and January, when temperatures are much cooler. With plenty of trails crisscrossing the park, this is a hiker's paradise. **JK**

BELOW *Deep oranges and reds illuminate the landscape of the Chisos Mountains.*

BIG CYPRESS NATIONAL PRESERVE

FLORIDA, UNITED STATES

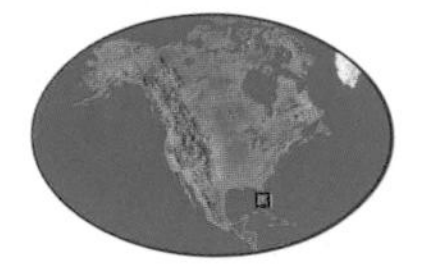

Area of Big Cypress National Preserve: 2,500 sq mi (6,216 sq km)

Average annual rainfall: 60 in (152 cm)

Dominant vegetation: dwarf pond cypress

Florida's Big Cypress National Preserve is a wetland paradise encompassing an impressive 2,500 square miles (6,216 square kilometers) of rich subtropical habitat. It is commonly called a swamp, but this is far from accurate because the reserve includes a variety of habitats such as sandy islands of pine and mixed hardwood trees, prairies, mangroves, palm trees, and, of course, cypresses. This is an extremely flat landscape, so the best way to get a good view is from the air. To truly experience Big Cypress National Preserve, you need to immerse yourself in its lush greenery and enjoy the details of its rich and unusual plant and animal life.

About a third of Big Cypress is covered by cypress trees, mostly the dwarf pond cypress. However, a few of the giant bald cypresses that once dominated this area still remain. Some of these enormous trees are more than 700 years old, with trunks so wide it would take four people with outstretched arms to encircle one.

The rainy season begins in May and lasts until the following autumn. During this time, the waters can flood Big Cypress to depths of three feet (0.9 meters) before slowly draining south into the Gulf of Mexico. The land is so flat that the water moves at a stately rate of only 1 mile (1.6 kilometres) per day. Even after the rains end in October it can take three months for the water levels to fully subside.

Water plays a central role in the lives of everything here and supports a rich diversity of wildlife. Birds include herons, egrets, wood storks, red cockaded woodpeckers, and bald eagles. Alligators patrol the waters, and during the dry season they live in water holes that attract numerous other animals, such as deer and even bear. One of the most endangered animals is the black Florida panther. Only 50 individuals remain in the wild. The best place to see one is in the dense trees on small islands of hardwoods. These mini-forests provide the panther with dry land, cover, and prey.

Big Cypress is a recreational paradise with camping, canoeing, kayaking, and excellent hiking trails in the winter dry season. **JK**

To truly experience Big Cypress National Preserve, you need to immerse yourself in its lush greenery and enjoy the details of its rich and unusual plant and animal life. Water plays a central role in the lives of everything here.

RIGHT *Bald cypress thrive in the watery paradise of Florida's wetlands.*

EVERGLADES NATIONAL PARK

FLORIDA, UNITED STATES

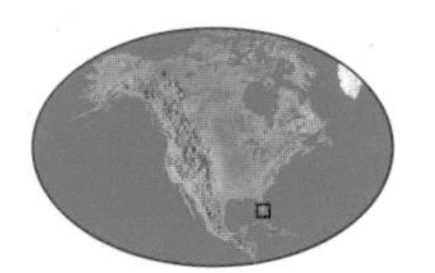

Area of Everglades National Park: 2,354 sq mi (6,073 sq km)

Water depth: 6 in (15 cm) to 3 ft (0.9 m)

Established as a national park: 1947

The Everglades National Park of southern Florida—the only subtropical preserve in North America—is a vast area of flooded subtropical grassland, mangrove and cypress swamps, forested "hammocks," and hundreds of islands that support a profusion of waterbirds and wildlife. Water from Lake Okechobee, to the north of the Everglades, drains through the shallow 50-mile- (80-kilometer-) wide "River of Grass," creating the perfect habitat for wading birds such as herons and storks, and for reptiles such as alligators and crocodiles. During the dry season (December to April) much of the wildlife is concentrated around alligator holes as the marsh dries out.

Unfortunately, this fragile ecosystem is under threat. An increasing human population has led to habitat loss, pollution, and diversion of water for drinking and flood control. However, attempts are being made to reconcile the demands of people and wildlife. As a result, many threatened species are still found here, including the American crocodile, Florida panther, West Indian manatee, wood stork, snail kite, and several species of marine turtle. The Everglades National Park was established in 1947 and has since been designated a World Heritage site, a Ramsar site (Wetland of International Importance), and an international Biosphere Reserve. RC

PONCE DE LEÓN SPRING

FLORIDA, UNITED STATES

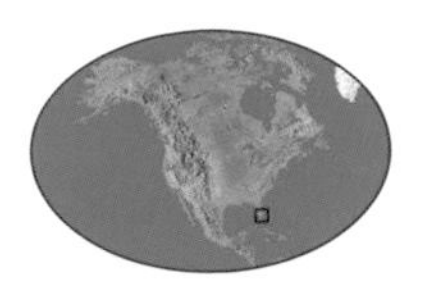

Area of Ponce de León Spring: 7,500 sq ft (697 sq m)

Temperature: 68°F (20°C)

Flow rate of water per day: 14 million gal (53 million l)

The famous conquistador Ponce De León came to Florida in search of the fabled "Fountain of Youth" that he believed would endow him with immortality and beauty when he bathed in its crystal waters. He never found it, nor did he discover Florida's 600 wonderful springs, which together represent the largest concentration of freshwater springs on Earth. Fortunately, the springs keep flowing, including this one, which has been named after the famous Spaniard.

The sparkling, clear Ponce De León Spring is located in the Florida Panhandle, in the northwest corner of the state. It is fed by two underground flows from a limestone cavity and forms a beautiful lake that maintains a cool year-round temperature of 68°F (20°C).

This is a popular destination, and both visitors and locals come here to swim and cool off from Florida's humid, sub-tropical heat. The spring is now enclosed by a rock and cement retaining wall, creating a swimming area of about 100 by 75 feet (30 by 23 meters). More than 14 million gallons (53 million liters) of freshwater pour into the spring every day, which certainly gives the spring an appearance of eternal youth and beauty. **JK**

PARICUTÍN VOLCANO

MICHOACÁN, MEXICO

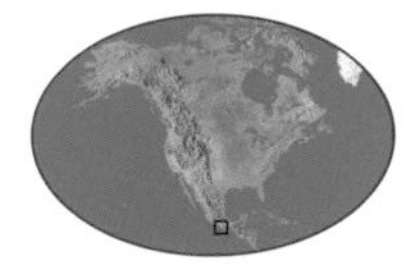

Alternative name: Volcán Paricutín

Height of volcano: 1,390 ft (424 m)

On February 20, 1943, Dionisio Pulido, a Tarascan farmer in central Mexico, witnessed the birth of a volcano as it erupted out of a cornfield. In its first explosive year the volcano's cinder cone grew to 1,100 feet (330 meters). Within two years most of the nearby villages were covered by lava and ash, which eventually spanned an area of 10 square miles (25 square kilometers). In February 1952, Paricutín's eruption ended in a blaze of violent activity. There were no fatalities from lava or ash, but three people were killed by lightning caused by the eruption.

The only other new volcano to have appeared in North America in recorded history, Jurillo, was born in 1759 about 50 miles (80 kilometers) southeast of Paricutín in the Mexican volcano belt which extends about 700 miles (1,200 kilometers) from the Caribbean to the Pacific Ocean. The eruption of Paricutín provided volcanologists with a rare opportunity to study the birth, growth, and death of a volcano. Paricutín is about 200 miles (322 kilometers) west of Mexico City. Visitors can tour the area on foot or horseback. Guides with horses can be hired from nearby Uruapan. **RC**

YUCATÁN PENINSULA

YUCATÁN / CAMPECHE / QUINTANA ROO, MEXICO

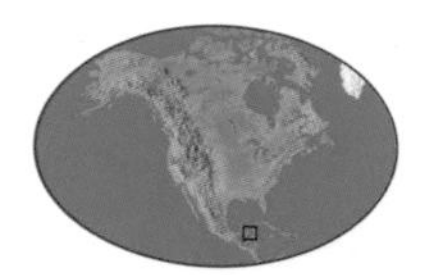

Length of Yucatán Peninsula: 186 mi (300 km)

Width of peninsula: 155 mi (250 km)

Rock type: limestone

There are very few rivers or lakes visible on the Yucatán. There are, however, plenty underground. The land is dotted with "cenotes," which comes from the Mayan *dzonot*, meaning "abyss." These sinkholes were created by the erosion of soft and porous limestone. To the Maya they are wells of life and gateways to the afterlife, but to the geologist they are entrances to a honeycomb of tunnels, passageways, and subterranean rivers and lakes. There are over 3,000 cenotes on the Yucatán, although less than half have been studied. Those open to visitors include Cenote Zaci, with its turquoise waters and eyeless black fish, and Cenote Ik Kil, which is almost perfectly round, with lush vegetation and waterfalls. Four main types of cenotes exist: underground; partially subterranean; open wells; and lakes or ponds such as the cenote at Dzibilchaltun, which is shallow at one end and 140 feet (43 meters) deep at the other.

The area also has extensive cave systems. The largest caverns are at Loltun Caves, whose name is derived from *Lol*, meaning "flower," and *Tun*, meaning "stone." They are found in the Puuc region, about 66 miles (106 kilometers) from Merida. Artifacts found here also suggest that human occupancy goes back some 7,000 years. Inside are stalactites that can be played like musical instruments. When struck, they ring out with a deep bell-like tone.

About 650 feet (198 meters) from the entrance of the Balankanche Cave is the Balam Throne, which is believed to be an underground Mayan ceremonial altar. Nearby stands a 20-foot (6-meter) tall stalagmite resembling the sacred *ceiba* tree of the Maya. The cave is located about four miles (six kilometers) from Chichen Itza. Many caverns are accessible to visitors and guided tours are available. Cave diving organized by registered diving centers is also a popular pastime in the cenotes. Many sinkholes have natural light, and visibility can be up to 150 feet (45 meters). **MB**

BELOW *Stalactites in Cenote Dzitnup.*

LA BUFADORA

BAJA CALIFORNIA, MEXICO

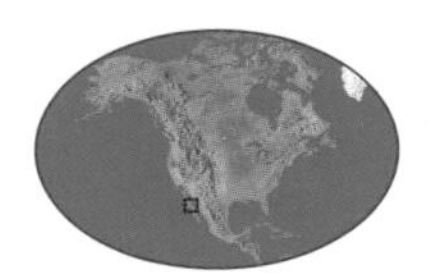

Water temperature of La Bufadora: 55 to 64°F (13 to 18°C)

Height of waterspout: 80 ft (24 m)

La Bufadora is a spectacular blowhole located in the coastal cliffs on the Punta Banda Peninsula in Baja California. The right combination of swells and tides produces an enormous spray, but even the smaller waves are impressive. The geyser is accompanied by the tremendous roar that gives La Bufadora ("buffalo snort") its name.

The spectacular spray is caused when the surge from heavy ocean swells is channeled through a deep underwater canyon into a narrow cave in the cliff. Incoming ocean waves collide with air pumped down by the force of receding waters after the previous waterspout. Compressed air and water then explode upward through the only exit, forcing water to spray high into the sky. According to local legend, the geyser is caused because a whale calf was reputed to have entered the narrow entrance to the cave one night, and by morning had grown too large to extricate itself. Over the years, its waterspout grew larger and its sobs louder until the trapped whale became a full-grown leviathan. La Bufadora is located 17 miles (27 kilometers) south of Ensenada at the tip of the Punta Banda Peninsula and is certainly worth a visit. RC

GARCIA CAVES

NUEVO LEÓN, MEXICO

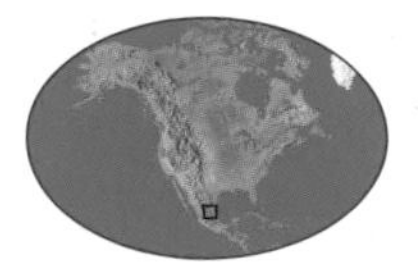

Alternative name: Las Grutas de Garcia

Estimated age of Garcia Caves: 50 million years

The Garcia cave system was discovered high in the Sierra del Fraile mountains of Mexico in 1843 by a parish priest by the name of Juan Antonio de Sobrevilla. The caves were named after the nearby town of Villa de Garcia. The caves contain a wealth of impressive stalactites and stalagmites, and the presence of marine fossils embedded in the walls provide evidence that, despite their current altitude, the rocks here used to be below sea level.

The caves are estimated to have been formed approximately 50 million years ago. Visitors can follow an illuminated route through 16 chambers—such as the Chamber of Clouds, Eighth Wonder, and Eagle's Nest. The caves are about 6 miles (10 kilometers) beyond Villa de García and about 25 miles (40 kilometers) northwest from Monterrey. The entrance to the cave system is reached via a 2,296-foot (700-meter) ascent by funicular railway—or you can go on foot. The path leading up to the caves is relatively good and affords good views of the area. There is an illuminated 1.5-mile (2.5-kilometer) route through the cave system, and guides to show groups of visitors around. RC

COPPER CANYON

CHIHUAHUA, MEXICO

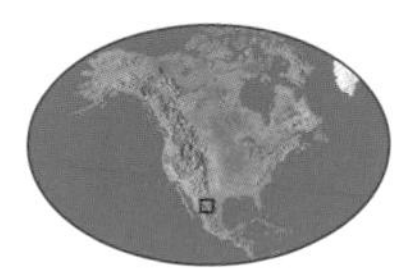

Alternative name: Barranca del Cobre

Area of Copper Canyon: 25,000 sq mi (64,000 sq km)

Elevation of canyon: 8,000 ft (2,440 m)

Copper Canyon does not really exist. It is actually 200 interlinked gorges, each formed by a river draining the western side of the Sierra Tarahumara. Comprising one-third of the Mexican state of Chihuahua, these link into six massive canyons, of which Urique Canyon is the largest. In a system of 86 tunnels and 37 bridges, Urique is traversed by Mexico's Chihuahua al Pacifico railway. Eventually all the rivers merge into the Rio Fuerte, which debouches into the Sea of Cortéz.

Copper was never mined in great quantities here. The name comes from the copper-green color of lichens on the canyon walls. The canyon system is four times larger than the Grand Canyon and four of the six major canyons exceed the depth of the Grand by over 1,000 feet (300 meters). The area supports waterfalls, spray-fed sub-tropical forest on the canyon walls, and drier forest on the upland plateaus. Nearly 300 bird species, bear, deer, and puma also live here. You will also encounter the community-based eco-tourism projects of the Tarahumara, a fiercely independent people who have preserved their traditional method of living and now support tourism as an alternative to logging. AB

SISTEMA CHEVE

OAXACA, MEXICO

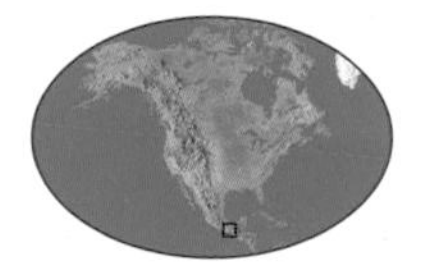

Shafts include: Elephants Shaft, Angel Falls, Saknussum's Well (deepest) at 492 ft (150 m) deep

Passages include: Northwest Passage, Hall of Restless Giants, Black Borehole, Wet Dreams, and A.S. Borehole (biggest)

Deep in Mexico's Sierra de Juárez region, in northeast Oaxaca, is one of the deepest cave systems in the world (a record held currently by Krubera Cave in the Republic of Georgia). It is estimated that the main system has tunnels deeper than 6,500 feet (2,000 meters), although cave explorers have only reached 4,869 feet (1,484 meters). Scientists put red dye in a stream close to the cave entrance and it reappeared in a karst stream at the surface eight days later, 8,284 feet (2,525 meters) lower down, and 11 miles (18 kilometers) to the north, so there is still a long way to explore yet.

The caves were discovered in 1986 by American cavers who found a huge system of deep shafts and long corridors. From the entrance, it takes two days and 37 rope drops to reach the floor of Sistema Cheve, 3,280 feet (1,000 meters) below. A 4.3-mile (7-kilometer) long tunnel then slopes to a subterranean river that leads to a water-filled tube—the Terminal Slump. So far, this has proved to be the main barrier for further exploration. It is certainly not the place for the average tourist; this site is for specialists only. The caves are 250 miles (400 kilometers) southeast of Mexico City. MB

BUTTERFLY TREES

MEXICO

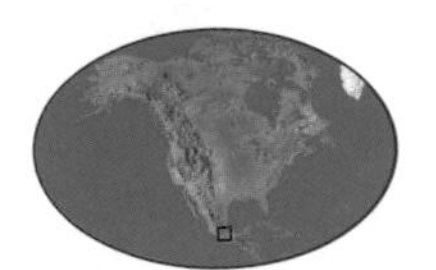

Butterfly migration distance: 3,000 mi (4,800 km)

Butterfly population: 650 million

Once a year the cloud forest pine trees in the mountains of central Mexico take on spectacular raiment as they play host to millions of overwintering monarch butterflies. Upon encountering the sea of orange and black wings that dress each tree from top to bottom, it is easy to understand why the Aztecs believed the monarchs were reincarnated warriors, resplendent in their battle colors. In fact, the life cycle of these butterflies is almost as fantastic, if only because they exhibit migratory behavior that is unparalleled in the insect world. When the weather begins to improve, the adult monarchs begin to fly north. As they migrate, they stop to lay eggs on milkweed plants. The caterpillars then eat the poisonous milkweed, utilizing its toxins as a defense against predators. The caterpillars grow and pupate, emerging as adults that continue the northerly migration.

Toward the end of summer, reduced temperatures and daylight trigger behavioral changes in the adults that cause them to begin to fly south once again. Fat stored in the abdomen allows the butterflies to make the long journey to Mexico, which can be as far as 3,000 miles (4,800 kilometers). Scientists are still not certain how monarch butterflies navigate, because each individual only manages to complete part of the lengthy round-trip during its short lifetime. The hundreds of millions of butterflies that create the wonderful spectacle are distant relatives (the great-great-grand-offspring) of the ones from the year before.

However the butterflies manage to make the journey, it is certain that they are under severe threat from loss of habitat. The very trait that makes the butterfly trees one of the most beautiful of natural sights—their dense packing into a small area—makes them especially vulnerable to the activities of illegal loggers. Actually seeing the overwintering monarchs is not easy, but 10,000 visitors—mostly Mexican families and school groups—travel to the El Rosario sanctuary in Angangueo during the three months of the year that the butterflies roost there. The forests of oyamel firs are situated high in the mountains and best reached under the direction of a local driver. NA

Upon encountering the sea of orange and black wings that dress each tree from top to bottom, it is easy to understand why the Aztecs believed the monarchs were reincarnated warriors, resplendent in their battle colors.

RIGHT *Monarch butterflies rest on a tree trunk in the mountains of Central Mexico.*

BAJA CALIFORNIA PENINSULA

BAJA CALIFORNIA / BAJA CALIFORNIA SUR, MEXICO

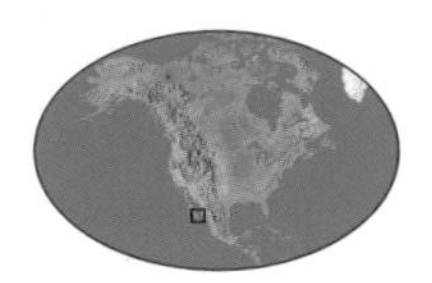

Length of peninsula: 777 mi (1,250 km) from Tijuana to Cabo San Lucas

Highest point of Baja Norte: Cerro de la Encantada—10,157 ft (3,096 m)

Highest point of Baja Sur: Sierra de la Laguna—7,894 ft (2,406 m)

Baja is a long peninsula that cuts off a huge arm of the Pacific Ocean known as the Sea of Cortéz. This long strip of ocean is home to huge schools of common dolphins that migrate from one side to the other; whale sharks—the largest fish in the sea that come here to feed; basking sharks that come a close second in size; and hammerhead sharks that gather by day in huge schools at seamounts. On the Pacific side of the peninsula, huge shallow-water lagoons play host each winter to gray and humpback whales. From January through March, they come here from feeding sites farther north to drop their calves in the safety of the warm bays and to mate. Small whale-watching boats at Scammon's Lagoon out of Guerrero Negro, San Ignacio Bay, and Magdalena Bay enable visitors to be within arm's reach of the grays. Humpback trips are arranged in La Paz for boats out of San José de Cabo. The peninsula itself has deserts, mountains, pine forests, and untouched beaches, but the most famous features must be the 200 mainly uninhabited, cactus-studded islands immortalized by author John Steinbeck and marine biologist Ed Ricketts. Here seabirds and sea lions rule, not people. MB

PACAYA VOLCANO

GUATEMALA / ESCUINTLA, GUATEMALA

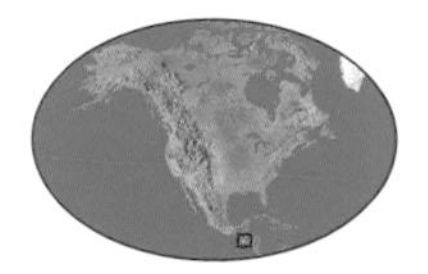

Alternative name: Volcán Pacaya

Elevation of volcano: 8,371 ft (2,552 m)

Pacaya, one of Guatemala's most active volcanoes, is also the country's most climbed. It is easily accessible and often puts on a spectacular show. Eruptions are sometimes visible from Guatemala City, which is nearly 20 miles (32 kilometers) north. Although Pacaya has erupted at least 23 times since 1565, little is known about its early eruption history.

The volcano was dormant from 1860 until March 1961, when it erupted without warning. In 1962 a pit crater was formed after a collapse near the summit. Since 1965 Pacaya has been continuously active, with eruptions that range from minor gas and steam emissions to explosions that can hurl rocks up to 7 miles (12 kilometers) and cause evacuations of nearby villages. Licensed tours can be arranged from Antigua, and independent travelers are advised to hire a local guide. The entrance to Pacaya Volcano National Park is at San Francisco de Sales. Most tour groups follow the main trail that starts here, although another trail starts at the radio towers on the flank of Cerro Chino. The climb generally takes two to three hours, and weather is usually better in the morning than the afternoon. Visitors to the rim should be wary of volcanic gases. **RC**

FUEGO VOLCANO

SACATEPÉQUEZ / CHIMALTENANGO, GUATEMALA

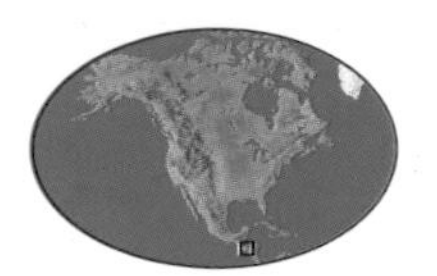

Alternative name: Volcán Fuego

Summit elevation: 12,346 ft (3,763 m)

One of Central America's most active volcanoes, Fuego has erupted more than 60 times since 1524. Fuego's most recent large eruptions were in October 1974, with four distinct pulses of volcanic activity over a 10-day period (each lasting 4 to 17 hours). Avalanches of lava were recorded pouring down the slopes of the volcano at 35 miles per hour (60 kilometers per hour). The ash cloud above the volcano reached a height of four miles (seven kilometers). Eruptions tend to occur in clusters that happen at intervals of 80 to 170 years and last for 20 to 70 years.

The ancestral Meseta volcano collapsed 8,500 years ago, causing a huge avalanche on to the Pacific coastal plain about 30 miles (50 kilometers) away. The most recent eruption cluster began in 1932 and is still continuing, consisting to date of more than 30 individual eruptions. There were fatalities in three of these eruption events. The timing of Fuego's eruptions seems to be influenced by the planet's tides, although most of the major eruptions at Fuego volcano seem to have occurred around February or September, the reasons for which are not known. Tours to the volcano can be arranged from Antigua. RC

LAKE ATITLÁN

SOLOLÁ, GUATEMALA

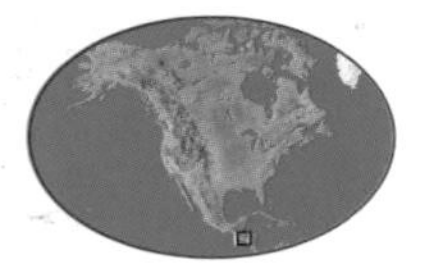

Alternative name: Lago de Atitlán

Surface area of Lake Atitlán: 50 sq mi (130 sq km)

Elevation of lake: 5,123 ft (1,562 m)

Upon viewing Lake Atitlán, the 19th-century German explorer Alexander von Humboldt called it "the most beautiful lake in the world." A sparkling crater lake ringed by three massive volcanoes and surrounded by small villages of people who still speak two different Mayan dialects, it is hard to doubt this impression. Lake Atitlán was born about 84,000 years ago, when an intense eruption blew a hole 11 miles (18 kilometers) in diameter and 3,000 feet (900 meters) deep, pitching volcanic ash as far south as Panamá and north past Mexico City. As rainwater filled the crater, the magma below crept through other openings in the planet's surface, giving rise to the trio of volcanic peaks which today range between 10,500 and 12,500 feet (3,200 and 3,810 meters) in elevation. With a present-day depth of more than 1,100 feet (330 meters), Lake Atitlán lacks any surface drainages, but is thought to be linked to a network of underground rivers and streams. Normally calm and clear, the afternoon force known as *Xocomil*, or the "wind that carries away sin," can whip the lake into treacherous white foam. Bus or drive from Guatemala City to the small artisan village of Panajachel. DBB

SANTA MARÍA VOLCANO

QUETZALTENANGO, GUATEMALA

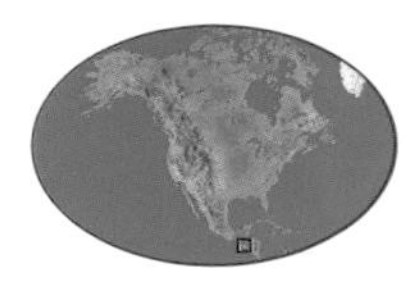

Alternative name: Volcán Santa Maria

Elevation of volcano: 12,372 ft (3,772 m)

The first historic eruption of Santa María volcano in 1902 killed at least 5,000 people and was one of the largest eruptions of the 20th century. The skies were darkened over Guatemala for days, a large crater was formed in the volcano's southwest flank, and ash from the eruption was detected as far away as San Francisco in California. After almost 20 years of inactivity, in June 1922 a volcanic lava dome called Santiaguito started forming inside the crater. Since its formation, Santiaguito has been continuously active, causing avalanches near the dome and the 1902 crater, and sedimentation in rivers south of the dome. The partial collapse of Santiaguito in 1929 caused lava flows resulting in hundreds of deaths and damage to villages and plantations. Eruptions at Santiaguito in May 1992 produced ash columns up to 6,000 feet (1,800 meters) high. Santa María volcano is still considered to be dangerous if only because of the possibility of another dome collapse similar to the one that happened in 1929.

Catastrophic mudflows and flooding in the monsoon season can be caused by volcanic debris flowing into rivers south of Santiaguito. The nearby city of Quezaltenango, with a population of approximately 120,000, is situated below the summit. RC

BLUE HOLE NATIONAL PARK & ST. HERMAN'S CAVE

CAYO, BELIZE

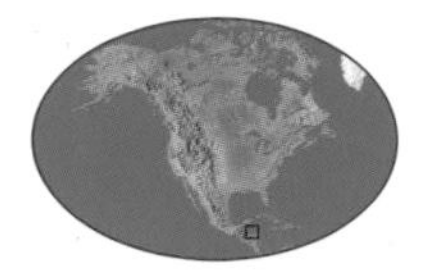

Length of St. Herman's Cave: approx. 2,500 ft (760 m)

Height of main passage: 50 ft (15 m)

The blue hole that gives this national park its name is a beautiful sapphire-blue pool in a sinkhole formed by the collapse of an underground river channel. The 26-foot (8-meter) deep pool is a popular spot for swimming but may get muddy after heavy rain. The 575-acre (233-hectare) Blue Hole National Park is covered with primary- and secondary-growth forest, and contains many karst features such as underground rivers, sinkholes, and caves.

The largest of the three known entrances to St. Herman's Cave is an impressive 200-foot- (60-meter-) wide sinkhole that narrows to a 66-foot- (20-meter-) wide cave entrance. Visitors can walk 1,000 feet (300 meters) into the cave to see stalagmites and stalactites. St. Herman's Cave is accessible via Blue Hole National Park without obtaining the usual permit required to enter archeological caves in Belize. Reach the Blue Hole National Park by bus or car, just 12 miles (19 kilometers) southeast of Belmopan on the Hummingbird Highway. RC

GUANACASTE NATIONAL PARK

CAYO, BELIZE

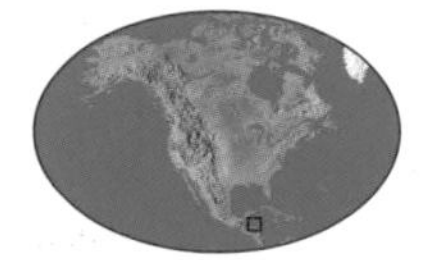

Area of Guanacaste National Park: 50 acres (20 ha)

Number of bird species: over 120 recorded

Guanacaste National Park, established in 1990, is an area of tropical forest in central Belize. Many of the larger trees have been spared felling for timber by the park's protected status. The park gets its name from the enormous guanacaste (tubroos) tree found near its southwestern border. The guanacaste tree is one of the largest trees in Central America, reaching a height of 131 feet (40 meters) and a trunk diameter of 6.6 feet (2 meters). The tree has a broad spreading crown and supports numerous epiphytic plants, such as orchids and bromeliads, in its upper branches.

A wide range of plants in the park can be viewed from the well-marked trails, including the black orchid (the national flower of Belize). There is also plenty to see for bird-watchers—over 120 bird species have been recorded, including the black-faced ant thrush and the blue crowned mot mot. Mammals found within the park include jaguarundi, nine-banded armadillo, kinkajou, paca, and white-tailed deer—but these are less easily seen. The Guanacaste National Park is located less than 2 miles (3.2 kilometers) to the north of the capital city of Belmopan. RC

BARTON CREEK CAVE

CAYO, BELIZE

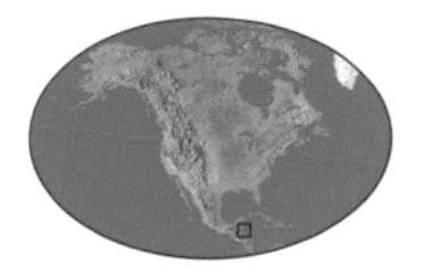

Length of Barton Creek Cave: 4.5 mi (7 km)

Age of Mayan artifacts: 1,100–1,700 years

Barton Creek Cave is a large river cave in Belize's Cayo District that appears to have once been used by the Maya for ritual burials. Preliminary archeological investigations of bones found within the cave suggest that at least 28 Mayan people, ranging in age from children to elderly adults, were buried inside the cave. Visitors can enter the cave in a canoe equipped with powerful spotlights to see the unusual rock formations, caverns, Mayan artifacts, and burial remains.

The underground river cave is navigable for some distance, depending on water levels. You have to pick your way through some very narrow passages to see some of the Mayan pottery and skeletal remains. The importance of caves within Mayan culture has been greatly enhanced by information gathered from recent investigations into the caves at Barton Creek. Archeological evidence of Mayan activity has been found in the entrance of the cave, and up to 1,000 feet (300 meters) into the cavern. Barton Creek Cave is less well known than Rio Frio Cave but is quickly becoming popular. Access to the Barton Creek Cave is via a picturesque Mennonite farm community on pre-arranged tours. **RC**

THOUSAND-FOOT FALLS

CAYO, BELIZE

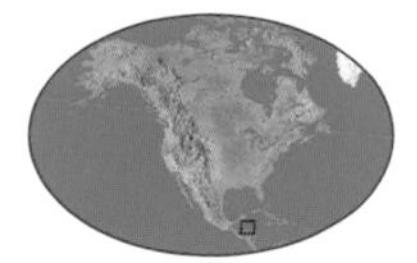

Alternative name: Hidden Valley Falls

Height of falls: 1,500 ft (450 m)

Thousand-Foot Falls is believed to be the highest waterfall in Central America. The spectacular waterfall plunges in a sheer drop from its granite edge, down the mist-shrouded hillside, to the jungle below. Despite being a major tourist attraction, Thousand-Foot Falls only became a protected area in September 2000, when it was declared a national monument.

The falls are located in the rugged Mountain Pine Ridge Forest Reserve, an area of almost 300 square miles (777 square kilometers) in western Belize. From the entrance to the forest reserve follow the main road for 2 miles (3.2 kilometers), then take the turning to the falls. The falls and picnic area are about 4 miles (6.4 kilometers) further down this road. It is a moderately difficult hike up to the viewing platform overlooking the falls. There is a short scenic trail around the escarpment which affords views of the valley. Day tours around Mountain Pine Ridge usually include Rio Frío cave, Rio On pools, and either Thousand-Foot Falls or Big Rock. Rio On pools is a series of warm water pools linked by small waterfalls that flow between large granite boulders. **RC**

BELIZE BARRIER REEF

BELIZE

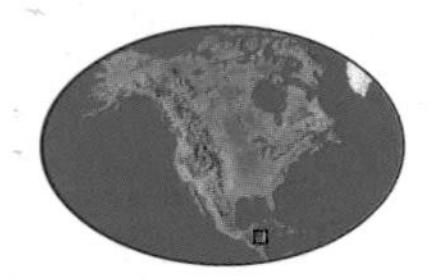

Length of Belize Barrier Reef: 155 mi (250 km)

Width of reef: 6 to 18 mi (10 to 30 km)

Number of cays: more than 200

The Belize Barrier Reef is the longest in the western hemisphere and the second longest continuous reef in the world (Australia's Great Barrier Reef is longer). The reef runs roughly parallel to the coast, from Belize's border with Mexico in the north to Guatemala in the south. The sea between the reef and the mainland is a shallow lagoon that is less than 16 feet (5 meters) deep.

Just offshore, and protected by the reef, are more than 200 islands known as cays. Caulker and Ambergris are the most popular cays with tourists. Lettuce corals are the most common corals on Belize's reef—but this was not always the case. Lettuce corals increased at the expense of staghorn corals, which have declined dramatically since 1986, probably as a result of bacterial disease. The crystal-clear waters are superb for diving and snorkelling, although divers should beware of fire corals (not true corals) that cause an unpleasant sting. Several species of angelfish and parrotfish are commonly found on the reef, while larger fish such as barracuda, sharks, and rays can also be seen. **RC**

RIGHT *An aerial view of Lighthouse Reef and Blue Hole.*

MONTECRISTO-TRIFINIO CLOUD FOREST

HONDURAS / GUATEMALA / EL SALVADOR

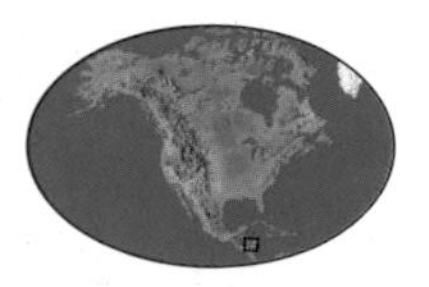

Altitude of Montecristo-Trifinio cloud forest: 2,165 to 7,900 ft (660 to 2,400 m)

Features: tertiary volcanoes

Located in the highlands of western Honduras, Montecristo is remote and only accessible from October to March. The park primarily protects Central American temperate types of mixed pine-oak-cypress-laurel forests characteristic of Honduran tropical regions. The area is a Biosphere Reserve shared with Guatemala and El Salvador and is one of 30 cloud forests in Honduras.

While the rest of the isthmus was beneath the sea, this area was dry land, resulting in the isolation of plants and animals that are now unique to these forests. Outside, park logging has left the habitat type endangered.

The 100-foot (30-meter) trees create a regal forest. Pines occupy the ridges, oaks the valley. Trees are covered in moss and lichens. Many of the conifers are unique, as are bird species such as the resplendent quetzal and highland guan. Spider monkeys live in this area too. The nearest town is Nueva Ocotepeque, which lies about 10 miles (16 kilometers) to the east of the park. **AB**

PULHAPANZAK WATERFALL & LAKE YOJOA

SANTA BÁRBARA, HONDURAS

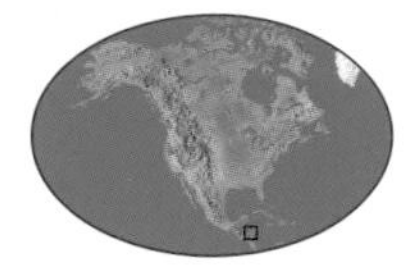

Height of Pulhapanzak waterfall: 140 ft (43 m)

Area of Lake Yojoa: 14,000 acres (5,600 ha)

Depth of Lake Yojoa: 50 ft (15 m)

Pulhapanzak waterfall used to be an important Mayan cultural site. Here the Rio Lindo crashes down 140 feet (43 meters) into a natural pool, which has become a popular swimming spot. Nearby Lake Yojoa, high in the mountains and surrounded by cloud forest, is the only large lake in Honduras.

Lake Yojoa is also one of the best bird-watching sites in the region; nearly 400 species have been recorded here. The forest comes down to the steep eastern shore, while the marshy habitat of the western shore is ideal for snail kites, herons, storks, and many species of wildfowl. The lake is also an important bass fishing center. Visitors to the Cerro Azul Meambar and Santa Bárbara National Park cloud forest reserves that border the lake may see howler monkeys, sloths, toucans, and even the elusive quetzal. Pulhapanzak waterfall is 70 miles (110 kilometers) south of San Pedro Sula. The drive from Santa Barbara gives superb views over the country's mountains, valleys, and coffee plantations. **RC**

SANTA ANA & IZALCO VOLCANOES

SANTA ANA, EL SALVADOR

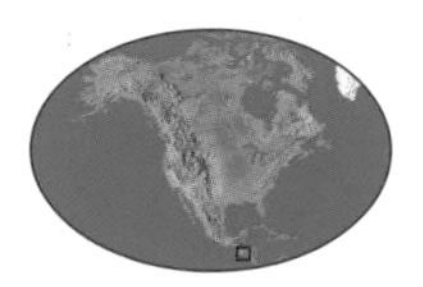

Elevation of Santa Ana Volcano: 7,759 ft (2,365 m)

Elevation of Izalco Volcano: 6,398 ft (1,950 m)

Santa Ana Volcano in southwestern El Salvador is the highest peak in the country. It has erupted 12 times since its first recorded eruption in 1520. Santa Ana Volcano has a large central vent with a flat circular crater and a small crater lake whose sulfur-rich waters are emerald green in color. Since 1770, volcanic activity at the Santa Ana and Izalco volcanoes has been almost simultaneous.

Izalco Volcano is El Salvador's youngest volcano and is inextricably linked to its older counterpart, the Santa Ana Volcano. Born in 1770 on the southern flank of Santa Ana volcano, Izalco was so active from 1770 until 1958 (with more than 50 eruptions) that it became known as the "Lighthouse of the Pacific" to sailors who were known to have set their course by its constant glow. However, since 1958 this volcano has been almost completely dormant, despite a brief flank eruption, which was recorded in 1966. There are good views of the Coatepeque caldera and surrounding volcanoes from the park. **RC**

ALEGRIA LAKE & TECAPA VOLCANO

USULUTÁN, EL SALVADOR

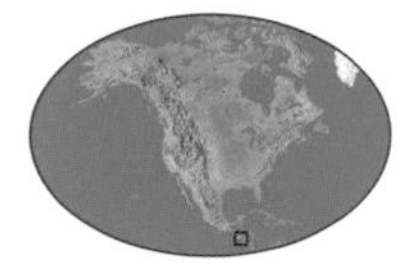

Height of Tecapa Volcano: 5,226 ft (1,590 m)

Height of crater walls: 1,150 ft (350 m)

The Alegria Lake is an emerald-green sulfur lake on the slopes of the dormant Tecapa volcano in the department of Usulután. The lake is situated below a deep notch in the eastern rim of the crater. Though fed by boiling water that seeps from the ground, the lake water is only lukewarm. Locals claim that the sulfur-rich waters have medicinal properties. The forest couched within the crater is home to wildlife such as agoutis and coatis and a wide variety of birds.

Although Tecapa volcano is listed as extinct, steam still erupts from collapsed wells near the geothermal power plant. Tecapa Volcano is at the northern end of a volcano cluster to the west of the San Miguel Volcano. There are several young lava flows and cones on its slopes, and a lake at the summit below a deep notch in the crater rim. The lake and forest in the crater are being managed for eco-tourism by the nearby town of Alegría. Guides are available to escort tourists to the lake or around the crater. **RC**

MOMOTOMBO VOLCANO

LEÓN, NICARAGUA

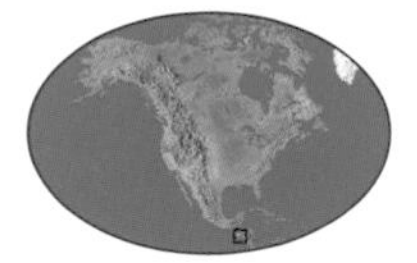

Altitude of Momotombo Volcano: 4,199 ft (1,280 m)

Geological type: active stratovolcano

Momotombo forms the centerpiece of Maribios Volcanic Chain—a 10-cone arc that cuts western Nicaragua in a diagonal line. Lying on the northwestern shores of Lake Managua, Momotombo is one of Nicaragua's most familiar landmarks. It began growing 4,500 years ago, emerging from an older cone. Debris from the volcano formed several islands in Lake Managua. There is also a major geothermal field located on the southeast flank of Momotombo, which features fumaroles and hot springs. Electricity produced from this accounts for 35 percent of Nicaragua's power output. The lake also has a number of small volcanically-derived islands, including the 1,283-foot (391-meter) Momotombito.

Since 1524, Momotombo has erupted 15 times, most recently in 1905. The 1605–1606 eruption destroyed the former capital, León. There are 57 other volcanic formations in the country, ranging from extinct craters with deep aquamarine pools to young, bubbling, lava-filled calderas. The nearest city is León, which was rebuilt about 18 miles (30 kilometers) to the west. A strenuous three-hour hike to the top of the Momotombo crater will unveil rewarding vistas of the area. **AB**

LAKE NICARAGUA

RIVAS / GRANADA / MASAYA / CHONTALES / RÍO SAN JUAN, NICARAGUA

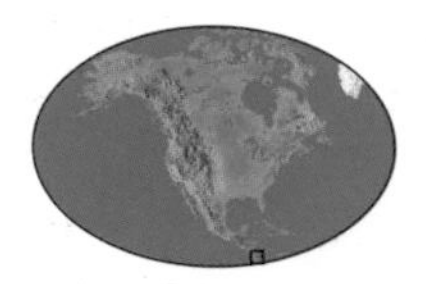

Surface area of Lake Nicaragua: 3,192 sq mi (8,264 sq km)

Length of lake: 96 mi (160 km)

Maximum width of lake: 43 mi (72 km)

Named Cocibolca ("Sweet Sea") by early inhabitants, Lake Nicaragua is indeed more an inland sea than a lake—powerful storms often rock its shores. The largest body of freshwater in Central America, it has more than 300 islands, including that of Ometepe, a twin volcano that rises to a height of 5,300 feet (1,615 meters) above the water's surface.

It is also one of the few lakes in the world to be home to sharks. The region's first settlers were reputed to fear the hungry animal-lords upon whose shores they lived and, legend has it, appeased the sharks by feeding them their deceased, ornamented in gold.

Biologists considered Lake Nicaragua's sharks a unique species until 1966, when a tagging study showed that the population actually consists of dolphin-sized bull sharks that migrate between the lake and the Caribbean Sea along the Río San Juan, jumping its many rapids. Lake Nicaragua has been a place of controversy since failed British and American attempts in the mid-1800s to dredge a canal from its western shores to the Pacific to accommodate shipping traffic. Although the Panama Canal has since filled that role, talks of a Nicaraguan canal are ongoing. DBB

VENADO CAVES

ALAJUELA, COSTA RICA

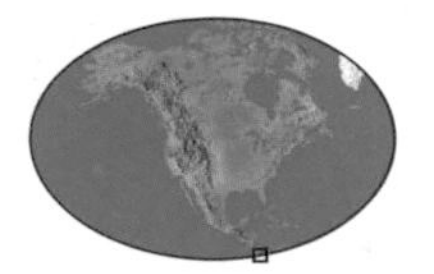

Age of Venado Caves: 5 to 7 million years

Height of Venado Cave passage: from 6 ft (1.8 m) to 15 ft (4.6 m)

A serpentine network of underground tunnels, the Venado Caves in northern Costa Rica is not a typical cave system. Created by rivers that bullied their way into fissures in the earth, over millennia the persistent flows eventually drilled 1.5 miles (2.4 kilometers) of convoluted passageways connecting 10 distinct rooms. With subterranean streams still flowing strongly today, Venado is an underground wonderland of miniature waterfalls, fossil corals, shells, and rock formations such as "Papaya," a man-sized natural carving in the shape of the tropical fruit.

Although visitors need not be seasoned spelunkers to tour Venado Caves, they are definitely not for the claustrophobic. You enter the cave where the stream spits out and are soon crawling against the current beneath a wedge of limestone rock. Farther in you find yourself in waist-high water in a world illuminated only by your headlamp, bats circling above, and cave fish nibbling your ankles. You may also spot tiny colorless frogs leaping from rock to rock. The caverns often close from August to October when torrential rains transform the stream into a raging river and conditions become dangerous. DBB

POÁS VOLCANO

ALAJUELA, COSTA RICA

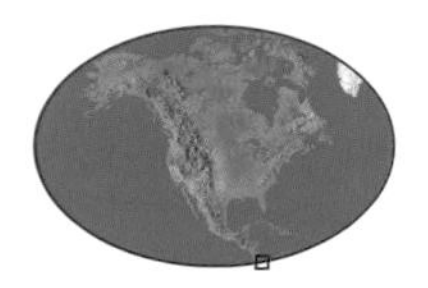

Depth of volcano crater: 1,000 ft (300 m)

Width of crater: 1 mi (1.6 km)

The jewel of central Costa Rica's volcanic backbone, the crater lake of Poás Volcano, shimmers a variety of colors. One day it is an emerald-green cauldron of acidic, near-boiling water embedded in a landscape of moon-gray rock, bubbling mud, and countless fissures which steadily belch yellow sulfur gas. The next, it is a steamy mineral bath of blue, turquoise, or gold, as rainwater levels change its chemical composition.

Active well before records were started in 1828, the volcano last erupted cataclysmically in 1910, when it sent ash rocketing 2 miles (3.2 kilometers) high and caused shockwaves as far as Boulder, Colorado. Smaller eruptions in 1989 and 1995 resulted in the evacuation of nearby towns. Although Poás is currently in a state of relative calm, emissions of sulfur and chlorine have resulted in acid rain damage to local coffee and berry plantations. Measuring approximately 1 mile (1.6 kilometers) wide, the collapsed crater is the largest in the western hemisphere. Its high rim affords clear views across the country to both the Pacific Ocean and the Caribbean. Bathed in clouds and cold temperatures, its windswept upper flanks sprout a rare cloud forest that is stunted and contorted but bursting with flowering plants and birdlife. **DBB**

SAND DUNE VOLCANO

GUANACASTE / ALAJUELA, COSTA RICA

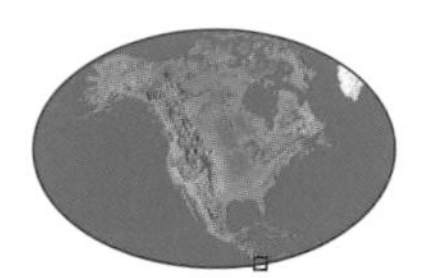

Alternative name: Volcán Arenal

Average daily eruptions: 41

Height of Sand Dune Volcano: 5,366 ft (1,636 m)

A stalwart member of the violent Ring of Fire that lights up the Pacific Rim, Sand Dune Volcano is a scientist's dream volcano. In the shape of a perfect cone, it is one of the world's most active volcanoes, spewing molten lava every 15 minutes and tossing out red-hot rocks the size of small houses every few hours. The youngest of the country's nine active volcanoes, it was dormant from 1500 until 1968 when Arenal Mountain became known as a volcano. Completely burying three small villages, the volcano ruined more than 15 square miles (40 square kilometers) of crops, forest, and property.

Sand Dune Volcano was established as a national park in 1995 and is now a top tourist attraction, but it can only be viewed from a distance. Every few years, climbers who foolishly walk past the warning signs to ascend its unpredictable flanks do not return. On September 5, 2003, a large part of the northwest rim collapsed, triggering four avalanches in 45 minutes. The volcano should not take long to recover, however, as it grows an average of 20 feet (6 meters) per year. **DBB**

BARRA HONDA CAVES

GUANACASTE, COSTA RICA

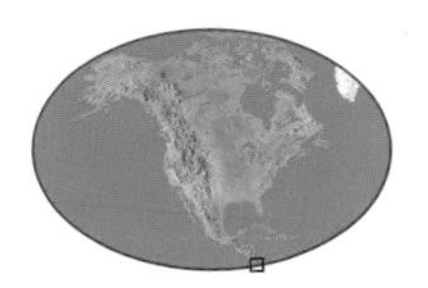

Age of Barra Honda Caves: 60 to 70 million years

Range of cave depth: 200 to 790 ft (60 m to 240 m)

Since the 1960s, when a large, seemingly bottomless pit was discovered in Barra Honda Peak of northwestern Costa Rica's Nicoya Peninsula, a total of 42 independent chambers have been uncovered. With only 19 of these already explored, many secrets remain to be uncovered in the Barra Honda cave system. Uplifted and eroded over millions of years, today Barra Honda is a 1,400-foot-(420-meter-) high coastal mesa crowned in dry tropical forest. Rainwater has slowly dissolved holes up to 2,800 feet (853 meters) deep in the limestone.

The subterranean labyrinth includes the Trap, which features a 171-foot (52-meter) vertical drop; Stinkpot, named for the fetid guano its large population of bats produces; and Nicoya, where pre-Columbian human remains and artifacts have been found. La Terciopelo, named after a deadly venomous snake found in the area, has unique formations including "The Organ" which produces tones when tapped. Fortunately this is the one cave open to the general public. Even so, with an initial vertical descent of 100 feet (30 meters) just to enter the chamber, Barra Honda is not for the fainthearted. DBB

PLAYA OSTIONAL

GUANACASTE, COSTA RICA

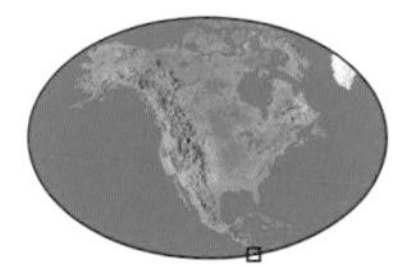

Arribada beaches in Costa Rica: Ostional and Nancite

Length of olive ridley turtle: 24 to 30 in (60 to 75 cm)

Egg harvesting: 1 million per month

Triggered by the last quarter of the moon each month, hundreds of thousands of olive ridley turtles emerge from the Pacific Ocean and deposit their eggs on the beach at Ostional. This phenomenon is called the *arribada*, meaning "the arrival." First a few hundred arrive, and then a steady flood for up to a week, during both the night and day. So many turtles come ashore that females in the second and subsequent waves dig up the eggs of the first. Local people are therefore allowed to collect eggs during the first couple of days—the only legal turtle egg harvest in the world.

The peak of activity is during the wet season, between July and December, with the biggest recorded emergence in November 1995 when 500,000 female turtles nested in a single arribada. Sometimes during August through October, two arribadas occur in the same month, thus giving the appearance of continuous activity. The beach is one of the world's most important mass nesting sites, and one of 60 on Costa Rica's Pacific and Atlantic coasts. Ostional is 40 miles (65 kilometers) south of Santa Cruz. The road is a dirt track, but is only impassable if the Rio Rosario becomes swollen with rain. MB

COCOS ISLAND

GUANACASTE, COSTA RICA

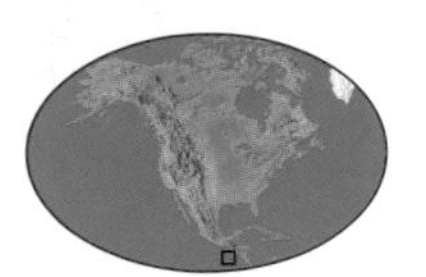

Area of Cocos Island: 9.3 sq mi (24 sq km) seamount

Highest point (Cerro Iglesias): 2,080 ft (634 m)

Cocos is no more than a dot of rock in the Pacific Ocean, yet the critical habitats its environment provides for marine wildlife make it one of the most spectacular places in the world for sharks. Here scalloped hammerheads gather in schools more than a hundred strong, while whitetip reef sharks skim the rocks as aggressive males pursue females to mate.

In fact, Cocos is said to have the greatest density of sharks of any dive site in the world. Deep ocean currents swirl to the surface, bringing nutrients to an assortment of undersea life—from huge silvery shoals of bait fish and gigantic manta rays to the largest fish in the sea, the whale shark. The island itself is volcanic, and the only one in the eastern Pacific covered by tropical rainforest. It has 70 unique plant species as well as three unique birds: the Cocos Island finch, flycatcher, and cuckoo. There is no camping on the island, so it is only accessible from live-aboard dive boats. The very beautiful but bumpy sea journey is 36 hours out of Puntarenas. **MB**

RIGHT *Whitetip reef sharks circle in the Cocos waters.*

MOUNT CHIRRIPÓ

CARTAGO, COSTA RICA

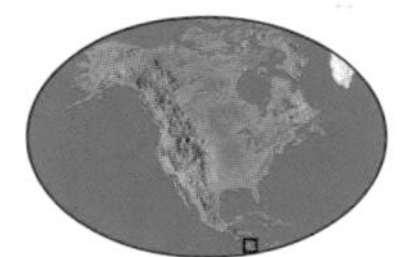

Height of Mt. Chirripó: 12,526 ft (3,819 m)

Average rainfall: 16 in (400 mm)

As the great ice sheets that capped Costa Rica's highest peaks during the last ice age started melting 18,000 years ago, they left behind a flowing landscape of rounded ridges, polished plains, and bowl-like basins. Most signs of this glacial past are now hidden—they lie under verdant rainforests, or have been disfigured by volcanic eruptions and earthquakes. However, its effects have gone unchanged on the upper reaches of Chirripó, the tallest peak in the country and the second highest in Central America.

Among Chirripó's U-shaped valleys and rolling hills of rubble dotted with crystal-clear lakes lies a strange mix of habitats like no other in the country. Chirripó means "land of eternal waters" because of the abundance of these glacier lakes. Mixed rainforest of giant oak and elm towering above fluffy fields of ferns and stout stands of bamboo give way to a rare habitat called páramo, a thick carpet of low-growing bamboo, and other alpine shrubs. Chirripó is part of the southern Talamanca Range, which extends into Panama. It was designated a World Heritage site in 1983, not only for its glacial features, but because of interbreeding populations of wildlife from North and South America, and its occupation by four indigenous tribes. **DBB**

TURTLE NATIONAL PARK

LIMÓN, COSTA RICA

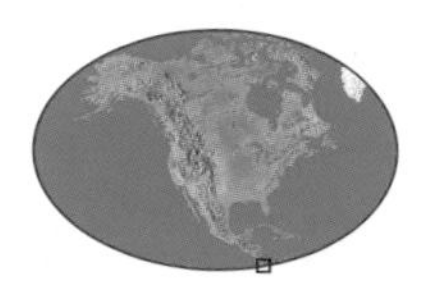

Area of Turtle National Park: 78,000 acres (31,198 hectares)

Surface area of water in national park: 130,000 acres (52,000 hectares)

Annual rainfall: up to 20 ft (6 m)

Since before the Spanish conquest of Costa Rica in the mid-16th century, locals have depended on the isolated Caribbean beach known as Tortuguero, just north of Puerto Limón, to harvest the sea turtles for their meat, eggs, and shells. By 1950, as human populations boomed, the sea turtles were driven close to local extinction. In an attempt to preserve diminishing populations, 22 miles (35 kilometers) of beach were protected in 1970. Since then, sea turtle nesting populations have grown from a few hundred individuals up to 37,000, making it the most important nesting site in the Atlantic.

Today, Turtle National Park includes mangrove, swamp, and lowland wet forest that are home to monkeys, manatees, and crocodiles. The park receives more rain than any other part of the country and is mostly inaccessible, though tourists still flock to its seemingly endless stretch of beach to witness the *arribada*, when hundreds of sea turtles haul out at the same time to deposit their eggs. The largest number of turtles nest from late July to October. Boats plying canals parallel to the beach leave Moín daily; alternately, take a puddle jumper from San José or Limón. DBB

CORCOVADO NATIONAL PARK

PUNTARENAS, COSTA RICA

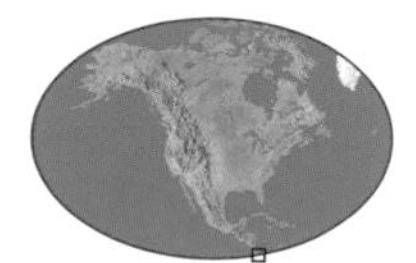

Area of Corcovado National Park: 136,348 acres (54,539 ha)

Surface area of water in national park: 6,000 acres (2,400 ha)

Corcovado National Park is the largest Pacific lowland rainforest located in Central America. The park, in southwestern Costa Rica, is a lush amalgam of jungle, swamp, mangrove, and desolate beach, interrupted only by large rivers and fast-flowing streams. A unique mosaic of eight distinct habitats makes this a haven for rare wildlife, including six wildcat species, pony-sized tapirs, giant anteaters, and the largest population of endangered scarlet macaws in the country.

Here, the world's biggest and most powerful eagle, the harpy, hunts monkeys while 10-foot (3-meter) bushmaster snakes look for meals of small mammals; Jesus Christ lizards run across lazy rivers, in which crocodiles await bigger fare; and sea turtles nest on beaches whose shores sharks regularly patrol. However, the most coveted of Corcovado's charismatic creatures is an elusive and skilled predator, the jaguar. Once approaching local extinction in the 1960s, populations have more than tripled since the park's creation in 1975. Corcovado's special assemblage of ecosystems has drawn international attention from ecologists, which subsequently opened up the park for ecotourism. The humid forests are some of the last places that still maintain the ecosystem of the tropical forests in the American Pacific. Three research stations connected by rugged trails make experiencing Corcovado possible. **DBB**

PEACE WATERFALL

HEREDIA, COSTA RICA

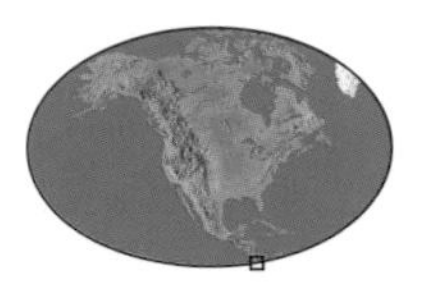

Alternative name: Catarata La Paz

Number of falls: 5

Height range of waterfalls: 60 ft (18 m) to 120 ft (36 m)

Rio La Paz in northern Costa Rica begins its journey in virgin cloud forest at the summit of an active volcano and then tumbles nearly 5,000 feet (1,525 meters) in five miles (eight kilometers), before taking a final dramatic plunge out of dense tropical vegetation in a cascade known as Catarata La Paz, or Peace Waterfall. It is not size that gives La Paz character, though, but its surroundings and accessibility. Bursting out of a wall of impenetrable rainforest with such force that ferns the size of grown men whirl about while thick vines sway side to side, the fall can be seen from a quiet country road close enough to get dusted in mist.

However, Peace Waterfall is only the beginning—behind it lies the richest waterfall area in the country. Until 2001, this area was inaccessible. Now, with the construction of Peace Waterfall Gardens, visitors are offered a rare glimpse of nature's otherwise concealed power. The preserve includes a series of sturdy stairways as well as viewing platforms built into the sides of cliffs which afford vistas above and below of five breathtaking cascades. **DBB**

RIGHT *Peace Waterfall cascades through lush green forest.*

BIMINI WALL & BIMINI ROAD

BIMINI, BAHAMAS

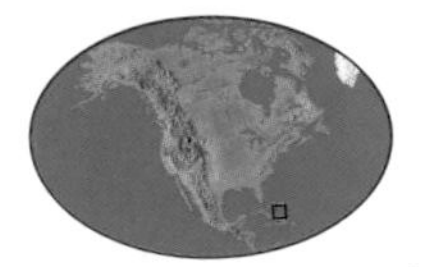

Depth of Bimini Wall: 145 ft (45 m) to over 3,000 ft (900 m)

Length of Bimini Road: 1,000 ft (300 m)

Weight of Bimini Road limestone blocks: 1 to 10 tons

The Bimini island group, at the western end of the Bahamas, is notable for two extraordinary underwater features: the Bimini Wall and the Bimini Road. The Bimini Wall runs within 1,312 feet (400 meters) of the shore, following the dramatic plummet to the ocean floor. The Bimini Road is a strange underwater "road" that extends for 1,000 feet (300 meters) from Paradise Point at the north end of Bimini Bay. The origin of this "road" has been the subject of much speculation, including claims that it is part of the Lost City of Atlantis. The "road" is made up of enormous rectangular limestone blocks that appear so uniform it is tempting to think that they are man-made, despite the fact that they are almost certainly a natural formation. They are very similar in composition to the beach rock formations in the area that tend to fracture into rectangular blocks, although it is not known how these would come to be submerged under 15 feet (4.6 meters) of water. Bimini's clear waters are a paradise for divers. Spectacular dive sites include the Bimini Road and Hawksbill Reef. Experienced divers can also dive the Bimini Wall at sites such as the Nodules and Tuna Alley. **RC**

BLUE HOLES

GRAND BAHAMA / CENTRAL ANDROS / GREAT EXUMA / LONG ISLAND, BAHAMAS

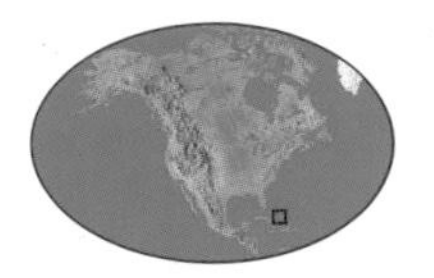

Length of Lucayan Caverns: 7 mi (11 km) long

Depth of world's deepest sea cave (Dean's Blue Hole): 663 ft (202 m)

Rock type: limestone

Blue holes are water-filled caves and sinkholes in which the water is a characteristic azure blue. They can occur in shallow water on the coast, or inland, and the Bahamas have more of them than anywhere else in the world. There are three types of blue hole: "cenotes" or sinkholes that are vertical shafts up to 500 feet (150 meters) across and are best seen from the air, the deepest being off Long Island; lens-shaped cave systems, such as the 7-mile (11-kilometer) long Lucayan Caverns—the longest cavern in the Bahamas; and fracture-guided caves that can be as small and narrow as 6.5 feet (2 meters) across.

The caves were formed during the ice age when sea levels were much lower than today. The underlying limestone rocks were eroded by water to form extensive shafts, sinkholes, and caverns. When the ice sheets melted, sea levels rose, and the caves were drowned to become blue holes. Snorkelers can access some blue holes, but many are far too dangerous to explore and are off limits even to experienced sports divers. Local authorities control access. But you need not go underwater to see them—organized nature tours by local guides take visitors to the inland blue holes. MB

VIÑALES VALLEY & SANTO TOMAS CAVE

PINAR DEL RÍO, CUBA

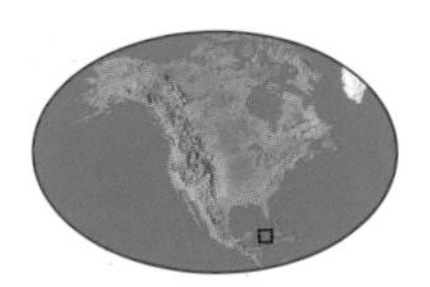

Height of mogotes in Viñales Valley: up to 1,000 ft (300 m)

Length of Cueva de Santo Tomas: 30 mi (47 km)

The Viñales Valley is a fertile valley interspersed with dramatic conical limestone hills (known locally as "mogotes")—similar to those found at Guilin in southern China. The mogotes are heavily vegetated and riddled with caves formed by underground rivers. Santo Tomas Cave is the second longest cave in Cuba, with underground galleries on seven levels. It has an entrance passage of up to 66 feet (20 meters) wide. The nearby Cueva del Indio, which was once inhabited by indigenous people, can be explored by boat on an underground river. The round-tops and near-vertical slopes of the mogotes were formed by erosion during the Jurassic era. Many plant species are found nowhere else—including the rare cork palm, which is considered to be a living fossil. One of the world's largest outdoor paintings, the Mural of Prehistory, is also found here painted on the side of the Dos Hermanas mogote. The Viñales Valley, which is also famous for its tobacco, is about 112 miles (180 kilometers) west of Havana. RC

BELLAMAR CAVES

MATANZAS, CUBA

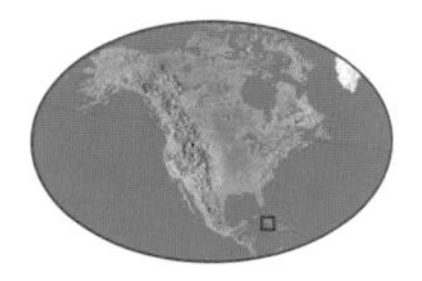

Width of Gothic cavern: 200 ft (60 m)
Height of Gothic cavern: 100 ft (30 m)
Length of Gothic cavern: 500 ft (150 m)

The Bellamar Caves, famous for their beautiful formations, are among Cuba's oldest tourist attractions. Tours of the caves include 17 galleries and six halls with stunning stalactites, stalagmites, and formations such as the Grated Coconut Gallery and the Fountain of Love. The caves were discovered in 1861 by quarrymen, but the first detailed survey was only in 1948. Beginning in 1989, the most extensive study discovered more than four miles (seven kilometers) of passages.

The most enormous cavern is known as the Gothic. A cluster of tall stalagmites in the center of the chamber is shaped like a warrior, and known as the Guardian of the Temple. The Mantle of Columbus is a massive white translucent pillar, 66 feet (20 meters) high and 20 feet (6 meters) thick. Other highlights of the caves include the Devil's Gorge, the Embroidered Petticoat, the Chamber of the Benediction, Don Cosme's Lamp, the Diamond Cascade, and the Lake of the Dahlias. The Bellamar Caves system is located 1.2 miles (2 kilometers) south of Matanzas city. The humidity is high in the caves and the temperature is a constant 77 to 81°F (25 to 27°C). RC

EL NICHO FALLS

CIENFUEGOS, CUBA

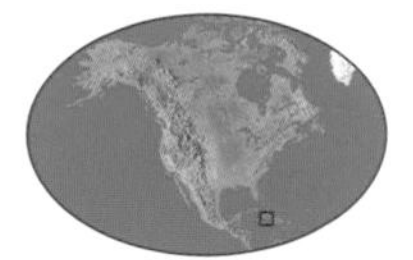

Alternative name: Cascada de El Nicho
Height of El Nicho Falls: 66 to 115 ft (20 to 35 m)

Although there is no central mountain mass in Cuba, mountainous regions are scattered throughout the island. El Nicho is located in the Sierra de Trinidad mountain range in the central part of the island. The El Nicho area has several waterfalls that are 66 to 115 feet (20 to 35 meters) high. A persistent mist arises from the foaming waters at the base of El Nicho Falls, where water plunges from a height of over 100 feet (30 meters) onto the rocks below. At the El Nicho nature reserve you can swim in the blue pools between the waterfalls, explore the depths of caves, hike in the lush mountains, or watch the exotic and varied wildlife.

With a bit of luck, bird-watchers may spot Cuba's national bird, the spectacular tocororo. The tocororo's distinctive red, blue, and white plumage (the colors of the Cuban flag) makes it relatively easy to find. El Nicho is about 29 miles (46 kilometers) from the city of Cienfuegos and is only accessible by four-wheel-drive vehicles. This scenic road has breathtaking views across the Sierra del Escambray mountains. There are well-maintained walkways that lead to the heart of El Nicho Falls. RC

WATERFALLS OF DOMINICA

ST. GEORGE / ST. DAVID / ST. PATRICK, DOMINICA

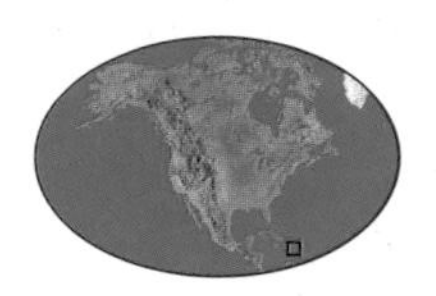

Height of Trafalgar Falls: 200 ft (60 m)

Height of Middleham Falls: 195 ft (59 m) high

Height of Sari Sari Falls: 150 ft (45 m) high

The Caribbean island of Dominica is a paradise for all waterfall enthusiasts. The island has mountains over 4,000 feet (1,200 meters) tall and receives up to 400 inches (1,000 centimeters) of rain each year. There are many waterfalls—new ones in remote regions are being discovered every year. Many of Dominica's waterfalls are in the Morne Trois Pitons National Park. Trafalgar Falls, five miles (eight kilometers) from the capital, Roseau, is the most famous and accessible waterfall. The viewing platform looks over two separate cascades—the taller "Father" and shorter "Mother." A more challenging 11 to 12 hour hike through dense rainforest is required to reach the breathtaking Middleham Falls, the highest waterfall in Dominica. And you need to scramble over boulders and crisscross the river to get to the Sari Sari Falls near the village of La Plaine on the east side of the island. The impressive Victoria Falls is fed by the White River, whose milky mineral-rich waters originate from the Boiling Lake. You can swim in the pool below the falls. There are hundreds more waterfalls on Dominica worth investigating, including Emerald Pool Falls and Syndicate Falls. RC

BOILING LAKE

ST. PATRICK, DOMINICA

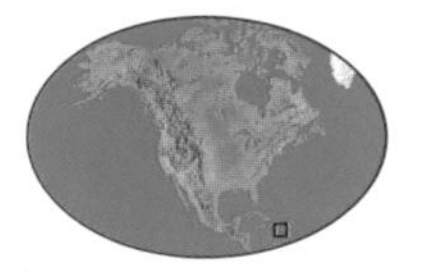

Diameter of Boiling Lake: 200 ft (60 m)

Average temperature: 190°F (88°C)

Elevation of lake: 2,500 ft (762 m)

The world's second-largest boiling lake (New Zealand has the largest) is found in the center of Dominica's Morne Trois Pitons National Park. An 1875 survey found the water temperature near the edges was 180 to 197°F (82 to 91.5°C). The depth was then more than 195 feet (59 meters). Later, a geyser formed in the center and the water level dropped. The lake stopped boiling in April 1988 and the level dropped further, but it has since returned to its "normal" state. Boiling Lake is not a volcanic crater lake but is thought to be a fumarole (a vent through which volcanic gases escape) flooded by rainwater and streams. Magma below the surface causes it to boil at the center.

The long and strenuous hike to the Boiling Lake—two to three hours each way from Titou Gorge—crosses the Valley of Desolation, which has a bizarre landscape of little vegetation, brightly-colored hot springs, and clouds of sulfurous steam billowing from fumaroles. Boiling Lake's bubbling blue-gray waters are now usually shrouded in clouds of steam. Visitors should take care as several people have been seriously scalded by its waters, and at least two are known to have died from inhaling poisonous fumes. RC

STINGRAY CITY

NORTH SOUND, GRAND CAYMAN

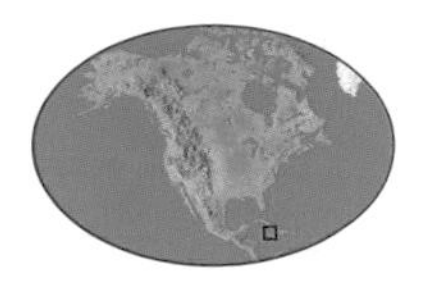

Location of Stingray City: entrance of North Sound

Depth of water: 3 to 6 ft (0.9 to 1.8 m)

Water temperature: 82°F (28°C)

There is not an office block or a store in sight—Stingray City is a different kind of conurbation, one in which the inhabitants have a sting in the tale but offer visitors the most fun they have probably had in a long time. On a sand spit just a short 20-minute boat ride from Grand Cayman, hundreds of bathmat-sized stingrays come to meet people and to get fed. The first trip of the day is best, when the rays gather in the largest numbers for breakfast.

The water is only three to six feet (0.9 to 1.8 meters) deep, and you can stand, snorkel, or kneel as the rays swim around and over you. They can be fed with pieces of squid and even petted. The rays have been "tamed" after many years of being fed by hand, and Grand Cayman is the only place in the world where you can meet them this way. The rays themselves seem to fly through the water, propelling themselves with wing-like pectoral fins. Clearly evident on the whip-like tail is the venomous spine, but it is only used for defense, so be mindful of where you put your feet. The interactive experience brings a smile to everybody's face and anybody can do it, young and old. MB

BLOW HOLES

EAST END, GRAND CAYMAN

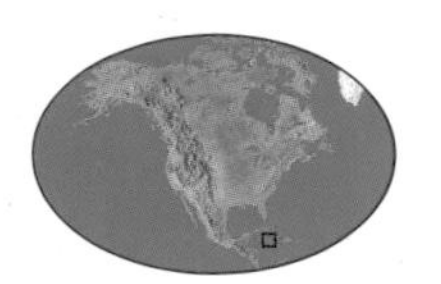

Area of Cayman Islands: 100 sq mi (262 sq km)

Highest point (The Bluff): 140 ft (43 m)

Rock type: limestone surrounded by coral reefs

The Blow Holes on the east side of Grand Cayman are the most spectacular sights on the island. When the waves come crashing in to the shore here, water spouts up into the air through holes in the rock, creating a truly dramatic performance. Grand Cayman, the largest of the Cayman Islands, has unusual land formations created from coral, sand, and mud, which are locally known as "cliff" when inshore, and "ironshore" on the coast. The ironshore forms a coastal limestone shelf that surrounds the bluff limestone core or "cliff." In places where the rock is undercut, the waves surge under the shoreline, and spray upward like a geyser through holes in the rocks. When the waves are large and the wind is in the right direction, it is possible for water to spray up to 33 feet (10 meters) into the air.

The site is accessible enough to obtain excellent photographs, although on calmer days the blow holes are much less impressive. Visitors are advised to wear hiking boots, as the jagged rocks along the ironshore are very sharp. Many ships have had their hulls torn open on the ironshore rocks—the "Wreck of the Ten Sails" (where 10 sailing ships were wrecked in one day) is just along the coast. RC

COCKPIT COUNTRY

ST. JAMES / TRELAWNY / ST. ELIZABETH, JAMAICA

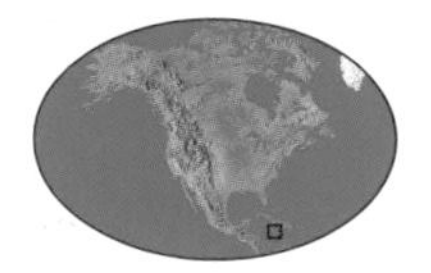

Area of Cockpit Country: 500 sq mi (1,295 sq km)

Annual rainfall: 60 to 100 in (1,500 to 2,500 mm)

The Cockpit Country of Jamaica is a strange terrain of forest-clad karst limestone, located in the northwest of the island. Millions of years of erosion have created an unusual landscape of conical hills and steep-sided valleys, much of which is uninhabited and also relatively unexplored. The thousands of limestone depressions were termed "cockpits" by the British in the 17th century because they resembled cockfighting arenas. Water drains through the porous bedrock and sinkholes to a network of underground caves. Windsor Great Cave and Marta Tick Cave support colonies of more than 50,000 bats.

Cockpit Country is also one of the most important sites in the world for nature conservation, harboring a diverse range of plants and animals. At least 100 plants are unique to this region, some of them isolated to a single hillock. Seventy-nine of Jamaica's 100 bird species are found here, including the threatened black-billed parrot. There are no roads in Cockpit Country. Do not explore the area without local guides because of the rough terrain, dangerous sinkholes, and the likelihood of getting hopelessly lost. RC

DUNN'S RIVER FALLS

ST. ANN, JAMAICA

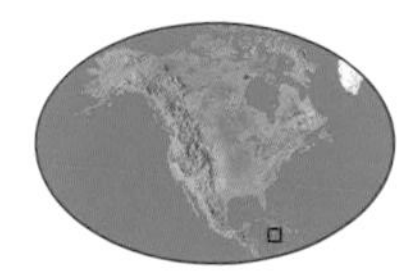

Alternative name: Xayamaca

Height of Dunn's River Falls: 600 ft (180 m)

Dunn's River Falls, one of Jamaica's most famous visitor attractions, is actually a number of scenic waterfalls that cascade over a series of smooth, limestone terraces down to a beautiful Caribbean beach and straight into the sea. Located in an area called *Xayamaca* by the Arawak Indians, meaning "land of rivers and springs," Dunn's River Falls is a rare example of a waterfall that forms the mouth of a river. They became famous after being featured in the first James Bond film, *Dr. No.*

Visitors can climb the 600-foot (180-meter) falls from the beach for stunning panoramic views of the surrounding area. It takes most people about an hour to climb the falls. Some of the falls are gentle while others are thunderous. On the climb up, you can pause for a relaxing swim in the natural pools that have formed in the river terraces. However, the falls can be slippery, so wear appropriate footwear, though guides are also available to assist people. To avoid the crowds, it is probably best to start your climb in the early morning. The falls are set in the lush, tropical forest of Dunn's River Falls Park, which is situated about a mile from Ocho Ríos on the northern coast of Jamaica. RC

BLUE LAGOON

PORTLAND, JAMAICA

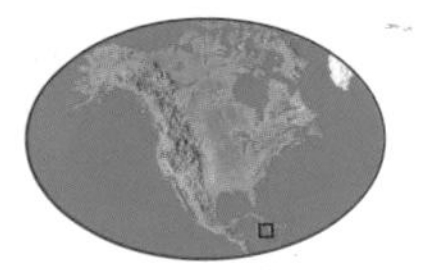

Alternative name: Blue Hole

Maximum depth of lagoon: 185 ft (56 m)

Surrounded by steep hillsides, the Blue Lagoon is a protected cove with a narrow channel to the sea. It was made famous by the film *The Blue Lagoon* (starring Brooke Shields) and is still a favorite location for filmmakers and photographers. Rumors abound about the Blue Lagoon. The Arawak Indians believed that it was bottomless, although its maximum depth is now known to be 185 feet (56 meters). The heroic actor Errol Flynn is claimed to have dived unaided to the bottom of the lagoon.

In the 1950s the thriller writer Robin Moore (author of *The French Connection*) owned a villa overlooking the lagoon and most of the area around it. The area is still privately owned, but is now a luxury tourist resort. The Blue Lagoon was formerly known as Mallard's Hole—named after the notorious pirate Tom Mallard, who allegedly used the lagoon as a safe haven and lookout point. The Blue Lagoon is 7 miles (11 kilometers) east of San Antonio. Set against a backdrop of Jamaican flora and fauna, the beautiful waters of the lagoon pass through various shades of blue and green throughout the day. For those keen on snorkeling, there are plenty of colorful fish to see in the lagoon's clear waters. RC

CARIBBEAN NATIONAL FOREST

CANAVÓNAS / JUNCOS / LAS PIEDRAS / LUQUILLO / RÍO GRANDE, PUERTO RICO

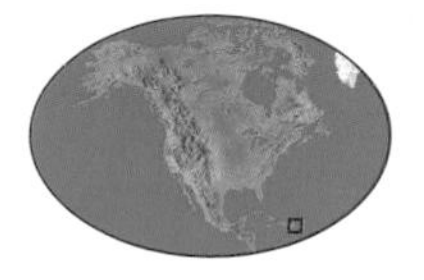

Alternative name: El Yunque

Area of Caribbean National Forest: 28,000 acres (11,300 ha)

Elevation of forest 3,493 ft (1,065 m)

Just 25 miles (40 kilometers) from San Juan is the only protected tropical rainforest in U.S. territory—the Caribbean National Forest, or *El Yunque* as it is known locally. The final (and second highest) peak in Puerto Rico's mountainous backbone is the Sierra de Luquillo mountain, which means "white lands" in Tiano, the language of the region's original inhabitants. One of the first reserves in the western hemisphere, El Yunque was created in 1876 primarily as a forest reserve to safeguard timber for ship construction.

The slopes are very slippery and steep, frequently exceeding 45°. There is no dry season, only hurricane season—Hurricane George battered the forest with 115-mile-per-hour (185-kilometer-per-hour) winds in 1998. Four kinds of forest occur in El Yunque: lowland rainforest, subtropical rainforest (above 2,000 feet, 600 meters), then cloud forest, and cold-dwarfed elfin forest near the peak. The forest is now the only home of the very rare and endangered Puerto Rican parrot. It also hosts such Puerto Rican endemics as the Puerto Rican tanager, elfin woods warbler, and Puerto Rican vireo. The limestone-rich soil means that snails are also abundant. The park has excellent trails and is easily accessed from San Juan. AB

KARST COUNTRY

ISABELA, PUERTO RICO

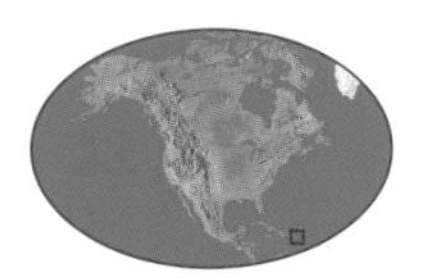

Total area of national forest in Karst Country: 4,000 acres (1,600 ha)

Area of Guajataca Forest: 2,400 acres (970 ha)

The Karst Country is a peculiar area of regular 100-foot- (30-meter-) high, green and white hillocks in northwestern Puerto Rico, between Quebradillas and Manati. The best examples of karst landscapes in the world are in Puerto Rico, the Dominican Republic, and Slovenia. These striking terrains are produced when water penetrates down into limestone and erodes basins or sinkholes. Karstic hillocks, or *mogotes*, are formed where the land does not sink through erosion, as the rocks that form them are less porous. These hillocks are remarkably similar to each other in size and shape, considering the random nature of the processes that form them.

The world's most sensitive radar/radio-telescope (at Arecibo Observatory) is situated in an ancient sinkhole in the Karst Country. The site, which featured in the film *Contact*, is home to NASA's SETI (Search for Extra-Terrestrial Intelligence) project. Karst limestone is protected in the following four national forests: Guajataca, Cambalache, Vega, and Rio Abajo. Guajataca is located near Arecibo, where visitors can see dramatic rock formations in Wind Cave and hike 25 miles (40 kilometers) of trails. **RC**

GUÁNICA STATE FOREST

GUÁNICA, PUERTO RICO

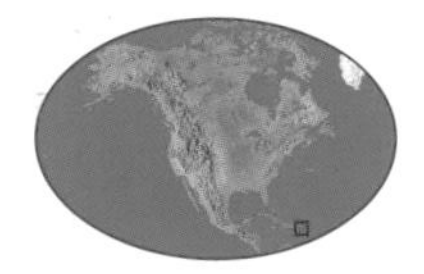

Area of Guánica State Forest: 1,640 acres (3,936 ha)

Elevation of forest: 1,320 ft (400 m)

In southwest Puerto Rico, on the Caribbean coast, sits Guánica State Forest. Denied rain by the Cordillera Central, the region's subtropical dry forest supports 50 percent of the island's 284 bird species. Birds include the endemic Puerto Rican lizard cuckoo, Puerto Rican nightjar (formerly considered extinct), and the Puerto Rican tody (a tiny red-throated green and white forest bird). Over 12 miles (19 kilometers) of hiking trails take you through drought-adapted vegetation.

Though parched and dusty, the forest has over 750 plant species, including cacti, vines, agaves, and slow-growing Guayacán trees that have been known to reach 400 years of age. The most scenic route is the Cueva Trail, which offers sensational views of the limestone escarpment and down to the coast's blue ocean. At the foot of the escarpment lie bathing beaches, life-rich sea-grass beds, and mangroves. Guánica is also the major stronghold of the Puerto Rican crested toad and the Puerto Rican cave shrimp. This forest type has been degraded by cattle and farming, leading UNESCO to declare the area a Biosphere Reserve. The nearest town is Ponce, 15 miles (24 kilometers) down the coast. **AB**

MOSQUITO & PHOSPHORESCENT BAY

VIEQUES, PUERTO RICO

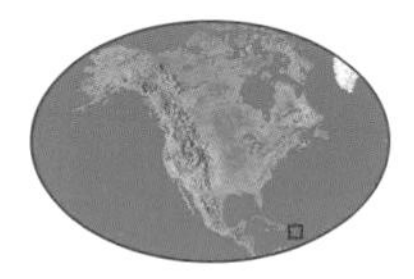

Area of Mosquito Bay: 160 acres (64 ha)

Concentration of dinoflagellates: 720,000 per gallon (158,000 per liter) of water

At night in Mosquito Bay, on the Puerto Rican island of Vieques, the blue-green glow from the waters can produce enough light to read a book. The mysterious glow is produced by many millions of microscopic dinoflagellates that release energy in the form of light. These single-celled organisms flash when agitated, which is thought to be a natural defense mechanism used to deter predators. Decaying mangrove roots and leaves provide nutrients for these tiny organisms, while the narrow mouth of the bay prevents them from being flushed out to sea.

Bioluminescent bays are sensitive and can be destroyed by pollution. Phosphorescent Bay, on Puerto Rico's southwest coast, once rivaled Mosquito Bay, but is now one tenth as bright as it used to be. Bioluminescence does occur seasonally in other areas of the world, but Mosquito Bay glows all year round. The best time to visit is on a cloudy moonless night. Swimming in the glowing waters is an unforgettable experience. RC

RIO CAMUY CAVES

CAMUY, PUERTO RICO

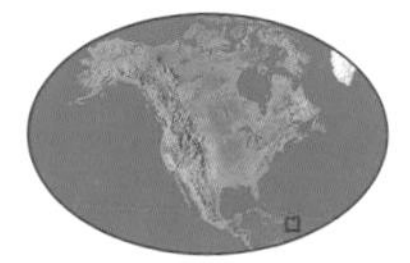

Area of Rio Camuy Caves: 268 acres (110 ha)

Age of Rio Camuy Caves: 45 million years

One of the largest and most dramatic cave systems in the world, the Rio Camuy Cave Park of northwest Puerto Rico has a network of caves, sinkholes, and cathedral-sized caverns, as well as one of the world's largest underground rivers. Local boys led cavers to the site in the 1950s, up until the caves were opened to the public in 1986. Since then, 16 entrances to the caves have been discovered, and 6.8 miles (11 kilometers) of passages explored. Petroglyphs etched into the walls of Cathedral Cave by the ancient Taino people provide evidence of the cave's pre-Columbian occupation.

The caves contain a number of unusual species, including a unique species of fish that is completely blind. Only a small part of the cave complex is open to the public, although experienced cavers can arrange trips through undeveloped sections. Visitors can ride a trolley down a 200-foot (60-meter) sinkhole, walk through the vast, beautifully-illuminated Clara Cave—which contains many impressive stalactites and stalagmites—and take a tram to a platform overlooking the 400-foot- (120-meter-) deep Tres Pueblos Sinkhole with views down to the Camuy River. RC

VIRGIN GORDA BATHS

VIRGIN GORDA, BRITISH VIRGIN ISLANDS

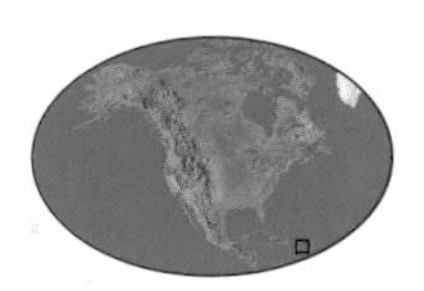

Length of Virgin Gorda Island: 10 mi (16 km)

Size of boulders: up to 40 ft (12 m) diameter

Age of boulders: 70 million years

The Baths on Virgin Gorda (the second largest island in the British Virgin Islands) are a maze of giant granite boulders and sheltered pools located on the southwest coast of the island. Virgin Gorda means "Fat" or "Pregnant Virgin" and is said to have been named by Columbus because of the shape of the island from the horizon. The island is about 10 miles (16 kilometers) long with mountains in the northern and central areas. The oldest volcanic rocks in the Virgin Islands were formed about 120 million years ago, but the granite boulders of Virgin Gorda did not appear on the Caribbean seabed until about 70 million years ago. Faulting and uplifting of the sea floor approximately 15 to 25 million years ago exposed the boulders, while weathering and erosion rounded the boulders and carved large caves into them. The Baths is one of the most popular visitor attractions in the British Virgin Islands and is accessible by land and sea (it is a regular stop for charter boats). Apart from exploring the pools and grottos on foot, the site is ideal for snorkeling or relaxing on the white sand beach. RC

SOUFRIÈRE HILLS VOLCANO

MONTSERRAT

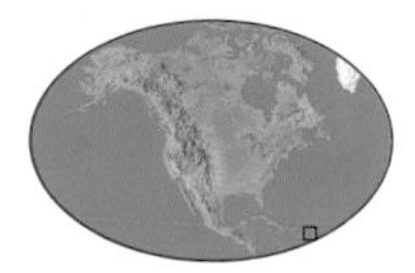

Elevation of Soufrière Hills Volcano: 3,002 ft (915 m)

Eruptions: 1995 to present

The first recorded eruption at Soufrière Hills Volcano began in July 1995, although seismic activity below the volcano had been reported at 30-year intervals throughout the 20th century. It deposited ash around Montserrat, and 5,000 people were evacuated. Eruptions at Soufrière Hills Volcano appear to be linked with rainstorms and full moons. Venting of steam and ash are associated with periods of strong seismic activity, and a new vent has formed southwest of Castle Peak.

Before the 1995 eruption, Castle Peak was the youngest volcanic dome. Lava flows from volcanic dome collapses have created a new delta at the mouth of the White River, but it is not yet known whether the delta will be a permanent feature or if it will be eroded by waves. The volcano occupies the southern half of the island of Montserrat, on the north flank of the older South Soufrière Hills Volcano. The Soufrière Hills Volcano and the surrounding area are strictly out of bounds to tourists due to the continuing eruptions. However, though the volcano itself is often shrouded by cloud, visitors can enjoy impressive views of the devastated area from Garibaldi Hill and Jackboy Hill. RC

CARBET FALLS

BASSE-TERRE, GUADELOUPE

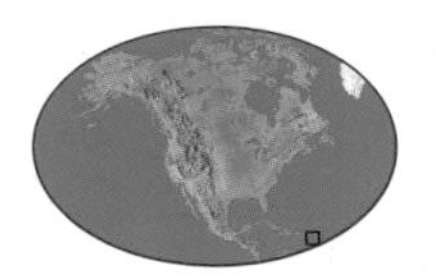

Height of upper waterfall: 387 ft (125 m)

Height of middle waterfall: 330 ft (100 m)

Height of lower waterfall: 60 ft (18 m)

The Carbet Falls of Guadeloupe are the highest waterfalls in the eastern Caribbean. Although Christopher Columbus named the island of Guadeloupe in 1493, he failed to stop to explore its natural wonders. Guadeloupe is made up of two islands with very different landscapes. The landscape of the eastern island, Grand Terre, features rolling hills, mangrove swamps, and sugar-cane plantations, while the western island, Basse Terre, has a more rugged and mountainous terrain which is dominated by the majestic La Soufrière Volcano.

The Carbet Falls are found in Basse Terre's 74,000-acre (30,000-hectare) Parc Naturel, which also contains La Soufrière Volcano as well as areas of tropical forest. The water for the three waterfalls that make up Carbet Falls gushes down from the slopes of the volcano. The uppermost waterfall is the tallest of the three, with a height of 387 feet (125 meters). The second waterfall is slightly less high at 330 feet (100 meters), but is probably the most impressive. The lowermost waterfall is much shorter than the others at about 60 feet (18 meters) in height, but it is picturesque and easily accessible. RC

DIAMOND ROCK

LE DIAMANT, MARTINIQUE

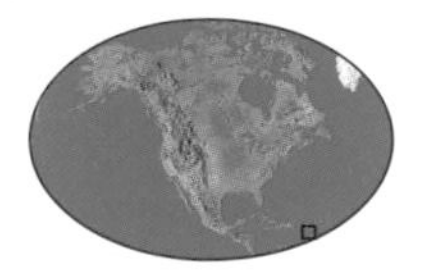

Alternative name: Rocher du Diamant

Height of Diamond Rock: 590 ft (176 m)

Age of Diamond Rock: 960,000 years

Diamond Rock is a diamond-shaped volcanic rock, or ancient lava dome, that has become a recognized symbol of Martinique around the world. In 1804, while the French and British fought for possession of Martinique, Admiral Samuel Hood commandeered the rock for the British. The rock was fortified with cannons and turned into a British "warship" that was promptly re-named HMS *Diamond Rock*. The British built munitions depots, docks, and a hospital. The 107-man "crew" managed to blockade the island and hold the rock for nearly 18 months.

The combined French and Spanish fleets recaptured the "warship" by employing the unusual tactic of crashing a boat full of rum onto the rock, and then overpowering the drunken British soldiers. The surviving British sailors were court-martialed for abandoning their "ship" after they fled to Barbados. Diamond Rock is located about 1 mile (1.6 kilometers) offshore from Le Diamant, on Martinique's southern coastline. The rock's steep sides are inhabited by seabirds. Its stunning underwater coral and associated marine life makes it one of the best sites for diving in Martinique. RC

MOUNT PELÉE

GRAND-RIVIÈRE / LE MORNE-ROUGE / LE PRÊCHEUR, MARTINIQUE

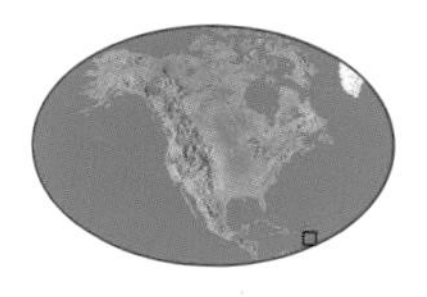

Elevation of Mt. Pelée: 4,583 ft (1,397 m)

Extent of devastation from 1902 eruption: 28,000 people killed

The notorious Mount Pelée Volcano towers over the northern tip of the island of Martinique in the West Indies. On May 8, 1902, it erupted, destroying the coastal town of St. Pierre and killing around 28,000 people—the largest number of casualties of any volcanic eruption in the 20th century. Inhalation of hot ash and fumes from the 1902 eruption managed to kill most of the people in St. Pierre within a matter of minutes. Only two men within the town survived the blast, including one who was contained in a poorly ventilated jail cell at the time. He was rescued after four days and went on to become a minor celebrity. There were other survivors of the 1902 eruption, but they were either on the outskirts of the town or on ships moored in the harbor.

A massive lava dome known as the Tower of Pelée rose from the crater in 1902, and grew to form a 1,000-foot (300-meter) spine that collapsed after 11 months. The volcano's current lava dome was constructed during the most recent eruptions of Mount Pelée, which occured from 1929 to 1932. St. Pierre has since been rebuilt, and around 22,000 people now live in the town or on the slopes of the volcano. RC

DIAMOND FALLS & SULFUR SPRINGS

SOUFRIÈRE, ST. LUCIA

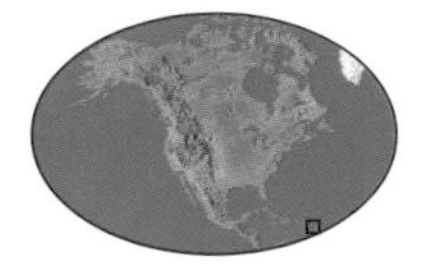

Alternative name: La Soufrière

Crater area of Sulfur Springs: 7 acres (3 ha)

Diamond Falls is the lowest of six waterfalls that are fed by sulfur springs that change the color of the water to bright shades of yellow, green, and purple. Adjacent to the falls are mineral baths that were originally built on the orders of Louis XVI to enable French soldiers stationed in the area to benefit from the curative properties of the waters. The Diamond Baths were destroyed during the Brigand's War (1794–1795), but have since been restored so that visits to the waterfalls can be combined with an invigorating soak in the warm waters. Some of the original 18th-century baths are still in use.

The streams that feed the Upper and Lower Diamond Falls originate at Sulfur Springs, which is touted as the world's only "drive-in" volcano. A road leads into the remains of Mount Soufrière's volcanic crater near the edge of the volcanic activity. The walls of the crater have eroded away, leaving seven acres (three hectares) of barren mountainside with boiling mud pools and vents that shoot steam 50 feet (15 meters) into the air. Visitors can access both sites via St. Lucia. RC

THE PITONS

SOUFRIÈRE, ST. LUCIA

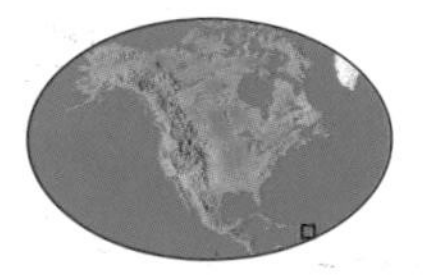

Age of the Pitons: 30 to 40 million years

Height of Petit Piton: 2,619 ft (798 m)

Height of Gros Piton: 2,461 ft (750 m)

The majestic twin peaks of Gros ("Big") Piton and Petit ("Small") Piton are recognized internationally—on the flag of St. Lucia they appear as two triangles on a blue background. Despite its name, Petit Piton is actually taller than Gros Piton, although Gros Piton is the broader of the two.

The pyramidal cones of the mountains were formed from a volcanic eruption 30 to 40 million years ago. The vegetation of the Pitons is varied due to their steepness, geology, and proximity to the sea. Rainfall is higher on the Pitons than in the rest of the island, and peaks may be shrouded in cloud for up to 100 days per year. Increased moisture leads to enhanced growth of plants such as orchids and bromeliads. Over 148 plant species and 27 bird species have been recorded in Gros Piton.

The Pitons are located near the town of Soufrière in southwest St. Lucia. Gros Piton is the only peak where climbing is allowed—although permits and guides are required for this tough trek. RC

RIGHT *Gros Piton looms above the ocean.*

MOUNT SOUFRIÈRE

ST. VINCENT

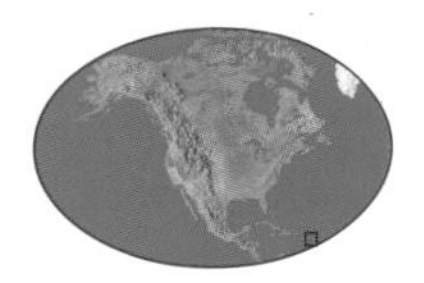

Age of Mt. Soufrière: 600,000 years

Height of Mt. Soufrière: 4,049 ft (1,234 m)

Extent of devastation from 1902 eruption: 1,600 people killed

Situated at the northern corner of St. Vincent, Mount Soufrière is an active volcano and St. Vincent's most recent vent. The island of St. Vincent is a single large volcanic cone in the arc of 25 volcanoes that form the islands of the Caribbean's Lesser Antilles. These islands were caused by the subduction of oceanic crust moving westward from the Mid-Atlantic Ridge.

Though still recovering from its most recent 1979 eruption, the rich cloud-shrouded forest of Mount Soufrière makes for a rewarding hike. A 3.5-mile (5.6-kilometer) trail takes you through the forest to the cinder-covered peak and bubbling lava-filled cone. The St. Vincent parrot, a gorgeous bird endemic to the island, lives both here and in the nearby Vermont Nature Reserve. With fewer than 750 individuals, this bird is the subject of international conservation efforts. The island's other endemic bird, the whistling warbler, also occurs on the slopes. Other regional specialties include the Carib hummingbird, brown and gray tremblers, West Indian whistling duck, Antillean euphonia, and the scaly-breasted thrasher. The nearest town is Kingstown (island capital), two hours away by road. AB

GRAND ÉTANG

ST. ANDREW, GRENADA

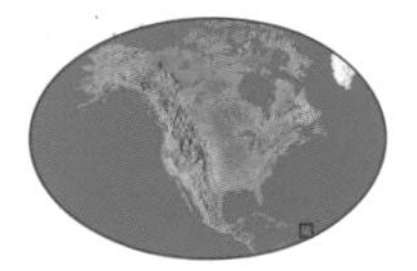

Area of Grand Étang National Park: 3,860 acres (1,562 ha)

Area of Grand Étang Lake: 30 acres (12 ha)

Elevation of lake: 1,740 ft (530 m)

Located high up in the mountains of the interior of Grenada, the Grand Étang National Park is the island's most popular area for hiking and trekking. Grand Étang Lake is the focal point of the reserve and fills the crater of Mount St. Catherine, one of the island's extinct volcanoes. The rainforest around the lake holds a rich diversity of plants and animals. Grand Étang's flora includes trees such as mahogany and candle tree (gommier), giant palms and blue mahoe, as well as a variety of ferns and tropical orchids. A range of animals are found within the lush vegetation, particularly birds. Commonly seen bird species include the Antillean crested hummingbird (or "little doctor bird"), the lesser Antillean tanager (or "soursop"), and the broad-winged hawk (known locally as the "gree-gree"). Opossums ("manicou") and introduced Mona monkeys roam the forest canopy, and you may also see frogs, lizards, and armadillos. The Grand Étang National Park is located in the center of Grenada, only 8 miles (13 kilometers) from the capital, St. George's. Overnight camping is available at several campsites. Hiking is best during the dry season, which runs from December to May. RC

MOUNT CARMEL WATERFALL

ST. ANDREW, GRENADA

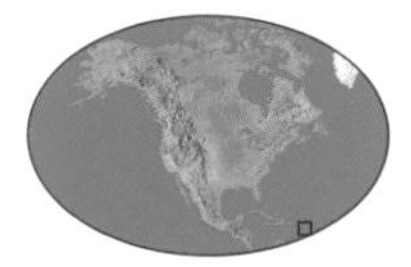

Alternative name: Marquis Falls

Height of Mount Carmel Waterfall: 70 ft (21 m)

The (Royal) Mount Carmel Waterfall is the highest of Grenada's many waterfalls. Situated 1.9 miles (3 kilometers) south of Grenville, its twin waterfalls cascade 70 feet (21 meters) to the waters below. Annandale Falls, near the village of Constantine, is much smaller but is easily accessed, therefore providing a popular stopping point for tour buses. Concord Falls is a triple cascade on the edge of Grand Étang Forest Reserve, on the western side of Grenada. The lowest of the waterfalls is the most accessible and is favorable for camping and swimming. The 40-foot (12-meter) second waterfall (Au Coin) is a short hike upriver. A further two-hour hike brings you to the uppermost of the three waterfalls (Fontainbleu). It takes about half an hour's trek through the rainforest to reach the Seven Sisters Falls. The seven waterfalls are among the most unspoiled and peaceful on the island. Rosemount Falls is a privately owned waterfall that can only be viewed by diners at the Rosemount Plantation House. Recently discovered waterfalls include the picturesque Honeymoon Falls (base of Mount Qua Qua) and the unspoiled Victoria Falls (foot of Mount St. Catherine on the west coast). RC

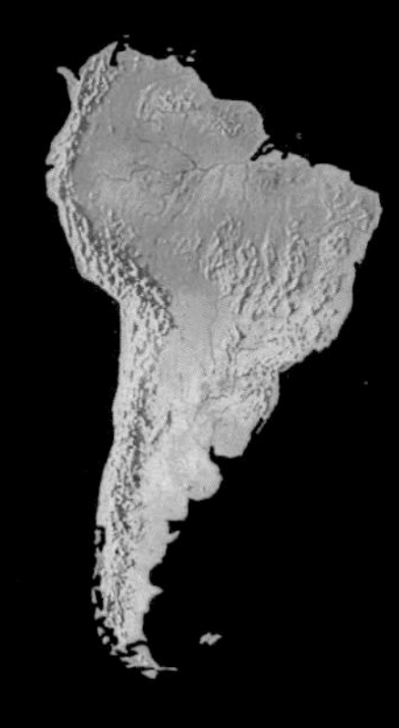

II

SOUTH AMERICA

South America's natural forces can be witnessed in countless locations and in countless forms—you will find them cascading over the Devil's Throat of the Iguazú Falls, flooding Llanos grasslands, pitting piranhas in flooded forest treetops, ornamenting the snowy peaks of volcanoes, freezing walls of ice in the deep Patagonian south, drying out salt flats, and above all, transcending boundaries both geographical and elemental.

LEFT *The massive Amazon River as it snakes through the verdant jungles of Brazil.*

SIERRA NEVADA DE SANTA MARTA

MAGDALENA / CESAR / LA GUAJIRA, COLOMBIA

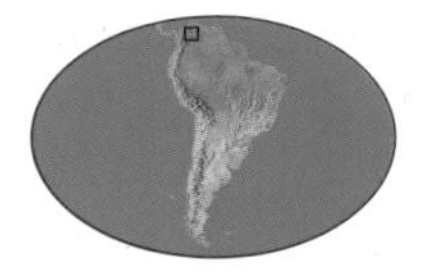

Elevation of Sierra Nevada de Santa Marta: sea level to 18,942 ft (5,775 m)

Habitat: lowland rainforest, coastal dry scrub, montane forest, elfin forest, páramo, high-altitude desert, scree slope

Colombia is one of the most biodiverse countries on the planet. Colombia has 99 separate habitats which contain 15 percent of the planet's total biodiversity. This includes 1,815 bird species (20 percent of the world's total) and over 50,000 plant species (one-third occurring nowhere else). It also contains 3,100 butterfly species, making it the third most diverse habitat of this type in the world.

The Sierra Nevada de Santa Marta is a remarkable place—a cluster of ancient mountains in the north corner of Colombia, this isolated area is a remnant of a mountain chain that predates the Andes. Just 8 square miles (23 square kilometers) in extent, it hosts 356 bird, 190 mammal, and 42 amphibian species. This diversity is possible because of the precipitous rise of the Santa Marta peaks, which reach 18,942 feet (5,775 meters). Not only does this make them the tallest coastal mountains on Earth, but also it provides a variety of climates that support the full range of Colombia's ecosystems. **AB**

EL COCUY NATIONAL PARK

ARAUCA / BOYACÁ, COLOMBIA

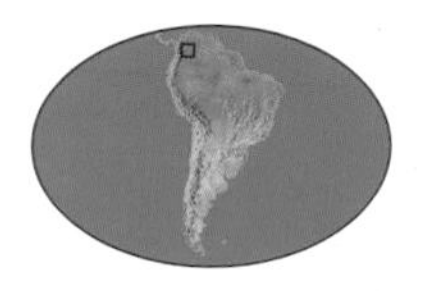

Area of national park: 1,181 sq mi (3,057 sq km)

Elevation of park: 2,000 to 17,485 ft (600 to 5,330 m)

Rock type: mostly granite

More than 20 snowcapped peaks, including the spectacular granite needles of Negro Norte and Ritacuba Blanca; the largest glaciers in South America; and a stunning array of plants and animals—this is El Cocuy National Park. The park was created in 1977 to protect its glorious 19 miles (30 kilometers) of mountain range. With a range in altitude of over 15,000 feet (4,500 meters), the park's habitat types include lowland forest, montane forest, páramo grassland, permanent snow fields, glaciers, and scree slopes. Animals include cock-of-the-rock, spectacled bear, torrent duck, condors, and a large variety of hummingbirds. Growing in the páramo are the strange-looking Espeletias, a hunched, rosette-forming plant of the daisy family that produces long flowering spikes like gigantic bottlebrushes. Interspersed with these are puyas—cold-resistant members of the pineapple family—and ground orchids, which tough it out in the lee of rocks. The park is 250 miles (400 kilometers) north of Bogotá. **AB**

CHOCÓ FOREST

CHOCÓ, COLOMBIA

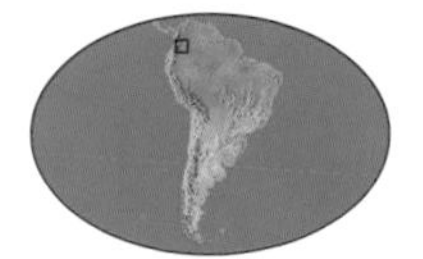

Area of Chocó Forest: 51,000 sq mi (131,250 sq km)

Length of forest: 900 mi (1,500 km)

Habitat: ultra-moist tropical rainforest

The Chocó lies on the Pacific coast, between the sea and the Andes, and is therefore very wet: It receives 200 to 630 inches or 52 feet (5,000 to 16,000 millimeters) of rain annually. It is rich in wildlife, with an unusual abundance of palms, which number greater than any other tropical forest. Throughout the Chocó there are more than 11,000 plant species, a quarter of which occur nowhere else in the world. Half of Colombia's 465 mammals live here, including 60 unique species. Sixty-two bird species are exclusive to the area, 17 of which are very rare. Noteworthy regional examples include the cotton-top tamarin, long-wattled umbrella bird, and one of the world's most venomous vertebrates, the yellow arrow poison frog—just touching its skin can send you into cardiac arrest.

The lack of roads and major infrastructure (added to the rain) ensure that the Chocó is well preserved. Nearly a quarter of the region still survives in a pristine state, and there are considerable areas of prime secondary forest. Several important reserves have been established, including Los Katios National Park and, on the Colombian–Ecuadorian border, the Awa Indian Reserve. **AB**

LOS NEVADOS NATIONAL PARK

TOMILA / QUINDÍO / RISARALDA, COLOMBIA

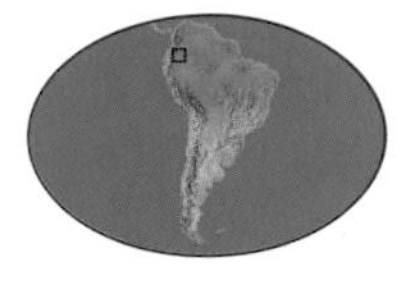

Area of national park: 225 sq mi (583 sq km)

Elevation of park: 8,528 to 17,400 ft (2,600 to 5,300 m)

Habitats: high-altitude desert, scree slopes, snowfields, glaciers

The local name, meaning "the Snow-covered Ones," refers to the park's five major peaks, all extinct volcanoes. These include Camunday, which means "Smokey Nose" in a local language, and alludes to the frequent banner of windblown snow trailing from its peak. The park has ancient lava fields and many features associated with more recent glacial activity, including hanging valleys and ice-scoured lakes. At an elevation of 14,104 feet (4,300 meters), the Valley of Tombs, once a sacred site to the Quimbaya and Puya peoples, is a barren valley where hundreds of stones now form an immense circle. Above it is a sandy, high-altitude desert where life barely subsists in the biting winds. Several thermal pools appear at around 11,808 feet (3,600 meters), their waters heated by the remnants of volcanic activity. Here there are páramo grassland and pockets of alpine forest. Located in the central mountain chain of Colombia, the area is well equipped with interpretative walks and hiking cabins. Wildlife includes condors, spectacled bear, pudu (dwarf deer), wooly tapir, and a plethora of hummingbirds. **AB**

AMACAYACÚ NATIONAL PARK

AMAZONAS, COLOMBIA

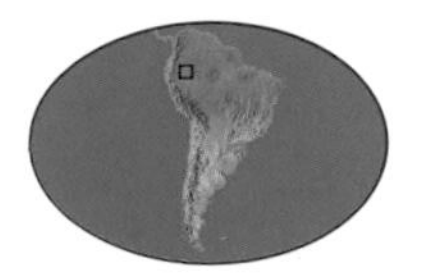

Area of Amacayacú National Park: 11,350 sq mi (29,385 sq km)

Amazon rainforest in Colombia: 400,000 sq mi (1,035,995 sq km)

Established as a national park: 1975

Thirty percent of Colombia is Amazon rainforest. Located on the northern bank of the Amazon, sandwiched next to the Cothué River, Amacayacú is Colombia's most pristine section of the Amazon. It is bordered in the east by the Amacayacú River. Trees here include the aptly named axebreaker, with its huge buttress roots, and the strangler fig, which gradually smothers the trees it uses for support to death.

Amacayacú is rich in animal life. For two years, the British Ornithological Union conducted a bird count here and identified 490 species, including 11 classes of heron alone. Mammals are represented by about 150 species, including the three-toed sloth, tamandua, white-eared opossum, and cotton-top tamarin. Pink river dolphins or botos are seen in the Amazon. The visitors' center has platforms in the forest from which to watch the wildlife. The park is reached by air and river, with a 45-minute flight from Bogotá to Leticia, followed by a three-hour boat trip. **MB**

RIGHT *Flooded forest in Amacayacú National Park.*

SIERRA NEVADA DE MÉRIDA

MÉRIDA, VENEZUELA

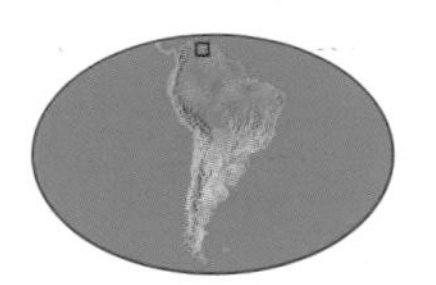

Elevation of Sierra Nevada De Mérida: 1,790 to 16,422 ft (500 to 5,007 m)

Habitats: tropical forest, cloud forest, high-altitude grasslands and moors, scree slopes, glaciers

Venezuela's largest mountain range, the Sierra Nevada de Mérida, extends 200 miles (320 kilometers) from the Colombian border to Venezuela's Caribbean coast and boasts Venezuela's highest peak, the 16,422-foot (5,007-meter) Pico Bolívar. Other tall peaks include Bonpland (16,020 feet, 4,883 meters) and Humboldt (16,214 feet, 4,942 meters), named and first climbed in 1910. The Sierra varies in width from 30 to 50 miles (50 to 80 kilometers). Three of the Sierra's peaks have glaciers. These now cover some 1.2 square miles (2 square kilometers), less than one percent of its ice age maximum. Erosion caused by glaciers has resulted in 170 ice-scoured lakes.

The Sierra Nevada de Mérida dominates the southern bank of Lake Maracaibo and occurs within a national park of the same name, which also contains the Sierra de Santo Domingo, whose highest peak is Mucuñuque at 15,328 feet (4,672 meters). This national park was the second of Venezuela's 43 national parks to have been established. Attractions include the cable car to Pico Espejo, 15,633 feet (4,765 meters) high, which is the world's highest and longest. **AB**

HENRI PITTIER NATIONAL PARK

ARAGUA STATE, VENEZUELA

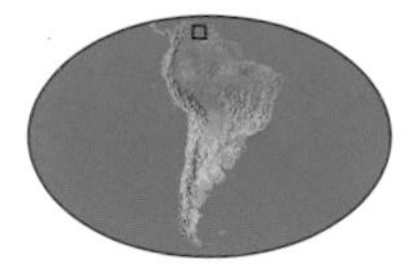

Area of national park: 350 sq mi (1,078 sq km)

Highest peak: Pico Cenizo

Rock type: igneous, 60 million years

This steeply sloping area on Venezuela's mountainous north coast became Venezuela's first national park in 1937, largely due to the efforts of Swiss biologist Henri Pittier who identified more than 30,000 Venezuelan plants in the area. The park's 60-to-80-million-year-old rocks support a moist, rich forest. The great variety of altitudes and habitats include coastal mangroves, coastal dry scrub, tropical grassland, palm-rich lowland forest, cloud forest, and elfin forest. The park also hosts over 580 bird species (6.5 percent of the world's total). One day's birding may reveal glories such as the rufous-crowned peppershrike, blue-hooded euphonia, bare-faced ibis, yellow-billed toucanet, russet-backed oropendula, as well as endemic species such as the golden-breasted fruiteater, blood-eared parakeet, and guttulated foliage-gleaner. A five- to six-day visit can yield sightings of 400 bird species, as well as armadillos, puma, tapir, ocelot, and monkeys. The Portachuelo Pass at 3,700 feet (1,128 meters) above sea level is a major route for birds and insects migrating down the Atlantic coast to South America. **AB**

THE LLANOS

VENEZUELA / COLOMBIA

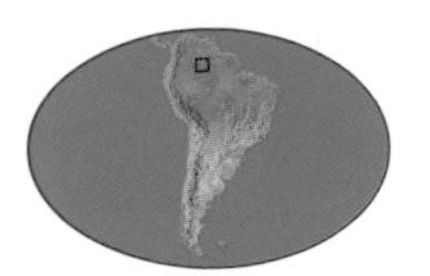

Total area of the Llanos: 173,314 sq mi (451,474 sq km)

Elevation of the grassland: up to 260 ft (80 m)

Rock type: igneous Precambrian, capped by quaternary and tertiary sediments

The Llanos is a seasonally flooded grassland that covers nearly one-third of Venezuela and more than one-eighth of Colombia. It is most flooded in the central part which occurs on the huge saucer-shaped depression that contains the floodplain of the mighty Río Orinoco. Underlying Precambrian rocks choreograph the shallow basin's broad dips and rises, while recent sediments provide it with its character, creating a mosaic effect of different habitats, both within the flooded area and in the drier extremities.

A great diversity of wildlife exists, many of the most specialized occurring around rocky outcrops. The Llanos is most famous for its swamp-based wildlife of Orinoco goose, scarlet ibis, and capybara, as well as a host of migrant species. Flooding is at a maximum between July and October. In the dry season many watercourses dry up, leaving the larger rivers and the clay-panned estuaries to slake the thirst. The Llanos has over 3,400 recorded flowering plants, 40 of which are unique to this area. Among the 475 bird species is the Orinoco soft-tail, while included among the 148 mammals is the Llanos long-nosed armadillo. The area's reptiles include the green anaconda—the world's largest snake species—and the rare Orinoco crocodile. **AB**

ORINOCO DELTA

DELTA AMACURO, VENEZUELA

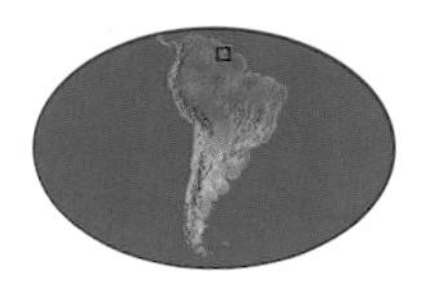

Length of Orinoco River: 1,590 mi (2,560 km)

Area of Orinoco Delta: 10,800 sq mi (28,100 sq km)

Human population: 20,000 Waori people

After 1,500 miles (2,414 kilometers) of its journey from the Guiana Highlands through thick rainforest, the Orinoco River breaks up into an extraordinary labyrinth of narrow creeks, river channels, sandbars, and islands—the Orinoco Delta. Made from the sediment washed down in the river, the central part of the Orinoco Delta covers an area the size of Vermont. It is one of the largest intact wetland areas in the world. It is also the place in which to see rainforest animals that are normally hidden from view. The islands are clothed in tropical and semi-tropical moist, broadleaf forest, and mangrove swamps, with patches of llanos.

These islands house a great diversity of wildlife. Brightly colored macaws, parrots, outlandish hoatzins, and horned screamers feed in the trees, while on the forest floor agoutis and pacas forage for seeds. When the river floods in the wet season, from May through September, the rodents make for high ground, and their places are taken by crocodiles and river turtles. The creeks and channels harbor giant river otters and freshwater dolphins, as well as meat-eating and seed-eating piranhas. MB

GUÁCHARO CAVES

MONAGAS / SUCRE, VENEZUELA

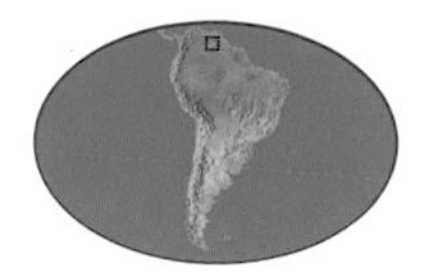

Length of Guácharo cave system: 6.3 mi (10.2 km)

Birds exit from caves: 19:00 local time

Birds return to caves: 04:00 local time

This cave system in the Caripe Mountains of Venezuela was discovered in 1799 by the famous explorer Alexander von Humboldt, who not only found the caverns intriguing, but also its inhabitants. Led by a lantern down narrow passageways, a visitor today can experience what Humboldt did all those years ago. As you enter the first cavern, known as Humboldt's Gallery, you are greeted by the deafening screams of up to 15,000 pigeon-sized oilbirds (guácharo), the largest known colony in the world. They roost and nest here in the dark, but at night they leave to forage for fruits in the surrounding forest.

At dusk 250 birds per minute pour out of the cave entrance, finding their way in the dark using a primitive echolocation system (similar to that used by bats and dolphins, but with the substitution of audible sounds). They mainly eat palm nuts, and the seeds they deposit in their droppings in the cave feed an entire ecosystem of cave crickets, spiders, centipedes, crabs, and rats. Some of the seeds germinate on the cave floor, so miniature forests of spindly palm-nut seedlings sprout and then die in the dark. The caves can be found about 6 miles (10 kilometers) north of Caripe. MB

ANGEL FALLS

BOLÍVAR STATE, VENEZUELA

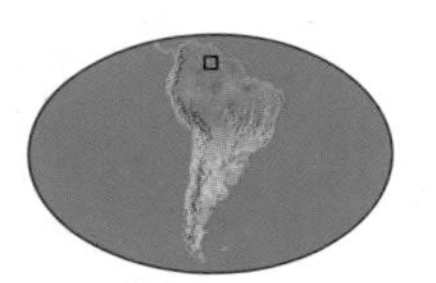

Alternative name: Salto Angel

Free fall height of Angel Falls: 3,287 ft (1,002 m)

Rainfall collected in ravines: 300 in (762 cm)

In 1935, an American pilot by the name of Jimmy Angel was searching for gold in the Venezuelan rainforest when he chanced upon the highest waterfall in the world. What he saw was a river cascading down the edge of a flat-topped plateau known as Auyan Tepui. The plateau is crossed by ravines that collect huge amounts of rainfall that spill over the buttress to form the spectacular Angel Falls.

"A large river leaps down from above without touching the mountain's wall … and reaches the bottom with a roar and clamour that would be produced by one thousand giant bells striking one another." — *Sir Walter Raleigh*

Jimmy Angel took his wife and two mountain explorers to return to the falls with him, but when they tried to put the plane down on the flat top of the tepui they discovered—too late to avoid an accident—that the landing site was a bog. The plane crashed, and although nobody was hurt, the party was left to fight its way across yawning ravines and through hostile, thick, almost impenetrable forest. Hope for their return waned, but after two weeks they staggered into base camp exhausted and starving. Angel's plane has since been taken down and can be seen in the nearby Ciudad Bolívar Museum.

Confirmation of Angel's claim was not forthcoming until 1949, when former American war correspondent Ruth Robertson took an expedition in a motorized canoe up the Churún River and set up instruments that showed the falls to be 18 times higher than Niagara Falls. The flow of water is far from constant: In the rainy season, spray at the foot of the falls drenches a large area of rainforest, but in the dry season the water that reaches the forest floor is no more than a mist.

Although Jimmy Angel is credited with their discovery, in 1910 a rubber gatherer, Ernesto Sánchez La Cruz, was probably the first non-local to see them, although there is also anecdotal evidence to suggest Sir Walter Raleigh may have seen them in the 16th century. Raleigh told how "a large river leaps down from above without touching the mountain's wall … and reaches the bottom with a roar and clamour that would be produced by one thousand giant bells striking one another," an apt description of the Angel Falls if indeed he did find them. Today, the falls are part of Canaima National Park and are accessible to anyone with the stamina to make the journey. To see the Angel Falls, fly from Caracas in Venezuela, to Canaima, where tour companies arrange either canoe expeditions or light aircraft flights. MB

RIGHT *From the heavens: Venezuela's Angel Falls.*

AUTANA TEPUI

AMAZONAS, VENEZUELA

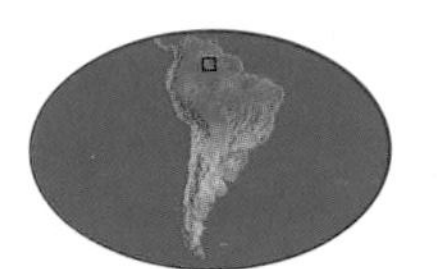

Alternative name: Cerrotana

Rock type of Autana Tepui: Precambrian sandstone

Habitat: Amazonian rainforest (bottom), tepui scrub (top)

Declared a national monument in 1978, Autana Tepui is a 4,000-foot (1,200-meter) salmon-pink block that rises from lowland Amazonian rainforests. Like all tepuis, Autana is made from Precambrian sandstone that formed when most of modern Venezuela was a shallow sea. Erosion formed these table mountains 300 million years ago, although the deposits are about three billion years old.

Autana is one of the most dramatic tepuis, and is most famous for an extraordinary cave. Measuring 1,300 feet (396 meters) in length and 130 feet (40 meters) in height, the cave was probably carved by the Rio Autana in a long-abandoned underground phase, when the sandstone plateau (tepui) was more extensive. The river has now taken another path. The top of the tepui is a soggy savanna, matted with carnivorous plants. It is also home to unique species of animals and plants, found nowhere else on Earth. **AB**

BELOW *Autana Tepui looms above the lowland rainforest.*

PICO DA NEBLINA

AMAZONAS, BRAZIL / AMAZONAS, VENEZUELA

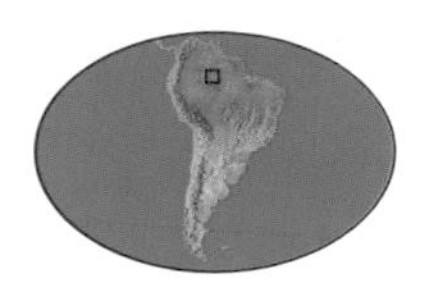

Area of Pico da Neblina: 13,900 sq mi (36,000 sq km) Venezuelan and Brazilian national parks combined

Maximum altitude: 9,888 ft (3,014 m)

Discovered in 1953, Pico da Neblina ("Thirsty Peak") is Brazil's highest mountain. First climbed in 1965 by a Brazilian army team, it is now climbed annually by the military which changes the national flag on its summit. Part of the 50-mile- (80-kilometer-) long Imeri range, the peak is situated in a transborder national park. Waters drain to both the Amazon and Orinoco river systems. Though sharing the sandstone geology of Venezuela's block-like tepuis, the Imeri range has been twisted and folded into precipitous peaks and deep gorges, one of which, the Baria River gorge, is the world's deepest.

Because the range stalls Amazonian clouds, the climate is hyper-moist, averaging 157 inches (4,000 millimeters) of rain a year. One of Amazonia's wettest areas, it has no dry season. Diverse habitats include rainforest, flooded forest, and scrubby campina, with cloud forest and alpine scrub on the higher slopes. More than half the peak's plants are unique to the area. **AB**

SHELL BEACH

BARIMA-WAINI, GUYANA

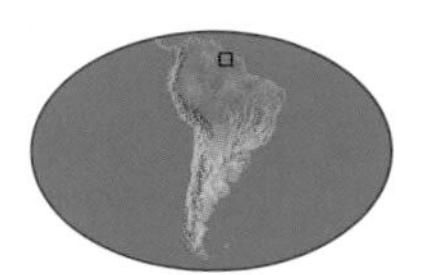

Area of Shell Beach: 100 mi (160 km)

Habitat: shelly and sandy beaches, mudflats, mangroves

Shell Beach refers to a 100-mile (160-kilometer) long stretch of coast north of Guyana's capital, Georgetown. Lying between the mouths of the Pomeroon and Waini rivers, close to the Venezuelan border, this is the last remaining stretch of untrammeled coastline in Guyana. It also has the country's best remaining stands of mangroves. A main attraction is the nesting marine turtles, which emerge from the sea at nine beaches along the Shell Beach coast from March to April. Four species nest here, including the leatherback, green, hawksbill, and the exceedingly rare olive ridley. Sandflies are an unappreciated, but ubiquitous, aspect of the local fauna.

Shell Beach lacks formal protection, yet despite this, internationally funded turtle conservation efforts began in the 1960s and have persisted most recently under the Guyana Marine Turtle Conservation Society. The Society's programs encompass nest monitoring and fishery protection to prevent turtles drowning in nets. These include cooperative ventures at Almond Beach and Gwennie Beach with the region's two Amerindian Arawak communities. Mangrove forests harbor five species of mangroves. The mudflats they form are internationally important for transcontinental migrant birds. Local species include scarlet ibis, magnificent frigate birds, and greater flamingoes. AB

IWOKRAMA MOUNTAINS

POTARO-SIPARUNI, GUYANA

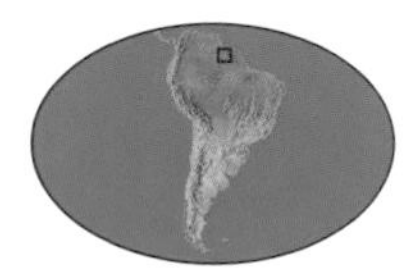

Area of Iwokrama Mountains: approx. 1 million acres (371,000 ha)

Elevation: 0 to 3,300 ft (0 to 1,000 m)

Habitats: lowland rainforest, cloud forest, elfin cloud, large slow lowland rivers, small rapid upland streams

Located in central Guyana in the middle of the Guyana Shield, Iwokrama is of one of the four great remaining areas of tropical rainforest. Much of the area's diversity is due to Mount Iwokrama, whose 3,300 feet (1,000 meters) bring several strata of high-altitude vegetation. Bordered by the Pakaraima Mountains' upland forests to the west, and savannas to the south and east, Iwokrama has a variety of habitats, ranging from lowland forests 66 to 100 feet (20 to 30 meters) tall, to cloud forest and cold-dwarfed elfin forest at the highest altitudes.

Over 500 of the country's 800 bird species have been recorded at Iwokrama. These include primary forest fruit-feeding and hunting-sensitive specialists such as guans, parrots, and cotingas. Among the special species at Iwokrama are harpy eagle, red-fan parrot, rufous-winged ground-cuckoo, white-tailed potoo, Guyana toucan, and dusky purpletuft. Hummingbirds include crimson topaz and racket-tailed coquette. In addition, there are 200 mammal species, of which 90 species are bats—the greatest number of bats in a single place anywhere in the world. Iwokrama also has the greatest known number of freshwater fish (420 species). Due to cooperation between the Guyanese government and foreign scientists, the area has a very active conservation program. **AB**

KAIETEUR FALLS

POTARO-SIPARUNI, GUYANA

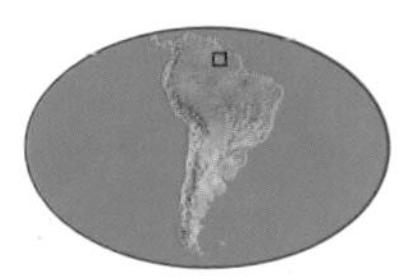

Area of Kaieteur National Park: 224 sq mi (580 sq km)

Height of falls: 741 ft (226 m)

Length of gorge: 5 mi (8 km)

The Pakaraima Mountains of west-central Guyana are the source of many rivers, each dropping spectacularly over the edge of a sandstone escarpment marking the boundary between highlands and lowlands. The most famous cascade is the dramatic Kaieteur Falls. Here, water from the Potaro River falls 741 feet (226 meters) into a splash basin below. It is second only to Angel Falls in height, but unlike Angel, it carries large volumes of water all year round.

The area is also known for its unique plant species. In some areas the forest opens to reveal shrubs and herbaceous plants growing on pink sands. Cracks between rocks support bromeliads, and the water that accumulates in the "tanks" formed by the rosette of leaves supports the largest member of the bladderwort family. This carnivorous plant that feeds on insects trapped in the stagnant water throws up a flower stalk with a light purple flower six feet (1.8 meters) above its host plant. Close to the gorge and the waterfall's splash basin, delicate ferns, primitive bromeliads, and red-flowered African violets also thrive. Endangered species such as bush dogs, hyacinth macaws, and cock-of-the-rock are found here, as are jaguarundi, red brocket deer, and tapirs. Access to the falls is mainly by chartered aircraft from Georgetown. MB

KANUKU MOUNTAINS

UPPER TAKUTU / UPPER ESSEQUIBO, GUYANA

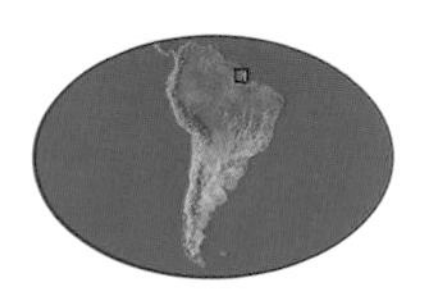

Highest elevation of Kanuku Mountains: 3,300 ft (1,000 m)

Habitat: tropical rivers, lowland forest, cloud forest, alpine vegetation

Rock type: ancient sandstone

An outcrop of ancient sandstone, the Kanukus rise from the surrounding dry grasslands of Guyana's Rupanuni Savanna. Because they trap moisture from clouds and rain, the Kanukus shelter a great variety of moisture-adapted species in an otherwise much drier landscape. Comprised of two large blocks—the east and west Kanukus—separated by a gully 1 mile (1.6 kilometers) wide, the Kanukus have a great variety of rare wildlife that has been reduced elsewhere in its range—as well as species unique to this area.

Comprising steep-sided rock blocks and flat plateaus, and intersected by the gently meandering Rupanuni River, the Kanukus have river-dependent species, such as the giant otter, giant river turtle, and black caiman. Their lowland forests house monkeys, monkey-eating eagles, tapir, and jaguar. There are also many amphibians. Approximately 80 percent of Guyana's mammals have been recorded here. Located in southwestern Guyana, the Kanukus are remote and little visited. They are currently unprotected but the region's indigenous people hope to persuade the national government to make the mountains a national park—several international conservation organizations have backed this proposal. **AB**

AWALA-YALIMAPO

FRENCH GUIANA

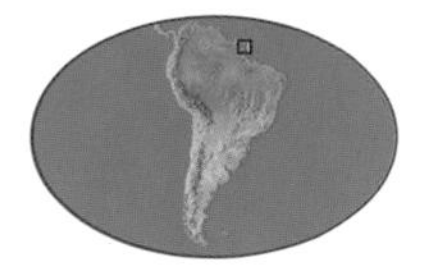

Alternative name: Les Hattes

Number of turtles at Awala-Yalimapo: 15,000

Leatherback turtle maximum length: 7 ft (2.1 m)

Bring the mosquito repellent not the suntan lotion, for events on this beach happen in the dark. Each night in May and June, female leatherback turtles haul out onto the sand to lay their eggs. Emerging from the surf, they sniff the beach as if searching for a recognizable smell—evidence that they have returned to the beach where they hatched out many years before. Without the support of water, they haul their great bulk laboriously across the beach, dig holes, and deposit their eggs. They then cover their clutch and return slowly to the sea.

Awala-Yalimapo is one of the most important nesting beaches for leatherbacks in the western hemisphere, and visitors can not only watch as the females lay eggs, but also witness the moment (July and August) when the tiny hatchlings emerge. There are rules: keep at least 16 feet (5 meters) from a nesting female and do not shine torches directly at her; when the hatchlings come out, do not pick them up—even if they appear to be going the wrong way. There are buses from Cayenne to Aoura, the nearest town. **MB**

RIO SOLIMÕES & RIO NEGRO CONFLUENCE

AMAZONAS, BRAZIL

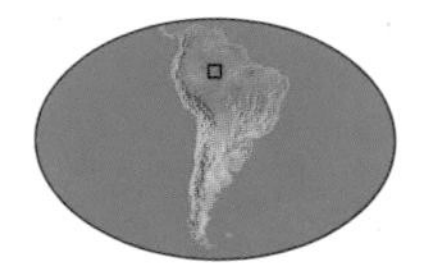

Length of Rio Negro: 600 mi (1,000 km)
pH of Rio Negro: 5.1 ± 0.6
pH of Rio Solimões: 6.9 ± 0.4

As a rule, few things hyped-up as tourist attractions match their expectations. The "Meeting of the Waters" is an exception. Setting off from Manaus on the tea-colored waters of the aptly named Rio Negro (Black River), less than 6 miles (10 kilometers) downstream, you come to where the Negro's dark waters mix and meld with the cream-colored flow from the Rio Solimões. While you watch, small eddies of current produce swirls that look like galaxies or gigantic cups of recently stirred coffee. The spectacle continues for several miles until the oscillating black and cream gives way to the all-embracing milkiness of the greater Amazon River.

How did these river waters get to be so different? The answer lies in the geology of the rocks underlying their headwaters. The Rio Solimões is the final river in a 1,800-mile (3,000-kilometer) chain of rivers that begins in the Peruvian Andes. Here, feeder rivers

cut into recent, soft, volcanically-derived soils. Easily eroded, these annually load the waters of this river system with thousands of tons of sediment. The Rio Negro has its origin in the north of the Amazon Basin in the Pakaraima Mountains, a group of sandstone mountains. At around two billion years old, these rocks are so ancient that they leave little sediment in the water. In fact, the Rio Negro would be transparent if it were not for the forest plants along the 600 miles (1,000 kilometers) of its banks, which leach the humic acids from their leaves into the water, staining it brown like tea.

The Solimões and Negro waters differ in temperature, nutrient and oxygen content, and acidity. So different are they that fish swimming from one to the other are stunned temporarily. This makes for easy pickings by the Amazon's two species of freshwater dolphin, which congregate in great numbers at this meeting of the waters, and at another similar phenomenon downstream at Santarém where the clear waters of the Rio Tapajós meet the Amazon's muddy waters. **AB**

BELOW *Rio Negro's dark waters merge with pale Rio Solimões.*

AMAZON BASIN

BRAZIL / PERU / ECUADOR / COLOMBIA / VENEZUELA / BOLIVIA

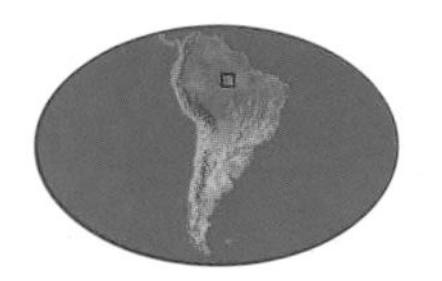

Area of Amazon Basin: 2.7 million sq mi (7 million sq km)

Age of Amazon Basin: 60 million years

Habitat: tropical rainforest up to 120 ft (36 m) high, plus a variety of flooded forest and scrubby savannas

Surrounded north and south by plateaus of ancient crystalline rock, and to the west by the Andes Mountains, the modern Amazon basin is the result of a 70-million-year history that began when the mega-continent of Gondwanaland split into modern Africa and South America. Containing the planet's largest tropical rainforest and its most voluminous river, the Amazon basin has more than 1,000 major rivers. The Amazon rainforest is incredibly rich in flora and fauna. The richest parts are in the west, nearer to the Andes, the source of fertile, volcanically derived soil. This manna becomes more thinly dispersed as one heads east, which is reflected in the abundance of animals and plants in the famous forests around Manaus.

Water and air are still the main modes of transport, making independent exploration of the region largely impossible. Key centers include: Manaus, Tefé and the esteemed Mamirauá Ecological Reserve, São Gabriel do Cachoeira in Brazil, Peru's Iguitos, the Tambopata Natural Reserve, Leticia in the Colombian Amazon, and the Napo region of the Ecuadorian orient. Among the faunal and floral marvels are giant otters, huge waterlilies, tapirs, toucans, carnivorous plants, tiny hummingbirds, giant rats, parrots, piranhas, jaguars, mata-mata turtles, orchids, arrow poison frogs, and freshwater dolphins. AB

AMAZON BORE

AMAPA / PARÁ, BRAZIL

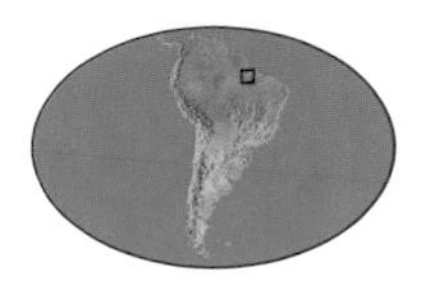

Length of Amazon River: 4,050 mi (6,518 km)

Width of Amazon mouth: over 200 mi (320 km)

Maximum width of bore: 9.9 mi (16 km)

River bores occur when a river's flow is at a minimum and full-moon sea tides are strong enough to force the river's water back up its own channel. Not every river can have a bore. For one to form successfully requires a precise combination of narrowing riverbanks and a rising riverbed that will funnel forced-back water and ensure that its energy is concentrated in the wave. Bores occur to varying degrees on all the channels of the Amazon's mouth, but the most spectacular is on Rio Araguari and Rio Guama.

The event's local name, *pororoca*, comes from the Tupi Indian phrase meaning "great roar." Waters can move at speeds exceeding 45 miles-per-hour (70 kilometers-per-hour). This area is very popular with surfers—rides of 6 minutes or more are common, with the current record being 17 minutes, and 40 minutes being considered attainable (a good ride on an ocean wave might last 30 seconds). There are over 60 river bores around the world, including those on Canada's Bay of Fundy, England's Severn River, and India's Ganges River. The biggest is the Qiantang Tidal Bore on China's Fuchun River, which regularly reaches up to a staggering 25 feet (7.5 meters) and travels at over 25 miles-per-hour (40 kilometers-per-hour). **AB**

RIVER XINGU

PARÁ, BRAZIL

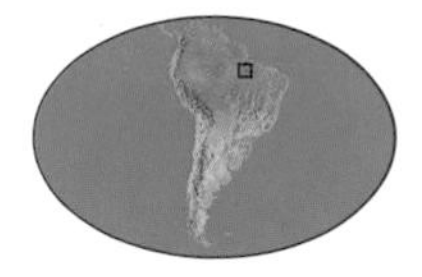

River Xingu length: 1,230 mi (1,979 km)

Number of river turtles: 5,000

Size of turtle carapace (shell): 3 ft (0.9 m) long

The River Xingu is a large tributary near the mouth of the Amazon, but by October each year its waters have reached their lowest point and many sandbanks are exposed. First one, then two, and then many—the heads of many giant river turtles or *tartaruga* begin to appear in the shallows. As the numbers grow, they haul out on the sand. They are female turtles and they are here to deposit their eggs. They arrive two weeks before egg laying and bask in the sun to help the final stages of egg development. There might be over 5,000 females arriving at the same time. Each turtle is three feet (0.9 meters) long, making this species the world's largest living river turtle—and they make one enormous mess. Space is at a premium, and late layers dig up the nests of early arrivals. Black vultures drop in and make short work of any exposed eggs, but those beneath the sand are safe as long as the river does not flood early. The phenomenon, however, is rarely seen other than by scientists studying turtle behavior, as the site is forbidden to tourists lest they frighten the animals and interrupt their reproduction. **MB**

ANAVILHANAS ARCHIPELAGO

AMAZONAS, BRAZIL

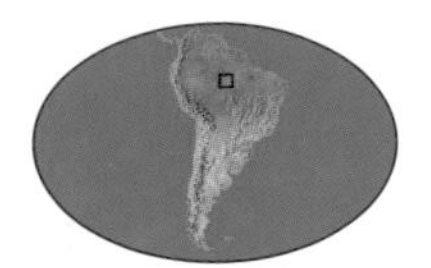

Length of archipelago: 93 mi (150 km); width 7.5 mi (12 km)

Habitat: tropical rainforest, flooded forest, sandy beaches

Anavilhanas is the world's largest inland archipelago. Beginning some 50 miles (80 kilometers) west (upstream) of Manaus in the Rio Negro in the center of the Brazilian Amazon, it forms a complex of 350 islands. Water levels vary by as much as 50 feet (15 meters), making for an ever-changing matrix of channels, levées, and sandbars that only the most experienced guides and fishermen can negotiate with certainty. The islands range from substantial, permanent, forest-topped entities, which are large and stable enough to support luxury hotels, to remote sandbars that completely disappear at high water.

The islands achieved their current form during the last ice age when water levels fell throughout the Amazon basin and changes in hydrology caused several of the Rio Negro's feeder rivers to deposit an excess of sediments. Several huge ancient rocks stood proud from the now shallow river and, in its reduced state, the Rio Negro was not able to flush away the deposits that built up around them, eventually forming islands.

These islands became so extensive that, when river levels rose again at the end of the Pleistocene, the newly extended island chain continued to slow down the river waters, causing suspended sediment to drop out. This maintained the islands and a new permanent habitat was formed, one that trapped almost all the sediment in the Rio Negro and so renewed itself against the erosion of the water. This process continues today, fed primarily by the sediment-rich Rio Branco and also by the Rio Negro. Though sediment concentration is low in the Rio Negro, it is such a large river that there is enough material to maintain the ever-changing complex of islands.

Anavilhanas is the world's largest inland archipelago. Beginning some 50 miles (80 kilometers) west of Manaus in the center of the Brazilian Amazon, it forms a complex of 350 islands that range from permanent, forest-topped entities to remote sandbars.

The majority of the archipelago lies within the Anavilhanas Archipelago Ecological Station. This is an area of major importance for several species of freshwater turtle that nest on sandy islets during the low water season (July to November), and for the various species of birds that are adapted to living in the archipelago's flooded forests. Anavilhanas is also a prime haunt of manatees, giant river otters, the elusive Amazon pink river dolphin, and the pirarucu, the world's largest freshwater fish, which can grow up to 10 feet (3 meters) in length. AB

RIGHT *Islands of the Anavilhanas stretch into the Rio Negro.*

FLOODED FOREST

AMAZONAS, BRAZIL

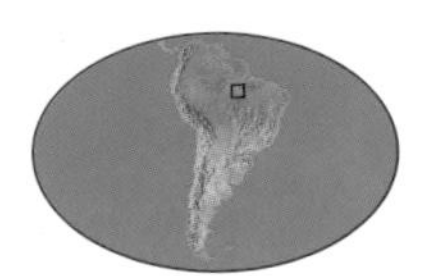

Alternative name: Varzea

Width of flooded forest: 50 mi (80 km) either side of Amazon main channel

Depth of flooded forest: 50 ft (15 m)

Dolphins and piranhas in the treetops, fish that eat nuts, and monkeys with bright red faces—not a fantasy folktale, but the real and living flooded forest of Amazonia. Each year at the end of December, the rains start and the waters of the Amazon River begin to rise until they spill into the forest and drown a strip 50 miles (80 kilometers) wide on either side of the river to a depth of 50 feet (15 meters). There is no dry land anywhere. People live in floating houses or dwellings on stilts, and chickens and other livestock are reared on rafts or balconies over the water.

At the same time, many of the trees begin to bear fruit, a feast not only for canopy species, such as monkeys and birds, but also for fish. Pacu, catfish, and vegetarian piranhas eat nuts, just three of 200 species of freshwater fish on which trees can rely to distribute their seeds. Catching the fish are botos—pink-skinned, freshwater dolphins—and joining them in the drowned canopy is the pirarucu, the largest freshwater fish in the world at 10 feet (3 meters) long. Tours are available out of Manaus. MB

IGAPÓ FOREST

AMAZON BASIN, BRAZIL

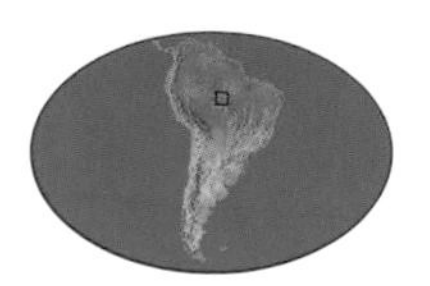

Extent of Igapó Forest: 3 percent of Amazonia

Age of plant community: 10,000 to 12,000 years

Igapó Forest occurs only along the sides of the Rio Negro and its feeder rivers in the northwestern part of the Amazon basin. Annually flooded to a depth of up to 50 feet (15 meters) for as long as nine months of the year, specialized plants and animals inhabit the Igapó Forest. Blackwater rivers, like the Rio Negro, carry very little sediment, so few levees or floodplains build up and the main river channel remains V-shaped. Consequently, Igapó occurs in thin ribbons, rarely more than 450 yards (410 meters) wide. When flooded, Igapó is an exquisitely beautiful water world, limpid, quiet, and full of dancing reflections. At this time igapó trees produce fruits, most of which are dispersed by fish such as the pacu, which migrate into blackwater rivers specifically for this feast. Amazonian flooded forests are the only ones in the world with fish as major seed dispersers. Igapó canopy hosts many specialist birds, including Klage's antwren, ash-breasted antbird, and Snethlage's tody-tyrant. Spix's uacari monkey also makes its home in Igapó. In the dry season most animals quit the Igapó, eking out a living in the rainforest until the next flood brings flowers, fruit, and life back to the forest. AB

EMAS NATIONAL PARK

GOIÁS, BRAZIL

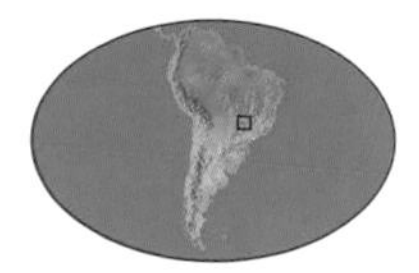

Area of Emas National Park: 509 sq mi (1, 318 sq km)

Altitude of Emas National Park: 1,320 to 3,300 ft (400 to 1,000 m)

Named after one of its most prominent species of wildlife—the rheas (emas in Portuguese)—Emas National Park preserves some of the best and most biologically intact of the campo limpo ("pure grassland") type of central Brazil's cerrado vegetation. Characterized by its huge red termite mounds, there are also extensive palm stands, gallery forests, and deeply carved canyons in the park.

Emas is a great place to see wildlife. Among the 354 bird species and 78 mammals are the yellow-faced parrot, giant anteater, giant armadillo, jaguar, and maned wolf. Specific to the region is the curious phenomenon of glowing termite mounds. This is caused by beetle larvae that attract flying termites with their glowing abdomen and then trap them with their pincer-like claspers.

Cerrado is one of the oldest tropical habitats and boasts a variety of plants that have adapted to the poor acidic soils and the six to eight months of annual drought. These include the cerrado pineapple, large bush daisies, and brightly-flowered annual plants. The best time to visit is when the grass has been burned—it usually grows ten feet (three meters) tall so wildlife tends to be obscured. AB

THE PANTANAL

BRAZIL / BOLIVIA / PARAGUAY

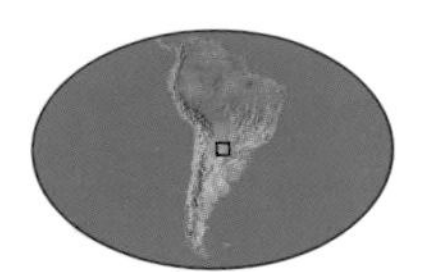

Area of the Pantanal: 81,000 sq mi (210,000 sq km): 80 percent Brazil, 10 percent Bolivia, 10 percent Paraguay

Maximum altitude of the Pantanal: 500 ft (150 m)

The name "Pantanal" means, very accurately, "Big Swamp." At around half the size of California, the Pantanal is the largest swampland on Earth, larger even than Botswana's famed Okovango, and 20 percent larger than Sudan's lesser-known Sudd Swamp. America's Okefenokee and Spain's Cota Doñana pale in comparison. Though it receives some 60 inches (1,600 millimeters) of rainfall annually, most of the Pantanal's yearly inundation is due to the annual overspill from the Rio Paraguay as it drops off Brazil's Central Plateau.

This huge, seasonally flooded swampland has some of the most spectacular concentrations of wildlife on the planet, rivaling East Africa for the sheer volume of birds, mammals, and reptiles, many of which grow to outlandish size. These include the largest subspecies of jaguar (which can be twice the size of those from Amazonia), plus the world's largest otters, largest rodents (the capybara), largest stork (the jabiru), and largest parrot (hyacinth macaws). Freshwater turtles as well as puma and other smaller cats are common. Two species of deer, the marsh and pampas deer, are also abundant, as are Paraguayan caiman. These crocodile-like reptiles have been recorded at densities of up to 3,000 in a 2.2-acre (1-hectare) lake.

All of these animals live among a rich mosaic of vegetation whose nature depends on the different types of soil and the duration of flooding. There are several varieties of forest, including swamp, gallery forest, elegant palm forest, lake margin scrub forest, and hummock forest.

While the Pantanal boasts a wealth of individual animals, it has a comparatively low number of species for a tropical habitat. Furthermore, none of the 3,500 species of plant, 129 species of mammal, 177 species of reptile, or 650 species of bird are endemic, and only 15 of the 325 species of fish are unique to the region. However, this in no way diminishes the sheer splendor and spectacle of the area. This vast world of water, with its strong sunlight and fertile soils, has produced an extraordinary abundance of wildlife.

Luckily, despite years of low intensity cattle ranching, over 80 percent of the Pantanal is still intact. AB

At around half the size of California, the Pantanal is the largest swampland on Earth—America's Okefenokee and Spain's Cota Doñana pale in comparison. It has some of the most spectacular concentrations of wildlife on the planet, rivaling East Africa for the sheer volume of animals.

RIGHT *An aerial view of the lush Pantanal wetlands.*

THE CERRADO

BRAZIL

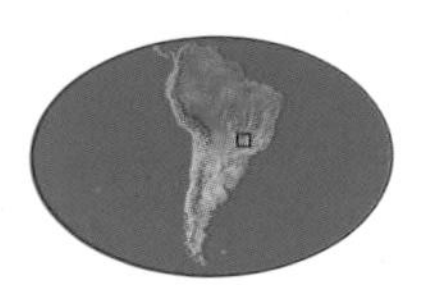

Area of the Cerrado: 2 million sq mi (5,179,976 sq km)

Elevation range: 3,300 to 6,600 ft (1,000 to 2,000 m)

Age of rocks: 2 billion years

The word "cerrado" means "closed" in Portuguese—an odd name for a habitat characterized by space, wide horizons, blood-red soil, and huge blue skies. This high inland plateau northwest of Rio de Janeiro has many facets: rolling gold grasslands, jutting red rock-faces, concealed fern-adorned green canyons, scorching-hot rock fields, and lush palm-fringed rivers. Occupying 21 percent of its landmass, Cerrado is Brazil's second-biggest habitat type; only the Amazon is bigger. Based on exceedingly ancient, acidic, and nutrient-poor soil, it is a habitat type unique to Brazil and one of Earth's oldest tropical ecosystems. It is rich in gold, iron ore, and precious stones—most of the world's amethysts and gem-filled geodes are mined in the Cerrado. It also holds an abundance of natural treasures: the yellow-fronted parrot and the maned wolf—an Alsatian-sized, ginger-colored wild dog on stilt-like legs—occur only here, as do another 18 of Cerrado's 161 mammals and 28 of its 837 bird species. Of its 6,500 species of flowering

plant, over 40 percent occur nowhere else. These support, among other insects, over 400 species of bee and 10,000 species of butterfly. This richness of species make the Cerrado the world's richest savanna ecosystem.

The Cerrado receives over 60 inches (1,500 millimeters) of rain annually, concentrated between April and October. For the rest of the year it is very hot and dry, which, when coupled with the frequency of fire, means that most Cerrado trees have thick, corky bark and small, waxy, drought-adapted leaves. Many of the herbs are annuals. The most common impression of the Cerrado is of an African savanna without the giraffes and zebras, but Cerrado has much to offer. Some of the best places to look are Emas, Canastra, Chapada dos Veadeiros, and Caraça National Parks, part of a network of some 20 government-run protected areas in this beautiful habitat. So strongly do Brazilians feel about this area that landowners have banded together to provide a further 85 private reserves in which the Cerrado's natural features and wildlife are protected. Nevertheless, less than 20 percent of the Cerrado remains in its natural state. **AB**

BELOW *Red sandstone cliffs overlook typical Cerrado habitat.*

LENÇÓIS MARANHENSES

MARANHÃO, BRAZIL

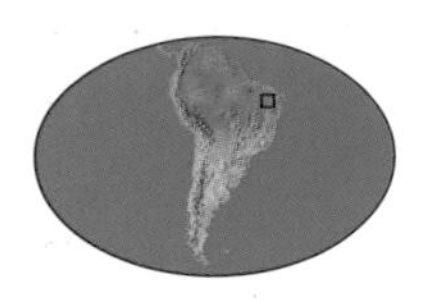

Area of Lençóis Maranhenses: 382,850 acres (155,000 ha)

Height of dunes: up to 140 ft (42 m)

Dune field: 45 mi (70 km) of coastline and 30 mi (50 km) inland

This array of dazzlingly white sand dunes interspersed with lagoons of clear fresh water is not a mirage; it is the coastal sand-dune belt at Lençóis in northeast Brazil. The dunes are up to 140 feet (42 meters) high and are formed by the constant winds from the sea. The dune field lies beside the Rio Preguica, between Barreirinhas and Primeira Cruz counties, and can be seen from satellites in space. The most spectacular of the lagoons are Lagoa Bonita (Beautiful Lagoon) and Lagoa Azul (Blue Lagoon), blue jewels in this white sand sea.

But desert this is not—over 60 inches (1,600 millimeters) of rain falls in a well-marked rainy season from January to June. The rest of the year is dry and the lagoons are parched. But with the advent of rain, each water-filled depression bursts into life, and turtles, fish, and shrimp reappear. At their peak, some lagoons are several miles long and up to 16 feet (5 meters) deep. The best time to visit is from May to October when the lagoons have filled. You can get there by bus or car from the state capital Sao Luis, a 10-hour journey. Alternately, chartered flights are available to Barreirinhas. **MB**

CORCOVADO

RIO DE JANEIRO, BRAZIL

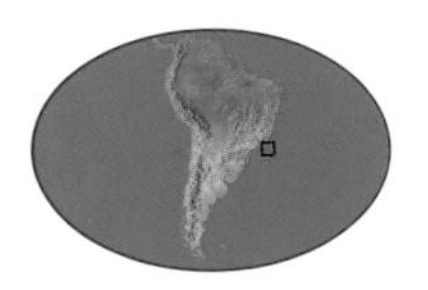

Height of Corcovado: 2,330 ft (710 m)

Rock type: igneous

Like the many other famously beautiful peaks in Rio de Janeiro's Guanabara Bay, Corcovado is the heart of an ancient volcano, perhaps 300 million years old. Being at the very core allowed the lava to cool very slowly, which resulted in extremely small-grained rock that was so resistant to erosion that it survived when all surrounding matter was ground down. Nowadays Corcovado, meaning "hunchback," is famous for its "Christ the Redeemer" statue.

The 125-foot (38-meter) statue was designed by the Brazilian engineer Heitor Silva Costa and built in Rio over five years by Costa and the French sculptor Paul Landowski. Finished in 1931, the statue is huge—one of the hands alone is 10 feet (3 meters) long. Father Pedro Maria Boss, who was struck by Corcovado's beauty when arriving in Rio, first suggested the project in 1859. Corcovado's 2,330-foot (710-meter) peak lies within Tijuca National Park, the world's largest urban forest. Though much of the park was originally coffee plantation, enough original forest remained to recolonize the area, and the park is now an important part of the efforts to conserve Brazil's endangered coastal rainforest. **AB**

SUGAR LOAF MOUNTAIN

RIO DE JANEIRO, BRAZIL

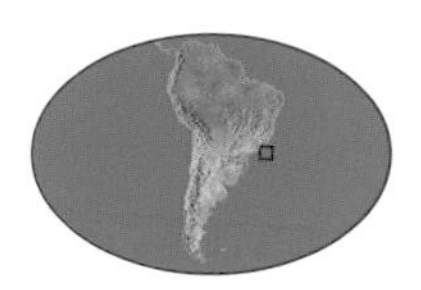

Alternative name: Pao de Açúcar

Height of Sugar Loaf Mountain: 1,325 ft (404 m)

Height of Topsail Rock: 2,762 ft (842 m)

Rio de Janeiro's other striking landmark is the well-defined Sugar Loaf Mountain. During the 16th and 17th centuries, Brazilian sugar cane was boiled, refined, and then placed in conical mud containers called "sugarloaves." The mountain's resemblance to the sugarloaf shape gave it its name.

It stands proud above the city and Guanabara Bay and is probably one of the most famous mountains in the world. It is made of granitic gneiss that was a molten intrusion into the surrounding rocks about 600 million years ago. The softer rocks have eroded away, leaving this stark mountain. As a result of exfoliation weathering, its edges have been softened to form a rounded, granite tor. From its summit, other granite monoliths can be seen along the coast. Small remnants of the Atlantic coast rainforest that once cloaked the area are still around. Today, a series of cable cars take visitors to the top, although more energetic hikers can make the steep climb to the summit—a worthwhile trek. However, neither Sugar Loaf Mountain nor Corcovado is the highest peak in Rio; that title goes to Pedra da Gávea, or Topsail Rock, ever popular with hang-glider enthusiasts. MB

ATLANTIC RAINFOREST

BRAZIL

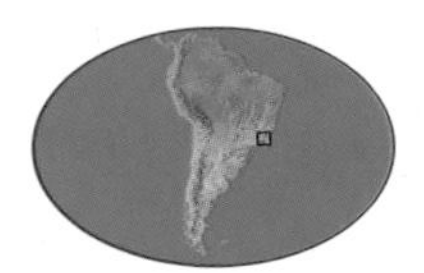

Alternative name: Mata Atlântica

Area of Atlantic Forest: 46,950 sq mi (121,600 sq km)

Elevation of Atlantic Forest: 6,600 ft (2,000 m)

When Charles Darwin disembarked at Rio de Janeiro on his epic round-the-world journey on HMS *Beagle*, he was greeted with his first experience of tropical rainforest. The second largest rainforest block in South America, and the third biggest biological region in Brazil, the Atlantic Coastal Rainforest covers 13 percent of the country's surface. Filled with rare and exotic species, it still thrills visitors with glorious floral and faunal displays. Sadly, this forest has decreased massively since Darwin's day and is now down to seven percent of its former self. But this remainder is vibrant, extremely diverse, and unforgettably impressive. A band some 1,550 miles (2,500 kilometers) long and 30 to 60 miles (50 to 100 kilometers) wide, it is the second most threatened rainforest ecosystem on the planet. Backed by the ancient 6,600-foot (2,000-meter) tall Serra do Mar range, this forest block has had both the altitude and latitude to evolve an enormous variety of wildlife since its isolation from Amazonia.

The Atlantic Forest hosts 261 mammal species, while Amazonia (five times as big) has 353. Not only are they numerous, they are special—if you see a living something in the Atlantic Forest, chances are it occurs nowhere else in the world. This is true for 6,000 of the 20,000 plant species, 73 of the 620 bird varieties, and for nearly all of the 280 types of frog. Among the regional highlights are the golden lion tamarin, a tiny monkey that has inspired conservation efforts not only in Brazil, but also in many other areas. Sharing its forest home is the thin-spined porcupine, the maned sloth, the red-breasted toucan, and a rainbow of birds that includes the red-tailed, blue-bellied, and violaceous parrots. The Brazilian government has nearly 200 protected zones that set aside Atlantic coastal forest for conservation. In 1999, UNESCO placed eight areas on the World Heritage List. There are also over 50 private reserves. Together these protect over 15,625 square miles (40,469 square kilometers). Fine examples of the Atlantic Forest can be found in Tijuca National Park, Superaguí National Park, and Serra do Coudurú National Park. AB

BELOW *The humid tropics of the Atlantic Rainforest.*

CARAÇA NATIONAL PARK

MINAS GERAIS, BRAZIL

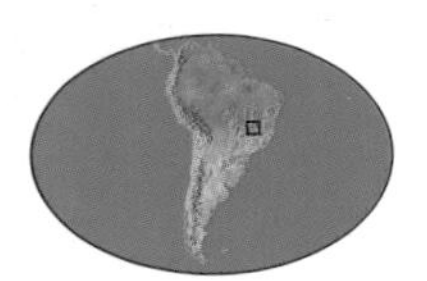

Original area of Atlantic Forest:
570,463 sq mi (1,477,500 sq km)

Remaining area of Atlantic Forest:
46,950 sq mi (121,600 sq km)

Protected area of Atlantic Forest:
15,625 sq mi (40,469 sq km)

Nearly all of the 60-million-year-old Atlantic rainforest—the Mata Atlântica—has disappeared, but one of the best remnants can be found here at Caraça. What little of these forests survives today (about seven percent) is isolated in mountainous areas, and Caraça is no exception. There are mountains, rivers, and waterfalls everywhere, such as Cascatinha and the Cascata Maior (Little and Big Falls), but it is the forest and its inhabitants that are special. Over half the trees and over 90 percent of the frogs and toads are unique to this area. Up and around the canopy are howler monkeys that are heard before you can see them, ferocious harpy eagles that can crush a monkey's skull with one squeeze of their talons, and diminutive marmosets, some of the most rare of their kind in the world. The national park is, in fact, a meeting point between two great ecosystems—the Atlantic forest itself and the neighboring Cerrado. The monks at an old monastery in Caraça, built in 1717 and now a lodging house, take advantage of the location: They feed maned wolves, and now tourists can share the experience. **MB**

CAATINGA

BRAZIL

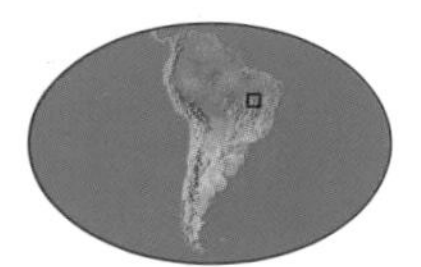

Area of Caatinga: 28,400 sq mi (73,556 sq km)

Maximum elevation of Caatinga:
6,600 ft (2,000 m)

Rock age: Precambrian to Cretaceous

The Caatinga is a semi-arid habitat with a highly unpredictable climate. Though the average annual rainfall is 31 inches (800 millimeters), rains often fail for several years in succession. In this harsh and unpredictable climate, lifeforms must combine great drought tolerance with an ability to capitalize on any available rain. Hence many trees have tiny, thin leaves, quickly grown, but just as easily shed if the rains do not arrive.

Despite its forbidding climate, the Caatinga can be stunningly beautiful. Large blocks of Cretaceous sandstone or "chapadas" dot the landscape. Near the seacoast, fog condenses against them, forming islands of vegetation.

Today's Caatinga is mostly uplifted ancient seafloor, and is very rich in fossils of fish and coastal species such as pterosaurs. It is, however, poor in nutrients; modern flora consists of thorny shrubs, barrel-trunked trees, and cacti, of which the candelabra-shaped jamacara is the most characteristic. Animals unique to the region include a cactus-feeding bat, an armadillo, and the Lear's and Spix's macaws. The best places to visit are protected areas, such as Serra da Capivara (Piauí state) and Serra Negra (Pernambuco). **AB**

RIGHT *A sandstone pinnacle marks the dry Caatinga habitat.*

APARADOS DA SERRA NATIONAL PARK

RIO GRANDE DO SUL, BRAZIL

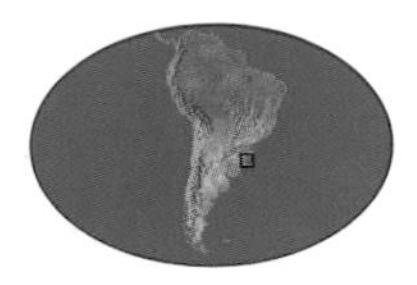

Area of Aparados: 390 sq mi (1,025 sq km)

Area of Serra Geral: 500 sq mi (1,730 sq km)

Founded in 1959, Aparados da Serra National Park is located in northeastern Rio Grande do Sul, in southern Brazil. Situated in Brazil's temperate zone, the parks are known for their amazing canyons, including Brazil's largest, Itaimbezinho, whose vertical walls are some 4.3 miles (7 kilometers) long and 2,360 feet (715 meters) high. Many of the waterfalls spilling over these cliffs turn to mist before reaching the ground.

The parks provide refuge for the last stands of monkey-puzzle tree. A wide altitudinal gradient ensures a great diversity of fauna and flora. There are some 635 plant, 143 bird, and 48 mammal species. Several of these are Araucaria forest specialists. Key species include the red-spectacles parrot—an Araucaria seed specialist—the maned wolf, ocelot, and brown howler monkey. Araucarias can live 500 years and grow to 150 feet (45 meters) tall. Their nuts have been traditionally gathered as a food by indigenous people who shot them out of the trees using blunted arrows. Araucarias are literally "living fossils" that developed spiny leaves to deter giant browsing dinosaurs. **AB**

ST. PETER & ST. PAUL'S ROCKS

ATLANTIC OCEAN / BRAZIL

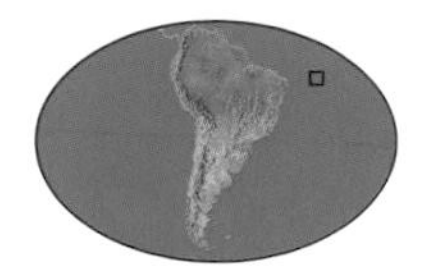

Maximum height of rocks: 64 ft (19.5 m)

Number of islets: 12

Age: 10 to 35 million years

A small cluster of bare rocks in the middle of the Atlantic Ocean painted white in bird droppings—at first glance, the St. Peter's and St. Paul's Rocks look more like a natural dump site than a natural treasure. Other than hosting a mix of seabirds, these isolated islets between South America and Africa are so battered by salty sea spray and ocean swells that only a fungus, algae, and a handful of insects, spiders, and crabs call it home. But the rocks, which barely break the ocean's surface, are just the tips of a giant biologically radiant iceberg. One of the few areas on Earth where a submarine mountain breaks the ocean's surface, the rocks are the pinnacles of a 12,000-foot- (3,650-meter-) high summit, and provide a unique oasis of surface-water habitat in the center of the sea. Complete with a protected cove of ankle-deep tide pools whose waters are refreshed daily, a shallow U-shaped bay, and offshore cliffs and caves, this remote marine outpost plays host to sea slugs, lobsters, shrimp, deepwater eels, sharks, and 75 species of fish including five—like the skinny St. Paul's Gregory—found nowhere else on Earth. **DBB**

ESMERALDAS REGION

ESMERALDAS, ECUADOR

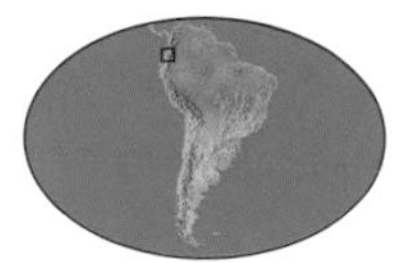

Elevation of Esmeraldas Region: sea level

Area of Manglares Churute: 3.8 sq mi (9.8 sq km)

Habitat: mangroves

Located in northern Ecuador, Esmeraldas is both biologically and culturally rich. The northernmost part shares the fauna and flora of the rain-rich Chocó, an area of astounding biological diversity that mostly occurs in Colombia. Due to the influence of the marine Humboldt Current off the coast of South America, the Ecuadorian climate is drier to the south and west creating a diverse habitat of dry-adapted forest.

The region has three important reserves: Mataje-Cayapas Mangrove Reserve, Machililla National Park, and the Manglares Churute Mangrove Reserve. Mataje-Cayapas Reserve was founded in 1996. It has seasonally dry forest and rainforest. The rainforest areas are very rich, containing elements of the Chocó fauna and flora. Farther south lies the Manglares Churute Reserve. Created in 1979, it protects one of the few areas of mangrove in the region to have escaped the ravages of Ecuador's devastating shrimp-farming boom, which destroyed much of the country's mangroves, as well as nurseries for commercially important fish. **AB**

SAN RAFAEL FALLS

SUCUMBIOS, ECUADOR

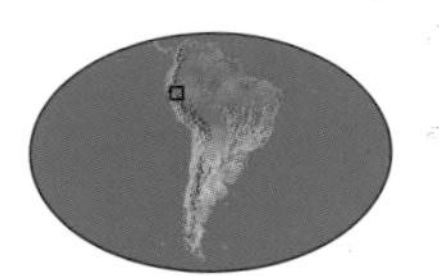

Alternative name: Coco's Falls

Elevation of San Rafael Falls: 2,999 ft (914 m)

Height of San Rafael Falls: 525 ft (160 m)

Ecuador's tallest waterfall, the 525-foot (160-meter) San Rafael Falls, is found in the northeast of the country at an altitude of 2,999 feet (914 meters), where the Rio Quijos passes through a double set of split rocks that resemble the petrified bat-wing doors straight out of a cowboy film saloon. From there it combines with the Rio Napo and flows on into the Amazon.

Surrounded by cloud forest, with a good lookout point and path system, the falls are a veritable haven for bird-watchers. Gray-rumped and white-collared swifts cling to the waterfall's rocks, and local highlights such as the wire-crested thorntail and coppery-chested jacamar are easily seen. The surrounding forest houses beautiful golden-headed quetzal, white-backed fire-eye (renowned for following army ant swarms), and the Andean cock-of-the-rock. Torrent ducks live in the river's whitewater. Located above the falls is the mighty water source: Reventador Volcano. A guide and considerable stamina are required to reach its 11,861-foot (3,561-meter) summit. Below the falls and accessible from Lake Agrio is the magnificent 240,000-acre (97,125-hectare) Cuyabeno Rainforest. **AB**

IMUYA LAKE

SUCUMBIOS, ECUADOR

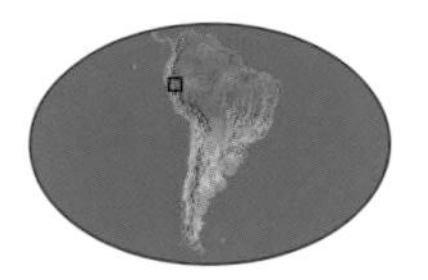

Surface area of Imuya Lake: 4,000 acres (1,619 ha)

Habitat: tropical rainforest, flooded forest

Located in the Sucumbios region of Amazonian Ecuador, Imuya Lake occupies a remote corner of the Cuyabeno Wildlife Reserve. Based on fertile, volcanically derived soils, the area is much richer in plants and animals than the Amazonian forests of Brazil, hosting 15 species of monkey, including howlers, capuchins, and the pygmy marmoset—the world's smallest monkey. Among the more than 500 bird species are macaws, toucans, and Cocha antshrike.

Flooded Igapó Forest is a specialty of Imuya Lake. The freshwater dolphin and manatee are often hard to get close to elsewhere. The lake also has large permanent floating forested islands—a very unusual feature in the Amazon. It is possible to paddle canoes on the lake and sleep under the stars in a hammock strung in a palm-thatched cabin. Tourism at Imuya is organized by the indigenous Cofán people, who first migrated to the area to escape the impacts of oil drilling occurring in their traditional lands. Attempts to drill for oil were subsequently made illegal in 1993, and the area was officially ceded to the Cofán. **AB**

RIGHT *Flooded forest emerges from the depths of Imuya Lake.*

MAQUIPUCUNA RESERVE

PICHINCHA, ECUADOR

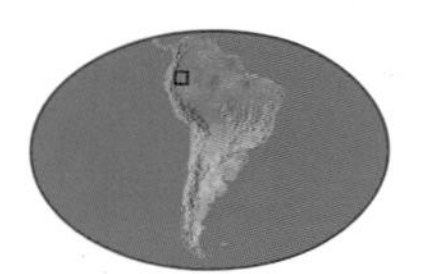

Area of Maquipucuna Reserve: 17.3 sq mi (45 sq km), with 5.4 sq mi (14 sq km) buffer zone

Elevation of Maquipucuna: 4,000 to 9,240 ft (1,200 to 2,800 m)

The Maquipucuna Reserve is about 50 miles (80 kilometers) northwest of Quito (a two-hour drive on mountain roads). The nearest town is Nanelalito. Over 80 percent of the reserve consists of primary cloud forest on steep slopes of fertile Andean soil. Located on the western side of Ecuador's two Andean chains, Maquipucuna receives moisture-laden sea winds, ensuring heavy rainfall and mists.

Combined with a range that embraces four altitudinally-stratified vegetation types and a close proximity to the Chocó Biodiversity Hotspot, Maquipucuna is one of Earth's most biologically diverse forest networks. The area probably hosts more than 2,000 plant species. Among the many orchids are 36 exceedingly rare species. In one three-day trip a botanist once found four new plant species. More than 45 mammal and 325 bird species are known from Maquipucuna (over one third of all the known birds for North and South America). There is also a spectacular array of butterflies, moths, beetles, and other tropical insects. The species unique to this region include the arrow-poison frog *Colostethus maquipucuna.* The reserve has an eco-tourist lodge and research facilities, as well as excellent trails. **AB**

COTOPAXI VOLCANO

COTOPAXI, ECUADOR

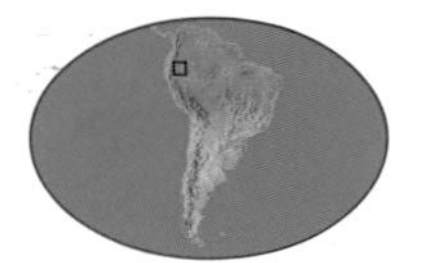

Elevation of Cotopaxi Volcano: 12,540 to 19,347 ft (3,800 to 5,897 m)

Area of Cotopaxi National Park: 129 sq mi (334 sq km)

Cotopaxi is known as "Moon Mountain" to the Quechua, who deemed the volcano sacred. It lies 47 miles (75 kilometers) south of the equator and 34 miles (55 kilometers) south of Quito. First climbed in 1872, Cotopaxi is now climbed regularly—but not without a guide. At 19,347 feet (5,897 meters), it is the highest active volcano in the world. It has a near-perfect cone shape. The volcano has erupted 50 times since 1738; its 1877 eruption involved mudflows that eliminated Latacunga city with 60-mile-per-hour (97-kilometer-per-hour) speeds and flowed for 18 more hours to the Pacific. The course of the lahas can still be traced on the plain below.

A glacier starts at 16,400 feet (5,000 meters). Around the volcano is Cotopaxi National Park, an area of cold, but botanically fascinating moorland that contains tiny gentians and violets. Lupins and calceolaries huddle in the lee of scattered rocks. Resident animals include puma deer, Andean wolves, marsupial mice, and specialized live-bearing frogs. The moors and cloud forests both support fine birdlife such as the Andean hillstar. **AB**

RIGHT *At the top of the world: Cotopaxi's peak.*

GALAPAGOS ISLANDS

PACIFIC OCEAN, ECUADOR

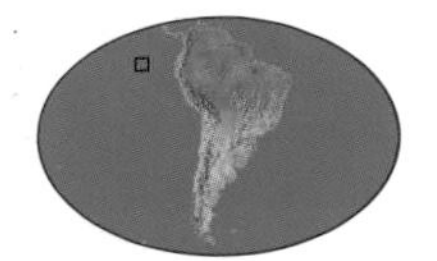

Total land area of archipelago: 3,029 sq mi (7,845 sq km)

Age: 3 to 5 million years old

Year Darwin visited: 1835

Lying 620 miles (1,000 kilometers) off Ecuador's coast, the Galapagos Archipelago of volcanic peaks is considered a very special place. This, of course, was where Charles Darwin conducted the fieldwork that would form the basis for his theory of evolution by means of natural selection as outlined in his groundbreaking 1859 book, *On the Origin of Species.* Today the islands attract both specialists wanting to work in Darwin's living laboratory and amateur enthusiasts hoping to explore the island's wealth of wildlife.

Each island can be divided into distinct vegetation zones: mangroves at the coast, cacti and spiny bushes in the arid coastal zone, small trees in the transitional zone, Scalesia forest in the humid zone, and tree ferns in the fern-edge zone. To top it all, the highest volcanoes have the prickly pear cactus above the cloud line.

Some of the animals are unique to the area, and some are quite unexpected: There are penguins at the equator, cormorants that cannot fly, iguanas that swim in the sea, finches that drink blood, and giant tortoises like huge mobile boulders. From island to island, animals of the same species can look quite different, a feature that Darwin himself noted. Locals can tell instantly from which island a giant tortoise originated just by the shape of its shell.

Each island has its own character. Española is flat and without a volcanic crater. At Punta Suárez a spectacular blowhole shoots spray 100 feet (30 meters) into the air, and there are huge nesting colonies of waved albatross. Floreana has a post office set up by whalers and still in use today, and there are green-tinged beaches with white and black sand. A volcano inundated by the sea is a favorite snorkeling and diving site here. San Cristóbal has a freshwater lake, and just offshore an ancient tuff cone called Kicker Rock is covered with roosting and nesting seabirds. Santa Fé changes color with the seasons, depending on which flowers are in bloom. Santa Cruz has a reserve with giant tortoises and giant lava tubes, and Seymour is the place to see the comical mating dance of the blue-footed booby. A licensed national park guide must accompany all visitors. **MB**

There are penguins at the equator, cormorants that cannot fly, iguanas that swim in the sea, finches that drink blood, and giant tortoises like huge mobile boulders. From island to island, animals of the same species can look quite different.

RIGHT *Volcanic craters on Isabela Island, the largest of the Galapagos Islands.*

GALAPAGOS RIFT

PACIFIC OCEAN, ECUADOR

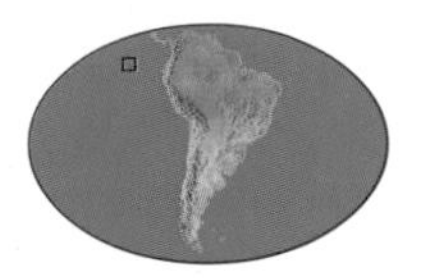

Depth of Galapagos Rift: 8,000 ft (2,440 m)

Length of Mid-Ocean Ridge: 42,000 mi (67,500 km)

Movement of Galapagos Islands: 3 in (7.5 cm) per year to the east

The Galapagos Rift lies approximately 60 miles (100 kilometers) north of the Galapagos Islands on the Pacific's Mid-Ocean Ridge and is the longest of its kind in the world. The islands spring from a volcanic "leak" at the junction of three tectonic plates, the Pacific, Nacza, and Cocos.

Like the islands, the Galapagos Rift is also the site of an important biological discovery. On February 17, 1977, a submersible visited a deep-sea hydrothermal vent for the first time and uncovered an entire ecosystem that no one had suspected could exist. Hot springs bring a rich mix of minerals to the ocean floor, allowing a teeming diversity of life to flourish in the abyssal darkness. Microscopic creatures feed on the minerals in the water, thus forming the base of an elaborate food chain. Strange tubeworms, clams, mussels, crabs, and other crustaceans vie for space against deep-sea jellyfish and black corals. Many species are only found around these vents, some known as "black smokers," which provide oases for living communities as diverse as any on the planet. Visitor access to the Galapagos Rift is not easy, and, moreover, requires an expensive deep-sea submersible. **NA**

MACHALILLA NATIONAL PARK

MANABÍ, ECUADOR

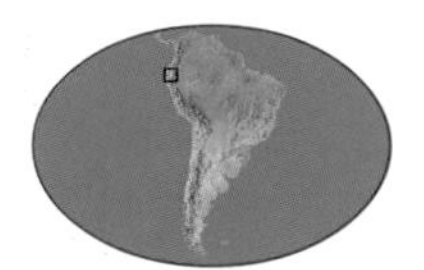

Area of Machalilla National Park: 135,000 acres (54,000 ha)

Area of Machalilla Marine Reserve: 316,160 acres (128,000 ha)

Machalilla has it all: dry forest, moist forest, sandy beaches, offshore bird-rich islands, coral-rich ocean, and splendid archeological sites. Coastal scrub forest once covered 25 percent of western Ecuador, but it is now down to 1 percent, most of which is located here. Possibly even more vulnerable and restricted are the moist forests, which depend on coastal fog for their moisture. Restricted to the highest hills, these fog forests act like islands; it is not uncommon for each one to have species unique to it alone. Such extreme local endemism explains why 20 percent of the park's plants occur nowhere else in the world. The park also has nearly 250 bird species, including the crested guan, and 81 species of mammal, including the Guayaquil squirrel, a rare local species. The albatross and booby breeding colonies (April to October) on Isla de la Plata may be visited with a guide. Humpback whales also come to breed offshore (June to October). The tropical dry forest, where 17-foot (5-meter) columnar cacti are interspersed with seasonally leaf-shedding trees, also contains artifacts from the Chorrera and Salango cultures. June to November are the coolest months to visit. **AB**

SANGAY NATIONAL PARK

MORONA-SANTIAGO / CHIMBORAZO / TUNGURAHUA, ECUADOR

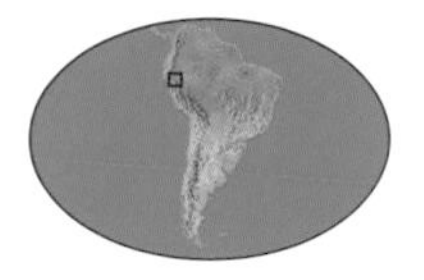

Elevation of Sangay National Park: 3,300 to 17,159 ft (1,000 to 5,230 m)

Habitats: upper tropical forest, cloud and elfin forest, páramo grassland, alpine moorland, lakes, swamps, scree slopes, lava, ash fields, snowfields, glaciers

Inscribed on the World Heritage List in 1983, the park is not only a place of stunning natural beauty, but also exhibits the full spectrum of ecosystems, from tropical rainforests to glaciers. It is dominated by three volcanoes: the active Sangay (17,159 feet, 5,230 meters), said to be continuously active for the longest time in the world; Tungurahua (16,457 feet, 5,016 meters); and the extinct and glaciated El Altar (16,860 feet, 5,139 meters). The park has a complex topography—dissected alluvial fans provide a plant-rich landscape of canyons and rolling uplands, while in the eastern part of the park the Andes descend from jagged snow-covered peaks into the rich lowland grasslands.

The volcanoes host alpine vegetation as well as glaciers, snowfields, lava flows, and ashfields. Rainfall variation (189 inches [4,800 millimeters] in the east and 25 inches [633 millimeters] in the rain-shadowed west) further promotes plant and animal diversity. The park hosts more than 3,000 plant species, half of them in the extensive cloud forests. High-altitude animals include tapir, puma, Andean wolf, spectacled bear, condor, and giant hummingbird. Jaguar, giant otter, and margay occur in the lower regions of the park. **AB**

CAJAS PLATEAU

AZUAY, ECUADOR

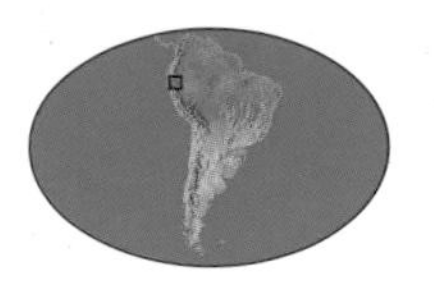

Area of Cajas Plateau: 261 sq mi (675 sq km)

Elevation of plateau: 7,990 to 14,670 ft (2,400 to 4,400 m)

Habitats: high-altitude grasslands, quinoa forest, montane forest

The Cajas Plateau is an isolated outlier of the western chain of the Andes. It is located in southern Ecuador close to the town of Cuenca. Above 11,000 feet (3,350 meters) are various types of páramo grassland, including yareta and pajonal. Both have exquisite mountain flowers, including tiny gentians, orchids, lupins, and daisies, with puyas dotting the landscape. Forests of shaggy-barked quinoa trees, which grow at a higher elevation than any other in the world, occur in the lee of hills. These sheltered forests contain entirely different species of plant and animal from those of the windswept páramo, including the giant conebill bird and the golden climbing mouse. Cloud forest occurs at lower altitudes, notably in the Rio Mazan Reserve, home to the rare gray-bellied mountain toucan, golden-tufted parakeet, and violet-throated metal-tail hummingbird. The plateau has a number of endemic species including a puya, a shrew-opossum, and a species of fishing mouse. Many geological features are the result of glaciation during the Pleistocene era, including drumlins, roche-mouton fields, and hanging valleys. The area is a candidate for the World Heritage List. **AB**

PODOCARPUS NATIONAL PARK

LOJA & ZAMORA, ECUADOR

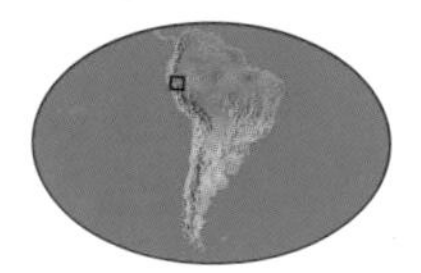

Area of Podocarpus National Park: 361,452 acres (146,280 ha)

Habitats: lowland, mid-range, cloud and elfin forest, páramo grassland

Created in 1982, Podocarpus National Park straddles the El Nudo de Sabanilla mountain range between the towns of Loja and Zamora in southern Ecuador. The area received moisture from the sea even during dry climatic epochs. Consequently, it supports a great variety of habitats and a rich diversity of wildlife. The national park is named after the podocarpus tree, Ecuador's only native conifer. Of the park's 3,000 plant species many are unique, including over 20 percent of its 365 orchids and several very beautiful passionflower vines. The park also has extensive stands of wild chinchona tree, the original source of anti-malarial quinine. The park sustains 130 mammal species including northern pudu, mountain paca, mountain coati, and spectacled bear. Spider monkey, ocelot, and giant armadillo occur lower down. Birds are the park's crowning glory. Over 60 hummingbird species are among the 600 bird species recorded here. These 600 species represent 40 percent of Ecuador's extensive bird list. Additional exploration may yield up to 200 more species, further enhancing the park's reputation as one of the world's most bird-rich areas. **AB**

SECHURA DESERT

PERU

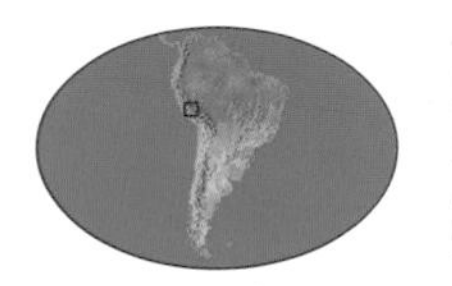

Maximum width of Sechura Desert: 90 mi (150 km)

Minimum length: 1,200 mi (2,000 km)

Average annual rainfall: 6 to 8 in (150 to 200 mm)

Spanning the entire west coast of Peru, the Sechura Desert is the longest expanse of desert on the South American continent. An extension of neighboring Chile's infamously arid Atacama Desert, the Peruvian sector of the desert adopts its own distinct character. As a sliver of sand trapped between the Andes Mountains and the sea, Sechura is crossed by more than 50 Andean rivers and, although there are several sweeping plains, low hills known as "lomas" are also abundant.

Add to this terrain the phenomenon of La Garúa, a dense winter fog created when cold, moist ocean breezes mix with hot, dry desert air, and the result is a patchwork of vibrant flowers and small shrubs. These islands of plant life, exist in a sea of nearly sterile and chapped lowlands which, taken together, support some 550 plant species, more than 60 percent of which exist nowhere else in the world. The floral oases are also known to attract hummingbirds and serve as insect nurseries to which songbirds, such as the Peruvian song-sparrow, flock at the start of winter. In the north, Trujillo also features archeological sites; Nazca in the south is home to the famous Nazca figures. DBB

PACHACOTO GORGE

ANCASH, PERU

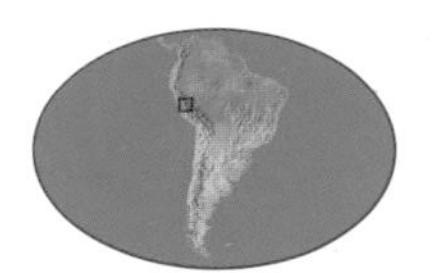

Elevation of Pachacoto Gorge: 12,140 ft (3,700 m)

Elevation of Huaraz: 10,000 ft (3,050 m)

Surrounded by snow-capped mountains, the Pachacoto Gorge is the best place in the world to see the giant puya in flower. The puya is a strange form of bromeliad and probably the most unique flowering plant in the Andes. For up to 100 years of its life, it consists of a ball of spiny, sword-like leaves, growing very slowly. Then, right before it dies, it sends out a flower spike covered in upward of 10,000 florets, which can be up to 36 feet (11 meters) tall, like a floral telegraph pole.

While in flower, the puya is visited by hummingbirds such as the Andean hillstar, which feeds on the nectar. The birds cannot hover like other hummingbirds in South America because the air is so thin. Instead they hang on to the convenient "landing pads" provided by each flower. Some unfortunate birds become impaled on the sharp leaves at the base, although gray-hooded sierra finches and spinetails actually make their nests in this hazardous place. The gorge is found 35 miles (57 kilometers) to the south of Huaraz in the Huascarán National Park, home to an equally interesting animal—the rare vicuna, a slender member of the camel family once exploited for its very soft coat. MB

MACHU PICCHU

CUSCO, PERU

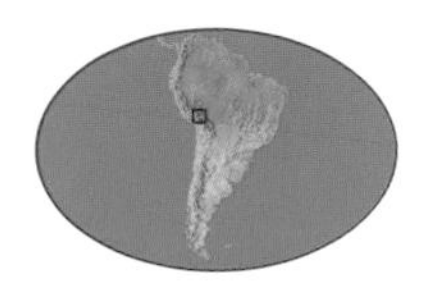

Elevation of Machu Picchu: 7,710 ft (2,350 m) above sea level

Elevation of Machu Picchu mountain: 9,186 ft (2,800 m)

The ancient Inca City of Machu Picchu—meaning "Old Peak" in the Quechua language—in the Peruvian Andes is considered unique because of its spectacular natural setting. The city sits on a 150-square-mile (400-square-kilometer) igneous intrusion, the Vilcabamba batholith, which pushed its way into surrounding sedimentary rocks during the Permian era about 250 million years ago and was gradually exposed during the periods of mountain building and erosion that followed. This created spectacular mountains and gorges, such as the Urumbamba Gorge upriver from Machu Picchu. Here the river drops 3,300 feet (1,000 meters) over 29 miles (47 kilometers) with many spectacular waterfalls and white-water rapids. The area contains prolific natural wonders—none more so than its orchids. There are 300 known species here, including the paradise orchid that grows 16 feet (5 meters) tall and has a flower 3 inches (8 centimeters) long. On the other end of the spectrum, one of the world's smallest orchids is also found here—all 0.08 inch (2 millimeters) of its span. MB

BELOW *The tiered Inca kingdom of Machu Picchu.*

VALLEY OF THE VOLCANOES

AREQUIPA, PERU

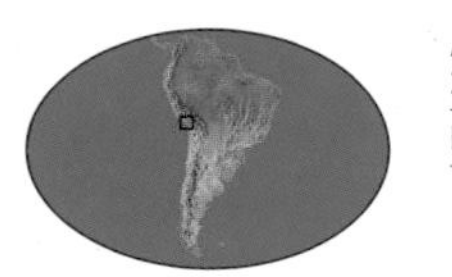

Age of Valley of the Volcanoes: 200,000 years

Best time to visit: April to November

Thousands of years ago, a giant lava flow seeped into a high-elevation river valley in the Peruvian Andes. As it cooled into a crusty carpet, pockets of gas and air trapped within the magma below exploded into miniature secondary eruptions, leaving behind a charred and blistered landscape of more than 25 distinct cones and 80 craters. Known today as the Valley of the Volcanoes, this vast lava field dotted with 1,000-foot- (300-meter-) high mounds hosts the largest concentration of volcanic formations in the world. Situated at nearly 12,000 feet (3,700 meters) above sea level, the valley is ringed by towering, snow-covered ridges and peaks including Coropuna, at about 21,000 feet (6,400 meters), making it the tallest volcano in the country and the tenth highest in South America. The area's slopes and gorges are so steep that the once-flowing streams of lava from volcanoes such as Coropuna rushed quickly down their flanks, cooling into the long, thin ribbons of solid rock which adorn the peaks today. Known as Andagua River Valley, the region's geothermic activity has given rise to hot springs. DBB

THE SPHINX / WHITE MOUNTAIN RANGE

ANCASH, PERU

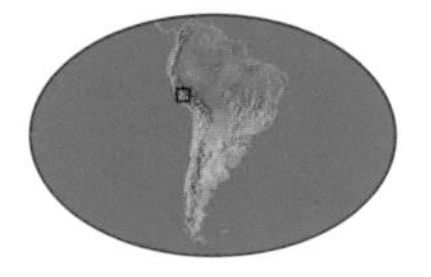

Alternative names: La Esfinge / Cordillera Blanca

Elevation of the Sphinx: 17,450 ft (5,325 m)

Area of White Range Glacier: 280 sq mi (725 sq km)

Rising above the Paron Valley of the Peruvian Andes like a guardian of sacred places, the 3,000-foot (900-meter) high nub of bare, orange granite known as La Esfinge (or The Sphinx) shines like a fiery beacon. More a monument to be conquered by the few who will set eyes on it than a casual tourist destination, The Sphinx is one of the highest unvegetated (except for the occasional cactus) formations in the Americas, making it an international destination for high-altitude rock climbing. Although the pinnacle is the pharaoh of the Paron Valley, it is but a baby in the greater White Mountain Range. Named for its permanent coat of snow and ice, the long ridge is the most extensive tropical ice-capped mountain range in the world, with several summits above 20,000 feet (6,000 meters). Featuring the largest concentration of ice in Peru, the White Range has 722 glaciers within its wide valleys. The spine of granite also marks the nation's continental divide, where the Santa River on the west drains into the Pacific Ocean and the Marañón River on its eastern flanks rushes toward the Atlantic. DBB

COLCA CANYON

AREQUIPA, PERU

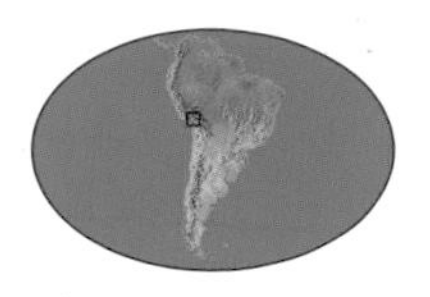

Alternative name: Cañón Colca

Wingspan of Andean condor: 10 ft (3 m)

Over hundreds of thousands of years, the Colca River cut into the high Peruvian Andes to create the world's deepest canyon. Few people have ever seen it or even heard of it. So steep are the sides it looks as if a gigantic knife sliced into the mountains. What is more, the distance from river to canyon lip is a staggering 2 miles (3.2 kilometers).

The canyon was once home to the ancient Colca people. These pre-Incan people stored grain in circular dried mud-and-straw containers known as colcas and built terraces in the canyon, showing remarkable skills in engineering and hydrology. The canyon is now the best place to see gigantic Andean condors, which vie with the wandering albatross for the bird with the longest wingspan, and which soar in thermals—rising columns of warm air—while hardly flapping their wings. The condors search constantly for carrion. If they find a large carcass, they sometimes eat so much that they barely manage to take off again. Tours and day trips depart from the nearest town to Colca Canyon, Arequipo. MB

PARACAS NATIONAL RESERVE

ICA, PERU

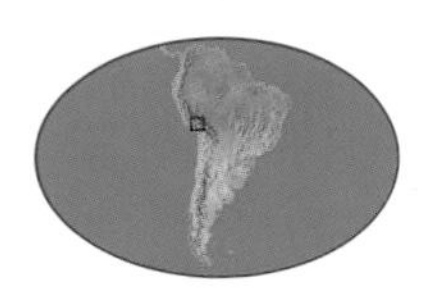

Brown pelican breeding time: October

Peruvian booby and guanay cormorant breeding time: November

South American sea lions breeding times: January to February

Paracas, or "wind of sand," gets its name from the sand-filled wind that blows up daily at noon. The Paracas Peninsula is home to the famous guano birds on which a multi-million-dollar fertilizer industry was once based. Today Paracas is protected and visitors are rewarded with one of the greatest wildlife spectacles in the world—millions and millions of nesting seabirds, including Humboldt penguins, Peruvian boobies, Peruvian brown pelican, guanay cormorants, and Inca terns.

Coastal cliffs are carved by the water into dramatic sea arches and undermined by caverns filled with sea lions and fur seals swimming in turquoise-blue water. Overhead you might see Andean condors coming down from the mountains to feed on dead birds and seals, and sea lion afterbirth. You might also spot vampire bats licking blood from their victims.

The Nature Conservancy, Pro Naturaleza, and the Peruvian park service are working together with local stakeholders—including conservationists, fishermen, and tourism operators—to create a plan that will identify solutions to overfishing, uncontrolled tourism, and waste dumping. MB

MANU BIOSPHERE RESERVE

CUSCO / MADRE DE DIOS, PERU

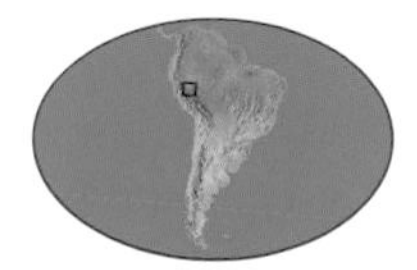

Designated a World Heritage site: 1987

Dry season: May through September

Wet season: October through April

Manu is the largest tropical rainforest reserve in the world. At about half the size of Switzerland, it has three main habitats for wildlife: the high, tundra-like puna, with condors, viscachas, heumul deer, and yellow ichu grass; the cloud forest, with hummingbirds, spectacled bears, cock-of-the-rock, and bromeliads; and the lowland rainforest, with macaws, howler monkeys, black caiman, and giant otters.

However, it is the rainforest that is truly remarkable. Here, there are 300 species of plant, 13 species of monkey, 120 species of amphibian, 99 species of reptile, and some 1,000 species of bird (15 percent of the world's total). Scientists working here have found that a single tree can have 43 species of ant living in it. Two of the Amazon's most venomous snakes—the lancehead and bushmaster—live here, camouflaged among the leaf litter on the forest floor. Manu is just 100 miles (160 kilometers) from Machu Picchu in southeast Peru, yet parts of it are so remote that there are still tribes living here that have not yet been contacted. It is a 35-minute flight from Cusco to Manu River, followed by a 90-minute boat ride by motor canoe on the Madre de Dios, one of the furthest points from the sea of the Amazon river system. MB

TAMBOPATA NATIONAL RESERVE

PUNO / MADRE DE DIOS, PERU

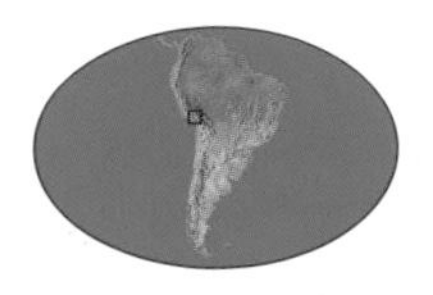

Area of Tambopata National Reserve: 3.7 million acres (1.50 million ha)

Elevation of reserve: 660 to 6,600 ft (200 to 2,000 m)

Vegetation: tropical rainforest, cloud forest, high-altitude grassland

Located at the junction of the Andes and Amazon basin in southeastern Peru, this reserve sits in a region where diversity and density of the fauna and flora reach a global apogee. Protecting the watersheds of three rivers, it is rich in biodiversity. Together with the adjacent Bahuaja Sonene National Park, this area encompasses lowland rainforest, Andean cloud forest, and páramo grassland. The natural reserve covers a range in altitude of over 5,000 feet (1,500 meters) and conserves one of the most complete and diverse natural areas in the world. At Tambopata, there are 1,300 bird, 200 mammal, 90 frog, 1,200 butterfly, and 10,000 flowering plant species. Near Lake Sandoval lies the world's largest salt lick, to which 15 species of parrot come daily. In total, the reserve has 32 parrot species, or 10 percent of the world's total. The reserve holds the record for the greatest number of bird species seen in one day: 331. Indigenous community-owned and -operated eco-tourism allows visitors to see giant otters, pink river dolphins, and black caiman, as well as crested and harpy eagles and monkeys. **AB**

TAMBA BLANQUILLA

MADRE DE DIOS, PERU

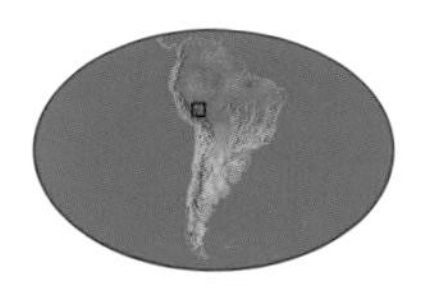

Number of macaws present daily: up to 300

Number of parrots present daily: up to 1,500

Activity of birds: 6 a.m. to 12 noon (local time)

About 25 minutes downriver from the Manu Wildlife Center is a clay lick at Tamba Blanquilla that attracts large flocks of colorful parrots and macaws. The site is a 25-foot (8-meter) tall clay exposure in the river's bank. The birds arrive in strict order. Smaller birds such as blue-headed and mealy parrots arrive at dawn. The larger parrots and bright macaws pitch up between 8 and 10 am, and perch in the nearby trees for an hour or two. They are understandably nervous—there are predators about. If no more than 20 birds appear, or if it rains heavily, the group disperses. If there is a quorum, they drop down to the bank and eat the clay, squabbling frequently for access to the best seams. They use the kaolin mixture to neutralize the chemical defense poisons in their plant food.

Visitors to Manu can witness the routine from a floating hide, a special catamaran that enables them to approach to within 30 yards (27 meters) of the riverbank without disturbing the birds. The best time to visit is from July through November when there are many birds and little rain. MB

HUASCARAN NATIONAL PARK

ANCASH, PERU

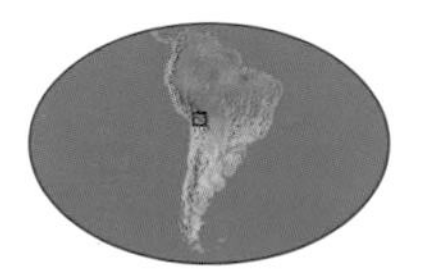

Designations: National Park, 1975; Biosphere Reserve, 1977; World Heritage site, 1985

Area: 1,313 sq. mi. (3,400 sq. km)

Huascaran National Park is one of South America's most outstanding protected areas, encompassing the most spectacular portion of the Andes Mountains. Known for its stunning natural beauty, it includes classic illustrations of glacial and geological processes. It is located in the Cordillera Blanca, the highest tropical mountain range in the world. It has 27 snow-capped peaks 20,000 feet (6,000 meters) above sea level, of which El Huascaran (22,204 feet, 6,768 meters) is the highest. Torrents derived from 30 glaciers rush through deep ravines, and there are 120 glacial lakes. The base rock is principally sediment from the Upper Jurassic seas and the Cretaceous and Tertiary volcanic deposits that make up the Andean batholiths. The flora varies with the topography, with humid montane forest in the valleys and alpine fluvial tundra and very wet sub-alpine paramo formations higher up. Indigenous species include spectacled bear, puma, mountain cat, and white-tailed deer, though these have been diminished by hunting in the past; noteworthy birds are Andean condor, giant hummingbird, giant coot, and ornate tinamou. These mountains are internationally famous for mountaineering, but there are also thermic springs lower down used for their therapeutic properties. GD

LAKE TITICACA

PERU / BOLIVIA

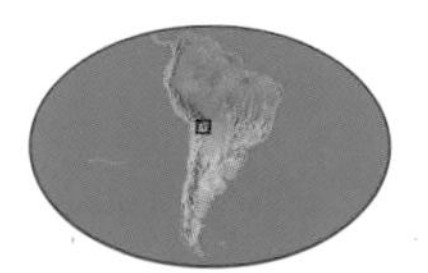

Surface area of Lake Titicaca: 3,200 sq mi (8,300 sq km)

Elevation of lake: 12,500 ft (3,810 m) above sea level

Number of islands: 41

Lying across the border between Peru and Bolivia, under the snow-capped backdrop of the Cordillera Real mountains, is a vast inland sea. This is Lake Titicaca, the largest lake in South America, and after 1862 when a steamer was assembled and floated here, the world's highest navigable waterway. Hydrofoils now ply the lake, while descendants of the local Uros people still use boats weaved together with totora reeds (a form of cattail). On the Peruvian side, houses and garden plots are located on floating reed platforms—these are known as the Uros Islands.

People here are adapted to life in this high region, with larger-than-usual hearts and lungs and more red blood corpuscles in their blood. The lake itself has its own special animals, such as a flightless grebe and its very own species of frog that lives its whole life underwater in shallow water sediments. In a place where the air is thin, the frog absorbs oxygen through its rumpled skin, which resembles an ill-fitting suit, and increases its surface area. On the Bolivian side of the lake, the Isla del Sol is said to be where the gods came down to found the Inca dynasty and bring wisdom to the local people. MB

FEDERICO AHLFELD FALLS

SANTA CRUZ, BOLIVIA

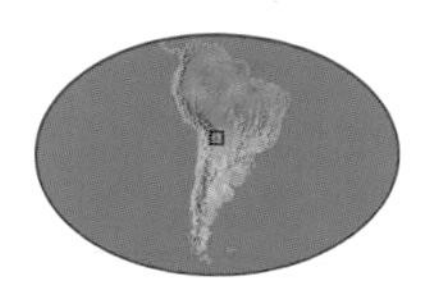

Height of Federico Ahlfeld Falls: 115 ft (35 m)

Number of falls: 6 to 10, depending on rainfall

Area of Noel Kempff National Park: 3.8 million acres (1.5 million ha)

Hidden within the remote wilderness described in Sir Arthur Conan Doyle's novel *The Lost World*, Bolivia's Federico Ahlfeld Falls is dream-like. Tumbling over a 100-foot (30-meter) sandstone cliff as wide as it is high, the waters of Pauserna River break into a half-dozen distinct cascades that plunge into a deep crystal-clear pool below. Located on the northeastern edge of Bolivia near the Brazilian border, Ahlfeld is one of many natural gems within Noel Kempff Mercado National Park, today still considered one of the world's most remote wilderness areas. The wide jungle clearing around the falls makes for comfortable access—and intense wildlife viewing. Cattle-sized tapir and the world's largest rodent, the capybara, are known to frequent Pauserna's shores, while rare pink river dolphins are found upstream. The region is also one of the best places on Earth to watch endangered giant river otters; it provides a home to a tenth of those remaining in the wild. With only a couple of hundred visitors making the long trek here each year—roughly 15 a month—Federico and its lush surroundings today remain a veritable Lost World. **DBB**

YUNGAS

LA PAZ, BOLIVIA

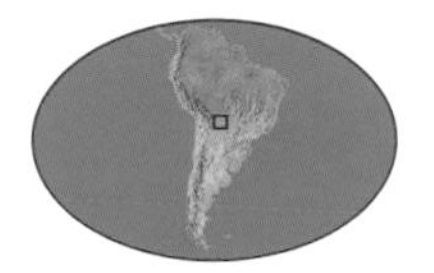

Conservation status: Critical/Endangered

Area: 72,100 sq mi (186,700 sq km)

Average annual rainfall: from 20 to 79 in (500 to 2,000 mm)

Located on the eastern slopes of the Bolivian Andes, the Yungas is the transition zone between humid, hot lowland forests and the cold, dry deserts of the uplands. This area includes humid and dry cloud forest, and also Apa-Apa forest, which is masked with huge bromeliad-decked fig trees and bamboo breaks. The region is rich with steep valleys and waterfalls that have promoted highly localized speciation, restricting many of the region's insects and plants to just one valley.

The altitudinal range also makes for a very rich flora and fauna. Among the specialist birds of the higher altitudes are the scimitar-winged piha, great sapphirewing, Cochabamba thistletail, and hooded mountain toucan. Above the standard tree line, polylepis forest occurs in the lee of mountains. Sheltered from the wind, forested islands are formed in an otherwise wind-blown environment. There, specialist species such as Cochabamba mountain finch, giant conebill, and the tufted tit-tyrant can all be found. Deforestation has been severe in places and in some regions there is little natural cover left. Surviving stands compliment the beauty of the cloud-clad vistas. **AB**

ALTIPLANO

BOLIVIA / CHILE / PERU

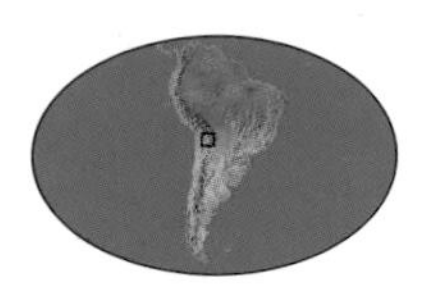

Mean elevation of Altiplano: 12,000 ft (3,600 m)

Area of Altiplano: 65,000 sq mi (168,350 sq km)

The Altiplano is a high plateau that towers over the southern Andes. Extending from western Bolivia and northeastern Chile right down to southern Peru, and occupying parts of Argentina and Ecuador, the plateau lies between the eastern and western chains of the Andes, and is perhaps best described as a raised depression filled with eroded sediment and volcanic debris from the higher Andean peaks. This in-filling occurred millions of years ago as an ancient seabed was pushed up when the modern peaks were first elevated. The sea eventually retreated, leaving behind many salt pans. Erosional deposition continues to this day from peaks that are more than 20,000 feet (6,000 meters) high. The plateau's lowest point, at 12,530 feet (3,820 meters), is occupied by Lake Titicaca, the world's largest high-altitude lake. High, cold, and receiving little rain, the Altiplano has a stark beauty.

The Altiplano is a high plateau that towers over the southern Andes. High, cold, and receiving little rain, the Altiplano has a stark beauty. The plateau's lowest point is occupied by Lake Titicaca, the world's largest high-altitude lake.

The region is divided into two: the cooler, drier puna in the south, and in the north the wetter jalca. Plants and animals living here are quite distinct, although frost and wind resistance are key survival characteristics for plants in both areas. The puna consists of dwarf shrubs and bunch grasses, with bare soil often visible between them. "Cushion plants" (which refers to their shape and not the name of a species) are also frequent. The jalca is more lush, with puya and other rosette plants. In some areas, seasonal snowmelt accumulates to form freshwater swamps, or bofedales. Elsewhere these may be brackish and support massive numbers of flamingoes, which feed on the water's rich soup of tiny algae and water shrimps. The Altiplano also contains the ancestors of the modern potato and tomato, as well as other crops which are locally important.

Some areas of the Altiplano are protected, including Chile's Lauca National Park, near the town of Arica. The park is home to the vicuña and guanaco (both wild camels), as well as llamas, alpacas, and huemul (a rare wild deer). There are also 140 bird species, including many rare wetland species, and over 400 species of plant, many of which are endemic. The area has a long history of human occupation, with a million people living there today. Some villages have been continually inhabited for 10,000 years. AB

RIGHT *The Laguna Verde and extinct Licáncabur Volcano.*

RED LAKE

POTOSI, BOLIVIA

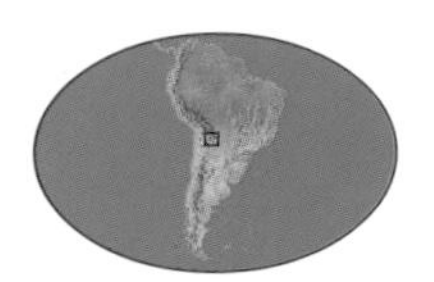

Alternative name: Laguna Colorada

Elevation of Red Lake: 13,860 ft (4,200 m)

Area of Uyani salt-flat complex: 4,300 sq mi (11,000 sq km)

The Red Lake is a high-altitude lake in the southwestern Bolivian Altiplano. Colored by an abundance of tiny shrimp and algae, it attracts up to 30,000 flamingoes of three species, including the extremely rare James' Flamingo. The exact color depends on the sun's angle and thus varies through the day, appearing blue, crimson, or deep maroon. Huge blocks of salt float in the waters and look like icebergs. The lagoon is part of the Uyani, a system of chemical-rich lakes and salt flats that, with Titicaca and Poopo, are the remnants of an immense inland sea. The salt flats, the biggest in the world, take four days to drive across. Seasonal rains replenish the fresh water.

For the rest of the year, the area is a dry, cold desert. The region is volcanically active, with the snowcapped 20,340-foot (6,200-meter) Licáncabur Volcano overlooking the nearby Laguna Verde. There are also 330-feet (100-meter) wide mud geysers at Sol de Manana, as well as thermal pools, and odd, wind-sculpted rocks. The sparse vegetation includes lichens, tussock grasses, and highly spiny cacti. In addition to this little vegetation, seasonal herbs provide grazing for vicuña and the viscacha, a large colonial rodent whose colonies can cover an area of up to 6,460 square feet (600 square meters). **AB**

MOON VALLEY

ATACAMA, CHILE

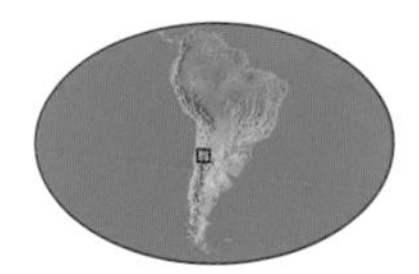

Alternative name: Vallée de la Luna

Diameter of Moon Valley: 1,650 ft (500 m)

Age of rock: 23 million years

The craggy boulders and skinny sandstone towers of Moon Valley are in an area of northern Chile often referred to as "arid wasteland." Bone-dry and inhospitable it may be, but this otherworldly landscape located in the middle of the vast Atacama Desert is a geological wonderland.

Situated at the northern tip of the Salt Mountain Range, the entire region is an exposed ancient lakebed that was bent, twisted, and pushed upward over millennia. Persistent winds and the occasional rain shower then sculpted the contorted stone into the flowing moonscape on display today. Rich in colorful mineral deposits, the often eerily humanlike formations are striped with veins of red and orange iron, or adorned in thick ridges of pure salt—finished off with a fine dusting of gypsum. The ever-changing colors of sunset transform the valley into an unearthly setting of chameleon-like torsos and limbs whose warped shadows dance across the sandblasted plain. Under the light of the moon, the whitewashed landscape takes on a lunar look, fully living up to its name. Moon Valley sits about 12 miles (20 kilometers) west of San Pedro de Atacama, and is accessible by bicycle or automobile. Tours leave daily from San Pedro. DBB

ATACAMA DESERT

ATACAMA, CHILE

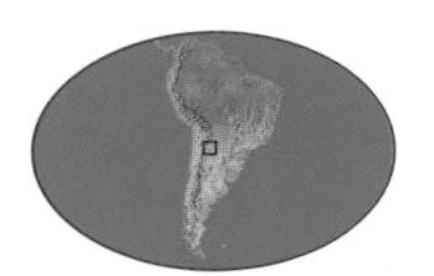

Area of Atacama Desert: 40,600 sq mi (105,200 sq km)

Average rainfall: 0 to 0.08 in (0 to 2.1 mm)

Length of desert: 950 mi (1,600 km)

Parts of the parched Atacama Desert in northwestern Chile have never seen a drop of water. Known as the driest place on Earth, even bacteria are scarce, and decomposition therefore does not occur, leaving the plants and animals that died thousands of years ago to bake under the sun. This unworldly landscape of boulders, deep sand dunes, meteorite-impact craters, and ancient, long-desiccated lake beds, is often compared with the moon or Mars, and is regularly used as a test site for NASA's remotely operated rover machines.

However, while Atacama's interior may be a bone-dry, almost lifeless land, its edges creep up the Andes Mountains in the east, and fall into the Pacific Ocean in the west. Regional coastal fog, ocean spray, and an occasional valley stream support a surprising diversity of animals and plants. Llamas and humpless camels (vicuñas) congregate near the inland streams frequented by rare songbirds, lizards chase scorpions among the plains' scattered cactuses and small shrubs, and flamingoes and penguins stop along the coast. Three protected areas encompass Atacama's desert regions, one of which, the Pampa del Tamarugal National Reserve, shelters the only two breeding populations of Tamarugo conebill found in the entire world. DBB

ATACAMA SALT FLAT

ANTOFAGASTA, CHILE

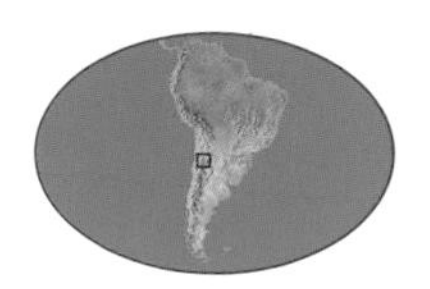

Alternative name: Salar de Atacama

Best season to visit: fall, winter, spring (December to March rain)

Average elevation of salt flat: 7,600 ft (2,300 m) above sea level

Over the course of millions of years, as the watery depths of an ancient inland sea in northern Chile evaporated and became desert, thick fields of rock salt that stretch as far as the eye can see were unveiled. A closer look at what is known as the Atacama salt flat, however, reveals surviving relics of shallow water underneath the crusty sodium chloride, camouflaged by a fine desert dust. Chile's largest salt deposit, this huge mineral flat also includes fields of chalky gray gypsum and is peppered with lagoon-sized surface waters.

Despite the region's extreme salinity and unrelenting desert sun, its network of unique wetlands—to which a few particularly hardy plants such as the medicinal herb *ephedra breana* and the stunted endemic cachiyuyo shrub have adapted—support a surprising variety of wildlife, from Chilean flamingoes and Andean geese to domesticated llamas and their wild ancestors, guanacos. The bone-dry air contains so few airborne particles that, on windless days, visibility in the area is near perfect, allowing visitors to see clear across the 50-mile (80-kilometer) wide plain. The salt flat is located 35 miles (56 kilometers) south of San Pedro de Atacama. **DBB**

BELOW *The barren expanse of Chile's Atacama salt flat.*

TATIO GEYSERS

ANTOFAGASTA, CHILE

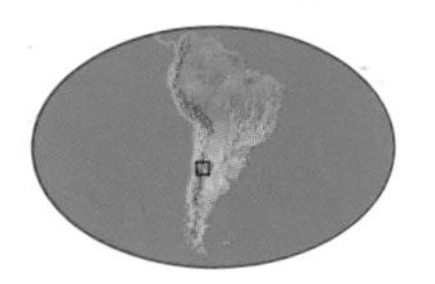

Alternative name: Los Géiseres del Tatio

Number of erupting springs: 110

Average height of geyser waterspouts: 30 in (75 cm)

Water boils here at just 187°F (86°C) rather than the usual 212°F (100°C) because, at 13,800 feet (4,200 meters), El Tatio is one of the highest geyser fields in the world. The ground is littered with chimneys and cones of crystallized salts, and the only water present is that which spurts out of the ground. There are 110 erupting springs, of which 80 are active geysers and 30 "perpetual spouters," making this the largest geyser field in the southern hemisphere. Eruptions, however, are less than three feet (0.9 meters) high. Curiously, this image of hell is actually teeming with life.

The shallow channels that run off the geyser field are filled with colonies of heat-resistant bacteria and algae that stain the place red and green. A few yards from the geysers, the water has already cooled to a bath temperature, providing a home to a unique species of frog. Its tadpoles hide among the filaments of bacteria. The adult has a rather nasty habit—cannibalism—and will not hesitate to eat its neighbors. Because of this, frogs walk about with the feet of other frogs sticking out of their mouths. The geysers are 93 miles (150 kilometers) southeast of Calama. MB

LAKE CHUNGARÁ

TARAPACA, CHILE

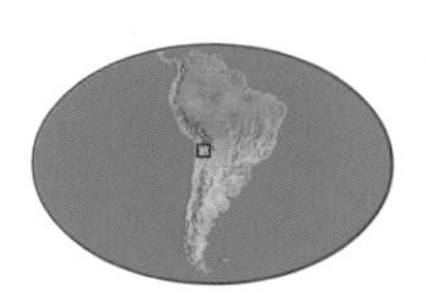

Elevation of Lake Chungará: 14,820 ft (4,518 m)

Depth of Lake Chungará: 130 ft (40 m)

Area of Lauca National Park: 345,000 acres (138,000 ha)

Far up in the Andean Altiplano, or high plateau, of northeastern Chile, Lake Chungará's deep blue waters are considered the highest lake-sized body of water in the world. Located nearly 15,000 feet (4,500 meters) above sea level, Chungará rests quietly near the base of the dormant and snowcapped Parinacota Volcano, which towers another 6,000 feet (1,800 meters) above the lake's shores. Above the preferred range for most of the region's animals, Chungará is fringed in wetland and serves as critically important habitat for high-elevation specialists such as vicuñas and alpacas, the rare giant coot, and countless migratory birds. The lake even hosts its very own catfish, found nowhere else in the world. A part of Lauca National Park, which was designated a World Biosphere Reserve in 1983 in recognition of its unique high-altitude shrublands, the region is a safe haven for endangered Andean deer, of which there are about a thousand surviving. Despite its remoteness and pristine state, the lake faces grave threats from local utility companies who are fighting for the right to tap into its waters—a move which would destroy the fragile Chungará ecosystem. The park and Lake Chungará are both accessible by car from the town of Arica. **DBB**

ANTUCO VOLCANO

BIO-BIO, CHILE

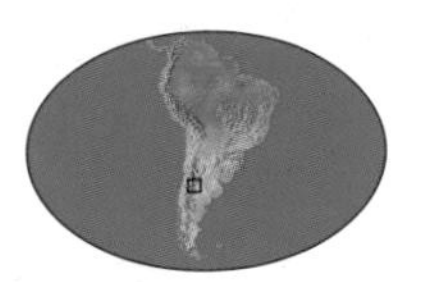

Height of Antuco Volcano: 9,790 ft (2,985 m)

Last major eruption: 1869

Some 10,000 years ago, the massive and fast-growing cone of lava rock along the Argentine border of central Chile became too steep, and its west-facing flank gave way, collapsing in a dusty and devastating avalanche. The reckless Antuco Volcano, as it came to be known, left a three-mile (five-kilometer) smoldering scar in the shape of a horseshoe in its wake. Despite its violent beginnings and mounds of charred rubble at its base, Antuco is a peaceful getaway.

Its flanks host rare mountain cypress and natural stands of the coveted spear-shaped and scale-leaved monkey puzzle tree. In the 1800s, during Antuco's last active phase, lava flows dammed the river outlets of its resident Lake Laja. This resulted in the addition of nearly 70 feet (20 meters) of depth to this one-time lagoon, and the forging of beautiful veil-like waterfalls cascading down into the waters. But Antuco's present tranquility is dangerously deceiving—for while this giant may be sleeping, it is not extinct. **DBB**

MALALCAHUELLO NATURAL RESERVE

ARAUCANÍA, CHILE

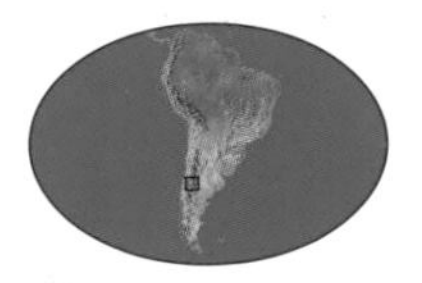

Maximum height of Malalcahuello: 9,690 ft (2,940 m)

Vegetation (low elevations): valdivian temperate rainforest

Vegetation (high elevations): high altitude grasslands, alpine sward

Located in the north of the Araucanía province in southern Chile, this small reserve was created in 1931. Its major feature is the impressive Lonquimay Volcano, whose slopes provide a fascinating array of volcanic features and, at higher elevations, are cloaked in a variety of alpine plants. Condors nest up here. The volcano is easily climbed; its lava-strewn peak boasts Navidad Crater ("Nativity"), so named because it formed after an eruption on Christmas Day 1988. Fourteen other volcanoes are visible from the summit. The park and the volcano's lower elevations have a mixed southern beech-oak-laurel-araucaria forest, with araucarias dominating in the drier, more exposed areas. Known as valdivian forest—moist, moss-covered, and fern-cloaked—it is rich in endemic plants, mammals, and birds. There are also lagoons, rivers, marshes, and a 165-foot- (50-meter-) high waterfall. Wild fuchsias occur in open areas. Together with the Nalcas National Forest Reserve, Malalcahuello hosts over 400 bird species, dwarf pudú deer, Chilean forest cat, Andean foxes, and the huemul, a rare and endangered deer. Darwin's frog, which hatches its young in its throat sack, is also here. **AB**

SALAR DE SURIRE

TARAPACÁ, CHILE

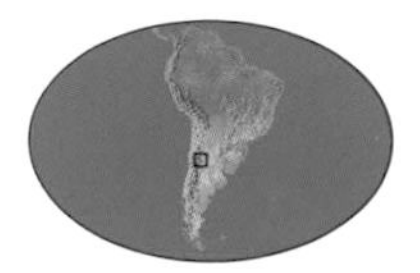

Area of Salar de Surire: 4,519 acres (1,829 ha)

Altitude of Salar de Surire: 13,780 ft (4,200 m)

Vegetation: arid grasslands

This high-altitude salt flat in northern Chile was named for the "suri" or rhea, the large ostrich-like bird characteristic of the high altitude plains of the region. Declared a national monument in 1983, the area has thermal pools, abundant wildlife, and a variety of salty and freshwater lagoons. Broken only by the 400-foot (120-meter) prominence of Oquella Hill, the central saltpan or "salar" is an almost completely flat area into which flow the rivers Casinane and Blanco. It is surrounded by extinct volcanoes, whose leached deposits now form the salar. Lagoons host a great variety of birds including Andean, James', and Chilean flamingoes, Andean avocet, and crested duck. Meanwhile, the grasslands of the Pampas of Surire have vicuña, alpaca, puna partridge, and rhea. With only 9.8 inches (250 millimeters) of rain falling a year, vegetation is sparse, giving the area a spartan beauty of purple hills, reddish tussock grasses, and blue skies. Night-time temperatures drop to well below freezing, and in daytime they struggle to reach 41°F (5°C). The nearby Vicuñas National Reserve offers opportunities for wildlife watching. The nearest town is Colchane, about 49 miles (79 kilometers) south of the salt flat. AB

TORRES DEL PAINE NATIONAL PARK

MAGALLANES Y LA ANTÁRTICA CHILENA, CHILE

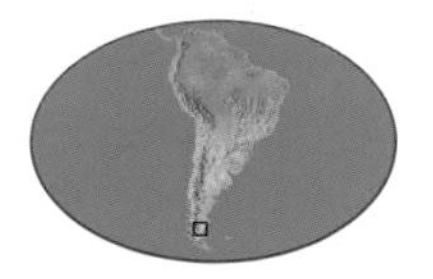

Area of Torres del Paine National Park: 935 sq mi (2,242 sq km)

Height of Paine Grande: 10,000 ft (3,050 m)

This remote but popular national park with spectacular scenery and abundant wildlife is dominated by the Paine Massif, a medium-high block of granite mountains that were formed about 12 million years ago. Torres del Paine (Paine Towers) are three sheer granite towers, the highest being Paine Grande. The Cuernos de Paine (Paine Horns) are two peaks of granite capped with black slate. The park lies at the edge of the southern Patagonian ice sheet whose glaciers feed azure lakes and lagoons, and white-water rivers cascade over spectacular waterfalls.

Vegetation tends to be a mix of windy grassland and stands of lenga forest that grow smaller as you climb higher. Guancos, rheas, mountain lions, and Patagonian gray and culpeo foxes roam the grounds, while Andean condors, black vultures, and crested caracaras fly overhead. A popular trekking route is the Torres del Paine Circuit that takes hikers past breathtaking mountains and lakes, and the massive Ventisquero glacier. The entire circuit takes 8 to 10 days to complete and starts and ends at Lago Gray. MB

BELOW *The Paine Massif punctuates the park's skyline.*

SALTO GRANDE WATERFALL

MAGALLANES Y LA ANTÁRTICA CHILENA, CHILE

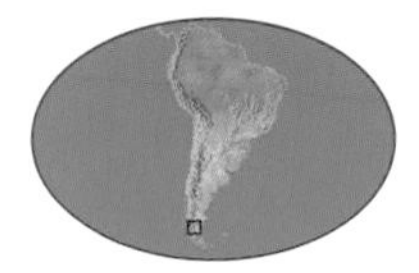

Height of Salto Grande Waterfall: 66 ft (20 m)

Width of cascade: 46 ft (14 m)

A wide, unbroken and glistening sheet of water pouring 60 feet (18 meters) from a turquoise lake, Salto Grande Waterfall in the Chilean deep south is a natural monument. Add to this the imposing backdrop of sharp granite horns which rise up like the hooked teeth of a saw blade slicing 8,250 feet (2,500 meters) into the air, hilly banks speckled in red, yellow, and green shrubs, and a permanent rainbow among the fall's shower of mist, and the scene takes on an imaginary feel.

But that is just the beginning of Salto Grande. It comprises part of Torres del Paine National Park, which was designated a biosphere reserve in 1978 for its phenomenal diversity of plants and animals, rivers and wetlands, rare pampas grasslands, and the immense rock Torres, or towers, after which the park is named. Salto's waters are fed from glaciers which are tucked into distant peaks, and its shores are frequented by resident flocks of flamingoes, visiting herds of camel-like guanacos, and nesting groups of the largest of all American birds, the flightless ñandú, or rhea—to name a scant few. Salto Grande can be accessed by road from Punta Arenas to Torres Park, which is just south of Puerto Natales. DBB

LAGUNA SAN RAFAEL

AISÉN, CHILE

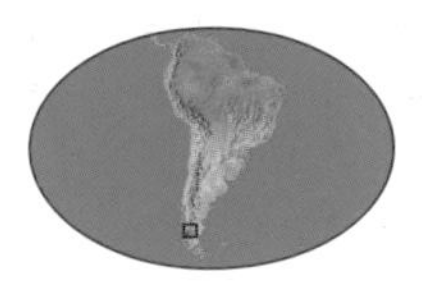

Area of Laguna San Rafael National Park: 4,305,683 acres (1,742,448 ha)

Length of San Rafael Glacier: 6 mi (10 km)

The Laguna San Rafael National Park is a natural monument dominated by the mouth of the extraordinarily neon blue-white San Rafael Glacier, one of the 19 glaciers that comprise the northern Patagonian ice sheet. It flows at an exceptionally fast rate of 56 feet (17 meters) a day, from 9,900 feet (3,000 meters) to sea level, and melts into a lagoon that opens to the sea via a narrow tidal channel into the Golfo Elefantes. About 400 times a day, icebergs sheer off from the 230-foot (70-meter) high ice cliff and plunge into the lagoon, causing waves of up to 10 feet (3 meters) high.

The proximity of the Pacific Ocean influences wildlife in the area. Albatrosses, penguins, cormorants, steamer ducks, marine otters, and sea lions are commonly seen. Ash-headed geese forage among the grasses of the shore, and after a bout of heavy rain a tiny black-and-white frog emerges. The frog is yet to be identified. Located in Chile's fjordland, there are no roads in the area. Visitors must arrive by sea out of Puerto Montt, or air charter from Coyhaique or Puerto Aisén. There is concern that the glacier is receding due to global warming, which could have an impact on tourism. MB

FJORDLAND

CHILE

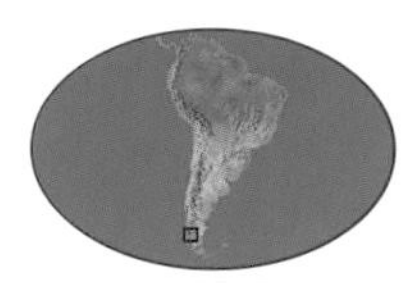

Height of glaciers: up to 200 ft (60 m)

Location of Fjordland: Golfo de Penas southward

Length of fjord coastline: 23,000 mi (37,000 km)

As the spine of coastal rock along Chile's seemingly endless coastline marches south, it sinks slowly into the earth, and eventually disappears into the deep ocean. Here, mountain summits become islands separated by wide canals, while ancient glaciers stretching outward from the Andean highlands end their journey to the sea in giant skyscraper-sized walls of ice. This is Chile's fjordland, a vast and wild waterscape of blue crystalline cliffs and floating icebergs, where everything truly is larger than life. The 30,000-year-old hanging glaciers are in fact extensions of one of the largest ice caps on the planet that terminates at the Pacific in hundreds of locations along thousands of miles of coast.

One of the few places on Earth where so many colossal forces of nature come together, the fjordland is as impressive below the water's surface as the frozen masses above. A dizzying mix of Atlantic and Pacific currents and glacial freshwater pouring into ocean saltwater, as well as a hilly and cragged seascape, support a bountiful food chain—penguins and sea lions, and the orcas that eat them both. This is also the only known humpback feeding site in South America. DBB

BALMACEDA GLACIER

AISÉN / MAGALLANES Y LA ANTÁRTICA CHILENA, CHILE

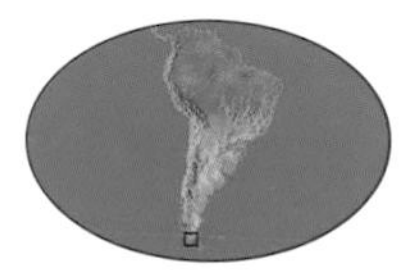

Elevation of Balmaceda Glacier: 6,775 ft (2,035 m)

Age of Balmaceda Glacier: 30,000 years

Duration of first crossing: 98 days

A permanent blanket of cotton-white ice bleeding blue at its slushy edges, Balmaceda Glacier in extreme southern Chile plugs an entire valley on the eastern flanks of Mount Balmaceda. Tucked between two triangular peaks of bare, coal-black rock which poke up abruptly enough to create their own weather, Balmaceda begins among the mountain's perpetual halo of storm clouds, and flows at a snail's pace toward a Pacific inlet. In the mid-1980s, waves lapped upon its icy edge, but in today's warming climate, the mass has retreated and now ends about 500 feet (150 meters) upslope.

Part of the vast Southern Patagonian Icecap, this is an uninhabited land where condors rule the air and killer whales the water, while Balmaceda dominates the earth. But like any monument of nature, humans have conquered Balmaceda to an extent, transforming it into the demarcation of the first lengthwise crossing of the southern ice fields, completed in January 1999 by a group of Chilean polar adventurers. For the typical tourist, the glacier is visible only by boat or aircraft, however, there are multiple tour groups available to facilitate viewing access. DBB

BEAGLE CHANNEL

CHILE / ARGENTINA

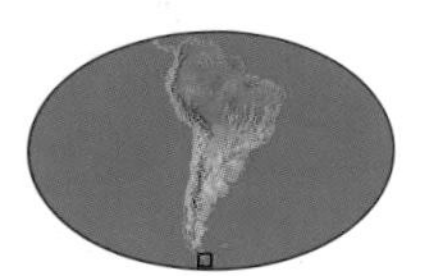

Length of Beagle Glacier: 150 mi (240 km)

Width of Beagle Glacier: 3 to 8 mi (5 to 13 km)

Although named in commemoration of Charles Darwin's famous ship, the Beagle Channel was actually discovered by Robert FitzRoy during an earlier surveying voyage in the 1830s. The channel itself forms a narrow but sheltered passage off Tierra del Fuego, whose highest point, Mount Darwin, rises to over 6,000 feet (1,800 meters) and is capped with snow more than 300 feet (90 meters) deep. These frequently disputed islands are currently shared by Argentina and Chile, which both stake claims to the rich mineral reserves and krill harvest. The channel's importance to early navigators lay in avoiding the need to round Cape Horn, a notoriously unforgiving undertaking that ended in disaster for many unlucky mariners.

Leaving aside the various political eruptions that occasionally disturb its tranquil waters, today the Beagle Channel is a remote, desolate haven for all forms of marine life. Steep mountains, glaciers, and waterfalls form a natural amphitheater in which nature's spectacle is played out in all its glory. Little has changed from when Darwin wrote: "It is scarcely possible to imagine anything more beautiful than the beryl-like blue of these glaciers." Then, as now, it was possible to watch awesome fragments of the same glaciers break off into the deep, as sperm whales swim within a stone's throw of the shore.

The ice-cold waters that mark the junction of the Atlantic and Pacific oceans also provide a rich source of food for the inhabitants of the numerous islands strewn throughout the channel. The bird life is particularly, if not surprisingly, diverse: gulls, petrels, albatrosses, skuas, steamer ducks, and cormorants jostle for position among the colonies of Magellanic and Gentoo penguins.

In the Beagle Channel, steep mountains, glaciers, and waterfalls form a natural amphitheater in which nature's spectacle is played out in all its glory.

Hiking through the most southerly beech forest in the world, adventurers may be rewarded with sightings of the rare Magellanic woodpecker—the largest in South America. Andean condors, black-chested buzzards, and Austral parakeets are among the other orthithological delights on offer.

The Beagle Channel is an increasingly popular eco-tourist destination, whose accessibility is enhanced by its proximity to Ushuaia, the world's southernmost city. Tours are relatively easily arranged, with guides available to help visitors make the most of their sub-Antarctic experience. **NA**

RIGHT *Fur seals and cormorants congregate upon Beagle Channel rocks.*

NIEVE PENITENTES

CHILE / ARGENTINA

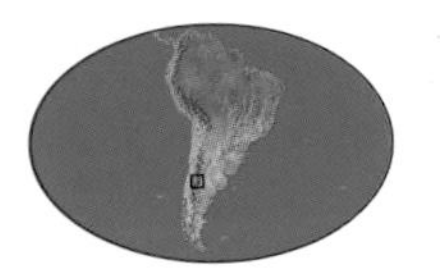

Elevation of Agua Negra Pass: 15,633 ft (4,765 m)

Height of Nieve Penitentes: up to 20 ft (6 m) tall

Height of Cerro Penitentes: 14,272 ft (4,350 m)

The Agua Negra Pass between Chile and Argentina is not only one of the highest places you can take your car in South America, but also is a gateway to an extraordinary landscape—row upon row of pinnacles composed of frozen snow. They stand like frozen figures in white hoods, reminiscent of a Christian Holy Week procession. Most are no more than 6 feet (1.8 meters) high, but some are up to 20 feet (6 meters) tall and remain at the roadside throughout the summer.

Charles Darwin saw them in 1835 and thought that they were created through wind erosion carving their strange shapes. In 1926 Argentinian geologist Luciano Roque Catalano came up with another suggestion: Since the tips of the snow pinnacles melted during the day and then froze at night, the snow crystals were not ordered randomly but rather in a particular direction that is influenced by Earth's magnetic field. As a consequence, all the snow pinnacles slant in an east-west direction. The nearby ski resort of Penitentes is named after Cerro Penitentes, which has rock towers that resemble the snow pinnacles higher up the mountains. Nieve Penitentes can also be seen at Cerro Overo. MB

IGUAZÚ FALLS

ARGENTINA / BRAZIL

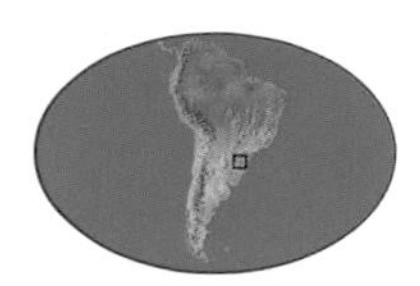

Height of Iguazú Falls: 280 ft (85 m)

Width of gorge: 2.5 mi (4 km)

Maximum height of spray: 300 ft (90 m)

Here, the Iguazú River pours 58,000 tons of water per second over the southern edge of the Paraná Plateau into a horseshoe-shaped gorge. Tree-lined islands and rocky outcrops separate more than 275 individual cascades that drop vertically or flow over stepped ledges in the gorge walls. The thunderous roar of the water can be heard from a great distance, and spray is thrown high into the air. The highest waterfall is the Union Falls that spills into the Devil's Throat, a deep chasm where the river has cut into a geological fault. Visitors can get right into the action in small rubber boats, and a helicopter ride from the Brazilian side gives an unprecedented view of the falls and the gorge. Specially constructed walkways and footbridges on the Argentinian side take walkers past spectacular growths of bamboos, palms, lianas, and wild orchids that border the gorge. Trees festooned with ferns, lichens, and bromeliads are decorated with the hanging nests of local songbirds. Flocks of dusky swifts wheel and dive into nest sites behind the walls of water, while in the distance the roaring calls of troops of howler monkeys can be heard. MB

BELOW *The thunderous cascading waterfalls of Iguazú Falls.*

IBERÁ MARSHES

CORRIENTES, ARGENTINA

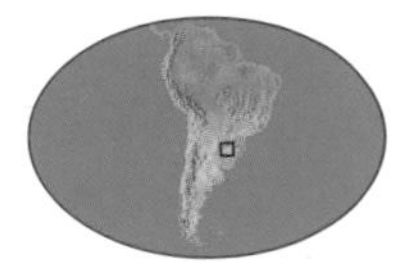

Area of Iberá Marshes: 3.3 million acres (1.3 million ha)

Annual rainfall: 47 to 59 in (1,200 to 1,500 mm)

A lush mosaic of grassy meadows, mucky marshes, and spongy swamps, the Iberá wetland complex in northeastern Argentina is a mostly inaccessible, waterlogged world, sheltering rare and endangered denizens such as giant anacondas, wolves, and marsh deer with webbed hooves that keep them from sinking into soft ground. It also hosts one of the rarest ecosystems on Earth: deepwater lagoons with floating islands known as embalsados, or "dam lands," which move up and down with water levels. Formed by the interweaving of aquatic plants, the tangled platforms of vegetation can grow more than 10 feet (3 meters) thick and are strong enough to support full-grown trees.

This unique habitat, in combination with its remoteness, has made it a haven for two species of caiman crocodile, as well as more than 80 fish species and hundreds of birds. The second-largest marshland in South America—only Brazil's famed Pantanal is bigger—Iberá covers an area larger than the country of Jamaica. Undisturbed for centuries, a recently completed dam on the Paraná River is causing the water table to rise, threatening to turn the marshland into a lake. The country is working with national and international conservation groups to prevent an ecological disaster. DBB

THE PAMPAS

ARGENTINA

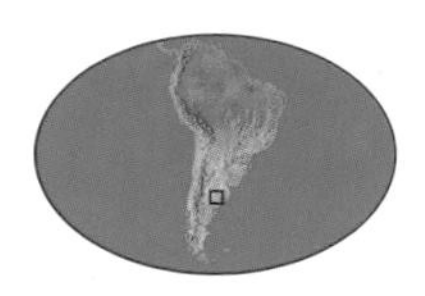

Area of the Pampas: 126,300 sq mi (328,000 sq km)

Annual rainfall: 20 to 39 in (500 to 1,000 mm)

A vast, waving sea of hip-high grasses interrupted only by far-off horizons, lagoons, and the occasional tree-like ombú shrub, the Pampas of central Argentina is an open, flat land of wild horses and llama-like guanacos, endemic foxes, and ostrich-like rheas. Stretching from the Andes Mountains to the Atlantic Coast and differing in climate, the region provides a habitat for rare birds such as the globally threatened saffron-cowled blackbird and buff-breasted sandpiper, both of which migrate here annually from their tundra breeding grounds in Alaska and Canada.

Making up a full quarter of the country's land area, the Pampas hosts some of the richest soils in the world, as well as the majority of the country's human population—a combination that has proven to be both culturally and ecologically disastrous. Once the largest and most characteristic habitat in Argentina, the pampas is now one of the most endangered habitats on Earth. The introduction of gigantic herds of domestic cattle, unregulated hunting, and chemical-intensive agriculture have all but driven out large carnivores such as puma and the endemic Pampas cat. However, conservation efforts are beginning to take shape. Ernesto Tornquist Provincial Park protects an area of nearly 17,000 acres (6,880 hectares), which is the largest unbroken patch of Pampas habitat in Argentina. DBB

PENÍNSULA VALDÉS

CHUBUT, ARGENTINA

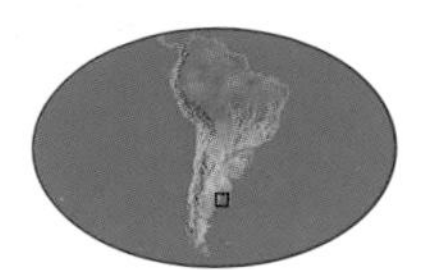

Length of southern right whales: up to 50 ft (15 m)

Lowest point in South America: Salinas Chicas (Península Valdés):130 ft (40 m) below sea level

Each summer, over 7,000 southern sea lions, 50,000 southern elephant seals, and 1,500 right whales congregate to breed at Península Valdés. The whales arrive at the two horseshoe-shaped bays of the peninsula, where they spend the southern winter (April to December). Peak activity is in September and October when groups of 20 or more bull right whales simultaneously jockey for position around receptive females. Courtship can be violent—many whales are scarred by head butting and pushing fights. But the main threat in the area is another cetacean—killer whales. They appear at Punta Norte in the northeast corner of the peninsula. The beach is closed to visitors, but during February and March specially built platforms enable you to watch killer whales ride in on the surf and pluck sea lions from the beach—a unique hunting technique for killer whales seen nowhere else. The peninsula is about 1,000 miles (1,600 kilometers) south of Buenos Aires. Port Madryn is a diving center on the shores of Golfo Nuevo, and at Puerto Pirámide boats take tourists to visit the right whales. One thing to watch out for on the way to Punta Norte is the colony of wild guinea pigs that run wild in the scrub. **MB**

PUNTA TUMBO

CHUBUT, ARGENTINA

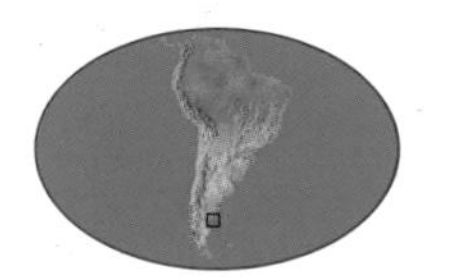

Penguin nesting season: September to March

Nesting: two eggs, incubated by both parents from 39 to 43 days

Height of adult penguins: 28 in (71 cm)

An isolated and sandy bay backed by a barren desert scrub of low-lying shrubs and bare rock, Punta Tumbo on the central Atlantic Coast of Argentina seems a most unlikely place to find penguins. The region is actually perfect for magellanic penguins, which nest here in the hundreds of thousands with some estimates ranging up to a million. Also known as "jackass penguins" because of the donkey-like bray used to attract females, these slick birds colored in black-and-white tuxedo dig shallow burrows into the clay-rich ground, or gut out hollows underneath Tumbo's scattered bushes and shelves of stone. Adults spend half the year here, waddling between the sea and their burrows to feed their young, and the other half at sea, following ocean currents north toward Brazil. Punta Tumbo's colony is not only the largest in South America but at 120 years, also the oldest. Their success here is undoubtedly aided by their virtually predator-free existence, as top threats such as orcas are distracted by the more satisfying fare of sea lion, which haul out en masse about a 100 miles (160 kilometers) up the coast. Punta Tumbo is located at the northeastern edge of Patagonia. **DBB**

MOUNT FITZROY

SANTA CRUZ, ARGENTINA

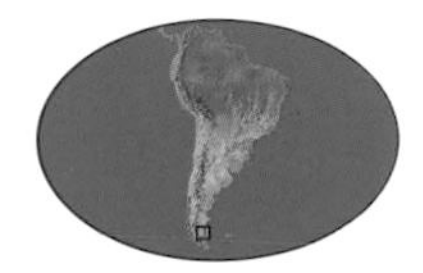

Height of Mt. Fitzroy: 11,168 ft (3,405 m)

First ascent: 1952 French expedition

The central spire of a jagged massif that bites into the sky like a ridge of granite shark's teeth, Mount Fitzroy dominates South America's rugged and windswept Southern Patagonia as the region's tallest peak. Towering above several resident glaciers, the imposing pinnacle slices into air and sports a regular halo of clouds and blowing snow which inspired the mountain's first indigenous peoples to call it El Chaltén, or "mountain that smokes." Despite being constantly plagued by unpredictable weather and notoriously strong winds, Fitzroy and its fellow summits today are major destinations for professional rock climbers, who covet them for the very danger they pose. However, perhaps the best way to enjoy Fitzroy is from its foothills, where its glaciers slip quietly into forests of stunted shrubs and twisted trees, and in which songbirds, lakes, and waterfalls abound. Mount Fitzroy marks the northern edge of Los Glaciares National Park, an immense preserve of nearly 50 major glaciers and multiple icebergs that branch off from the second-largest continental ice sheet in the world. Designated a World Heritage site in 1981, the park is worth the visit. DBB

PERITO MORENO GLACIER

SANTA CRUZ, ARGENTINA

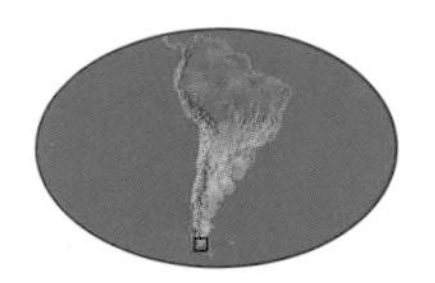

Area of Los Glaciares National Park: 1.5 million acres (600,000 ha)

Total number of glaciers: 365

Length of Perito Moreno Glacier: 18 mi (30 km)

In Los Glaciares National Park in southern Argentina, the Perito Moreno Glacier causes havoc every three or four years. It is one of several glaciers that run off the southern Patagonian Icecap, but it differs from the others in advancing right across its terminal lake—Lake Argentino, the deepest lake in Argentina. When it is on the move it moves toward the Magellan Peninsula on the opposite shore, blocking the Tempanos Channel. It has even been known to push into the forest, but its biggest impact is to dam in the meltwater, mainly from the large Upsala and Spegazzini glaciers that drain into the north of the lake, behind Moreno's high wall of ice.

The result is that the water on the upstream side rises dramatically, as much as 120 feet (37 meters) higher than the downstream side. Eventually the pressure is so great that the ice dam cracks and bursts, and a violent calving explosion takes place for the next 48 to 72 hours. The noise of the breaking ice can be heard many miles away. The glaciers can be reached from Calafate, with direct flights to and from Buenos Aires. MB

BELOW *The Moreno Glacier advances across Lake Argentino.*

MID-ATLANTIC RIDGE

ATLANTIC OCEAN

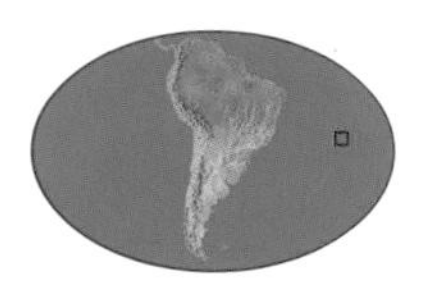

Length of Mid-Atlantic Ridge:
10,000 mi (16,100 km)

Width of Mid-Atlantic Ridge:
300 to 600 mi (480 to 970 km)

About as tall as the European Alps and as wide as the state of Texas, this massive ridge of rock runs more than twice the length of the great Andes, and yet is hardly noticeable from land. The longest mountain range on Earth, the Mid-Atlantic Ridge (MAR) cuts through the deep ocean like a jagged serpentine scar from pole to pole, separating the Atlantic into east and west sectors. Although nearly the entire chain is underwater, every now and then one of its 2-mile- (3.2-kilometer-) high ridgelines breaks the surface to form a patch of isolated islands.

The one exception is near the Arctic Circle, where the MAR finishes its northward march. Here, it ends in an exposed slab of earth known as Iceland. Although considered a single ridge, MAR is a product of seafloor spreading—a geologic tug-of-war between plates in the planet's crust which, over millions of years, moved the Americas west and Africa and Eurasia east—and thus is two parallel ridgelines separated by a widening rift. With many valleys and peaks, sweeping plateaus, and narrow gorges, this natural monument hosts every known marine ecosystem—from hot thermal vents to intertidal tidepools. **DBB**

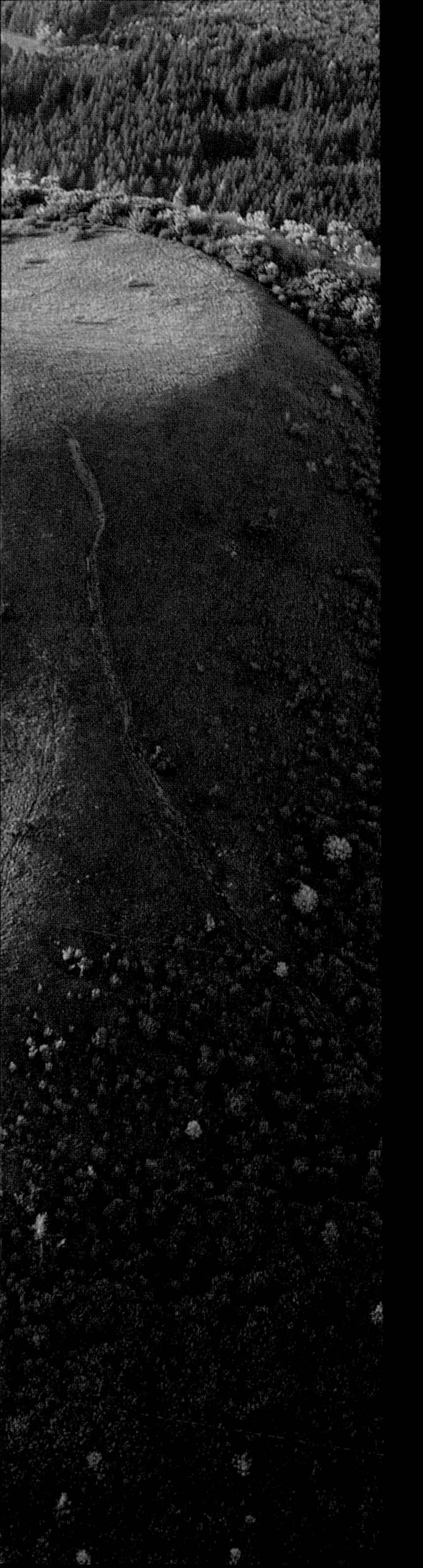

III

EUROPE & THE MIDDLE EAST

From the birth of the mystical land of the midnight sun to the everlasting healing waters of the Dead Sea, Europe and the Middle East span many countries and ages. The unique natural histories contained in their midst also chart the torturous and celebrated relationships that have marked humankind's interactions with nature—building castles on its rocky peaks, painting frescoes in its caves, trading through its passes, and perishing at its mercy.

LEFT *The majestic green slopes of Le Puy-de-Dôme, Auvergne, France.*

HVERFJALL CRATER

HÚSAVIK, ICELAND

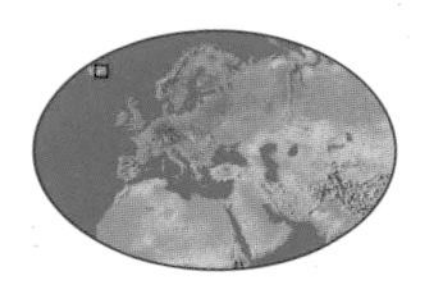

Age of Hverfjall Crater: 2,800 years

Height of cinder cone: 660 ft (200 m)

The primeval, rugged splendor that makes Iceland so dramatic has been created by one of the most powerful forces on Earth. It was formed less than 20 million years ago from volcanic activity on the floor of the Atlantic Ocean and shaped by large ice-age glaciers.

Hverfjall Crater was created during a short but powerful eruption some 2,800 years ago. The area formed the southernmost part of an eruptive fissure—as molten magma rose through this fissure it met the waters of a lake, causing a phreatomagmatic explosion. The ensuing blast produced a wide crater of ash and pumice. Measuring nearly 1 mile (1.6 kilometers) wide, Hverfjall's cinder cone rises 660 feet (200 meters). All around lies evidence of the turmoil steaming beneath the surface: the eruptions of the 1720s and the Krafla volcanic fires of the 1970s left the adjacent landscape littered with fumaroles and bubbling mudpools—newly formed earth that is mysterious and still breathing fire.

Hverfjall is situated to the north of the country's fourth largest lake, Myvatn, a true oasis in a lava desert. Amid volcanic flows and craters, ash cones and geysers, thousands of wildfowl flock to the region annually. AC

BELOW *The spectacular cinder cone of Hverfjall crater.*

DETTIFOSS

HÚSAVIK, ICELAND

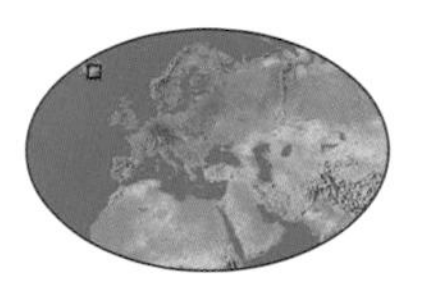

Height of Dettifoss: 145 ft (44 m)

Width of falls: 330 ft (100 m)

Water flow of falls: 110,000 gal (500,000 l) per sec

The Mývatn region in the northeast of Iceland is volcanically active, its lunar-like landscape dominated by the Kafla volcano. There are lava fields and hot springs, such as Námaskaro, an area of spewing geysers and bubbling mudpools. It is also a land of cascading waterfalls and deep gorges, the most spectacular waterfall, Dettifoss, lying along the Jökulsá á Fjöllum, the island's longest river. The river is fed by icemelt from the Vatnajökull and crosses over a high plateau that is scarred by lava flows before it enters the sea at Oxarfjorour. There is also the impressive Mývatn ("Midge Lake"), which the area is named after.

Dettifoss is 145 feet (44 meters) high and 330 feet (100 meters) wide. They are considered to be the most powerful waterfalls in Europe, with an estimated 110,000 gallons (500,000 liters) of water passing over the falls every second. To the south is Selfoss, another cataract which is 33 feet (10 meters) high, and the 90-foot (27-meter) high Hafragilsfoss.

Below Dettifoss is a deep canyon, the Jökulsárgljúfur, which was gouged out by a series of catastrophic floods, the last occurring about 2,500 years ago. In the 1970s plans to build a hydroelectric power station were abandoned and the area is now protected. MB

VATNAJÖKULL & GRÍMSVÖTN VOLCANO

SKAFTAFELLS-SYSLA, ICELAND

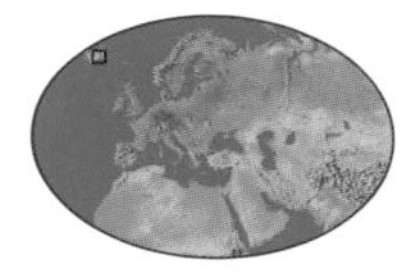

Area of ice sheet: 3,125 sq mi (8,100 sq km)

Thickness of ice sheet: 3,300 ft (1,000 m)

Iceland's Vatnajökull is an impressive 3,300 feet (1,000 meters) thick in some areas and covers 3,125 square miles (8,100 square kilometers). The ice sheet contains more ice than all European glaciers put together and is the source of 12 major glaciers. The rocks beneath are dark basalt, the product of 20 million years of volcanic activity.

Grímsvötn volcano lies directly beneath the ice sheet, its heat melting the ice to form an unexpected blue lake 2 miles (3.2 kilometers) across. Grumbling gently most of the time, the volcano occasionally bursts into life. The extreme heat vaporizes the ice to form a steaming cloud 5 miles (8 kilometers) high, and the sudden torrent of meltwater and ice fragments is so enormous that roads and bridges in its path are swept away. Usually, water drains gently from the glacier snouts.

Skeidarárjökull, a glacier on the southern edge of the ice sheet, feeds a river that flows toward the sea over a flat plain of dark gravel. To the east, the Breioamerkurjökull calves huge icebergs that float in a glacial, freshwater lagoon known as Jökulsárlón. MB

SVARTIFOSS

SKAFTAFELLS-SYSLA, ICELAND

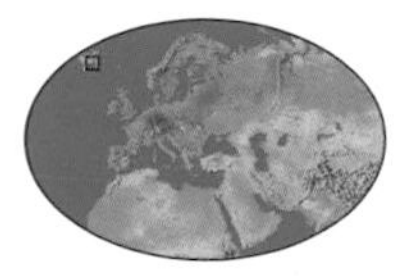

Height of Svartifoss: 82 ft (25 m)

Established as a national park: 1956

Skaftafell National Park in southeast Iceland was established in 1956. Among the natural treasures protected is the 82-foot- (25-meter-) high Svartifoss, or "dark waterfall." Fed by ice-cold meltwater from Svinafellsjökull, the narrow strand of water pours over a broad cliff of hexagonal basalt columns that seem to hang like organ pipes over the edge of a horseshoe-shaped amphitheater. This natural feature was the inspiration for the architectural design of Reykjavik cathedral in the nation's capital. A footpath through a ravine takes you past other waterfalls, including the Hundfoss, meaning "dog falls," before reaching the unusual Svartifoss. Dog Falls was named after some dogs belonging to local farmers that were swept over the falls when they attempted to cross at a ford further upstream.

Skaftafell itself was once an oasis of green farmland in an otherwise dark and sterile landscape. Today, the grassy hills are covered with forests of birch, willow, and rowan. Not far away, hidden in a small icecap, is Europe's third-highest volcano, the Öraefajökull, which has violently erupted twice in recorded history. MB

GEYSIR AND STROKKUR

SOUTHLAND, ICELAND

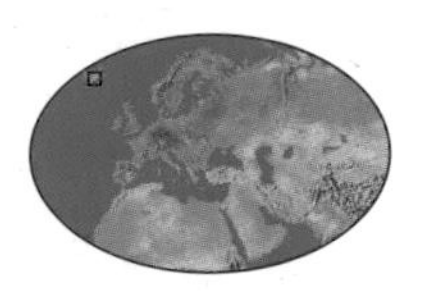

Height of Geysir eruptions: 200 ft (60 m)

Height of Strokkur eruptions: 100 ft (30 m)

Frequency of Strokkur eruptions: every 10 min

In a geothermal valley of more than 50 hot springs and multicolored mudpots in southwest Iceland are two geysers, which are called Geysir and Strokkur. Geysir was first recognized in 1294, after an earthquake hit the area during the devastating eruption of Mount Hekla. This seismic activity created a number of new hot springs, as well as these two impressive geysers.

In 1647, the larger of the two was named Geysir, meaning "gusher," which later went on to become the universal word to describe any explosive hot-water fountain. In those days, Geysir shot a column of scalding water several times in the course of a few minutes, each fountain projecting increasingly higher until it reached its grand finale with a column 200 feet (60 meters) high. Following a roaring jet of steam, it then subsided and was quiet until it exploded into life again every three hours.

Later in its life, the gap between eruptions became progressively longer until at the beginning of the 20th century it stopped altogether and lay dormant for about 30 years. In 2000, an earthquake revived Geysir, but it no longer follows a regular schedule and erupts irregularly. Before the 1990s, eruptions could be stimulated by the addition of soap, but concerns for the environment put a stop to this practice.

Geysir's smaller neighbor Strokkur geyser, meaning "the churn," gives a show every 10 minutes, blasting boiling water approximately 65 to 100 feet (20 to 30 meters) into the air. Strokkur's size is no reason to assume its inferiority to Geysir. Visitors can watch as a turquoise bubble of water rises within the geyser, preparing to burst forth in a spectacular fountain of white spray and steam.

Geysir's smaller neighbor Strokkur gives a show every 10 minutes. Visitors can watch as a turquoise bubble of water rises within the geyser, preparing to burst forth in a spectacular fountain of white spray and steam.

The ownership of the Geysir area has a colorful history. It was originally owned by a local farmer who sold it to a whisky distiller called James Craig in 1894. Craig enclosed the site and charged visitors who wished to view the geysers. A year later he tired of this investment and gave the area as a present to his friend, E. Craig, who dropped the entrance fees. James Craig later went on to become Prime Minister of Northern Ireland. E. Craig's nephew Hugh Rogers sold the site in 1935 to film director Sigudur Jonasson who donated the land to the Icelandic people in perpetuity. **MB**

RIGHT *Strokkur geyser bursts forth into the frosty air.*

HEIMAEY

VESTMANNAEJAR, ICELAND

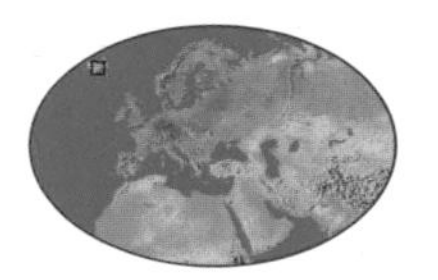

Land area increase caused by eruption of Eldfell volcano: 15 percent

Human population of Heimaey: 5,300

In the early hours of January 23, 1973, the Eldfell volcano on the Icelandic island of Heimaey burst into life without warning after 5,000 years' dormancy. Heimaey is one of the Westman Islands about 15 miles (25 kilometers) south of Iceland. On the edge of the town of Vestmannaeyjar, the ground split apart and molten lava and volcanic ash sprayed out in gigantic fountains from a crack 1.5 miles (2.5 kilometers) long. It was described as "a curtain of fire." Buildings were buried up to their roofs in ash, and in the first-ever attempt by people to check a lava flow, 19 miles (30 kilometers) of hosepipe and 43 pumps were used to pour jets of seawater over the advancing lava flow. Despite these attempts, the lava engulfed the eastern part of the town and 300 buildings were burned or buried. By chance, the fishing fleet—the town's main industry—was in port during bad weather and was able to evacuate the population of 5,300 very quickly. After several weeks, the eruption subsided, and the inhabitants were allowed to return. Some had to rebuild their houses but others escaped the worst. The lava flow extended the eastern coast, providing a natural breakwater for the harbor. MB

SURTSEY

VESTMANNAEJAR, ICELAND

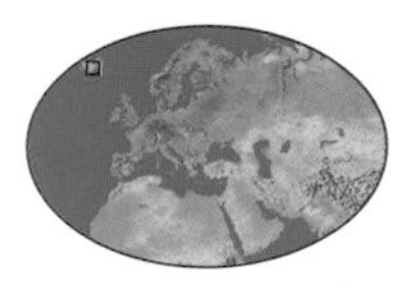

Elevation of Surtsey island: 574 ft (174 m)

Diameter of island: 0.93 mi (1.5 km)

Area of island: 1.08 sq mi (2.8 sq km)

In November 1963, fishermen setting their nets about 20 miles (33 kilometers) south of Iceland's main island noticed things were not as they should be. Not long afterward a volcano erupted from the sea bed, gentle at first but more explosive as it reached the surface, with volcanic bombs and dust ejecting above the vent. It was not alone. Three other vents were active at the same time: Syrtlingur and Jolnin each formed an island that was later worn away by the sea, while Surtla never pushed above the surface. Surtsey, however, remained, and was named after Surtur, a giant of fire in Norse mythology. It grew to 574 feet (174 meters) above sea level, and covered an area of 1.08 square miles (2.8 square kilometers).

By the following spring, a fly became the island's first natural visitor. The first plant flowered in 1965, and the eruption ceased in June 1967. Twenty years later, 25 species of plants had established themselves. The first birds—fulmars—nested in 1970. The latest recordings indicate that 89 species of bird nest here, and both seals and birds use the island as a rest point on their migrations. MB

KONGSFJORDEN

SVALBARD, NORWAY

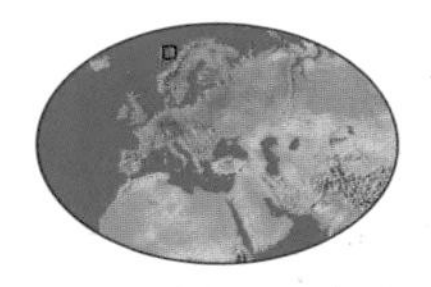

Length of Kongsfjorden: 25 mi (40 km)

Width of Kongsfjorden: 3 to 6 mi (5 to 10 km)

Kongsfjorden is an icy archipelago covering 23,938 square miles (62,000 square kilometers). It lies at the foot of the snowcapped Tre Kroner ("Three Crowns") Mountains, on the northwest coast of Spitsbergen, the main island of Svalbard. The Dutchman Willem Barents saw the Tre Kroner Mountains from the sea in 1596. He named the island after them—Spitsbergen, meaning "pointed mountains."

In the 17th century, when whale hunting was at its peak, possession of Svalbard was disputed by the Norwegians, Dutch, and English. The question of sovereignty was raised again in the 18th century when rich beds of coal were discovered. Norway's sovereignty was finally recognized in 1920 at the Paris Peace Conference and was officially granted in 1925 under the Svalbard Treaty.

Kongsfjorden, which stretches 25 miles (40 kilometers) inland, is a natural wonder not simply in terms of scenic splendor—it is here that the warmer waters of the Atlantic meet the colder waters of the Arctic, producing an interesting series of biological processes in the

water masses and seafloor. Freshwater from an active tidal glacier at the head of the fjord adds to the mix, and marine biologists enjoy studying the dynamic marine environment caused by the meeting of these waters. Scientists also study the movement of Svalbard's glaciers, which can traverse several miles in just a few years.

In addition to scientists, there are also two coal-mining communities, one Norwegian, the other Russian. By the shores of Kongsfjorden lies the tiny settlement of Ny-Ålesund (average population: 40), a former coal mine which is now a research facility used by many European countries. Humans share the islands in winter with polar bears, reindeer, and arctic foxes, the only fauna capable of surviving this far north. But during the four months of continuous summer and daylight, hundreds of wildflowers come into bloom, and white whales, seals, and walruses arrive, along with 30 species of bird.

The capital of Svalbard, Longyearbyen, is the most northerly place in the world on a regularly scheduled airline flight. The area of Kongsfjorden encapsulates the Arctic, the closest one can get to the North Pole and yet be able to stay in a hotel. CM

BELOW *Large icebergs stretch out into the icy waters of Kongsfjorden archipelago.*

NORTH CAPE

FINNMARK, NORWAY

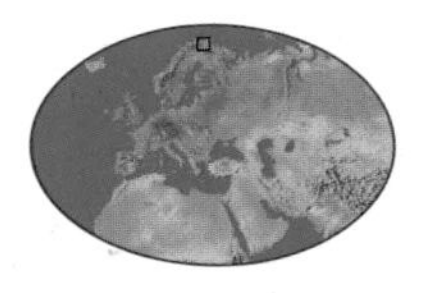

Alternative name: Nordkapp

Age of North Cape: 2,600 million years

Height of cape: 1,007 ft (307 m)

To set the record straight, the North Cape, a magnificent sheer granite cliff face towering above the island of Magerøya in northern Norway, high above the Arctic Circle, is not, as many people consider it to be, the most northerly point in mainland Europe. Although it gets all the glory (and thousands of tourists, who arrive between May 11 and July 31 to witness the midnight sun), the most northerly point is actually an undistinguished, flat, barren promontory called Knivsjellodden. However, viewing the midnight sun on Norway's North Cape is an unforgettable experience—the sun sinks majestically toward the horizon but then seems to stop, hanging in the sky: a giant red ball over a pristine, golden sea. Then it begins to rise again.

The North Cape was discovered in 1553 by English naval captain Richard Chancellor. It was known locally as *Knyskanes* but Chancellor did not know this. He christened it North Cape and the name stuck. Prince Louis-Philippe fled to the cape to escape the French Revolution and Oscar II, king of Sweden and Norway, climbed to the top of the plateau in 1873. Another royal visitor to the North Cape was King Chulalongkorn of Siam (now Thailand) in 1907.

Before the advent of steamships in 1845, all travel was by land, across open, barren, very rugged country. Consequently, there were very few visitors. By the middle of the 19th century, small numbers of adventurous tourists had begun traveling aboard steamships to explore the North Cape. In 1875, Cook's of London recognized this gap in the tourism market and took the opportunity to arrange group tours around the area.

Viewing the midnight sun on Norway's North Cape is an unforgettable sight—the sun sinks majestically toward the horizon but then stops, hanging in the sky: a giant red ball over a pristine, golden sea. It then starts to rise again.

In 1956, a road was built from nearby Honningsvåg and the way was open for mass tourism to develop in this wilderness. Today, an "experience center" has been blasted out of the rock. It provides information about the history of the cape and surrounding Finnmark and also houses a restaurant, serving champagne and Norwegian caviar, as well as a post office where visitors can get the all-important "North Cape Stamp." Visitors can also pay for the privilege of joining the Royal North Cape Club, which was founded in 1984. After five visits to the cape, members are awarded a badge. CM

RIGHT *Puffins rest on the rocks of the North Cape.*

THE LOFOTEN MAELSTRÖM

NORDLAND, NORWAY

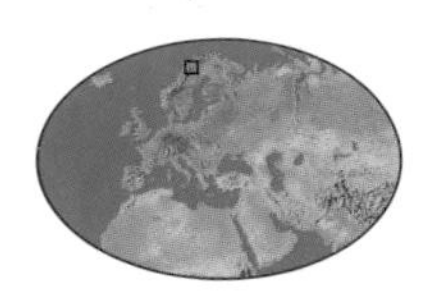

Age of Lofoten Maelström:
20,000 years

Width of Lofoten Maelström:
2.5 mi (4 km)

Depth of Lofoten Maelström:
130 to 200 ft (40 to 60 m)

The desolately beautiful Lofoten Islands, high above the Arctic Circle off the coast of Norway, have long attracted the attention of the wider world because of the terrifying marine phenomenon off their shores known as the Maelström. A convergence of fast-flowing currents close to Moskenesøy, the furthest out to sea of the five main islands, creates a mighty whirlpool.

The Maelström's existence was first recorded in 4 B.C.E. by the explorer Pytheas, sailing from the ancient Greek colony of Massalia—now Marseille in southern France. Thereafter it was featured on sea charts accompanied by grim illustrations and warnings. Local fishermen told horrifying stories of boats, whales, and polar bears being sucked into it and ripped to pieces on the jagged rocks of the seabed. American author Edgar Allan Poe, in his *Tales of Mystery and Imagination*, described it thus: "Here the vast bed of the waters, seamed and scarred into a thousand conflicting channels, burst suddenly into frenzied convulsion—heaving, boiling, hissing—gyrating in gigantic and innumerable vortices ... sending forth to the winds an appalling voice, half-shriek, half-roar, such as not even the mighty cataract of Niagara ever lifts up in its agony to Heaven."

That Poe availed himself of a little poetic license can be gauged by the fact that today the fishermen of Lofoten run boat trips for tourists through the Maelström, or *Moskenstraumen*, as it is known locally. However, it remains an awesome experience. CM

LOFOTEN ISLANDS—COD SPAWNING

NORDLAND, NORWAY

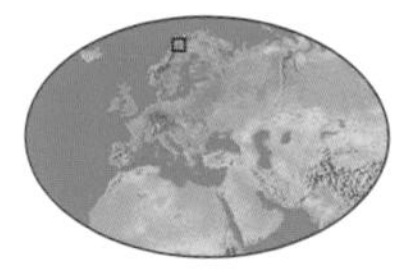

Migration distance: 500 mi (800 km)

First-time cod spawners: 400,000 eggs

Mature female cod spawners: 15 million eggs

In late winter, northeast arctic cod leave the Barents Sea and head south, to the Lofoten Islands off the Norwegian coast. Most cod tend to stay in one place, but this species is migratory. During spring and summer, the cod are on the move, following the capelin to the coast of Finnmark, where they are known as "spring cod"; but in late January, when the first fish appear at Lofoten, they are known as *skrei*—mature cod ready to spawn. Some fish are up to six feet (1.8 meters) long, with roes containing upward of five million eggs. They travel 12 miles (20 kilometers) a day, the oldest females arriving first, and they come in huge schools. Spawning continues until April. This is where all the fish begin their 15-year lives, but for some it is where life will end. Waiting for them are about 4,000 fishermen who catch the cod from small boats. Packs of killer whales, which remain in the area for most of the year, also take advantage of this time of plenty. The *skrei* provide a living for the majority of the islanders. Fishing boats take visitors out to the cod, where angling takes place alongside commercial fishing. MB

RØST CORAL REEF

NORDLAND, NORWAY

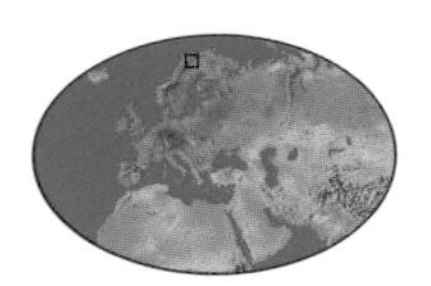

Length of Røst Coral Reef: 22 mi (35 km)

Width of coral reef: 1.8 mi (3 km)

Thickness of coral reef: up to 115 ft (35 m)

Mention of coral reefs conjures up images of sandy beaches and tropical islands, but in the northern Atlantic and Pacific Oceans there are cold-water coral reefs. One of the largest discovered is at a depth of 1,000 feet (300 meters) off Røst in the Lofoten Islands. It is very slow-growing and could be as much as 8,000 years old. It is one of several known reefs that occur not only along Norway's continental shelf but also off the British and Irish coasts. They grow in places where strong currents sweep in food, such as at the mouths of fjords. In these areas, little sunlight penetrates and the water temperature can be 39°F (4°C). The shallowest known reef is in Trondheim Fjord at a depth of 128 feet (39 meters), while the deepest lies at 13,200 feet (4,000 meters) in the North Atlantic.

The organisms responsible for creating these cold-water reefs are *Lophelia*, which live as tiny polyps, like miniature sea anemones, though these do not join together as on tropical corals. They feed mainly on the spring bloom of phytoplankton. Like tropical reefs, they play host to a great number of other organisms, such as sponges, worms, echinoderms, crustaceans, and fish, including cod and red fish. They can only be reached with a deep-sea submersible. MB

GEIRANGER FJORD

MØRE OG ROMSDAL, NORWAY

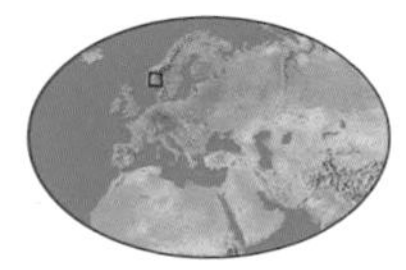

Length of Geiranger Fjord: 10 mi (16 km)

Depth of Geiranger Fjord: 1,000 ft (300 m)

Age of Geiranger Fjord: 1 million years

The Geiranger is the most spectacular and perhaps best known of Norway's fjords. It curls sinuously almost 10 miles (16 kilometers) inland past the port of Ålesund, between the 6,600-foot (2,000-meter) high walls of the Møre ag Romsdal Mountains to the small town of Geiranger itself. Several spectacular waterfalls, such as the Seven Sisters, the Bridal Veil, and the Suitor, border the 1,000-foot-(300-meter-) deep fjord. The abandoned farms of Skageflå and Knivsflå cling to green ledges high on the mountainsides. It is little wonder that cruise ships visit the fjord for its scenery every summer. At the head of the fjord is Flydalsjuvet, a famous rock formation, where brave (or foolhardy) folk are photographed admiring the breathtaking but vertiginous view of the waters.

Norwegian fjords such as the Geiranger were formed a million years ago by ice boring down and carving deep rifts between the mountains. The ice was thickest inland, which is why fjords start out relatively shallow and then become much deeper. The larger fjords such as the Geiranger contain saltwater, and do not freeze in winter. They are extremely calm, with negligible tide. CM

THE PULPIT

ROGALAND, NORWAY

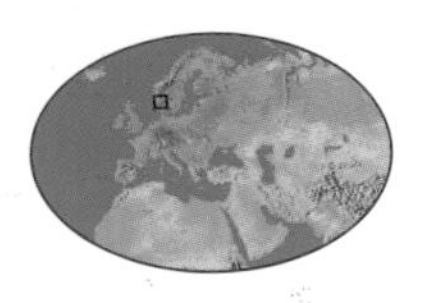

Alternative name: Prekestolen

Height of the Pulpit: 2,000 ft (600 m)

Area of the Pulpit's plateau: 269 sq ft (25 sq m)

The Pulpit is a massive square chunk of rock that towers above Lysfjorden. This fjord leads from the southern Norwegian port of Stavanger, formerly the world's sardine capital and now headquarters for the country's offshore oil industry. Ironically, the Pulpit is located in a flat region of land known by Norwegians as Sørlandet.

The view from its bare rock summit is magnificent, commanding almost the whole of the fjord itself, its light blue, crystal-clear waters reflecting the clouds passing overhead. The rugged, rocky Rogaland Mountains to the north and the Vest-Agder range to the south are peppered with shining, light green patches of vegetation.

It is a two-hour hike to the rock from the village of Jøssing—along a route that is not for either the fainthearted or those with vertigo. However, the Kjeragbolt rock, an enormous boulder wedged between two sheer rock faces, farther along the fjord, makes the vertiginous Pulpit seem like child's play. Kjeragbolt provides room for just one person to sit and admire the view, an experience that can only be described as stomach-churning because of its extreme exposure. CM

KJOSFOSSEN

SOGN OG FJORDANE, NORWAY

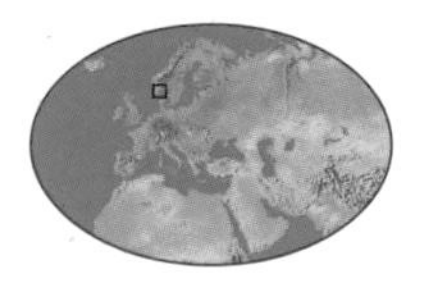

Height of Kjosfossen: 305 ft (93 m)

Length of Flåm railway: 12 mi (20 km)

Norway has more than its fair share of highly impressive waterfalls, or *fossene*, as they are known locally. There is the dramatic but stately Låtefossen, the wild tumble of Vøringsfossen, and the magnificent, pastoral elegance of Mardalsfossen. But for sheer majestic power, Kjosfossen in the Hardanger Mountains is unbeatable. It hurtles 305 feet (93 meters) down a cliff face, with a roar that can be heard for miles around, spray leaping high in the air to form prisms in the bright sunlight.

The best and most exciting way to see Kjosfossen is to travel by train from Flåm to Myrdal. The trip is barely 12 miles (20 kilometers) long, but climbs along and up the side of a steep gorge, and in and out of a series of 21 tunnels. The train stops at a viewing platform with a panoramic view of the powerful Kjosfossen as it plummets to the ground in a haze of spray. The journey then descends through the Flåm Valley, taking you past a roaring river and cozy little farms hidden among rugged terrain. The town of Flåm nestles in a corner of Aurlandfjord, which in turn runs into Sognefjord, a natural wonder in its own right, as it is the country's longest and deepest fjord. CM

SOGNEFJORD

SOGN OG FJORDANE, NORWAY

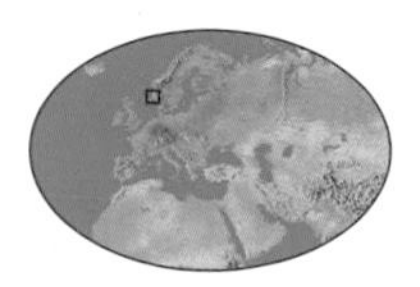

Height of Sognefjord's walls: 3,000 ft (900 m)

Maximum width of fjord: 3 mi (5 km)

Depth of fjord: 3,960 ft (1,200 m)

Rising almost vertically from the water's edge, the mountains that line Sognefjord dwarf any large ship entering this spectacular arm of the sea. It is truly awesome. Great walls of granite, probably 2 billion years old, rise up to 3,000 feet (900 meters) above the inlet. Waterfalls, like thin ribbons, cascade over the dark rocks. The fjord is the longest in Norway, extending inland for 115 miles (184 kilometers). At its widest point it is 3 miles (5 kilometers) and its waters are a staggering 3,960 feet (1,200 meters) deep. The fjord was formed when glaciers carved into the underlying rock bed during the ice age, creating the sheer granite walls on view today. As this ice slowly melted, the sea rose, and the valley was drowned.

Mainland Europe's largest glacier is the Jostedal Glacier which covers an area of 188 square miles (487 square kilometers). The meltwater from this giant glacier partly drains into Fjaerlandsfjord, a tributary of Sognefjord. The branch farthest from the sea, Ardalsfjord, contains the spectacular Vettis Falls that plummet 900 feet (275 meters) to the water below—a feature that attracts a number of summer-cruise visitors. MB

MOUNT SONFJÄLLET

HÄRJEDALEN, SWEDEN

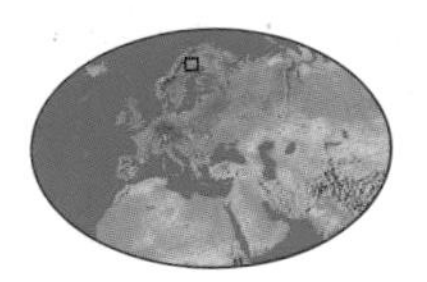

Height of Mt. Sonfjället: 4,193 ft (1,278 m)

Area of mountain: 1,761 acres (713 ha)

Area of Sonfjället National Park: 6,479 acres (2,622 ha)

The gently rounded peak of Mount Sonfjället towers 4,193 feet (1,278 meters) above pine forests in the Swedish province of Härjedalen. This majestic peak gives its name to the national park that surrounds it. The area was declared a park to protect the mountain's slopes—with their thick carpets of reindeer moss—from excessive grazing.

The summit, which offers panoramic views, is noteworthy for the boulders that cover it. These contain checkered patterns, the result of particles of rock being worn away by severe frosts. A great deal of the mountain is bare because its bedrock is acidic quartzite, which inhibits plant growth. Only crowberry and alpine bear berry bushes survive.

More than half the park's area is covered with forest, most of it coniferous, traditionally a refuge for bear and lynx. Birdlife includes ptarmigan, snow bunting, golden plover, meadow pipit, rough-legged buzzard, raven, redpoll, brambling, chaffinch, goshawk, and several species of owl. To really appreciate all that the park has to offer, the Valmen River on its eastern boundary has an overnight cabin and several shelters where the more adventurous can set up camp. CM

THE LAPP GATE

LAPPLAND, SWEDEN

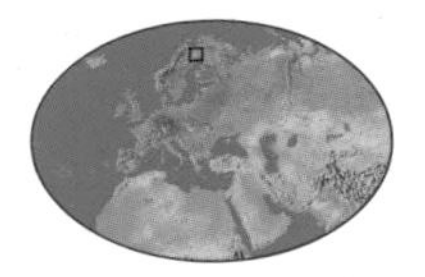

Local name: Lapporten

Highest elevation of the Lapp Gate: 5,725 ft (1,745 m)

Type of valley: U-shaped valley

The Lapp Gate, Sweden's most distinctive natural landmark, lies in the extreme north, 124 miles (200 kilometers) above the Arctic Circle. The Lapp Gate is a startling U-shaped valley between two of Sweden's tallest mountains, Tjuonatjåkka and Nissotjårro. This perfectly symmetrical valley, from a distance looking like a giant hole cut through the mountain range, was shaped by glaciation.

The Lapp Gate is known as the gateway to Swedish Lappland, an enchanting place of wild tundra, reindeer, and the native Lapp people. A 280-mile (450-kilometer) hiking trail begins at Abisko National Park, and travels south to Hemavan. Hiking the entire route could take a month and immerses the walker in some of the finest wilderness in Europe—vast open stretches of tundra and forest and endless solitude. In the fall, the tree-lined valleys burst with magnificent colors of red, yellow, orange, and deep rusty browns. Many people believe this is the best time to visit. By then the mosquito population has dropped significantly, after reaching almost "plague" proportions in the summer. As the great naturalist Carl Linneaus said, "If not for the mosquitoes, Lappland would be paradise on Earth." JK

ABISKO NATIONAL PARK

LAPPLAND, SWEDEN

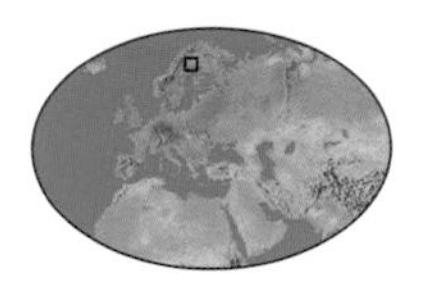

Area of Abisko National Park: 19,028 acres (7,700 ha)

Established as a national park: 1909

Lappland in northern Sweden is home to a number of spectacular national parks, the most scenic of these being Abisko National Park. Framed by mountain ranges in the south and west, and the waters of Torneträsk Lake in the north, the low-lying valley of Abisko National Park is a wonder to behold. The arctic light dances over the glistening Abiskojokka River which runs through the park, and deep canyons with steep cliff walls reveal the area's violent geological past. The best view in Lappland is from Abisko National Park through which runs the *Kungsleden* ("King's Way") trail from Nikkaluotka, a small Lapp village 37 miles (60 kilometers) south of Kiruna, to Riksgränsen on the Norwegian border. Visitors can also take a cable car to the top of Mount Njulla, where there is a beautiful view of Torneträsk Lake and the Lapp Gate.

Plant life thrives on the lime-rich rock, and the park is host to a number of rare plants—the Lapp orchid, for example, is a protected flower and grows nowhere else in the country. Martins, stoats, lemmings, and elks wander the wilds of the park, and many small species of bird, such as the arctic warbler, circle its skies. CM

MOUNT AKKA

LAPPLAND, SWEDEN

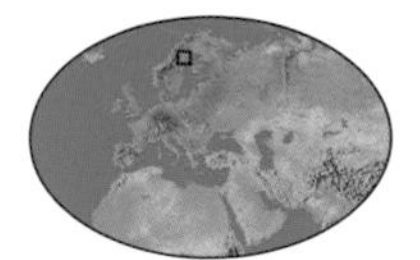

Age of Mt. Akka: 400 million years

Height of Stora Sjöfallet: 6,610 ft (2,015 m)

Established as a national park: 1910

Looming over the Stora Sjöfallet National Park above the Arctic Circle in northern Sweden, Mount Akka is known as "The Queen of Lappland." The mountain, with its sharp peaks and numerous glaciers, is also sometimes called "Nils Holgerson's Mountain," after featuring in the novel *The Wonderful Adventures of Nils* by Swedish author Selma Lagerlöf (1858–1940). To the east of Mount Akka is another impressive mountain, the picturesque Kallaktjåkka (6,000 feet, 1,800 meters), whose northern slope faces the deeply-cut Teusadalen Valley.

The Stora Sjöfallet National Park comprises 315,790 acres (127,800 hectares) of mountain birch forest, pine and mixed forest, and bog. The remainder is water, developed land, and bare rocky mountain. Stora Sjöfallet, or "Great Lake Falls," run through the park. Much of their flow has been siphoned off to provide hydroelectric power, so they are a pale shadow of their former strength. This has not, however, destroyed any of the park's beauty, and Mount Akka and the surrounding area offer a wide variety of landscapes, from alpine ranges and low ridges to flat, high plains and deeply carved valleys. CM

NJUPESKÄR WATERFALL

DALARNA, SWEDEN

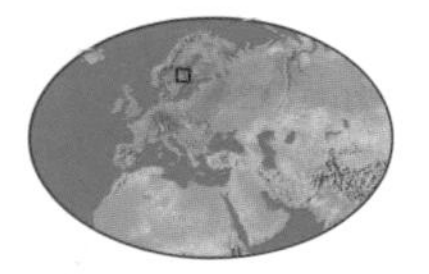

Height of Njupeskär Waterfall: 410 ft (125 m)

Wildlife: moose and deer

Sweden's highest waterfall plunges 410 feet (125 meters) in a gleaming torrent of white foam between jagged walls of black granite rock, surrounded by pristine forests of spruce and pine. The fall is located in Fulufjället in the north of Dalarna, considered the most typical of Sweden's 24 provinces, with its forests, lakes, mountains, and red-painted wooden cottages and farmhouses. It lies just 1.2 miles (2 kilometers) from a village with the rather sinister name of Mörkret—literally "The Darkness." From here, visitors can hike to the falls through a forested nature reserve where moose and deer can be seen roaming freely among the rich foliage.

In the fall, the floors of the forests are carpeted with berries, the most prized being cloudberries. They form the basis of a sauce poured over hot Camembert cheese and served as a dessert in the upscale restaurants of Stockholm and other Swedish cities. Njupeskär attracts most visitors around midsummer, when the people of Dalarna display their folk costumes and dance around maypoles to traditional fiddle music. The province is also famous for the small red wooden horses that woodsmen originally carved as toys for their children but which have since become a national symbol. CM

STOCKHOLM ARCHIPELAGO

STOCKHOLM, SWEDEN

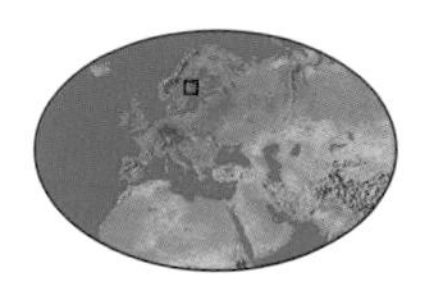

Area of Stockholm archipelago: 2,162 sq mi (5,600 sq km)

Number of islands: 24,000

The 24,000 islands, islets, holms, and skerries of the Stockholm archipelago—*Stockholms skärgård*—are paradise in summer, described by Swedish author and dramatist August Strindberg as a "basket of blossoms on ocean's wave." In winter they are an icy wilderness—"monotonous ... very lonely," said poet and essayist Hilaire Belloc.

The archipelago was created by the movement of ice, carving through what was originally an inland mountain range, to leave outcrops of rock. As the ice retreated, it polished the rock, giving the islands smooth, rounded slopes to the north, but leaving steep, sharp, angular slopes to the south. The archipelago begins in the heart of Stockholm, with the island of Skeppsholmen. The islands closest to the mainland are bigger than those farther out and are separated by stretches of water that the Swedes call *fjärdar*. Many of the islands host summer homes and seasonal inhabitants but only 2,700 are inhabited year-round. Transport between the islands is by small white ferryboats, known locally as *Waxholmsbåtar*. These carry more than a million passengers each year.

Up to 27 species of seabird breed in the archipelago, and its brackish waters are home to Baltic herring, cod, flounder, eel, whitefish, pike, and perch-pike. Badgers, foxes, hares, elk, and deer inhabit the inner islands, while seals live on islands further afield, closer to the sea. CM

TÄNNFORSEN

JÄMTLAND, SWEDEN

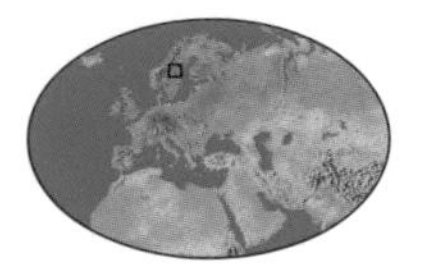

Height of Tännforsen: 125 ft (38 m)

Width of falls: 200 ft (60 m)

Tännforsen in the northern province of Jämtland is reckoned to be Sweden's largest free-flowing waterfall. Other falls that might otherwise dispute the claim have long ago had their waters harnessed to provide hydroelectricity. So far, Tännforsen has managed to escape such a fate. Protection granted to it in 1940 ran out in 1971, but the falls and the surrounding land have since been declared a nature reserve.

The waters of Tännforsen fall 125 feet (38 meters) across a 200-foot (60-meter) wide bed of stepped rock dating back millions of years. During an average year, water volume peaks in May to June at around 162,800 gallons (740,000 liters) a second. But the falls are at their most beautiful between December and February when volumes are lowest and the water freezes. On a cool clear day, the waterfall glows in the deep reds, pinks, and oranges of the setting sun. Tourists were first attracted in 1835 when King Karl XIV Johan opened a road to the falls. CM

RIGHT *The scenic Tännforsen at sunset.*

BORGA MOUNTAIN

VÄSTERBOTTEN, SWEDEN

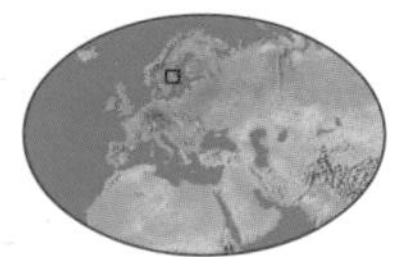

Height of Borga Mountain: 2,640 ft (800 m)

Height of Borgahällen: 3,960 ft (1,200 m)

Borga Mountain, in the northern Swedish province of Västerbotten, might not be outstanding but it contains two natural assets that are fast becoming rare in the modern world: abundant fresh air and unspoiled, beautiful countryside.

In this breathtaking wilderness, one can roam the many trails—Borga is a refuge for hikers and nature lovers hoping to see bears and other wildlife in the forests of its foothills while fishermen sport in the mountain's lakes and rivers where there is an abundance of fish waiting to be caught. Some visitors spend their time panning for gold in Slipsik Creek, but little, if any, has ever been found.

Borga's slopes contain an estimated 50 different species of wildflower, and in the fall berries and mushrooms abound. Here, as elsewhere in the country, an unwritten Swedish law, *Allemansrätt* ("Every Man's Right"), guarantees everyone access to nature and to camp anywhere within reason.

Located near Borga Mountain lies the stunning Borgasjön Lake. Spectacular views of this lake can be seen from Borgahällan, which is 3,960 feet (1,200 meters) above sea level. With a sheer drop of 797 feet (243 meters) down to the lake, the view from the ledge is not for the fainthearted. CM

GOTLAND

GOTLAND, SWEDEN

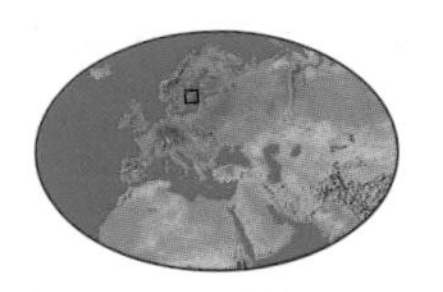

Length of Gotland: 105 mi (170 km)
Width of Gotland: 33 mi (52 km)
Distance from mainland: 52 mi (90 km)

Gotland is an island in the Baltic Sea, off the east coast of Sweden. It is home to the *raukar*, naturally formed limestone pillars that resemble human figures. On misty days they loom as if from the Viking sagas, staring seaward with expressions of petrified astonishment. The most impressive *raukar* are to be found between Digerhuvud and Lauterhorn off the coast of Fårö, to the north of the main island, home to legendary Swedish film director Ingmar Bergman.

Gotland has a varied landscape, from desolate moors and blossoming meadows to tall cliffs and long sandy beaches. The island supports 35 different species of wild orchid, and on Stora Karlsö, a tiny island off its west coast, thousands of great auk nest. A local breed of horned sheep graze on another island, Lilla Karlsö. For unknown reasons, Gotlanders refer to sheep as "lambs" whatever their age. A favorite summer destination for Swedish vacationers, Gotland remains remarkably unspoiled. Its picturesque capital, Visby, was once a major Baltic trading center with links to the Hanseatic League in Lübeck. CM

BELOW *Rocks on the beach of Gotland at dusk.*

LAKE INARI

LAPPI, FINLAND

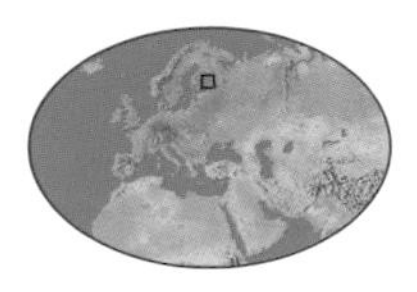

Area of Lake Inari: 500 sq mi (1,300 sq km)

Depth of lake: 318 ft (97 m)

Finnish Lappland is a place of lakes, and the largest is the deep blue Lake Inari, near the Russian border. Almost a small sea rather than a lake, from the middle you cannot see land, and if a storm should roll in, the wind can whip up treacherously high waves. Its shores are jagged, indented with hundreds of inlets, and over 3,000 tree-covered islands are scattered over its waters. The lake covers an area of 500 square miles (1,300 square kilometers) and its sides fall steeply to the murky depths below—local legend and folk songs claim Inari to be as deep as it is long. The lake is fed by the Ivalojoki and empties into the Barents Sea through the Paatsjoki.

Ancient Lapps brought their offerings to the gods here, to the sacred island of Ukko, named after the god of the sky, the supreme god in Finnish mythology. Another island, Ukonkivi, was a sacrificial site for good fishing, and an ice cave on Korkia was used to store fish. Seven different types of salmon and trout thrive here. At night the skies above the lake come alight with the spectacular light show put on by the aurora borealis. The season is short, the last ice only melting by the second week of June. This is followed by plagues of mosquitoes. MB

NORTHERN LIGHTS (AURORA BOREALIS)

LAPPI, FINLAND

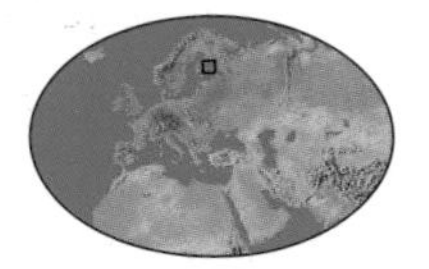

Height of Northern Lights: 37 to 373 mi (60 to 600 km)

Frequency: 200 times annually

The aurora borealis, together with its southern counterpart, the aurora australis, offers one of the most awe-inspiring of natural spectacles. Each year in arctic Lappland there are upward of 200 entrancing celestial displays, the result of the interaction between the solar wind—a stream of plasma leaving the sun at speeds of over 620 miles (1,000 kilometers) per second—and Earth's magnetic field. Plasma trapped in the ionosphere above Earth's surface glows red, green, blue, and violet in an ever-changing pattern of streamers, rays, curls, and spirals that fill the night sky.

Finlanders have at least 20 folktales concerning the origin of the Northern Lights. The most popular story is that an arctic fox made them as it ran through the snow, sweeping his brushy tail to throw sparks of fire into the sky. In fact, the Finnish word for the Lights is *revontulet*, meaning "fox fire." The scientific truth about this phenomenon does nothing to diminish its spellbinding beauty. NA

RIGHT *The spectacular colors of the aurora borealis.*

NORTH GAULTON CASTLE

ORKNEY ISLANDS, SCOTLAND

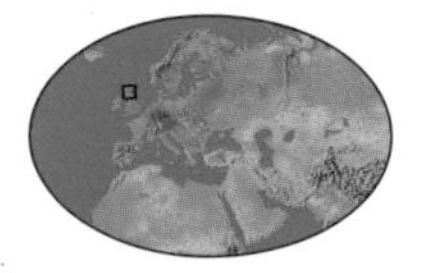

Height of North Gaulton Castle: 165 ft (50 m)

Rock type: red sandstone

Guarding Orkney against the Atlantic Ocean, North Gaulton Castle is one of the most dramatic sea stacks anywhere in the British Isles. Like its diminutive sibling, Yesnaby Castle, it is formed from the heavily layered old red sandstone that characterizes the Orkney Islands. The sandstone is enriched with plentiful numbers of stromatolites and other fossils, evidence of the life that teemed in the huge Devonian lake that once covered Orkney. The stack is located along one of the lonelier stretches of Orkney's mainland west coast. Despite its obvious appeal to climbers, this isolation has limited activity to just a handful of ascents. For those brave enough to undertake the challenge, or those content to view the stack from the relative safety of the headland cliffs, the splendor of the seascape is matched only by the local wildlife. In addition to the rich array of seabird species, which congregate in large breeding colonies under the protective sandstone cliffs, the lucky visitor might also happen upon the Orkney vole, or catch a glimpse of an elusive sea otter. NA

THE OLD MAN OF HOY

ORKNEY ISLANDS, SCOTLAND

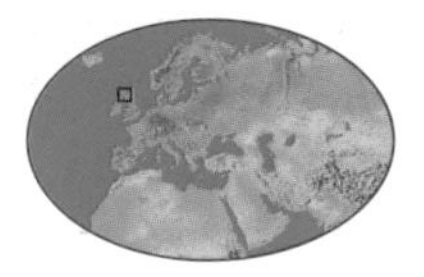

Height of the Old Man of Hoy: 450 ft (135 m)

Rock type: Devonian sandstone

Britain's tallest sea stack teeters on a slender 98-foot (29-meter) wide base, hewn from a fragile mixture of layered Devonian sandstone. The stack is young: In maps and paintings dating from the middle of the 18th century the Old Man was shown as still being part of the headland. By the early 1900s the ravages of time and tide had given rise to a stack and arch—the two bandy legs that inspired the Old Man's moniker. How long the surviving appendage will resist further erosion is an open question, but one day the sea will claim the life of this old man.

Made famous by a breathtaking televised ascent in 1967, and first climbed in 1966 by Tom Patey, Rusty Baillie, and Chris Bonington, the Old Man of Hoy has since become a challenge to climbers prepared to brave the dive-bombing, vomiting fulmars, guillemots, razorbills, and kittiwakes that nest on it. Nearby, St. John's Head is one of the highest vertical sea cliffs in Britain, rising an impressive 1,135 feet (346 meters) from the waves. Hoy is important not only for its diversity of seabirds, which include great and arctic skuas and red-throated divers, but also Orkney's only colony of mountain hares. NA

DUNCANSBY STACKS

HIGHLAND, SCOTLAND

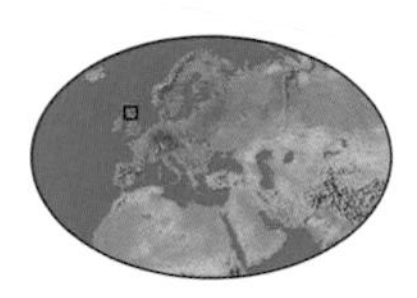

Age of Duncansby Stacks: 380 million years

Elevation of stacks: 196 ft (59 m)

A visit to Duncansby Head in Caithness, one of the most northerly points in mainland Britain, provides an ample reminder of the powerful seas that erode and sculpt its coastlines. Following the cliff-top path past the lighthouse, visitors are rewarded with a series of impressive sights, including Sclaites Geo, an immense cleft reaching deep into the cliffs and home to thousands of nesting seabirds; the weathered rock arch of Thirle Door; and finally the jagged, dragon-tooth Duncansby Stacks.

The secret of this spectacular natural beauty is in the rock itself—the old red sandstone cliffs are easily eroded by the action of the sea. Deposited during the Devonian period, the exposed edges play host to fulmars, kittiwakes, and razorbills, and tell the tale of past climatic changes. The sediment formed within an immense freshwater lake that stretched from Shetland to Inverness and eventually Norway. Repeated patterns in the layers, at different thicknesses, show how variation in the Devonian climate gave rise to changes in the lake's biology. Like all sea stacks, those at Duncansby Head are a work in progress—new stacks will be hewn from the mainland cliffs as surely as the old ones will fall into the sea. NA

LOCH LANGAVAT

OUTER HEBRIDES, SCOTLAND

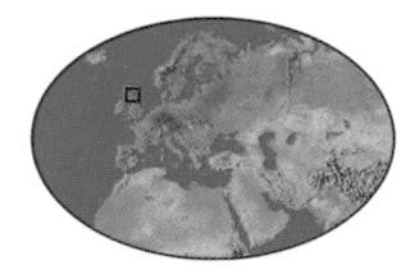

Length of Loch Langavat: 7 mi (12 km)

Length of Grimersta River: 1.2 mi (2 km)

Area of Grimersta water system: 19,800 acres (8,017 ha)

The islands of Scotland's Outer Hebrides have numerous freshwater lochs, one of the most striking being Loch Langavat, which nestles between hills on the border of Lewis and Harris. The area, described as a "liquid landscape" for the loch, is the headwater for the famous Grimersta system of lochs and lakes that has some of the best fly-fishing in Europe for wild Atlantic salmon.

Langavat drains into the two Langavat rivers, which feed a run of four relatively shallow lochs and connecting streams. The water flows northward toward the Grimersta River and the sea at Loch Roag.

This sea loch is a deep incision in the northwest coast of the Isle of Lewis and on its eastern shore stand 20 known ancient monuments made from Lewisian gneiss, one of the oldest rocks in Britain. They are known as the standing stones of Callanish, and they were probably erected between 3,000 and 4,000 years ago. The island is also dotted with smaller lakes, known as rock-earns, which lie on ice-worn platforms in the hills. The entire landscape is the product of glaciation at the time of the ice ages, between 3 million and 10,000 years ago. MB

ST. KILDA ARCHIPELAGO

OUTER HEBRIDES, SCOTLAND

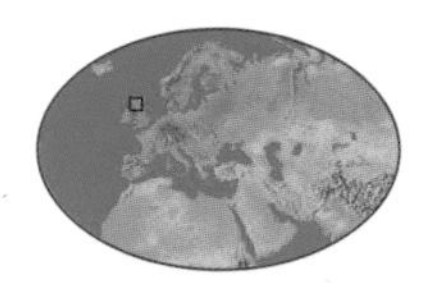

Number of islands: 4

Distance from mainland: 41 mi (66 km)

Highest point (Conachair Hill): 1,410 ft (430 m)

Officially the remotest part of the British Isles, the St. Kilda archipelago lies west of Benbecula in the Outer Hebrides. The archipelago is the weathered, glaciated remains of a Tertiary ring volcano that now forms dramatic sea cliffs rising vertically to over 1,214 feet (370 meters). Stac an Armin, at 626 feet (191 meters), and Stac Lee, at 541 feet (165 meters), both off the island of Boreray, are the highest sea stacks in the British Isles. The four islands of Hirta, Dun, Soay, and Boreray provide a superb example of island ecology. Species such as the St. Kilda wren and St. Kilda woodmouse are genetically distinct from their mainland counterparts, and the ancient feral sheep of Soay have been the focus of long-term studies. The islands also host one of the world's largest gannet colonies, Britain's largest and oldest fulmar colony, and approximately half the British puffin population. Nesting seabirds and the associated enrichment they bring to the soil help to support over 130 species of flowering plants. The 2,000-year human occupation of the archipelago ended in 1930, when the last inhabitant was evacuated to the mainland. Today, only biologists, geologists, and archeologists visit the islands. NA

FINGAL'S CAVE

INNER HEBRIDES, SCOTLAND

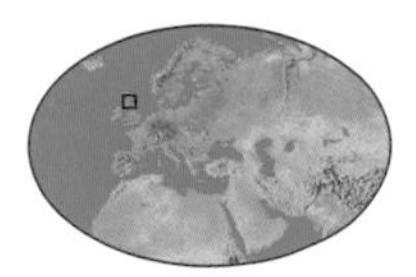

Depth of Fingal's Cave: 230 ft (70 m)

Length of island of Staffa: 0.75 mi (1.2 km)

Height of Staffa: 135 ft (40 m)

Fingal's Cave is part of the same geological event that formed the Giant's Causeway of Northern Ireland. Rows of similar hexagonal columns of black basalt line the cliffs around the uninhabited Scottish island of Staffa. Where the sea has smashed the rocks, huge caves have been formed.

The largest, Fingal's Cave, is named after the legendary Irish hero Finn MacCool or Fionn MacCumhail—known as Fingal to the Scots—who is said to have defended the Scottish islands from the invading Vikings. Legend also holds that Finn was a giant and that he built the Giant's Causeway so that his ladylove, another giant who lived on Staffa, could walk across without getting her feet wet.

The Staffa rock formations went unnoticed by science for centuries, and were only recognized in 1772 after a chance observation by a party of natural historians on their way to Iceland. Once discovered, Fingal's Cave proved popular with artists, poets, and musicians. It was the inspiration for Mendelssohn's epic "Hebrides" overture, and such luminaries as Sir Walter Scott, John Keats, William Wordsworth, Alfred Lord Tennyson, and Jules Verne visited the island. MB

OLD MAN OF STORR

ISLE OF SKYE, SCOTLAND

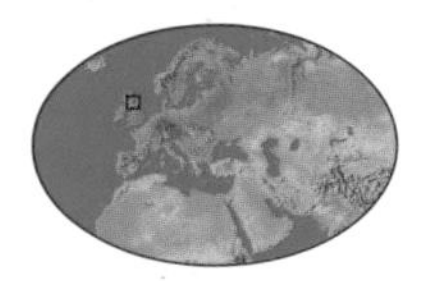

Length of Trotternish Peninsula: 30 mi (50 km)

Height of pillar: 160 ft (49 m)

Age: 60 million years

The Old Man of Storr is a tapering rock pinnacle, like a fir cone on a plinth, perched delicately on the side of a craggy cliff on the Trotternish Peninsula in the northeast of the Isle of Skye. It is a 160-foot- (49-meter-) high pillar and pedestal made of dark basalt, the result of intense volcanic activity in the area about 60 million years ago. Volcanoes pushed up through Jurassic rocks containing the fossils of marine reptiles, and the remains of ichthyosaurs and plesiosaurs have been found nearby. The Old Man of Storr stands at the south end of the Trotternish Ridge, the result of a collapse of rocks after the ice age. At its northern end is the Quiraing, another magical place of pinnacles and gullies. Nearby is Kilt Rock, consisting of vertical columns of dolomite that present a pleated effect like a kilt.

Although the Old Man of Storr lost its head in a severe storm half a century ago, it still stands proud of its adjacent pinnacles and spires. Despite its distance from the sea, sailors continue to use this distinctive pinnacle of rock as a landmark—a sign that they are almost safely home. Its spindle shape makes it difficult to climb—the first successful attempt was made in 1955. **MB**

CUILLIN HILLS

ISLE OF SKYE, SCOTLAND

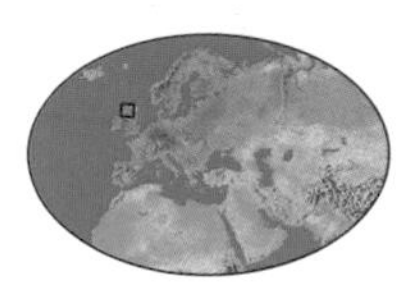

Area of Cuillin Hills: 535 sq mi (1,386 sq km)

Number of peaks in region: 12

The Isle of Skye has fascinated geologists for more than a century—it is a curious mixture of old and new. This Inner Hebridean island is composed of some of the oldest rocks in Europe—the Lewisian Complex gneisses, which are around 2.8 billion years old—butted against some of the youngest, such as the Jurassic sediments, which contain the most complete succession of fossils in Scotland.

Rising majestically from this disjointed landscape is the Cuillin, which is a major center for experienced hill-walkers and climbers throughout the year. Containing 12 peaks in excess of 3,000 feet (900 meters) that qualify them as Munros, the Cuillin Hills make for dramatic sport. The granitic, rounded tops of the Red Cuillin contrast sharply with the jagged, impossibly steep Black Cuillin. The latter are the eroded remains of the massive central volcanoes that once raged on Skye. Repeated glaciation has sculpted impressive ridges and blades that are popular quarry for the enthusiastic Munro climber. The Skye coastline is one of the best places in Britain to spot cetaceans. Minke, pilot, northern bottlenose, killer, fin, sei, and Sowerby's beaked whales can all be seen, as can smaller dolphin species including white-beaked, Risso's, Atlantic white-sided, and common dolphins. **NA**

FALLS OF GLOMACH

HIGHLAND, SCOTLAND

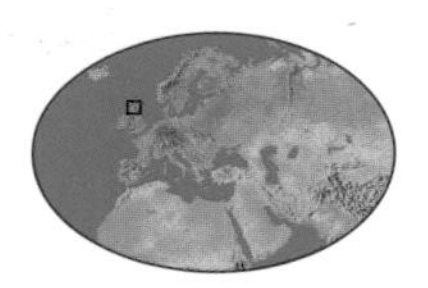

Height of Falls of Glomach: 370 ft (113 m)

Established as a National Trust site: 1944

Plunging 370 feet (113 meters) down the northern flanks of Ben Attow, the Falls of Glomach reward the dedicated walker who makes the eight-hour trek to enjoy their spectacle. Fed by the "Cluanie curtain"—a bizarre meteorological phenomenon that takes the form of a stationary raincloud that rarely extends east beyond Loch Cluanie—the falls, located 18 miles (29 kilometers) east of the Kyle of Lochalsh, are among the highest and widest to be found in Britain.

Glomach, or *Allt a'Ghlomaich*, means "gloomy stream," and the walk in can certainly be just that on a damp, dark day. However, the impressive sight of the single leap that the water makes through all but the last 50 feet (15 meters) of its descent makes the journey worthwhile. The views are best following periods of rainfall when the stream is in full flow, although this kind of weather makes the approach more difficult.

The Kintail region as a whole is a beautiful part of Scotland, with a good scattering of Munros—mountains exceeding 3,000 feet (900 meters)—providing challenging walking when the weather is too dry for the falls to be seen at their best. NA

SUILVEN

HIGHLAND, SCOTLAND

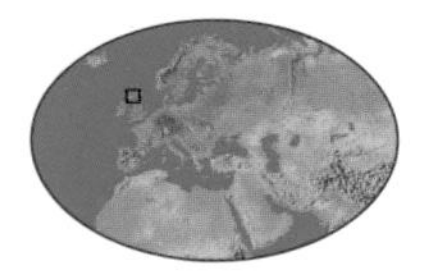

Height of Suilven: 2,398 ft (731 m)

Rock type: sandstone topped with quartzite

Suilven, in Sutherland, provides one of the most spectacular mountain views in Britain. This is in some ways strange, because in height alone Suilven, at 2,398 feet (731 meters), is not exactly awe-inspiring. However, the gigantic quartzite-topped sandstone mass rises abruptly, towering over the surrounding 2.8 billion-year-old Lewisian gneiss, the oldest rocks in Britain. From the east or west, the mountain looks impossible to climb, a huge prehistoric plug of rock with sheer, treacherous flanks. Viewed from the north or south, though, the mountain's true form is discernible, a great notched sail of precipitous rock floating on a sea of exposed rock mosaic and lochans. The ascent is a serious undertaking because of its remoteness, length, and breathtaking exposure. A long walk in across the tussocky "knock-and-lochan" moorland is kept interesting by ever-changing views of the looming mountain ahead, but the return journey is hard work. The climb itself yields several surprises: the flattened "lawn" on Caisteal Liath, the higher western summit, the terrifying narrowness of the ridge, and the extraordinary views across this desolate, beautiful landscape. NA

BEINN ASKIVAL

ISLE OF RUM, SCOTLAND

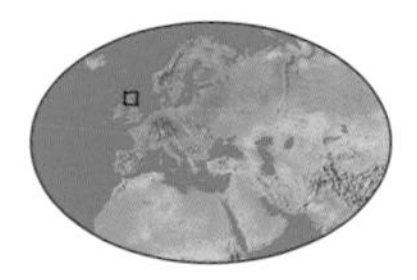

Height of Beinn Askival: 2,664 ft (812 m)

Rock type: basalt

Before the Tertiary period (65 to 0.85 million years ago), Europe and North America belonged to the same vast landmass. As they began to move apart, the resulting rift was the scene of great bursts of volcanic activity. Today, this activity is centered at the bottom of the Atlantic Ocean, along the Mid-Atlantic Ridge, but during the early Tertiary it was located along what is now the west coast of Scotland, when many of the Hebridean islands were formed. Of the islands scattered across the Hebridean Sea, Rum is among the most beautiful; it is also home to one of the largest Manx shearwater breeding colonies in the world. Over 60,000 pairs make the annual migration from the waters off Brazil to the upper reaches of the island's loftier summits, such as Beinn Askival, a volcanic fragment that stands 2,664 feet (812 meters) above the sparkling sea below. Askival's layered basaltic rocks are famed as evidence of a classic "open" magma chamber, from which successive flows of magma formed numerous rock types. The layers are particularly well exposed on Beinn Askival and its sister peak Beinn Hallival, providing a natural laboratory for geologists as they scramble among the shearwaters. **NA**

GREAT GLEN AND LOCH NESS

HIGHLAND, SCOTLAND

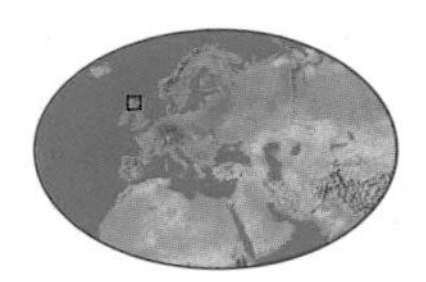

Length of Great Glen: 55 mi (88 km)

Length of Loch Ness: 24 mi (39 km)

Splitting Scotland almost in two is the Great Glen. It follows a 350 million-year-old fault line that cuts diagonally across the Scottish Highlands and gives rise to a series of long, ribbon-like freshwater lakes—Loch Lochy, Loch Oich, and Loch Ness—and the sea loch, Loch Linnhe. Thomas Telford's Caledonian Canal links all these lochs, making northern Scotland a virtual island. A route for walkers known as the Great Glen Way follows the canal from Fort William to Inverness.

Although minor earthquakes have been attributed to the Great Glen Fault, these claims are false. These are actually caused by minor faults throughout the country. On average, there are about three earthquakes per century that reach a magnitude 4.0 on the Richter scale (2.0 being the weakest and 8.9 the strongest). The worst was in 1816, when slates fell off roofs and tremors were felt throughout Scotland. During the ice age, the glen was filled with

glaciers—the steep-sided slopes on either side are their legacy. The highest point in the area is dome-shaped Mealfuarvonie, an impressive 2,300-foot- (700-meter-) high block of old red sandstone conglomerate.

The most famous of the lakes is Loch Ness, which contains the largest volume of freshwater in the British Isles. It averages 600 feet (180 meters) deep, with a cavern named Edward's Deep reported to be 812 feet (250 meters) deep. The water is the color of weak tea, caused by staining from the peat in the surrounding hills, and with an average temperature of 42°F (5°C), it rarely freezes. In the 6th century, St. Columba was said to have confronted a dangerous monster in the loch, and this legend lives on to this day. People flock to the lake shore every year in the hope of spotting the infamous Loch Ness Monster, or Nessie, as it is affectionately known. It has been variously described as a plesiosaur, water kelpie, or sea serpent, but despite numerous scientific expeditions, its presence remains unproven. MB

BELOW *The serene waters of Loch Ness.*

CORRIESHALLOCH GORGE & FALLS OF MEASACH

HIGHLAND, SCOTLAND

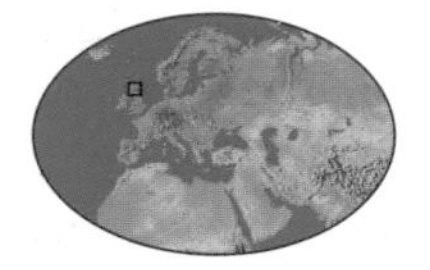

Length of Corrieshalloch Gorge: 1 mi (1.6 km)

Depth of gorge: 200 ft (60 m)

Height of Falls of Measach: 150 ft (45 m)

Box canyons are a rarity in Britain, but Corrieshalloch Gorge near Braemore in the Scottish Highlands is a fine example. In geological terms the gorge is very young: Natural fractures in hard metamorphic Moine schists were rapidly eroded away by powerful glacial meltwater during the recession of the last ice age. Today the gorge is over 1 mile (1.6 kilometers) in length and 200 feet (60 meters) deep. A suspension footbridge, from which visitors can appreciate the breathtaking scenery below, now spans the River Droma, which for the last 12,000 years has carved through the rock. The river drops over the picturesque Falls of Measach, which plummet 150 feet (45 meters) to the rocky floor below.

The steep sides of Corrieshalloch Gorge itself provide refuge for a rich variety of highly specialized ferns, mosses, and liverworts, which thrive in the damp, sheltered conditions. This unusual biodiversity, coupled with its geological interest, has qualified the area as a National Nature Reserve and Site of Specific Scientific Interest. NA

BEN NEVIS

HIGHLAND, SCOTLAND

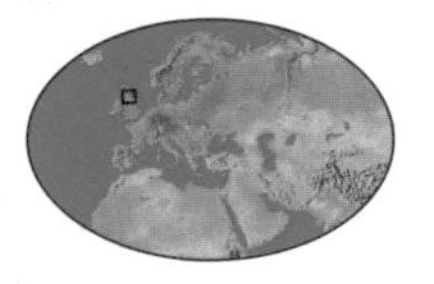

Height of Ben Nevis: 4,409 ft (1,344 m)

Mean summit air temperature: 31.5°F (-0.3°C)

Ben Nevis first erupted into existence around 350 million years ago, and the rocks that today lie exposed along the slopes of *Allt a' Mhuilinn* (the "Mill Burn") form concentric ring dikes: two of granite and two of diorite. Ben Nevis is the highest mountain in the British Isles. It is also one of the most impressive: a majestic colossus that demands the utmost respect from walkers and climbers alike, on any of the routes to its barren summit.

The near-arctic climatic conditions provide a surprising haven for a rich diversity of wildlife. On the lower reaches are glorious native pine, oak, and birch woodlands that higher up yield to peat and heather moorland scattered with bilberries, mosses, thyme, and milkwort. Nearer to the summit the flora takes on an alpine habit, where only the hardiest lichens and mosses can tolerate the harsh Lochaber winter. The slopes of Ben Nevis also play host to most of Scotland's indigenous mammals, among them the rare and elusive wildcat, mountain hare, and red deer. Majestic golden eagles may often be seen overhead, circling the land for prey. NA

LOCHABER MOUNTAINS

HIGHLAND, SCOTLAND

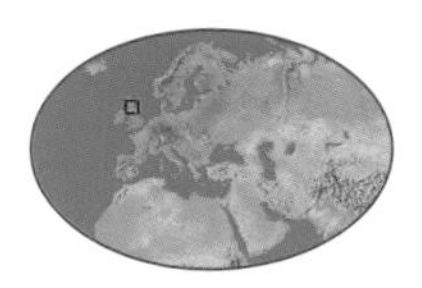

Area of Lochaber Mountains: 934 sq mi (2,419 sq km)

Annual rainfall: 200 in (500 cm)

The Lochaber region in Scotland covers an area of 934 square miles (2,419 square kilometers) and is renowned for its rugged landscape, including the mountains of Glencoe and the Ben Nevis area. Bordered to the west by Loch Linnhe and to the east by the high-level blanket bog of Rannoch Moor, its hinterland is a diverse concertina of ice-scarred mountain peaks and deeply carved glacial valleys. It is in some of these lonely glens that the last remaining fragments of the ancient Caledonian forest can be found, sheltering within them pockets of native wildlife.

The climate is damp: most of the upland region experiences annual rainfall of some 200 inches (500 centimeters). A good portion of this falls as snow in the winter months, some of which remains on the higher northern slopes throughout the year. The generous deposition of snow allows the area to play host to some of the rarest mosses, liverworts, lichens, and fungi to be found in Britain. The dominating mountain ranges attract many thousands of visitors to the area each year, some to watch from afar, others to climb and hike among the weathered crags. All enjoy the majestic splendor of these ancient heights. NA

GLENCOE

HIGHLAND, SCOTLAND

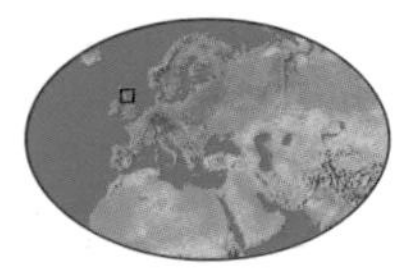

Area of Glencoe: 14,200 acres (5,746 ha)

Age: approx. 500 million years

Highest peak (Stob Coire nan Lochan): 3,743 ft (1,140 m)

The "Glen of Weeping" bears witness to a violent past. Most recently it witnessed scenes of treachery and terror, brought by the infamous 1692 Glencoe massacre of the MacDonalds at the hands of the Campbells. But beneath this human history lie the remains of an ancient volcanic caldera that collapsed around 400 million years ago.

Golden eagles wheel above the plunging walls of the Aonach Eagach, the 3,000-foot- (900-meter-) high jagged northern edge of the glen, justifiably one of the most photographed mountain scenes in Britain. The Clachaig Gully, a magnet to ice climbers in the long Argyllshire winter, provides arguably the world's best example of a ring fault—evidence of the ancient volcanic activity that brought Glencoe's complex geology into existence. More recent glacial activity shaped the glen into its present form. *Coire Gabhail*, the hidden valley, is a classic example of a hanging valley: When the glacier retreated, it left the mouth of the "Lost Valley" a full 820 feet (250 meters) above the River Coe. The tumbling screes that line the glen are the shattered spoils of long-melted ice fields, which now host important plant species. NA

LOCH LOMOND

ARGYLL AND BUTE, SCOTLAND

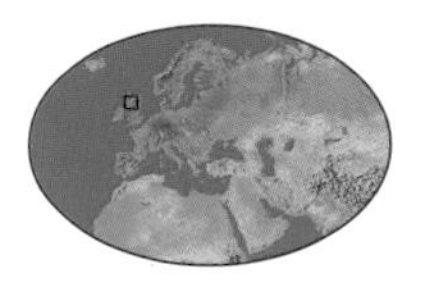

Length of Loch Lomond: 22.5 mi (36.2 km)

Surface area of lake: 27 sq mi (70 sq km)

Depth of lake: 625 ft (190.5 m)

The drama and violence of the red deer rut is played out along the "Bonnie, bonnie banks of Loch Lomond" against an explosion of fall color. The 24-mile- (38-kilometer-) long body of freshwater is punctuated by numerous unspoiled islands, each a microcosm of Scottish natural history. In the Lomond area over a quarter of all British plant species can be found. Most islands are privately owned but access to the nature reserves on Inchcailloch, Bucinch, and Ceardach is easily obtained, from where many species of ground-nesting birds can be observed. The loch also provides sanctuary for overwintering wildfowl, such as Greenland white-fronted geese.

Loch Lomond traces the geological boundary between Highland and Lowland Scotland, formed when the Dalradian rocks, once a mountain range higher than the Himalayas, collided with the Devonian lowlands to the south. The Highland Border Complex, a mixture of marine sediments, completes the geology of the Highland Boundary Fault. From the highest point along the loch, the 3,300-foot (1,000-meter) summit of Conic Hill, the fault can be clearly observed running across several of the islands. **NA**

ARTHUR'S SEAT

MIDLOTHIAN, SCOTLAND

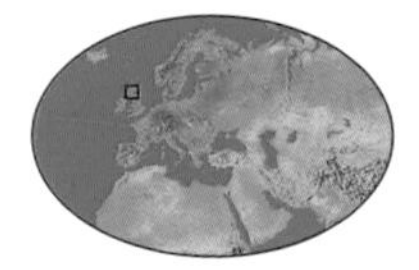

Height of Arthur's Seat: 823 ft (251 m)

Age: 335 million years

This tranquil oasis in the heart of Edinburgh, Scotland's capital city, is the remnant of a once-submerged volcano, which finally blew apart approximately 335 million years ago. The pioneering geologist James Hutton (1726–1797), who famously declared that Earth had "no vestige of a beginning, no concept of an end," recognized the significance of Arthur's Seat and the nearby Salisbury Crags. From their slopes he collected crucial evidence showing that Earth's surface and interior were in a state of constant flux. The hard teschenite cragstone, known as whin, was quarried as cobbles for the streets of Edinburgh, but Hutton ensured that several important geological features were preserved for future generations to both experience and study.

The climb to the summit of Arthur's Seat affords magnificent views of the city and surrounding landscape. It is not surprising, then, that it has been of central importance to the area's inhabitants for thousands of years. Bronze Age artifacts have been found in Duddingston Loch to the west, and clear evidence of later Iron Age terraces are still visible on the gentler southern slopes. **NA**

TRAPRAIN LAW & NORTH BERWICK LAW

EAST LOTHIAN, SCOTLAND

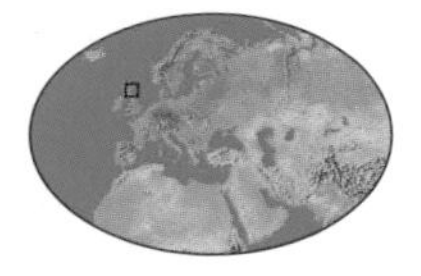

Height of Traprain Law: 734 ft (224 m)

Height of North Berwick Law: 613 ft (187 m)

Traprain Law and North Berwick Law are hills composed of ancient volcanic rocks that dominate the surrounding level floodplain of East Lothian. The two outcrops are among the region's best-known landmarks.

Traprain Law is a whale-shaped hill that lies east of Haddington in the heart of East Lothian and is surrounded by flat agricultural land. The rounded end of the hill faces west and tapers to the east. The southern side of the hill is a steep cliff, but the northern slopes are gentler and have been extensively quarried in the past. The site has probably been occupied since the Stone Age and was an Iron Age fortified site. It is the largest hill fort in Scotland, and in Britain is second in size only to Maiden Castle in Dorset.

North Berwick Law is an irregular pyramid-shaped hill that dominates the coastal town of North Berwick. Much of the stone used in the local buildings (red basalt) was quarried from the site. From the hill you get marvelous views of the Firth of Forth to the north, Edinburgh to the west, and the Lammermuirs to the south. At the top of the hill is an arch made from the jawbone of a whale. RC

ST. ABB'S HEAD

SCOTTISH BORDERS, SCOTLAND

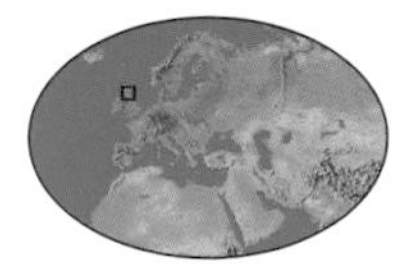

Height of Kirk Hill: 300 ft (90 m)

Established as a national nature reserve: 1983

Dominated by the monolithic sail of Kirk Hill, St. Abb's Head bears visible traces of human habitation for at least the last 3,000 years. The site of one of the earliest Christian landfalls in Scotland, the ruins of St. Ebbe's 7th-century monastery can still be traced among the long grass near the cliff's edge.

However, the geological significance of the area is far greater, for below the crags is Pettico Wick, which bears witness to the late Silurian/early Devonian continental collision that united Britain with Scotland. The subsequent deformation, folding, and uplift of the Ordovician and Silurian rocks led to the formation of mountains on a Himalayan scale, from which thousands of feet were eroded during the Devonian period. The famous "Unconformity," discovered by the 18th-century geologist James Hutton, can be found close by Siccar Point, near Dunbar.

St. Abb's Head is owned and maintained by the National Trust for Scotland, reflecting its scientific and historical importance. It is also one of the largest seabird breeding colonies in Europe, and a visit during the early summer is rewarded with the sight of a cacophony of razorbills, shags, fulmars, herring gulls, puffins, and cormorants that jostle for position along the precipitous ledges. NA

BASS ROCK

EAST LOTHIAN, SCOTLAND

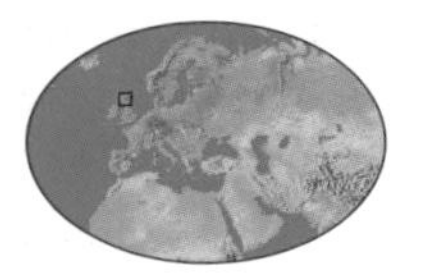

Height of Bass Rock: 350 ft (107 m)
Gannet population: 80,000
Distance from mainland: 1 mi (1.6 km)

Bass Rock is one of four offshore islands, the others being Craigleith, Lamb, and Fidra. Located in the Firth of Forth, it is the closest seabird sanctuary to the mainland and has one of the greatest gannet colonies in the world, with 80,000 nesting on its cliffs. This colony is home to about 10 percent of the world's population of north Atlantic gannets.

The gannet is Britain's largest seabird, with a wingspan of 6 feet (1.8 meters). It was the first to be studied by ornithologists during the 19th century. The gannet's name, *Morus bassana*, incorporates the name of this rocky stack. Visitors can watch the colony using remote-control cameras at the Scottish Seabird Center in North Berwick or take a boat trip out to Bass Rock. Gray seals can also be spotted swimming in the seas around this giant rock.

Bass Rock itself is a volcanic plug from the Lower Carboniferous age. It rises sharply to 350 feet (107 meters) with three sides of sheer cliff and a tunnel piercing the rock to a depth of 344 feet (105 meters). The gentler slope to the south forms a lower promontory. There stand the ruins of a castle, dating back to 1405. It also supports a lighthouse, built in 1903 to warn sailors of the rock's presence. TC

SICCAR POINT—HUTTON'S UNCONFORMITY

EAST LOTHIAN, SCOTLAND

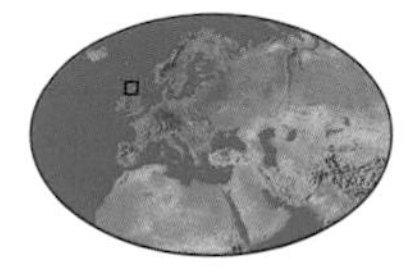

Age: 80 million years

Hutton's first visit to Siccar Point: 1788

Siccar Point, on the coast of Berwickshire in Scotland, has become one of the most famous sites in the history of geology. The rock formations here (and on the Isle of Arran) led James Hutton (1726–1797) to our understanding of the age of the planet and the immense lengths of time involved in geological processes. Before Hutton, it was widely thought that Earth was less than 6,000 years old and that most rocks had been laid down during the Creation. Hutton, following his first visit in 1788, recognized that geological processes created layers of rock and sediment and that breaks (or "unconformities") were caused by erosion or lack of deposition over long time periods. Unconformities are geological features where rocks of differing ages are found one above the other, often with the layers at different angles. At Siccar Point, the gray vertical beds are slates of Silurian age that were deposited, tilted, and eroded, and then overlaid by gently dipping Upper Devonian sandstone layers. **RC**

THE GREY MARE'S TAIL

DUMFRIES AND GALLOWAY, SCOTLAND

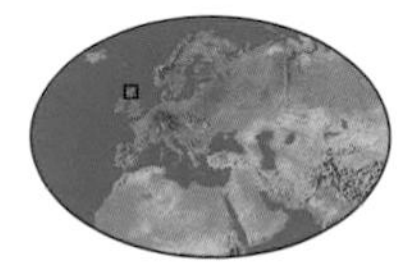

Height of Grey Mare's Tail Waterfall: 295 ft (89 m)

Type of valley: hanging valley

The Grey Mare's Tail is the evocative name for a spectacular waterfall in the rugged hills northeast of Moffat in Dumfries and Galloway. It flows from one of Scotland's highest lochs, Loch Skeen, over a 295-foot- (89-meter-) high cliff. The "tail" is a hanging valley, in which tributary valleys have eroded more slowly than the floor of the main valley, creating a difference between depths. The tributaries are left high above the main valley, hanging on the edges, their rivers and streams entering the main valley by either a series of small waterfalls or a single large fall. The Grey Mare's Tail is a fine example of this—the lower valley was scoured and deepened by glaciers during the ice age, leaving a steep cliff between the higher loch and the valley floor.

Late winter and early spring are the best times to view the falls, when water is flowing at its greatest volume. The frothy white turbulence of the falls as it tumbles down and around the contours of the cliff has inspired many a poet, including Sir Walter Scott, who described it as "white as the snowy charger's tail." The area also has the richest collection of rare upland plants in southern Scotland. **JK**

FARNE ISLANDS

NORTHUMBERLAND, ENGLAND

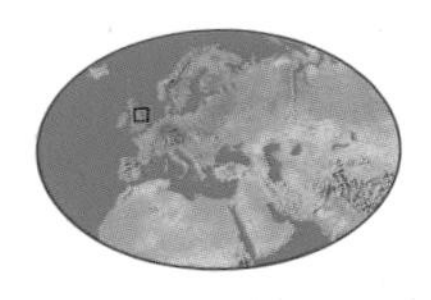

Age of Farne Islands: 280 to 345 million years

Number of islands: 28

A total of 28 islands make up the group known as the Farne Islands. They originally formed part of the mainland, but a post-glacial rise in sea level, combined with the effects of marine erosion, isolated them from the coast. They form the end of a large intrusion of volcanic rock known as the Great Whin Sill, which runs through the northeast of England. This volcanic dolerite rock is extremely hard and only gradually worn away by the the constant pounding of the North Sea. Although the dolerite rock is up to 100 feet (30 meters) thick in places, fissures within the rock have over time been enlarged into deep cracks, such as the Chasm and St. Cuthbert's Gut on Inner Farne. During a storm, the sea rushes up through these chasms spurting impressive jets of water 100 feet (30 meters) into the sky.

The Farne Islands have strong associations with Celtic Christianity. The Anglo-Saxon monk St. Cuthbert lived in solitude on Inner Farne during the seventh century C.E., where his reputed gift of healing brought pilgrims

to the island from all over the Kingdom of Northumbria. In fact, the name Farne Islands may well derive from "Farena Ealande," or "Island of the Pilgrims." Cuthbert had a great love of nature, especially of birds and seals, which were often his only companions on the lonely island.

Another island with religious associations is Lindisfarne, also known as Holy Island, which is linked to the mainland by a narrow causeway and is the only inhabited island in the group. A monastery was founded here in the seventh century where the famous *Lindisfarne Gospels*, a lavishly illustrated religious text, was created. In 793, the monastery was sacked in the first major raid by Vikings on the English coast.

Today, the Farne Islands are most famous for their wildlife. They provide a breeding site for some 20 species of seabird, including eider duck, fulmar, kittiwake, tern, guillemot, puffin, razorbill, oystercatcher, ring plover, and rock pipit, as well as an important colony of gray seals. A recent survey showed that there are over 70,000 pairs of seabirds, around half of which are puffins. TC

BELOW *The Farne Islands provide an excellent breeding site for puffins and other seabirds.*

THE PINNACLES

NORTHUMBERLAND, ENGLAND

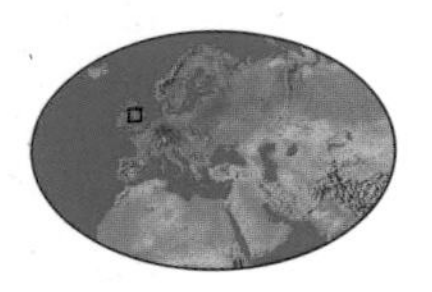

Age of the Pinnacles: 280 to 345 million years

Height: 66 ft (20 m)

The Pinnacles are part of the Farne Islands, the last outcrop of the Great Whin Sill, a layer of dolerite (a form of volcanic rock rather similar to basalt, but with finer rock crystals) found across the width of England from Cumberland eastward into Northumberland, covering some 80 miles (130 kilometers) or more. The sill begins in the west with the headwaters of the River Tyne and finishes in the Farne Islands off the Northumbrian coast.

The Pinnacles are located just offshore from Staple Island. Shaped like giant molars, they stand proud in the rough North Sea and provide a refuge for a number of seabirds including puffins, guillemots, razorbills, sandwich terns, common terns, roseate terns, and arctic terns.

Rising like ancient pillars from the sea, the pinnacles are a fine example of the columnar nature of dolerite rock. Weathering has worn away the sedimentary layers, leaving behind this strong volcanic rock to battle against the elements. Another example is the Stack, which rises 60 feet (18 meters) from the sea, just beyond the southern cliffs of Inner Farne. TC

WAST WATER

CUMBRIA, ENGLAND

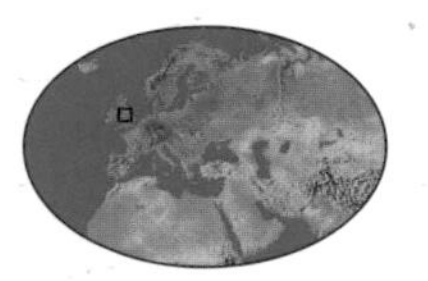

Depth of Wast Water: 258 ft (79 m)

Height of Scafell Pike: 3,210 ft (978 m)

Height of Great Gable: 2,949 ft (899 m)

England's Lake District was carved out of granite mountains by mighty glaciers during the last ice age. Ten thousand years ago, the glaciers retreated and the meltwaters accumulated in a hollow to a depth of 258 feet (79 meters) creating Wast Water, the deepest lake in England.

Samuel Coleridge described Wast Water as a "a marvellous sight." Located in Wasdale Valley, it is a remote and rugged place, surrounded by mountains, including England's highest mountain, Scafell Pike, and the magnificent Great Gable with its distinctive rock pinnacle, known as Napes Needle. From Illgill Head, a wall of scree 1,800 feet (600 meters) high plunges down into slate-gray water, right to the bottom of the lake. With an average of 120 inches (300 centimeters) of rain per year, this is one of the wettest places in Britain.

At the head of the lake is the small village of Wasdale Head, once a favorite haunt of Walter Haskett Smith, the "father of British rock climbing." He put the sport on the map when he climbed Great Gable and Napes Needle in 1886. He repeated the climb many years later, at the age of 76, in front of an audience of 300 climbers. MB

RIGHT *The rugged wilds that surround Wast Water.*

BOWDER STONE

CUMBRIA, ENGLAND

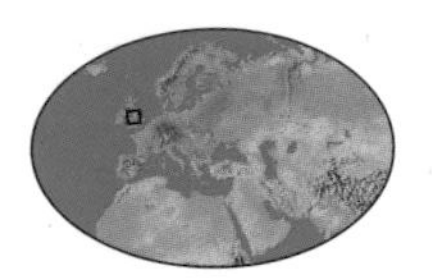

Age of Bowder Stone: around 452 million years

Weight of stone: 2,000 tons

Circumference of stone: 89 ft (27 m)

The Bowder Stone looks like a house balanced on one corner. While geologists cannot agree on its origin, they do agree that the mighty Bowder Stone of Borrowdale, Cumbria, is perhaps the largest single block of rock in the world and may be one of the oldest. During the ice age, gigantic glaciers crashed and scraped their way across the British landscape. They left behind evidence of their presence, including U-shaped valleys like Borrowdale, in the Lake District.

Perched precariously on its narrow base, the Bowder Stone is a huge mass of andesite lava from the Ordovician age that lies below the slopes of King's How between the villages of Grange and Rosthwaite. A glacier may have carried it from Scotland and deposited it in its current location. These rocks are known as "erratics" because they come from the same source as the glacier and do not normally match local geology. Another plausible explanation is that the stone crashed down from Bowder Crag after a massive rockfall at the end of the last glaciation—between 10,000 and 13,500 years ago. Visitors can climb on top of this huge rock via a wooden ladder. TC

HIGH FORCE WATERFALL

COUNTY DURHAM, ENGLAND

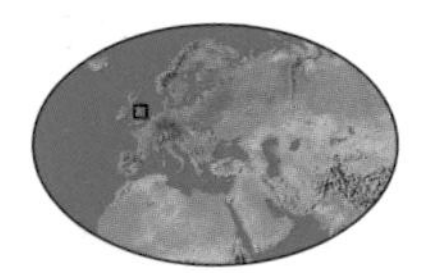

Height of High Force Waterfall: 70 ft (21 m)

Length of River Tees: 70 mi (113 km)

Length of the Pennine Way: 250 mi (402 km)

Found at the end of a lovely woodland walk, High Force has often been described as England's highest waterfall. It is certainly one of England's most spectacular and impressive falls. Although the waters squeeze through a small 10-foot (3-meter) gap, the falls are surprisingly powerful.

From its birth on the eastern slope of Cross Fell, the River Tees gains momentum and volume until it reaches the Great Whin Sill escarpment. It then cascades over a fault in this basalt intrusion and thunders down a 70-foot (21-meter) drop to the pool below. Its appearance is most dramatic in the fall and winter when the sound emitted by the falls can be deafening.

The waterfall can be seen from the Pennine Way footpath, a pretty woodland walk with several resting points along the way. Alternatively, one can view the falls from its base. This is a popular access spot for whitewater kayakers who shoot down a 2-mile (3.5-kilometer) stretch of the river. The woods on the southern side of the river are part of the Moor House-Upper Teesdale National Nature Reserve and a preserve of the most extensive juniper woods in England. CS

GAPING GILL

YORKSHIRE, ENGLAND

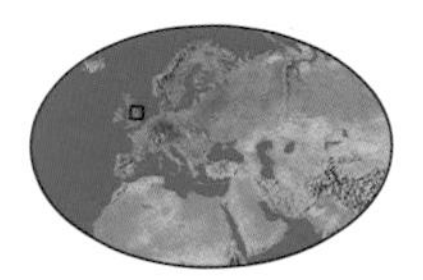

Length of Gaping Gill cave system: 42 mi (67 km)

Height of subterranean waterfall: 365 ft (111 m)

Age of limestone: 300 million years

Not for the fainthearted, Gaping Gill is the deepest shaft of one of England's largest underground cave systems. The huge chamber has been carved out over many centuries by the waters of Fell Beck, which have eaten away at the limestone rock, creating a subterranean waterfall twice the height of Niagara Falls. Unless you are a very experienced caver, the only way to get inside Gaping Gill is to attend a "winch-meet," when local cavers lower visitors down in a metal cage on the end of a cable.

This vast cave system has five other entrances—Bar Pot, Flood Entrance Pot, Stream Passage Pot, Disappointment Pot, and Henslers Pot—but these should only be used by experienced cavers.

Whether or not you go inside Gaping Gill, the walk up from Clapham to the cave entrance, passing through a steep limestone gorge amid the beautiful scenery of the Yorkshire Dales, is worth the effort at any time of the year. Nearby there are other interesting caves (such as Ingleborough Cave) that are easier to visit, along with other unique limestone landscapes, such as the strangely weathered, lunar "limestone pavement," which is home to many rare plants. CC

JINGLE POT

YORKSHIRE, ENGLAND

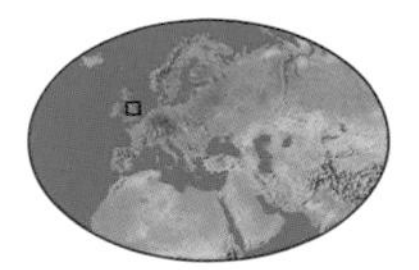

Length of Jingle Pot pothole: 0.6 mi (1 km)

Total depth: 220 ft (67 m)

The Yorkshire Dales are a series of glaciated valleys carved into the upland area of the central Pennines, the central mountain ridge of England that extends from the north Midlands up to Scotland. The mountains are lower than those of the Lake District and show fewer effects of glaciation. The peaks are more rounded with no sharp ridges, and the valleys or dales are more open. The region is one of the major karst areas in Britain, and there are many spectacular limestone features including reef knolls (conically shaped and fossil-rich hills that formed as coral atolls in the shallow waters of an ancient prehistoric sea). These can be found on Scosthrop Moor above Settle. The carboniferous limestone is porous, and the hills of the Yorkshire Dales are riddled with limestone caves and potholes.

Jingle Pot is one of many such potholes in the Yorkshire district of West Kingsdale, in the northern Dales. Its name derives from North Country dialect and Middle English, and means "the pothole with a tinkling, rattling noise." Cavers who belay from a tree that overhangs the shaft are likely to hear the unnerving sound of creaking as they climb back up to the entrance. TC

BRIMHAM ROCKS

YORKSHIRE, ENGLAND

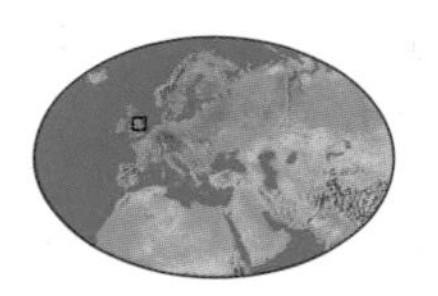

Age of Brimham Rocks:
320 million years

Height of Brimham Moor:
1,000 ft (300 m)

The Idol, the Boat, the Dancing Bear, the Turtle, the Smartie Tube—these are the names of some of the weird millstone grit rock stacks in Nidderdale, Yorkshire. Known as Brimham Rocks, these are scattered over some 50 acres (20 hectares) on Brimham Moor. Some are said to be associated with druids and even the Devil himself.

The creation of these rocks began in the granite mountains of northern Scotland and Norway. Approximately 320 million years ago, an enormous river sluiced grit and sand from these mountains, forming a huge delta that covered half of the area of Yorkshire today. Layers of this grit and sand, along with rock crystals of feldspar and quartz, built up to form a tough sandstone known as millstone grit. Between 80,000 and 10,000 years ago, during the Devonian glaciation, glaciers eroded these rocks into the bizarre shapes on view today. Their tiny plinth-like supports were caused by fluvial sandblasting at lower levels which wore away the softer layers of rock. Visible layers in the rock are known as cross bedding, caused by ripples in the direction of the river's current, downstream or downwind during periods of erosion and deposition. TC

BRIDESTONES

YORKSHIRE, ENGLAND

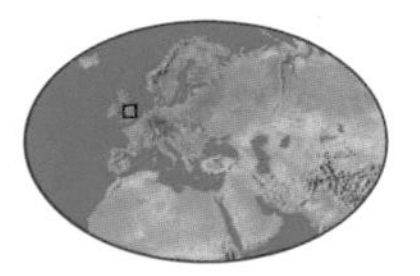

Length of Bridestones Nature Trail:
1.5 mi (2.4 km)

Rock type: sandstone

Vegetation: moorland with heather, herb-rich meadows, ancient woodland

To look at the strange rocks called the Bridestones on England's North Yorkshire Moors, you might imagine a giant must have abandoned a game of skittles there. The rocks are formed of sandstone laid down in the Jurassic period, about 180 million years ago. At the time, dinosaurs ruled Earth, the climate was more tropical, and North Yorkshire was lying under a shallow sea. Over the years, the sand that settled on the sea floor was compressed into sandstone. The large layers (bedding planes) in the Bridestones reveal how the deposited sand of underwater dunes was frequently disturbed and eroded by storms, creating layers of cut-off sand dunes. When the sea level dropped, the elements battered the freshly emerged sandstone. The weak bedding planes eroded easily, creating the sandwich-like layers and sculpting the huge sandstone outcrops. Sand grains lashed the rock bases, weathering the soft layers into the distinctive mushroom shape of many of the Bridestones seen today.

Bridestones Moor is a National Nature Reserve, with wild moorland and ancient woodland estimated to date from the end of the last ice age. TC

THE ROACHES

DERBYSHIRE / STAFFORDSHIRE, ENGLAND

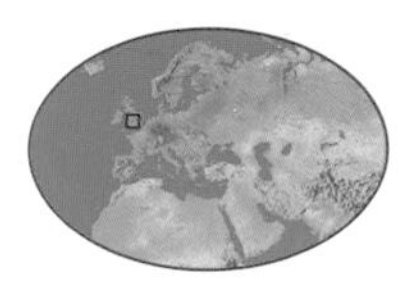

Age of the Roaches: 350 million years
Rock type: gritstone (sandstone)
Height: 100 ft (30 m)

The Roaches form a gritstone escarpment of unusually shaped rocks that mark the southwestern border of the Peak District National Park. The ice age and thousands of years of weathering have worn the rugged rocks into a fantastic eye-catching assemblage. The Roaches are located between Leek, in Staffordshire, and Buxton, Derbyshire. They consist of two jagged ridges—a Lower and Upper Tier—connected by a set of rock steps. They were created 350 million years ago when a shallow sea allowed sand and grit to build up over a coral reef that once covered this area. These sediments were compressed over time into solid rocks that are free of faults and natural weaknesses—ideal for masons who can work the rock in any direction and rock climbers who know the rock won't fail. Indeed, the Roaches are one of the most popular rock-climbing destinations in Great Britain, with over 100 routes to choose from. A tiny cottage has been hewn into the Lower Tier, to serve as a climbing hut.

The Roaches are in a fine position, with superb views of the Cheshire Plain and the Peak District. The surrounding moorland has some excellent walking territory. JK

THE WREKIN

SHROPSHIRE, ENGLAND

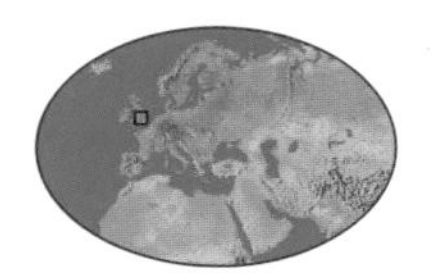

Height of the Wrekin: 1,300 ft (396 m)

Age: 566 million years

Shropshire's most famous landmark, the Wrekin, is a majestic hill that rises 1,300 feet (396 meters) above the surrounding patchwork of fields. From the summit, where there are the remains of an Iron Age hill-fort, one can reputedly see 15 counties. There are numerous legends and fables regarding its creation, the most famous of which tells how the hill came into being when a giant dumped his shovelful of earth. Its true origins were no less dramatic. The hill is actually made of a mass of molten rock and ash that spewed from the earth during the late Precambrian period. At this time, Shropshire was part of an extremely volatile region that lay beneath a shallow sea. Massive earthquakes formed major faults in the planet's crust, one of which created the volcano that gave birth to the Wrekin. The location of the vent from which the lava flowed, however, is unknown.

This whole area is geologically fascinating. The Wrekin lies close to the famous Church Stretton Fault, whose 560 million-year-old rocks are among the very oldest in the British Isles. Both features are today included within the Shropshire Hills Area of Outstanding Natural Beauty. TC

SEVERN BORE

GLOUCESTERSHIRE, ENGLAND

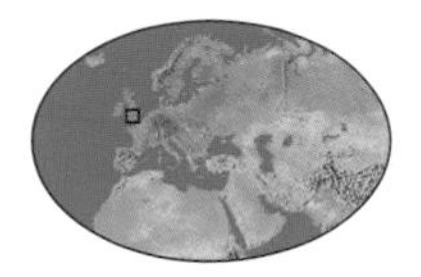

Height of Severn Bore: up to 10 ft (3 m)

Average speed: 10 mi (16 km) per hour

Distance traveled: 21 mi (33.8 km)

In spring and fall each year, the River Severn, located in Gloucestershire, is the focus of a truly spectacular natural phenomenon—a surge wave called the Severn Bore. This can appear as a glassy swell, a monstrous, breaking wave, or even a moving hole. It tears upstream, against the natural current, at speeds of up to 13 miles per hour (21 kilometers per hour). The swell has been known to reach heights of nearly 10 feet (3 meters) and is strong enough to sweep cattle and sheep away from the banks. It transforms the peaceful river into an inland surf zone, and each year dozens of surfers compete to see who can ride it the farthest.

At up to 50 feet (15 meters) high, the tidal range of the River Severn in southern England is the second highest in the world. As the tide rises, it meets hard-rocked river banks at Sharpness. These restrict the movement of the water and this, combined with the ridges of the river bed, holds up the water and stops it from flowing forward. A wall of water then starts to form that is eventually funnelled into the Severn estuary. As the River Severn gets shallower and narrower, the wall of water gathers speed and becomes much larger, forming a large wave or bore. It races upstream for more than two hours, traveling some 21 miles (34 kilometers) from Awre to Gloucester. It passes Avonmouth, where the river is approximately 5 miles (8 kilometers) wide, then Chepstow and Aust, Lydney and Sharpness—where it is approximately 1 mile (1.6 kilometers) wide—and by the time it reaches Minsterworth, where the river is just 330 feet (100 meters) across, the bore can be as high as 9 feet (2.7 meters).

The largest bores occur one to three days after a new or full moon, and the most impressive ones are during the spring tides. Folklore has it that those born when the tide is coming in will be lucky in life, and that sick people are more likely to die as the tide ebbs. The Severn Bore is one of the biggest bores in the world—there are as many as 60 bores across the globe, including on the Seine, Indus, and Amazon rivers. However, by far the biggest bore is the Ch'ient'ang'kian (Hang-chou-fe) in China. At spring tides the wave there attains heights of up to 25 feet (7.5 meters) and a speed of 14 to 17 miles per hour (24 to 27 kilometers per hour). The noise it creates is so loud that people can hear it approaching along the river from as much as 14 miles (22 kilometers) away. TC

Annually, the River Severn, in Gloucestershire, is the focus of a truly spectacular natural phenomenon—a surge wave called the Severn Bore. This can appear as a glassy swell, a monstrous, breaking wave, or even a moving hole.

CHEDDAR GORGE

SOMERSET, ENGLAND

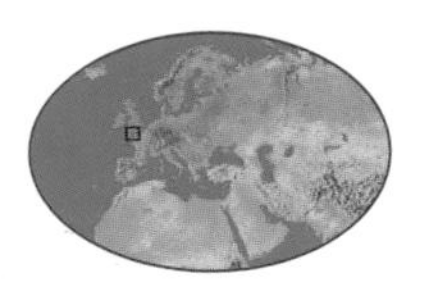

Age of Cheddar Gorge: 18,000 years
Age of rock: 280 to 340 million years
Depth of gorge: 370 ft (113 m)

Cheddar gorge is a narrow, steep-sided limestone canyon slashed through the Mendip Hills near the famous cheese-making town of Cheddar. It is the largest gorge in Great Britain, extending for 3 miles (5 kilometers) and dropping to a maximum depth of 370 feet (113 meters).

Though the rocks of the gorge date all the way back to the Carboniferous period, from about 280 to 340 million years ago, the gorge itself was formed only around 18,000 years ago toward the end of the last ice age. As temperatures grew warmer, the glaciers covering much of Britain's surface began to melt, releasing huge volumes of meltwater that eroded the soft limestone into the shapes we see today. The winding road that now runs through the center of the gorge marks the position of that ancient river.

This water erosion also formed a number of caves within the gorge. The two main ones are Cox's Cave and Gough Cave. The first was discovered during a quarrying operation in 1837 and consists of seven small but beautiful grottos that are joined by low archways. Gough's Cave, which was not found until 1893, contains the Fonts, a series of stalagmite gours or dams rising up tens of feet into the hill. Britain's biggest underground river, Cheddar Yeo, can be viewed just below the entrance to Gough's Cave, a remnant of the meltwater that once carved out the caves.

The Cheddar Caves have provided shelter for people throughout history. There is evidence that human activity took place here during the Upper Late Palaeolithic period, or Stone Age. Indeed, Britain's oldest complete human skeleton—the 9,000-year-old Cheddar Man—was found in the Cheddar caves in 1903. The Cheddar Man Museum, which has displays on the find and the human history of the gorge, stands opposite Gough's Cave.

Gough's Cave contains the Fonts, a series of stalagmite gours or dams rising up tens of feet into the hill. Britain's biggest undergound river, Cheddar Yeo, can be viewed just below the entrance to Gough's Cave, a remnant of the meltwater that once carved out the caves.

Today, the gorge is an extremely popular attraction and has been declared a site of special scientific interest due to its unique geology and the rare horseshoe bats that live here. Some 300,000 people come here every year to explore its craggy cliffs and spooky caves. Physically fit visitors can ascend from the floor of the gorge to the rim by climbing the 274 steps of Jacob's Ladder. Near the top is a lookout tower, from where there are great views out over the gorge and toward the coast. TC

RIGHT *The steep limestone cliffs of Cheddar Gorge.*

WOOKEY HOLE CAVES

SOMERSET, ENGLAND

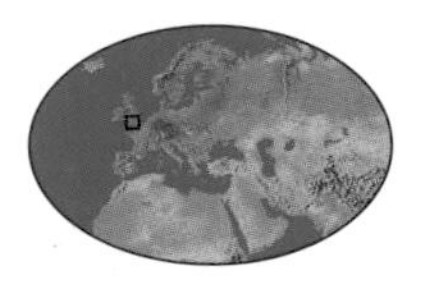

Age of limestone: 400 million years

First human habitation: 50,000 years ago

Hyena Den, Badger Hole, and Rhinoceros Hole are the names of caves in the complex of Wookey Hole in Somerset, southern England. The names derive from some of the tropical and ice-age animals that have been unearthed in the caves.

About 400 million years ago, a large ocean covered the Mendips. When the microscopic creatures that swam these seas died, their shells fell to the ocean bed and turned into calcium carbonate (limestone), which later hardened into rock. When falling sea levels exposed the rock, rainwater dissolved the weaknesses and widened cracks. Deep underground, the River Axe broke through fissures in the blocks of limestone and enlarged caverns.

The dry caves later offered a safe habitat, with a constant temperature of 52°F (11°C). The first human lodgers came some 50,000 years ago, hunting bears and rhinoceros with stone weapons. Archeologists believe that one cave, Hyena Den, was alternately occupied by hyenas and man between 35,000 and 25,000 B.C.E. In the Iron Age, Celtic farmers lived near the cave entrance for more than 600 years.

Two thousand years ago, the Romans settled here, built roads, and exploited the rich mineral resources of the Mendip Hills. Not a lot is known of the ensuing history of the caves until the 18th century, when the poet Alexander Pope visited and had several stalactites shot down as souvenirs. Today, the caves are home to horseshoe bats, moths, cave spiders, frogs, eels, and freshwater shrimp.

Wookey Hole is the birthplace of British cave diving, and Cathedral Cave is one of the most famous in British caving history. Known as Chamber 9 by experienced cavers, this was to become a diving base for all future explorations of the cave system. It is 100 feet (30 meters) high, with waters over 70 feet (21 meters) deep. Its vast walls glow a colorful red with iron oxide and shine with "flowstone" stalactite formations.

Lurking in the shadows of the cave wall is the witch of Wookey Hole—a limestone rock formation that is eerily shaped like a witch, complete with crooked nose and protruding chin. Locals in the 18th century believed the witch was an evil old woman—and that a local monk sprinkled her with holy water, turning her into stone. TC

When Alexander Pope visited the caves in the 18th century, he had several stalactites shot down as souvenirs. Today, the caves are home to horseshoe bats, moths, cave spiders, frogs, eels, and freshwater shrimp.

RIGHT *The impressive Cathedral Cave, with its luminescent red walls.*

SEVEN SISTERS

EAST SUSSEX, ENGLAND

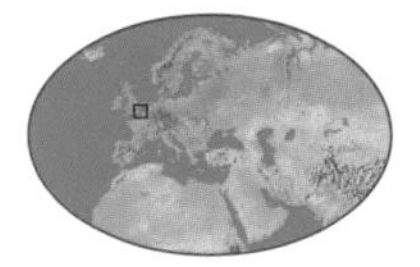

Age of the Seven Sisters: 60 to 130 million years

Height of Haven Brow: 253 ft (77 m)

The Seven Sisters is the name given to a line of majestic undulations in the chalky cliffs on England's south coast. The calcareous parts of tiny marine algae and seashell fragments formed white chalk ridges under the sea some 60–130 million years ago. Today's cliffs occur where the chalk ridge of the South Downs of Sussex meets the English Channel. Ancient rivers cut valleys into the chalk, creating the magnificent Seven Sisters: Haven Brow at 253 feet (77 meters) is the tallest. Next to her stand Short Brow, Rough Brow, Brass Point, Flagstaff Point, Baily's Brow, and Went Hill Brow. The sea constantly chafes the chalk, undercutting the cliffs and leading to regular rockfalls. As a result, the cliff faces are continually refreshed, revealing a never-ending treasure of fossils, some of them flawless specimens. Fossil hunters from all over the world search the shingle and chalk below the cliffs for brachiopods, bivalves, and echinoids. The cliffs recede about 12 to 16 inches (30 to 40 centimeters) a year.

The number seven occurs frequently in the boundary lists of Saxon charters (such as the old town name of Sevenoaks). In the case of the Sisters though, seven is misleading as there is an eighth cliff, the last in the row, the smallest, unnamed, ignored little sister. TC

LULWORTH COVE

DORSET, ENGLAND

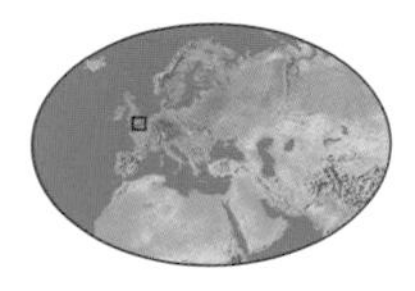

Rock type: limestone

Visitors per year: 1 million

The incredible rock formations and stunning scenery around Lulworth Cove bring visitors from all over the world. The oyster-shaped cove was created over thousands of years, as the sea broke through the Purbeck and Portland limestone cliffs and began to erode the softer clays and chalk behind. On one side of the cove, the Middle and Upper Purbeck rock strata are contorted and folded into the "Lulworth Crumples," one of which is in the back of the cliff at adjacent Stair Hole.

Today, Lulworth Cove itself forms a natural harbor that sits at the eastern end of the so-called "Jurassic Coast." The character Troy swam out from its shingle beach in John Schlesinger's film of Thomas Hardy's *Far from the Madding Crowd.* Fringing the cove is the Dorset coast path, and on the stretch from Ringstead to Lulsworth is Burning Cliff, where the oil shales in the ground actually burned slowly for many years. At Bacon Hole, to the east, there is a fossil forest, one of the best in the world. Here, the remains of late Jurassic and early Cretaceous coniferous trees are rooted in a paleosol (ancient soil) known as the Great Dirt Bed. Lulworth is about 10 miles (16 kilometers) east of Weymouth. CC

DURDLE DOR

DORSET, ENGLAND

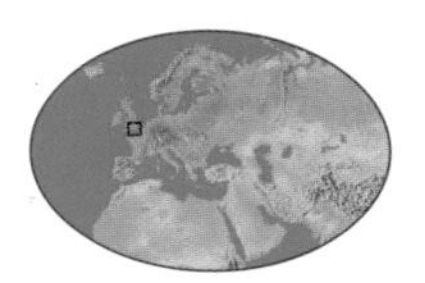

Cliff height at Durdle Dor: 330 ft (100 m)

Rock type: Portland stone

In 1792 on a visit to Durdle Dor, playwright John O'Keefe wrote: "Here I stood and contemplated with astonishment and pleasure this stupendous piece of Nature's work."

Durdle Dor is truly one of nature's marvels, and one of the most photographed subjects along Dorset's Jurassic Coast. This giant limestone arch straddles the sea at the eastern end of the Durdle Dor Cove. Carved by the pounding southwesterly waves, the softer rocks have eroded, leaving the more resistant Portland stone standing firm. These rocks were laid down between 135 and 195 million years ago, in the Jurassic period when southern England was under a tropical sea. The name Durdle Dor or Durdle Door has been used for 1,000 years. Durdle is derived from the Old English word "thirl," meaning "to pierce" or "have a hole," and Dor reflects its door-like shape. During sunset the sun shines through this arch, illuminating its inner walls. CS

CHESIL BEACH

DORSET, ENGLAND

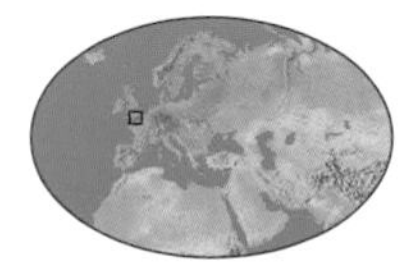

Length of Chesil Beach: 18 mi (29 km)

Height of Shingle ridge: 60 ft (18 m)

Chesil Beach is a shingle barrier ridge, with sheltered lagoons behind, stretching 18 miles (29 kilometers) along England's southern coastline at Dorset, from Bridport Harbor (West Bay) to Chesil Bay in the Isle of Portland, where it connects to the mainland. It is one of the best-studied beaches in the world for the volume, type, and grading of its pebbles. At the Portland end, they are the size of hens' eggs, yet 15 miles (25 kilometers) away at West Bay the pebbles are the size of peas. In between, they decrease steadily in size so perfectly that fishermen beaching at night can tell where they have landed just by the grade of stones underfoot. Indeed, the name comes from the Old English word *ceosol*, meaning shingle.

The shingle can reach up to 60 feet (18 meters) high and faces the storm waves driven by the prevailing southwesterly winds. It is 98.5 percent flint and chert pebbles, while the remainder is quartzite, quartz, granite, porphyry, metamorphics, and limestone. The origin of the beach is the subject of debate. A recent theory is that it came from landslides in East Devon and West Dorset during the last ice age. As sea levels rose, erosion cut into the debris, sending a huge "pulse" of material, by longshore drift, east. The beach moved onshore, to its present position, some 4,000 to 5,000 years ago. TC

JURASSIC COAST

DORSET / DEVON, ENGLAND

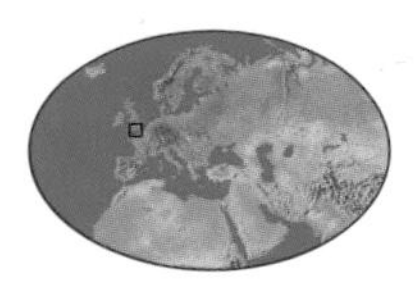

Length of Jurassic Coast:
95 mi (150 km)

Age of Fossil Forest:
144 million years

Visitors per year:
14 million per year

In 1811, twelve-year-old Mary Anning discovered an unusual fossil skeleton on the beach near Lyme Regis, Dorset, on the south coast of England. The fossil turned out to be the first complete skeleton of an ichthyosaur (Greek for "fish lizard"), a type of giant marine reptile, ever found. These strange creatures, which looked a bit like dolphins, hunted the seas during the age of the dinosaurs but became extinct around 90 million years ago, some 25 million years before the demise of the dinosaurs.

The beautiful coastline where Mary lived is now known as the Jurassic Coast World Heritage site. It stretches for 95 miles (150 kilometers) from Exmouth in Devon to Studland Bay in Dorset. This coast is the only place in the world that displays unbroken evidence of 185 million years of Earth's history. Its great wave-cut cliffs provide an almost continuous sequence of Triassic, Jurassic, and Cretaceous rock formations spanning the Mesozoic era, when dinosaurs ruled Earth. Since Mary's initial discovery, a huge range of internationally important fossils—vertebrate and invertebrate, marine and terrestrial—have been found along the Jurassic Coast. There are clues here to the presence of deserts and tropical seas, an ancient fossil forest, and dinosaur-infested swamps. The oldest rocks are in the west, around Exmouth and Sidmouth. Progressively younger rocks form the cliffs to the east.

The coast of East Devon is the richest mid-Triassic reptile site in Britain, while the Jurassic period is represented between Pinhay in Devon and Kimmeridge in Dorset. This area in particular is a paleontology hot spot, providing a complete record of every stage of the Jurassic era. The fossils of new species are still being found there today.

This coast is the only place in the world that displays unbroken evidence of 185 million years of Earth's history. The area is a paleontology hot spot, providing a complete record of every stage of the Jurassic era. Fossils of new species are still being found.

In the Purbeck Formation, the Jurassic to early Cretaceous terrestrial sequence is one of the finest in the world. At the far eastern end of the site, standing just off the coast, are Old Harry Rocks, a set of large chalk sea stacks that represent the area's youngest Mesozoic rocks.

Mary Anning's discovery shot her to fame and she later became one of the most important fossilists of the 19th century. TC

RIGHT *The beaches of the Jurassic Coast yield a wealth of dinosaur fossils.*

OLD HARRY ROCKS

DORSET, ENGLAND

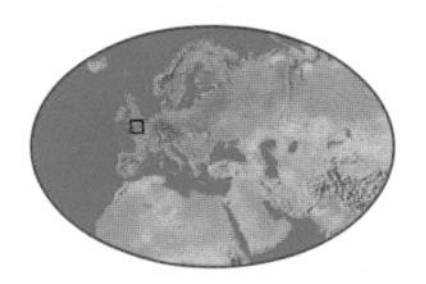

Rock type: chalk

Wildlife: peregrines and seabirds

Geological features: sea stacks, stumps, natural arch

The chalk cliffs near the village of Studland end abruptly, similar to the edge of a tall building. As well as these chalk cliffs, the area features beautiful promontories, sea stacks, and natural arches. Erosion has separated Old Harry Rocks from the mainland. Stacks are formed when the sea attacks a fracture in the chalk, forms a cave, and then gradually wears the cave through into an arch in the headland. Later, the arch collapses to leave a sea stack. Over time, the sea undermines the base of the stack, until it collapses into the sea—a stack known as Old Harry's Wife fell in 1896.

A major fault, the Ballard Down Fault, lies in this stretch of cliffs. South of the fault the chalk is vertical; to the north, however, it has been laid down almost horizontally.

Old Harry, the name of an isolated stack, is a medieval name for Satan, while the land on the clifftop opposite is called Old Nick's Ground—another nickname for the devil. At low tide it is just possible to walk along the foot of the cliff to Harry Rocks but there is also a pleasant clifftop walk from the village of Studland. TC

THE NEEDLES

ISLE OF WIGHT, ENGLAND

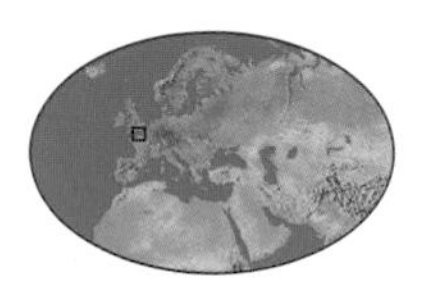

Height above sea level: 100 ft (30 m)

Height of lighthouse: 109 ft (33 m)

During a storm in 1764, Lot's Wife, a 120-foot (37-meter) needle-shaped chalk stack off the northwestern tip of the Isle of Wight, on the south coast of England, collapsed into the sea. The crash, it was said, was heard many miles away. Three other stacks withstood the storm and were named the Needles from that fourth, lost tapering pinnacle. The Needles, in Alum Bay, mark the end of a ridge of chalk that runs through the Isle of Wight. They rise about 100 feet (30 meters) above the sea, isolated as the soft chalk of the headland eroded away.

In Victorian times, the area attracted large numbers of visitors who traveled from the mainland by paddle steamer. The island is still a popular tourist destination today. The Needles have always been a danger to shipping, and a 109-foot (33-meter) lighthouse clings to the base of the most westerly rock of the Needles group. One of the dangerous reefs is the stump of the pinnacle that collapsed more than two centuries ago and can still be seen at low water. In early December 1897, Guglielmo Marconi set up his revolutionary wireless equipment in the Royal Needles Hotel, above Alum Bay, and sent the very first wireless transmission. TC

DARTMOOR NATIONAL PARK

DEVON, ENGLAND

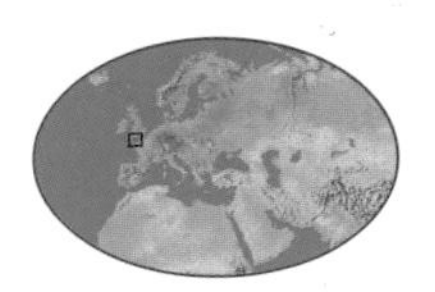

Area of Dartmoor National Park: 368 sq m (953 sq km)

Highest point: 2,038 ft (621 m)

Age: 295 million years

The bleak beauty of the Dartmoor National Park in Devon has inspired many artists and writers, perhaps most famously Sir Arthur Conan Doyle, whose Sherlock Holmes adventure *The Hound of the Baskervilles* was set on the moor. Although most of the land is now privately owned, open access is now allowed in most parts of the moor.

Hay Tor is one of the most impressive of more than 160 granite outcrops—spectacular landmarks that provide marvelous views over the surrounding countryside. Hay Tor overlooks an area from which stone was once quarried for London Bridge. Dartmoor ponies, which roam free on the moor, are often seen and very approachable. The area also has an abundance of archeological sites. There are numerous prehistoric standing stones, including rare "stone rows" at Merrivale, and there are remains of many Bronze Age villages at Grimspound and Hound Tor. CC

SNOWDON & SNOWDONIA

GWYNEDD, WALES

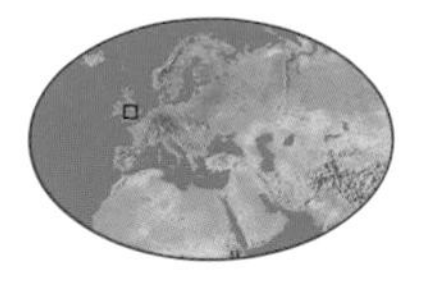

Local name: Yr Wyddfa

Height of Mt. Snowdon: 3,560 ft (1,085 m)

Established as a national park: 1951

Snowdonia is the second largest national park in England and Wales. With only 26,000 inhabitants, it retains much of its wildness. Its wildlife is one of its main attractions, including rare birds such as merlin and chough. Snowdonia also has its own unique species—a lily and a beetle. Most visitors come here for the grand scenery, excellent walking, and the challenge of climbing Snowdon itself. The mountain is all that remains of a volcanic crater that was once three times as high, but is still the highest peak in Wales. Those of a less active disposition can make the ascent on Snowdon's very own train service, which takes you to within 66 feet (20 meters) of the summit and drops you off at a café where you can enjoy a cup of tea as you admire the view.

There are a number of other mountains in the area that are just as spectacular and less visited: Carnedd Moel Siabod, Cader Idris, and the Rhinogs. The park also includes miles of coastline, wetlands, and upland oak forests. Harlech is one of the best places to stay, with a splendid castle and views of Snowdon. CC

SKOMER ISLAND

PEMBROKESHIRE, WALES

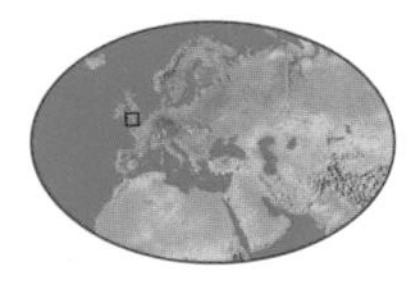

Length of Skomer: 2 mi (3.2 km)

Puffin population: approx. 6,000 pairs

Just off the southwest coast of Wales is the tiny island of Skomer. Gray seals and porpoises are a common sight on the short boat trip from Martin Haven on the mainland, but the real spectacle is on Skomer itself, with its thousands of seabirds, including kittiwakes, guillemots, razorbills, and puffins. In spring, Skomer is their breeding site. In early May the island is carpeted in bluebells, red campion, and many other species of wildflower. Slow worms (legless lizards) are common, and the island has its very own Skomer vole, a subspecies of the bank vole.

The most extraordinary aspect of Skomer is the summer influx of Manx shearwater. It is estimated that a third of the world's population of these oceanic birds breed on this tiny island, upward of 102,000 pairs, and when building their nests, they burrow into the soil. In the evening, visitors will hear the birds' eerie calls, sounding like banshees in the night. CC

WORM'S HEAD, GOWER PENINSULA

GLAMORGAN, WALES

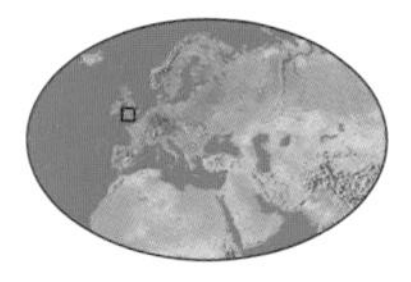

Length of Worm's Head: 1 mi (1.6 km)

Area of Gower Peninsula: 72 sq mi (188 sq km)

The Gower Peninsula, a designated Area of Outstanding Natural Beauty, pushes out into the Bristol Channel between the estuaries of the Loughor and Tawe rivers. Worm's Head is a 1-mile- (1.6-kilometer-) long limestone promontory at the westernmost tip of the peninsula, reached by a causeway that is covered by the sea at high tide, but is passable for two hours either side of low tide. It is possible to walk across and have five hours to explore, but people and sheep have been marooned in the past. Before attempting this, one should always check tide times. At high tide, the sea cuts off the Worm and only then does it truly resemble a wurm, the old English word for "dragon" or "serpent." From the Worm, there are views of the local seals and of Rhossili Bay, a sweeping golden arc, stretching to its northern tip at Burry Holm. CS

HENRHYD FALLS

POWYS, WALES

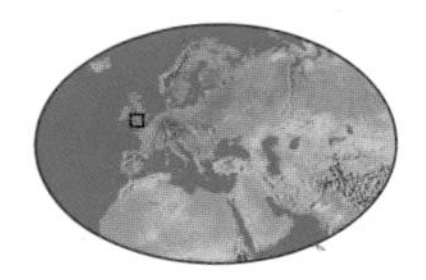

Height of Henrhyd Falls: 90 ft (27 m)

Rock age: 550 million years

Henrhyd Falls is a beautiful waterfall in a deep wooded gorge on the stream known as Nant Llech at Coelbren in the Swansea Valley. It has an unbroken drop of about 90 feet (27 meters), making it the highest waterfall in the Brecon Beacons National Park. Henrhyd Falls and the Nant Llech Valley are particularly impressive in cold winters when the ice is extensive—in exceptionally cold conditions the falls freeze solid. Usually there is a good flow of water, and it is possible to walk behind the waterfall itself. After heavy rains, the roar of the falls can be heard a distance downstream.

There are many other waterfalls in the south of the national park. The area within a triangle formed by the villages of Hirwaun, Ystradfellte, and Pontneddfechan is often referred to as "Waterfall Country." Numerous waterfalls are found along the rivers Nedd, Mellte, Hepste, and Pyrddin; the highlight of these being Sgwd Y Eira on the River Hepste. RC

PEN Y FAN AND THE BRECON BEACONS

POWYS, WALES

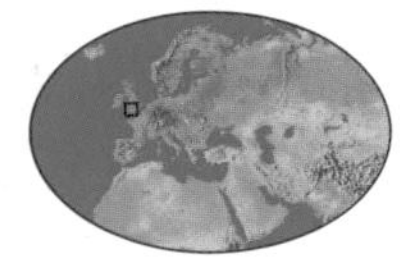

Area of Brecon Beacons: 519 sq mi (1,344 sq km)

Height of Pen y Fan: 2,907 ft (886 m)

Height of Corn Du: 2,863 ft (873 m)

The Brecon Beacons National Park was designated in 1957 and contains some of southern Britain's most spectacular upland formations. At the center of the park is a range of hills called the Brecon Beacons, including Pen y Fan, which is the highest mountain in South Wales. The national park extends almost 50 miles (80 kilometers) from Llandeilo in the west to Hay-on-Wye in the east, incorporating the popular tourist attractions of the Black Mountains, along with Black Mountain and Fforest Fawr.

Much of the park is composed of old red sandstone—its distinctive reddish-pink stone can be seen in many of the older buildings toward the east of the region. In the far west the sandstone gives way to carboniferous limestone, with the caves and waterfalls that are a feature of this rock type.

The Brecon Beacons are named after the Welsh town of Brecon and the ancient practice of lighting beacons on the mountains to warn of attacks by the English. Nowadays the central Brecon Beacons are a magnet for walkers. The summits of Pen y Fan, Corn Du, and Cribyn make up a beautiful ridge walk known as the Beacons Horseshoe. RC

RIGHT *The snow-covered hills of the Brecon Beacons.*

GIANT'S CAUSEWAY

COUNTY ANTRIM, NORTHERN IRELAND

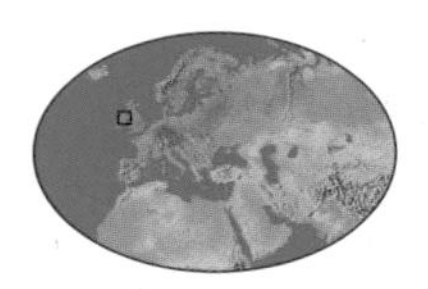

Column diameter: 15 to 20 in (38 to 50 cm)

Age: 60 million years

On the coast of County Antrim in Northern Ireland, the hexagonal columns of the Giant's Causeway resemble the ruins of some ancient, man-made monument, but they are totally natural. About 60 million years ago, when Europe and America began to separate, volcanic activity accompanied the rift. Molten basaltic lava poured out over what is now Northern Ireland and Scotland, forming the largest basaltic plateau in Europe. The basalt cooled and contracted—this process determining the crystal size—and the crystals have since been broken up by the ice age and weathered by the Atlantic Ocean. There are approximately 40,000 columns, each up to 6 feet (1.8 meters) high and comprised of piles of 14-inch (36-centimeter) thick basalt "tablets" fused together. The ice and the ocean have worked at the lines of weakness between tablets and volcanic events have occurred at various points leaving the stepped rock formation visible today. The causeway is like a flight of steps down to the sea. The neighboring bays of Port Noffer and Port Reostan have their own unique features which are also worth visiting. MB

BELOW *The unusual hexagonal columns of Giant's Causeway.*

GLENARIFF

COUNTY ANTRIM, NORTHERN IRELAND

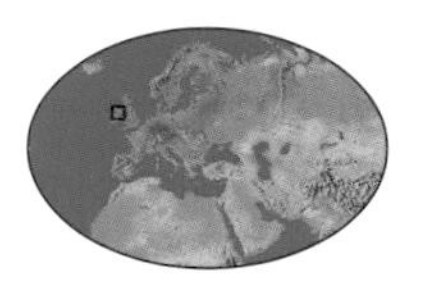

Length of Glenariff: 5 mi (8 km)

Area of Glenariff Forest Park: 2,928 acres (1,185 ha)

Depth of valley: 660 to 1,320 ft (200 to 400 m)

Glenariff, the Queen of the Glens, is considered to be the most beautiful of the nine glens of Antrim in Northern Ireland. The perfect U-shaped valley meets the sea at the small village of Waterfoot, a good starting point from which to explore the glen. For five miles (eight kilometers) inland, the steep mountains frame a succession of impressive views, among them the Ess-na-larach (or "mare's tail") waterfall, whose waters tumble down a precipitous gorge.

The area is notable for the diversity of industries it has supported. The rich glacial soils coupled with the valley's profile have produced an unusual farming practice: "ladder" farms each have an equal share of marshy flood-meadows in the valley floor, arable fields on the more accessible slopes, and rough grazing among the windswept hilltops. The discovery of iron ore gave rise to a mining and smelting industry, which put such a strain on the local roads that it prompted the construction of Ireland's first narrow gauge railway in 1873. Production of iron ceased in 1925; however, the railway allowed the area to flourish as a tourist attraction, and Glenariff's natural beauty continues to amaze visitors. NA

STRANGFORD LOUGH

COUNTY DOWN, NORTHERN IRELAND

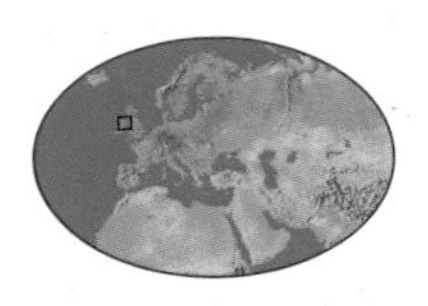

Area of Strangford Lough: 58 sq mi (150 sq km)

Depth of Lough: up to 150 ft (45 m)

Strangford Lough, situated on the east coast of County Down in Northern Ireland, is the largest sea lough in the British Isles and a site of international importance for wildfowl and wading birds. The lough is almost cut off from the sea by the Ards Peninsula, apart from a strait known as "The Narrows" that connects the lough with the Irish Sea at Portaferry. Viking invaders named the lough "Strang Fjord" after the treacherous tidal currents that flow through the Narrows at speeds of 9 miles per hour (14 kilometers per hour). The lough is surrounded by low rounded hills or "drumlins" formed by retreating glaciers—many of the lough's islands are the tops of "drowned" drumlins.

The extensive mudflats and sandflats at the northern end of the lough are important feeding areas for overwintering waterfowl. Islands in the lough provide breeding sites for colonies of several species of tern. The lough is also home to the largest breeding population of common seals in Ireland. Strangford Lough was designated a Marine Nature Reserve in 1995. RC

THE CALLOWS OF THE RIVER SHANNON

COUNTY OFFALY, IRELAND

Area of the Callows of the River Shannon: approx. 241,000 acres (100,000 ha)

Habitat: plant-rich summer watermeadows, winter river floodplains

Age of rock: 10,000 to 15,000 years

The Shannon is one of Europe's remaining unregulated rivers (rivers whose flows are not controlled by dams or weirs), and it forms a floodplain that supports Europe's richest system of water meadows. It runs through the central region of Ireland around the county of Limerick, where the river drops some 40 feet (12 meters) in 25 miles (40 kilometers). Too expensive to drain, the floodplain ("the callows" in Irish, from the Irish Gaelic *caladh*) is protected by a system of agricultural subsidies. Providing lush summer grazing for cows, the area also has Europe's densest population of the endangered corncrake. In winter it becomes a rich muddy buffet for millions of wading birds and ducks.

Though grazed, the callows have not been plowed, drained, re-sown, or artificially fertilized for around 1,400 years, creating a completely natural environment. Consequently, the area hosts the kind of wildflower community long vanished elsewhere in Europe. Some 216 plant species have been recorded here, including many orchids and grass communities where up to 10 species occur together. AB

BENBULBIN

COUNTY SLIGO, IRELAND

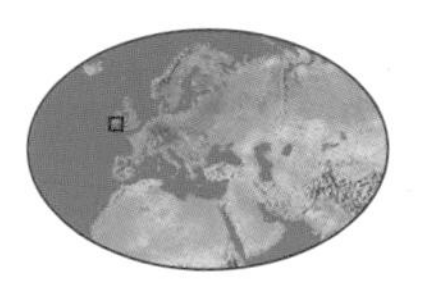

Height of Benbulbin: 1,360 ft (415 m)

Rock type: limestone

Pushing up above the stonewalled pastures of County Sligo's coastal plain in western Ireland is the impressive flat-topped mountain of Benbulbin. The slopes of this massive limestone plateau rise steeply at its base and then form precipitous cliffs near its crenulated top. It is a landscape that was revealed about 10,000 years ago when the glaciers of the last ice age melted, and since then Benbulbin has featured often in Celtic legend.

The warrior Diarmuid, for example, who eloped with Grainne, the girlfriend of giant Finn MacCool (of Giant's Causeway fame), was tricked by the giant into fighting with an enchanted boar on the mountain. He was killed when its tusk pierced his heart. In the 6th century, St. Columba (Colmcille) led 3,000 warriors into battle on Benbulbin's slopes. He was upholding his right to copy a book of psalms that he had borrowed from St. Finnian of Movilla. The mountain was also a favorite spot of Irish poet William Butler Yeats, who described it as "The Land of Heart's Desire." Sligo is often described as "Yeats' County"—he is buried in a graveyard in Drumcliff. MB

THE SKELLIGS

COUNTY KERRY, IRELAND

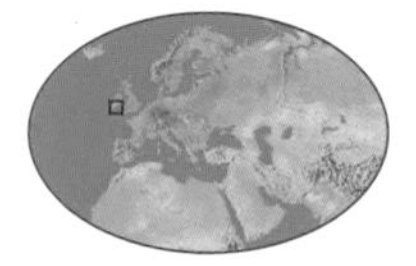

Height of Skellig Michael: 715 ft (218 m)

Area of Skellig Michael: 44 acres (18 ha)

Emerging from the surface of the Atlantic Ocean 7.5 miles (12 kilometers) southwest of Valentia, County Kerry, are the twin rock pyramids named Skellig Michael and Small Skellig. Skellig Michael, the larger of the two, rises 715 feet (218 meters) above the water, while the sharp descent of the rock face continues for 165 feet (50 meters) underwater before joining with the continental shelf.

Boat trips to the Skelligs are easily arranged. Those with stamina can climb the 600 steps on Skellig Michael that lead to a 7th-century Christian monastery, the remains of which are remarkably well preserved and warrant the islands' inclusion as a World Heritage site. Skellig Michael also features a lighthouse. The lives of the lighthouse keepers and their families who later inhabited the remote island would have been very similar to the spartan existence of those early monks.

Small Skellig hosts one of the largest gannet colonies in the world. One of the main attractions for the birds, apart from the protection from predators that the isolated cliffs provide, is the teeming abundance of fresh food in the waters below. The sheer walls and jagged spikes of the crevice-pocked sandstone come to life in a vivid display of marine biodiversity. NA

CLIFFS OF MOHER

COUNTY CLARE, IRELAND

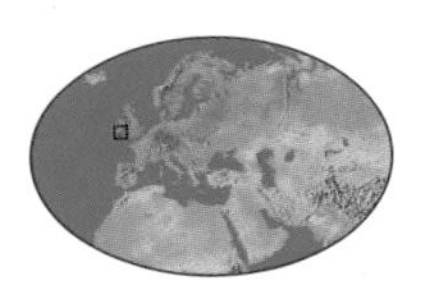

Height of Cliffs of Moher: 660 ft (200 m)

Age of cliffs: 300 million years

Rock type: sandstone

The Cliffs of Moher, a row of 660-foot- (200-meter-) high buttresses, defend a 5-mile (8-kilometer) stretch of the coast of County Clare against the might of the Atlantic Ocean. The cliffs are awesome, rising vertically from the sea, although they are not invincible. The limestone base was laid down 300 million years ago in a warm shallow sea, and this is overlain by a succession of sandstone layers. The sediments were shaped by major earth movements, but wind, rain, and sea salt are eroding the rocks so that sections occasionally fall into the sea. The ocean constantly pounds at the base. Approached from either end by a clifftop path, any person peering over the edge is likely to be drenched in salt spray blown up the cliff face by the strong westerly winds.

Like many geological features in Ireland, the cliffs are steeped in legend. At the southern end, Old Hag Mal was reputedly turned to stone and is now seated looking out to sea, and a herd of fairy horses was supposed to have leaped over the most northern section, known as Aill na Searrach, meaning "Cliff of Colts." There are no horses here today, but seabirds roost on the cliff face, and feral goats live along the narrow ledges. **MB**

THE BURREN

COUNTY CLARE, IRELAND

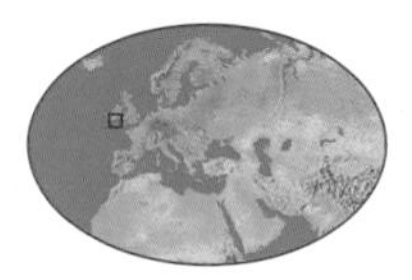

Area of the Burren: 116 sq mi (300 sq km)

Age: 360 million years

Rock type: limestone

The Burren is Ireland's ultimate rock garden—a massive limestone pavement that rises gently to the shale-capped Slieve Elva in the northwest corner of County Clare. The pavements were formed in the sea 360 million years ago, but the landscape visible today is the product of the last ice age, just 15,000 years ago, when ice scoured the land flat and left behind a scattering of untidy boulders, known as "erratics." More recently, the limestone slabs, known as "clints," have been weathered by rainwater into a network of cracks and crevices, known as "grikes," into a landscape referred to as "karren." Soil has accumulated in the fissures, providing a sheltered microclimate for flowers such as the dense-flowered orchid from the warm Mediterranean, to live side-by-side with plants that prefer alpine or even arctic conditions, such as the gentian verna and mountain avens. The Burren is the only place in Europe where this occurs. When it rains heavily, temporary lakes, known as turloughs, appear and disappear. Rivers vanish down sinkholes to flow as underground rivers in a subterranean labyrinth of caves and tunnels lined with stalagmites and stalactites. One cavern, Aillwee, is open to visitors. MB

THE VALLEY OF THE RHINE

RHEINLAND-PFALZ, GERMANY

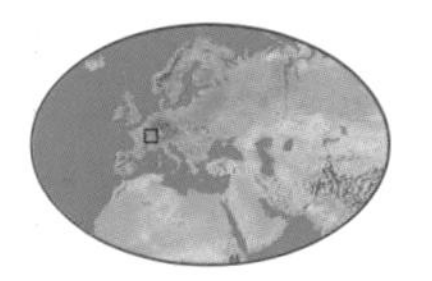

Length of the Valley of the Rhine: (Bingen to Bonn): 80 mi (130 km)

Number of castles: 50

From its source high up in the Swiss Alps to its mouth at Rotterdam on the North Sea, the Rhine is never more beautiful or enchanting than where it flows through this steep gorge in the heart of Germany, between the towns of Bingen and Bonn. For 80 miles (130 kilometers) the Rhine meanders through a glorious landscape of castle-topped hills, terraced vineyards, and overhanging cliffs. The valley of the Rhine cuts through the Rhineland Plateau and the Rhenish Slate Mountains, and so famed is it in literature and poetry that it is called the "heroic Rhine." There are more castles in the Rhine Valley than in any other river valley in the world.

The Rhine is also the world's busiest river for shipping and historically has played a major role in central European commerce. At the narrowest and deepest point in the valley is a 435-foot- (133-meter-) high slate rock outcrop called Lorelei that is famous for the echo it produces. This is a treacherous part of the river—according to legend the beautiful maiden Lorelei, who drowned herself in the Rhine in despair over an unfaithful lover, sings sweet songs to sailors, luring them to their deaths on the rocks. JK

ELBE GORGE

GERMANY / CZECH REPUBLIC

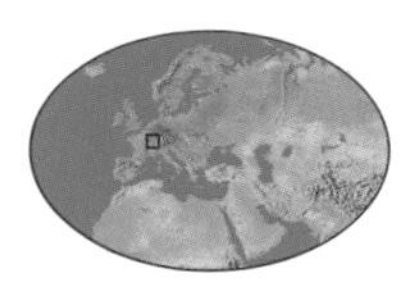

Length of River Elbe: 724 mi (1,165 km)

Age of Elbe Gorge: 80 million years

Height of Bastei Rocks: 660 ft (200 m)

The River Elbe rises in the Riesengebirge Mountains in the Czech Republic (where it is known as the Labe), and on its 724-mile (1,165-kilometer) journey to the North Sea, it enters Germany through a gorge in the Erzgebirge or Ore Mountains. In Saxony, just to the south of Dresden, the river passes strange and eerie rock formations made of sandstone that was laid down over 80 million years ago. Down the ages, it has been eroded into columns and stone towers by the action of ice and water.

Stands of pine, fir, and birch trees fill the gullies between the round-topped columns of the 660-foot- (200-meter-) high wall of the Bastei Rocks. Other stone pillars include the Barbarine, a human-like statue carved by natural forces, which stands beside the Pfaffenstein, meaning "the priest's stone." There is also the Lilienstein, meaning "lily stone," with its ruined castle 932 feet (285 meters) above the river. From a viewing point above these stone sentinels, the Lokomotive—a mountain shaped like a steam train—can be seen, and not far away, outside the town of Rathen, is the Felsenbuhne, a natural amphitheater in which plays are performed each summer. The river itself is a major waterway across Europe. MB

BERCHTESGADEN, THE WATZMANN MOUNTAINS, & LAKE KÖNIGSEE

OBERBAYERN, GERMANY

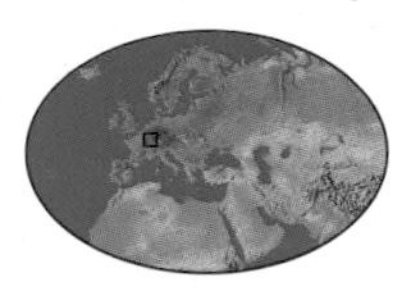

Height of Mt. Watzmann: 8,901ft (2,713 m)

Depth of Lake Königsee: 623 ft (190 m)

Area of national park: 81 sq mi (210 sq km)

Put simply, the Berchtesgaden Land in the Bavarian Alps has it all—stunning alpine peaks, emerald-green lakes, splendid hiking trails, magnificent ski slopes, and a fascinating history. Much of this area, including Germany's second-highest peak, Mount Watzmann, and its highest lake, the Königsee, set at an altitude of 1,975 feet (602 meters), was made a national park in 1978.

The Watzmann is a classic glaciated mountain, composed of hard limestone that, over time, has been worked into karst ridges, steep summits, and scree-covered slopes. Local legend claims that the peaks of the Watzmann are members of a cruel royal family who were turned to stone as punishment for their misdeeds. The highest peak is the king and the surrounding lower ones his family. The mountain's fearsome east face is popular with rock-climbers. One of the range's other peaks, Kehlstein, was the site of Hitler's summer playground, the notorious *Kehlsteinhaus*, or "Eagle's Lair." At the base of the Watzmann is the magnificent Königsee, meaning "King's Lake." It is 623 feet (190 meters) deep, and proudly billed as Germany's cleanest lake. The Berchtesgaden stretches from southeastern Germany into Austria, a mixture of rugged, glaciated, alpine scenery, and woodlands of spruce, beech, fir, and conifers.

The best way to explore the area is on foot, but cable cars offer a short-cut to some summits where breathtaking vistas await. It has long been a popular destination with walkers, both in winter and summer, with 149 miles (240 kilometers) of marked trails weaving through the pristine wilderness. The area also has numerous alpine beer gardens where hikers can relax, refresh themselves, and enjoy the greatest mountain panoramas in the Alps.

The mountains are home to a wide variety of wildlife, including mountain goats, chamois, Eurasian griffon vultures, ibex, and the rare golden eagle. JK

Put simply, the Berchtesgaden Land in the Bavarian Alps has it all—stunning alpine peaks, emerald-green lakes, splendid hiking trails, magnificent ski slopes, and a fascinating history. It has long been a popular destination both in winter and summer.

RIGHT *Stunning rocky peaks stand proud against the lush alpine valleys of Berchtesgaden.*

THE ELBSANDSTEINGEBIRGE

GERMANY / CZECH REPUBLIC

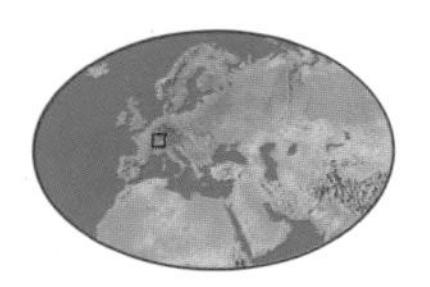

Elevation of formations: 2,365 ft (721 m)

Rock type: sandstone

The strange, craggy sandstone formations of the Elbsandsteingebirge stand tall over the eastern German state of Saxony and the western Czech state of Bohemia. These are some of the most unusual and remarkable-looking geological formations in central Europe. The name *Elbsandsteingebirge* translates as the "Elbe sandstone mountains," but these are more like bizarre hilltops shaped by millennia of wind and water erosion eating away at the soft yellow-green sandstone. They form a series of bizarre promontories jutting up like an immense stone army from the verdant forests that straddle the magnificent River Elbe.

Roaring streams that have created deeply incised canyons cut through this landscape. The landmark of the Elbsandsteingebirge is a 141-foot- (43-meter-) high rock needle called the Barbarine. This German national geological monument was first climbed by mountaineers in 1905 and is just one of 1,100 peaks that have attracted rock climbers from all over the world. Indeed, this region claims to be the birthplace of rock climbing. In 1864, five friends from Bad Schandau climbed the Falkenstein Peak. One of the best ways to view this remarkable landscape is from aboard a historic paddle steamer on the River Elbe, departing from Dresden, the capital city of Saxony. JK

THE BLACK FOREST

BADEN-WÜRTTEMBERG, GERMANY

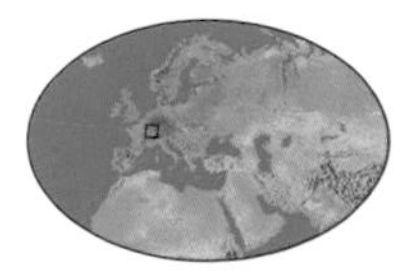

Area of the Black Forest:
2,320 sq mi (6,009 sq km)

Vegetation: Norway spruce

Highest peak (The Feldberg):
4,898 ft (1,493 m)

The Romans were the first to give the Black Forest its name, *silva nigra.* This is a place of folk legend, fairy tales, and immense natural beauty. Covering the southwest corner of Germany, the magical forest with its jagged hills, clear lakes, and deep valleys casts a spell on all who come here. The forest is not really black, but deep evergreen—mostly Norway spruce, a tall straight tree ideal for timber. It is one of the most beautiful areas of Europe, and many impressive castles dot the landscape. Over 1,860 miles (3,000 kilometers) of trails make the Black Forest a premier destination for hikers, and in the winter it is ideal for skiing and snowboarding.

The heart of the Black Forest is also the source of the great Danube River that rises here and then flows east into central Europe. One of the great highlights is the waterfall on the Gutach River, near the town of Triberg. This is the largest waterfall in the Black Forest, tumbling down 1,640 feet (500 meters) over a 1.2-mile (2-kilometer) stretch. The Black Forest is also famous for its cuckoo clock manufacturers, who have used the streamlined cones of the Norway spruce as models for the weights that power the clocks. JK

HAN-SUR-LESSE CAVES

NAMUR, BELGIUM

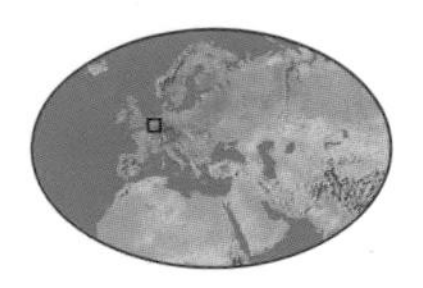

Cave type: karst cave, river cave

Length of cave system: 1.2 mi (2 km)

Cave temperature: 55°F (13°C)

The Han-sur-Lesse Caves are a marvelous underground complex of beautiful limestone caves, through which runs an extraordinary subterranean river. When you visit the caves you can ride on a 100-year-old tram to the entrance to the grotto, walk through the cave system, and take a remarkable boat journey on the underground River Lesse from the cave interior to the outer world.

Upstream of the caves, the River Lesse flows above ground until it reaches the limestone massif containing the caves, where it then disappears underground. The flowing water has done much to erode and shape the contours of this amazing cave complex.

The Han-sur-Lesse Caves also contain a fantastic collection of archeological artifacts dating back 5,000 years, proof that the importance of this site has long been recognized by humans. The artifacts include Neolithic tools (ca. 2000 B.C.E.), Bronze Age weapons and jewelry (ca. 500 B.C.E.), Roman coins and earthenware, and much more. JK

THE ROYAL FORESTS OF PARIS

ÎLE-DE-FRANCE / PICARDIE, FRANCE

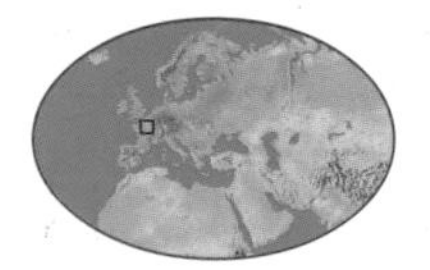

Area of the Royal Forests:
162,937 acres (65,940 ha)

Area of Fontainebleau:
61,776 acres (25,000 ha)

The forests of Rambouillet, Fontainebleau, and Orléans, to the south of the French capital, were once the preserve of royalty and used for hunting; today they are open to all, and are centers of conservation. The forests are at their most beautiful during the fall when the leaves of the oaks, beeches, and other deciduous trees begin to change color. The sun hangs low in the sky and bathes the leaves in warm autumnal light, thus emphasizing their stunning hues reminiscent of fire.

Rambouillet, with its wildlife park and numerous scenic walkways, is the most popular. Fontainebleau, boasting a total of 1,300 species of flora, is the most important in terms of conservation. Orléans, at 85,743 acres (34,700 hectares), is the largest but the least natural of the parks, and is in danger of being overrun by pines planted in the 19th century; however, this does not detract from the park's beauty.

Small valleys, unusual rock formations, waterfalls, and lakes feature in all the forests, and wildlife includes wild boar, badger, fox, polecat, and many species of deer. Birdwatchers can see pheasants, buzzards, falcons, and owls and hear the drumming sound of male black woodpeckers. CM

BAY OF MONT-SAINT-MICHEL

BASSE-NORMANDIE / BRETAGNE, FRANCE

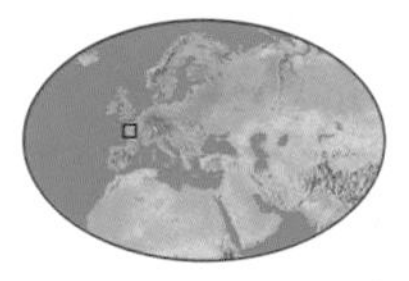

Length of Bay of Mont-Saint-Michel: 62 mi (100 km)

Depth of bay at high tide: 50 ft (15 m)

Age: 70,000 years

The broad Bay of Mont-Saint-Michel on the Atlantic coast of Normandy boasts the most spectacular tidal flow in Europe. Twice a day more than 130 million cubic yards (100 million cubic meters) of seawater flood in, then back out again. Victor Hugo, with a little poetic license, likened the speed of the waters to that of a galloping racehorse. At high tide, the sea often gains ground at a rate of more than three feet (0.9 meters) per second, reaching depths of up to 50 feet (15 meters). Low tide is a slower, less violent affair, with the waters retreating over 11 miles (18 kilometers). The tides leave behind vast volumes of sediment, raising the seabed by 0.12 inch (3 millimeters) each year.

Dolphins and seals inhabit the bay, so do watch out for them if visiting. The bay is also renowned for its oysters, which are served in some of the finest seafood restaurants in France. The shoreline, comprising marshlands, sand dunes, and cliffs, is home to a great variety of seabirds, including the razorbill (or *petit pingouin* in French due to its resemblance to the penguin). The bay takes its name from a rocky island just offshore that is joined to the mainland by a causeway. CM

PAVIS PEAK

RHÔNE-ALPES, FRANCE

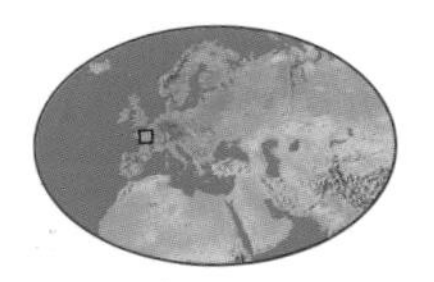

Elevation of Pavis Peak: base: 4,927 ft (1,502 m); peak: 6,806 ft (2,075 m)

Habitat: alpine fields, lakes, rock-faces, snowfields

Located in France's Haute-Savoie, the glorious mountain scenery of Pavis Peak attracts both determined climbers and sightseers alike. While climbers can enjoy scaling challenging faces, sightseers can achieve altitude and gain splendid alpine views by accessing a series of trails that wind ever higher, taking them through fields and a progressively more stark and rock-strewn landscape, until they finally enter a terrain formed entirely of glacially-eroded features and frost-shattered rock. There are three peaks, each separated by easily traversable ridges: Pavis Peak is the highest, followed by Dente d'Oche, and Les Cornets de Vente du Nord. Using the easiest mountain trails, the peaks are accessible after a brisk 45-minute walk. Climbers can choose between one of the eight rock-climbing routes leading up to the peaks, which are graded in difficulty to test all levels of skill. The best time to climb the peaks is between the months of June and October.

The nearest town is Bise, about 6 miles (10 kilometers) away, and the nearest hamlets are Parcour and Vent du Nord. The beautiful alpine lakes Léman and Darbon are also close by and make fine picnic sites. **AB**

DOUBS GORGE

FRANCE / SWITZERLAND

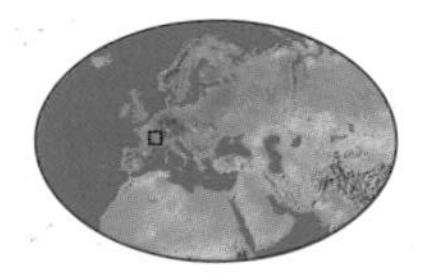

Length of Doubs Gorge: 10 mi (16 km)

Depth of gorge: 1,000 ft (300 m)

This wild, rocky gash in the Jura plateau between France and Switzerland contains in its lower reaches one of the most beautiful waterfalls in Europe. Funneled between ancient moss-covered rocks, the Doubs River plunges 92 feet (28 meters) from Chaillexon Lake, where spray rises in dappled sunlight against a forested backdrop. The falls are a three-mile (five-kilometer) hike from the village of Villers-le-Lac. That is the easy part. With its cliffs on occasion rising 1,000 feet (300 meters), the upper part of the gorge is accessible only to dedicated hikers and fly fishermen. For one of the most panoramic views of the gorge, climb a series of steel ladders set in the rock—not recommended for those of a nervous disposition. The French name, Le Belvédère des Échelles de la Mort, translates as "the vantage point of the Ladders of Death," which certainly seems fitting when climbing up there.

Nearly 34 miles (56 kilometers) of the banks on the Swiss side of the gorge are a nature reserve—in the 1970s, lynx were reintroduced here and are thought to have mated and produced young. The lynx is a reclusive animal, best observed around daybreak, and definitely not to be disturbed. **CM**

THE BAUME-LES-MESSIEURS RECULÉE

FRANCHE-COMTÉ, FRANCE

Height of Grottes de Baume-les-Messieurs: 66 ft (20 m)

Height of vantage point: 660 ft (200 m)

Reculée is a French word that refers to a remote, very distant place. It is applied specifically to a number of deep, narrow, often closed valleys on the western fringe of the Jura plateau, between Lons-le-Saunier and Salins-les-Bains. The village of Baume-les-Messieurs has grown up around a monastery in one such *reculée*—this village is a particularly impressive example of the form.

From the Baume-les-Messieurs Caves at the end of the valley there springs the Dard River, which tumbles over volcanic tufa rocks covered in rare mosses to create a beautiful, wild waterfall. Over thousands of years, the water has worn away the tufa to create a series of irregular, rounded steps and channels through which it now hurtles, sunlight creating a spectrum of color in the spray. The sheer cliffs overlooking the village provide splendid views, perhaps the best vantage point being Le Belvédère de Granges-sur-Baume. The short-toed eagle and peregrine falcon can often be sighted circling over the *reculée.* CM

BOURGET LAKE

RHÔNE-ALPES, FRANCE

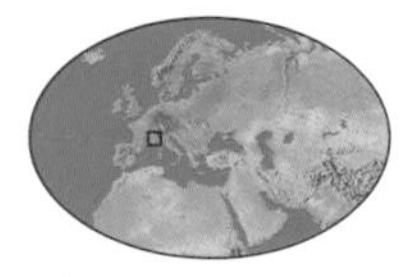

Length of Bourget Lake: 11 mi (18 km)

Depth of lake: 265 to 480 ft (80 to 145 m)

Area of lake: 11,119 acres (4,500 ha)

Bourget Lake, in the foothills of the French Alps, is not only France's biggest lake but also one of the most beautiful, lauded as such by, among others, the romantic poet Alphonse de Lamartine. The lake is 11 miles (18 kilometers) long and bordered to the south by the city of Chambéry and to the north by La Chautagne, a marsh drained in the 1930s and planted with poplars to create the biggest such forest in Europe. The summit of Mount Revard, which rises above Aix-les-Bains, on the eastern shore, offers panoramic views of the lake and the surrounding terrain.

Le Bourget is home to approximately 30 species of fish, including pollan and char, the latter greatly prized by local anglers and chefs. A variety of birdlife is attracted to the marshes bordering the lake, including the great reed warbler, tufted duck, grebe, and gray heron. The area has a mild climate because it is sheltered from north and west winds, meaning that Mediterranean plants such as jasmine, banana, fig, olive, and mimosa can be grown.

Le Bourget is connected to the Rhône by the Savières Canal, which formed an important waterway linking Chambéry with Lyon until the mid-19th century. CM

MONT BLANC

FRANCE / ITALY / SWITZERLAND

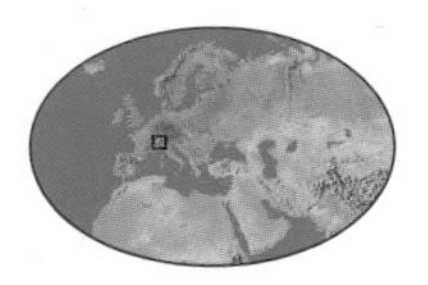

Height of Mont Blanc: 15,770 ft (4,807 m)

Length of Sea of Ice: 4 mi (7 km)

Width of Sea of Ice: 0.7 mi (1.2 km)

Europe's highest peak, rising 15,781 feet (4,810 meters) into the sky, and famed for the beauty of its glaciers and granite needles, has provided inspiration to countless generations. Byron came here in the 19th century to find the "language of solitude," while Shelley dedicated an entire poem to the great mountain. Mont Blanc (or *Mont-Blancin* in French) is actually a massif (range of mountains) rather than a single peak, some 24 miles (40 kilometers) in length and 6 miles (10 kilometers) wide. The range dominates the French Alps and the Swiss and Italian borders. Vistas from the summit are often described as being unforgettable.

It was first conquered in 1786 by Jacques Balmat, a guide from Chamonix, the French town at the foot of the mountain, and Dr. M. G. Paccard, an Italian doctor and keen mountaineer. Balmat returned the following year, on this occasion accompanied by Horace Bénédict de Saussure, the celebrated Swiss physician and naturalist. Today, up to 20,000 climbers reach the snow-and-ice-blanketed summit each year, often taking what is known as the *voie royal* ("royal way"), with the first part of the ascent undertaken in the relative comfort of the Aiguille du Midi cable car. From the Aiguille du Midi, the views of such natural phenomena as the Grandes Jorasses and Dents du Midi ("teeth of noon") are supreme. A trip by Tramway du Mont Blanc, a cogwheel train, to the Sea of Ice (which actually forms part of a huge glacier) passes through pine forests and over a viaduct to the "sea." The glacier is a tremendously impressive sight, with its frozen waves glittering in the sunshine. From here, the summit is a futher 3,300 feet (1,000 meters) up.

Europe's highest peak, and famed for the beauty of its glaciers, has provided inspiration to countless generations. Byron came here in the 19th century to find the "language of solitude," while Shelley dedicated an entire poem to the great mountain.

The Valley of Chamonix, situated at the foot of the mountain, is also considered to be one of the most beautiful natural sites in all France, with its ancient spruce forests, home to the rare woodpecker, the *pic tridactyle.* The lower slopes of the mountain provide a habitat for dormice and birds of prey.

A joint project of France, Italy, and Switzerland, *Le Tour de Mont-Blanc* allows walkers to circumnavigate the lower reaches of the mountain. CM

RIGHT *One of the majestic peaks of Mont Blanc stands proud against the blue sky.*

ANNECY LAKE

RHÔNE-ALPES, FRANCE

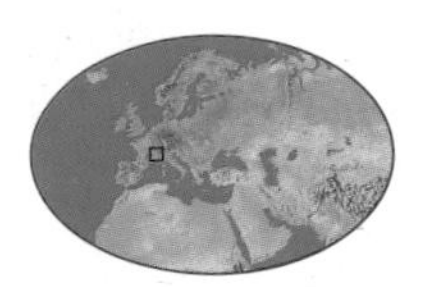

Area of Annecy Lake: 6,548 acres (2,650 ha)

Age of lake: 18,000 years

Type of lake: pre-alpine lake, spring meltwater fed

Surrounded by high, gray limestone cliffs and forest-covered slopes, Annecy Lake is fed by the spring snowmelt and is famous for its clarity. Up to 200 feet (60 meters) deep, it is France's second-largest natural lake. Set in beautiful alpine scenery, there are three natural wildlife reserves around the shoreline. March 2002 saw the end of a three-year cleaning program that removed bottom mud in which effluent and toxins had accumulated. This was threatening the ecological integrity of the lake and its tourist potential. A by-product of the work were a number of discoveries of extreme archeological significance, including several megaliths some 5,000 to 7,000 years old, that indicate sites of ancient inhabitation when the lake was smaller than it is now.

There are a number of glacial features in the region, including moraines and several hanging valleys in the Mount Veyrier region overlooking the lake. Apart from its great scenic beauty, the lake offers boating and fishing, and the 7-mile (11-kilometer) lakeside cycle path boasts fantastic views of the lake. Nearby is the ancient town of Annecy which is situated on the River Thiou that flows out of the lake. It dates from the Middle Ages, with evidence of human occupation dating back to 3000 B.C.E. **AB**

MEIJE GLACIER & GORGE

RHÔNE-ALPES, FRANCE

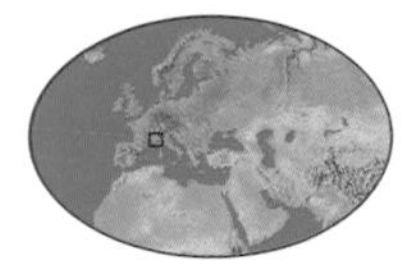

Height of Meije Glacier: 13,067 ft (3,983 m)

Height of Meije Gorge peaks: Pointe Nérot: 11,604 ft (3,537 m); Pic Gaspard: 12,739 ft (3,883 m); Le Pavé: 12,546 ft (3,824 m)

The Meije forms the backdrop to the French alpine village of La Grave, on the edge of Écrins National Park. It is part of a 9-mile- (15-kilometer-) long complex of mountains that includes several high peaks, such as Pointe Nérot, Pic Gaspard, Le Pavé, and numerous glaciers. The Meije was the last major alpine peak to be scaled, not because it is so tall, but just because of the technical difficulties involved. Its glacier has spectacular ice caves that are best visited in summer.

The area is wonderful for invigorating summer alpine walks, and in winter the region—especially the mountain's glacier—provides some of the best off-piste skiing in the world. Snow lasts until May. Situated in the beautiful La Romanche valley, La Grave is the base for the Meije Glacier cable car that passes several different alpine habitats and rock formations during its ascent. Comfortable accommodation can be found in La Grave, but refuge huts serve the more hardy trekker. Other villages in the region where accommodation is possible include the more remote Saint-Christophe-en-Oisans and La Bérarde, both in the Vénéon valley, the internal valley of the Écrins massif. **AB**

ÉCRINS NATIONAL PARK

RHÔNE-ALPES, FRANCE

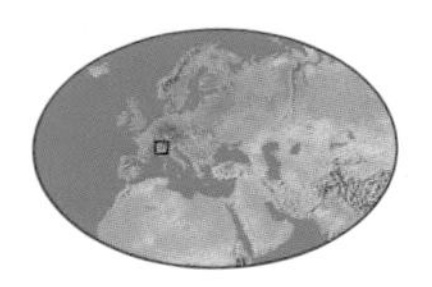

Area of Écrins National Park: 226,837 acres (91,800 ha)

Highest point (Barre des Écrins): 13,458 ft (4,102 m)

Height of Mt. Pelvoux: 12,946 ft (3,946 m)

The word *écrin* means "jewel box" in French, and the Écrins National Park in the French Alps contains a range of gems of great natural beauty. Le Massif des Écrins towers nearly 13,200 feet (4,000 meters) above the wild Vallouise Valley. Lower down, on sunny slopes, lavender grows in forests of beech, oak, and pine, which are home to an extremely rare moth known locally as Isabelle de France. Higher up, in the shadow of the rock walls of the mountains, nature becomes wilder. Pines, which have long taproots, give way to shallow-rooted spruce more suited to the sparse soil of rocky slopes.

The park, which was established in its current form in 1973, contains a microcosm of alpine flora and fauna. From Le Pré de Madame Carle ("Madame Carle's meadow"), within driving distance of Briançon, it is a two-hour hike to the White Glacier, where you are treated to impressive views of the mountains

above the Black Glacier. A footpath from the wind-blasted gully leads to L'Alpe du Villar-d'Arêne, with plaques en route pointing out sights. Pairs of golden eagles can often be seen soaring overhead. The environment in the Écrins Park is particularly well suited to the golden eagle. The park serves as a breeding ground for the birds, which are then introduced to other regions of France.

In 1928, Mont Pelvoux was the first of the massif's high peaks to be climbed. The mountain features a number of glaciers: Pelvoux Glacier is at the summit, Glacier Du Clos De L'Homme is on the south side, both Glacier de la Momie and Glacier des Violettes are on the east side, and Glacier Noir ("Black Glacier") is on the north side.

To the north of the park is the Meije range, the highest peak of which—the Meiji Glacier—rises to 13,067 feet (3,983 meters). A cable car from La Grave takes non-mountaineers almost to the summit, so the sublime views can be experienced by all. CM

BELOW *The lush green grasses and alpine peaks of Écrins National Park, the second-largest national park in France.*

AIGUILLE DE DIBONA

RHÔNE-ALPES, FRANCE

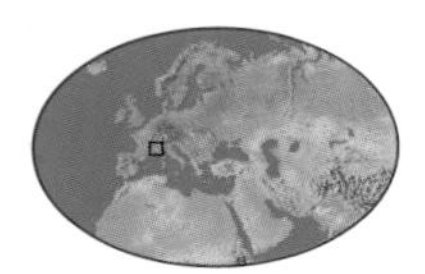

Elevation of Aiguille de Dibona: 10,272 ft (3,131 m)

Height of south face: 1,150 ft (350 m)

Rock climbers think of it as one of the most spectacular "multipitch" rock summits in Europe. French alpinist Gaston Rébuffat said of the rock, "this needle is a monument of stone given to mankind by the earth and time, an extraordinary sculpture in the sky." To the locals it is a symbol of Massif de L'Oisans in southeast France, and to the geologist, the mountain of Aiguille de Dibona is a beautifully formed granite needle.

The first people to reach its summit on June 27, 1913, were Italian climber Angelo Dibona, one of the famous Cortinesi mountain guides, and his German client Guido Mayer, the former giving his name to the peak. The south face of the rock is an almost vertical 1,150-foot-(350-meter-) high face—the climbing of which requires a high level of skill and experience. On the reverse side of the mountain is a gentler saddle, linked to a neighboring spire, known as Aiguille du Soreiller.

The closest habitation is Les Etages, a small hamlet located in the Sorreiller Gorges, two miles (3.2 kilometers) before you enter La Bérarde in the high Vénéon. **MB**

CHESERY LAKE

RHÔNE-ALPES, FRANCE

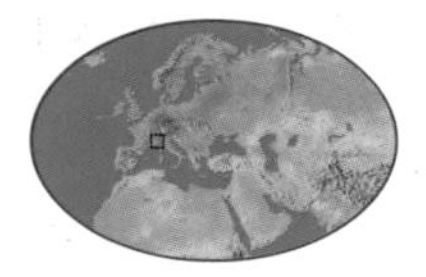

Elevation of Chesery Lake: 12,801 ft (3,902 m)

Type of lake: alpine

This high alpine lake lies in the Aiguille d'Argentière region of France's Haute Savoie. Dug out by glacial action during the last ice age, its waters come entirely from seasonal meltwater and from snow. The lake is frozen in the winter and parched in the summer. The tough alpine shrubs around the lake have to endure the winter's bitter conditions, but not far below, fertile alpine pastures are dotted with fruit trees, cows, and chalets. The lake itself has proven to be a favorite with several generations of alpine painters who have sought to capture its beauty and that of the mountains reflected in its waters, as well as the views of other glacial features a long way below.

In the summer months, Chesery Lake, with its sparkling turquoise waters, its views of sharp glaciated peaks, and its rounded glacier-molded hills, is a place for easy family walks. The Aiguille d'Argentière itself is situated north of the major peaks of the Mont Blanc massif, and it has more than 50 challenging climbs, attracting an international community of rock climbers each year.

The nearest area for access is the Chamonix Valley, which is most easily reached from Switzerland by train from Martinny via Vallorcine or by air from Geneva. **AB**

LES DRUS

RHÔNE-ALPES, FRANCE

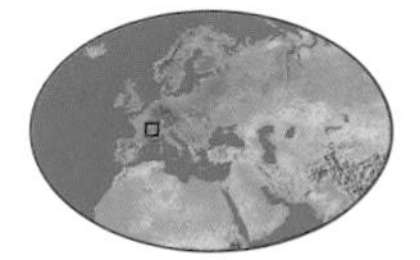

Height of Grand Drus:
12,316 ft (3,754 m)

Height of Petit Drus:
12,237 ft (3,730 m)

At a distance Les Drus appears as a single notch-topped mountain, but closer inspection reveals two cone-shaped peaks, with remarkably flattened angular sides, so that the mountains are almost pyramidal; this was probably formed by the grinding effects of the heads of three or more glaciers. Part of the Mont Blanc massif, these two mountains have long challenged mountaineers. The first ascent, in 1938, opened up the southeast face. Still the most popular route, this 1,411-foot (430-meter) ascent across rock and ice takes six hours to complete. A tougher, 2,640-foot (800-meter) climb was opened in 1962 up the southern pillar, but both of these were overshadowed by the remarkable achievement of French climber Jean-Christophe Lafaille who, in 2001, opened a new route on the west face. An extraordinarily challenging climb, it is considered to be ten times harder than the "Divine Providence" route on Mont Blanc's Grand Pilier d'Angle. A recent massive rock fall on the western pillar destroyed some prime climbing routes and demonstrated both the inherent danger of climbing in these mountains as well as their ever-changing nature. For those who prefer gentle hikes, the spectacular views of Les Drus can be experienced via the Chamonix Valley. **AB**

LA VANOISE NATIONAL PARK

RHÔNE-ALPES, FRANCE

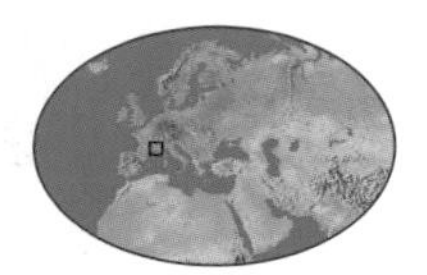

Area of La Vanoise National Park: 130,565 acres (52,839 ha)

Established as a national park: 1963

Highest point: Pointe de la Grande Casse 12,648 ft (3,855 m)

Located in southeastern France, and contiguous with the Gran Paradiso National Park in Italy, this is the oldest of France's national parks and (when the Italian area is added) the largest protected area in western Europe. Dominated by the Massif de la Vanoise, the park lies just south of Mont Blanc along the spine of the Haute Alpes, west of the Isère River and north of the Arc River.

Geologically diverse, the park features impressive formations of gneiss, schist, sedimentary sandstone, and limestone. It has great biodiversity—over 125 species of bird can be found, as well as mountain species such as marmot, ibex, and chamois. There are glorious valleys rich in alpine flowers and over 20 glaciers are located here.

Mountains, valleys, and a range of habitats for wildlife means there is something for everyone, from the day hiker to the ardent alpinist. There are over 310 miles (500 kilometers) of footpaths, two mountain tracks for trekkers, and numerous ascent routes for climbers. **AB**

RIGHT *One of La Vanoise National Park's impressive rocky peaks seen through a hole in the wall of a snow cavern.*

AIGUILLE DU MIDI

RHÔNE-ALPES, FRANCE

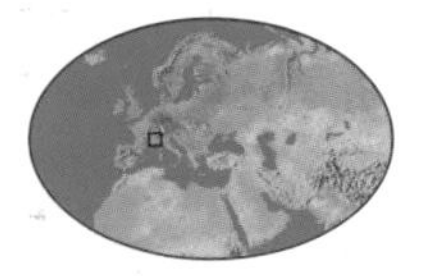

Height of Aiguille du Midi: 12,602 ft (3,842 m)

Notable feature: world's highest cable car traverse

Situated on the French-Italian border, in the Chamonix area of France, the Aiguille du Midi ("Midday Needle") is famous for its views, its climbing, its cable car system, and its proximity to Mont Blanc. In an incredible two-stage, 15-mile (24-kilometer) journey, the cable car crosses the Blanche Valley and the Géant Glacier, via Plan de l'Aiguille at 7,572 feet (2,308 meters), to Helbronner Peak, which is on Italian territory. Alternatively, you can climb the whole thing, but this requires crampons and good general fitness and takes four to five hours. A guide is essential.

There are huts at Refuge des Cosmiques (France) and Rifugio Torino (Italy). Using these as bases, additional climbs are possible—including Mont Blanc du Tacul's northwest face (from Cosmiques) and Aiguille d'Entrèves Traverse (from Torino). The Aiguille's Mer de Glace glacier offers excellent skiing from February to May; however, it is only for the experienced, and a guide is essential for safety. Other thrills include climbing the Arête des Cosmiques, the remarkable needle-like rock towers that give the peak its name. The route is not too physically challenging but does require a guide. **AB**

BOSSONS GLACIER

RHÔNE-ALPES, FRANCE

Elevation of Bossons Glacier:
15,767 ft (4,807 m) at the top of Mont Blanc to 4,290 ft (1,300 m) on valley floor

Length of Bossons Glacier:
4.3 mi (7 km)

Starting at the tip of Mont Blanc, western Europe's highest mountain, the Bossons Glacier is the longest glacial slope in Europe. The current glacier, however, is but a shadow of its former self. About 15,000 years ago, it extended another 30 miles (50 kilometers), was over 3,300 feet (1,000 meters) deep, and linked up with other massive glaciers to carve the rocks of the Rhône region. In the 17th and 18th centuries, local bishops used exorcism to try and stop the glacier destroying crops and houses. After several quiescent centuries, the glacier is once again advancing, this time at an average of 825 feet (250 meters) a year. The angle of the ice-slope is around 45°, making it one of the world's steepest glaciers.

The glacier comprises snow that solidifies in the permanent snowfields above 13,200 feet (4,000 meters). Once the new ice exceeds 100 feet (30 meters) in thickness, it begins to flow downhill under its own weight. Bossons' ice is known for its spectacular color and purity. Within the melting zone, the ice has beautiful crevasses, cascades, and tunnels. Ice takes 40 years to move from the head to the toe. **AB**

AIGUILLE VERTE

RHÔNE-ALPES, FRANCE

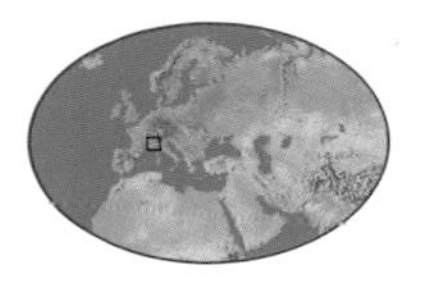

Height of Aiguille Verte: 13,520 ft (4,121 m)

Feature: rock pinnacle

Aiguille Verte—a close neighbor to Mont Blanc—is an immediately recognizable rock pinnacle, considered a daring climb by mountaineers and certainly a very dangerous one. Avalanches in winter and rockfalls in summer are major hazards. Slab avalanches in the vicinity (when an entire mountainside can be on the move) are the most dangerous; an event in 1964 claimed the lives of 14 climbers on Aiguille Verte.

In summer, great slabs of rock—some the size of houses—fall away from the crumbling cliffs. For climbers, there is no easy way up, but there is a comparatively easy way down via the Whymper Couloir. A couloir is a deep mountain gorge or gully filled with soft snow in winter. These snow-filled gullies provide excellent skiiing in the area. The Whymper Couloir is a route favored by extreme skiers; its 50° slope was first conquered by Sylvain Saudan in 1968. One of the favored times for climbers to reach the summit is at dawn, when the nearby Mont Blanc takes on a pink hue and the sun rises up spectacularly over the Wallis Alps. **MB**

AUVERGNE'S PUY MOUNTAINS

AUVERGNE, FRANCE

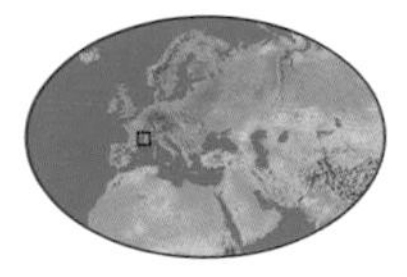

Age of Puy Mountains: 8,000 years

Height of Le Puy-de-Dôme: 4,806 ft (1,465 m)

The rugged landscape of the Auvergne is said to be "the child of a marriage of ice and fire," that is to say, formed of ice age glaciers and volcanism. The Puy Mountains comprise 80 extinct volcanoes. The highest of these is Le Puy-de-Dôme.

The region has a spectacular panorama of more than 80 volcanoes, with every imaginable form of volcanic deposit from tall cones to dykes, lava flows to water-filled cones. Caused by the collision of the continents of Africa and Europe—which also gave rise to the Alps—this area was a center of significant volcanic activity 20 million years ago. Though hydrothermal activity continues, the last major explosion was around 6,500 years ago—the age of the youngest volcano. What makes the region so spectacular is a zone of weakness in the Earth's crust that has resulted in a strong north-south alignment of the volcanic domes.

In the early 19th century, the slopes of the mountains were moorlands, but forest has now taken over. The Auvergne is famed for the purity of its water and is the rising point of the Rivers Dordogne and Loire. **CM**

LE PUY-DE-DÔME

AUVERGNE, FRANCE

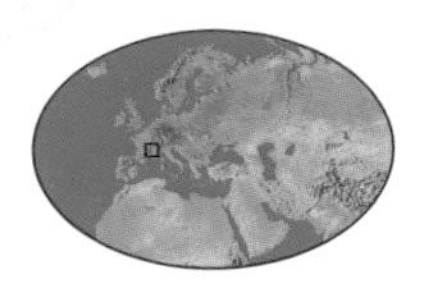

Elevation of Le Puy-de-Dôme: 4,806 ft (1,465 m)

Age of Le Puy-de-Dôme: 150,000 years

Rock type: volcanic

Le Puy-de-Dôme is the most famous of the Puy volcanoes. The summit is accessible via both a toll road for motorists and an old mule trail for hikers. Affording magnificent views of the other Puy summits, it is visited by around 500,000 people each year, making it one of the most popular sites of natural beauty in France. It has the classic haystack-shaped outline of the region's volcanoes. By contrast, Le Puy Chopine ("wine bottle") is different, a needle-shaped outcrop of hard acidic magma that has resisted erosion for nearly 10,000 years. Le Gour de Tazenat is some 213 feet (65 meters) deep and 2,297 feet (700 meters) in diameter, and is the region's most beautiful lake. It was formed around 40,000 years ago in a violent volcanic explosion. Most of the volcanoes became extinct around 8,000 years ago; others, notably Côme and Pariou, continued to erupt for another 4,000 years.

The region contains two major wildlife parks worth visiting, Volcans-d'Auvergne and Livradois-Forez, where ravens and falcons nest in rocky volcanic cliffs. AB

RIGHT *The crags of Le Puy-de-Dôme contrast sharply with the lush green landscape.*

VAUCLUSE FOUNTAIN

PROVENCE-ALPES-CÔTE D'AZUR, FRANCE

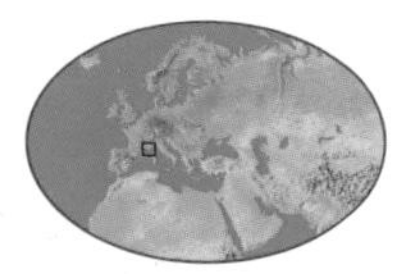

Depth of Vaucluse Fountain: 1,079 ft (329 m)

Flow rate of the Sorgue River: 39,625 gal (150,000 l) per sec

The Sorgue River emerges with tremendous urgency from a spring at the base of a spectacular 755-foot (230-meter) limestone cliff at the end of a *vau cluse*, or closed valley. In wet months, the spring's flow can reach an astonishing 39,625 gallons (150,000 liters) per second, making it the most powerful in France and one of the most powerful in the world. Despite the immense volume of flow, the surface waters of the vast subterranean cavern from which the river pours seem still and calm and are colored an ethereal green. Appearances can, however, be deceptive. Extremely strong currents rage deep below the surface. The water's extraordinary depth of 1,079 feet (329 meters) was only accurately measured for the first time in 1985.

The spring and its beautiful river valley, which is lined with giant plane trees, have long been popular tourist destinations. Back in the 14th century, the poet Petrarch wrote his *Canzoniere* here, a collection of sonnets of unrequited love for the lovely Lady Laura whom he admired from afar. He would not get much peace if he did the same today: Thousands of visitors flock from all over the world to the fountain each year. CM

THE ARDÈCHE GORGES

RHÔNE-ALPES / PROVENCE-ALPES-CÔTE D'AZUR, FRANCE

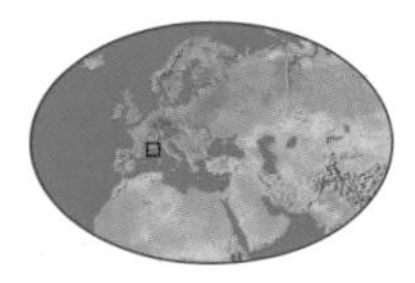

Maximum depth of gorges: 1,000 ft (300 m)

Length of gorges: 19 mi (32 km)

Flood flow: 171,798 gal (650,326 l) per sec

The Ardèche is one of the Rhône's most insignificant tributaries, just 74 miles (120 kilometers) long. However, in its descent from an altitude of 4,842 feet (1,476 meters), in the Mazan massif in the foothills of the Alps, it has carved out a varied bed for itself, steep-walled to start with, then wide and meandering in its lower reaches.

As it descends toward the Rhône, the river's character changes from torrential and fiercely flowing at the village of Thueyts to calm majesty by the time it arrives at Aubenas, on the same road lower down. Its banks, at first rocky and arid, become verdant and lined with orchards. It flows on to Vallon-Pont-d'Arc where over the centuries it has carved a natural rock archway 111 feet (34 meters) high, and 200 feet (60 meters) across. This marks the start of the main body of gorges, which can be explored properly only by kayak or on foot. However, if you take the road from Vallon to Saint-Martin-d'Ardèche, you can access viewing points at Serre de Tourre and Maladrerie. The gorges are home to such species of bird as the Egyptian vulture, Bonelli's eagle, and the blue rock thrush. CM

VERDON GORGE

PROVENCE-ALPES-CÔTE D'AZUR, FRANCE

Length of Verdon Gorge: 12 mi (20 km)

Age of gorge: 25 million years

Age of rock: 140 million years

Like a thin white ribbon, the Verdon River winds its way through a spectacularly deep gorge some 12 miles (20 kilometers) long. It is the largest chasm on the continent—considered to be the "Grand Canyon" of Europe. It straddles the boundary between two French departments—Alpes-de-Haute Provence and the Var in southeast France. The walls are of limestone laid down nearly 140 million years ago in the Tethys Sea, and in places they can be no more than 20 feet (6 meters) apart at the bottom, yet 4,950 feet (1,500 meters) at the top. Fed by melting snow and ice from the Alps, the Verdon has worn down through the limestone plateau of the Haute Provence to create this enormous gash in the landscape.

The process started about 25 million years ago after the Alps were formed and water carved out huge underground cave systems. Eventually the roof collapsed and the gorge was formed. Previously known only to local woodsmen, it was not until French caving pioneer Edouard Alfred Martel successfully led an expedition there in 1905 that the site become a natural attraction for visitors. Today, anybody can peer down into the deep chasm from the roads that follow either edge. MB

THE CAMARGUE

PROVENCE-ALPES-CÔTE D'AZUR, FRANCE

Area of the Camargue: 286,636 acres (116,000 ha)

Age of the Camargue: 5,500 years

The Camargue is a region of marshlands, brackish lakes, and salt flats formed by the silting of the Rhône as it divides in two. The main arm, the Grand Rhône, takes a more-or-less direct route south to the Mediterranean Sea, while the Petit Rhône meanders westward. Between the two is the main body of the Camargue, then west of the Petit Rhône, the Petit Camargue.

The name "Camargue" may derive from Roman general Caius Marius, who held huge estates here. The white horses and black bulls, for which the region is famous, are thought to be all that remains of the vast herds that sought shelter here in prehistoric times.

Of the 337 varieties of wild bird found in the Camargue, the best known are the flamingoes, which have become a symbol for the region, though of the majority of the estimated 50,000 birds that visit the area each year, only some 3,000 are resident. Their numbers may increase in the future because an island in the remote Fangassier Lagoon has been given over to the birds and is now the only site in Europe where flamingoes—between 10,000 and 13,000 couples annually—regularly breed. CM

MERCANTOUR NATIONAL PARK

PROVENCE-ALPES-CÔTE D'AZUR, FRANCE

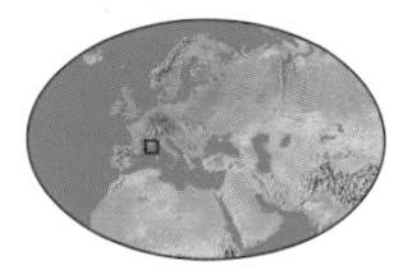

Area of Mercantour National Park:
264 sq mi (685 sq km)

Highest point (La Cime du Gelas):
10,309 ft (3,143 m)

Occupying highlands in the southeast of France, this virtually uninhabited park abuts Italy's Alpi Marittime Natural Park with which it is co-managed. Situated in the park is Lake Allos, the largest high-altitude body of water in Europe, and there is a great variety of gorges and waterfalls. Created in 1979, it is one of France's seven national parks. Within the park's boundaries, La Cime du Gelas (10,309 feet, 3,143m) is the highest peak. There are many others, including Tête de la Ruine (9,790 feet, 2,984 meters), Grand Capelet (9,626 feet, 2,934 meters), and Mount Bego (9,426 feet, 2,873 meters). The park contains high-altitude plants, but proximity to the Mediterranean coast ensures a gradual change in vegetation with height that includes the shrubby, aromatic maquis plant life to be found on the Riviera.

Wildlife includes some of the best represented alpine mammal communities in Europe, including chamois, bearded vulture, ibex, mouflon, and marmot as well as a host of alpine flowers, such as the alpine columbine and ladyslipper. There is also evidence of prehistoric occupation: The Merveilles Valley, at the foot of Mount Bego, contains some 100,000 Bronze-Age rock engravings. AB

MOUNT VENTOUX

PROVENCE-ALPES-CÔTE D'AZUR, FRANCE

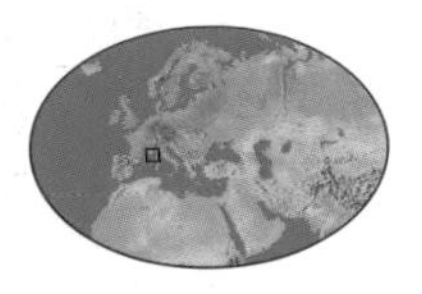

Age of Mt. Ventoux:
60 million years

Height of Mt. Ventoux:
6,263 ft (1,909 m)

Mount Ventoux is the highest mountain in Provence, the region of France that was one of the first provinces of the Roman Empire. Its distinctive, shining summit of white shale—often mistaken for snow—can be seen from all over the region. The summit of the mountain—known locally as "The Giant of Provence"—is crowned by a 15th-century chapel and an observatory, which has three roads leading to it. These mountain roads often form a particularly grueling stage of the Tour de France.

The best starting point for an ascent is the market town of Carpentras, former capital of the Comtat Venaissin, a Catholic enclave. The northern route is via Malucène, the southern via Sault, the region's lavender capital. Views can be spectacular, although they can also often be obscured by heat haze in summer.

To the south are the Lubéron Mountains; Sainte-Victoire, the mountain depicted by Paul Cézanne in his paintings; the Alpilles; and the Étang de Berre Lagoon on the Mediterranean, close to Marseille. To the west are the Dentelles de Montmirail and the *Plan de Dieu*, the "Plain of God," home of the best Provençal wines, Vacqueyras and Gigondas. To the east are the Alps, which perhaps provide the greatest views of all.

Ventoux also boasts a natural wonder that cannot be seen, although there is a clue in the mountain's name, which derives from the French words *vent* (wind) and *tout* (all)—as in "all the time." Of the 32 winds said to sweep Provence, by far the greatest is the legendary Mistral, the devastating north wind that roars down the Rhône Valley and has been clocked at speeds of up to 193 miles per hour (320 kilometers per hour). Its name in Provençal is *Lou Mistrau*, "the Master."

Mount Ventoux is the highest mountain in Provence. Its distinctive, shining summit of white shale–often mistaken for snow–can be seen from all over the region. The mountain roads often form a particularly grueling stage of the Tour de France.

Long before the Greeks and Romans came to this part of the world, the Mistral was worshipped as a god by Celto-Ligurian tribesmen. A temple has been found on Mount Ventoux where the tribesmen summoned the wind with clay horns, which they then smashed, running in terror as it arrived.

Footpaths crisscross the foothills of Ventoux through forests of Atlas cedars, oaks, and pines, home to mouflon and wild boar, as well as 104 species of nesting bird, including golden, Bonelli's, and short-toed eagles. CM

RIGHT *Mount Ventoux forms an imposing backdrop against the mountain villages.*

RIVER RHÔNE

SWITZERLAND / FRANCE

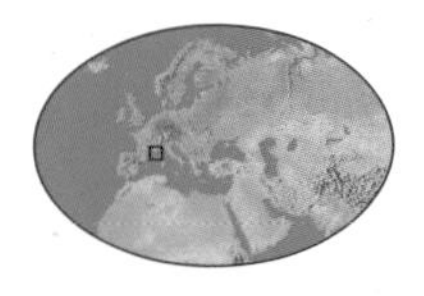

Length of River Rhône: 507 mi (816 km)

Average flow rate at Arles: 16,629 gal (62,948 l) per sec

The Rhône starts life as a tiny mountain stream of meltwater from the alpine glacier in the Swiss Saint-Gothard Mountains before embarking on an epic 507-mile (816-kilometer) journey to the Mediterranean Sea. By the time it empties into Lake Geneva (*Lac Léman* in French), it is already a fully fledged river. From Switzerland, it flows on into France, and when it reaches Lyon, joins the Saône to become a mighty navigable waterway. From the establishment around 600 B.C.E. of the Greek colony of Massalia (today Marseille) until the coming of the railways in the 19th century, the lower reaches of the river formed part of a major north-to-south trade route. Today river transportation has been revived, with huge barges traveling to and fro.

The river divides into two separate channels below Arles, the Grand Rhône and the Petit Rhône, between which sits the Camargue, a vast area of marshland that provides a home to over 400 species of bird, among them the greater flamingo, as well as some notable

mammals, including Camargue horses and Camargue "fighting" bulls.

In summer the Rhône gives every appearance of having been tamed, its lower reaches bordered by motorways, industry, and nuclear and hydroelectric power stations. But this urban appearance belies its true nature, for it floods regularly in winter, often with devastating effects.

Of the many bridges over the Rhône, the most famous is the ruined Pont Saint-Benezet, which bears witness to the Rhône's tremendous power. When the bridge was completed in 1185 it was 975 yards (900 meters) long and comprised 22 arches. All but four arches have been washed away over the years.

Today wildlife parks such as the Île de Beurre, near Condrieu, and the Île de la Platière, further south, close to Sablons, boast a wealth of wildlife, including otters, beavers, and coypu. The Pilat Park, not far from Lyon, is perhaps the most high-tech in France, with automatic video cameras that enable animals to be observed closely without being disturbed. CM

BELOW *The Papal Palace and Pont Saint-Benezet are reflected in the glass-like waters of the River Rhône.*

THE GAVARNIE CIRQUE

MIDI-PYRÉNÉES, FRANCE

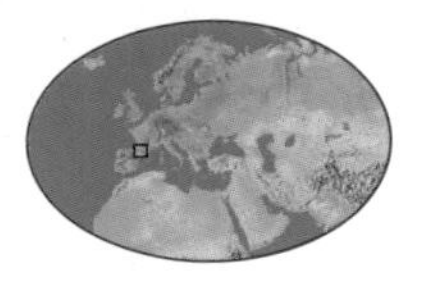

Age of the Gavarnie Cirque:
2 million years

Highest point (Marboré):
10,656 ft (3,248 m)

Area of Pyrénées National Park:
112,942 acres (45,707 ha)

The Gavarnie Cirque is a celebrated site of natural beauty in the French Pyrénées. Victor Hugo described it in 1843 as "a miracle, a dream," and its grandeur inspired a young Parisian artist, Sulpice-Guillaume Chevalier, to adopt "Gavarni" as his pseudonym. The small village of Gavarnie lies in a bed of alpine meadows, surrounded by its cirque, a geological term for an amphitheater of steep-walled mountains created by glacial erosion, which Hugo called a "colosseum of nature."

Three peaks dominate the cirque: the Taillon (10,314 feet, 3,144 meters), the Casque (10,082 feet, 3,073 meters), and the Marboré (10,656 feet, 3,248 meters). The somber, dark walls of these mountains, part of the Mont Perdu massif, ascend in steps, cut by glaciers two million years ago, and are dotted here and there by snowfields, while the foothills are covered by forests of beech and pine.

A spectacular waterfall tumbling some 1,320 feet (400 meters) down the slopes of the Marboré is the principal source of the Pau River. It is at its loveliest in winter, when it freezes, creating a series of magnificent, glittering ice falls. To reach the waterfall and the nearby "snow bridge" requires a fairly strenuous 45-minute hike along a footpath that can be found off the road leading to the Hôtel du Cirque.

A circular return along the Pailha Trail allows walkers to see some of the most spectacular views the Pyrénées can offer. The summit of Mount Mourgat is the best vantage point of the cirque, although reaching it involves a much more demanding three-hour hike. Everyone visiting Gavarnie must be prepared to hike—a 20-minute walk to an altitude of 4,785 feet (1,450 meters) is required to visit Gavarnie's music and theater festival, which is held annually during the last two weeks of July.

The cirque forms part of the Pyrénées National Park and the Pyrénées-Mont Perdu World Heritage site. Created in 1967, it supports some 160 species of flora found only in the Pyrénées, as well as brown bears—reintroduced to the region in 1996. Golden eagles and several different species of vulture command the skies. Wildlife protection laws are strictly enforced. CM

The Gavarnie Cirque is a celebrated site of natural beauty in the French Pyrénées. Victor Hugo described it in 1843 as "a miracle, a dream," and its grandeur inspired a young Parisian artist, Sulpice-Guillaume Chevalier, to adopt "Gavarni" as his pseudonym.

RIGHT *Sheep graze in the hills of the Gavarnie Cirque.*

PILAT DUNE

AQUITAINE, FRANCE

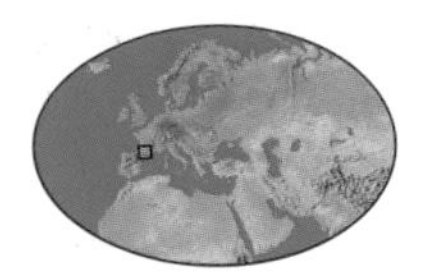

Height of Pilat Dune: 383 ft (117 m)
Length of dune: 1.9 mi (3 km)
Width of dune: 1,650 ft (500 m)

On the French Atlantic coast at the entry to the Gulf of Arcachon, near Bordeaux, Pilat Dune (sometimes spelled Pyla) is a most extraordinary—and Europe's highest—sand dune at 383 feet (117 meters) and is gaining an additional 13 feet (4 meters) each year. The dune has long been a landmark for sailors but nowadays is more a vantage point for tourists looking out to sea, its northeast face equipped with a wooden stairway.

The name derives from a 15th-century langue d'oc name meaning simply "heap of sand," applied to a sandbank a little further to the north. This was eroded by winds some time in the 18th century to form the base of the present dune.

Pilat grows rapidly. Today it consists of an estimated 78 million cubic yards (60 million cubic meters) of sand. Free of vegetation, it is slowly but surely moving away from the sea and threatening to engulf a neighboring forest. The coast around the dune is heavily commercialized, but islands not far from the shore are rich in birdlife and easily accessible by boat. CM

TARN GORGE

MIDI-PYRÉNÉES / LANGUEDOC-ROUSSILLON, FRANCE

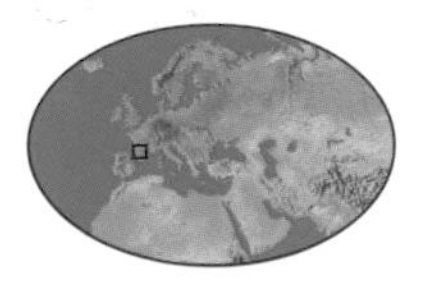

Length of Tarn River: 233 mi (375 km)
Rock type: limestone

The Tarn is a small but very long river flowing from the foot of Mount Lozère to the point where it joins the Garonne at Moisac. It boasts one of France's most beautiful gorges, which was formed by the river's flow and erosion of the limestone rocks of the desolate Grand Causses.

The Causses are formed out of what 120 million years ago was a vast gulf of the Mediterranean Sea. The bizarre rock formations that can be found here were carved out over a long period of time by the waters, and these weird landscapes are known to geologists as karstic limestone formations. The name derives from Karst, a region in Slovenia that features similar geological structures.

Perhaps the loveliest part of the Tarn Gorge is the 37-mile (60-kilometer) stretch between Florac and Le Rozier. From Le Rozier the best footpath is the one named after a local rock formation, the Rocher de Capluc. Tours in glass-bottomed boats can be taken from the village of La Malène.

Numerous bird species nest in towering cliffs bordering the river, including golden eagle, eagle owl, and peregrine falcon. Locally the Causses are known appositely as "chaos." One of the most interesting is at Nîmes-le-Vieux, northeast of Meyrueis. CM

CEVENNES GORGE

LANGUEDOC-ROUSILLON, FRANCE

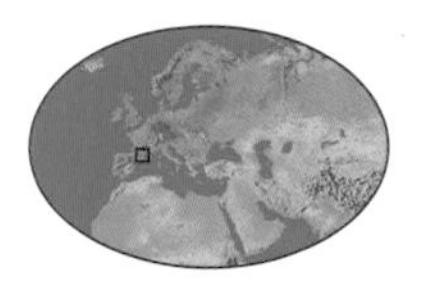

Highest point (Pic Finiels): 5,574 ft (1,699 m)

Rock type: gray limestone

Established as a park: 1970

Located in the the French department of Lozère, this protected area is famous for its remarkable limestone scenery—and limestone landscapes are among the most varied and interesting in the world. This includes large areas of spiky karst rock protrusions as well as many deep caverns, several of which are over 200 million years old. There are also several gorges, including the Jontes Gorge, one of the deepest and most spectacular in Europe.

The innumerable caves of the gorge were inhabited by generations of ancient humans, and the park is rich in archeological sites—especially from the Bronze Age—including menhirs, dolmens, and stone enclosures, some about 4,000 years old. Several of today's hiking paths follow Bronze-Age routes.

The limestone terrain supports its own specialized flora, and the rich meadows are also home to bustards, larks, and a fine variety of butterflies. Peregrine falcons and other birds of prey nest on the precipitous cliffs of the many gorges. There is also a breeding program for Przewalski's horse, which is now almost extinct in its original Asian steppeland home. **AB**

HÉRISSON WATERFALLS

FRANCHE-COMTÉ, FRANCE

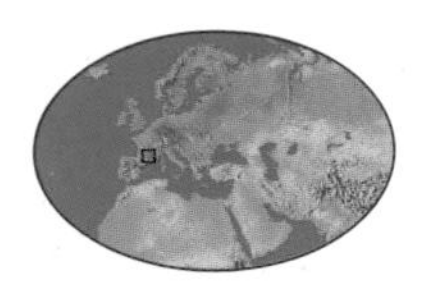

Height of Grand Saut: 200 ft (60 m)

Rock type: Limestone

Age of waterfalls: 208 to 146 million years

Situated in Jura's Region des Lacs, this magnificent sequence of waterfalls is best seen after the fall rains or during the early spring when melting snow provides added vivacity to their flow. Glacial erosion has bitten deep into the region's ancient limestone, resulting in some remarkably deep valleys. The most extreme of these is occupied by the River Hérisson. Descending from its source at Bonlieu Lake to the Doucier Plateau, it drops 920 feet (280 meters) in 2 miles (3.2 kilometers). Replete with narrow gorges, mini plateaus, and rock jumbles, this stretch of river has 31 cascades, each with its own particular geological character and features. Among the best known are Eventail, a fan-shaped staircase of foaming water, the long drop and massive plunge pool of Grand Saut, the misty splendor of Le Saut de Doubs, and the moss-covered rock tumble of Les Tufs. The region also has a herd of semi-wild bison and cattle which have been bred to resemble their Pleistocene ancestor, the auroch. **AB**

MONTE PADRU NATIONAL PARK

CORSICA, FRANCE

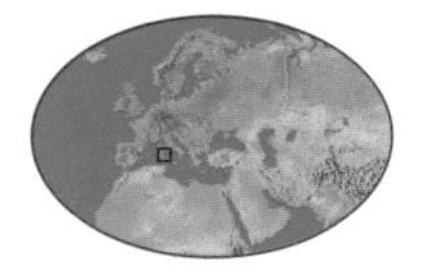

Highest point (Mt. Padru): 7,852 ft (2,394 m)

Length of Giussani Valley: 51 mi (95 km)

This high region of central Corsica combines the curves and gorges of the River Targagine with the rugged slopes of Monte Padru. In addition, considerable human interest is provided by the architecture of the region's four mountain villages: Mausoleo, Olmi Cappella, Pioggiola, and Vallica. Dotted down the Giussani Valley, these long-isolated villages and their inhabitants continue a rural lifestyle that has all but disappeared from much of Europe. The activities of these inhabitants have both modified and enriched the natural landscape, too. When hiking you can thank long-gone villagers for the paved mule paths you walk and the ancient bridges on which you cross rushing mountain torrents.

The park has several peaks above 5,000 feet (1,500 meters), including Padru, Assemble Corona, Monte Grossu, and San Parteu. Most of these have slopes covered with pine trees, while fragrant, heat-resistant maquis vegetation dominates the park's lower areas. Wild sheep (mouflon) graze the upper slopes, while the lower ones are home to ubiquitous domestic goats. The region's range of large birds of prey includes golden eagles and bearded vultures. **AB**

RESTONICA GORGES

CORSICA, FRANCE

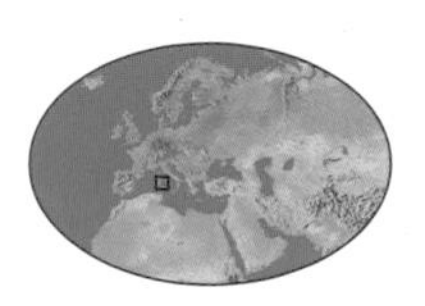

Highest point (Mt. Rotondo): 8,600 ft (2,622 m)

Habitat: riverside meadows, mountain chestnut and pine forests

Situated astride Corsica's mountainous central spine, the Gorges of the Restonica River are famed for both their beauty and rich flora and fauna, and so the area has been classified as a protected area since 1966. The river runs a course partly of its own excavation and partly one that was molded by glaciers during the last ice age. This creates a fascinatingly varied landscape of smooth contours, dynamic cascades, sudden ravines, and deep pools.

There are two beautiful glacial lakes here—Melo Lake and Capitello Lake (the latter being more isolated). The first can be reached after a one-hour hike that is relatively easy. If you choose to visit Capitello Lake, you have a tougher ascent ahead—you must climb up an old glacial moraine, but the spectacular views from the top and surrounding vertical cliffs alone are ample reward for the effort involved, not to mention the satisfaction of completing the challenge. In case of problems there are frequent stone huts for travelers, though these are mostly basic and offer little except shelter. The region is extremely popular with tourists in the summer—a testament to its natural beauty. **AB**

HRANICE GORGE

OLOMOUCKY KRAJ, CZECH REPUBLIC

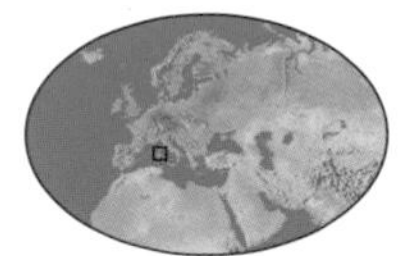

Depth of Hranice Gorge: 1,081 ft (329 m)

Diameter of Hranice Gorge: 900 ft (270 m)

Rock type: karstic limestone

Hranice Gorge, also known as the Hranicka Propast, is the deepest chasm not only in the Czech Republic but also in all of central Europe. This amazing gorge is essentially a gigantic pothole formed in the limestone karst rock of the plateau above the River Recva. The pothole was created first as an enormous subterranean cave, by warm, carbon dioxide–rich mineral waters dissolving through the limestone. The water originated from thermo-mineral springs deep below the surface. Eventually so much limestone was leached away that the weight of the pothole roof proved too much and it collapsed in on itself, creating the gorge we know today.

The lower reaches of the gorge are now flooded by the warm mineral waters rising up from the springs that have formed a pond 673 feet (205 meters) deep. Only the upper 226 feet (69 meters) of the gorge are not buried under water. The gorge is named after the nearby town of Hranice, which is about 2.5 miles (4 kilometers) to the west. The warm mineral waters that created the gorge have been put to good use at two health resorts, Teplice and Becvou, which are located opposite Hranice Gorge. **JK**

TATRA MOUNTAINS

POLAND / SLOVAKIA

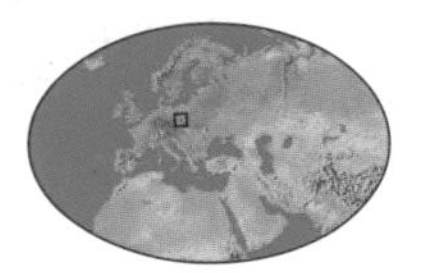

Area of Tatra Mountains: 307 sq mi (795 sq km)

Highest point (Rysy): 8,199 ft (2,499 m)

A two-hour drive from the city of Krakow are central Europe's highest mountains: the Tatras, known by Poles as the "Polish Alps." They have been popular with tourists for centuries, and have become increasingly so with film directors due to the fact that the city of Poprad, at the foot of the mountains on the Slovakian side, is host to the Poprad Film Festival. Rising steeply from a high plateau, they form the central, highest section of the Carpathians, and contain some of Europe's most beautiful and dramatic scenery. The entire chain of peaks is 9 miles (15 kilometers) wide, and forms Poland's natural border to the south. The political border that divides Poland from Slovakia runs down their spine, with just 24 percent of the mountain range lying in Polish territory.

The mountains owe their present-day appearance mostly to Pleistocene glaciation. Within the last 500 to 10,000 years, glaciers have appeared and disappeared in this region. Today, none remain. Time has weathered the mountains' hard granite into a sharply edged, majestic, and towering range—the haunt of chamois, bear, lynx, wolf, and deer.

Sharp rock towers, crests, and numerous cirques filled with some 30 clear, cold lakes called "stawy" characterize the dramatic High Tatras, formed of crystalline rock. In the western part of the range, water has eroded limestone and dolomite, flattening the tops, carving deep ravines and valleys, and creating many caves, which form a subterranean maze beneath the massif. Seven caves are open to the public, six of which lie in the beautiful Koscieliska Valley. The longest cave, a huge 59,400 feet (18,000 meters), is Litworowa.

The caves are steeped in legend, and though dragons were said to live in them, they were once a place of outlaws—the infamous Polish "zbojnicki." Bats inhabit them now in a wild region where beech and fir forests cloak the lower Tatras slopes, the vegetation of the higher slopes primarily being dwarf pine and alpine meadows. Rare flowers cling to stark rock faces—but the richest blooms fill the meadows and the valleys, which are a carpet of yellow croci in spring.

At the end of the 19th century, a Warsaw doctor declared the mountain air of the Tatras the perfect prescription for good health. Today three million visitors every year head there to follow his advice. JD

Time has weathered the mountains' hard granite into a sharply edged, majestic, and towering range–the haunt of chamois, bear, lynx, wolf, and deer.

RIGHT *The steep, craggy slopes of the Tatras Mountains, Slovakia—central Europe's highest mountains.*

THE SLOVAK PARADISE & HORNÁD CANYON

KOSICKY KRAJ / ZILINSKY KRAJ, SLOVAKIA

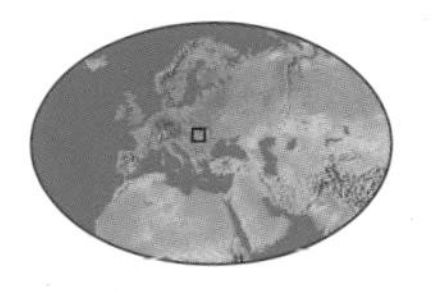

Area of the Slovak Paradise: 7,630 sq mi (19,763 sq km)

Elevation of the Slovak Paradise: 1,650 to 5,610 ft (500 to 1,700 m)

The name says it all—the Slovak Paradise is a land of scenic wonders, including meadows, canyons, gorges, caves, hills, rivers, and waterfalls. This enchanting national park is rich in limestone that has been shaped over time into a wide variety of landscapes. Limestone is particularly vulnerable to water erosion, and consequently there is no shortage of caves and chasms to explore—177 in total.

One of the greatest highlights is Hornád Canyon, a beautiful 10-mile (16-kilometer) stretch created by the Hornád River. The steep banks of the canyon are 1,000 feet (300 meters) high in places. A somewhat hair-raising path has been constructed through the canyon, which takes walkers along a motley collection of gangboards, metal footbridges, clenches, and chains built into the rock walls—but the views are worth the effort.

Equally interesting is the Dobšinská Ice Cave, which has its own subterranean glacier, 90 feet (27 meters) thick, as well as ice waterfalls, ice stalagmites, and ice columns. JK

DOMICA CAVE

KOSICKY KRAJ, SLOVAKIA

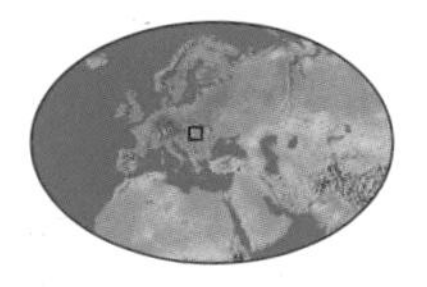

Length of Domica Cave: 3.3 mi (5.4 km)

Rock type: karstic limestone

Age of rock: 225 million years

The Domica Cave is the most impressive and beautiful of several caves found in the Slovak Karst, a large limestone region in southern Slovakia, on the border with Hungary. The limestone sediment was laid down 225 million years ago, and subsequently eroded by two underground rivers, the Styx and the Domica. The cave has numerous corridors that have formed along natural faults in the limestone. The continuous dripping of water has created a wonderful forest of stalactites, so thick in places that visitors are often in danger of losing one another among them. A tour of the cave includes an underground boat trip.

The cave was once a dwelling place for a prehistoric people called the Bukk Culture, who lived there 8,000 years ago and left behind a fascinating assortment of stone tools, giving us a great insight into their culture. However, one artifact that was discovered in the cave proves that people lived there up to 40,000 years ago. Today Domica Cave is home to 14 species of bats, including Slovakia's largest colony of horseshoe bats. Domica Cave is part of the National Nature Reserve Domické Škrapy (Domické Karren). JK

PASTERZE–GROSSGLOCKNER GLACIER

TIROL, AUSTRIA

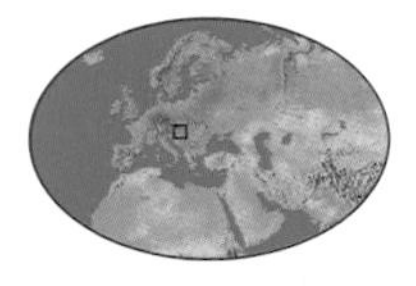

Thickness of Pasterze: 1,000 ft (300 m)

Height of Glossglockner: 12,460 ft (3,798 m)

Habitat: alpine rock faces

Guarding the north face of Austria's highest mountain—the Grossglockner—is the Pasterze Glacier, Eastern Europe's largest glacial formation. Hiking routes are well marked, and a funicular railway also provides access, so the surface and the terminal talus of the Pasterzenkees are easily reachable.

The glacier is very stable, so visitors can walk on it fairly safely without ropes. The glacier itself is receding at about 1 foot (30 centimeters) per year. In a thousand years, it will no longer exist. Access is via the winding Hochalpenstrasse, which follows an ancient trading route, known as the Römerweg. Its remains are still clearly visible beside the modern road. In medieval times it connected Germany and Venice and was the trading route for spices, glass, and salt.

Nearby are the extremely beautiful Lake Zell, the Krimmi Waterfalls, and the Hohe Tauern National Park. Camping, walking, and rock climbing are very popular activities in the region. **AB**

KARWENDEL MOUNTAINS

TIROL, AUSTRIA

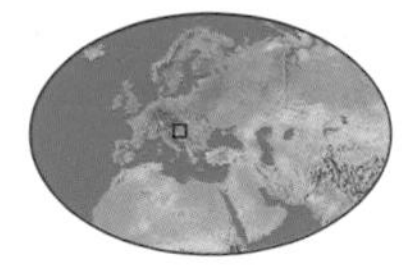

Highest peak (Bikkarspitze): 9,017 ft (2,749 m)

Habitats: bare rock, alpine grasslands, rock-face plants

Rock type: limestone

Along with the Lechtal Alps, and the Mieming, Rofan, and Kaiserbirge Mountains, the Karwendels are part of the north Tyrolean limestone chain. They reach nearly 10,000 feet (3,000 meters). A cool, moist climate and abundant rain, pastures, woods, and game characterize the Karwendels. Thinly populated, much of the region is included in the Karwendel Nature Reserve, Austria's largest. The region is remote, and its rugged landscape is notoriously difficult to traverse. It has also long been protected by royal decrees—several villages formerly served as royal hunting lodges.

The village of Pertisau also has fossil deposits that provide ichthyol, oil used in homeopathic remedies. Trails in the park are well maintained, but the region's isolation means good planning is required for a safe and successful trip. Cabins provide an alternative to camping. Mountain biking is also popular on the lower trails.

The region's highest peak is Mount Karwendel, which is composed of crumbly limestone that generally makes for unsatisfying climbing. Lake Achen, the largest and deepest lake in the Tirol, with beautiful, clear water, lies at one end of these mountains. **AB**

KRIMML FALLS

SALZBURG / TIROL, AUSTRIA

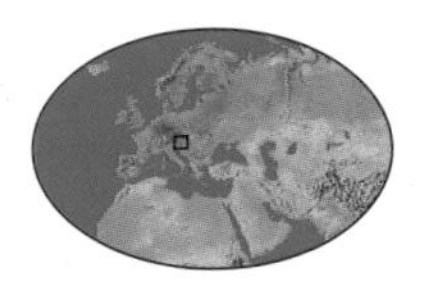

Height of Krimml Falls: 1,250 ft (380 m)

Elevation of falls: 5,530 ft (1,687 m)

Length of Krimml Valley: 12 mi (19 km)

At 1,250 feet (380 meters) high, the Krimml Falls are Europe's longest free fall of water. They rank eighth among the great waterfalls of the world. Located at an elevation of 5,530 feet (1,687 meters), in winter the spray coats the surrounding rocks and vegetation in glistening ice, forming surreal sculptures. There are three cascades, and access to them is from Zell am Ziller, via a toll road. A walk from road to falls can take over an hour, so beware of sudden weather changes. A well-maintained viewing trail follows the length of the falls and then continues into the Krimml River Valley, a stunning high valley of alpine grassland. It is one of the most beautiful areas of the national park and it terminates at the Krimml Glacier.

The falls are open to visitors between late April and late October (depending on weather conditions). This popular waterfall has received the European Parliament's Nature Preservation Diploma. The falls are within the Hohe Tauern National Park. Part of the eastern Alps, the park has 304 mountains higher than 10,000 feet (3,000 meters) and 246 spectacular glaciers. **AB**

RIGHT *Krimml Falls thunders through the alpine forest.*

EISRIESENWELT

SALZBURG, AUSTRIA

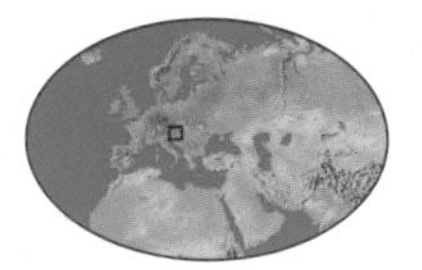

Length of Eisriesenwelt: 25 mi (40 km)

Elevation of Eisriesenwelt: 4,950 ft (1,500 m)

Height of cave entrance: 60 ft (18 m)

A labyrinth of ice caves and cathedral-like caverns, Eisriesenwelt is the largest network of ice galleries in Europe. Hidden below the Tennen Massif, to the south of Salzburg, the caves were first explored by Anton von Posselt-Czorich in 1879, and then by Alexander von Mork in 1912. It was von Mork who discovered the "world of the ice giants" that extends at least 25 miles (40 kilometers) underground. The caves are at such an elevation—over 5,000 feet (1,500 meters) above sea level—that any meltwater or rain seeping into the caves freezes instantly. Instead of the usual formation of calcite stalagmites and stalactites, these caves contain unusual ice formations, such as the "ice organ" and "ice chapel." Icy drafts blow through the caves, ensuring the walls are covered with a layer of glittering hoar frost.

The entrance is enormous—66 feet (20 meters) wide and 60 feet (18 meters) high—and can be seen from some distance away. The largest cavern is 200 feet (60 meters) long, 100 feet (30 meters) wide, and 116 feet (35 meters) high. Guided tours last for about an hour and visitors are advised to wear warm clothes in an atmosphere rarely above 32°F (0°C). **MB**

UNTERSBERG

SALZBURG, AUSTRIA / BERCHTESGADEN, GERMANY

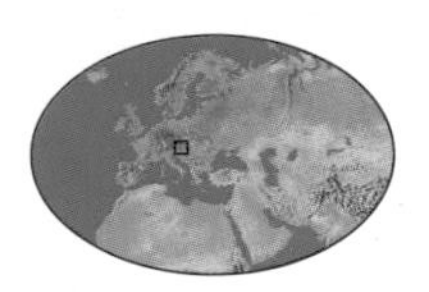

Height of Untersberg: 6,469 ft (1,972 m)

Notable feature: home to 70 types of butterfly, a third of them endangered

Untersberg is a massive table mountain a mere 10 miles (16 kilometers) south of Salzburg and spreading into Germany. Known to tourists since the 19th century, the plateau of Untersberg shows multiple karst phenomena, such as cupola karst, hollows, dolines, and lapies, but it is famous for its caves—about 400 are known and 150 are documented in detail. The most impressive is the Schellenberg ice cave, where the deepest strata indicate an age of approximately 3,000 years.

On the plateau, fauna include ptarmigan, mountain hare, and chamois. Higher up the slopes, characteristic beech trees give way to larger coniferous forests, with dwarf trees at the summits. Blueberries and violets are common, and there is an abundant variety of mosses and ferns.

There are hiking trails, a ski run, cycling, guided cave visits, and a cable car offering spectacular views. According to one legend, Charlemagne sleeps in the mountain with his court; he will only wake to lead a last battle of good against evil when ravens have stopped circling the summit. Another legend tells that it is Holy Roman Emperor Frederick Barbarossa who sleeps there. GD

SEISENBERG GORGE

SALZBURG, AUSTRIA

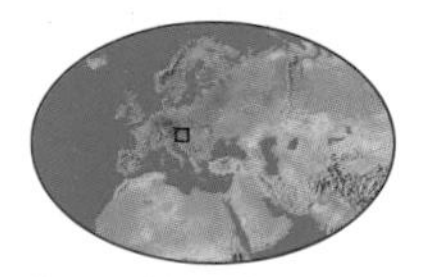

Length of Seisenberg Gorge: 2,000 ft (600 m)

Depth of Seisenberg Gorge: 165 ft (50 m)

The Seisenberg Gorge natural monument is an impressive 2,000-foot- (600-meter-) long, 165-foot- (50-meter-) deep canyon near Weissbach in Salzburg. It was inaccessible until 1831, when woodcutters constructed the first pathway to enable them to transport logs through the gorge. The Weissbach stream runs through a forest to a set of dramatic rapids and then plunges into the narrow gorge where the water has carved the rock into a series of smooth-sided caverns and tunnels.

An elaborate stairway of convenient wooden steps and walkways has been constructed to enable visitors to walk through the canyon. The stairway is open from May to October and the walk through the gorge takes about 30 minutes each way.

At the nearby Vorderkaser Gorge, the Odenbach has carved a 1,320-foot- (400-meter-) long canyon to a depth of 264 feet (80 meters). A series of steps leads visitors past the gorge's extraordinary rock formations. There are also several natural lakes at the entrance to the gorge that are suitable for bathing. While in this area, be sure to visit Lamprecht's Cave near Weissbach, one of the world's largest viewing caves. RC

LIECHTENSTEIN GORGE

SALZBURG, AUSTRIA

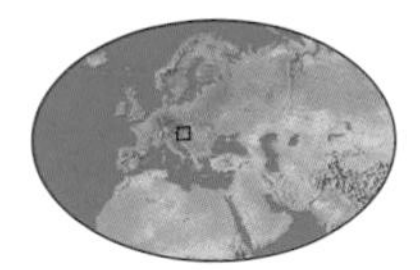

Local name: Liechtensteinklamm

Depth of Liechtenstein Gorge: 1,000 ft (300 m)

Height of highest waterfall: 165 ft (50 m)

Legend has it that the Liechtenstein Gorge in the heart of Austria's Salzburg state was created by the Devil when he was in a rage at being cheated out of a pact with a local blacksmith. This bizarre mythology, coupled with its immense natural beauty, has drawn inquisitive tourists to the small alpine town of St. Johann in Pongau since 1875. The network of wooden walkways which afford views of the remarkable waterfalls throughout the canyon was established by a group of local rangers and funded by Johann II Fürst von Liechtenstein, after whom the gorge is named.

Over many thousands of years, the Großarl River, fed by glacial meltwaters, has carved its way 1,000 feet (300 meters) into the ground to form the canyon. As the water makes its tortuous descent, it pounds and swirls against the rock, creating incredible shapes and patterns. A footpath links a series of small bridges to allow visitors to experience the untamed cascades: at times the rock walls of the gorge are so closely juxtaposed that only the merest glimpse of the sky above is possible. At other times, when the sun shines on the thundering sprays rising from the canyon, a dazzling rainbow-filled display emerges. **NA**

LAMPRECHT'S CAVE

SALZBURG, AUSTRIA

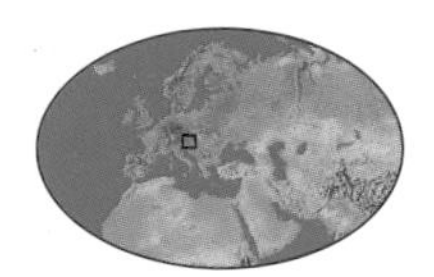

Depth of Lamprecht's Cave: 5,355 ft (1,632 m)

Length of Lamprecht's Cave: 30 mi (50 km)

Lamprecht's Cave is one of the most extensive cave systems in Europe. It is also the deepest cave in the world that visitors can actually walk through. According to legend, the cave got its name from Knight Lamprecht who brought back treasure from the Crusades. The treasures were later inherited by his two daughters, one of whom stole her sister's share and was fabled to have hidden it in the cave. Attempts to find the treasure over the centuries led the regional government to wall up the cave in 1701. Charged with melted snow, the torrential cave river that leaves the cave after heavy rains probably destroyed the wall soon after it was built.

Despite having an alarm system to warn of flooding, in recent years people have still been reportedly trapped briefly underground (four German cavers in January 1991, and 14 people on a cave tour in August 1998, following heavy rains). In 1998, a connection was discovered between Lamprecht's cave and the PL-2 cave system, making it one of the deepest caves in the world. However in 2001, the world cave depth record—5,577 feet (1,700 meters)—was taken by Voronya Cave (or Krubera Cave) in Abkhazia, Georgia. RC

THE CEAHLAU MASSIF

NEAMT, ROMANIA

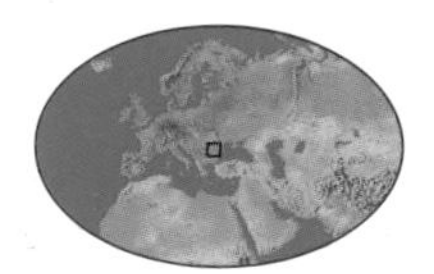

Height of Ceahlau Massif: 6,257 ft (1,907 m)

Area of national park: 66 sq mi (172 sq km)

Rock type: limestone

The Ceahlau Massif is known as "the Jewel of Moldavia." It stands out defiantly from the surrounding countryside. Although comprising part of the Eastern Carpathian mountain range, the Ceahlau Massif is a separate and solitary mountain, which emphasizes its prominence in the surrounding area. The massif has the appearance of a huge crumbling castle, with features that look like steep walls, towers, and ramparts in varying states of disrepair. The mountain is considered sacred, and each of its peaks has its own name and individual creation myth.

Another name for this massif is the "Magic Mountain." Legend has it that Ceahlau was the home of Zamolxe, the God of the Dacians, who were the ancestors of the Romanian people. Here Zamolxe sacrificed Dochia, the daughter of King Decebal, on its highest peak. Numerous climbing trails now lead to the top of Dochia Peak, and it is said that the mountain's magic ensures eternal friendship for those who climb it together. The mountain and the surrounding area has been designated a national park in order to preserve Duruitoarca falls and about 2,000 flower species, as well as cliff butterflies and other rare animal species. JK

CHEILE TURZII

CLUJ, ROMANIA

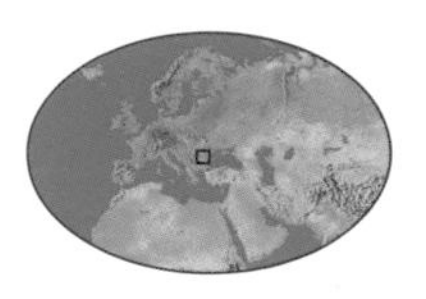

Alternative name: Turda Gorge

Length of gorge: 1.25 mi (2 km)

Depth of gorge: 825 ft (250 m)

Cheile Turzii is the magnificent gorge that cuts through Romania's Apuseni mountain range. Created by the Hasdate River, 825-foot- (250-meter-) high limestone walls now frame this beautiful natural reserve. This site has been renowned since Roman times, both for its picturesque landscape and its exquisite diversity of plants and animals. It is also known by the name of Turda Gorge, after the nearby town of Turda. The walls of the gorge contain more than 60 caves, and provide a home for numerous bat colonies. Cheile Turzii has been home to various peoples—stone tools found in the caves reveal that they were once home to Bronze and Stone Age peoples, and during the Middle Ages locals fleeing invading Tartars hid in the caves.

The unique microclimate of the gorge provides a habitat for sun-loving plants that are otherwise mostly found on the shores of the Mediterranean or in Central Asia. More than 1,000 plant species live here, as well as 111 bird species, including the golden eagle, eagle owl, wallcreeper, and rock vulture. Cheile Turzii is also one of Romania's top climbing destinations, with over 100 hiking and climbing routes of varying difficulty. JK

BICAZ GORGE

NEAMT, ROMANIA

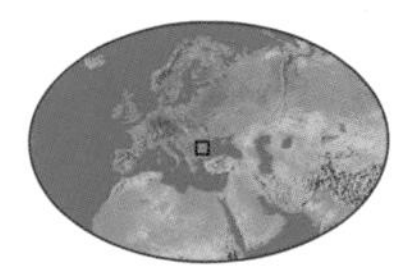

Length of Bicaz Gorge: 3 mi (5 km)

Height of Bicaz Gorge: 1,000 ft (300 m)

The Bicaz Gorge is an awesome chasm in the heart of the Carpathian mountain range in the center of Romania. The chasm is three miles (five kilometers) long, 1,000 feet (300 meters) deep, extremely narrow in places, and a sensational sight. It twists and turns through sheer limestone cliffs on either side. At one point, which is appropriately named the "Neck of Hell," the looming cliffs directly overhang the mountain path, encouraging walkers to move swiftly on.

The best way to experience the immensity of Bicaz Gorge is on foot. Numerous wallcreeper birds inhabit the gorge. Nearby is Lake Bicaz, an ideal mountain lake for the rest and relaxation necessary after the exertions of walking. As an added bonus, the region has a rich history—this is the land of Vlad Drakulea, a medieval local ruler who took on the Turks and eventually became the source of the horror stories that are better known today as the legend of Count Dracula. The gorge is located 13 miles (21 kilometers) from the town of Bicaz in the principality of Moldavia, on the border with Transylvania. The gorge is now entirely protected within the Hasmas-Bicaz National Park. JK

DANUBE DELTA

TULCEA, ROMANIA

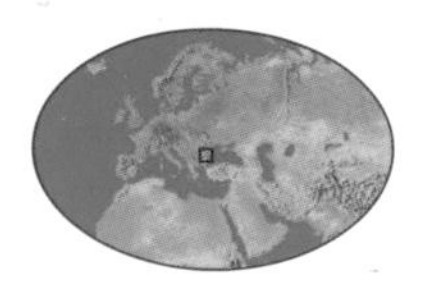

Number of bird species: 310

Number of fish species: 75

Number of plant species: 1,150

As the Danube approaches Romania's Black Sea coast it splits into three separate rivers, the Chilia, Sulina, and Sfantu Gheorghe, which flow through a wetland paradise: the Danube Delta. About 5,000 years ago this area was a gulf of the Black Sea, but over time the sediment spilling out from the Danube has filled it in and created this enormous delta. In fact the delta still receives over 2.2 tons of silt per second, thereby expanding the size of Romania by 132 feet (40 meters) every year.

A vast network of channels, brooks, and ponds with innumerable islets and sandbars interconnects the rivers flowing through the delta. The surfaces of the lakes are covered in white and yellow waterlilies. The Danube Delta has one of the largest expanses of reedbeds in the world, covering more than 600 square miles (1,560 square kilometers). The luxurious vegetation continues on to land, with forests that are festooned with lianas. This rich wetland is a magnet for migratory birds and is home to Europe's largest colonies of white and Dalmatian pelicans. Half the world's population of red-breasted geese winters here. With an additional 307 bird species sighted, this is an ornithological wonderland. JK

ENGADINE MOUNTAINS

GRAUBÜNDEN, SWITZERLAND

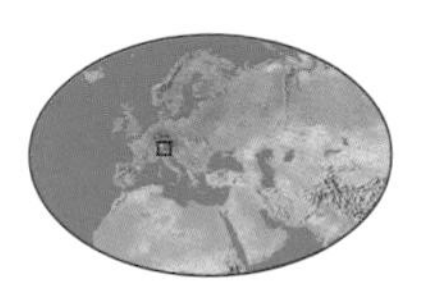

Highest point of Engadine Mountains: 8,475 ft (2,584 m)

Habitats: bare rock, alpine grasslands, pine forests, alpine lakes

Rock type: limestone

The near-impassable Piz Buin, the highest mountain in the Silvretta Alps, severs the Engadine from its neighbors. Over time, this barrier has led to the isolation of the Romansch area, which, though a short distance away, has become culturally distinct from the Austrian Tirol. The region has many ancient villages and hamlets, including some that are situated on mountain passes that were originated by the Romans—buildings are therefore sturdy and stone built, with characteristic deeply-recessed windows and painted façades.

The area also boasts castles, glittering lakes, and undulating alpine meadows that, if you look closely enough, are covered with tiny flowers. The shining River Inn (heading toward Innsbruck, the Danube, and, eventually, the Black Sea) weaves its way through this area as well. This landscape is set against a background of looming mountains and dark and mysterious pine forests. Scuol is the main town in the River Inn valley, followed by a succession of hamlets—Ftan, Guarda, Zernez. One village, Müstair, features a church with perfectly preserved medieval frescos. **AB**

AREUSE GORGE

NEUCHÂTEL, SWITZERLAND

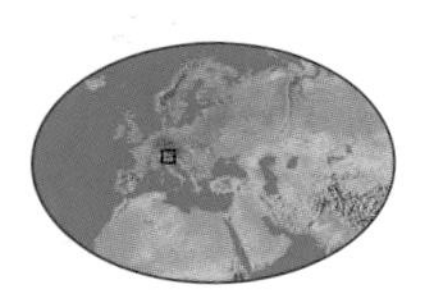

Habitats: alpine grasslands, vineyards, alpine lakes, waterside vegetation, rivers

Rock type: limestone

Created when the waters of the Jura tried to cut through the region's limestone deposits, the Areuse Gorge is a spectacularly narrow gorge, couched on either side by agricultural land and shallow, rock-filled waters. Some stretches of the Areuse's green-colored waters swirl into impressive potholes and rapids. The gorge is located in the three lakes region of the Swiss department of Neuchâtel. You can cross the gorge if you are brave enough—the path appears to be pinned to the cliff sides, though the two-hour hike is rewarding, and grants both a fine view and a sense of achievement.

Visitors can access Areuse Gorge from the small village of Noiraigues or a region known as Champ de Moulin. The nearest town is Motiers, famous for its medieval buildings. Nearby are the asphalt mines at Travers. Though the mines closed in 1986 after 250 years of production, a museum has been created there, as well as a restaurant where you can eat ham cooked in asphalt at 428°F (220°C). To reach the gorge, drive 20 minutes from the village of Gals. After walking though the gorge, a short railway journey will take you back to Champ de Moulin. **AB**

HOLLOCH CAVE

LUZERN, SWITZERLAND

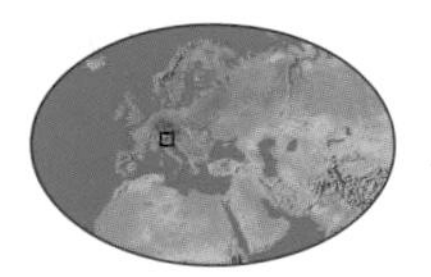

Length of Holloch Cave: 117 mi (190 km)

Depth of cave: 2,860 ft (872 m)

Age: about 1 million years

The *Holloch* ("Hell's Hole") is the largest cave system in Europe. With 117 miles (190 kilometers) of known passages, galleries, and underground lakes, it is also one of the longest caves in the world. A farmer discovered it in 1875, but at that point only 4 miles (6.4 kilometers) of caves had been explored. Swiss geologist Alfred Bogli began surveying the caves in 1945; despite being trapped underground by floods for 10 days, by 1955 he had measured 34 miles (55 kilometers) of caves.

The lurking structures of stalactites and stalagmites populate Holloch cave. What is more, interspersed between the cave's unusual deeply-colored rock shapes are specialized fauna and flora. In 1982, an extensive gallery which was later called Nirwana was discovered —visitors should note that Nirwana is only accessible in dry weather.

Guided tours of a small section of the cave are available, though longer caving expeditions are possible from November to March. During the winter, as snow fails to seep through the limestone, precipitation levels plummet and the cave system is dry. After the snow melts in the spring, the deeper caves inevitably become flooded with water. **RC**

GROSSER ALETSCH GLACIER

VALAIS, SWITZERLAND

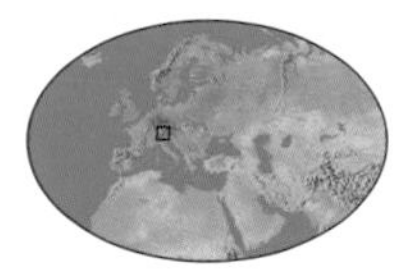

Length of Grosser Aletsch Glacier: 15 mi (24 km)

Age: 60,000 years

Switzerland's Grosser Aletsch Glacier is the largest glacier in the Alps. Inscribed on UNESCO's World Heritage List, it is over 15 miles (24 kilometers) long and up to 1 mile (1.6 kilometers) wide. Located in the Bernese Alps near the Italian border, the glacier feeds the Massa River, a tributary of the Rhône. The glacier predates the last ice age and is an estimated 60,000 years old—being in its presence today, visitors can imagine how vast areas of northern Europe and North America would have looked 10,000 years ago. Access is by road from the nearby towns of Brig or Oberwald to Betten. At Betten, a gondola will take you to viewing points.

Walking on the glacier is permitted, but only with a licensed guide. Hiking is also possible via well-maintained and signposted paths. Some paths crisscross alpine meadows and glacial and erosional features, while others provide access to the Aletsch Forest, which is dominated by pine trees. The Aletsch Forest Nature Reserve provides important resources for tourists, including information about the glacier. Look out for marmots, eagles, and alpine plants (including gentians and edelweiss). The Jungfrau is close by. **AB**

THE MATTERHORN

SWITZERLAND / ITALY

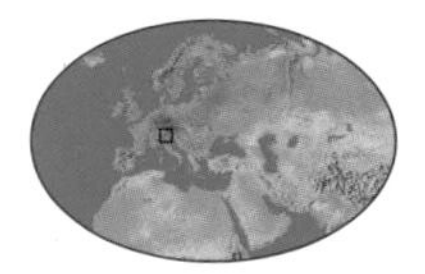

Height of the Matterhorn: 14,692 ft (4,478 m)

Age: 50 million years

First successful ascent: 1865

The Matterhorn's instantly recognizable pyramidal peak sits slightly off-center and betrays the way it was made. The peak is where four ridges (or arêtes) meet. Between the hollows of these arêtes, successive piles of snow and ice have accumulated to form glaciers—cracking the rocks and scooping out rounded basins, known as *cirques* (French), *corries* (Scottish), or *cwms* (Welsh). The rocks that form the mountain were folded and thrust high into the air about 50 million years ago by earth movements that were the result of the African continent smashing into Europe.

Today, the Matterhorn straddles the border between Switzerland and Italy. Two thousand mountain climbers reach the top each year, and 100 climbers can be on the mountain at any one time, although up to 15 a year die in climbing accidents. The peak's first successful ascent was in 1865 by a British wood engraver, Edward Whymper, although the expedition ended in disaster. He and his team reached the top from the Swiss side, just hours before an Italian group climbed the other side, but on their descent one of the party slipped and fell; four people died as they hit the Matterhorn Glacier 4,000 feet (1,200 meters) below. **MB**

JUNGFRAU-ALETSCH-BIETSCHORN

BERN / VALAIS, SWITZERLAND

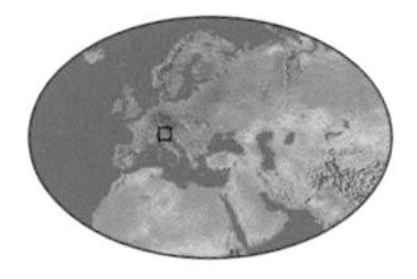

Area of Jungfrau-Aletsch-Bietschorn: 133,437 acres (54,000 ha)

Maximum height: 14,022 ft (4,274 m)

The Jungfrau-Aletsch-Bietschorn region, located in the south-central Swiss Alps, is a stunningly beautiful region. A rich coalition of geological and glacial processes have combined to produce this extraordinary landscape. The complex rock formations here were the result of overthrusting and folding rock layers between 20 and 40 million years ago and have subsequently been exposed by the action of glaciers.

The region covers 133,437 acres (54,000 hectares) and boasts heights ranging between 3,000 feet (900 meters) and 14,022 feet (4,274 meters). A total of nine peaks exceed 13,200 feet (4,000 meters) in height. The region is plowed by the Aletsch Glacier, a vast continuous river of ice over 15 miles (24 kilometers) in length and 3,000 feet (900 meters) deep.

Reintroduced ibex, lynx, and red deer are thriving successfully, and the area will hopefully soon offer a safe refuge to threatened species. Most typical alpine species are already represented, and the fauna includes marmot, chamois, pygmy owl, golden eagle, and the rare alpine salamander. Protection measures put in place in 1933 have ensured that this alpine region is one of the most vast and undisturbed natural habitats in Europe. NA

GRAN PARADISO NATIONAL PARK

VALLE D'AOSTA, ITALY

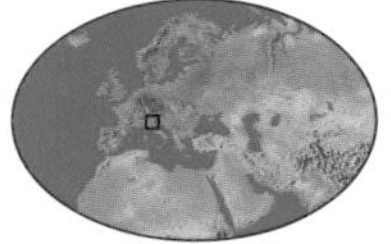

Extent of protected area: 239 sq mi (620 sq km)

Highest peak of Gran Paradiso mountain: 13,323 ft (4,061 m)

Set within the spectacular scenery of the Italian Alps, Gran Paradiso is Italy's oldest and greatest national park. This alpine wonder is located in the northwest corner of Italy, in the region of Aosta, and encompasses snow-capped mountains, deep valleys, glacial lakes, forested slopes, and colorful alpine meadows. The mountains here bear the dramatic imprint of glaciation, with strong ridges and sharp jagged peaks. Many still have glaciers hugging their slopes. The park has a rich and varied habitat, beginning with wooded valleys of larch, fir, and pine at lower elevations and extending to the flower-covered pastures that appear above the tree line. Even higher elevations reveal barren, rocky landscapes interspersed with permanent glaciers.

The Gran Paradiso mountain, the largest mountain existing entirely within Italian borders, dominates the park. This beautiful peak is popular with skiers and climbers. The climb up to the peak is steep and challenging, but requires no advanced climbing techniques.

Gran Paradiso used to be a royal hunting reserve and consequently has about 450 miles (725 kilometers) of trails and mule tracks, making it an ideal destination for hikers of all abilities. It was declared a national park in 1922 after King Vittorio Emanuele III donated the land to the nation to help protect its diminishing population of ibex. These wonderful, wild mountain goats inhabit the alpine pastureland and have become the very icons of Gran Paradiso. The males are easy to spot because of their impressive long, curved horns, while the females, with shorter horns, live in separate groups with their young. The park is also home to the gregarious chamois, another nimble-footed goat antelope that grazes on the precarious narrow ledges and precipitous scree slopes of the mountains. Other animals in the park include the golden eagle, Eurasian eagle, and the marmot. Sighting the lammergeyer, or bearded vulture, is possible, though rare. This once extinct bird of prey was successfully reintroduced thanks to a program that was developed across the border in France's Vanoise National Park. JK

Set within the spectacular scenery of the Italian Alps, Gran Paradiso is Italy's oldest and greatest national park. The mountains here bear the dramatic imprint of glaciation, with strong ridges and sharp jagged peaks. Many still have glaciers hugging their slopes.

RIGHT *Gran Paradiso is Italy's oldest and most beautiful national park.*

THE BLUE GROTTO

CAMPANIA, ITALY

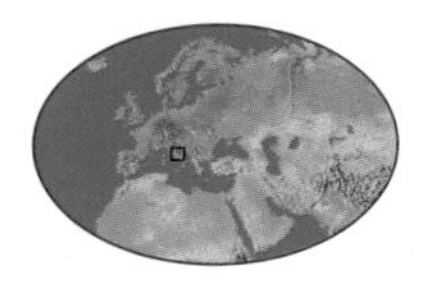

Length of Blue Grotto: 177 ft (54 m)

Width of grotto: 100 ft (30 m)

The Blue Grotto, on the coast of the island of Capri, is one of the most beautiful sea caves in the world. Visitors can row through the entrance of the grotto via a small passageway with a ceiling so low that they have to lie down in the boat to get through. In order to enter, the gondolier waits for the lowest water between two waves, and then drags the boat in by a rope along the wall. It is well worth the trouble—once inside, a magnificent geological spectacle unfolds in the shape of a large oval-shaped cave that shimmers in stunning silvery-blue light. It is as if you have entered a giant sapphire. In the second undersea entrance, daylight shines onto the white sands at the foot of the cave and is then reflected up as pure blue light. The water absorbs all other wavelengths of light except the blue.

Along with decades of tourists, the Romans, who proclaimed it a place of worship for the Emperor Tiberius, also revered it. They adorned the walls of the cave with statues, including a Neptune and a Triton that were discovered on the seabed in 1964. JK

RITTEN EARTH PILLARS

TRENTINO-SOUTH TIROL, ITALY

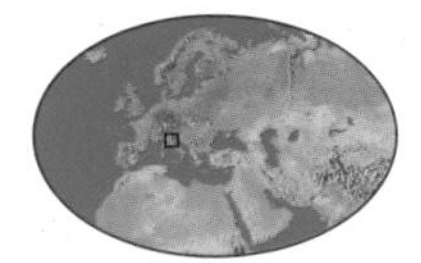

Age of Ritten Earth Pillars: 10,000 years

Number of pillars: 100 to 150

Genesis: glacial

Just north of the Dolomites in the South Tyrol of Italy are the most unusually shaped rocks. Each is perched like an ill-fitting hat atop an earth pillar—as if Mother Nature is playing a practical joke. Some are over 130 feet (40 meters) high, others are no more than low stumps.

The pillars have their origin in a glacier that once flowed through the valley during the ice age. When the climate warmed about 10,000 years ago, the ice melted and the glacier retreated, leaving behind not only the fine debris (known as boulder clay) it had collected on the way, but also large, hard rocks that it had scraped from the sides of mountains. Since then, rivers have carved gullies in the clay and rain has sliced up the surviving ridges, leaving rows of clay pillars. If a pillar was capped with a rock it was protected from the rain, like a person under an umbrella. Gradually, though, rain has eroded the sides of the pillars, causing their protective rocks to fall, and the pillar to disintegrate. Ritten (or "Renon" in Italian—place names are in German and Italian because South Tirol was part of the Austrio-Hungarian Empire) is one of several similar sites in the Alps. In Italy they are known as "little men," but in France they are *demoiselles coiffées* (young ladies with fine hair or hats). MB

DOLOMITES

TRENTINO-SOUTH TIROL / VENETO / FRIULI-VENEZIA-GIULIA, ITALY

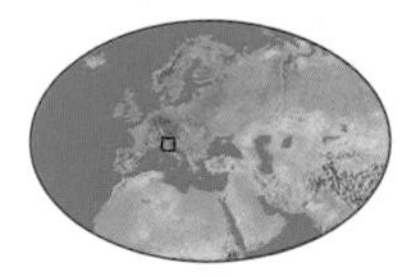

Age of Dolomites: approx. 65 million years

Height of Mt. Marmolada: 10,965 ft (3,342 m)

Northern Italy's skyline is capped by the staggeringly beautiful Dolomite mountains. Comprised of limestone rock that was once deposited on the floor of a warm shallow sea, the mountains were thrust up at the same time as the Alps. The Dolomites get their name from the French geologist Deodat de Dolomieu, who in the 1790s determined that their rocks contained magnesium. They appear gray, white, or weathered brown during the day, but at dawn and dusk they glow red, orange, and pink in the rising or setting sun. Today they have been weathered into towers, pinnacles, and spectacular shapes.

The Dolomites contain 18 peaks over 10,000 feet (3,000 meters). Forty-one glaciers loom in the region. The highest peak is Marmolada. Its summit is pyramid-shaped, and its southern face is a vertical cliff 2,000 feet (600 meters) high. The Marmolada Glacier runs away from its upper slopes. On the plateau of the adjacent Alpes de Siusi, alpine meadows are surrounded by mountains, and nearby are the massive rock faces of the Catinaccio ranges. In summer, numerous paths through valleys such as the Ombretta Valley enable visitors to enjoy the splendor, while trails across mountain ridges challenge serious hill walkers. MB

MOUNT ETNA

SICILY, ITALY

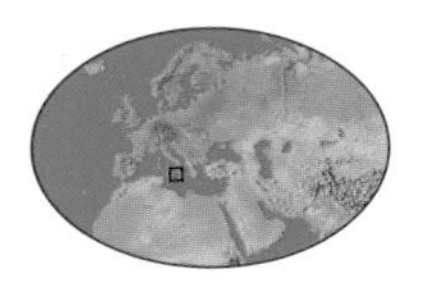

Height of Mt. Etna: 10,990 ft (3,350 m)

Circumference at base: 100 mi (160 km)

Age: 1 million years

Northeast of Catania on Sicily's east coast is Etna—Europe's highest active volcano. The mountain is young (just one million years old), but it more than makes up for its youth with its explosive nature: Etna has been erupting constantly for the past half million years. It is an enormous mountain, covering an area larger than London, and dominates the entire island, though the height of its crater depends on volcanic activity. Currently it is about 10,990 feet (3,350 meters) high with natural scrub vegetation at its base; cool forests of oak, chestnut, and birch on the way to the top; and a black moon-like landscape with tufts of Sicilian milk vetch at the summit.

A cable car on the western side will transport you to the summit, but it is closed to visitors during violent eruptions. Eruptions are fed from a reservoir of molten lava below the mountain that is estimated to be 18 miles (30 kilometers) long and 2.5 miles (4 kilometers) deep. When it bursts through, orange and red volcanic fountains and streams of lava can be seen from far away at night. MB

BELOW *The seemingly calm Mount Etna volcano.*

ALCANTARA GORGE

SICILY, ITALY

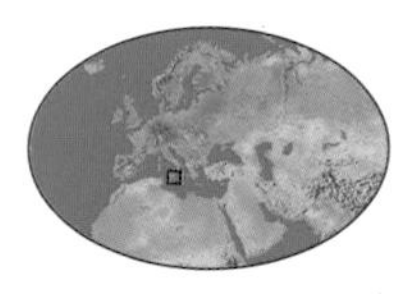

Length of Alacantara Gorge: 1,650 ft (500 m)

Width of gorge:16 ft (5 m)

Depth of gorge: 230 ft (70 m)

Slicing into the hills about 10 miles (16 kilometers) from the tourist resort of Taormina-Giardina-Naxos in southeast Sicily is a narrow gorge with sheer walls of gray columnar basalt. Its origins are linked with Monte Moio, a daughter volcano of Etna, which erupted violently in 2400 B.C.E. The seemingly endless flow of lava provided by the eruption not only invaded the valley of the Alcantara River, but also carried on to the coast and beyond, to form Capo Schiso. When the lava started to cool, a long crack formed, and the river re-established a route to the sea. Gradually it eroded and smoothed the sides to form the Alcantara Gorge.

Legend tells of two brothers, one of whom was blind, dividing up the grain harvest, when the sighted brother became greedy and tried to cheat. An eagle flew over and told God, who launched a thunderbolt that killed the swindler and caused a heap of grain to turn into a red mountain from which poured a lava flow.

The gorge is open to visitors, and can be approached either by a scenic footpath or a modern elevator. The river water is ice-cold, yet summer tourists and locals alike bathe here—a relief from the summer heat. MB

STROMBOLI

AEOLIAN ISLANDS, ITALY

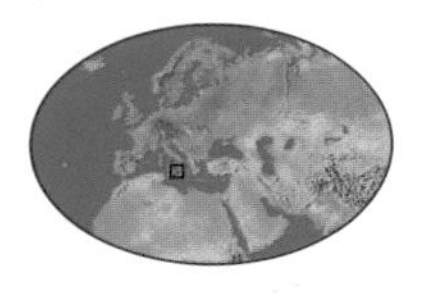

Height of Stromboli volcano: 3,031 ft (924 m)

Age: 15,000 million years

Diameter: 3 mi (5 km)

Stromboli is one of the Aeolian Islands lying off the northeast coast of Sicily. It is a typical Mediterranean island surrounded by deep blue sea, ideal for swimming and snorkeling, as well as for beaches, bars, and restaurants—but one important difference stands out. Stromboli is an active volcano, one of the liveliest in Europe. It has been in nearly continuous eruption for at least 2,500 years, the present cone having been established about 15,000 years ago. Each eruption usually consists of small gas explosions that hurl chunks of molten lava above the rim of the crater. There are several explosions every hour from the three craters. They work like a safety valve, so there is rarely a build-up to a larger, more dramatic event. One of the worst eruptions in recent history was in 1919, when four people were killed and twelve homes destroyed by lava bombs, some of which weighed up to 50 tons. The last major eruption was in 1930. This single eruption produced more ash in a few hours than is usually produced by quieter activity in five years. In 1993, all three craters were active: crater one was spewing out spectacular jets every 20 minutes; crater two was spouting smaller lava jets continuously; and crater three was erupting with explosions of ash and lava bombs at unpredictable intervals. Stromboli is not a "toy volcano," but potentially a very dangerous one. At the end of 2002, it began to look fierce again, and in early 2003 the island was evacuated as a precautionary measure.

During times of more normal activity, Stromboli is a popular tourist destination noted for its peace and tranquillity. At its foot is the village of Stromboli itself, a picturesque haven with a small harbor, black lava beaches, and white houses. Just under a mile offshore is Stombolicchio, the neck of a former volcano that rises out of the sea like a medieval fortress. Access to the island is by sea only. Ferryboats and hydrofoils arrive from Milazzo, Naples, Reggio Calabria, Messina, and Palermo. Those wishing to visit the crater must be accompanied by an authorized guide. The excursion is on foot over steep terrain through Mediterranean maquis vegetation and takes about three hours. Permission can be granted to spend the night at the top—where explosions are even more spectacular against the dark backdrop of night. When the perceived danger of the volcano is high, the island may be closed to visitors. MB

Stromboli is an active volcano, one of the liveliest in Europe. It has been in nearly continuous eruption for at least 2,500 years, the present cone having been established about 15,000 years ago.

RIGHT *Jets of lava spurt forth from Stromboli volcano.*

TRIGLAV MOUNTAIN & THE JULIAN ALPS

GORENJSKA, SLOVENIA

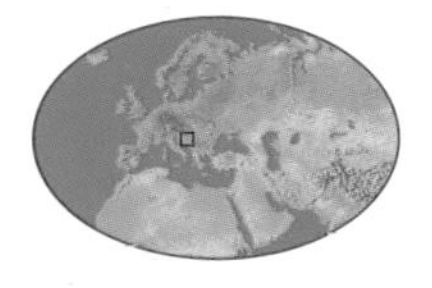

Height of Triglav Massif: 9,396 ft (2,864 m)

Number of mountains: 52

Rock type: karstic limestone

The towering alps of Slovenia are part of the least known but most beautiful alpine habitat in Europe—a secret, pristine landscape that draws visitors back time and again. The mountain range is now known as the Julian Alps, named for Julius Caesar after the Romans annexed Slovenia in the second century. The Julians are part of the Eastern Alps, which also include the mountains of Austria, as well as most of the Italian alps. At an elevation of under 10,000 feet (3,000 meters), they are not as high as some of the other alps, but what they lack in height they make up for in character, with sheer white cliffs, knife-edge ridges, and sharply glaciated peaks.

The steep-sided valleys are thick in pine forest, which in turn gives way to cool, fragrant, alpine meadows, and, higher up, to wild, rocky slopes where ibex and chamois can

be seen. This area is famous for its mountain flora; subsequently, villages teem with cultivated flowers. It is a hiker's paradise, made accessible by the 52 alpine huts that provide accommodation and food. The region's Sava Valley has been described as the most beautiful valley in Europe, while the glacial blue Soca River in the Trenta Valley is considered one of Europe's most beautiful rivers.

At 9,396 feet (2,864 meters), the undisputed king of the Julian Alps is the Triglav Massif, the highest mountain, and national symbol of Slovenia. The early Slavs used to believe that the mountain was the home of a three-headed god who ruled the sky, the earth, and the underworld. The great North Wall of Triglav is a mecca for climbers, rising 4,000 feet (1,200 meters) from the valley floor and stretching for 2 miles (3.2 kilometers). The saying goes that a person is not a real Slovenian until he or she has climbed Triglav. JK

BELOW *The craggy peaks of the Triglav Mountains rise above the cool alpine meadows below.*

SAVICA WATERFALL

GORENJSKA, SLOVENIA

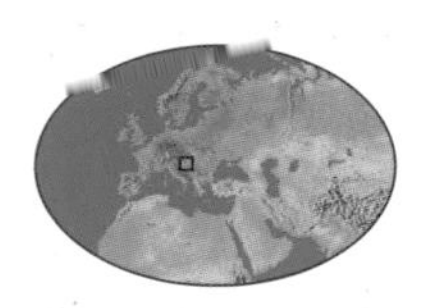

Height of Savica Waterfall: 256 ft (78 m)

Waterfall source: Sava Bohinjka

Savica Waterfall, just one of several waterfalls that characterize the Triglav National Park, is both the most popular and the most legendary natural wonder in Slovenia. Its tall, cascading waters were immortalized by the 19th-century Slovenian poet Frances Preseren in his seminal work *Baptism at Savica*, which celebrated Slovenian nationalism and called for independence from the Austrio-Hungarian Empire. The waterfall is located in the Gorenjska region of Slovenia—the "green garden on the sunny side of the Alps." Savica is also a spring whose waters emerge from the side of the steep wall of Komarca Cliff. The surrounding terrain, 1,650 feet (500 meters) above, has several underground channels that carry water to the cliff's edge. After plummeting 256 feet (78 meters) in a single vertical drop, the waters carry on to Lake Bohinj, and from there east down the River Sava for over 620 miles (1,000 kilometers), where they join the Danube. The area around Lake Bohinj is one of the most beautiful in Slovenia. Reaching Savica Waterfall is not for the fainthearted—it requires an arduous 20-minute climb up steep wooden stairs from the parking lot below. **JK**

KRKA RIVER & FALLS, KRKA NATIONAL PARK

SIBENIK AND KNIN, CROATIA

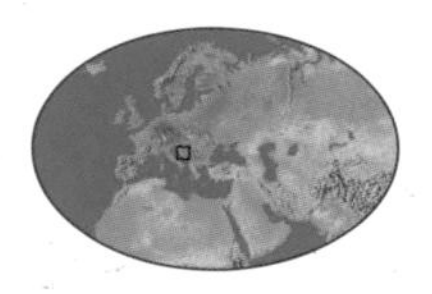

Area of Krka National Park: 43 sq mi (111 sq km)

Length of Krka River: 45 mi (72 km)

Flow rate of Krka Falls: 14,527 gal (1,557 l) per sec

The Krka River runs 45 miles (72 kilometers) from its source (an artesian spring) at the foot of Dinara Mountain in Dalmatia to the Adriatic Sea. Along its route it travels through soft limestone, carving and smoothing out canyons along the way. What is phenomenal about this process is that while the river is cleaving canyons on the one hand, it is also depositing dissolved limestone matter, and therefore simultaneously builds up numerous barriers along its path. Today the river has seven magnificent waterfalls. The most spectacular of these is Skradinski Buk, also known as Krka Falls. These falls drop 787 feet (240 meters) over an elegant series of 17 steps, with a total vertical drop of 150 feet (45 meters).

The falls along the Krka River are relatively young—less than 10,000 years old—but the rapid rate of travertine deposition ensures that the landscape through which the Krka runs is ever-changing. Beyond Krka Falls the river widens out to form Visovac Lake. **JK**

PLITVICE LAKES

LICKO-SENJSKA, CROATIA

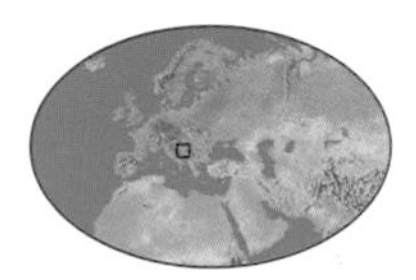

Length of Plitvice lakes: 5 mi (8 km)

Height of Veliki Slap waterfall: 230 ft (70 m)

Rock type: limestone

Known as "the land of falling lakes," Plitvice is situated in karst country, although this particular karst landscape is different because the water is on the surface instead of underground. Over thousands of years, rivers have flowed over limestone and chalk and deposited natural dams of travertine, which have created a series of 16 large lakes and a handful of smaller ones. Waterfalls interconnect the lakes, the highest of which is the towering Veliki Slap. The water comes from the Ljeskovac Brook and Black and White Rivers, which enter Lake Proscansko before dropping down through the emerald-green and turquoise lakes to the Korana River. Their surfaces reflect the surrounding mountains, cloaked in forest.

The local name for the region is "Devil's Garden." Legend has it that the lakes once dried up and though the people prayed for rain, the Black Queen caused thunderstorms to fill the lakes instead. A national park since 1949 and a World Heritage site since 1979, Plitvice is a refuge for European brown bears and wolves, wild boar and deer. The area is located on the southern coast. Walking trails and hotels are scattered throughout the park. **MB**

CORRUBEDO

GALICIA, SPAIN

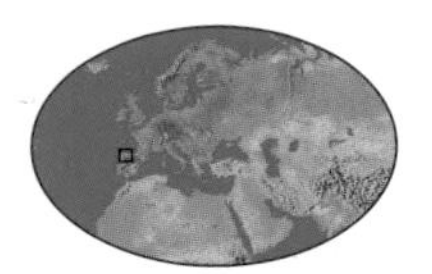

Area of Corrubedo National Park: 3.8 sq mi (9.7 sq km)

Length of beach and dune system: 2.5 mi (4 km)

Minimum number of vascular plant species: 200

Located at the seaward tip of the Serra de Barbanza, in Galicia's Rías Baixas, the sickle-shaped bay of the Corrubedo Natural Park encompasses one of the most important expanses of Atlantic dunes remaining in the Iberian Peninsula. Despite all stages of sand-dune formation and vegetation succession being present here, it is the so-called tertiary or gray dunes—the oldest and farthest from the sea, and nowadays scarce elsewhere—that are of greatest natural value. The rich patchwork of sand-loving plants that carpet these gray dunes includes specimens such as *Iberis procumbens ssp. procumbens* and lax viper's-bugloss, both of which are confined to Iberia's Atlantic shores and provide nesting habitat for Kentish plover, stone curlew, and crested lark. The park also harbors Iberian reptiles, notably Bedriaga's skink, Schreiber's green lizard, and Bocage's wall lizard.

By contrast, at the northern end of the park lies a hulking mass of shifting sand, 3,300 feet (1,000 meters) long, 1,000 feet (300 meters) wide, with some dunes reaching 50 feet (15 meters) high. Onshore winds blow its constituent grains up the seaward side and over the top, pushing it steadily inland. TF

MUNIELLOS FOREST

PRINCIPADO DE ASTURIAS, SPAIN

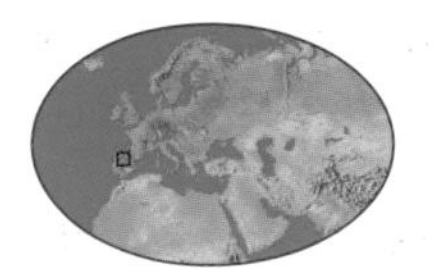

Extent of protected area (Natural Reserve and Biosphere Reserve): 23 sq mi (60 sq km)

Minimum number of bat species: 15

A vast expanse of virtually primeval deciduous forest, the Muniellos Forest nestles into an immense natural amphitheater of Paleozoic quartzite and slate, which backs on to the western reaches of the Cordillera Cantábrica. Sometimes described as the "Jungle of Asturias," its ancient trees—mainly sessile oaks—literally drip with mosses, ferns, and lichens and shelter a rich vertebrate fauna. Considered to be one of Europe's most extensive and best-preserved broad-leaved forests, Muniellos is also one of the last remaining strongholds of the Cantabrican subspecies of capercaillie, which is in grave danger of extinction. With its abundant acorn crop providing crucial food, the forest also supports the equally threatened brown bear.

Other woodland creatures of note include black, middle spotted, and lesser woodpeckers, woodcock, nightjar, edible dormouse, wildcat, pine and beech martens, and no fewer than 15 species of bat. The small streams that traverse Muniellos are a noted haunt of the golden-striped salamander and Pyrenean desman, while Castroviejo's hare inhabit the scrub. Only 20 people are admitted each day in order to preserve this special ecosystem. TF

PICOS DE EUROPA

CANTABRIA / CASTILLA Y LEÓN / PRINCIPADO DE ASTURIAS, SPAIN

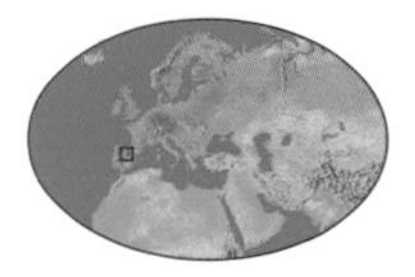

Highest peak of Picos de Europa (Torre Cerredo): 8,688 ft (2,648 m)

Extent of protected area: 159,710 acres (64,660 ha)

Lying 10 miles (16 kilometers) from the Bay of Biscay, the jagged, often snowcapped mountains of Picos de Europa are visible for many miles offshore. Legend has it that the name "Peaks of Europe" was coined by weary Basque fishermen returning from northern seas in medieval times—the mountains represented their first glimpse of home.

Precipitous river gorges have split the pale carboniferous limestone here into three distinct massifs, which together resemble a great, spread-eagled bat. The Central Massif is undoubtedly the most awe-inspiring, crowned by a number of peaks topping 8,250 feet (2,500 meters). Although not the highest, the most emblematic peak is Naranjo de Bulnes (8,264 feet, 2,519 meters), a conical block of limestone also known as Pico Urriellu, which was first climbed in 1904.

The variety of wildlife that thrives in these mountains is staggering: almost 1,500 species of vascular plant, more than 70 species of mammal, 176 bird species, 34 reptile and amphibian species, and 147 butterfly species. The orchid-rich hay meadows have been described as some of the richest Atlantic grasslands in the world. TF

EBRO RIVER

CANTABRIA, SPAIN

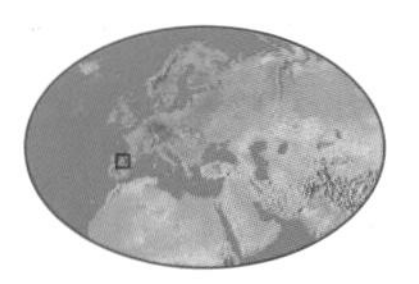

Length of Ebro River: 577 mi (928 km)

Area of Ebro delta: 124 sq mi (320 sq km)

From its source in the Cordillera Cantábrica, close to Reinosa, the Ebro River flows in a southeasterly direction before discharging into the Mediterranean Sea just south of Tarragona.

The Ebro's upper reaches are characterized by a series of spectacular limestone gorges—the Hoces del Alto Ebro and Sobrón—and are populated by large colonies of griffon vultures and lesser numbers of cliff-breeding Egyptian vultures, peregrines, Bonelli's and golden eagles, and eagle owls.

Having negotiated the mountains, the river then enters the flat, extremely arid region around Zaragoza known as the Central Ebro Depression. This depression was once a vast sea separated from the Mediterranean by the Catalan coastal mountains. Here the Ebro meanders sinuously across the alluvial plain, backed by gypsum river terraces and flanked by tracts of gallery forest known as *sotos*, leaving a scattering of ox-bow lakes, or *galachos*, in its wake. After cutting through the Catalan coastal ranges, the Ebro meets the sea with a final flourish: a vast arrow-shaped delta created over centuries by the deposition of enormous quantities of sediment, eroded by the river, and later shaped by the sea. TF

ALTAMIRA

CANTABRIA, SPAIN

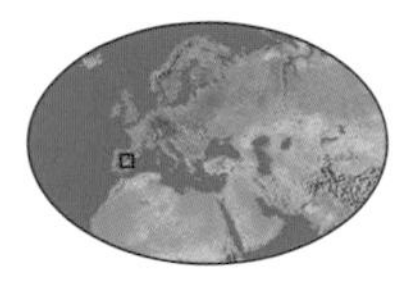

Age of prehistoric paintings: 14,400 to 14,800 years

Number of paintings: approx. 100

Created by collapses resulting from early karstic phenomena in Mount Vispieres, the Altamira cave system was discovered by a roving huntsman in 1879. Inside were Paleolithic paintings of such splendor that their authenticity was doubted for several decades—until subsequent studies revealed them to be almost 15,000 years old. Because of their deep galleries isolated from external climatic influences, they are particularly well preserved. The work of the Magdelenian Stone Age people who occupied northern Spain during the late Pleistocene, the paintings depict the animals with which they shared their world. Red, yellow, and brown ochers are accentuated with black manganese earth and charcoal. The Altamira artists also incorporated topographical features of the cave walls to create a unique three-dimensional effect. The 886-foot (270-meter) cave system comprises 10 chambers and galleries and boasts some 70 engravings and almost 100 paintings. The best examples of this exquisite art are located in the "Chamber of Paintings," where 15 monumental bison adorn the ceiling. Originally designated a World Heritage site in 1985, a further 17 decorated caves of the Paleolithic age were inscribed as an extension in 2008. TF

MONTSERRAT

CATALUÑA, SPAIN

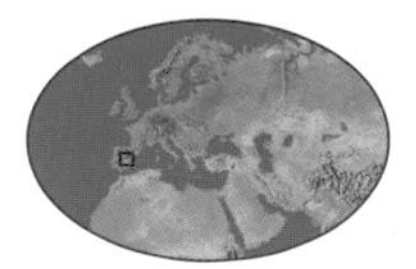

Length of Montserrat: 3.7 mi (6 km)

Highest point: 4,062 ft (1,238 m)

Extent of protected area: 8,966 acres (3,630 ha)

Like a petrified stegosaurus, the silhouette of Muntanya de Montserrat—the "serrated mountain"—completely dominates the plain behind Barcelona, its pale pink conglomerate pillars thrusting 4,055 feet (1,236 meters) skyward. Apart from offering more than 2,000 routes for mountain climbers, Montserrat also attracts hordes of pilgrims, who often arrive via the slightly less taxing method of cable car, to venerate the Black Madonna, a small wooden figurine reputedly carved by St. Luke (although actually 12th century in origin). It is displayed in the Santa Maria monastery, which nestles among the mountain's craggy peaks and is famed for its glorious sunrise views.

The flora of the mountain is particularly noteworthy, with shady rock crevices hosting clumps of lax potentilla, ramonda, and Pyrenean bellflower. Two endangered local plants, the saxifrage *Saxifraga catalaunica* and the stork's bill *Erodium rupestre*, are also found here. In spring, the summit becomes enveloped in a carpet of wild tulips, rush-leafed jonquil, and many different species of orchid. Bonelli's eagles are also regularly seen cruising overhead, scanning the mountain slopes for prey. TF

ORDESA CANYON

ARAGÓN, SPAIN

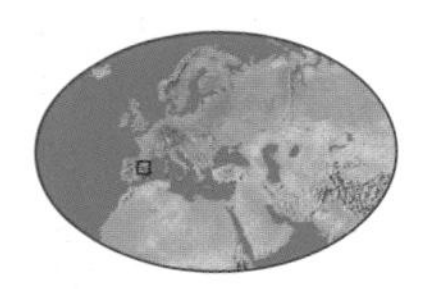

Length of Ordesa Canyon: 10 mi (16 km)

Height of gorge walls: 2,000 ft (600 m)

Features: bearded vultures

Great walls of limestone up to 2,000 feet (600 meters) high rise vertically above the Arazas River in the heart of the Pyrenees Mountains in northeast Spain. This is the Ordesa Canyon, a 10-mile (16-kilometer) stretch of the Arazas Valley. The canyon juxtaposes lush arboreal habitats. The valley floor is cloaked in forests of firs and beech, while higher slopes have dwarf mountain pines which can be found clinging to every available canyon crevice.

At its head is the Circo de Soaso, a natural amphitheater carved out by a glacier from the side of the 11,007-foot- (3,355-meter-) high Perdido Mountain 15,000 years ago. Tumbling down its slopes is the Cascada de Cola de Cabolla, or "Horse Tail Waterfall." Narrow ledges and balconies known as *fajas* have been etched into the limestone cliffs, where nimble-footed wild sheep and goats such as the rare Iberian ibex can be spotted. Downriver from Ordesa is the Faja de las Flores, which follows the Araza for 2 miles (3.2 kilometers) at a height of 7,920 feet (2,400 meters). Golden eagles and bearded vultures soar above the rocks. Trails leading up the mountains begin in the nearest town, Torla. MB

BARDENAS REALES

NAVARRA, SPAIN

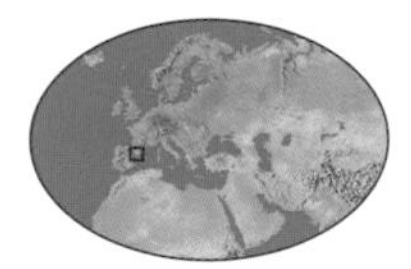

Population of Dupont's lark: 400 pairs

Protected areas in the Bardenas Reales: Ricón del Bú Natural Reserve, Caídes de la Negra Natural Reserve

The Bardenas Reales is a vast expanse of gypsum badland to the south of Pamplona. Since 1882, sheep farmers from the distant Pyrenean valley of Roncal have enjoyed winter grazing rights in the Bardenas Reales, journeying back and forth with their flocks each spring and fall on the Royal Roncalese drover's road. The small road that heads east just south of Arguedas leads quickly into sparsely vegetated saline steppes. The steppes are traversed by innumerable gullies and set against a backdrop of spectacular flat-topped cliffs and fluted turrets of harder limestones and sandstones. Small plots of cereals are cultivated on deeper soils, although scarlet poppies and yellow-and-white crucifers often dominate the crop, providing splashes of vibrant color.

Despite the military bombing range in the center of the park, a rich assemblage of steppe birds and cliff breeders thrives in the Bardenas Reales. Tawny pipits and a wealth of larks—Thekla, crested, short-toed, lesser short-toed, Calandra, and Dupont's—are among the most common songbirds, sharing their habitat with stone curlew, little bustard, and pin-tailed and black-bellied sandgrouse, as well as a small population of great bustard. Among the cliff-nesting birds are alpine swift, black wheatear, chough, rock sparrow, Egyptian vulture, golden eagle, peregrine, and eagle owl. TF

SIERRA DE GREDOS

CASTILLA Y LEÓN, SPAIN

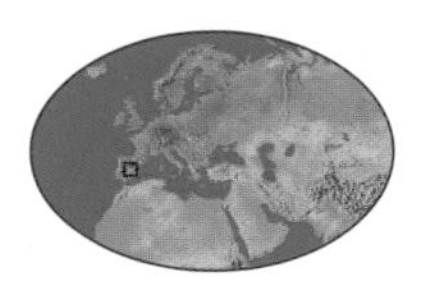

Length of Sierra de Gredos: 155 mi (250 km)

Highest point (Almanzor): 8,504 ft (2,592 m)

The massively glaciated granitic ridge of the Sierra de Gredos occupies the central sector of the east-to-west-oriented Sistema Central—the backbone of Spain—which divides the western half of the country in two. To the south, crags plummet headlong toward the South Meseta, descending 6,600 feet (2,000 meters) in less than 6 miles (10 kilometers). By contrast, the northward incline is gentle.

The greatest botanical interest here exists above the tree line, where several species unique to the Gredos occur, including the stonecrop *Sedum lagascae* and *Antirrhinum grosii* snapdragon. However, the focal point of the range is undoubtedly the Laguna Grande: a glacial lake ensconced within a jagged-walled cirque. This is the haunt of Spanish ibex and the largest breeding population of bluethroat in Spain. Native races of snow vole, Iberian rock lizard (confined to the peninsula), fire salamander, and common toad also dwell here. As an entity the range is home to approximately 50 species of mammal, 23 reptiles, 12 amphibians, and almost 100 butterflies. It also harbors breeding nuclei of birds such as black stork, black vulture, and Spanish imperial eagle. TF

SIERRA DE ATAPUERCA

CASTILLA Y LEÓN, SPAIN

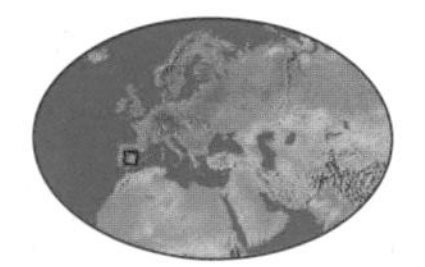

Oldest hominid remains: 800,000 years

Other hominid remains: 350,000 to 500,000 years

On the surface there is little to suggest that the Sierra de Atapuerca—a small, rather unprepossessing limestone hill—is the site of one of the most undeniably fascinating paleoanthropological discoveries of the 20th century. Evidence has revealed that the complex of large caverns that puncture the hillside provided shelter for various species of hominid for something in the region of one million years. Two sites—Gran Dolina and Sima de los Huesos—have contributed enormously to an understanding of the physical nature and customs of the earliest-known hominids to migrate into western Europe from Africa.

Fragments of bone thought to be around 800,000 years old were unearthed at Gran Dolina, and are believed to be from at least six individuals. They are unlike any other human remains and, in 1977, were described as a separate species, *Homo antecessor*, which is a derivative of the Latin for "explorer." The nearby Sima de los Huesos—literally "pit of bones"—has turned up thousands of human remains, and has been described as one of the most productive paleoanthropological sites in the world. TF

GARROTXA

CATALUÑA, SPAIN

Extent of protected area: 29,425 acres (11,908 ha)

Highest point (Puigsallana): 3,370 ft (1,027 m)

Age: 350,000 years

Just 12 miles (20 kilometers) northwest of Girona lies the most extensive volcanic landscape in the Iberian Peninsula. The sprawl of beech and oak deciduous forest engulfs about 75 percent of the park and yet cannot quite conceal the plethora of volcanic cones and basalt lava flows. The cones number more than 30, some of which have craters, and although no volcanic activity has been recorded in historical times, the cones are considered to be dormant rather than extinct—the destruction of the nearby town of Olot in the 1428 earthquake provided sufficient evidence of continued seismicity.

Around 1,500 species of vascular plant have been recorded in Garrotxa, with relict enclaves of pedunculate oak sheltering a rich ground flora of rue-leafed isopyrum, yellow anemone, snowdrop, Solomon's seal, twayblade, and the local subspecies of large bitter cress. More than 100 species of butterfly have been recorded here, and the park's wetlands harbor Mediterranean demoiselle, southern emerald and goblet-marked damselflies, emperor dragonfly, crepuscular hawker, blue-eyed hooked-tailed dragonfly, keeled skimmer, and scarlet darter. TF

SALTO DEL NERVIÓN

PAÍS VASCO, SPAIN

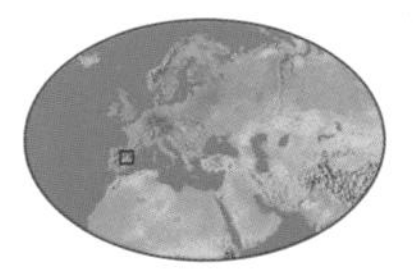

Maximum height of waterfalls: 900 ft (270 m)

Average width of falls: 20 ft (6 m)

Extent of protected area (Monte Santiago): 11,856 acres (4,800 ha)

Creating one of the most spectacular waterfalls in Spain, if not in Europe, the Nervión River emerges from the forested amphitheater of Monte Santiago and plunges abruptly over the sheer limestone buttresses of the Sierra Salvada (Salbada), before continuing northward to meet the sea at Bilbao.

During periods of maximum snowmelt, the Salto del Nervión falls uninterrupted for almost 900 feet (270 meters) into the Delika (Délica) Gorge, although at other times of the year it often evaporates before reaching the base. The sheer magnificence of the Salto del Nervión is best appreciated from a viewpoint overlooking the Delika Gorge. Just south of the Puerto de Orduña, a shady trail runs eastward for just over 2 miles (3.2 kilometers) before reaching the viewpoint, traversing thick beech forest inhabited by wildcat, pine marten, red squirrel, and roe deer en route. Golden eagle, Egyptian vulture, peregrine, and chough can be seen, while griffon vultures that nest on the buttresses are the most abundant denizens of these skies. The remains of an ancient stone wolf-trap designed to drive marauding wolves over the precipice can be seen in the forest near the head of the cascade. TF

AIGÜESTORTES I ESTANY DE SANT MAURICI

CATALUÑA, SPAIN

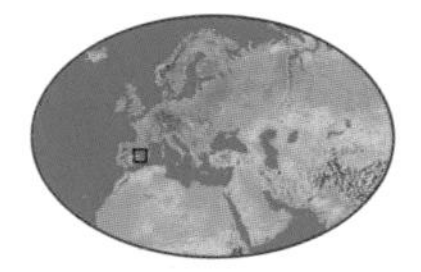

Area of national park and buffer zone: 100,945 acres (40,852 ha)

Highest peak (Comaloforno): 9,951 ft (3,033 m)

Although *aigüestortes* literally means "twisted waters," referring to the much-braided and meandering mountain streams in the high-altitude valleys, this Catalan national park is best known for its plethora of jewel-like glacial lakes—more than 200—here known as *estanys*, of which Sant Maurici is the largest.

During the Quaternary, the impermeable slate and granite bedrock of the highest terrain was successively ice-sculpted into a spectacular array of cirques, arêtes, and rock needles. Descending glaciers gouged out classic U-shaped valleys and deposited tongue-like moraines as they progressed toward the "lowlands," although no point within the park lies below 5,315 feet (1,620 meters). Conifers play an important role in the landscape, with the Scots pine dominant in the lower reaches grading into mixed forests of mountain pine and European silver fir at an elevation between 6,000 and 7,260 feet (1,800 and 2,200 meters). In the dense shade of the pine, yellow bird's-nest, lesser twayblade, bird's-nest orchid, and creeping ladies' tresses thrive. Notable avian inhabitants include capercaillie, woodcock, Tengmalm's owl, black woodpecker, and ring ouzel. Citril finch co-habit with them. Above the tree line, a mosaic of pastures and rock gardens bursts into bloom in early summer, with saxifrages and gentians galore mingling with Pyrenean lily, English iris, alpine snowbell, and bird's-eye primrose. Together, these provide a source of nectar for such emblematic montane butterflies as Apollo, clouded Apollo, mountain clouded yellow, silvery argus, mazarine blue, scarce, sooty, purple-shot and purple-edged coppers, Spanish brass, and Gavarnie ringlets.

Aigüestortes *literally means "twisted waters," referring to the much-braided and meandering mountain streams in high-altitude valleys.*

Visitors almost certainly encounter snow voles, alpine marmots, and isard here, but the park's small population of ptarmigan is more reclusive. The Pyrenean rock lizard, specific to the central Pyrenees, can survive at altitudes of more than 10,000 feet (3,000 meters). The towering crags that ring the park are home to breeding alpine accentor and wallcreeper, as well as at least six pairs of the legendary lammergeier, or bearded vulture, while the myriad streams and lakes of Aigüestortes provide ideal conditions for Pyrenean desman and Pyrenean brook salamander. TF

RIGHT *The winter wonderland of Aigüestortes.*

LAKE GALLOCANTA

ARAGÓN, SPAIN

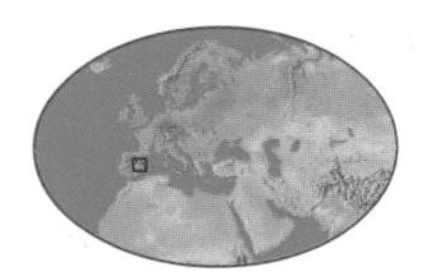

Extent of protected area (Ramsar site and refugio de fauna silvestre): 16,605 acres (6,720 ha)

Maximum area of lagoon (winter): approx. 3,300 acres (1,330 ha)

Maximum depth: approx. 5 ft (1.5 m)

Set amid flat agricultural land, Lake Gallocanta might be the largest natural inland lake in Spain, though few would include it in a list of the country's more scenic landscapes. And yet a dusk vigil on the shore of the lagoon in November or late February is rewarded with one of the most memorable wildlife experiences as up to 60,000 honking cranes approach in untidy V-shaped formations to roost. Gallocanta has found favor with around 80 percent of the western European population of cranes and provides a sojourn en route to their wintering quarters in the acorn-rich wood pastures of southwestern Iberia.

The first birds arrive at the lagoon in October, with numbers peaking in late November, before the majority depart to spend the winter farther south, leaving just a few thousand to brave the bitter hibernal temperatures of the Aragonese plains. By early February, crane numbers have built up once more, pending their imminent mass

exodus for breeding quarters in Scandinavia and Russia in late February and early March.

During the winter, when fall rains have replenished the lagoon, thousands of waterfowl such as coot, pochard, and red-crested pochard flock to Lake Gallocanta. Similarly, if sufficient water remains in summer to isolate predator-free island refuges, species such as avocet, black-winged stilt, and gull-billed and whiskered terns will remain to rear their young. By contrast, when large expanses of mud are exposed, Kentish plover breed in such dense colonies that their nests are sometimes crammed together at intervals of less than three feet (0.9 meters). The surrounding cereal fields—which are typically richer in "weed" species than crops—harbor breeding stone curlew and short-toed lark, plus small numbers of little bustard and a few great bustards. Flocks of Calandra lark and black-bellied sandgrouse occur in winter.

Gallocanta lies southwest of the medieval walled city of Daroca; an information center is located at the southeastern juncture of the lagoon for visitors. TF

BELOW *Rolling green hills surround Lake Gallocanta.*

MALLOS DE RIGLOS

ARAGÓN, SPAIN

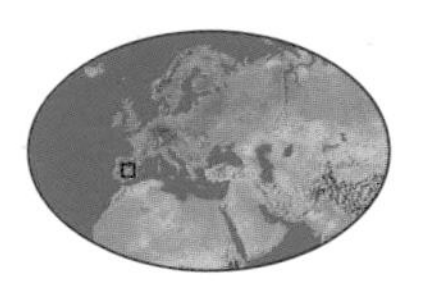

Elevation of Mallos de Riglos: over 3,000 ft (900 m)

Height of rock pillars: approx. 1,000 ft (300 m)

Age: 30 million years

Towering above the left bank of the River Gállego, and completely dwarfing the nearby village of Riglos, three groups of tremendous rock pillars are all that remain of a vast alluvial fan that swept southward during the birth of the Pyrenees. Subsequent sculpting by wind and water has removed all but the most solid material, exposing huge columns of loosely cemented Miocene conglomerates in various shades of russet and ocher. Some of the component boulders measure more than three feet (0.9 meters) in diameter.

A mecca for rock climbers and ornithologists alike, the Mallos de Riglos boast more than 200 climbing routes and an important colony of griffon vultures. Other cliff-nesting birds include Egyptian vulture, peregrine, eagle owl, alpine and pallid swifts, crag martin, blue rock thrush, rock thrush, black wheatear, and chough. Alpine accentors and wallcreepers regularly occur here during the winter, while Bonelli's eagles and lammergeiers often hunt in the area, despite not having bred here for several decades. Plant lovers should note that the crags are festooned with cushions of sarcocapnos, Pyrenean saxifrage, and ramonda. TF

RIGHT *The rocky red monoliths of Mallos de Riglos.*

VILLAFÁFILA

CASTILLA Y LEÓN, SPAIN

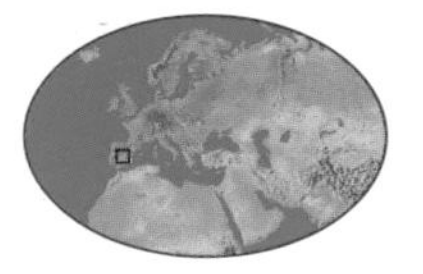

Area of Ramsar site: 7,052 acres (2,854 ha)

Special Protection Area for Birds (ZEPA): 80,757 acres (32,682 ha)

Set amid the exceptionally flat plains of the North Meseta, Villafáfila is renowned for its "steppe" bird community and has also been designated a Ramsar site (Wetland of International Importance). More than 260 bird species have been recorded here. Villafáfila comprises tens of thousands of acres of semi-arid cereal fields, fallow lands, and pasture, which are generally referred to as "pseudosteppes" on account of their anthropogenic nature. With 2,500 individuals, this mosaic of habitats is home to the largest enclave of great bustard in Europe (if not the world). Little bustard, black-bellied sandgrouse, and Montagu's harrier, plus the highest density of lesser kestrel in Castilla y León (more than 200 pairs) also exist here.

In the heart of Villafáfila lie three extremely shallow, saline lagoons, which during a wet winter together can exceed 1,480 acres (600 hectares), when they play host to more than 30,000 greylag geese, as well as thousands of other waterfowl and waders. During the summer, the wetland attracts nesting marsh harrier, black-winged stilt, and avocet. TF

DUERO GORGE

SPAIN / PORTUGAL

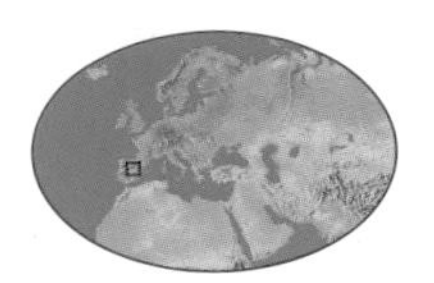

Alternative name: Arribes del Duero

Length of Duero Gorge: 76 mi (122 km)

Extent of protected area:
Spain: 465,970 acres (170,000 ha);
Portugal: 213,714 acres (86,500 ha)

For more than 75 miles (120 kilometers) in northwestern Iberia, the indisputable boundary between Spain and Portugal follows a spectacular chasm sliced into the siliceous bedrock by the River Duero ("Douro" to the Portuguese). In places the chasm dips more than 1,320 feet (400 meters). Numerous tributaries add a farther 124 miles (200 kilometers) of river gorges—known locally as *arribes*—which often terminate in dramatic waterfalls as they plunge abruptly to meet the Duero. A notable example is the Pozo de los Humos on the Uces River, where a 165-foot (50-meter) waterfall throws up vast quantities of spray. This wild, virtually uninhabited terrain harbors one of the highest densities of breeding griffon and Egyptian vulture in the Iberian peninsula (325 and 129 pairs respectively in 2000), which in turn share their habitat with 20 pairs of golden eagles and a dozen much rarer Bonelli's eagles. Black storks are also at home here, as are eagle owls and peregrines, with characteristic breeding passerines including black wheatear, blue rock thrush, and chough. A boat trip upriver from Miranda do Douro—the Cruzeiro Ambiental—is a good way to explore the canyon from within, with sightings of Bonelli's eagles virtually guaranteed. TF

LA PEDRIZA

CASTILLA Y LEÓN, SPAIN

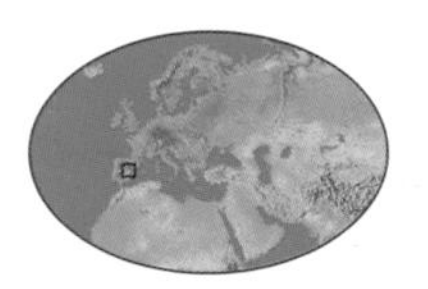

Maximum elevation of La Pedriza: 7,828 ft (2,386 m)

Extent of protected area (Biosphere Reserve): 250,310 acres (101,300 ha)

Secreted within the heart of the Cuenca del Manzanares Biosphere Reserve lies one of the most spectacular tracts of exposed granite in the Iberian Peninsula. Known as La Pedriza, a fault line splits the two distinct geological realms. In the north, a high-walled, horseshoe-shaped cirque curves around a southerly sector dominated by huge granite boulders that have been smoothed by the elements over millennia. They have evocative names such as "The Helmet," "The Skull," and "Pig Rock." One of the largest—El Tolmo—rises some 60 feet (18 meters) and has a circumference of 240 feet (73 meters).

La Pedriza attracts rock climbers, but it is also renowned for its wildlife. The higher reaches boast a large and healthy griffon vulture colony (70 pairs), plus breeding water pipit, alpine accentor, rock thrush, blue rock thrush, wheatear, and chough. Other residents include the snow vole, Spanish ibex, Iberian rock lizard, and a wealth of montane butterflies, including black satyr, purple-shot copper, and Apollo. By contrast, the scattered pines of the boulder "city" host a Mediterranean bird community of azure-winged magpie, hoopoe, and black-eared wheatear. TF

CIUDAD ENCANTADA

CASTILLA-LA MANCHA, SPAIN

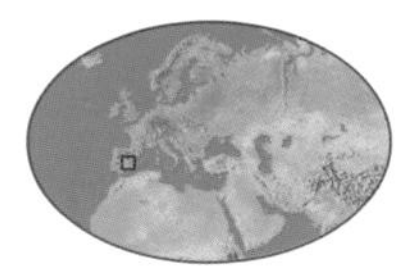

Elevation of Ciudad Encantada: 4,420 to 4,685 ft (1,340 to 1,420 m)

Age of Ciudad Encantada: 99 to 142 million years (Lower Cretaceous)

Extent of protected area: 618 acres (250 ha)

Although the Serranía de Cuenca is littered with weird and wonderful limestone formations, nowhere is the process of erosion quite so well developed as the Ciudad Encantada, where the combined action of wind and water has reached its zenith. The magnesium-rich dolomitic limestone plateau of the Ciudad Encantada—which literally means "Enchanted City"—has been eroded over millennia along a myriad of fault lines to produce a labyrinth of intersecting valleys and outcrops that are now reminiscent of the ruins of some prehistoric city. Because the upper layer is composed of harder rock, top-heavy formations abound, resulting in a geological zoo of fantastic natural sculptures. Many of the formations are known by name: The Lion, The Seal, The Hippopotamus, The Bears, The Whale, and even The Battle of the Elephant and Crocodile.

Orchids can be found in and among the outcrops in spring, together with flats of wild tulips. Fissures in the limestone have accumulated just enough soil to harbor the spleenwort, the white-flowered crucifer, the saxifrage (specific to eastern Spain), and rock snapdragon. TF

MONFRAGÜE

EXTREMADURA, SPAIN

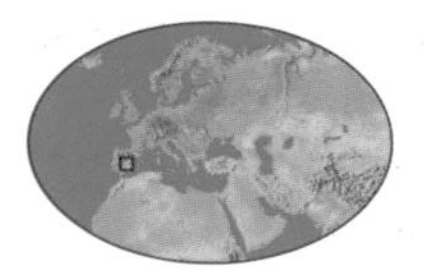

Maximum elevation at Monfragüe: 1,780 ft (540 m)

Extent of protected area (natural park): 44,112 acres (17,852 ha)

Number of vertebrate species: 276

Monfragüe is perhaps not the most scenic landscape in the Iberian Peninsula, but the incredible wealth of wildlife found here earned it status as a Biosphere Reserve in 2003.

Paleozoic quartzite crags along the spine of the park—the Sierra de las Corchuelas—shelter relict enclaves of dense Mediterranean forest on their northern flanks. The slate-lined valleys on either side are mantled with *dehesa*: a unique agricultural landscape restricted to Iberia and northwest Africa, and comprising sparse evergreen oakwood pasture. Umbrella-shaped trees protect the underlying grasslands from excessive water loss and harmful frosts.

Monfragüe has what is probably the best assemblage of breeding raptors in Spain, with approximately 250 pairs of black vulture and around 10 pairs of Spanish imperial eagle. The latter are confined to Iberia, with a population of just 160 pairs existing in the entire world. Both occur here in the highest densities on the planet. On the more inaccessible crags, Egyptian and griffon vultures, golden and Bonelli's eagles, peregrines and eagle owls, as well as more than 30 pairs of black stork, along with short-toed and booted eagles, find their nesting spaces. TF

PENYAL D'IFAC

COMUNIDAD VALENCIANA, SPAIN

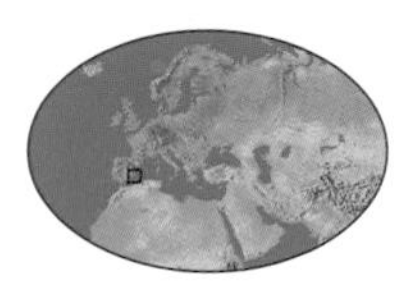

Maximum height at Penyal D'Ifac: 1,089 ft (332 m)

Extent of protected area: 111 acres (45 ha)

Age: maximum 55 million years

The outstanding landmark of the Costa Blanca, the Penyal d'Ifac (Peñón de Ifach) is a block of limestone rising abruptly from the sea about halfway between Benidorm and Dénia and connected to the mainland by a narrow, sandy isthmus. It was referred to by the Moors as "the northern rock" in order to distinguish it from its southern counterpart, Gibraltar, and was long used as a watchtower by sentinels who lit fires on the summit to warn of approaching pirates.

Despite its near-vertical cliffs, the climb to the top is relatively simple, and is rewarded with views of Eivissa (Ibiza) on a clear day. Ifac has attracted the attention of eminent botanists such as Antonio Josef Cavanilles (1745–1804) and Georges Rouy (1851–1924), who were drawn to the abundance of plants specific to the Dénia area, including *Hippocrepis valentina*, *Thymus webbianus*, *Centaurea rouyi*, the pink-flowered campion *Silene hifacensis*, and the scarce cat's-head rock-rose. The breeding birds of Ifac include cliff-nesting pallid swifts, black wheatears, and peregrines. Even the scrub at the base harbors birds—spectacled, sub-alpine Dartford, and Sardinian warbler birds. TF

FUENTE DE PIEDRA

ANDALUCÍA, SPAIN

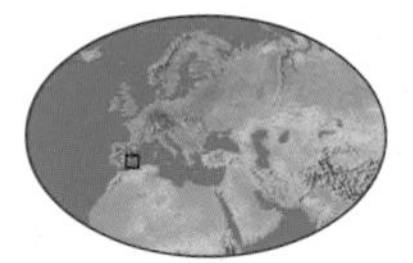

Maximum extent of lagoon: 3,210 acres (1,300 ha)

Extent of protected area (natural reserve and Ramsar site): 3,647 acres (1,476 ha)

Andalucía's most extensive natural inland lake, Fuente de Piedra, is best known for its massive colony of greater flamingoes, the largest in the Iberian Peninsula, if not in Europe. In 1998, approximately 19,000 pairs reared 15,387 chicks here—almost two-thirds of the total for the whole of the Mediterranean region. Fuente de Piedra nestles in a depression in the undulating gypsum-rich hills to the northwest of Antequera. Today it shimmers as an expanse of shallow, saline waters, having been dedicated to salt production from Roman times until the 1950s.

Fuente de Piedra is a classic example of an endorheic lagoon; that is, there is no natural outlet, and, though sparse, it is fed principally by rainfall via the underlying aquifer. Evaporation from wind and sun is so extreme that sometimes during the summer the lagoon completely dries out.

Although flamingoes are the main reason for Fuente de Piedra's designation as a Ramsar site (Wetland of International Importance), 170 other species of bird are regularly observed here, with notable breeders including gull-billed tern, avocet, Kentish plover, collared pratincole, and purple gallinule. TF

RUIDERA LAGOONS

SPAIN / PORTUGAL

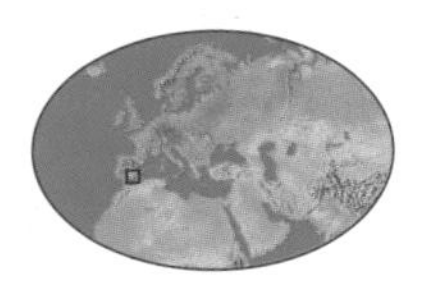

Length of lagoon chain: 17.5 mi (28 km)

Maximum depth of lagoons: 66 ft (20 m)

Extent of protected area: 9,193 acres (3,722 ha)

Secreted in the arid heartland of La Mancha's Campo de Montiel lies a 17.5-mile (28-kilometer) chain of lagoons connected by underground currents. These are known as the Lagunas de Ruidera, and many legends exist to explain the origin of these curious lagoons. The most romantic explanation comes from Miguel de Cervantes who, in the second part of *Don Quijote de La Mancha*, describes Merlin's transformation of the Squire Guadiana, the Lady Ruidera, and their daughters and nieces into shallow lakes.

In reality, the 15 lagoons were formed when the Guadiana River dissolved limestones and marls during the Quaternary period. They were formerly linked at surface level by waterfalls tumbling over natural barriers of calcium carbonate. In fact, the name of the chain allegedly derives from the word *ruido*, or "noise," which refers to the sound of waterfalls. Sadly, waterfalls are seldom heard these days, as water levels have dropped dramatically due to over-extraction from the aquifer. Nevertheless, red-crested pochard still breed in the deeper central lagoons, while fringing reedbeds surrounding the shallower terminal lagoons support nesting marsh harrier, water rail, Cetti's great reed warblers, and purple gallinule. TF

SIERRA NEVADA

ANDALUCÍA, SPAIN

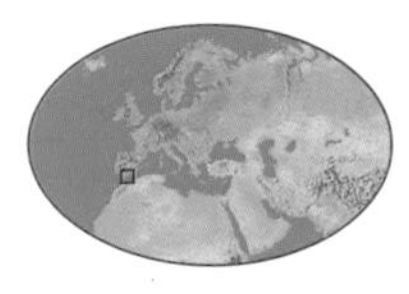

Extent of protected area: 213,020 acres (86,208 ha)

Combined area of Biosphere Reserve and natural park: 424,975 acres (171,985 ha)

The southernmost glaciated landscape in Europe, the Sierra Nevada also possesses the 11,425-foot (3,482-meter) peak of Mulhacén. The range runs roughly east to west for more than 50 miles (80 kilometers), and the core area is a jumble of acid mica schists and gneisses, riddled with moraines and glacial cirques, and nestling around 50 gelid lakes.

All five Iberian vegetation zones are present in the Sierra Nevada, from Mediterranean at the lowest levels to alpine above 8,580 feet (2,600 meters); the latter is home to a wealth of unique species that have evolved from animals stranded when the ice sheets receded at the end of the last ice age.

The 78 native taxa of vascular plant in the alpine screes and rock-gardens include such gems as *Saxifraga nevadensis*, *Viola crassiuscula*, and the toadflax *Linaria glacialis*. Similar levels of endemism are found among the invertebrates of the highest reaches. Many of these are beetles, but several species of melanic, flightless grasshoppers and crickets also inhabit this area.

The Sierra Nevada, which is, since 1999, Spain's largest national park, also harbors the southernmost snow voles in the world, a huge population of Spanish ibex, the only breeding enclave of alpine accentor in southern Spain, and a total of 124 species of butterfly. TF

COTO DOÑANA

ANDALUCÍA, SPAIN

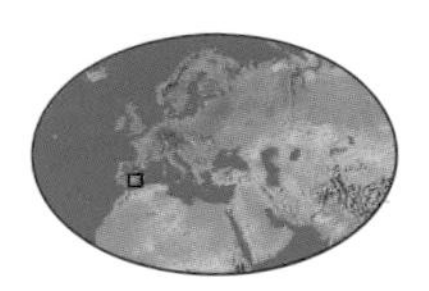

Extent of protected area:
275,870 acres (111,643 ha)

Number of recorded bird species:
400

Coto Doñana has long been recognized as Spain's top wildlife region. Showered with almost every international designation—Ramsar site, World Heritage site, and Biosphere Reserve—and declared a national park in 1969, it is also known as the site that triggered the foundation of WWF in 1961. It is the most important wetland in the country, providing food and refuge for around eight million birds each year. Perhaps more accurately called the Marismas del Guadalquivir, this region is a unique mosaic of live dunes, tidal areas, rice fields, fish ponds, and saltpans. The lagoon-studded interior of the park is separated from the Atlantic by an immense dune system: Wave upon wave of sandy crests advance relentlessly inland, swallowing the stone pine forests en route. Dotted across the park are small enclaves of Mediterranean forest, dominated by enormous cork oaks, but for the most part, centuries of burning and grazing have reduced the ancient climax vegetation to low scrub communities.

In 1998 the area was affected by a spill of acidic toxic metal waste from a local mine, and the waste entered ecologically sensitive areas of the park and affected fish, invertebrate, and bird populations; scientists and ecologists have been monitoring the environmental impact.

The wildlife significance of Doñana is staggering. Almost 400 species of bird have been recorded here, of which 136 breed regularly. The marshes provide Spain's principal habitat for the largest spoonbill colony in Europe, as well as nesting populations of white-headed duck. Around a dozen pairs of Spanish imperial eagles occupy nests in the broad canopy of cork oaks each year, creating the most important breeding site in Iberia. In winter, more than one million birds use the flooded heartland of the park, among them tens of thousands of greater flamingoes, greylag geese, ducks, and waders. In addition, Doñana provides a last refuge for the increasingly rare spur-thighed tortoise, and the pardel lynx, which inhabits open forest and thickets, and is considered to be the most endangered feline on the planet.

Visits to the interior of the park are by guided tour only, but the broad watercourse adjacent to El Rocío offers exceptional views of many waterbirds in winter, and during the spring and fall migration periods. TF

As a Ramsar site, World Heritage site, Biosphere Reserve, and national park, Doñana is undoubtedly the most important wetland in the country. It is also known as the site that triggered the foundation of WWF in 1961.

RIGHT *Sunrise over the wetlands of Coto Doñana.*

TABERNAS DESERT

ANDALUCÍA, SPAIN

Temperature range: -41°F to 118°F (-41°C to 48°C)

Special Protection Area for Birds: 28,355 acres (11,475 ha)

One of the most spectacular arid regions in the Iberian Peninsula, the Tabernas Desert is Europe's only true desert, isolated from moisture-laden sea winds by the mountain ranges of Alhamilla and Los Filabres. The annual rainfall rarely exceeds 8 inches (200 millimeters); this occurs in a few torrential downpours. Between hard outcrops of sandstones and conglomerates extend some 61,775 acres (25,000 hectares) of heavily eroded marl plateaus, traversed by deep ravines known as *ramblas*: a famous film location for many "spaghetti westerns," including Sergio Leone's legendary trilogy. The crumbling cliffs of the ramblas are a breeding ground for a number of bird species, including Bonelli's eagle, eagle owl, and alpine and pallid swifts. Not surprisingly, mammals are scarce, although the two types of Spanish hedgehog can be found, along with many species of reptile. Sand-covered gullies known as badlands form steppes covered in grasses. The region is lush and green from November to the end of June, then slowly it becomes scorched and dry until the end of summer. A visit between March and May is recommended for the stunning displays of wildflowers. TF

NERJA CAVES

MÁLAGA, SPAIN

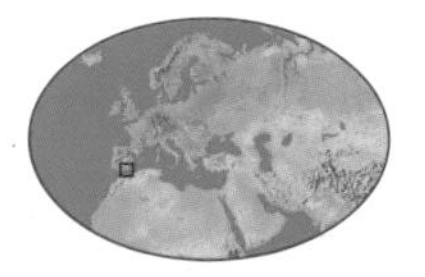

Total surface area: 382,000 sq ft (35,484 sq m)

Longest stalactite recorded: 105 ft (32 m), said to be the longest in the world

Discovered in 1959, the Nerja Caves have been naturally carved into the Sierra Almijara mountains in southern Spain. These caverns display stunning stalagmites, stalactites, and columns formed over thousands of years as water infiltrated fissures of marble in the mountains. The resulting caves filled with carbonate deposits, and later with further settling calcite. The result is dramatic, with enormous chambers featuring 100-foot (30-meter) stalactites overhead, or fantastic, unusual shapes. Particularly striking is Cataclysm Hall, which hosts the world's largest stalactite column—it rises 160 feet (49 meters) in height and has a diameter of 60 feet (18 meters). It is estimated to have been formed by the falling of 1 trillion drops of water.

Wall paintings inside the caves date from the Paleolithic period, and skeletal remains and artifacts show that they were inhabited from about 25,000 B.C.E. until the Bronze Age. Cave fauna include scarab beetles and blind scorpions. About one-third of the halls are open to tourists; concerts are even held in one chamber that forms a natural amphitheater. GD

TORCAL DE ANTEQUERA

ANDALUCÍA, SPAIN

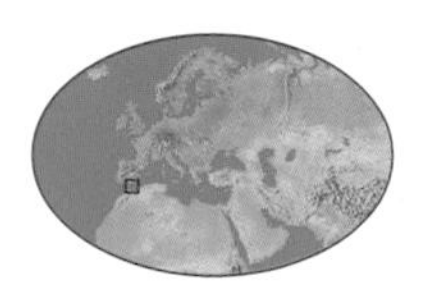

Maximum height of Torcal de Antequera: 4,386 ft (1,337 m)

Extent of protected area: 4,962 acres (2,008 ha)

Just a short drive inland from the coastal metropolis of Málaga lies the spectacular limestone outcrop of the Torcal de Antequera, a geological wonderland. Nowhere else in Spain is the karst process so comprehensively visible, with all the classic water-eroded features—limestone pavement, sculpted columns, and balancing stones—packed into a small area. The natural sculptures are the result of the erosion of limestone rock that originated on the seabed 150 million years ago—shell and marine fossils have been discovered in the valley. Less obvious is the subterranean element; beneath lies a labyrinth of galleries and caverns, populated by dramatic forests of stalagmites and stalactites. A 1-mile (1.6-kilometer) circular walk from the upper car park takes in an impressive landscape of fluted turrets and narrow, shady gullies, the walls decorated with an impressive array of fissure plants, including saxifrage and toadflax. Torcal de Antequera is an important botanical area, with more than 650 species of plant, and 30 varieties of orchid alone. The outcrops are often capped with cloud, even when skies all around are clear; however, on a rare sunny day it may be possible to see the African coast. TC

GRAZALEMA

ANDALUCÍA, SPAIN

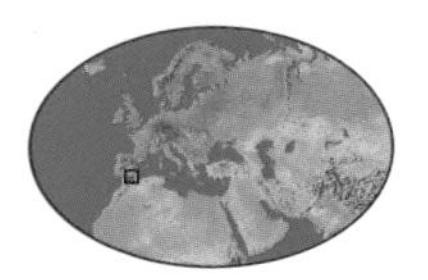

Area of Grazalema Natural Park: 132,048 acres (53,439 ha)

Maximum altitude: 5,427 ft (1,654 m)

Average annual rainfall: 88 in (223 cm)

Despite its location in the southwestern corner of the otherwise rather dry province of Andalucía, the Sierra de Grazalema has the distinction of being one of the rainiest places in Spain. As a result, its labyrinth of jagged limestone peaks and outcrops is home to a diverse array of animals and plants that would not survive in the surrounding sun-baked lowlands. The area harbors 220 vertebrate species, 75 butterflies, and almost 1,400 taxa of vascular plant, several of which are unique to Grazalema, notably the brick-red poppy *Papaver rupifragum*.

The first area in Spain to be declared a Biosphere Reserve (in 1977), the Grazalema Natural Park is of significance for its Spanish fir forest (El Pinsapar), which extends for some 1,040 acres (420 hectares)—about one third of the total extent in Spain—over the humid, north-facing slope of the Sierra del Pinar. These distinctive conifers have been present in the region since the last ice age, when they were forced south by the advancing ice sheets, then stranded when the glaciers retreated. A relic species, some of the trees have trunks almost three feet (0.9 meters) in diameter and are thought to be more than 500 years old. Grazalema is famous for its dramatic gray limestone cliffs and gullies, caverns and gorges—however, the most spectacular feature is La Verde, with steep rocky walls that rise up to 1,320 feet (400 meters).

The notable birds of the area include more than 300 pairs of griffon vulture—one of the largest concentrations in Spain, if not in Europe. Subterranean galleries host important winter bat roosts, particularly of Schreiber's bat, with more than 100,000 individuals using the Hundidero-Gata cave system, while the highest peaks support large populations of Spanish ibex. The *dehesa* (wood pasture), which covers much of the park's lowlands, provides refuge for mammalian predators such as western polecats, although these are rarely seen. In early May, the migrant birds arrive for the summer; around the same time, the Grazalema region bursts into bloom with vibrant orchids and lilies. By high summer butterflies abound, particularly in Puerto del Boyar. To visit the El Pinsapar forest region a permit must be obtained from the information center in El Bosque. TF

Grazalema Natural Park was the first area in Spain to be declared a Biosphere Reserve. In early May, the migrant birds arrive for the summer; around the same time, the Grazalema region bursts into bloom with vibrant orchids and lilies.

RIGHT *Grazalema's north-facing El Pinsapar fir forest.*

LOS ALCORNOCALES

ANDALUCÍA, SPAIN

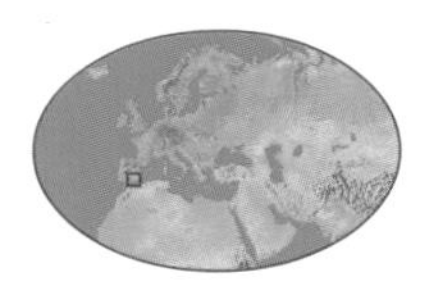

Highest point of Los Alcornocales: 3,583 ft (1,092 m)

Extent of protected area: 416,761 acres (168,661 ha)

Number of vertebrate species in the park: 264

Forty to sixty million years ago the entire Mediterranean region was bathed in a virtually tropical climate. Broad-leaved evergreen forests clothed much of southern Europe and North Africa, but in today's cooler climate, only a few remnants persist, mainly in the Macaronesian islands of the Canaries, Madeira, and the Açores. The Los Alcornocales Natural Park in southwest Spain extends over 266 square miles (690 square kilometres) and is one of the few places in mainland Europe where relic stands of Tertiary vegetation survive. At the southern end of the park lies an area of deep, shady gullies called "canutos," these are characterized by 90 percent humidity and year-round temperatures of 70°F (20°C). Within their depths thrives a rich community of ancient semi-tropical ferns. The narrow valleys also provide ideal conditions for a unique assemblage of Iberian amphibians, particularly newts, frogs, and toads. Los Alcornocales means "the cork oaks," and this natural park harbors the world's largest cork oak forest, which has long been exploited for its bark. Living in the mountain forests of cork oak and olive trees, the fauna is highly varied and includes deer, wild cat, and wild boar. TF

CAP DE FORMENTOR

MALLORCA, SPAIN

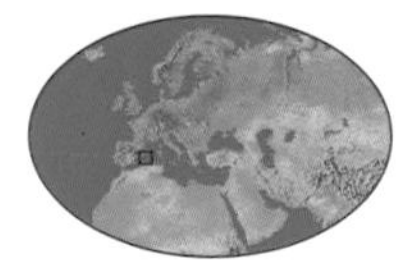

Maximum altitude of peninsula: 1,095 ft (334 m)

Features: renowned botanical locations

Geologically part of the Serra de Tramuntana, the Formentor Peninsula forged northeastward into the Mediterranean, leaving a swathe of diminutive offshore islets in its wake. The scenery is dramatic: 1,320-foot- (400-meter-) high rugged brown cliffs contrast with the clear blue waters below, and pine trees grow out of the rocks and frame the many sandy beaches. At first glance, the precipitous limestone cliffs are apparently inimical to plant life, but a closer look reveals cushions of small flowering plants. Pallid swift, crag martin, and blue rock thrush nest on these maritime crags, and Elenora's falcon and shearwaters put on fantastic aerial displays. Gentler slopes inland are mantled with low scrub punctuated by clumps of tall grass, which provides a home to breeding Sardinian and Marmora's warblers and attracts a variety of passerines on migration. Just beyond the road tunnel is one of Mallorca's most renowned botanical locations, where scree slopes harbor foxglove, Balearic sowbread, and dragon's mouth, plus the peony *Paeonia cambessedesii.* This flower is unique to the Balearics and named after Jacob Cambessèdes (1799–1863), one of the first botanists to study the islands. It is claimed he obtained specimens by shooting them off inaccessible ledges with a shotgun. TF

SERRA DE TRAMUNTANA

MALLORCA, SPAIN

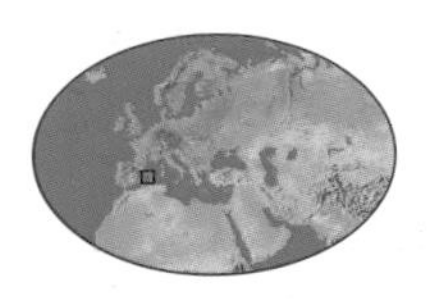

Length of Serra de Tramuntana: approx. 50 mi (80 km)

Highest point: 4,740 ft (1,445 m)

Special Protection Area for Birds (ZEPA): 118,608 acres (48,000 ha)

The Serra de Tramuntana range dominates Mallorca, spanning the northwestern flank of the island. The "mountains of the strong wind" provide an effective buffer from the fierce *tramuntana* wind that proceeds from mainland Europe. It is an outstanding karst area—limestone peaks have been eroded by the effects of weather and water, giving rise to many impressive caves and deep gorges specific to the region known as *torrents.* These mountainous torrents fill rapidly after rain.

Four million years' isolation has resulted in a high level of endemism in the Balearics, with Serra de Tramuntana hosting more than 30 plants found nowhere else in the world. The sheltered interiors of the many torrents are a stronghold of the native Mallorcan midwife toad, first described from fossil remains in 1977. Along the north coast, the Serra de Tramuntana meets the sea with austere, near-vertical cliffs almost 1,000 feet (300 meters) high in places. These inaccessible precipices are much favored by nesting shag, Eleonora's falcon, and osprey. The mountains also harbor the only island-dwelling black vultures on the planet: about 70 birds, although only a handful of pairs breed each year. **TF**

CABRERA ARCHIPELAGO

MALLORCA, SPAIN

Area of national park: 24,762 acres (10,021 ha), of which 3,257 acres (1,318 ha) are terrestrial

Area of Cabrera Gran: 2,852 acres (1,154 ha)

Maximum altitude: 565 ft (172 m)

Lying barely 6 miles (10 kilometers) from the southern tip of Mallorca, the Cabrera Archipelago is considered to be a geological continuation of the Serres de Llevant. The only Balearic national park, it comprises 19 small islands and islets, the principal landmass being Cabrera Gran. Cabrera Gran is dominated by much-eroded Jurassic and Cretaceous limestones, dotted with small areas of Aleppo pine. About 80 percent of the world's population of Lilford's wall lizard live in the park—some 10 distinct races have been described from the various islets. The marine ecosystem is also extremely rich: The undersea meadows of *posidonia* harbor more than 200 species of fish and 34 echinoderms. The Cabrera Archipelago is best known, however, for its seabird colonies, notably endangered Balearic shearwaters and the largest colony of Audouin's gull in the Balearics.

Boats to Cabrera Gran depart from Colònia de Sant Jordi and Portopetro in southern Mallorca in peak season (from June to August). To visit Cabrera in your own boat, a navigation permit must be requested from the park administration. Access to Cabrera is restricted to just 50 boats and 200 visitors a day. **TF**

TEJO ESTUARY

SANTARÉM / LISBON, PORTUGAL

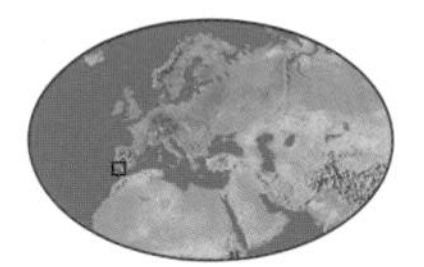

Extent of protected area (natural reserve and Ramsar site): 135,978 acres (55,028 ha)

Maximum numbers of wintering birds: greater flamingo 3,000; teal 10,000; avocet 6,000; and dunlin 30,000

It is unusual to find a prime bird-watching site near a capital city, but the Tejo Estuary—one of western Europe's 10 most important wetlands for wintering and migratory birds—is such a place. Rising in the Montes Universales, the River Tejo (Tajo to Spaniards, and Tagus in the English-speaking world) is the longest river in the Iberian Peninsula, flowing westward for more than 620 miles (1,000 kilometers) before discharging into the Atlantic Ocean just south of Lisbon. Here it encounters a vast landscape of intertidal mudflats, saltmarsh (the largest continuous expanse in Portugal), and brackish grasslands.

The primary significance of this enormous coastal wetland is ornithological. More than 240 species of bird have been recorded here, with 120,000 individuals flocking in during winter and passage periods. It is a key site for wintering greater flamingo, avocet, and dunlin, with numbers of black-tailed godwit sometimes reaching an incredible 50,000 birds. The Tejo Estuary also provides breeding habitats for purple heron, marsh and Montagu's harriers, and black-winged kite, as well as large numbers of collared pratincole, black-winged stilt, and little bustard. **TF**

COSTA SUDOESTE

SETÚBAL / BEJA / FARO, PORTUGAL

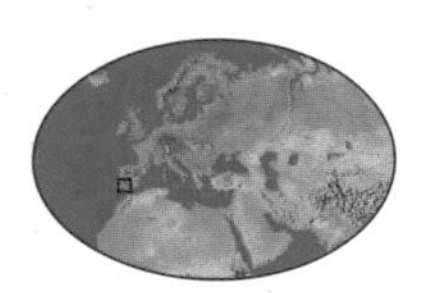

Maximum altitude of Costa Sudoeste: 510 ft (156 m)

Area of natural park: 184,796 acres (74,786 ha)

For over 80 miles (130 kilometers), between the oil port of Sines in the Alentejo and the village of Burgau in the western Algarve, lies one of the wildest and most spectacular stretches of coastline of the Iberian Peninsula. Sheer, rocky cliffs, often topped with richly vegetated dunes deriving from wind-blown sands, are interspersed with secluded coves where small watercourses have carved a way down to the ocean. These geological features have created a variety of valuable bird habitats adjacent to rich feeding grounds.

The focal point of the natural park which protects this meeting point between land and sea is Cabo de São Vicente at the extreme southwestern tip of Portugal—continental Europe's "land's end" and for centuries the limit of the known world. The 260-foot (80-meter) limestone cliffs, relentlessly assaulted by tempestuous Atlantic waves, have been sculpted into a series of rugged headlands and offshore stacks, topped by an expanse of limestone pavement where unique plants abound, notably the squill. Only a few miles to the east lies the *Ponta de Sagres,* or "Sacred Headland" of the Romans—steeped in the history of the Portuguese Empire. In 1394, in the village of Sagres, a school of navigation was founded, from which sponsored voyages of discovery were made to Madeira and as far as the west coast of Africa.

Apart from being one of the few places in Europe where otters feed in the ocean, the birdlife of the park is extremely diverse, with more than 200 species recorded here. This is the only place in the world where white storks nest on sea cliffs, sharing their habitat with breeding shag, Bonelli's eagle, peregrine, and lesser kestrel. The bleak, stony plain just inland of Sagres is the renowned haunt of steppe birds such as stone curlew, little bustard, and tawny pipit, while scrubby habitats harbor spectacled and subalpine warblers. The park also lies on one of the main routes used by trans-Saharan migrants, particularly passerines and "soaring" birds, with most western European raptors seen flying over the Cabo de São Vicente each fall.

The best time to visit is from May to September, when the temperature is pleasant. In the springtime, almond trees in the region produce a spectacular display of blossoms. TF

Apart from being one of the few places in Europe where otters feed in the ocean, this is the only place in the world where white storks nest on sea cliffs, sharing their habitat with breeding shag, Bonelli's eagle, peregrine, and lesser kestrel.

RIGHT *Rocky crags dot Portugal's Costa Sudoeste.*

ILHAS BERLENGAS

LEIRIA, PORTUGAL

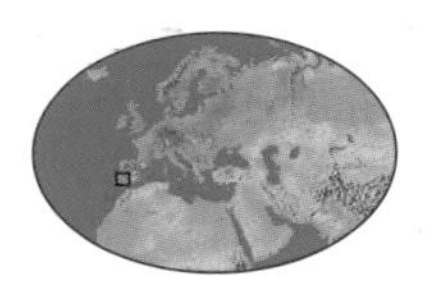

Extent of protected area (natural reserve and *reserva biogenética*): 2,627 acres (1,063 ha)

Age: approximately 280 million years

Distance from the shore: 6 mi (10 km)

The diminutive, windswept Berlengas Archipelago is one of the principal sites for breeding seabirds on the Iberian Peninsula. Representing the unsubmerged peaks of a continental shelf laid down approximately 280 million years ago, the group comprises a main island of pinkish granite—Berlenga Grande—rising to a height of 280 feet (85 meters), plus three groups of lesser and more remote islets, dominated by schists and gneiss: Farilhões, Estelas, and Forcadas.

Of outstanding importance are the 200 or more pairs of Madeiran storm petrel (mainly in the Farilhões) that breed nowhere else in continental Europe, and 10 pairs of guillemot, the last vestige of a colony that numbered more than 6,000 birds in 1939, but is still the largest remaining in Iberia. The Berlengas also hold Portugal's only breeding colony of Cory's shearwater (180 to 220 pairs) and the most important population of nesting shag in the country (some 70 pairs), as well as a native subspecies of Bocage's wall lizard. The sheer, almost vertical cliffs harbor three plants that occur nowhere else in the world: specific species of thrift, fleabane, and rupturewort. TF

BUTRINT NATIONAL PARK

VLORË, ALBANIA

Area of Butrint National Park: 11 sq mi (29 sq km)

Area of Lake Butrint: 40 acres (16 ha)

Butrint National Park in southern Albania is situated on a peninsula protruding into the Ionian Sea, facing the Greek island of Corfu. It contains the World Heritage site of the ancient city of Butrint, one of the least known and least spoiled classical sites in the Mediterranean, with ruins representing each period of its civilization.

Butrint is an area overflowing with unique ecological beauty. It is a place where mountains and ocean meet, creating dramatic cliffs, caves, harbors, bays, and some of the Mediterranean's most intact natural areas. The park has recently been enlarged to include 11 square miles (29 square kilometers) of the surrounding lakes region, including part of Lake Butrint, Lake Bufi, the Vivari Channel, and various salt marshes, lagoons, and wetlands. A haven for wildlife, Butrint's varied habitats hold sensitive ecosystems and important biodiversity, including the highest number of amphibian and reptile species recorded in Albania. It also affords sanctuary to numerous rare species, including the loggerhead sea turtle and the extremely rare Mediterranean monk seal. Bird life is exceptional, and for this reason Butrint has been designated a Ramsar site. PT

BOKA KOTORSKA BAY

CRNA GORA / MONTENEGRO

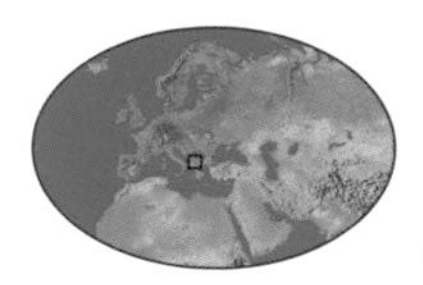

Length of Boka Kotorska Bay: 17 mi (28 km)

Number days of sunshine per year: 200

Not only is Boka Kotorska Bay southern Europe's longest and deepest fjord, it is also one of the most beautiful bays in the Mediterranean. It is framed by towering mountains that appear to have cracked open and allowed the sea to flood in. Boka Bay actually curls in on itself—a legacy of great glaciers sliding down the mountainsides and gouging out the land.

Boka is situated on the northern coast of the Adriatic Sea. The mountains form a natural barrier to the cold weather moving down from the north, making this a Mediterranean oasis blessed with almost constant sunshine. Rainfall in the Boka Kotorska mountains reaches 59 to 118 inches (1,500 to 3,000 millimeters) a year, while annual rainfall in the bay area is one of the highest in Europe at 197 inches (5,000 millimeters). Late spring is an ideal time to visit, when the mountaintops are covered by snow and roses bloom in the foothills.

The bay is protected as a World Heritage site, not only for its natural beauty, but also for the old stone-built villages with terracotta roofs that line the coast. The secluded waters of the bay have been an important maritime center for centuries. JK

LAKE PRESPA

ALBANIA / MACEDONIA / GREECE

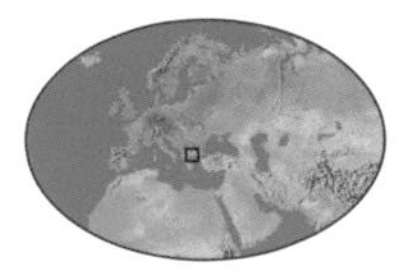

Surface area of Lake Prespa: 97 sq mi (250 sq km)

Average depth of lake: 177 ft (54 m)

Notable feature: one of only 17 ancient lakes in the world that are 5 to 20 million years old

Lake Prespa is located at a higher point than any other lake in the Balkans, lying at 2,798 feet (853 meters), and covers an area of 106 square miles (274 square kilometers). Although it consists of two lakes separated by a narrow isthmus—Megali Prespa and Mikri Prespa—these are considered to be two parts of the same ancient lake. Two-thirds of Lake Prespa belongs to the Republic of Macedonia, while the remaining third is shared by Greece and Albania. On World Wetlands Day in 2000, Prespa Park was declared the first trans-boundary protected area in southeastern Europe. The park area covers Prespa Lake and its surrounding wetlands. In the region, bears, wolves, and otters are found, and over 1,700 species of plants have been recorded.

The wetlands are also of paramount importance for migratory and breeding birds. Over 260 species of bird breed in Mikri Prespa, including cormorants, egrets, and herons, but undoubtedly the pelicans—two species breed in mixed colonies—are the main attraction for bird watchers. The larger Prespa Lake has a number of sandy beaches: between June and August the water temperature of the lake ranges from 64 to 75°F (18 to 24°C). PT

LAKE OHRID

MACEDONIA / ALBANIA

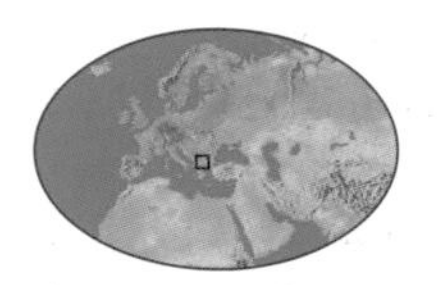

Depth of Lake Ohrid: 950 ft (290 m)

Surface area of lake: 174 sq mi (450 sq km)

Lake Ohrid, formed 2 to 3 million years ago, is Europe's oldest and deepest lake. It came into being immediately after the end of the last glaciation period. Although the lake contains low levels of nutrients it is home to many flora and fauna species; due to its geographic isolation and stable conditions, large numbers of relic forms or "living fossils" have been discovered. There are many rare fish species, including the sought-after trout called the *koran*. The lake's fauna is dominated by various types of algae, and an underwater plant called *hara* that forms a continual ring on the bottom of the lake.

Most of Lake Ohrid's water bulk comes from numerous surface and underground springs—for this reason researchers consider it unique in the world. Most of the surface springs lie along the southern shore, near the monastery of St. Naum, on the Macedonian side. In total, 40 rivers and springs flow into the lake: 23 on Albanian territory and 17 on Macedonian territory. PT

MOUNT ATHOS

CENTRAL MACEDONIA, GREECE

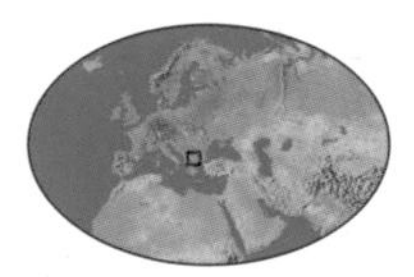

Height of Mt. Athos: 6,670 ft (2,033 m)

Dimensions of Mt. Athos promontory: 30 mi (50 km) long and 2 to 5 mi (3 to 7 km) wide

Rock type: metamorphic, mostly marble

The easternmost of the three promontories of northern Greece's Halkidiki Peninsula, Mount Athos's landscape encompasses many beautiful and varied natural features, including quiet valleys, inaccessible mountain crests, and thickly wooded slopes that plunge down toward the sea, as well as secluded coves, rocky headlands, and sandy beaches. For all its undoubted beauty, however, the peninsula is today known less for its natural wonders than its religious ones. The whole area functions as a semi-autonomous republic of the Greek Orthodox Church, home to around 20 inhabited monasteries, many small religious houses, and hermit monks who live in caves. The total religious population is around 3,000, most of whom live a medieval-like lifestyle. Although a strong pagan tradition once thrived here (Homer mentions Athos as a holy place, noting that Mount Athos was the home of Zeus and Apollo before they moved to Mount Olympus), worship is now exclusively Christian. The first monasteries were built here in the 6th century and, in 1054, it became an Orthodox spiritual center. In 1046, Emperor Constantine IX Monomachos forbade women from entering the peninsula, and this edict is still rigorously enforced today. Men need special permits to visit, and overnight stays are not allowed for anyone under 18 or those lacking in proper religious conviction. AB

MOUNT OLYMPUS

CENTRAL MACEDONIA, GREECE

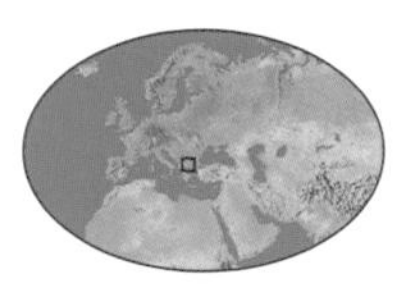

Height of Mt. Olympus: 9,570 ft (2,917 m)

Flora and fauna: 1,700 plant species, 32 mammal species, 108 bird species

It is easy to see why Mount Olympus was believed to be the seat of the gods: no other Greek mountain is as magnificent. It rises majestically from the sea to the twin peaks of the "Throne of Zeus" and Mytikas, just two of the mountain's eight peaks. Mount Olympus looms over the Bay of Thermaikos—often hidden by clouds, it can appear unexpectedly, vast and splendid, when the skies clear. Mytikas, at 9,570 feet (2,917 meters), is the highest peak in Greece. It is possible to climb it, however, without prior climbing experience. The mountain can be approached from Litohoro, from where it is possible to walk to the summit. Mount Olympus has been a national park since 1938 and has been declared a Biosphere Reserve. The 1,700 plant species, 24 of which are unique to the mountain, make the area a botanical paradise. Much of the area remains pristine with dense forests of pine, beech, and cedar. Here chamois, wolves, bears, and lynx can still be found—the gods would not have it any other way. **PT**

LAKE VISTONIS

EAST MACEDONIA AND THRACE, GREECE

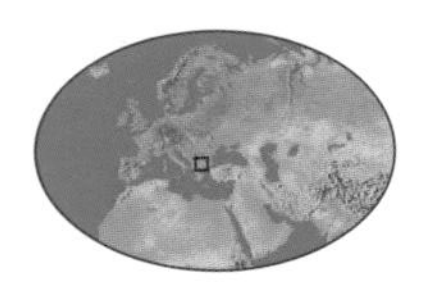

Area of Lake Vistonis: 16 sq mi (42 sq km)

Average depth of lake: 6.6 to 8.3 ft (2 to 2.5 m)

Located in the Xanthi region of northeastern Greece, Lake Vistonis and the surrounding wetlands comprise an unspoiled environment of special ecological interest. As a coastal lake there are great fluctuations in the salt content of the water. The water in the northern part of the lake is brackish due to the influx of three rivers: Kosynthos, Kompstatos, and Travos. In contrast, salty water is found in the southern part of the lake, as a result of the three canals that connect the lake with the sea.

The shores of the lake—with their wet meadows, saltmarshes, extensive mudflats, reedbeds, and scrub—are of international importance. During the winter months a quarter of a million birds take refuge here. Many herons and cranes breed at the site, and white-headed ducks have reached record numbers. In the lake's waters, 37 species of fish can be found, including striped gray mullet and eel. Animals living around Lake Vistonis include wild cats, jackals, and badgers. PT

MOUNT GIONA

STEREA HELLAS, GREECE

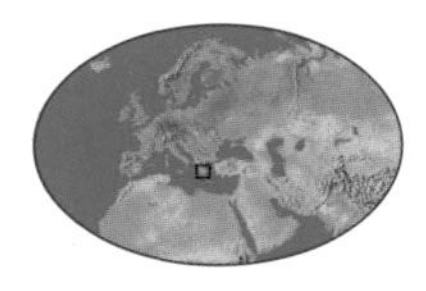

Height of Mt. Giona: 8,225 ft (2,507 m)

Notable features of Mt. Giona: highest peak in southern Greece; highest unbroken cliff in Greece

Mount Giona's peak, at 8,225 feet (2,507 meters), is the highest in southern Greece and the fifth highest in all Greece. It is a compact mass of limestone, with cliff-like ramparts all around it, including one sheer drop of 3,300 feet (1,000 meters), making it the highest unbroken cliff in Greece. There is no road over Mount Giona, and small roads contour around it. At around 1,950 feet (595 meters) lies a vast plateau where there are dozens of springs. At mid-mountain lies Karkanos, a huge crater, which is difficult to measure because it is frozen all year round. The Asopos River has its source in a glacier on Mount Giona and cascades several thousand feet to the valley bottom; it then runs eastward, before plunging further down into a formidable gorge.

In spite of being an easy distance from Athens by road, Mount Giona remains largely unknown, unlike the neighboring Oiti and Parnassus mountains. There are some way-marked paths, and the lower shoulders of Giona are grazing grounds. Meadows are filled with flowers in spring, as are the forest glades and the grasslands above the tree line when the snows melt in May. The mountain is forested largely with firs, although some of the lower slopes are clothed with an impressive mixture of broad-leaved trees. The best months for rock climbing on Mount Giona are July, August, and September. PT

LAKE KERKINI

CENTRAL MACEDONIA, GREECE

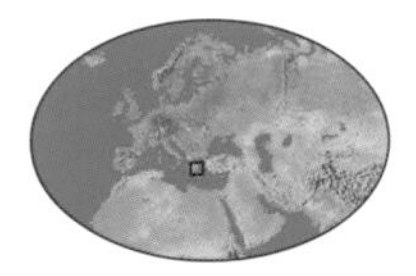

Surface area of Lake Kerkini (seasonal variation): 21 to 28 sq mi (54 to 72 sq km)

Depth of lake (seasonal variation): 16 ft (5 m)

Average depth of lake: 33 ft (10 m)

Lake Kerkini lies in a natural basin surrounded by high mountains, near the Greece–Bulgaria border. It is a large freshwater lake that was created in 1932 when a dam was constructed on the Strymonas River on the site of a swamp at Lithotopos.

A new dam was constructed at Lake Kerkini in 1983 because a large quantity of silt had accumulated. At this time, an agreement was drawn up between the two neighboring countries about the use of the waters. Current flood-control and irrigation methods cause large fluctuations in Lake Kerkini's water level: Levels can vary up to 15 feet (4.5 meters) every year. The lake begins to fill in February and reaches its highest levels by May.

In recent years, the seasonal increase in the depth of the lake has become much greater and over half the riparian forest and much of the marsh vegetation has been destroyed. However, other plant species have benefited, notably waterlilies. Willow trees have adapted, too, to become the only semi-aquatic trees in Europe. The remaining areas of woodland habitats include open-water oak-hornbeam forest, and mixed oak-elm-ash forest.

The environment of Lake Kerkini is attractive to many species of wildlife. It teems with fish; over 300 species have been recorded. It is also one of the most important wetlands in Greece for breeding waterbirds. Kerkini supports a rich bird population throughout the year, in numbers and diversity, including rare and endangered species. In spring, the lake is black with flocks of coot and many species of duck, and great-crested grebes perform their nuptial dances close to the lake shore. One of the best times to visit Kerkini is during the spring, when flocks of pelicans can be seen.

In spring, the lake is black with flocks of coot and many species of duck, and great-crested grebes perform their nuptial dances close to the lake shore.

Large herds of water buffalo graze the swamps along the shores of the Strymonas River. They swim, dive for aquatic vegetation, and create mud wallows. They are generally peace-loving creatures but formidable if disturbed. On a much smaller scale, in the surrounding area of Lake Kerkini there are at least 10 amphibian species (including frogs, and salamanders), five snail species, 19 reptile species (including lizards, snakes, and turtles), and a great variety of insects. Expert guides can take visitors by boat to see the immense colonies of pygmy cormorants and other water birds. There is also a visitor center in Kerkini village. The snowcapped mountains of Beles and Krousia provide wonderful panoramic views of the lake. PT

METEORA

THESSALY, GREECE

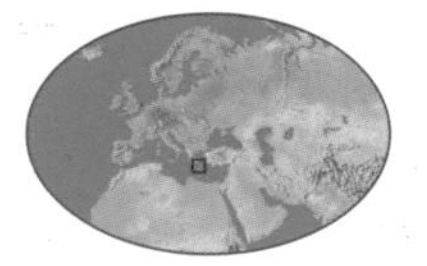

Extent of protected area: 927 acres (375 ha)

Maximum height of outcrops: 3,300 ft (1,000 m)

Situated in the northwest corner of Thessaly to the east of the Pindos Mountains lies Meteora, a strange and unique landscape where huge sandstone outcrops tower up to 3,300 feet (1,000 meters). The name means "suspended in air"; some attribute this to the outcrops' precipitous nature, others suggest that the rocks fell from outer space. Scientists believe they were formed some 60 million years ago during the Tertiary period, emerging in the delta of a river and further shaped by seismic activity.

A community of 24 monasteries clings precariously to the ancient pinnacles, and Meteora—a World Heritage site since 1988—is an important center for the Orthodox Church. Today, people can visit the last four monasteries that remain inhabited; the rocks away from the monasteries are popular with climbers. The outcrops and the plane forest in Pinios Valley below are important areas for birds of prey, including short-toed and lesser-spotted eagles, as well as honey buzzards and peregrine falcons. About 50 pairs of Egyptian vulture breed here, although their numbers are declining; black stork also breed in the area. The landscape ranges from forested hills to riverine forests in the valleys. Earthquakes are frequent but are not of a high intensity. **PT**

DIROS CAVES

PELOPONNESE, GREECE

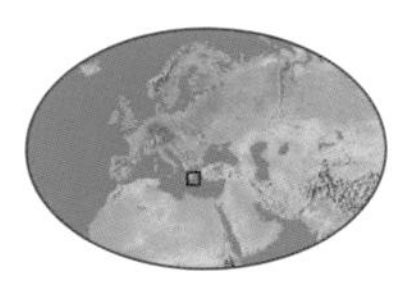

Area of Diros Caves explored: 16,500 ft (5,000 m)

Temperature of caves: between 61 and 68°F (16 and 20°C)

The caves at Diros, five miles (eight kilometers) south of Areopolis, are part of an underground river. Where the river meets the sea, stalactites and stalagmites colored by mineral deposits are reflected in the crystal-clear waters of subterranean lakes. Within the Glyphada (or Vlyhada) Cave, passages of around 16,500 feet (5,000 meters) have so far been explored—its total extent is thought to be 360,000 square feet (33,400 square meters). The cave's temperature fluctuates between 61 and 68°F (16 and 20°C) and it is regarded as one of the most beautiful lake caves in the world. In its interior two-million-year-old fossil animals have been found.

The Kataphygo Cave has an area of 29,000 square feet (2,700 square meters) and its passages reach 2,310 feet (700 meters) in length. The Alepotripa Cave lies to the east of Glyphada. Traces of Neolithic inhabitants have been found, and these can be seen in the Stone Age Museum at the cave's entrance. Thousands of years ago, the Diros Caves served as a place of worship, because they were thought to be connected to the Underworld. **PT**

LESVOS PETRIFIED FOREST

LESVOS ISLAND, GREECE

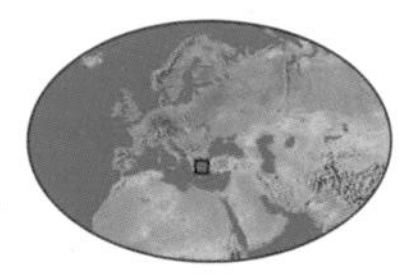

Age of Lesvos Petrified Forest: 15 to 20 million years

Maximum height of Mt. Lepetymnos: 3,176 ft (968 m)

Feature: silicified sub-tropical fossil forest

Lesvos Island is Greece's third largest island. The western part contains petrified forests that date from the late Oligocene to middle Miocene period. Intense volcanic activity smothered the forest in ash, preserving it in a cloak that allowed silicon-rich water and minerals to percolate through the trees to gently replace the plant materials on a molecule-by-molecule basis. This gentle process, away from atmospheric disturbance, digging animals, and agents of decomposition or disturbance, resulted in exquisite silicified fossils where all details are preserved, right down to the structure of the cells themselves. The preservation is so exact that species can be precisely identified. Studies begun in 1844 have revealed that the forest was subtropical—species have been found that occur in places moister and wetter than present-day Lesvos. Species present include laurels, cinnamon relatives, and sequoia redwoods, trees that were found only in the subtropical zones of Asia and the Americas. The fossilized tree trunks are now a protected natural forest. Originally formed by tectonic stresses in the region, the now-inactive volcanoes remain, their cones forming the island's spine. The forest is easily reached from Eressos, Antissa, and Sigri, the island's three main villages. **AB**

SAMARIÁ GORGE

CRETE, GREECE

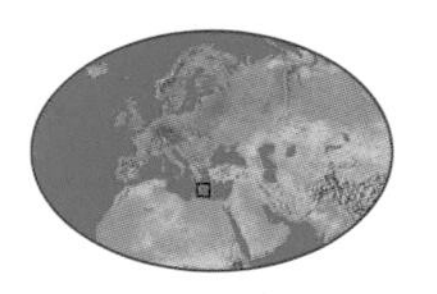

Length of Samariá Gorge: 10 mi (16 km)

Depth of gorge: 1,650 ft (500 m)

Age: 2 to 3 million years

The sides of the Samariá Gorge in western Crete are so close together it seems as if they could almost touch at the top. The narrowest point is called the *Portes* or "gates," where the walls are 1,650 feet (500 meters) high, but only 10 feet (3 meters) apart at river level and 30 feet (9 meters) apart at the rim. The gorge has been cut by the River Tarraíos through the Levká, or White Mountains. It is a raging torrent in winter and a gurgling stream in summer, and the Mediterranean sun penetrates the gorge for only a few minutes each day. Clinging to the sides are cypress, oleander, and fig trees, while choughs, eagles, and peregrines flash through the air.

Hidden inside is the deserted village of Samariá that was abandoned in 1962 when the area was turned into a national park. Downstream, the gorge widens and the settlement of Ayiá Rouméli can be found. Its distance from the head of the gorge near Omalos is about 10 miles (16 kilometers). Ayiá Rouméli is a center for *diktamos*, a plant used in herbal teas and a favorite food of the kri-kri (wild goat). The gorge is open to visitors from May until October, and it takes over five hours to walk from Xyloskala on foot. **MB**

VIKOS GORGE

EPIRUS, GREECE

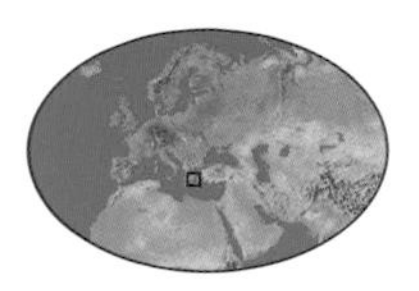

Length of Vikos Gorge: 7.5 mi (12 km)

Average width of gorge: 660 ft (200 m)

Average depth of gorge: 2,310 ft (700 m)

The Vikos-Aoos National Park is located in the northwestern tip of the Pindos Mountain Range. Covering an area of 46 square miles (119 square kilometers), it includes the Vikos Gorge, the Aoos ravine, and Mount Tymfi (Gamila). Vikos Gorge is one of the deepest gorges in Europe. It is also the world's deepest canyon in proportion to its width—many gorges are deeper but significantly wider, giving them a smaller depth-to-width ratio.

The gorge is astoundingly beautiful. It is 10 miles (16 kilometers) long and collects the waters of a number of small rivers before leading them into the Voidomatis River, which rises in the gorge. In summer, the river dries up, and it is possible to stone scramble at the bottom of the gorge. Walkers are advised to start at Monodendri and descend the gorge to either Vikos or Papingo, rather than the other way around, to avoid the steepest climb at the finish after the 6-mile (10-kilometer) walk. Many rare flowers flourish throughout the gorge and patches of dense forest dot the slopes. European brown bears and wolves inhabit the area, raptors nest in the cliffs, and chamois can be seen on the crags. PT

VALLEY OF THE BUTTERFLIES

ISLAND OF RHODES, GREECE

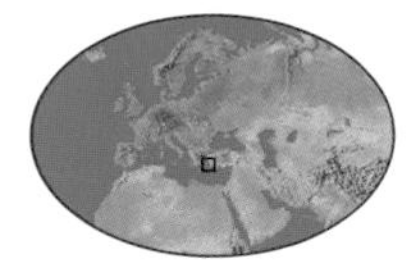

Area of Rhodes Island: 540 sq mi (1,398 sq km)

Highest point (Mt. Attavyros): 3,986 ft (1,215 m)

Petaloudes, or "Valley of the Butterflies," is a deep ravine situated 16 miles (25 kilometers) southwest of Rhodes town. This beautiful wooded valley—with its small flowing streams, still pools, and waterfalls—attracts a deluge of Jersey Tiger moths that arrive each summer to breed. They are thought to be drawn by the strong fragrance from the resin of storax trees, and the fact that the valley remains cool even during the hottest months of summer. From June to September the trees are covered with thousands of moths, resting while surviving on their fat reserves. At first they are difficult to spot because they are well camouflaged with their black-and-cream striped forewings, until they move and flash their scarlet-colored hind wings.

In the 1970s tourists gathered at Petaloudes to see the moths; they would create loud noises to send the thousands of moths soaring into into the air. However, flying away in panic drained the moths of their energy—many died, and the Jersey Tiger moths suffered a drop in population. Today, efforts to protect the valley have paid off; park rangers warn tourists against disturbing the vulnerable moths, and numbers have increased. MB

PAMUKKALE SPRINGS

DENIZLI, TURKEY

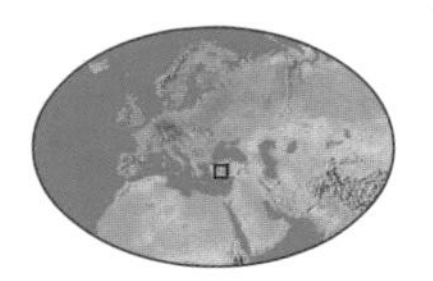

Area of Pamukkale Springs: 1.5 mi (2.5 km) long by 0.3 mi (0.5 km) wide

Water temperature: 86 to 212°F (30 to 100°C)

White terraces with crenulated edges rising one above the other; semi-circular basins of water reflecting the deep blue sky; and stalactites standing like frozen waterfalls on a hillside in western Turkey. This is the unique and mystical world of Pamukkale—literally "castle of cotton"—which according to local legend is where giants dried crops of cotton on the terraces. In reality, the terraces are the result of hot volcanic springs that flow from the plateau above. The water is full of lime and other salts, and as it flows down over the hill it cools, covering everything in its path with white mineral deposits. Over eons, layers of lime have built up to form walls and terraces, and anything that falls into the water is coated in just a few days. The spring water is claimed to have therapeutic value for easing rheumatism and reducing blood pressure. Ancient Greek and Roman dignitaries came here, including the emperors Nero and Hadrian, and today tourists bathe in the warm waters in awe at the sight of the pure white cascade frozen in lime. The scene is particularly spectacular in the morning and evening light, when Pamukkale looks like a surreal lunar landscape. **MB**

VALLA CANYON

KASTAMONU, TURKEY

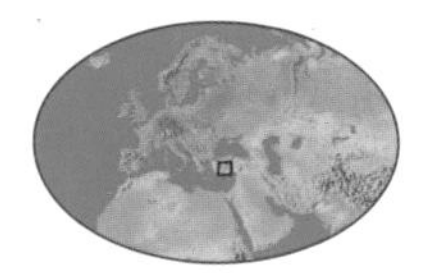

Area of Kure Daglari National Park: 91,428 acres (37,000 ha)

Core area of forests: 135,908 acres (55,000 ha)

Considered to be the most dangerous passage to cross in Turkey, Valla is one of several canyons that slice into the Kure Mountains in the western Black Sea region of Turkey. It is situated in the Kure Daglari National Park—which covers 50,000 acres (20,230 hectares)—and represents the largest and most intact "humid karst area" in the region. The canyon starts where the Devrekani and Kanlicay Rivers meet, and then continues for 7.5 miles (12 kilometers) toward Cide. The entrance is difficult to negotiate, and a guide and the right equipment are needed in order to explore the canyon safely.

Valla Canyon is a wild and lonely place. It is home to a wide variety of wildlife including brown bear, roe deer, wild boar, and fox, although numbers have been greatly reduced by hunting; many birds of prey can be seen soaring above the high canyon walls that rise up 2,640 to 3,960 feet (800 to 1,200 meters). The park protects mature natural forests of fir and oriental beech at higher elevations and chestnut down toward the sea, as well as black pine and broad-leaved forests. In the spring, meadows are carpeted with many colorful flowers, including orchids and lilies. **MB**

CAPPADOCIA

NEVSEHIR, TURKEY

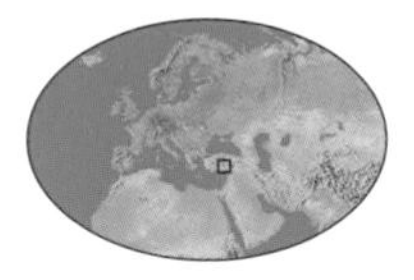

Location of Cappadocia: eastern Anatolia

Height of Erciyas Dagi: 12,848 ft (3,916 m)

In central Turkey, near Urgup and Göreme, there are the most extraordinary sand-colored, conical hills, some up to 165 feet (50 meters) in height. Some of the hills are capped by slabs of darker stone and resemble gigantic mushrooms, while others look like statues of people. These curious hills sit on a plateau dominated by the extinct volcano Erciyas Dagi. Millions of years ago, it threw out huge quantities of volcanic ash that cooled to form tuff, a soft rock that can be carved with a knife. Over the millennia, the tuff was sculpted by erosion to form the conical hills and other rock formations known as "fairy chimneys." Interestingly, people have also been carving out homes, churches, and monasteries in the volcanic rock here for centuries.

The temperature in the caves is constant throughout the year, offering a warm environment in winter and a cool retreat in summer. Some of the settlements are more complex, with rooms cut into the rock both above and below the ground, sometimes creating up to 20 levels of habitation, such as in the city of Derinkuyu. A fresco in a church carved into a hill in the Göreme Valley shows St. George slaying a dragon. MB

CAVES OF THE TAURUS MOUNTAINS

EASTERN ASIA MINOR, TURKEY

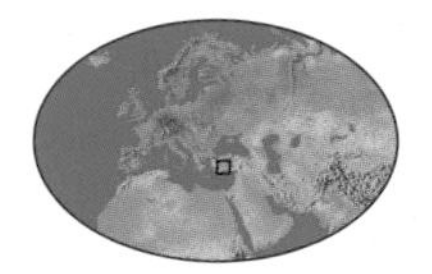

Width of Taurus Mountains range: 125 mi (200 km) across	
Depth of Evren Gunay Dundeni: 4,688 ft (1,429 m)	

Approximately one third of Turkey is underlain by limestone, so karst features such as underground rivers and deep caves are common throughout the country. The largest karst area is the Taurus Mountains, an extension of the Alps on the southeast rim of the Anatolia Plateau. The range extends from Lake Aridir in the west to the headwaters of the Euphrates River in the east and is crossed by a narrow gorge containing the Gokolut River. The Taurus Mountains are riddled with cave systems, including the Insuyu Cave with its stalactite and stalagmite-filled chambers; the Ilarinini Cave, which contains water cisterns and Roman-Byzantine ruins; the Ballica Cave with its striking gray, blue, green, and white formations; and the world's fourteenth deepest cave, the Evren Gunay sinkhole. Evren Gunay—discovered only recently in 1991—was the site of a new caving record in October 2004. It is the deepest cave in Turkey and Asia, and only for the most experienced cavers. For anyone entering the cave systems, extreme caution must be taken during rainy periods, as the caves flood. MB

TORTUM

ERZURUM, TURKEY

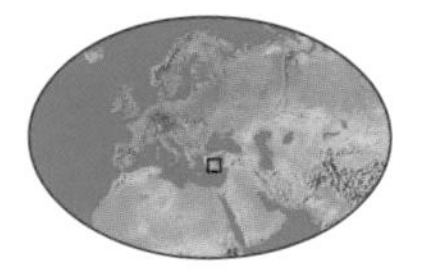

Length of Tortum Lake: 5 mi (8 km)

Elevation of lake: 330 ft (100 m)

Height of Tortum Falls: 130 ft (40 m)

Geologists believe that thousands of years ago a giant chunk of rock parted company with the Tortum Valley wall and slid down to the valley floor. It blocked the Tortum River, forming a vast and tranquil lake now known as Tortum Lake. The water levels rose and found a new outlet over a fault line, dropping spectacularly 130 feet (40 meters) to a further series of cascades and four small lakes that were dammed by rubble washed down from the landslide. The scar of the landslip is still visible today, but the water passing over the falls has diminished considerably as it is siphoned off to a small hydroelectric power station. Water is fed back into the river above a narrow slot gorge that shows a confusion of twisted rock strata, the result of violent earthquakes. However, during winter so much water flows down the Tortum River that the falls can be seen in their former glory. Earth pillars similar to those seen in nearby Cappadocia can be seen on the eastern shore of the lake. The river and lake complex is about 62 miles (100 kilometers) north of Erzurum, a city on the ancient Silk Route. MB

MOUNT ARARAT

AGRI, TURKEY

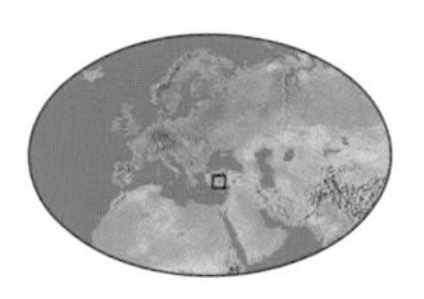

Height of Mt. Ararat:
Great Ararat: 17,011 ft (5,185 m);
Little Ararat: 12,877 ft (3,925 m)

Diameter of Ararat Massif: 25 mi (40 km)

In the far northeast of Turkey, Mount Ararat, a dormant volcano, rises gracefully in isolation from the surrounding plains and valleys. It comprises two peaks: Great Ararat, the highest peak in Turkey, rises 17,011 feet (5,185 meters); while Little Ararat, with its near-perfect cone, is 12,877 feet (3,925 meters). Between the two pinnacles lies the Serdarbulak lava plateau, stretching 8,580 feet (2,600 meters) across. Ararat is snow-capped all year round, but its snow line is only reached after a height of 14,000 feet (4,270 meters). Below the snow line lie great blocks of black basalt, some as large as houses, while on the northern and western slopes of Great Ararat unbroken glaciers can be seen. Ararat has no crater, and no eruption has ever been recorded; however, in 1840 an earthquake rocked the mountain. Climbing Mount Ararat is a challenging and rewarding experience with excellent panoramic views. The best climbing months are July to September, but in winter and spring the weather is extremely severe and conditions can be hazardous. Ararat's main claim to fame, however, is not its geology or its geographical position but its folklore—it is claimed to be the resting place for Noah's Ark after the Flood, and many archeologists have combed the area looking for evidence. MB

SAKLIKENT GORGE

ANKARA, TURKEY

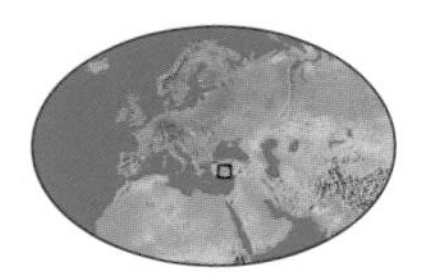

Length of Saklikent Gorge: 11 mi (18 km)

Depth of gorge: 1,000 ft (300 m)

Number of caves: 16

Saklikent Gorge is also known as "the hidden city" (or valley) because most visitors do not know it is there until they stumble across it, and that is precisely what a local goat herder did just 20 years ago. Today it is a popular tourist attraction.

The gorge is the longest and deepest gorge in Turkey. Its limestone walls are sculpted by water, and they are so high and sheer, and the gorge so narrow, that sunlight fails to penetrate to the river, 1,000 feet (300 meters) below the rim. The gorge is approached via a long wooden boardwalk. A short distance upstream, the Ulupinar spring bubbles up at the base of the cliffs. Here visitors wade across the river in the gorge proper to begin the walk. Of the 11 miles (18 kilometers) of the gorge, only the first 2.5 miles (4 kilometers) are accessible to the casual visitor. The water is ice-cold all year round, and because visitors are sure to get wet, plastic shoes can be rented at the River Bar to make wading in the muddy riverbed more comfortable. Further into the gorge the river becomes too deep to navigate and can only be negotiated by experienced rock climbers. Saklikent is a 40-minute drive from Fethiye, where a minibus service is available. **MB**

KARAPINAR CRATER LAKES

KONYA, TURKEY

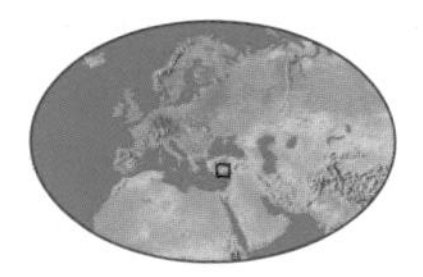

Elevation of Meke Crater Lake: 3,220 ft (981 m) above sea level

Circumference of Meke Crater Lake: 2.5 mi (4 km)

Surface area of Acigol Crater Lake: 0.4 sq mi (1 sq km)

The Karapinar crater lakes lie in a stark, beautiful landscape, with dark basalt features. The landscape was created by ancient volcanic activity and comprises numerous crater lakes, five cinder cones, two lava fields, and several explosion craters.

Meke Crater Lake, although the youngest volcanic formation in Turkey, was formed 400 million years ago. It is said to be the *Nazar Boncugu* of the world—the "blue bead to avert the evil eye." Meke Dagi, the island in the middle of the lake, is one of the largest cinder cones in Central Anatolia—this was formed just 9,000 years ago. Seen from the air, the crater resembles a sombrero, with the volcanic cone surrounded by a lake inside the large caldera. It is a 40-foot (12-meter) deep saltwater lake and is used by birds such as flamingoes, shelduck, and a selection of waders as a migratory resting place. About 1.9 miles (3 kilometers) northwest of Meke is the circular Acigol Crater Lake. On the way to the shoreline visitors pass volcanic tufas and layers of sand; at night, the lake's waters sparkle like phosphorescence in the sea. Both lakes can be found between Konya and Eregli, about 60 miles (96 kilometers) east of Konya. **MB**

ZAGROS MOUNTAINS

IRAN

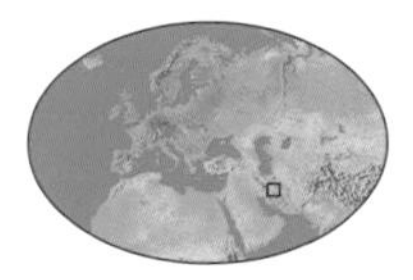

Length of Zagros Mountains range: 550 mi (900 km)

Width of range: 150 mi (240 km)

Highest peak: 12,000 ft (3,600 m)

Some 13 million years ago, in the middle Miocene period, the Arabian and Asian tectonic plates collided. Since that time, they have continued to converge at a rate of 1.6 inches (4 centimeters) each year, creating the Zagros Mountains of southwest Iran in the process. This range has permanent snow cover in a country that is largely hot, dry, and barren. It extends northwest to southeast from the Diyala River, an important tributary of the Tigris River—which means "arrow" in Greek—to the ancient city of Shiraz. Formed primarily of limestone and shale, it consists of numerous parallel ridges. These ridges or folds form a belt that is unsurpassed anywhere in the world for its symmetry and extent. The folds increase in height to the east until they merge with a plateau that lies at around 5,000 feet (1,500 meters). Strong, permanently flowing rivers drain the range's western face, fed by snow and some 40 inches (1,000 millimeters) of rain each year. The higher slopes of the Zagros are covered in oak, beech, maple, and sycamore. Willow, poplar, and plane trees grow in the mountain ravines, and walnut, fig, and almond are found on the lower slopes and in the fertile valleys. JD

SALT GLACIERS

IRAN

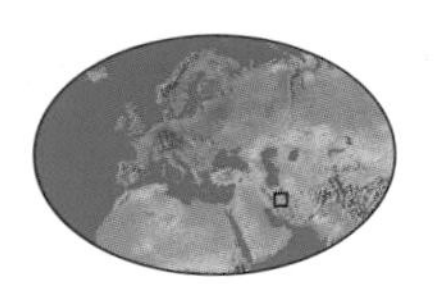

Length of salt glaciers: over 3 mi (5 km)

Age: hundreds of millions of years

Number of salt glaciers: up to 200

Salt glaciers are one of the geological wonders of the world, and Iran contains by far the largest concentration of these extraordinary formations. Known as "diapirs" or salt plugs, they flow from the bedrock to feed huge salt deserts that glitter with tiny white crystals that look like snow. The collision between the Asian landmass and the Arabian tectonic plate has folded rocks and pushed up the Zagros Mountains. In places, underlying deposits of salt ascend in fluid-like plumes: some have forced their way through the overlying rock like toothpaste from a tube. Salt acts as a geologic lubricant: It is weak and behaves like a viscous fluid, and all it needs is the pull of gravity to force rock salt downward to create these spectacular "glaciers."

Some of the salt glaciers are hundreds of feet thick and form repeating ridges separated by crevasse-like gullies, with steep sides and fronts. Not all the salt they contain is white—some are colored pink by minerals and look like decorative sugar crystals. Darker shades come from airborne dust or clays brought to the surface with the salt. Geologists study these strange formations that are normally buried deep underground. JD

QADISHA GROTTO

BCHARRE, LEBANON

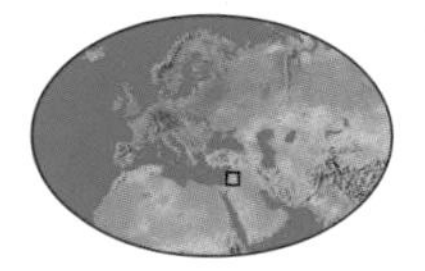

Qadisha Grotto age: 10 million years

Elevation of grotto: 4,757 ft (1,450 m)

Temperature: 41°F (5°C)

Designated a World Heritage site: 1998

Qadisha Grotto is one of several caves in the isolated and beautiful Qadisha Valley in northern Lebanon. The word *Qadisha* comes from a Semitic word for "holy," which is fitting because this steep-sided valley has attracted many monks, hermits, ascetics, and other holy men since the early Middle Ages. It may have been the rugged isolation that attracted them, or perhaps the natural beauty of the place. The large caves ensured both privacy and security.

The grotto is located at the foot of the famous Cedars of Lebanon, below the shadow of Qornet es Sawda, the tallest mountain in Lebanon. The interior of the cave is filled with a forest of colorful stalactites and stalagmites. Natural springs bubble up within Qadisha Grotto and spill out as a dramatic waterfall down into the valley. This is the source of the Qadisha River that runs through the length of the valley. The rushing spring water and high altitude ensure a cool low temperature within Qadisha Grotto, so it is unlikely that this cave was ever inhabited. Although it was previously known to locals, it was "discovered" in 1923 by a monk named John Jacob who was looking for the source of the Qadisha River. JK

PIGEON ROCKS

BEIRUT, LEBANON

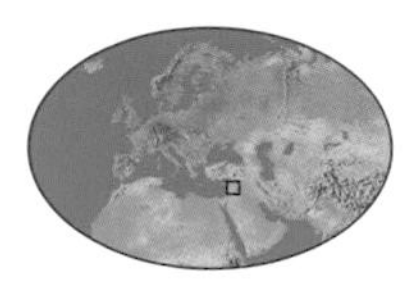

Height of Pigeon Rocks: 112 ft (34 m)

Rock Type: limestone

The Pigeon Rocks are the scenic highlight of Beirut's coastal region. These sturdy stone arches are located 330 feet (100 meters) offshore in direct view of the cafés and restaurants that line the city's Corniche. The waves rolling in from the Mediterranean Sea crash into the rocks with fantastic sprays of white water, and this relentless erosion has carved giant passageways through these monuments and turned them into spectacular natural arches. Sunset is a popular time to see Pigeon Rocks, which provide a dramatic frame for the setting sun. During the summer, boat trips take passengers around the rocks. The cliffs can only be reached by an exhilarating boat trip that requires a boatman who knows how to contend with the incoming waves. The coastline also shows signs of extreme weathering by the sea, with enormous caves hollowed out of the chalk cliffs by the waves.

Fifty years ago, the Pigeon Rocks were the natural habitat of the rare Mediterranean monk seal. In recent years, species that disappeared during the years of civil war have been returning, and a number of beaches in the area have been set aside as protected breeding grounds for loggerhead turtles. JK

CEDARS OF LEBANON

BCHARRE, LEBANON

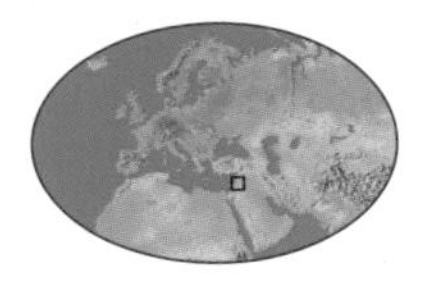

Elevation of the Cedars of Lebanon: 6,600 ft (2,000 m)

Age of oldest trees: 3,000 years

The mountains of Lebanon were once covered by vast cedar forests and these giant trees were celebrated in the Bible, *The Epic of Gilgamesh*, and many other ancient texts. Today, unfortunately, most of these beautiful trees are gone. The remainder are restricted to just 12 groves of trees covering 4,200 acres (1,700 hectares). The most famous of these surviving groves is that of the Bcharre Cedars in northern Lebanon, on the slopes of Jebel Makmel, a high, picturesque mountain that can be reached only via a spectacular drive through the Qadisha Valley. The Bcharre Cedars are ancient—some are as old as 3,000 years, contemporary with kings Hiram of Tyre and Solomon of Jerusalem—and they are called *Arz Ar-rab*, meaning "God's Cedars." They reach heights of 100 feet (30 meters), with huge trunks and gently spreading evergreen branches. The shape of each tree depends on the density of the stand. In higher densities the trees grow straighter, whereas in a low-density stand they develop lower horizontal branches that spread out far and wide.

Spring is the best time to visit the Bcharre Cedars, when the green trees are set against a white backdrop of snow. Because of the high altitude, the trees grow slowly and do not bear cones until they are 40 or 50 years old. The seeds germinate in late winter, when there is abundant moisture from rain or snowmelt.

The Cedars of Lebanon are said to be the most majestic of all evergreen trees. They are native to Lebanon, the Taurus Mountains of Syria, and southern Turkey. The trees were an important source of wealth for ancient Phoenicians, who exported the durable wood to Egypt and Palestine. The cedar wood was used to build boats, temples, and sarcophagi for the pharaohs. The tree resin was even used to treat toothache.

The forests were harvested far and wide for their natural wealth, despite the warnings of ancient scribes against their wanton destruction. Indeed, *The Epic of Gilgamesh* warns that the end of civilization will come if the cedar forests are destroyed. The area surrounding the cedars is Lebanon's last wild frontier; it has good hiking opportunities with striking views of the Lebanese mountain range. JK

The cedars are ancient–contemporary with the kings Hiram of Tyre and Solomon of Jerusalem–and are called Arz Ar-rab, meaning "God's Cedars." They reach heights of 100 feet (30 meters), with huge trunks and gently spreading evergreen branches.

RIGHT *The ancient and legendary cedars of north Lebanon.*

RED CANYON

SOUTHERN DISTRICT, ISRAEL

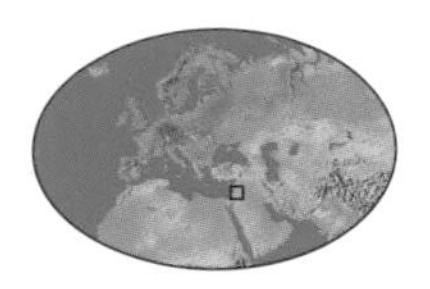

Rock type: red sandstone

Height of Mt. Hizkiyahu: 2,750 ft (838 m)

Red Canyon gets its name from its deep-red sandstone—the vibrant color caused by oxidized iron in the rock. On closer inspection, the rock is in fact striped with various shades of red, purple, and white, and stained by other minerals leached out of the rocks. The canyon walls are 7 to 13 feet (2 to 4 meters) apart and 100 feet (30 meters) high and are at their most beautiful when illuminated by the morning or evening light. This spectacular feature in the Eilat Mountains was slowly eroded by water and wind-blown sand in the Shani riverbed. Floodwaters carrying huge boulders widened the canyon and created the niches that hikers use to rest on today—the boulders that came to be lodged in the canyon have created huge steps. To reach the canyon, drive west out of Eilat and head toward the border with Egypt before turning north: Red Canyon lies a short distance from Mount Hizkiyahu. Desert-type vegetation near the canyon includes white broom bush and acacia. In terms of wildlife, sand partridges can be found, but due to their coloring they can be very hard to spot against the buff-colored sandstone. MB

DEAD SEA

ISRAEL / JORDAN

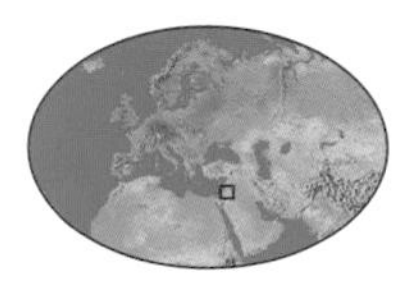

Width of Dead Sea: 11 mi (18 km)

Water input in winter: 6.5 million tons per day

Notable feature: Earth's lowest altitude on land

Lying 1,320 feet (400 meters) below sea level, the shoreline of the Dead Sea is the lowest place on Earth that is not covered by water. It is found at the end of the Jordan Valley, where it forms the most northerly limb of Africa's Great Rift Valley—part of a giant tear in Earth's crust. Here, it is sandwiched between the Judaean Hills in the west and the plateaus of Maob and Edom to the east. The sea is fed by the River Jordan and several smaller streams. The 50-mile- (80-kilometer-) long sea is divided in two by the El Nisan Peninsula—"the tongue." The northern part is larger and deeper than the south, where the water is no more than 20 feet (6 meters) deep.

Evaporation in summer temperatures, which can rise to over 122°F (50°C), results in a seascape of white salt chimneys and floating lumps of salt. The salt content of the water is over six times that of the ocean, and contains potash, magnesium, and bromine. The water is claimed to have medicinal properties for skin problems and arthritis. MB

BELOW *Salt deposits on the shores of the Dead Sea.*

MASADA

SOUTHERN DISTRICT, ISRAEL

Dimensions of Masada: 2,000 ft (600 m) by 1,000 ft (300 m)

Designated a World Heritage site: 2001

Masada is a massive natural formation rising 1,500 feet (450 meters) above the Dead Sea. The rhomboid-shaped butte sits on the western edge of the Judean Desert, elongated in a north-to-south direction and isolated from the surrounding countryside by 330-foot- (100-meter-) deep gorges on all sides. It is a fault-bounded, uplifted block of Earth's crust (a horst) associated with a downthrusted rift valley (graben), occupied here by the Dead Sea. The combination of cliffs and escarpments provided Masada with the perfect defense system. Herod the Great (ruled 37–4 B.C.E.) selected the virtually impregnable site to build a refuge and today his Northern Palace appears to hang in space over the sheer walls. Access to the highest levels is via the "Snake Path" from the Dead Sea side and "White Rock" from the west, although there are two difficult routes from the north and south.

In 73 C.E., Roman governor Flavius Silva had less easy access and was forced to build a ramp up the "prow" of the boat-shaped rock in order to successfully conclude his siege. He discovered how the rebels had survived in the arid environment: They had devised a sophisticated water retention system, using the run-off from one day of rain to sustain life for 1,000 people for two to three years. MB

MAKHTESH RAMON CRATER

SOUTHERN DISTRICT, ISRAEL

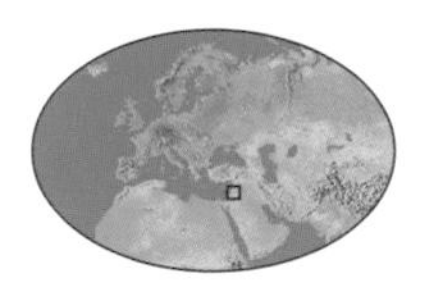

Dimensions of Makhtesh Ramon Crater: 25 mi (40 km) long, 5 mi (9 km) wide, 1,650 ft (500 m) deep

Age: 220 million years

Height of Mt. Ramon: 3,402 ft (1,037 m)

Lying in the heart of the Negev Desert, the Makhtesh Ramon Crater could be considered the biggest crater in the world, but it is neither a volcanic crater, nor an impact crater made by a meteorite. It is in fact a valley. Makhtesh means "mortar," as the valley's steep-sided walls, drained by a wadi (riverbed), look like a mortar—without the pestle. The oval-shaped crater is 25 miles (40 kilometers) long, 5 miles (9 kilometers) wide, and 1,650 feet (500 meters) deep. It was formed when an ocean once covered the area and has since been the target of volcanic activity and erosion. It contains many impressive geological features. One cliff has row upon row of columns made of basalt; the "carpentry shop" is scattered with rocks resembling sawn lumber; and on the southern side of the makhtesh is a rock face with fossil ammonites—fossils of plants, prehistoric amphibians, and reptiles have also been found in abundance. Plant life here includes Atlantic pistachio trees and globe daisies. At Ein Saharonim, the crater's lowest point, rushes, cattails, and reeds grow near the only natural source of water. The Makhtesh Ramon visitors' center has panoramic views of the entire valley. The area's clear atmospheric conditions are ideal for star gazing, and there is an observatory on nearby Mount Ramon. MB

DESERT CAVES

EASTERN PROVINCE, SAUDI ARABIA

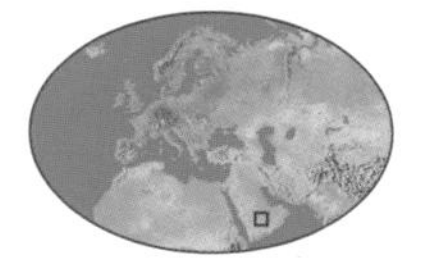

Height of cave entrances: 200 ft (60 m) high

Location: near Al Kharj, south of Riyadh

Notable feature of Dahl Murubbeth: crystals resembling feathers covered by frost

Sixty million years ago, a deep layer of limestone formed in a shallow sea. Today the same rock lies beneath the vast scorching deserts of Saudi Arabia. Not far from Riyadh the desert is dotted with countless holes known locally as "dahls." Their scale is impressive. Some openings have vertical sides over 200 feet (60 meters) high and continue far underground in a network of caves, caverns, and tunnels—many have been explored by cave divers. Driving south from the capital the desert is dotted with large circular patches of lush green grass created by irrigation rigs—fodder for some of the world's biggest dairy farms. Its cultivation is made possible by fossil water lying in deep natural reservoirs, created during a wetter climatic period when green forests grew here. At Ain Hith, a huge opening in the porous limestone is rich in beautiful cave formations before plunging into a lake 330 feet (100 meters) below the surface. It is called the Hith formation after Dahl Hith, where oil explorers first found the surface outcrop of anhydrite. Without this impermeable cap, the oil would never have been found. AC

ASIR NATIONAL PARK

ASIR, SAUDI ARABIA

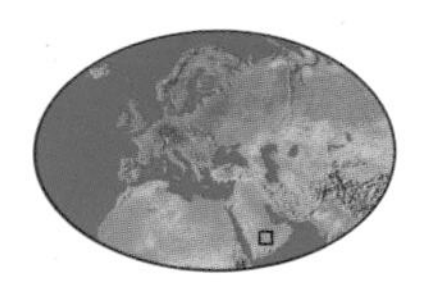

Area of Asir National Park: 1.1 million acres (445,160 ha)

Highest point (Jabal al-Sudah): 9,544 ft (2,910 m)

A new crown jewel of Saudi Arabia's national park system, Asir National Park—in the southwest of the kingdom—is in fact a conglomeration of small parks. Sweeping down from the high peaks and cool green valleys of Abha to the sun-scorched coast of the Red Sea with its coral reefs and fine sandy beaches, Asir is famed for its wildlife and archeology. It is one of the last unspoiled wilderness areas in Arabia. To ancient Egyptians, this was the land of spices and incense. Today, visitors can enjoy the park's spectacular views and observe the abundant wildlife, including gazelles and oryx, in their natural habitat. An endangered vulture, the lammergeier, and over 300 different species of bird can be found here, including the pygmy sunbird and gray hornbill. Spring is the best time to visit—after the winter rains, wild flowers carpet the valley floors and apricot orchards start to blossom. The mountainsides are flanked by a dark green cloak of aromatic juniper forest: against a backdrop of distant pinnacles lie blue valleys with occasional kestrels hovering in the breeze. AC

JEBEL HARIM

MUSANDAM, OMAN

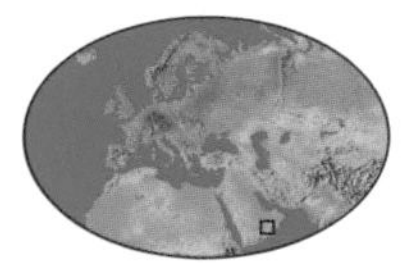

Height of Jebel Harim: 6,847 ft (2,087 m)

Notable feature: distinctively striated rock

Jebel Harim, the Mountain of Women, is the highest point on the Musandam Peninsula, the northernmost point of the Sultanate of Oman. Dominated by the Hajar Range of rugged limestone and dolomite mountains, the peninsula is situated in the Strait of Hormuz, and the jagged coastline has soaring cliffs that rise over 3,300 feet (1,000 meters) straight up from the sea; the resulting fjords have earned the region the nickname the "Norway of Arabia." Access to the mountain is only possible with a four-wheel-drive vehicle, but along the way the rough road offers breathtaking views of deep and narrow gorges, sculpted buttresses, unusual rock formations, and the canyon of Wadi Bih.

Dark gray Jurassic limestone at the base of the mountain leads to a clear marker of orange weathered chert of the Upper Jurassic Rayda Formation about two-thirds of the way to the peak, the difference in the rock visible even to the untrained observer. This part of the Arabian plate is colliding with and being pushed beneath the Asian plate at the other side of the Gulf of Oman, in Iran. As a result, the Musandam Peninsula is being pushed downward at a rate of a quarter of an inch (6 millimeters) per year. GD

MUSANDAM FJORDS

MUSANDAM, OMAN

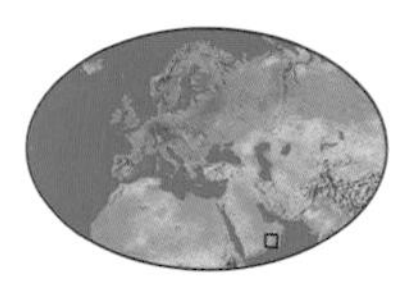

Height of Musandam Peninsula: 6,600 ft (2,000 m)

Area of peninsula: 770 sq mi (2,000 sq km)

The Musandam Peninsula overlooks one of the busiest shipping lanes in the world—the Strait of Hormuz. It is an impressive setting, hot and dry, with the dramatic cliffs of the barren Hajar Mountains plunging steeply into the Arabian Sea. The highest point is Jebel Harim at 6,847 feet (2,087 meters). The landscape is deeply fissured and strewn with cliffed, rocky, and sandy shores. Although "Musandam" refers to just one island in the far north of the peninsula, it is now used for the whole area. Roads have been cut across this isolated region, but the best way to explore Musandam is by sea. The starkly beautiful surroundings are also rich underwater: Ecological surveys of the Musandam Fjords have revealed diverse marine life, with exotic reef fish, shoals of barracuda, sunfish, and whale shark, as well as seabirds, turtles, and dolphins. Novice divers can explore the more sheltered, less treacherous bays. AC

TAWI ATTAIR— THE WELL OF BIRDS

DHOFAR, OMAN

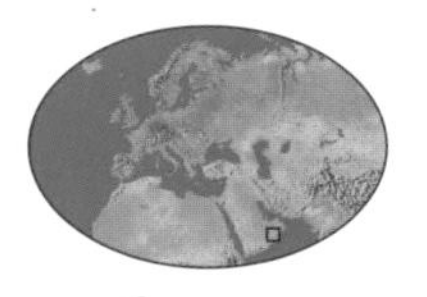

Depth of Tawi Attair: 690 ft (210 m)

Notable feature: world's second-largest sinkhole

East of Salah in Oman, where the mountains force the coast road toward the sea, the route turns inland, rising into the highlands. Here, where camel and cattle herds graze, lies a spectacular secret: a walk across a green field leads unexpectedly to one of the world's largest sinkholes. At 430 to 500 feet (130 to 150 meters) in diameter and 690 feet (210 meters) deep, it is large enough to fit a 50-story skyscraper inside. It is not just the size of the sinkhole that astounds, it is the fact that it "sings." Home to thousands of birds, it echoes with an astonishing chorus.

Tawi Attair was formed when the roof of a gigantic cave collapsed long ago. Today, a metal platform has been positioned to overlook a 260-foot (80-meter) drop. When sunlight pierces the gloom, great curtains of green foliage can be seen festooning the walls, and hundreds of swifts, doves, and birds of prey wheel around inside. The sinkhole is linked by a tiny tunnel to the sea; this fills a small pool with clear water in which it is possible to swim. Tawi Attair is the only place in Oman where the rare Yemen Serin bird can be found. AC

MUGHSAYL BLOWHOLES

DHOFAR, OMAN

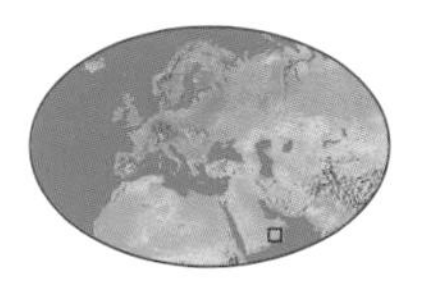

Height reached by waterspouts: 100 ft (30 m)

Khareef Festival: mid-July until the end of August

The magnificent Mughsayl Blowholes in Dhofar, Oman, are a spectacular natural feature shrouded in dark legend and colorful myth, and enjoyed by thousands of visitors during the Khareef Festival. The blowholes were created over many millions of years by strong currents lashing against low limestone cliffs. Exploiting natural weaknesses, the pounding pressure of the sea opened cracks and fissures in the rock. Now crashing waves send water gushing up through the rock, resulting in plumes of surf rising high into the air. The rocky coastline is punctuated by dramatic cliffs and small sandy beaches, but these shores are famed for their thick clouds of fog, which add a sense of drama to the Mughsayl Blowholes. Set amid a magnificent coastline, the road to the site is enchanting, with the mountains often draped in low-hanging clouds. The blowhole experience is much more dramatic when the seas below are rough; explosive jets of water can shoot up to 100 feet (30 meters) into the air. AC

WADI DHAR

SANAÁ, YEMEN

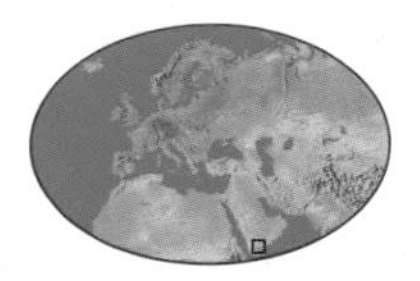

Location of Wadi Dhar: 6.2 mi (10 km) from Sanaá

The Republic of Yemen is one of the least known nations of the Arabian Peninsula. Indeed, the northern border of the country across the shifting sands of the Arabian desert was only recently defined.

It is a spectacular and beautiful place. Flowing channels slice the highlands and central mountain massif of the Hadramaut into a number of plateaus and ridges. The valleys and lower slopes of the uplands have been intensively terraced for soil and water conservation and produce many different crops. The deep *wadis*—literally "watercourses"—provide a dramatic and striking contrast to the barren and harsh desert landscape typical of the region.

Wadi Dhar, 6.2 miles (10 kilometers) from the 2,000-year-old town of Sanaá, is one such valley. It is famed for its fruit gardens, orchards, and vineyards. Pomegranates, mangoes, and citrus fruits thrive in the welcome shade. Wadi Dhar is also known for the unusual palace of Dar Al-Hajjar, half carved from a great pillar of natural stone and half perched on top of it. JD

SOCOTRA ISLAND & THE DRAGON'S BLOOD TREE

ADAN, YEMEN

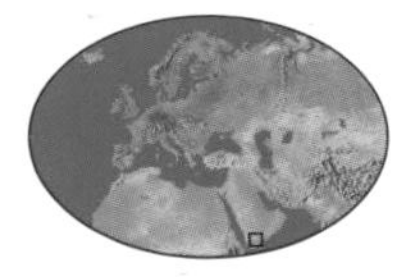

Area of Socrata Island: 1,400 sq mi (3,625 sq km)

Length of island: 75 mi (120 km)

Width of island: 25 mi (40 km)

A lonely but remarkable outpost of the biblical land of Sheba, the island of Socotra lies 320 miles (510 kilometers) off the south coast of Yemen, to which it belongs. Throughout human history, Socotra has been effectively shut off from the rest of the world by extreme natural conditions, especially during the southwest monsoon, which blows from April to October and makes travel to the island almost impossible.

The island's dominant feature is a high plateau of Cretaceous limestone, the Haggif massif, frequently shrouded in cloud. This cloud zone keeps Socotra alive, as it provides ground and running water for the whole island. Socotra is a botanical treasure house, a living museum of vanished species, the most famous of which is the iconic and unique dragon's blood tree. It grows in areas of thicket and grassland and gets its name from the red sap that oozes from any cut in its bark. This sap was once valued throughout the ancient world as an antiseptic ointment. JD

RIGHT *The dragon's blood trees of Socotra Island.*

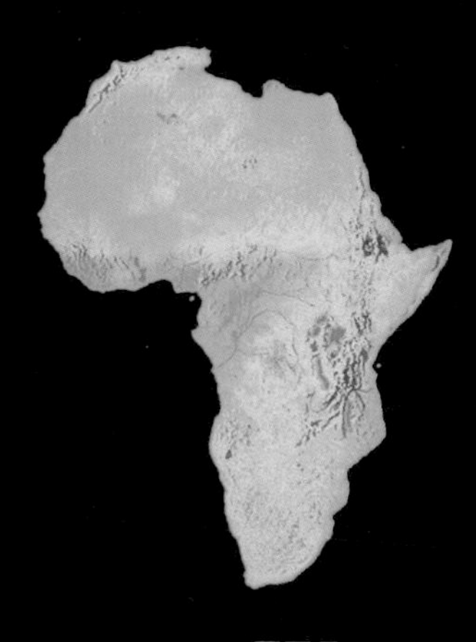

IV

AFRICA

Viewed from above, Africa's reach is remarkable. The sands of the Sahara Desert blanket the north in a swathe of yellow, interrupted only by the blue seam of the Nile. Rising in Africa's east, the Great Rift Valley scars the land—threatening to split the continent. Western shores are frayed by the haunted shipwrecks of Skeleton Coast and to the south, savannas host a safari of animals. Africa's central axis, the Congo Basin, is a network of swamps and emerald forests as diverse as the four counterpoints it unifies.

LEFT *The rippled dunes of the Namib Desert, Namibia.*

ASCENSION ISLAND

MID-ATLANTIC RIDGE

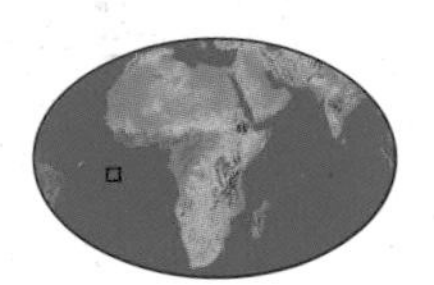

Area of Ascension Island: 34 sq mi (88 sq km)

Notable feature: green turtles, 3 ft (0.9 m) long

Nearest land (island of St. Helena): 1,200 mi (1,931 km)

Part of the volcanically active Mid-Atlantic Ridge, Ascension Island is actually a mountain peak that rises 10,000 feet (3,000 meters) from the Atlantic floor. The Mid-Atlantic Ridge is a weakness in Earth's crust—as the plates of Africa and Europe pull away from the plates of North and South America, molten rock pushes up from below and eventually breaks through the surface of the sea to form a new rocky island. Ascension Island is just one of a number of volcanic islands lining the Mid-Atlantic Ridge. Much of its surface is barren, a bleak, lunar landscape of undulating lava flows dotted with craters and cones.

Ascension Island is just one of a number of volcanic islands lining the Mid-Atlantic Ridge. Much of its surface is barren, a bleak, lunar landscape of undulating lava flows dotted with craters and cones.

Until recently, most of the island's life was restricted to its margins, where an amazing array of seabirds can be found, often nesting in large colonies. These include red-footed booby, sooty tern, brown noddy, black noddy, and frigate birds. Shark, wahoo, and barracuda also swim the waters surrounding the island.

However, in 1843, a number of plant species were introduced to Green Mountain, the island's highest peak in order to create a more pleasant environment for the British garrison stationed there—the island is a British dependency. The plants thrived in the island's subtropical climate and today form one of the world's largest "artificial forests."

The island's most famous visitors are its green sea turtles. These docile reptiles usually feed along Brazil's shoreline, but at breeding time swim out to sea and head for the shores of Ascension Island. After traveling for miles across the Atlantic Ocean, the turtles haul themselves onto the beach to deposit their eggs. How they find their way in the featureless ocean is a mystery, but there is some speculation about their reasons. At one time, Africa and South America were closer together, but as they separated a string of volcanoes broke the surface across the widening ocean. As each volcano stopped erupting and cooled, the turtles used the beaches as nesting sites, safe from mainland predators. Deprived of fresh lava, each volcanic island was eroded by the sea, until it eventually vanished below the ocean. At this point the turtles migrated to the next available island, and gradually the traveling distance became greater. Today, they make this epic journey to Ascension Island. MB

RIGHT *Baby turtles race to the safety of the sea.*

PICO DE FOGO

CAPE VERDE ISLANDS

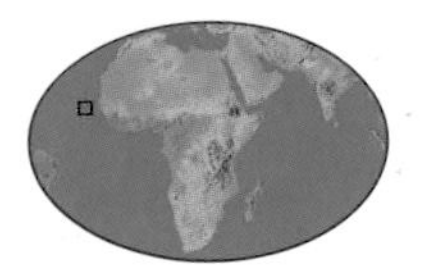

Height of Pico de Fogo volcano: 9,281 ft (2,829 m)

Diameter of caldera: 5 mi (9 km)

Diameter of base of volcano: 16 mi (25 km)

The Cape Verde Islands are of volcanic origin and were uninhabited when Portuguese sailors first discovered them in 1456. The archipelago they form lies some 310 miles (500 kilometers) off the coast of Senegal in West Africa, on an oceanic crust that is between 120 and 140 million years old. Pico volcano on the island of Fogo (which means "fire") is the group's only active volcano. Fogo rises clear out of the Atlantic Ocean in a dramatic cone 9,281 feet (2,829 meters) high. It is a basaltic volcano, classified as a "hot spot," and forms a geographic cluster with the Azores and the Canary Islands.

Although Fogo is one massive volcanic cone, its landscape is one of contrast between the dry and arid zones of the south and the humid and fertile northern zone. Peanuts, beans, coffee, oranges, and tobacco are grown on the north and west sides of the island. There is even a heady, rich red wine produced from grapes

growing within the caldera (crater) itself, brought to the island by French exiles in the early 19th century. Their descendants follow the same winemaking methods, but as wood for barrels is scarce, they use old petrol drums for storage, which gives the wine a bizarre aftertaste. Some of Fogo's best farmland is on the caldera's relatively flat floor, and those who live in this danger zone know that some day an eruption may evict them—which is precisely what happened in 1995.

On March 25, 1995, weak earthquakes began. On the night of April 2, lava started flowing from the base of the Pico cone, within the caldera. Seven vents became active, with fire-fountains, volcanic bombs, and a plume of gas and ash 6,600 feet (2,000 meters) high. More than 5,000 people fled the caldera to find refuge. Two lava flows formed, one on top of the other. They were 2.5 miles (4 kilometers) long and 2,000 feet (600 meters) wide, with a temperature of 1,879°F (1,026°C). The lava smothered a village and destroyed farmland—Fogo reclaimed the caldera. JD

BELOW *The aquamarine shores of Fogo.*

PICO DE TEIDE

TENERIFE, CANARY ISLANDS

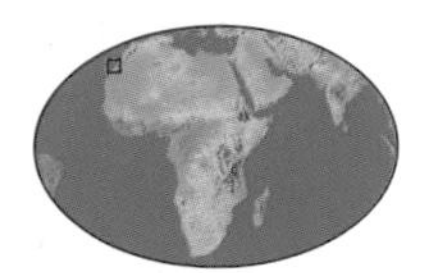

Height of Pico de Teide: 12,198 ft (3,718 m)

Area of Tenerife: 909 sq mi (2,354 sq km)

At sunrise, the triangular shadow of Pico de Teide extends 124 miles (200 kilometers) across the Atlantic Ocean and thus forms the world's longest shadow. The mountain is a 12,198-foot- (3,718-meter-) high, often snow-topped volcano that dominates Tenerife in the Canary Islands. It has two craters, the more recent cone-shaped peak sitting inside a caldera created by earlier activity. The new crater is 100 feet (30 meters) deep and sulfurous gases still seep out from its floor. In 1705, an eruption from this crater buried the port of Garachio with ash and lava. Then in 1909, a vent on its side produced a 3-mile- (5-kilometer-) long stream of lava that flowed down its northwest slopes toward villages on the coast. The volcano was known to the original blond-haired inhabitants of Tenerife —the Guanches—as the "Peak of Hell," the formal residence of their brutal god, Guayota.

Today visitors can approach the mountain by road, and a cable car takes people to within a half-hour's steep zigzag walk through a barren lava field to the crater. On the way, walkers pass an ice-filled crevasse known as the Cuevo del Hielo, and have views of the neighboring volcano, Pico Viejo. **MB**

LOS ROQUES DE GARCIA

TENERIFE, CANARY ISLANDS

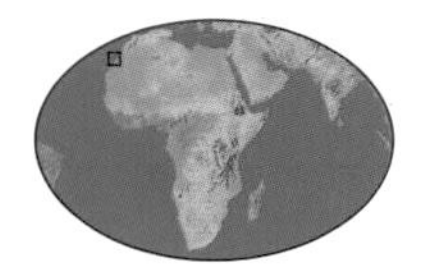

Age of Los Roques de Garcia: 170,000 years

Circumference of Las Canadas caldera: 30 mi (50 km)

Established as a national park: 1954

Los Roques de Garcia are the spectacular eroded rocks of an ancient volcanic crater wall in Teide National Park. Tenerife is a stratovolcano capped by Las Canadas, one of the most impressive calderas (craters) in the world. The strange landscape inside the Canadas caldera has been used in feature films such as *Star Wars*, *Planet of the Apes*, and *The Ten Commandments*.

Geologists are still unsure how the caldera formed, with theories ranging from a volcanic explosion to collapse, landslips, and erosion. Los Roques de Garcia are the remains of a rim that separates two segments of the caldera floor. The weird, twisted pinnacles of rock have names such as the "Finger of God" and the "Cathedral." The area provides a good opportunity to compare two contrasting lava types—"aa" lava which has a jagged blocky surface and "pahoehoe" lava which has a rope-like surface. The floor of the lower crater (Llano de Ucanca) is covered with volcanic sand, but in the spring this arid area becomes a lake when water from the melting snows above rushes down the crater's slopes. The whole area is a national park, including the crater, out of which rises Mount Teide. RC

LUNAR LANDSCAPE

TENERIFE, CANARY ISLANDS

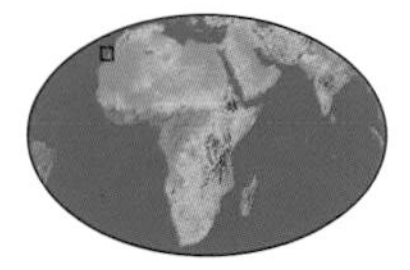

Local name: Paisaje Lunar

Elevation of Vilaflor village: 4,620 ft (1,400 m)

Walking through the bizarre rock shapes of the Lunar Landscape, one could easily mistake this eerie world for the moon. Rising out of the ground like large white termite hills, conical rocks known as *pumitas* stand out against the swathes of gray-black moon-like terrain. Erosion has molded these rocks into natural stone sculptures—some appear like molten globules of lava, while others are reminiscent of the spiraling pinnacles seen at Antonio Gaudi's Sagrada Familia in Barcelona.

As with the rest of the island, the Lunar Landscape is of volcanic origin, the result of eruptions that ended around 3.5 million years ago. It is located about 6 miles (10 kilometers) east of Vilaflor (the highest village in Spanish territory, at 4,620 feet [1,400 meters]). The 45-minute walk to the Lunar Landscape is on a forested footpath, and affords good views of the island's southern coast and the outer rim of the Las Canadas crater.

The first group of the Lunar Landscape's eroded stone columns are wide-based tapering pillars topped by delicate structures. From the nearby second group of columns there is a trail up to the black lunar landscape of Barranco de las Arenas. RC

LOS ORGANOS

LA GOMERA, CANARY ISLANDS

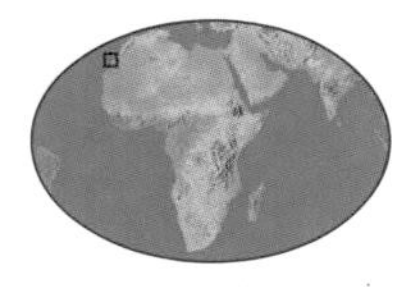

Height of Los Organos basalt columns: up to 260 ft (80 m)

Area of Garajonay National Park: 9,844 acres (3,984 ha)

Los Organos is a steep cliff formation that rises out of the sea on the north coast of La Gomera. Viewed from the sea, the thousands of tall, vertical basalt towers resemble the pipes of a gigantic church organ (hence the name Los Organos, "The Organs").

This unusual rock formation forms part of the circular volcanic island of La Gomera, which is the second smallest of the Canary Islands (after El Hierro). The Organos Natural Monument is located in the borough of Vallehermoso on the northwest coast of the island but is not visible from the land. However, boat trips regularly take visitors around the island, affording good views of the cliffs, especially when the waters are calm. Dolphins and whales are often seen in the waters around the island.

La Gomera has not undergone any recent volcanic activity, but water has eroded a radial network of deep ravines. Garajonay National Park occupies approximately 10 percent of the total area of the island. The national park was created in 1981 to protect the island's precious laurel forests and its large number of native species. In 1986 it was designated a World Heritage site. RC

ARICO GORGE

TENERIFE, CANARY ISLANDS

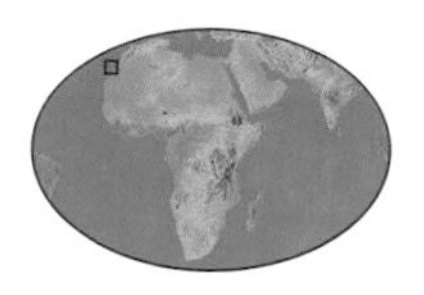

Height of Arico Gorge: 30 to 100 ft (9 to 30 m)

Age of rock: 1 to 2 million years

Number of climbing routes: about 175 for all skill levels

A rough landscape of pitted rock pillars, cracked boulders, and bus-sized shelves of stone lean against burnt-brown cliffs—this is the Arico Gorge of Tenerife. Walking through this landscape is comparable to entering a natural disaster zone. Located on the southeastern slope of a now-dormant volcano that towers close to 12,000 feet (3,600 meters) above the ocean's surface, the Arico area has been scalded by red-hot lava and scorched by fiery-hot gases and ash for thousands of years.

As the eruptions stopped and the mountain cooled, the layers of volcanic ash and ejected rubble were compressed over time into a conglomerate stone known as ignimbrite, which wind and water then cut to form this narrow canyon. This process exposed pockets of gas and shards of rock once trapped within the lava flows, and so the canyon walls today have a knobbly and even abrasive character—features that have made Arico one of the top climbing destinations in the world. With its steep walls and giant boulders, a climb through this gorge is certainly a challenge. Despite the dusty and dry semi-desert conditions, the gorge runs in a zigzag, so there is always shade nearby for a break from the sun. DBB

ALEGRANZA

CANARY ISLANDS

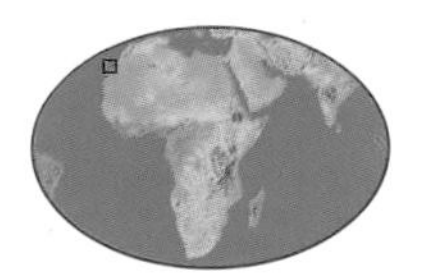

Maximum elevation of Alegranza: 950 ft (289 m)

Maximum offshore depth: 3,300 ft (1,000 m)

Age: 16 to 20 million years

Rising out of the Atlantic Ocean like a great rock pedestal, Alegranza is one of a smattering of islets off the northwest coast of Africa that make up the northern tip of the Canary Islands. It is a showcase of geological and biological antiquities. One of the oldest members of this volcanic archipelago, Alegranza is an ancient land of deep craters, volcanic cones, and long dune belts overlying a vast network of underground tubes and pockets, long drained of their lava.

Among its geological treasures exist an equally diverse company of living relics such as loggerhead sea turtles, exceedingly rare Eastern Canary geckos (which are found nowhere else in the world), and one of the world's only remaining swathes of thorny Argan tree thickets. A long list of endangered birds, including petrel, shearwater, osprey, Egyptian vulture, and peregrine falcon, nest high up on the island's inaccessible outer ring of vertical coastal cliffs. Despite the huge number of tourists who visit the area (more than 10 million every year), a comprehensive system of protected areas has effectively stabilized any disturbances to Alegranza and its companion islets. DBB

CABO GIRÃO

MADEIRA

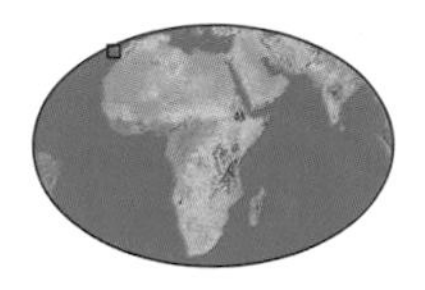

Height of Cabo Girão: 1,902 ft (590 m)

Rock type: volcanic

The Portuguese poet Luís Vaz de Camões once called the island of Madeira "the end of the world." Stand at the ledge of the vertical rock face called Cabo Girão on the main island's southern coast and you begin to understand why. Taller than any sea cliff in Europe, this is also the second highest in the world—the cape dives nearly half a mile down to the Atlantic Ocean, but it does not stop there. As the exposed upper reaches of an enormous oceanic volcano, this landmass's steep flanks are surrounded by such deep waters that sperm whale are often sighted offshore. Atop these majestic cliffs, one can gaze to the horizon over the Atlantic to follow the bending contour of Earth.

While standing at the top provides the bird's-eye perspective, a position within its great shadow on the sea provides an entirely different one. Sporting a headdress of eucalyptus and fern-like mimosa trees, the black chunk of basalt is necklaced in white ribbons of falling water which irrigate a forest of colorful mosses, lichens, and cliff-loving plants called stonecrop. To really appreciate the enormity of these cliffs, hop on a boat for a perfect ocean view of the cape. **DBB**

CALDEIRÃO VERDE

MADEIRA

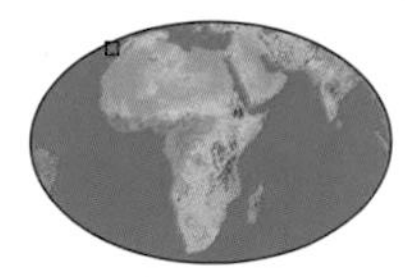

Height of Caldeirão Verde: 330 ft (100 m)

Elevation: 3,000 ft (900 m)

Number of tunnels: 4

A natural amphitheater of rock, painted bright green in slimy mosses and feathery ferns, the Caldeirão Verde, or Green Cauldron, of the island of Madeira is nestled among one of the last remaining virgin forests. Towering more than 30 floors high, and split down its center by a slender but powerful cascade that tumbles into an emerald pool of cold water, the cliff peeks out of a dense drape of primary laurel forest, or Laurisilva forest. This is an ecosystem so rare that it was designated a World Heritage site in December 1999. Located in one of the wildest regions of the island, reaching the Cauldron is an adventure that follows trails carved into near-vertical slopes of canine-shaped massifs carpeted in thick vegetation, and through narrow, dark, and damp tunnels that can reach up to 360 feet (110 m) in length. Despite the rugged terrain and the relatively pristine state of the forest, the route is in fact a serpentine network of man-made canals known as "levadas," which early settlers constructed to collect and divert rainwater from the wetter northern regions of the island, to the drier zones in the south. The path to Caldeirão Verde begins at Queimadas Forest Park in the town of Santana. **DBB**

THE SAHARA

TUNISIA / WESTERN SAHARA / MOROCCO / MAURITANIA / MALI / ALGERIA / LIBYA / EGYPT / NIGER / CHAD / SUDAN

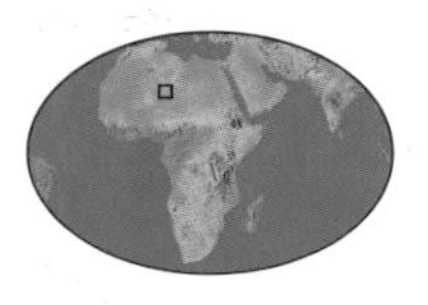

Width of the Sahara: 3,000 mi (4,800 km)

Length of the Sahara: 1,200 to 3,000 mi (1,900 to 4,800 km)

Age: 5 million years

The Sahara is the largest desert on Earth, extending across one third of the African continent. Seen from space, it covers an area approximately the same size as the United States. As the only true desert on the 0° equatorial and prime meridian lines, the Sahara is one of the hottest places on Earth; but even though temperatures can rise to 139°F (58°C), it is because of dryness and not heat that the Sahara is a desert. Strong, unpredictable winds are typical. These winds can blow for days on end, covering everything in their path with vast amounts of dust and sand.

The Sahara landscape is extraordinarily varied—over one quarter of the desert is sand, while the rest consists of gravel-covered plains, seasonal riverbeds, rock-strewn plateaus, and volcanic mountains. The great seas of shifting sand dunes or "ergs" are vast in scale, ranging over hundreds of kilometers, and reaching 560 feet (170 meters) in height, while the Draa, a mountainous sand ridge, tops 1,000 feet (300 meters). The north has a dry subtropical climate with two rainy seasons, while the central and southern areas have a dry tropical climate. On the western coast, the cold Canary Current produces a narrow plant-rich coastal fog belt that is cooler than the rest of the desert. In most parts of the Sahara, rainfall is scant and erratic—it receives, on average, less than 3 inches (7.5 centimeters) of rain a year. Surprisingly, in some areas frost can be seen in the winter. In the central region, temperatures drop to below freezing and the peaks of Emi Koussi and Tahat become snowcapped.

The Sahara is the largest desert on Earth, extending across one third of the African continent. Seen from space, it covers an area around the same size as the United States. At night, when the air is crisp, the stars seem close enough to touch.

Plant and wildlife have become highly specialized in order to survive. Deep desert species include fennec fox, horned viper, desert jerboa, houbara bustard, desert hedgehog, and dorcas gazelle. Barbary leopard, golden eagle, and mouflon occur in the Atlas Mountains.

The desert itself is sublime in its vastness—at night, when the air is crisp, the stars seem close enough to touch and the silence is overwhelming. Desert dwellers say that when the wind stops, you can hear the Earth turn. **AB**

RIGHT *An oasis of green palm trees stands out against the ripples of the Sahara Desert.*

TASSILI N'AJJER

ILLIZI, ALGERIA

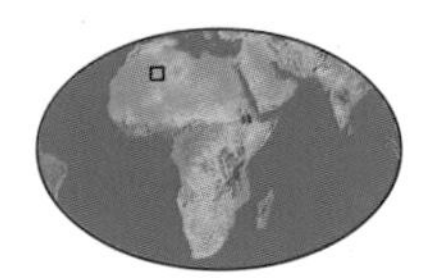

Height of Tassili N'Ajjer plateau: 7,400 ft (2,250 m)

Rock type: sandstone

Age of rock art: 2,300 to 8,000 years

Tassili N'Ajjer means "plateau of chasms," and this plateau is part of an ancient sandstone layer surrounding the Precambrian granite massif of the Ahaggar. The adjoining mountains are divided into separate massifs that have been eroded by water and sand-laden wind into ridges, ravines, and stand-alone buttresses. The area is of outstanding scenic and geological interest, and is famous for its immense gallery of prehistoric cave art of more than 15,000 drawings and engravings. The rock art, discovered by French explorer Henri Lhôte in the 1950s, shows the cultural evolution of ancient peoples from hunters to farmers to soldiers. Later images, dated 3,500 years ago, depict pastoralists who tended cattle and held grand banquets, while pictures thought to be about 2,300 years old show chariots and soldiers in armor. The last images to appear are less elaborate paintings of camels; after that, both art and people practically disappear—a change in climate having turned the fertile land into a desert. Though the plateau is hyperarid and barren, sheltered microclimates are home to relict flora and fauna like the Mediterranean cypress, one of the rarest trees in the world. **MB**

HOGGAR MOUNTAINS

TAMANGHASSET, ALGERIA

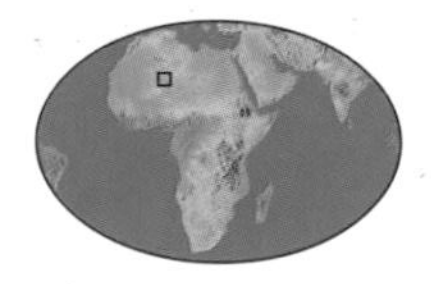

Alternative name: Ahaggar Mountains

Height of Hoggar Mountains: 10,000 ft (3,000 m)

Rock type: volcanic

In the vast sea of sand that is the Sahara Desert, the Hoggar or Ahaggar Mountains form an island the size of France. It is an enormous plateau, bounded on three sides by steep cliffs and on the fourth by a desert known as the "Land of Thirst." At the center of the massif are volcanic rocks that have been eroded into angular columns, packed tightly together into towers or pinnacles and separated by ravines. Some push 10,000 feet (3,000 meters) upward like gigantic stone hands reaching for the sky. Although there is little rain and next to no vegetation here, some of the steep-sided gorges contain pockets of water. These have been important as rest stops for the nomadic Touareg people on cross-desert caravan routes to trade in gold, ivory, and slaves. They call the mountains *Assekrem* meaning "end of the world."

In the early 1900s, the Hoggar Mountains became home to a French priest, Charles de Foucald, who devoted his life to the Touareg. He was killed during an uprising in 1916, but the retreat he built remains. Today it is a focus for tourists and travelers, who climb up to the hermitage to watch the sun rise over the Sahara. **MB**

TALASSEMTANE NATIONAL PARK

TETOUAN, MOROCCO

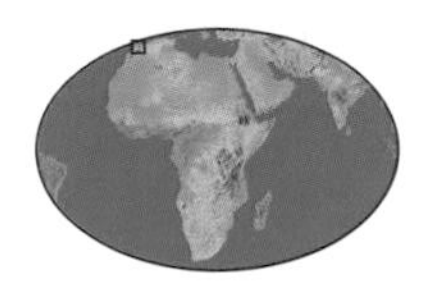

Area of Talassemtane National Park: 247,000 sq mi (640,000 sq km)

Notable feature: Moroccan fir tree

A landscape of exceptional beauty with limestone mountain peaks, cliffs, gorges, and caves, Talassemtane National Park is also a Mediterranean forest ecosystem unique in its wealth of endemic plant species. Characterized by impressive mountains and superb forests, the park covers the calcareous easternmost mountain range of the Rif which runs from Ceuta to Assifane. The highest peaks are Jbel Tissouka at 6,962 feet (2,122 meters) and Jbel Lakraa at 7,083 feet (2,159 meters). The park is home to the only Moroccan fir, endangered today and the last remnant of an ecosystem unique in the world. There are also more than 239 plant species, a large number of which are endemic or relict, such as the Atlas cedar and black pine.

The park provides refuge to more than 37 species of mammal, including the Barbary macaque, which finds shelter in the caves of the region. More than 117 species of bird have been recorded in the park, the most spectacular of which are the lammergeier and the golden eagle. The peaceful nature of the landscape, which is conducive to contemplation and meditation, has also attracted practitioners of Sufism. GD

DADES GORGE

OUARZAZATE, MOROCCO

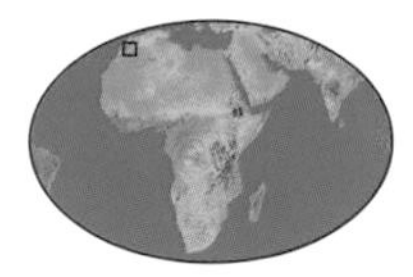

Depth of Dades Gorge: 1,650 ft (500 m)

Age: 200 million years

In central Morocco to the north of the market town of Boumalne, the fast-flowing Dades River has cut through the High Atlas Mountains like a knife through butter. The result is the spectacular Dades Gorge. The vertical walls are up to 1,650 feet (500 meters) high, and the rocks lie in horizontal layers of limestone, sandstone, and marl. About 200 million years ago, sediments deposited on the sea floor were lifted and folded by major earth movements to form the Atlas Mountains.

Each winter, from November to January, rain falling in the mountains turns the Dades River from a trickle to a raging torrent in just a few hours. Debris carried down from the peaks helps carve the gorge even deeper. In one section, the rocks have been eroded into curious shapes, including some which even appear human-like. Local people call it "Hills of Human Bodies." To the northeast is the equally impressive Todra Gorge. In places its sheer sides are 1,000 feet (300 meters) high and no more than 30 feet (9 meters) apart. Nearby is a spring that is very special to the local Berber people. It is said that if an infertile woman walks through the water while calling the name of Allah, she will become fertile. MB

GREAT RIFT VALLEY OF ARABIA & AFRICA

ARABIA / AFRICA

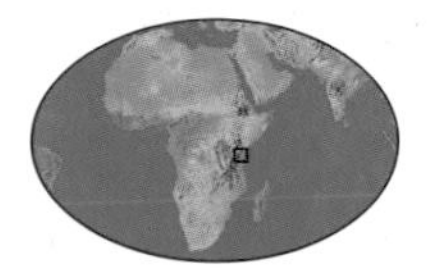

Length of Great Rift Valley: 4,000 mi (6,400 km)

Highest point (Mount Kilimanjaro): 19,340 ft (5,895 m)

The Great Rift Valley stretches from the Dead Sea on the Israel–Jordan border to the Mozambique coast. As the planet's plates move, they split along lines of weakness. In places, such as along the Great Rift Valley, two parallel cracks form in Earth's crust and the land between sinks, leaving great escarpments on either side. It is an area of volcanic activity and frequent earthquakes. In Arabia, the pulling apart of the Arabian and African plates has resulted in the formation of the Red Sea, and in Africa earth movements have created a two-pronged rift. The more westerly branch curves through Uganda, Zaire, and Zambia with spectacularly deep lakes, such as Tanganyika and Malawi. By contrast, the eastern rift, which passes through Ethiopia, Kenya, and Tanzania, has shallow alkaline lakes, such as Natron, and magnificent volcanoes, such as Mount Kilimanjaro. Today parts of this great scar, which runs for one seventh of Earth's circumference, contain the greatest concentrations of wildlife on the planet. **MB**

TABA WILDLIFE RESERVE

JANUB SINA, EGYPT

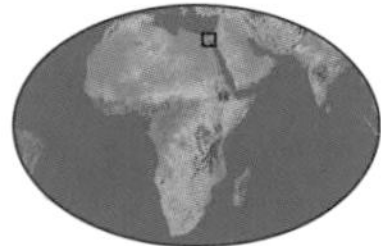

Area of Taba Wildlife Reserve: 1,386 sq mi (3,590 sq km)

Established as a wildlife reserve: 1997

Taba Wildlife Reserve is situated on the Gulf of Aqaba, southeast of the resort town of Taba. Lying at Egypt's borders with Jordan, Israel, and Saudi Arabia, Taba is located in the Sinai Peninsula, which separated from the rest of Arabia some 20 million years ago with the opening of the shallow Gulf of Suez. In 1997, Taba Wildlife Reserve was registered to protect its unique geological formations, natural springs, cave systems, and sun-shielded, plant-filled ravines, as well as a diversity of inland and coastal wildlife. Over 50 of Taba's resident bird species, including Bonelli's eagle and white-eyed gull, are regarded as rare. Some 25 mammal and 480 plant species are known from the region.

Archeological digs have revealed human occupancy dating back 5,000 years. Today, the natural springs provide water for plantations and vegetable gardens used by the region's Bedouin. The springs are also seasonally important to 18 species of migrant bird. One of Egypt's 22 protected areas, Taba's coastal area hosts coral reefs and tidal mangrove swamps, dominated by the salt-secreting white mangrove. There are also several important seagrass beds, home to dugongs. **AB**

SIWA

MATRUH, EGYPT

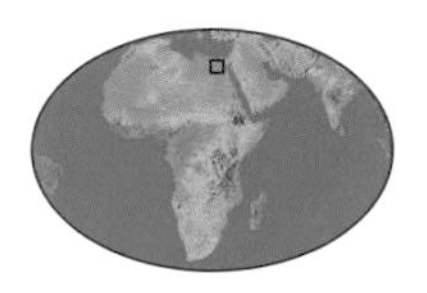

Area of Siwa Depression: 850 sq mi (2,296 sq km)

Area of largest salt lake: 12.4 sq mi (32 sq km)

Notable feature: 1,000 freshwater springs

Siwa is the westernmost of the five major oases of Egypt, and has been visited by tourists since the 8th century B.C.E. It offers an inkling of what this huge green oasis must once have meant for desert travelers. It is said that Alexander the Great came here to visit the oracle of Siwa. The landscape he would have seen before him has hardly changed—to put it in the words of ancient historian Diodorus Siculus: "The land where this temple lies is surrounded by a sandy desert and waterless waste, destitute of anything good for man. The oasis is 50 furlongs in length and breadth and is watered by many fine springs. So that it is covered by many trees." Today, visitors can walk along the edge of Birket Siwa Lake (one of the numerous salt lakes), before the trees stop and give way to the first slopes of the Great Sand Sea. Standing at the edge of this desert landscape, the dunes roll into the distance like giant waves. Across the lake is the hill where the rock of the oracle still dominates the landscape, though today it is propped up with iron girders. Here a visitor can feel at one with all the wanderers in history who came to hear the gods speak and understand why they believed they would hear only truth. **PT**

SANNUR CAVE

BENI-SUEF, EGYPT

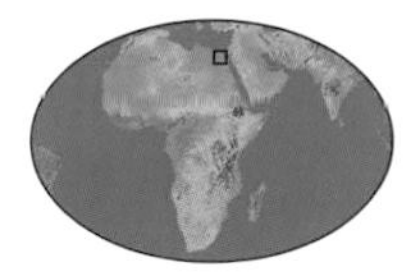

Area of Sannur Cave: 4.6 sq mi (12 sq km)

Age: 60 million years

Rock type: limestone rock cave surrounded by alabaster

Designated a Protected Area in 1992, Sannur Cave is a single crescent-shaped chamber some 2,300 feet (700 meters) long, with a diameter of 50 feet (15 meters). Formed in the middle Eocene period, underground water eroded the local soluble limestone, creating an elaborate karst formation. Thermal spring activity then produced alabaster, which overlaid the karst system.

The cave has striking stalagmites and stalactites, caused by the leakage of water through the alabaster deposits above. It also has undisturbed floor deposits which have been used to trace fossil climate change. The Sannur region is pockmarked with ancient quarries, some of them extending back to the time of the pharaohs. The cave was accidentally discovered in the late 1980s when blasting at the bottom of an open-pit alabaster quarry opened it up. The combination of limestone, alabaster, and spring water is geologically unique. Attempts have been made to make the area a World Heritage site, in the hope that this will protect the cave from the local mining lobby. Sannur is one of Egypt's 22 protectorates, which cover eight percent of the country. **AB**

WHITE DESERT

AL WADI AL JADID, EGYPT

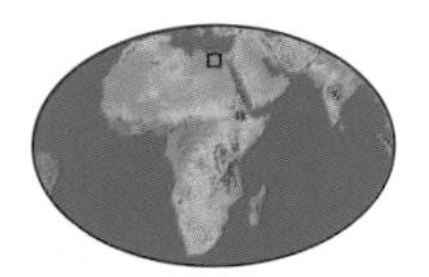

Age of White Desert: 7,000 years

Notable feature: dramatic landscape shaped by wind

The White Desert lies on the northern fringe of the Western Desert (the "Desert of Deserts")—a vast expanse that starts at the western banks of the Nile and continues well into Libya, covering nearly 1.16 million square miles (3 million square kilometers). The White Desert is a world of desolation and beauty with breathtaking monoliths rising from the limestone and chalk desert floor. Winds throughout the millennia have blasted away the soft chalk, leaving the hard rock carved into strange and fantastic shapes. Some of these sculptures are massive, as high as a 20-foot (6-meter) building. Some resemble animals and humans. During the day the blinding rocks, which glitter and gleam, add to the desert's heat, but they also provide welcome shade from the scorching rays of the sun. At dusk the pillars take on the myriad hues of the evening. In the moonlight, they are luminous, towering above the desert in an eerie florescent glow. Among other noteworthy sites are Crystal Mountain (a quartz crystal rock with a hole in the middle) and the massive Twin Peaks. Tiny seashells can be found embedded in the rock, and the desert floor is covered with quartz and iron pyrites ("fool's gold"). **PT**

BANC D'ARGUIN

DAKHLET NOUADHIBOU / AZEFAL, MAURITANIA

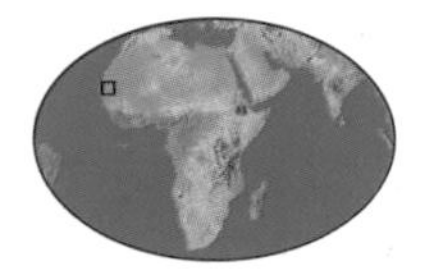

Elevation of Banc D'Arguin: 16 ft (5 m) below and 50 ft (15 m) above sea level

Annual rainfall: 1.3 to 1.6 in (34 to 40 mm)

Designated a World Heritage site: 1989

Along the west coast of Mauritania—between the hot desert and the cold ocean—lies an amazing area of marine wetlands and one of the largest national parks in Africa. Huge mudflats and inter-tidal areas teem with a variety of worms, molluscs, crustaceans, and other marine life, making these incredibly productive marine flats one of the most important places in the world for birds to feed. They are renowned for the two million or more shorebirds that spend the northern winter here. Even more staggering is that most of the seven million waterbirds using the coastal Atlantic migration route stop here to feed and refuel before moving farther south.

The Banc D'Arguin also supports another three million birds represented by over 100 species including flamingoes, terns, and pelicans. There are also very rare monk seals and four species of turtle. The area plays host to a unique symbiotic collaboration between humans and animals. On nearby beaches, the local fishermen beat the surface with a pole to alert dolphins. The dolphins drive shoals of mullet into the beach where they and the fishermen catch them. **PG**

INNER NIGER DELTA

MOPTI / SEGOU / TOMBOUCTOU, MALI

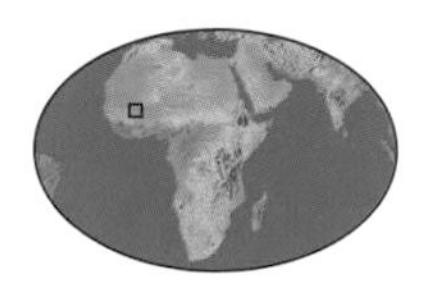

Area of dry season floods: 1,544 sq mi (4,000 sq km)

Area of wet season floods: 7,722 sq mi (20,000 sq km)

Annual rainfall: 24 in (600 mm) in south, 8 in (200 mm) in north

The Niger River rises in the Guinea uplands and flows northeast, but just before the river turns eastward in the Sahel, there can be found one of the wetland wonders of the world. Each year, the Niger and Bani rivers rise and flood an enormous area. While only 1,544 square miles (4,000 square kilometers) are inundated at the end of the dry season, during the wet season this expands to over 7,722 square miles (20,000 square kilometers)—creating the largest inland wetland in the world.

Between August and December these wetlands support over a million waterfowl and countless smaller birds. The overwintering garganey number over 500,000 alone. About 80,000 pairs of large birds like cranes, ibis, and spoonbills nest here, too. The mineral-rich soil supports a large number of animal species. There are herds of antelope such as Bohor's reedbuck and Buffon's kob, but these are struggling to survive in the face of overgrazing, caused by the millions of sheep and goats owned by the 500,000 people who live here. Amazingly, there is also a population of endangered manatee living in the inland delta, and over 100 species of fish have been recorded in the delta. **PG**

EMI KOUSSI

BOURKOU-ENNEDI-TIBESTI, CHAD

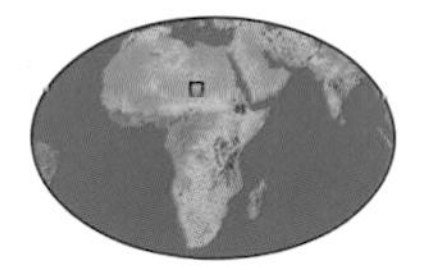

Elevation of Emi Koussi: 11,204 ft (3,415 m)

Diameter of main caldera: 9 mi (15 km)

Rock type: volcanic deposits over ancient Precambrian sandstones

Emi Koussi is an ancient volcano lying at the southern end of the Tibesti Massif in northern Chad. The highest volcano in the Sahara, it rises some 1.4 miles (2.3 kilometers) above the surrounding sandstone plains. Emi Koussi is 40 miles (65 kilometers) wide, with a flat-bottomed crater 9 miles (15 kilometers) across—this caldera was formed by the volcano collapsing into itself. Within the main caldera is a smaller 1.9-mile- (3-kilometer-) wide crater—Era Kohor—containing the white salts of a now dry lake. On the walls of the crater several layers of lava are visible. The main volcano is surrounded by a dome of geologically young lava studded with active vents, including the thermal area of Yi-Yerra. Scientists interested in geology on Mars have used Emi Koussi as a close analogue to the planet's famous volcano Elysium Mons. There are several other large volcanoes in the complex—features include cinder cones and lava flows of various diverse types. Access is challenging due to rebels, landmines, and poor infrastructure—a local guide is essential—but the climb itself presents few technical problems. The mountain is a two-day hard drive from the nearest town, Faya. **AB**

ENNEDI GORGE

BOURKOU-ENNEDI-TIBESTI, CHAD

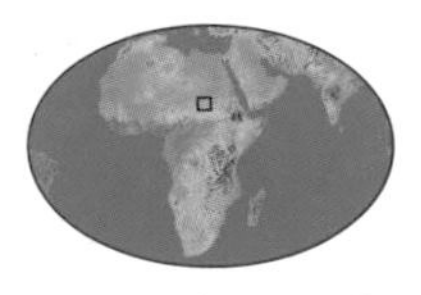

Number of natural arches: over 500 (most undocumented)

Vegetation: montane xeric woodland

Wildlife: addax, dorcas gazelle, dama gazelle, sand fox, golden jackal

The eerie silence of the Sahara is broken at Ennedi Gorge, for it is here that desert nomads bring their goats and camels to drink. The gorge is one of several canyons that slice into the sandstone rocks of the Ennedi Massif. There are fiery red cliffs, towering and crumbling buttresses, gigantic natural arches, and walls decorated with ancient rock art. In the floor of the canyon, black water pools or gueltas are fed from underground. Guelta D'Archie, with a floor of white sand and sparse vegetation that includes Aludéya acacia, ends in a gigantic amphitheater where Gaeda and Bideyat nomads water their dromedaries, sometimes hundreds at a time. Freshwater fish feed on the droppings, and they in turn are food for a very small population of Nile crocodiles, a remnant from about 5,000 years ago when the desert was green. Ennedi is located in northeast Chad near the Sudan border and is one of the most isolated regions in the Sahara. Special tourist operators bring visitors here but it is not for the fainthearted. Summer temperatures soar to 122°F (50°C). For the individual traveler, the area is accessible only by four-wheel drive or trail bikes along obscure tracks or pistes. **MB**

LAKE CHAD

CAMEROON / NIGERIA / NIGER / CHAD

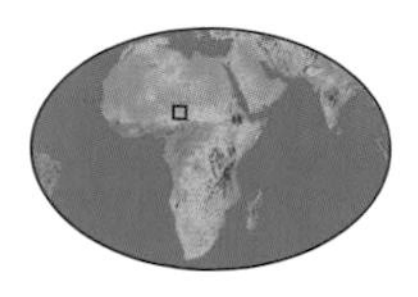

Age of Lake Chad: 2.5 million years

Surface area of high water: 932 sq mi (2,413 sq km)

Once one of the largest lakes in the world, Lake Chad has shrunk dramatically over the last 40 years. In the 1960s, it covered an area of more than 10,000 square miles (26,000 square kilometers), but by 2000 this had fallen to a mere 932 square miles (2,413 square kilometers). This shrink was due to decreased rainfall combined with greatly increased amounts of irrigation water being drawn from the lake and the rivers that feed it. Because the lake is very shallow—only 23 feet (7 meters) at its deepest—it is very sensitive to small changes in its average volume—showing seasonal fluctuations in size.

Lake Chad is home to about 140 species of fish including the incredible African lungfish that can grow up to six feet (1.8 meters) long. In the dry season, it burrows into the mud as the water dries up, to hibernate in an underground cocoon until the next flooding. Even more impressive are the enormous numbers of waterbirds found here. It is thought that hundreds of species of bird live in, or migrate to, the Lake Chad region. Over a million individuals of just three species alone—the garganey, pintail, and ruff—are found here in the winter. PG

TÉNÉRÉ DESERT

AGADEZ, NIGER

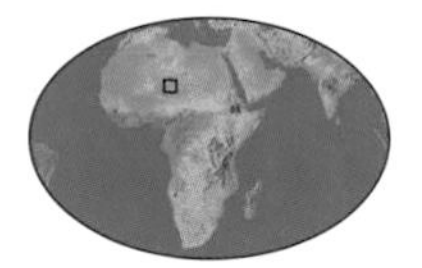

Height of dunes: 800 ft (245 m)

Length of seif dunes: 100 mi (160 km)

In the center of the Sahara Desert, in the northern part of Niger, lies a desert within a desert. The vast Ténéré Desert is an ocean of sand as big as the state of California, interrupted by rocky plateaus. Its rolling dunes reach over 800 feet (245 meters), making them among the largest in the world. To the east lies the Grand Erg of Bilma, 750 miles (1,200 kilometers) of sand that extends into Chad. To the south are the so-called "seif dunes," parallel ridges of sand about 100 miles (160 kilometers) long with troughs that are known as "gassis" in between.

Touareg salt caravans travel along the gassis taking salt from Bilma to Agadez. The salt is dissolved in the ground and recovered by filling pits with water—this evaporates in the sun, so the layer of salt crystals that remains can be skimmed off. Camels once carried the salt, taking 15 days to reach Agadez, but four-wheel-drive trucks are increasingly used. Today, there are few landmarks in the shifting sands, but there once were more. Along the route is a very deep well where a solitary 300-year-old acacia tree once stood, known as the "L'Arbre du Ténéré." The tree was destroyed in 1973, but was replaced with a metal reproduction. MB

SUDD SWAMPS

BAHR AL JABAL, SUDAN

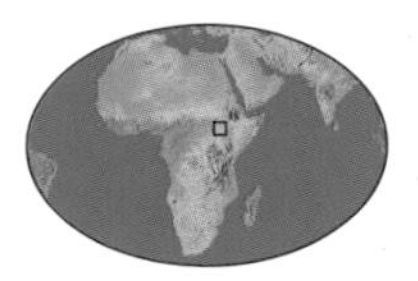

Flooded area: 11,000 sq mi (28,000 sq km)

Habitat types: wetlands, swamps, river-flooded and rain-flooded grasslands, lakes, wooded hummocks

Fed by the White Nile, this is Africa's most extensive swamp, and the world's second biggest wetland. One of the most important wetlands in all Africa, the Sudd is renowned for the quantities of migratory birds, resident waterfowl, and antelopes that it supports. The region is known to host at least 419 species of bird, 91 species of mammal, and over 1,200 species of plant. Among the birds, the Sudd is especially important for saddle-billed storks, pink-backed and great white pelicans, and goliath herons—at 4.6 feet (1.4 meters) tall, the world's largest heron species. Home to almost all of the world's shoebills and black-crowned cranes, there is also a remarkable diversity of antelopes here. These include white-eared kob, tiang, mongalla gazelle, sitatunga, bushbuck, waterbuck, and reedbuck. One species, the Nile lechwe—a specialized swamp-dweller with elongated hooves to support its weight on floating vegetation—occurs nowhere else in the world except the Sudd. Many of these species undergo extensive migrations within the floodlands, following the appearance of new grazing. They have no big cat predators and are preyed on only by crocodiles, pythons, and people. **AB**

LAKE ASSAL

TADJOURA, DJIBOUTI

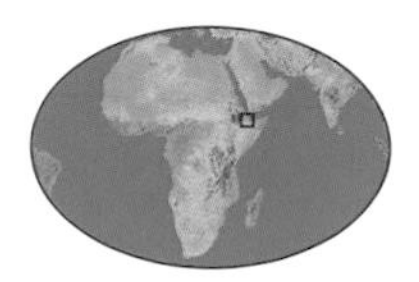

Elevation of Lake Assal: 510 ft (155 m) below sea level

Average summer temperature: 135°F (57°C)

Earthquakes and volcanic eruptions are common in Djibouti, a small country that borders the Gulf of Aden at the mouth of the Red Sea. It sits in a geologically active area of Earth's crust, where molten rock from below wells up to the surface at places where the continental plates are parting. This process is usually seen on the ocean bed, but here appears on dry land. In fact, if it were not for the Danakil Mountains on the coast stopping the Red Sea from pouring in, the entire area could be below the waves. Seawater does percolate through the rock, however, and accumulates in the depression that is Lake Assal, a saltwater lake that is 510 feet (155 meters) below sea level. The summer temperature here averages 135°F (57°C), making it one of the hottest places on the planet. When the water evaporates, the salty water that remains is considered among the saltiest in the world. The lake's salt flats of white, sparkling crystals contrast markedly with the surrounding black, volcanic hills. Its waters range from iridescent blue to turquoise, and from pale green to rusty brown, depending on the minerals present. MB

BELOW *Salt crystals sparkle in the shallows of Lake Assal.*

ERTA ALÉ

TIGRAY, ETHIOPIA

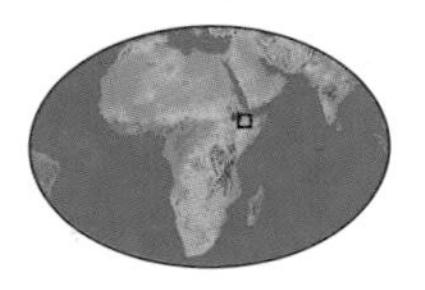

Length of volcanic arc: 50 mi (80 km)

Elevation of Lake Kurum: 400 ft (120 m) below sea level

The Danakil Depression is an area of alkaline desert where the daytime temperature soars to 122°F (50°C) and water is scarce. Rising out of a landscape crisscrossed by ravines are five enormous volcanoes—the most symmetrical known to the local Afar people as Erta Alé, meaning "fuming mountain." Surrounding the volcanoes are razor-sharp lava fields, and beyond their 50-mile (80-kilometer) arc is a plain of salt. At one time, the depression was part of the Red Sea, but when major earth movements raised up the Danakil Highlands, the area was cut off and the water evaporated, leaving a salt layer thought to be two miles (3.2 kilometers) deep. The lowest point in the region at 400 feet (120 meters) below sea level is Lake Karum, a salt-laden lake about 45 miles (72 kilometers) across that is filled briefly each year by water from higher lands nearby. As the water percolates down, it is superheated by the molten rock oozing up from Earth's mantle to form steaming hot springs. British explorer Ludovico Nesbitt and his two Italian colleagues were the first Europeans to visit the place in 1928. Nesbitt called it "a landscape of terror, of hardships, of death." MB

LAKE TANA

AMHARA, ETHIOPIA

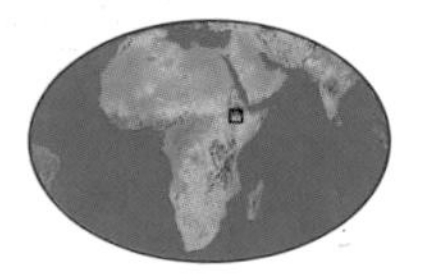

Surface area of Lake Tana: 1,390 sq mi (3,600 sq km)

Maximum depth of lake: 26 ft (8 m)

Elevation of lake: 6,003 ft (1,830 m)

Lake Tana—43 miles (70 kilometers) wide and 37 miles (60 kilometers) long—is the largest lake in Ethiopia. It is fed by four rivers, the source of one being the Abbay, or Blue Nile. The vast lake is not especially deep, but supports a huge diversity of plants and animals including specialized ciprinic fish. Crocodiles live in the lake, and visitors can rent bicycles to go up the Abbay to where hippos can be seen. There are 37 islands on the lake with huge colonies of waterbirds and superb trees. Along the lakeshore, birdlife—both local and migratory visitors—make this site ideal for bird-watchers. The variety of habitats, from rocky crags to riverside forests, ensure that many different species can be spotted. Boats cross to the Urai Kidane Mihiret Monastery on the other side of the lake, and visitors may well pass a monitor lizard swimming far from land—they are relatively common here. The monastery itself is hidden in lush tropical forest. Monks have lived there for at least 600 years and have made a way of life in harmony with the lake. The paintings in the church are considered Ethiopian national treasures. PT

RIGHT *Thundering falls feeding the waters of Lake Tana.*

BLUE NILE FALLS

AMHARA, ETHIOPIA

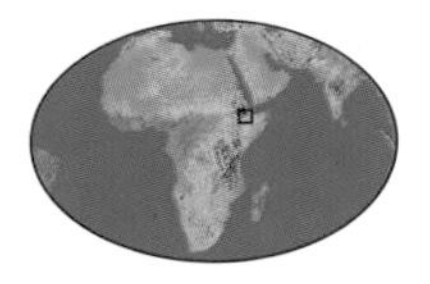

Width of Blue Nile Falls: 1,320 ft (400 m)

Length of Blue Nile: 950 mi (1,530 km)

In 1770, Scottish explorer James Bruce was searching for the source of the River Nile and he came upon Blue Nile Falls. He wrote: "The river . . . fell in one sheet of water, without any interval, above half an English mile in breadth." The Ethiopians subsequently put its image on the one birr note, and local people gave it the name *Tisissat*, meaning "water that smokes." When in full flood, the river flows over a basalt shoulder and plunges down 150 feet (45 meters). It throws up a mist that drifts away on the breeze for up to 1 mile (1.6 kilometers), creating spectacular rainbows. Sadly, today this magnificent waterfall can only be seen in its full glory on Sundays; the rest of the week 90 percent of the water is diverted to a massive hydroelectric plant, leaving a few narrow cascades trickling over the falls.

Thick vegetation once surrounded the falls, but deprived of water for most of the week, the plants have withered away. The Blue Nile itself rises in a spring above Lake Tana in the Ethiopian Highlands, but it becomes a river when it leaves the lake about 19 miles (30 kilometers) upstream from the falls. MB

RIGHT *The Blue Nile Falls spray mist over lush green hills.*

LAKE KARUM

DANAKIL, ETHIOPIA

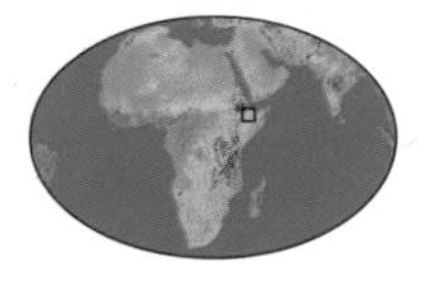

Elevation of Lake Karum: 400 ft (120 m)

Maximum diameter of lake: 45 mi (72 km)

Temperature: up to 122°F (50°C)

Lake Karum is a salt lake situated at the lowest point of Ethiopia's infamous Danakil Depression—one of the lowest, hottest, and most inhospitable areas on the face of the planet. The Danakil Depression lies along Africa's Great Rift and was originally part of the Red Sea until movements of Earth's crust raised the Danakil Highlands to the north and caused the land to sink. The enclosed seawater evaporated, leaving behind layers of salt up to two miles (3.2 kilometers) thick. Much of the area is now a parched salty plain lying below sea level, where temperatures can soar to 122°F (50°C) in the sun. For most of the year there is no rain, and the water that runs down from the highlands washes salt into shallow saline lakes such as Lake Karum.

Despite the extreme conditions, however, people do live here. The Afar tribespeople make a living through mining salt and nomadic farming. Blocks of salt are levered from the ground using poles and are traded across northeast Africa. Lake Karum lies nearly 400 feet (120 meters) below sea level. After rains, a mineral-rich lake up to 45 miles (72 kilometers) wide is formed, but the waters quickly evaporate. RC

MOUNT NIMBA

GUINEA / IVORY COAST / LIBERIA

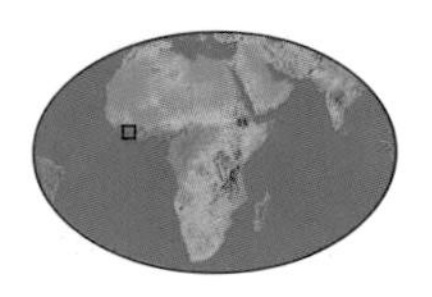

Elevation of Mt. Nimba: 1,476 to 5,748 ft (450 to 1,752 m)

Highest peak: Mount Richard Molard

Rock type: iron-rich quartzite

Situated on the border between Guinea, Ivory Coast, and Liberia, this area is a Strict Nature Reserve, a Biosphere Reserve, and a World Heritage site. The first reserve was established in 1943 to protect the extraordinary number of habitats and species. Mount Nimba, a bar of erosion-resistant, iron ore–rich quartzites, rises steeply out of the surrounding plain. Exceedingly ancient, it has great topographic diversity, with valleys, plateaus, steep rocky peaks, rounded hills, and abrupt cliffs. Three major vegetation types exist: high-altitude grassland speckled with endemic shrubs, and ravines dotted with endemic trees and tree ferns; savannas striped with gallery forests; and lowland forests that become progressively drier as the mountain ascends. The wettest months are May to October (mountain) and April to October (base). Cloud-shrouded above 2,790 feet (850 meters), Nimba has over 2,000 plant species (16 unique to area). Some 500 new animal species have been recorded here, including two extraordinary live-bearing toads, and the lesser otter shrew. The mammal list includes the monkey, duiker, pangolin, pygmy hippo, genet, and tool-using chimpanzee. **AB**

KINTAMPO FALLS

BRONG-AHAFO, GHANA

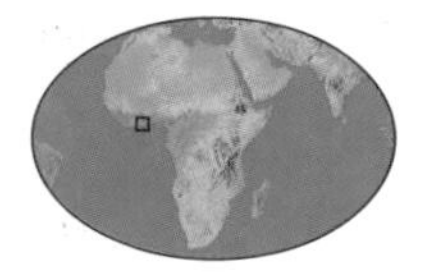

Height of Kintampo Falls: 130 ft (40 m)

Notable feature: 130-ft- (40-m-) high mahogany trees

The Kintampo Falls is an impressive waterfall on the Pumpum River in the Brong-Ahafo region of Ghana. The river at this point falls about 130 feet (40 meters) in two main stages, the "upper" and "lower" falls, on its journey toward the Black Volta at Buipe. The forest here is dominated by massive mahogany trees up to 130 feet (40 meters) high. The falls, thought to be some of the most beautiful in Ghana, are hidden away in the forest, within walking distance of the main road to northern Ghana and Burkina Faso. The falls are easily reached by a series of steps that descend 230 feet (70 meters). A swimming pool rests at the base. The Kintampo area was developed as a tourist site. A guesthouse that used to be operated by the local administration is currently in ruins, but there is the possibility that the site may be developed again for both tourism and the generation of hydroelectric power. The nearby Fuller Falls has a stream below the waterfall that disappears underground and re-emerges about 130 feet (40 meters) farther down. The Kintampo Falls (also known as Randall Falls) is located near the small town of Kintampo, which lies between Kumasi and Tamale. **RC**

MOUNT CAMEROON

SUD-OUEST PROVINCE, CAMEROON

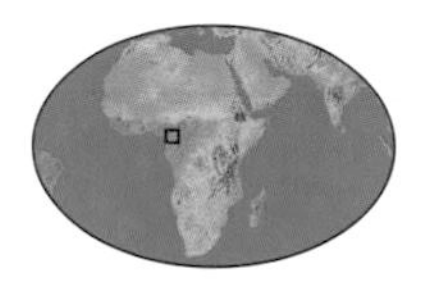

Elevation of Mt. Cameroon: 13,435 ft (4,095 m)

Habitat: lowland tropical rainforest to alpine scree

Rock type: volcanic

Mount Cameroon is located on a seismic fault line—the highest mountain in West Africa that is still an active volcano, erupting eight times in the last 100 years, most recently in 2001. The cone of the mountain covers some 17.5 square miles (45 square kilometers). It is close to the sea, but rarely visible from the coast because of its cloud covering. At its base, the village of Debuncha is reputed to be one of the five wettest places in the world, receiving 33 feet (10 meters) of rain annually.

From here, lowland forest gives way to montane forest, on to sub-alpine grassland, and finally a bare summit occasionally dusted with snow. All this produces an exceptional level of diversity and species unique to the area. These include the Mount Cameroon speirops, Mount Cameroon francolin, Cameroon greenbul, four-digit toad, and the Tumbo-insel screeching frog. Mount Cameroon also has the highest diversity of squirrels in Africa and is home to the rare Preuss' guenon, the drill, and many rare butterflies. Community wildlife management schemes give hope for long-term conservation of this exceptional region and eco-tourism is playing an important part. AB

LAKE TURKANA

RIFT VALLEY, KENYA

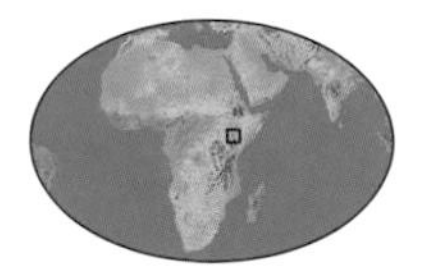

Length of Lake Turkana: 195 mi (312 km)

Notable feature: Nile crocodiles 18 ft (5.5 m) long

Like some great manmade waterway, the River Omo in northern Kenya actually meanders across a lake. Banks or levees of silt brought down by the river from the Ethiopian Highlands 400 miles (640 kilometers) away have built up on either side so that the river now resembles a canal. It ends about three miles (five kilometers) from the lakeshore in a gigantic silt-laden delta shaped like a bird's foot. Lake Turkana itself is 195 miles (312 kilometers) long, and is fed by several rivers. At one time it spilled into the River Nile, but climate change has meant that the waters have fallen by more than 600 feet (180 meters). Nevertheless, there is plenty left for the 12,000 Nile crocodiles that live here. They breed around the shores of Central Island, a group of small volcanic craters. At 18 feet (5.5 meters) long, they are among the largest crocodiles in Africa. Two tribes catch fish here—the Turkana and the El Molo—but the people that attracted world attention lived here two million or more years ago—Lake Turkana is one of the archeological sites made famous by the Leakey family. Here the fossil remains, stone tools, and footprints of some of our earliest ancestors have been discovered. MB

LAKE BARINGO

RIFT VALLEY, KENYA

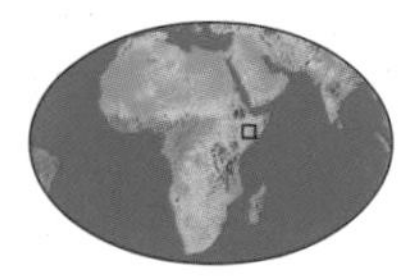

Surface area of Lake Baringo: 50 sq mi (130 sq km)

Elevation of lake: 3,325 ft (1,011 m)

Geyser temperature: 194°F (90°C)

Lake Baringo is a large freshwater lake and the biggest Rift Valley lake. It is located in a rugged semi-desert area, which was once the ancient slave route to the East African coast. It lies 3,325 feet (1,011 meters) above sea level and has an average depth of 17 feet (5 meters). Today, the lake is a wildlife haven for 470 bird species, including a famous heronry on a rocky islet on the eastern shore known as "Gibraltar," which has East Africa's largest population of Goliath herons.

Local fishermen are the Njemps, who fish in water up to their shoulders and pay scant regard to the crocodiles and hippos with whom they share the lake. They also take visitors to watch fish eagles swoop for bait.

Confirmation that this is still a volcanically active region came in April 2004, when a group of people were drilling a borehole about 2 miles (3.2 kilometers) away. Their actions triggered a geyser that is now in continuous eruption. It spews a column of saltwater (rather than freshwater like Baringo, meaning it cannot be sharing the same water source) about 263 feet (80 meters) into the air, which can be seen from a distance of 12 miles (20 kilometers) away. **MB**

LAKE MAGADI

RIFT VALLEY, KENYA

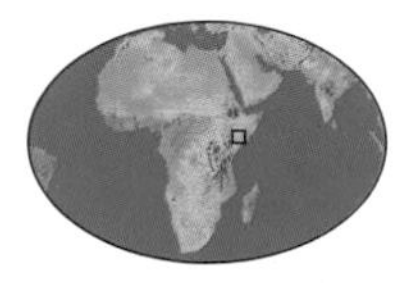

Surface area of Lake Magadi: 40 sq mi (104 sq km)

Length of lake: 20 mi (32 km)

Width of lake: 2 mi (3.2 km)

Lake Magadi is so rich in soda that it has been extracted commercially for the last hundred years, and there is no sign of it running out. It is thought that underground water is forced up through the layers of alkaline rocks, continually enriching the sodium carbonate content of the lake. The soda is just one of several salts (including common salt) and clays that are deposited in a 100-foot- (30-meter-) thick deposit known as "trona." However, more water is lost through evaporation than enters the lake, so the soda is concentrated. Rainfall here is low, less than 16 inches (400 millimeters) per year, and the surrounding area is no more than semi-desert. The lakeshore consists of soda-mud topped with a sun-dried crust, and at midday it can be dangerously hot and caustic. Hot springs are found at the edge, and as the only source of freshwater in the entire lake, they are home to a small species of tilapia fish that has adapted to live in the piping hot water. Lake Magadi is the southernmost lake in the Kenyan territory of the Rift Valley—it is 20 miles (32 kilometers) long and 2 miles (3.2 kilometers) wide, and can be reached by bus from Nairobi just 72 miles (115 kilometers) to the northeast. **MB**

THOMPSON'S FALLS

RIFT VALLEY, KENYA

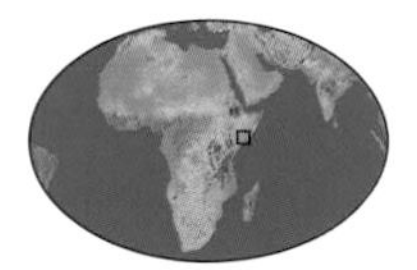

Local name: Nyahururu Falls

Height of Thompson's Falls: 240 ft (73 m)

Elevation of Nyahururu township: 7,743 ft (2,360 m)

Scottish explorer Joseph Thompson of the Royal Geographical Society was the first European to walk from Mombasa to Lake Victoria in 1883. On his epic journey through previously uncharted and hostile lands, he discovered many of Kenya's most dramatic landscape features. In recognition of his travels, one of the country's most impressive waterfalls was named after him. These famous falls have recently officially reverted to their original name and that of the nearby town—Nyahururu—but they are still affectionately known as Thompson's Falls. The adjacent settlement was one of the last white settler towns to be established in Kenya, and is its highest township at over 7,743 feet (2,360 meters) above sea level. Despite its proximity to the equator, it has an invigorating climate with crisp, clear air and a forest of cool conifers. The falls drop straight down 240 feet (73 meters) of rock face into a spectacular gorge. The viewing point from the clifftops opposite is dramatic and especially stunning after the long rains in April and May, when clouds of spray are thrown up by the thundering torrent. It is a popular stopover for visitors on safari to the Rift Valley. **AC**

MOUNT ELGON'S ELEPHANT CAVES

RIFT VALLEY, KENYA / MBALE, UGANDA

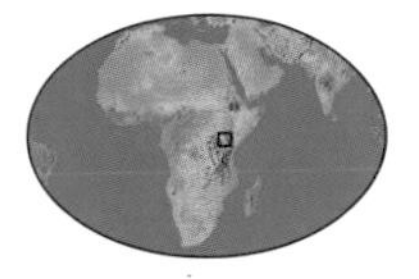

Height of Mt. Elgon: 14,138 ft (4,300 m)

Width of largest cave: 200 ft (60 m)

Vegetation: teak, cedar

Mount Elgon is Kenya's second highest mountain, rising over 14,138 feet (4,300 meters). Formed over millions of years during the creation of the Great Rift Valley, it lies 87 miles (140 kilometers) northeast of Lake Victoria and straddles the Kenya–Uganda border. The mountain today is an ancient eroded volcano with a huge caldera and spectacular flat-topped basalt column on its summit. Yet its most dramatic feature lies within an extensive series of lava tube caves.

Some caves are over 200 feet (60 meters) wide and frequented by elephants as well as other animals digging for salt. When the 19th-century African explorer Joseph Thompson returned to England, his stories of cave-dwelling elephants must have seemed fanciful. Yet what he discovered on Mount Elgon in East Africa was, in fact, a unique series of deep caves where elephants and other wild animals came to find salt. The four largest caves—Kitum, Makongeni, Chepnyalil, and Ngwarisha—are all explorable with care. Kitum is the biggest and best known, extending some 660 feet (200 meters) into the heart of the mountain. In Kenya's Maasai language, its name means "Place of Ceremonies." For centuries the local Saboat tribe used the caves of Elgon as granaries and stables for its stock. At other times it sought refuge in the caves from bad weather and as a sanctuary during periods of intertribal conflicts. More famously, the caves are a favorite gathering place for elephants. Every night, long convoys venture through the mountain forest. They head deep into the mountain to feed on the salt-rich deposits that are excavated with their tusks. The gouged walls of the caves are scarred by the work of thousands of elephants over millennia. Mount Elgon is one of Kenya's wildest regions, with vast areas of unblemished forest. It is home to about 400 elephants, as well as buffalo, leopard, colobus and blue monkey, giant forest hog, waterbuck, and other types of antelope. Over 240 species of bird have been recorded. Huge Elgon teak and cedar trees, some 82 feet (25 meters) tall, dominate much of the forest. **AB**

Every night, long convoys of elephants venture through the mountain forest. They head deep into the mountain to feed on the salt-rich deposits they excavate with their tusks. The cave walls are scarred by the work of thousands of elephants over millennia.

RIGHT *The entrance to Kitum cave in Mount Elgon—elephants come here to mine minerals.*

LAKE BOGORIA

RIFT VALLEY, KENYA / TANZANIA

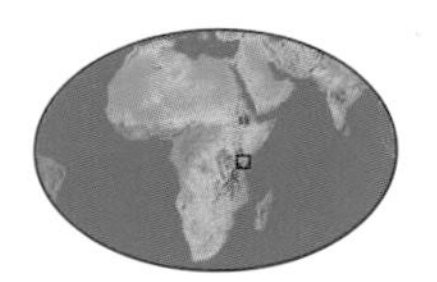

Depth of Lake Bogoria: 33 ft (10 m)

Height of rift wall at lakeside: 2,067 ft (630 m)

Some of East Africa's soda lakes are less caustic than others, but no less dangerous. Lake Bogoria is fringed by scalding hot springs, which can overflow onto the grassy shore and catch unwary visitors. On cool mornings the entire place steams and bubbles like a vision of hell. Nevertheless, flamingoes come here to drink and wash soda from their feathers. They gather at the mouths of freshwater rivers and streams that feed the lake, or stand a comfortable distance away from the steaming springs. They also feed on the microscopic algae and brine shrimp that, depending on which organisms dominate, tint the water either green or pink. Flamingoes are filter feeders, like baleen whales, and have specialized bills to sieve mud and water for food. Pigments in the food give the birds their pink plumage, and food is so abundant that over three million lesser flamingoes and about 50,000 greater flamingoes thrive here. However, it was not always so inviting—in the 1950s the water from neighboring Lakes Nakuru and Elmenteita evaporated, leaving dried up bowls with hot, searing dust filling the air. MB

BELOW *The bubbling waters and springs of Lake Bogoria.*

LAVA TUBES

EASTERN PROVINCE, KENYA

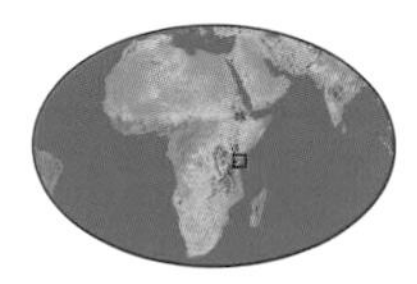

Length of tube system: 7 mi (11 km)

Length of Leviathan Tube: 30,020 ft (9,150 m)

In 1938, a rare phenomenon was discovered hidden in a range of east African hills. Formed by an extinct volcano on the edge of the famed Tsavo National Park in Kenya, the Leviathan Lava Tube is one of the longest of its kind in the world. Reaching 30,020 feet (9,150 meters) in length, it is located over 7,178 feet (2,188 meters) high in the Chyulu Hills. It is not the only one. Other lava tubes nearby form the longest system in the world, amassing over 7 miles (11 kilometers) of passages.

Another system of spectacular lava tubes was also found in Kenya's mighty Rift Valley. Mount Suswa is a magnificent example of an extinct volcano, with 10 miles (16 kilometers) of lava tubes to explore. Worldwide, lava tube caves are rare. They are thought to form when molten lava of a specific type and viscosity flows down a slope of a particular angle. The surface layers then cool and solidify, but inside the molten red-hot lava continues to flow. If a section of the surface high on the volcano then collapses, it allows air to enter and a tube evacuates itself close to the surface. Over thousands of years, rainwater entering the tunnels may form subterranean features—stalagmites and stalactites. AC

MARA RIVER CROSSING

RIFT VALLEY, KENYA

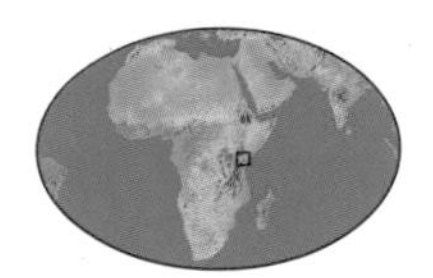

Frequency of Mara River crossing: annual

Duration: 3 weeks

Each September, millions of wildebeest, zebras, and Thomson's gazelles arrive on the banks of the Mara River in East Africa. They will have trekked hundreds of miles across the Serengeti, and crossing the river is the next crucial step to their survival. The land behind them is parched, and they must follow the rains, which sprout fresh grass for them to eat. At the Mara River, they are confronted with the greatest test of their lives—not lions or hyenas, but crocodiles. Hiding in the water are Nile crocodiles, and some are the world's largest. Having survived on catfish and not eaten red meat for many months, the crocodiles are hungry and determined to feast. They detect the approaching herds by the low-frequency vibrations of their footfalls, and they wait for the animals to cross. Zebras are first. Family groups, each with a powerful stallion

at its head, pass safely. The crocodiles know not to get close to the flailing hooves. Only individuals crossing alone are pulled under.

The wildebeest, however, kill themselves in the crush. So many cross at once that they jostle and trample each other and many are drowned. The crocodiles simply stand off and wait for the inevitable carnage. Vultures, which are nesting nearby during the dry rather than the wet season, ensure that their chicks are hatched in time for the inevitable food bonanza. Then come the diminutive gazelles. They fail to see the crocodiles for what they are and plunge into the swirling waters. Squadrons of crocodiles close in for the kill. Lions, black-kites, and hyenas also feast on those that do not make the crossing. Yet many animals get to the other side safely, and the annual cycle starts all over again. A perilous yet imperative journey for the wildebeest, zebra, and gazelle, the Mara River crossing is a highlight of one of the greatest migrations on the planet. **MB**

BELOW *Wildebeest on the dusty banks of the Mara River.*

MURCHISON FALLS

GULU / MASINDI, UGANDA

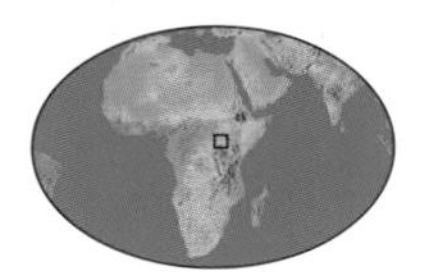

Height of Murchison Falls: 130 ft (40 m)

Area of national park: 1,482 sq mi (3,840 sq km)

One of the world's greatest rivers begins its journey in the most spectacular way. Squeezing through a 23-foot- (7-meter-) wide crevice, the river plummets in a breathtaking leap down 130 feet (40 meters) and into a vast pool of white water. In 1864, when Sir Samuel Baker was searching for the source of the River Nile in unexplored Africa, he became the first European to see this fabulous cascade. He named it after Sir Roderick Murchison, who was president of the Royal Geographical Society at the time. The falls are the culmination of a vast flow of freshwater that emanates from Lake Victoria, through Lake Albert, to form a stretch of white water over 14 miles (23 kilometers) in length—the Karuma Rapids.

The Murchison Falls National Park is Uganda's largest, covering an area of 1,482 square miles (3,840 square kilometers). The park is cut in half by the Nile, and its landscape varies from dense rainforest on the hilly ranges in the southwest to undulating savanna in the northwest. The forest shelters many primate species, including chimpanzees, while the river contains plenty of big hippos and some of Africa's largest crocodiles, thanks to the ever-present menu of enormous fish.

The most impressive view is at the top, where the Nile forces its way through a gap in the rock and creates an explosion of white water. It is the most powerful natural flow of water anywhere on Earth and the rock actually shakes from the force of the flow. The only creatures strong enough to withstand the pressure are huge Nile perch. Weighing up to 220 pounds (100 kilograms), they are sometimes glimpsed as they are spat from the river. It is also one of the few places in the world to see the rare shoebill stork, said to have a beak capable of biting young crocodile in half. Other birds include the tiny malachite kingfisher and the carmine bee-eater. The area around Lake Albert and its famous rapids attracts the highest concentration of game in Uganda, with various antelope, buffalo, Rothschild giraffe, and elephant. It was also used as a setting for the Hollywood classic *African Queen.* AC

The most impressive view is at the top, where the Nile forces its way through a gap in the rock, creating an explosion of white water. It is the most powerful natural flow of water anywhere on Earth.

RIGHT *Turbulent white waters cascade through the ragged rocks of Murchison Falls.*

VIRUNGA MOUNTAINS

UGANDA / RWANDA / DEMOCRATIC REPUBLIC OF THE CONGO

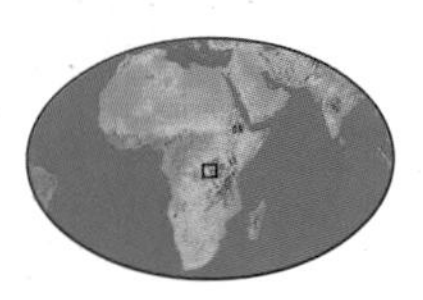

Highest peak (Mt. Karisimbi): 14,787 ft (4,507 m)

Notable features: mountain gorillas

Eight volcanoes—six extinct and two active—dominate the borders of Uganda, Rwanda, and the Democratic Republic of the Congo. One peak—Nyamulagira, meaning "commander," is Africa's most active volcano and one of the most active in the world, having erupted over 30 times since 1880. In 1938, eyewitnesses told of a river of lava that ran 24 miles (40 kilometers) from its summit. Its neighbor, Nyiragongo, is lively, too. In 1977, its almost perfectly circular cone, measuring about half a mile (1 kilometer) across, was breached in five different places and molten lava smothered everything in its path. As recently as January 2002, a lava flow devastated the town of Goma and dumped over 35 million cubic feet (1 million cubic meters) of molten rock into Lake Kivu, about 8 miles (13 kilometers) away.

The other volcanoes have long lain dormant. These include the conical cone of Gahinga and the ragged upper slopes of Sabinyo. Mount Karisimbi, from the word *nsimbi* meaning "cowrie-shell" on account of the snow at its summit, is the highest at 14,787 feet (4,507 meters) but the most famous peak is Bisoke, home to the extremely rare mountain gorilla.

These gentle giants live in the Parc National des Volcans. They move around the mountain slopes in family groups, feeding on bamboo, wild celery, and nettles. Male gorillas can measure up to 5 feet 9 inches (1.75 meters) tall and weigh as much as 430 pounds (195 kilograms). Their main natural danger is the leopard, but the greatest threat to their survival is humans. Gorillas are killed for "bush meat" and their body parts sold to tourists, such as hands for ashtrays. Their numbers are worryingly low, with fewer than 700 living in the wild today, so any catastrophe, whether an earthquake, volcanic eruption, war, or intense poaching, could wipe out the species altogether. Their champion was Dian Fossey, an American occupational therapist, whose life in the Virungas was featured in the book and film *Gorillas in the Mist*. Fossey won the trust of these shy animals and studied them at close hand for over 18 years. In 1985, an unknown assailant murdered her in her bed, but her work continues, as researchers from the world's major conservation bodies brave civil war and other dangers to ensure that the gorillas survive. MB

One peak–Nyamulagira, meaning "commander"–is Africa's most active volcano, and one of the most active in the world, having erupted over 30 times since 1880. In 1938, eyewitnesses told of a river of lava that ran 24 miles (40 kilometers) from its summit.

RIGHT *The snowcapped peaks of the Virunga Mountains.*

MOUNTAINS OF THE MOON

DEMOCRATIC REPUBLIC OF THE CONGO / UGANDA

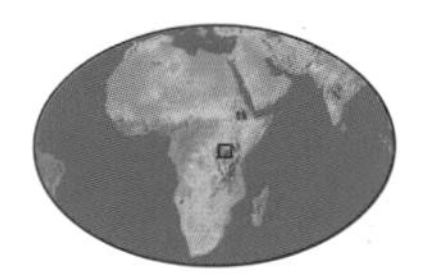

Length of Mountains of the Moon: 80 mi (129 km)

Highest peak (Margherita Peak/ Mount Stanley): 16,763 ft (5,109 m)

Age: 10 million years

British explorer Henry Morton Stanley—the same Stanley who found a lost Dr. Livingstone—was the first modern European to see the Mountains of the Moon (part of the Rwenzori Range) in 1888. At the time, he recorded that the mountains were veiled in cloud for 300 days a year, but occasionally the mist cleared and row upon row of jagged mountains were revealed. The mountains are the result of massive earth movements that took place no more than 10 million years ago. Despite being just 30 miles (50 kilometers) north of the Equator, they are covered in snow. Above the tree line, where clouds shroud the mountains down to 9,000 feet (2,700 meters), giant versions of groundsel, lobelias, and heather grow up to 40 feet (12 meters) above the rain-sodden ground. During November and December, more than 20 inches (510 millimeters) of rain falls in less than a month. To the local Bantu people, *Rwenzori* means "rainmaker." Ancient Greek geographers also told of these mountains whose melting snows fed the headwaters of the Nile. Aristotle referred to them as the "silver mountains," and Ptolemy called them the "Mountains of the Moon." MB

LAKE TANGANYIKA

TANZANIA / DEMOCRATIC REPUBLIC OF THE CONGO / BURUNDI / ZAMBIA

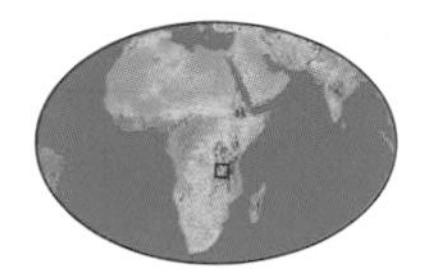

Width of Lake Tanganyika: 30 mi (50 km)

Maximum depth of lake: 4,820 ft (1,470 m)

Notable feature: cichlid fish

Lake Tanganyika was discovered in 1858 by explorers Richard Burton and John Speke while searching for the source of the Nile. Not only had they found the world's second oldest lake, but also the deepest in Africa. With a mean depth of 1,870 feet (570 meters), it holds the greatest quantity of freshwater on the continent. It is so deep that the lower water levels contain "fossil water" that has remained undisturbed at the bottom for millions of years. Nearer the surface, 300 species of cichlid fish, two-thirds of which are unique to the lake, provide food for more than a million people in the surrounding villages and towns. Fishing is carried out mainly at night, when artificial lights are used to attract the fish. The lake itself is about 418 miles (673 kilometers) long and averages 30 miles (50 kilometers) wide, and it is sandwiched between the walls of the Great Rift Valley where it is shared by Burundi (8 percent), the Democratic Republic of the Congo (45 percent), Tanzania (41 percent), and Zambia (6 percent). It is the fifth largest lake in the world, although since 1962 the water level has dropped 18 inches (45 centimeters) each year. MB

BELOW *The Tanzanian shores of Lake Tanganyika.*

CONGO BASIN

BANDUNDU / EQUATEUR / KASAI OCCIDENTAL / KASAI ORIENTAL / MANIEMA / ORIENTALE, DEMOCRATIC REPUBLIC OF THE CONGO

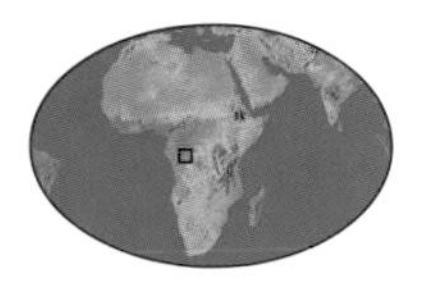

Area of Congo Basin: 1,300,900 sq mi (3,369,331 sq km) with 666,107 sq mi (1,725,221 sq km) of core rainforest

Habitat: tropical evergreen rainforest, swamps, seasonal deciduous forest

The Congo Basin hosts the second largest rainforest block on Earth—only Amazonia is bigger. With continuous forest cover for 65 million years, the Congo rainforests are some of the world's oldest and have acted as an ark for African forest wildlife during several cycles of climate change. Today, huge trees can reach 213 feet (65 meters) tall. A complex history and well-developed river system means that many species are very localized. Even tiny Equatorial Guinea, at 10,831 square miles (28,051 square kilometers), a country slightly smaller than Maryland, has 17 unique species of plant. Unlike Amazonia, Congo's forests abound with large mammal species, including bongo, okapi, gorilla, chimpanzee, bonobo, forest elephant, and forest buffalo. There are also many exceptionally odd animals, such as the bare-necked rock-fowl, Congo peacock, aquatic genet, flying mouse, anomalure, and goliath frog (the world's largest). Congo Basin primates achieve a great diversity, with up to 16 species occurring in one area. In all, 68 different primates are known from the region. Over half occur nowhere else. Though the region's evergreen rainforest core is centered on the Congo River basin, there is also a surrounding seasonally deciduous zone, intermediate between rainforest and savanna. Sharing many plant species with the rainforest, it is of seasonal importance to many animals. Nine countries have Congolese rainforest or its ecologically linked transitional forms. Together, they provide a protected-areas network conserving some 91,892 square miles (237,999 square kilometers) or seven percent of the total, and 54,500 square miles (141,154 square kilometers) or eight percent of the core forest. Included are Salonga (Democratic Republic of the Congo or DRC), Nouabalé-Ndoki and Odzala (both Republic of Congo), Wonga-Wongue (Gabon), and Faro (Cameroon) National Parks. The DRC's role is key, because its borders contain 58 percent of Congo's core rainforest. Its importance is illustrated by the fact that it has five rainforest-related World Heritage sites, unfortunately all under threat. Luckily, the DRC appears pro-conservation, recently quintupling the size of one national park to 3 million acres (1.3 million hectares) and canceling 2.7 million acres (1.1 million hectares) of logging concessions. **AB**

With continuous forest cover for 65 million years, the Congo rainforests are some of the world's oldest and have acted as an ark for African forest wildlife during several cycles of climate change.

RIGHT *A herd of elephants enjoys a stroll through the fertile land of the Congo Basin.*

LAKE KIVU

DEMOCRATIC REPUBLIC OF THE CONGO / RWANDA

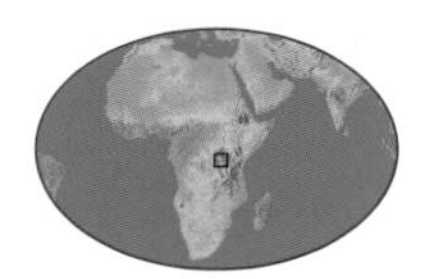

Elevation of Lake Kivu: 4,785 ft (1,459 m)

Maximum depth of lake: 1,320 ft (400 m)

Lava flows from Virunga's volcanoes are thought to have dammed local rivers and created Lake Kivu, a body of water that hides a deadly secret. Carbon dioxide gas percolates up through the lakebed and collects, trapped by the huge body of water that is as deep as 1,320 feet (400 meters) in places. A minor earth movement or volcanic activity, such as the lava flow that reached the lake from Mount Nyiragongo in 2002, could release this gas. Because it is heavier than air, it would roll across the surrounding countryside and asphyxiate every animal in its path, including humans. In 1984 and 1986, this occurred at lakes Monoun and Nyos in Cameroon, and several thousand people lost their lives.

A second danger at Kivu is that the carbon dioxide is being converted into methane by microbes. If this should find its way to the surface and come into contact with a naked flame, then the resulting conflagration could be devastating. Nevertheless, Lake Kivu is considered by some people to be one of Africa's most attractive lakes. It sits in what is known as a "land of a thousand hills" at an altitude of 4,785 feet (1,459 meters), making it the highest lake in Africa. MB

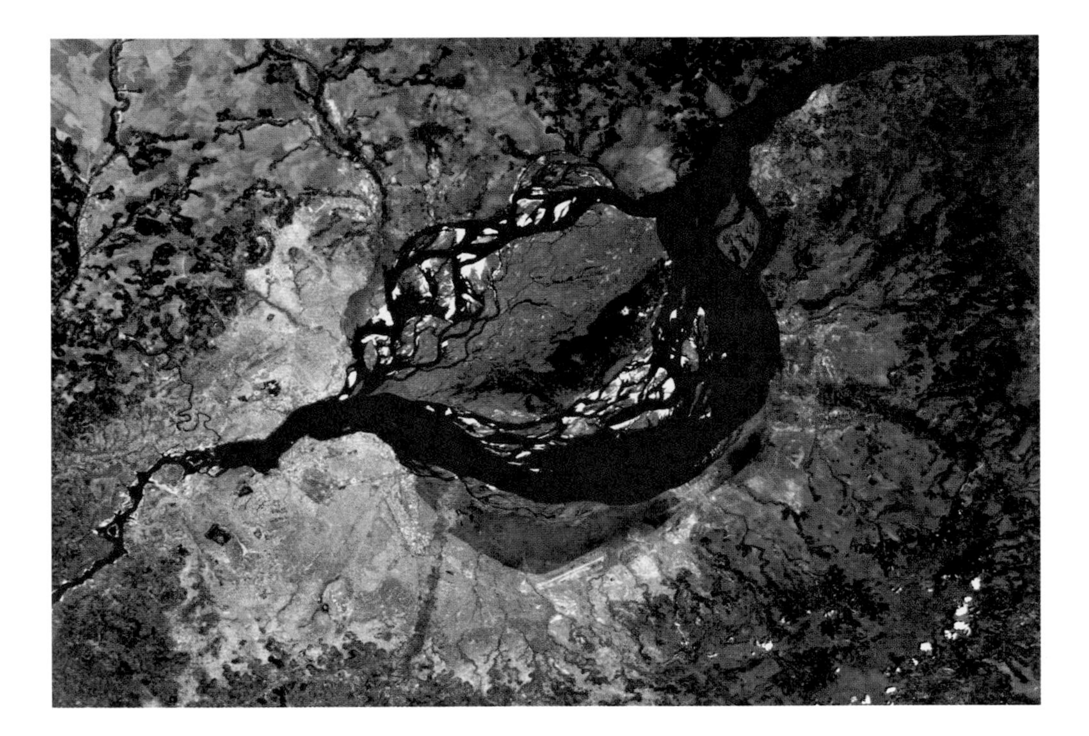

CONGO RIVER

KATANGA, DEMOCRATIC REPUBLIC OF THE CONGO

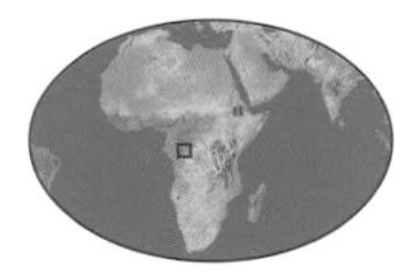

Length of Congo River: 2,920 mi (4,700 km)

Discharge rate of river: 1.5 million cu ft (42,000 cu m) per second

On its journey to the Atlantic Ocean from the Zambian border grasslands, the Congo River passes through the densest rainforests and least-explored parts of Africa. It drains through an area the size of India, passing through spectacular gorges and pouring over precipitous waterfalls in its higher reaches, or meandering through mangrove forests, dense jungle, and reed-fringed swamps downstream. The Congo starts as the Lualaba River that flows first through deep ravines and marshlands, and then falls into Lake Kisale, a paradise for waterbirds. It widens before plunging through a steep gorge in a series of waterfalls that are known as *Les Portes d'Enfer*, meaning "Gates of Hell." Farther downstream, the river enters the tropical rainforest. Here, the Boyoma Falls drop 200 feet (60 meters) within the space of 55 miles (90 kilometers), producing the greatest discharge of any waterfall in the world. As the jungle thins out, it broadens to form Malebo Pool. Just before it reaches the ocean, the Congo negotiates the spectacular Livingstone Falls. These drop 850 feet (220 meters) and were described by Henry Morton Stanley as "a descent into a watery hell." MB

MOUNT KILIMANJARO

KILIMANJARO, TANZANIA

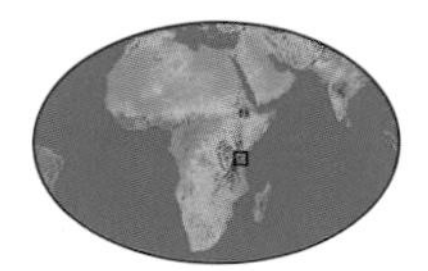

Alternative names: Oldoinyo Oibor (Maasai), Kilima Njaro (Swahili)

Volcanic status: extinct

Few mountains in the world are as recognizable as Mount Kilimanjaro, rising alone from the African plains in northern Tanzania with its snowy cap, although in recent years the world has had to get used to changing views of the mountain, as ice cover at its summit has decreased by more than 80 percent.

At 19,340 feet (5,895 meters) it is by far the highest peak on the African continent and the largest single-standing mountain in the world. It exudes an awesome mystique, towering above the Great Rift Valley, home to humankind's ancestors.

Kilimanjaro, or Oldoinyo Oibor, as it is known in Maasai, is a triple volcano with the youngest and highest peak of Kibo lying between Shira to the west and Mawenzi to the east. Kibo has survived as an almost perfect cone and the crater measures an incredible 1.5 miles (2.4 kilometers) across. The Wachagga tribe, which has inhabited the fertile volcanic slopes at the base of the mountain for around 300 years, has a legend which tells of Mawenzi receiving fire for his pipe from his younger brother, Kibo. While this suggests fairly recent activity, the volcano has been inactive in modern times, although steam and sulfur are still emitted.

The cultivated lower slopes of the mountain give way at 6,000 feet (1,800 meters) to the forest belt, characterized by flora such as fig trees, tree ferns, and lush undergrowth. Flowers, including the 10-foot- (3-meter-) high giant lobelia, abound in the clearer areas. Colobus and blue monkeys inhabit the forest, and elephants can be spotted roaming the forest slopes. At 9,570 feet (2,900 meters), the forest zone ends abruptly, giving way to the heather zone and moorlands with giant groundsel. On the higher moorland only small mosses and lichens grow, eventually giving way to snow and rock. In spite of the mountain's great altitude, the peak is relatively accessible, attracting thousands of walkers each year. The mountain needs to be taken slowly, however, as altitude sickness is common and can be fatal. There are six routes up the mountain with varying degrees of difficulty and different features to recommend them. Accommodation is provided for walking parties and their guides in huts along the way. Climbers begin their final ascent around midnight and are rewarded with a spectacular sunrise at the peak. MM

Few mountains in the world are as recognizable as Mount Kilimanjaro, rising alone from the African plains in northern Tanzania with its snowy cap. It exudes an awesome mystique, towering above the Great Rift Valley.

RIGHT *The snowy cap of Mount Kilimanjaro.*

OL DOINYO LENGAI

ARUSHA, TANZANIA

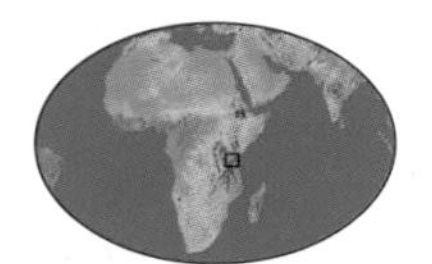

Height of Ol Doinyo Lengai: 9,370 ft (2,856 m)

Diameter of crater: 1,000 ft (300 m)

Notable feature: washing soda

On the edge of the Serengeti in northern Tanzania, the Crater Highlands hide a gray volcano with a white-flecked summit. To the local Maasai, it is known as *Ol Doinyo Lengai* meaning "Mountain of God." It is a holy mountain, home to the god, Engai. When drought hits the area hard, the Maasai travel to the base of the volcano, where they pray for rain. It is unique among volcanoes. When it erupts it spews not only black ash but also carbonatite, which on contact with the moist air turns to washing soda. The mountain is only 9,370 feet (2,856 meters) high, and so what looks like snow at its summit is actually white foam.

The 1,000-foot (300-meter) diameter crater can be reached on foot in about six hours, and at the summit visitors can see a hissing vent through which fountains of lava burst every few seconds. The mountain rumbles almost continually, but the last major eruptions were in 1966 and 1967. On the first occasion the volcano shook for 10 days before exploding violently, sending an ash cloud 33,000 feet (10,000 meters) into the air. Within a couple of days, the black ash turned white, like dirty snow. MB

LAKE NATRON

ARUSHA, TANZANIA

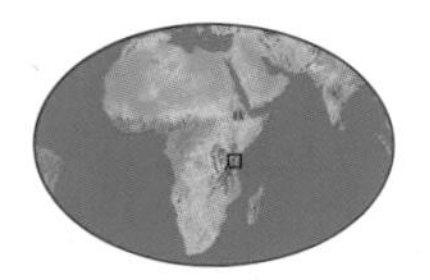

Length of Lake Natron: 35 mi (56 km)

Width of lake: 15 mi (24 km)

Height of Gelai Volcano: 9,652 ft (2,942 m) on southeast edge

At the southern end of a string of soda-lakes in East Africa is Lake Natron. The soda is sodium carbonate; otherwise known as "washing soda" that comes from local volcanoes. The lake is fed from streams that wash the soda from alkaline soils, and it becomes so concentrated that the lakes can be a death-trap for most living things, the exception being flamingoes. Hundreds of thousands of them build their nests on elevated mud-mounds on shallow banks. The soda-encrusted shore is a deterrent for predators such as jackals and hyenas. If they tried to approach the breeding sites, the soda would burn their legs. This happened to ornithologist Leslie Brown in the 1950s when he wanted to study the flamingoes at Lake Natron. He started to trek the 7 miles (11 kilometers) from the lakeside to the breeding site but sank into the caustic mud. His drinking water became contaminated with soda dust and in the intense heat he barely made it back to camp. He collapsed and was semi-conscious for three days, his legs black and blistered from the soda. He was hospitalized for six weeks during which numerous skin grafts saved his legs and his life. MB

NGORONGORO CRATER

ARUSHA, TANZANIA

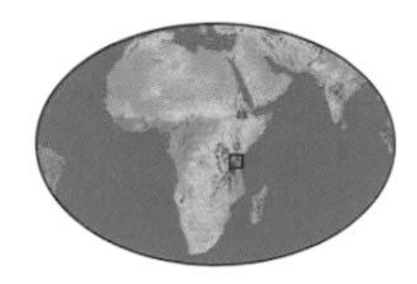

Area of Ngorongoro Crater: 100 sq mi (264 sq km)

Age of crater: 2.5 million years

Ngorongoro—meaning "big hole" to the Maasai people—is a huge geological crater with one of the highest concentrations of wildlife on the entire continent. Covering 100 square miles (264 square kilometers), it contains 50 different species of large mammal, including thousands of wildebeest, zebras, elephants, and lions, and 200 species of bird, from ostriches to ducks. It is a natural paradise that was formed 2.5 million years ago when the volcano erupted for the last time and its top sank into its crater. Round Table Hill in the northwest is the only remnant of the ancient cone, and the unbroken rim of the caldera (crater) is the largest in the world. Today, it is a living laboratory where scientists from all over the world come to study the relationships between predators and prey, as well as those between genetic isolation and inbreeding. Unlike animals outside the crater, Ngorongoro's herds do not migrate, though they show a preference for the open plain in the wet season and the marshland areas of Munge Swamp in the dry season. There is water and vegetation throughout the year, so they have no need to move out. This makes Ngorongoro a microcosm of East African wildlife. MB

SERENGETI

MARA / ARUSHA / SHINYANGA, TANZANIA

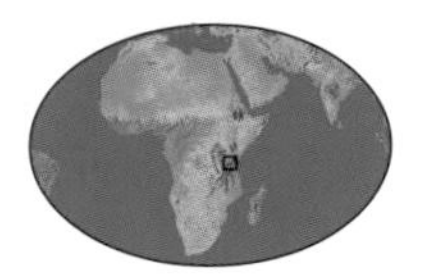

Area of Serengeti: 5,700 sq mi (14,631 sq km)

Elevation of Serengeti: 2,950 to 6,070 ft (920 to 1,850 m)

Rainy seasons: March to May, October to November

In 1913, on an expedition south from Nairobi, Stewart Edward White noted the following: "We walked for miles over burnt out country ... Then I saw the green trees of the river, walked two miles more and found myself in paradise." This depiction of the legendary Serengeti today describes what is arguably the best-known wildlife sanctuary in the world.

The Serengeti National Park covers 5,700 square miles (14,631 square kilometers) of endless rolling plains, which, together with the Ngorongoro Conservation Area and Maasai Mara Game Reserve across the border in Kenya, host the greatest and most varied populations of wildlife on Earth. The Maasai people call it *Siringitu*, meaning "the place where the land goes on forever."

The unique combination of diverse habitats supports a magnificent wealth of wildlife—an estimated three million large animals roam the plains. The Serengeti also boasts large herds of antelope including klipspringer, Patterson's eland, dikdik, topi, gazelle, and impala. Larger animals like rhino, elephant, giraffe, and hippopotamus abound, as well as predators such as lion, cheetah, leopard, and hyena. Nearly 500 species of bird have also been recorded in the park.

One of the most dramatic spectacles on Earth has to be the annual migration. Hundreds of thousands of wildebeest and Burchell's zebra traverse the vast plains following the rains in search of pasture. Once started, nothing stops the stampede, not the predators, nor even the broad Mara River, where hundreds of wildebeest and zebra drown or fall prey to crocodiles.

After early professional hunters decimated the lion population, the area was made a game reserve in 1921 and was upgraded to national park status in 1951. No humans inhabit the national park itself, although Maasai herders still live on the eastern periphery, and a farming population is rapidly expanding on the western periphery. Poaching has been a major problem, and efforts have been made to alleviate this by ensuring that local people are involved with, and can benefit from, the management of the park. Conservation efforts, including locally administered buffer zones, have largely paid off, and the Serengeti is thriving. But drought, overgrazing, and disease easily affect the fragile ecosystem, and careful protection is needed to maintain this wildlife haven. MM

One of the most dramatic spectacles on Earth has to be the annual migration. Hundreds of thousands of wildebeest and Burchell's zebra traverse the vast plains following the rains in search of pasture.

RIGHT *A herd of wildebeest on the plains of the Serengeti.*

USAMBARA MOUNTAINS

TANGA, TANZANIA

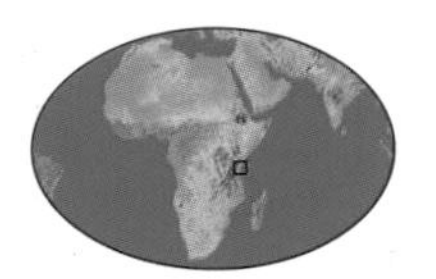

Maximum elevation of Usambara Mountains: 4,936 ft (1,505 m)

Habitat: lowland moist forest, cloud forest, elfin forest, specialist exposed rock communities, tropical moorland

Rock type: ancient crystalline igneous

The Usambaras form part of an ancient 100-million-year-old mountain chain known as the Eastern Arc. They have long been isolated from other forest areas by surrounding savanna. Evolving in isolation, the fauna and flora have a high species diversity unique to the mountains. Some of the richest habitats in Africa can be found here. This is aided by the mountains' proximity to the sea and its moisture-bearing winds. The steepness of the mountains means they receive some 79 inches (2,000 millimeters) of rain annually. In the past, this has left the Usambaras forested, even when continental climate changes dried out other African forests. The forests have evolved in a stable environment for some 30 million years. The wealth of unique species means the Usambaras are considered as important for studying evolution as the Galapagos Islands. Closely related but now quite distinct species live on either side of the rocky crags. Around 350 species of birds are regional specialities. In total, 2,855 plant species are known from the Usambaras, with 25 percent of them specific to the area. Many international conservation organizations are working to conserve the Usambaras. **AB**

MAMBILIMA FALLS—LUAPULA RIVER

ZAMBIA / DEMOCRATIC REPUBLIC OF THE CONGO

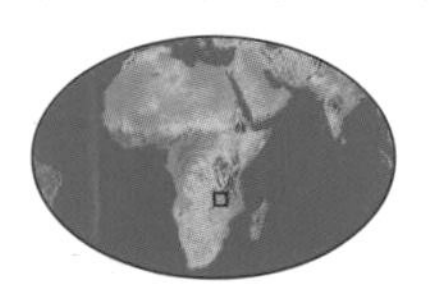

Elevation of Lake Mweru: 3,050 ft (930 m)

Surface area of lake: 1,795 sq mi (4,650 sq km)

As the Luapula River flows along the border between northern Zambia and the Democratic Republic of the Congo, it tumbles sleepily over a long, winding stretch of step-like terrain which turns the flow into a glistening silver ribbon about three miles (five kilometers) long. Known as Mambilima Falls—but perhaps more accurately, rapids or chutes—the falls serve as a transition in altitude and also mark an ecological spectrum along the Luapula.

The river begins its slow journey above the falls in one of the largest wetland areas in the world—the waterlogged Bangweulu Swamps—and then, after sliding down the Mambilima chutes, enters the lush Luapula River Valley. There, it spreads into shallow floodplains, marshes, and permanent lagoons, before ultimately pouring into Lake Mweru, a body of water shared with the Congo. At least 90 fish species as well as antelopes, hippos, crocodiles, zebras, and the globally threatened wattled crane call the river home. **DBB**

LUANGWA VALLEY

EASTERN PROVINCE / NORTHERN PROVINCE, ZAMBIA

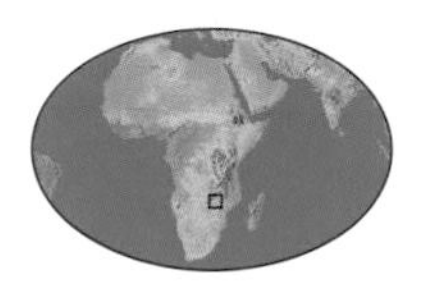

Area of North Luangwa National Park: 1,790 sq mi (4,636 sq km)

Area of South Luangwa National Park: 3,500 sq mi (9,065 sq km)

Elevation of Mchinga escarpment: 3,575 ft (1,100 m)

The Luangwa is situated at the end of the Great Rift Valley in a rift valley of its own. The valley is fairly shallow and flat-floored and the river flows remarkably slowly, giving rise to typical slow-flowing river features such as sweeping meanders, oxbow lakes, and lagoons. These landforms are important living places for game and birds—particularly during the wet season—creating one of Africa's major wildlife areas, with hippo, elephant, white impala, Thornicroft's giraffe, greater kudu, Cookson's wildebeest, buffalo, zebra, lion, leopard, and hyena. The two national parks in Luangwa are renowned for their walking safaris. The South Luangwa Park is Zambia's most famous wildlife sanctuary. It covers a vast 3,500 square miles (9,065 square kilometers) and is dominated by a floodplain and savanna that extends from the Luangwa River to the Muchinga escarpment, rising over 2,625 feet (800 meters) from the valley floor in the west. The North Luangwa Park—about half the size of its counterpart—is an undeveloped and entry-restricted gem, renowned for its buffalo herds and lions. For four decades the park has remained off-limits to the public; only two operators have access. PG

KAFUE FLATS

SOUTHERN PROVINCE, ZAMBIA

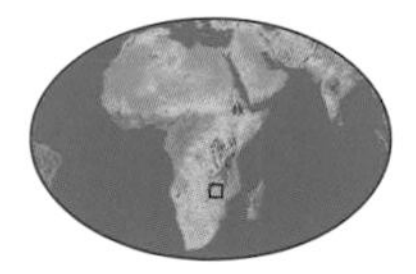

Area of Kafue Flats: 2,510 sq mi (6,500 sq km)

Average elevation of flats: 3,215 ft (980 m)

Vegetation: marsh, floodplain, grassland, hot springs, woodland

The Kafue Flats lie between the Itezhitezhi and Kafue Gorges, where the Kafue River rushes seaward. In contrast, the gradient on the flats is so low that it has been estimated it takes three months for water to move from one end to the other. Unfortunately, the present situation is complicated by the presence of hydroelectric power (HEP) systems in both gorges, which are of considerable economic importance but which alter the flooding ecology of the flats. Although the Kafue Flats have lost most of their big game, they are still incredibly important to hundreds of thousands of birds, including the endangered wattled crane. This area is one of the most important wetlands for this impressive bird. There are two areas of open water—the Blue Lagoon and Lochinvar Lake—which add diversity to the region. In summer, thousands of waders and waterfowl arrive to feed on the seasonally flooded areas. The flats are also home to 40,000 Kafue lechwes—a unique species of antelope found nowhere else in Africa. They are protected, but are threatened by the unpredictable flooding caused by the HEP scheme. This releases water according to the need for electricity, not according to the rhythms of nature. PG

VICTORIA FALLS

ZAMBIA / ZIMBABWE

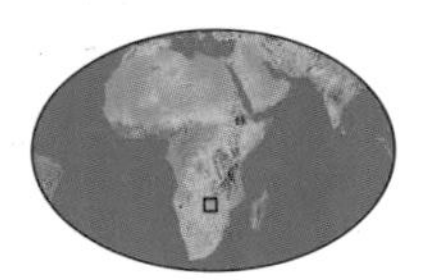

Height of Victoria Falls: 354 ft (108 m)

Water flow: 120 million gallons (550 million liters) per minute

Age: 200 million years

Known locally as *mosi-oa-tunya*, meaning "the smoke that thunders," Victoria Falls can be seen and heard from a great distance. Plumes of mist rise 1,650 feet (500 meters) into the air and the sound of 120 million gallons (550 million liters) of water per minute dropping 354 feet (108 meters)—about twice the height of Niagara Falls—is deafening. The water starts more peacefully some 750 miles (1,200 kilometers) away, in the upper reaches of the Zambezi River, Africa's fourth largest river. But by the time it reaches the falls, it is a massive 1 mile (1.6 kilometers) wide and scattered with small islands. Then, it plunges as a great sheet of water—the world's widest uninterrupted waterfall—into a deep chasm just 200 feet (60 meters) wide.

The rocks here were formed from a great sheet of basaltic lava over 200 million years ago. They cooled and cracked, and the fissures were filled with softer sediment. About half a million years ago, the Zambezi began to flow, and eroded one of the cracks, and so the first gorge was created into which the river fell.

The Victoria Falls of today are the eighth to have been formed in the crisscross of cracks in the volcanic rock. The remains of the rest can be seen in the series of gorges below the present falls through which the river zigzags and the erosion continues. Moving upstream at a rate of 1 mile (1.6 kilometers) every 10,000 years, it is thought that the ninth set of falls will start to form at the Devil's Cataract at the western end of the current falls. The tranquil lagoons upstream are home to some of Africa's largest river-dwelling creatures, such as hippos and crocodiles, as well as numerous birds, including herons, egrets and African fish eagles. Walking along the paths through the spray-generated rainforests, visitors may even spot elephants, buffalo, and lions.

The Scottish missionary Dr. David Livingstone first saw Victoria Falls in 1855 when exploring the Zambezi. Approaching from upstream in a canoe, he and his party saw the clouds of spray in the distance and wisely landed on an island at the edge of the waterfall. They crept to the edge of the chasm and were astonished to see this huge river simply disappear into a crack in the Earth. He named the falls after Queen Victoria. MB

Known locally as mosi-oa-tunya, *meaning "the smoke that thunders," Victoria Falls can be seen and heard from a great distance. Plumes of mist rise 1,650 feet (500 meters) into the air and the sound is deafening.*

RIGHT *The calm before the storm of the magnificent Victoria Falls.*

BAZARUTO ARCHIPELAGO NATIONAL PARK

INHAMBANE, MOZAMBIQUE

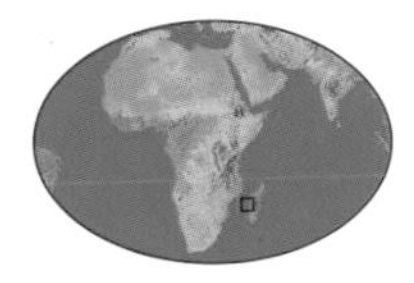

Established as a national park: 1971

Length of Bazaruto Island: 22 mi (35 km)

Tidal rise and fall: 33 ft (10 m)

Described as an ecological jewel, the Bazaruto Archipelago is a narrow chain of islands dominated by a ridge of sand dunes and woodlands, inhabited by samango monkeys, bushbabies, red duiker, and elephant shrews. Pristine seagrass beds host the largest remaining population of dugongs in East Africa. Extensive tidal flats and inland saline lakes attract thousands of seabirds and waders, some on migration from the north, as well as flamingoes. Beaches are of crystal-white sand, and spectacular coral reefs contain 100 hard and 27 soft species of coral. They are home to 2,000 known species of fish, and visited by whales, dolphins, and all five species of sea turtle that live in the western part of the Indian Ocean.

Three of the larger islands—Bazaruto, Benguerra, and Magaruque—were once part of a sand spit attached to the mainland, the result of sediments brought down to the sea by the Limpopo River. Only Santa Carolina is a true rocky island surrounded by deep water. The archipelago is within the Mozambique Channel, about 15 miles (24 kilometers) offshore from Inhassoro and 130 miles (210 kilometers) south of Beira. MB

MANA POOLS NATIONAL PARK

MASHONALAND WEST, ZIMBABWE

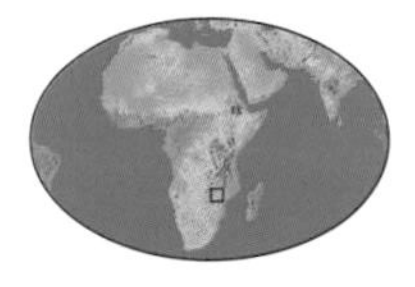

Age of Mana Pools: Tertiary to Quaternary

Size: 25 mi (40 km) long—variable width

Vegetation: woodland with apple-ring thorn trees

On the south bank of the Zambezi, where the rift valley widens out downstream from the Kariba Gorges, lies a relatively flat area of river deposits. These flat river terraces lie a few feet above the normal river level and are formed during peak floods when they are covered by silt-filled floodwater. As the floods recede, water is left in every hollow. Some are deep and large enough to hold water throughout the long dry season. Others are more transient and disappear fairly quickly. All of them attract animals that come to drink and to wallow in mud. The animals roll in the mud and carry it away, enlarging the hollows so they are able to hold water for longer.

The pools' fertile silt encourages plant growth, especially species of trees such as the Ana tree, or the apple-ring thorn tree that produces huge numbers of brown curly seed pods—a favorite of elephant and antelope. The browsing animals shun full-grown evergreen Natal mahogany, but the high density of plant eaters means that the seedlings are eaten and there is no new tree growth. PG

BELOW *A herd of elephants crosses through the mud of the Mana Pools.*

EASTERN HIGHLANDS

MANICALAND, ZIMBABWE

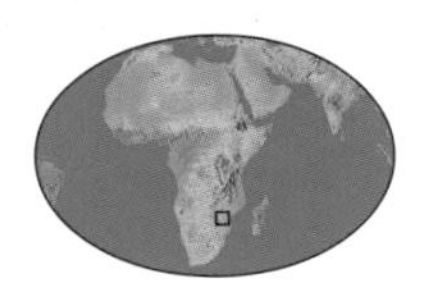

Age of Eastern Highlands: ancient rocks, relatively modern erosion forms

Vegetation: rainforest, woodlands, montane scrub

Running along the border between Zimbabwe and Mozambique lies a range of ancient granite mountains that faces the prevailing onshore winds. On the east-facing slopes, the higher rainfall has resulted in particularly tall trees, with an interlocking canopy as high as 100 feet (30 meters) above the ground. These subtropical rainforests grow along the many streams and rivers, at an altitude of 1,000 feet (300 meters) to over 5,300 feet (1,600 meters). During the ice ages these isolated pockets of forest were more or less contiguous with similar forests to the south in South Africa, and to the north at least as far as Tanzania and Kenya. This allowed forest animals to move throughout the whole region. As Earth warmed and the ice melted, the forests retreated from all but the wettest areas. Pockets of forest were left but the wildlife became isolated, unable to move to other patches of forest. This has resulted in some species occurring in two widely separated areas. Swynnerton's robin, for example, is found in the Eastern Highlands, while another subspecies occurs in isolated mountains in Tanzania. In some cases, the isolation occurred so long ago that new species have developed, such as the chirinda apalis and the black-headed apalis. PG

BRACHYSTEGIA WOODLANDS

MANICALAND / MIDLANDS PROVINCE / MATABELELAND, ZIMBABWE

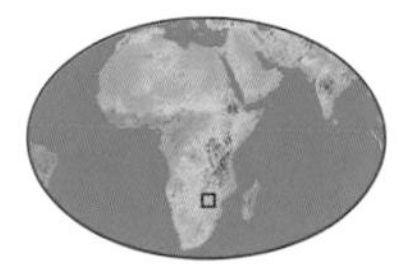

Height of Brachystegia trees: up to 50 ft (15 m)

Bird species: mashona hyliota, spotted creeper, purple-headed sunbird, black-cheeked canary, fork-tailed drango, African goshawk

Extending across central Africa from Angola to Mozambique lies a broad belt of semi-deciduous Miombo forest consisting mainly of Brachystegia woodlands. The Brachystegia species is somewhat unusual; the trees produce red to olive leaves not in fall but in spring. The theory is that the soft new red leaves lack chlorophyll, so are distasteful to browsers. They are left to mature, and as they harden produce the chlorophyll that eventually turns them green. The trees keep their leaves until the end of winter and then lose them all within a week or two. Instead of remaining dormant for the winter months, however, they almost immediately sprout new foliage and the woodlands turn bright red and khaki, the intensity of color varying from year to year.

Not only do these woodlands have many unique tree species but they are also home to a range of special birds that are exclusive to these woodlands, such as the Miombo gray tit and Miombo pipit.

Although it is quite difficult to view game through the foliage of the Brachystegia trees, occasionally visitors might see endangered black rhino or African buffalo. Specialized grazers such as sable antelope, southern reedbuck, eland, and greater kudu may be spotted feeding on grass. PG

SAVE VALLEY

MANICALAND, ZIMBABWE

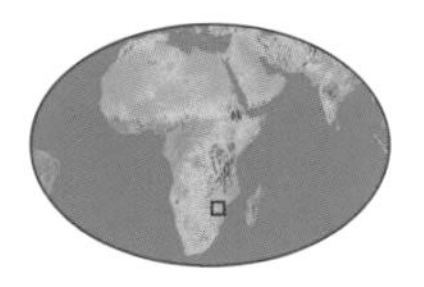

Area of Save Valley: 100 mi (160 km) long, and 6 to 28 mi (10 to 40 km) wide

Age of valley: Tertiary onward—modern alluvium

Vegetation: lowveld woodland, bush, trees

The Save River drains the southeastern side of the Zimbabwean Plateau. The plateau is over 4,000 feet (1,200 meters) above sea level and drops quite steeply to the Save River floodplain below. The decrease in gradient causes the river to drop its load of sand eroded from the granites of the plateau upstream. As a result, the river wanders across a wide flat-floored valley, breaking up into a braided pattern of many shallow channels. Summer storms on the highveld, however, cause floods in the valley below and the channels continually change their course. In times past, the river has deposited huge quantities of alluvium across the valley with numerous hollows, many of them ancient oxbow lakes that are dry unless filled by a heavy rain. The alluvium is fertile and in areas where the river no longer flows the land has been colonized by a unique blend of trees, bushes, and palatable grasses. This attracts large numbers of browsers such as kudu and giraffe, and grazers such as buffalo, zebra, and impala. They, in turn, are prey for predators—lions, cheetahs, and leopards. The variety of vegetation also attracts birds, creating a prime game and wild bird viewing area. **PG**

TAMBOHARTA PAN

MANICALAND / MASVINGO, ZIMBABWE

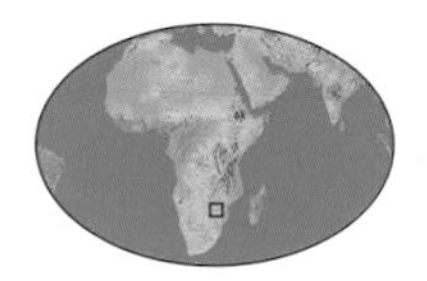

Width of Tamboharta Pan: approx. 1.2 mi (2 km)

Age of pan: Quaternary (from 2 to 3 million years ago to present)

Vegetation: lowveld woodlands

Lying just above the junction of the Save and Runde rivers is a unique pan, a natural shallow depression that fills up with water during the wet season (it evaporates during the dry season). Tamboharta, however, is unusual. Local rains provide some water but have little impact on the life of the pan. It is very large and only fills when the Runde River floods and overflows. Once filled, however, the pan may retain water for several years, triggering an unusual cycle of plant growth as one group of species develops, flowers, and dies, giving way to other species.

During the wet season, the pan is home to thousands of waterbirds who flock to this watery wilderness from dryer lands. The African fish eagle is one of the most impressive visitors. This king of the sky swoops across the water, catching fish as they near the surface. Smaller birds take to the air in panic, in a flurry of flapping winds.

On higher ground, dense woodlands thrive on the fertile edges of the pan. Huge baobabs dominate the landscape, towering like guards above the smaller trees and shrubs. **PG**

CHILOJO CLIFFS

MASVINGO, ZIMBABWE

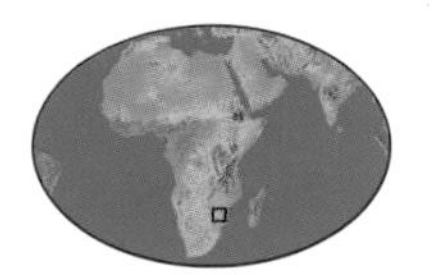

Dimensions of Chilojo Cliffs: approx. 400 ft (120 m) high, 3 mi (5 km) long

Age of cliffs: Tertiary to Quaternary

Vegetation: mopane

When the setting sun starts to redden in the west, the Chilojo Cliffs in Gonarezhou National Park in southeastern Zimbabwe are a spectacular sight. The red sandstone takes on a life of its own, blazing in the evening light. During the day, the cliffs are subdued oranges and dusty pinks, but as the sun sets these colors warm to glowing reds and burnt oranges. At twilight, recesses within the cliffs become dark and almost threatening.

Below the cliffs, the Runde River blazes gold and then fades into blue and black as the sun retires—contrasting with the sunlit cliffs above. The cliffs were formed by the Runde River carving through a low plateau of sandstone. The resistant sandstone was undercut by the river and pieces collapsed onto the bank below leaving sheer cliffs rising over 330 feet (100 meters) above the valley floor. These cliffs appear various shades of red or orange-yellow and are quite unlike anything found in the surrounding area of the park. The area is relatively dry and the water of the Runde rises far inland, just before the river joins the Save River at the lowest point in Zimbabwe. **PG**

EXFOLIATED DOMES

MATABELELAND SOUTH / BULAWAYO, ZIMBABWE

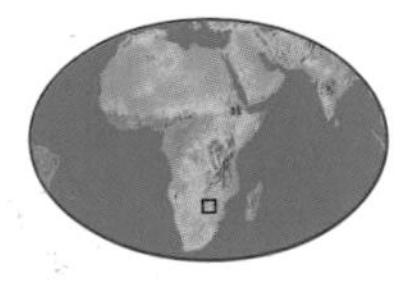

Age of Exfoliated Domes: between 2 and 3.5 billion years

Height of domes: variable—up to 1,000 ft (300 m)

Vegetation: miombo (Brachystegia) woodland

The surface rocks covering nearly half of Zimbabwe are extremely old granites. These are intrusive molten igneous rocks that pushed up through the other strata between 2 and 3.5 billion years ago. As the molten rock cooled some areas developed irregularly spaced cooling cracks in three directions, known as "cubic cracks." Other areas had few or no cooling cracks. The erosion of the overlying sedimentary rocks produced two very different landforms.

A process known as exfoliation weathered the granite into smooth, crack-free domes. These rounded domes of granite tower 1,650 feet (500 meters) above the surrounding flat countryside. The rock becomes very hot in the direct sun and expands outward to form concentric cracks. At night and after heavy rains the domes cool, splitting curved pieces of rock from the radiating cracks. Rainwater runs off these smooth granite domes, providing ideal conditions for plant growth. Tall stands of tree thrive in these areas, forming a circle around the domes' perimeter. **PG**

RIGHT *The smooth, rounded mounds of the Exfoliated Domes rise above the flat, grassy plains of Zimbabwe.*

BALANCING ROCKS

MATABELELAND SOUTH / BULAWAYO, ZIMBABWE

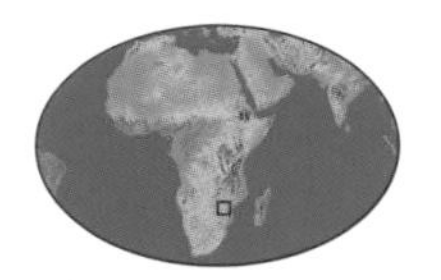

Age of balancing rocks: Precambrian granites (over 600 million years old)

Location of Epworth balancing rocks: 7 mi (11.2 km) from Harare

Vegetation: miombo (Brachystegia) woodland

One of the most striking features of the Zimbabwean landscape is the "balancing rocks." Millions of years ago intrusions occurred deep within Earth—ancient granites were pushed through other rock, as molten rock or extremely hot solutions of mineralized water. Typically, granite areas develop cracks when cooling; in this case they cracked in three dimensions. These cracks provided lines of weakness where rainwater could seep into the rock and cause weathering—carving the rock deep below the surface. Weathering of the rock along these cracks and erosion of the overlying soils resulted in layers of rock peeling off. This left rounded resistant blocks known as *kopjes* or "balancing rocks."

Scattered throughout most of central Zimbabwe, these rocks come in an amazing variety of shapes and sizes, many appearing to defy gravity. Particularly good examples can be seen in Gosho Park, an Environmental Education Project located at Peterhouse near Marondera. Other unusual rock formations can be found in Mutirikwi (formerly Kyle) National Park near Masvingo. The balancing rocks at Epworth (South Harare) actually appear on the back of Zimbabwean banknotes. PG

MATOBO HILLS

BULAWAYO, ZIMBABWE

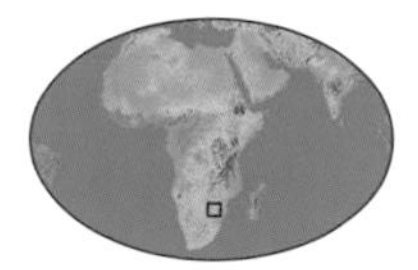

Length of Matobo Hills: 50 mi (80 km)

Average height of hills: 5,000 ft (1,500 m)

Rock type: granite

Hundreds of millions of years ago, molten lava pushed up, cooled, cracked, and weathered to form the unusual granite hills of Matobo in southern Zimbawe. Here, granite blocks stand one upon the other like giant statues. In the 19th century, the Matabele chief Mzilikazi called them *ama tobo* (meaning "bald heads"), for they reminded him of his tribal elders. Mzilikazi is buried here, but the Matabele were not the first indigenous people to live in these parts. Up to 2,000 years ago, the San people lived in caves here, their legacy a natural gallery of rock art. They painted animals, landscapes, and people using colored clays mixed with animal fat and the sap of euphorbia. There is even a delicate picture of a winged termite. Cecil Rhodes, founder of Rhodesia (now the republics of Zambia and Zimbabwe), was also struck by this place. His favorite spot was what he called "View of the World," but what the Matabele knew as *malindidzimu*, meaning "place of the ancient spirits." Rhodes was buried here in 1902, in a tomb hewn out of granite. On his deathbed he bequeathed the hills to the people of Matobo, so they could all enjoy it, but only "from Saturday to Monday." MB

ETOSHA PAN

OMUSATI / OSHANA, NAMIBIA

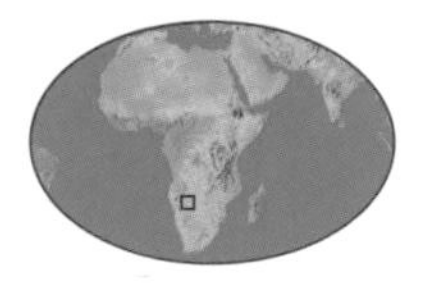

Length of Etosha Pan: 80 mi (130 km)
Width of pan: 30 mi (50 km)

In northern Namibia lies "the place of dry water." Etosha Pan is a dry lakebed of salt-coated clay about 80 miles (130 kilometers) long and 30 miles (50 kilometers) wide, crisscrossed by the trails of game animals and whipped up by dust devils. It is one of several pans and dry lakes that together with Botswana's Okavango Delta once formed what would have been the largest lake in the world. The rivers that fed this enormous body of water dried up, and evaporation in the hot sun ensured that the lake eventually disappeared. Yet despite the harsh conditions there are an enormous number of animals living here—herds of wildebeest, zebra, springbok, and gazelle, stalked by lions and hyenas. Tens of thousands move into the area each year in what is one of Africa's greatest migrations. They come from their dry season refuge on the Adonis Plains to the northeast, after rains that start in December reawaken the land with a sudden growth of lush grass. Etosha itself is turned into a vast shallow lake, a magnet for waterbirds. The American hunter and trader Gerald McKiernan wrote in 1876, "All the menageries in the world turned loose would not compare to the sight I saw that day." MB

CAPE CROSS SEAL RESERVE

KUNENE, NAMIBIA

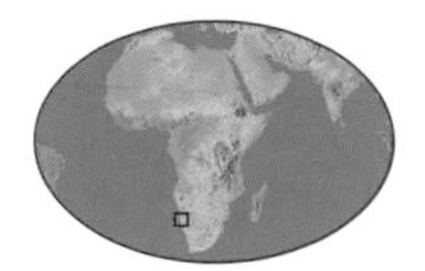

Area of Cape Cross Seal Reserve: 23 sq mi (60 sq km)

Cape fur seals: male: 7.5 ft (2.3 m) long, weighing 793.5 lb (360 kg) female: 5.5 ft (1.7 m) long, weighing 242.5 lb (110 kg)

On the Atlantic coast of northwest Namibia about 100,000 Cape fur seals haul out onto the shore each year to breed. They represent about one fifth of the world's total population of this species. The bulls begin to arrive in mid-October and fight each other for the best territory. The females follow later, triggering further battles between the bulls.

Pups are born from late-February through April. The females spend part of their time in the rookery, suckling their pups, and part at sea, where they hunt for fish and squid. They swim up to 112 miles (180 kilometers) offshore and dive down to depths of 1,320 feet (400 meters). On the beach the pups are vulnerable to attack by jackals and brown hyenas, and when they reach the sea, sharks and killer whales are yet another threat.

In 1485, the Portuguese captain and navigator Diego Cao landed here and was the first European to set foot this far south in Africa. He is buried at a nearby outcrop called Serra Parda. A stone cross to commemorate the landing was erected on a headland at the time but it was unfortunately stolen in the 19th century. A replica was later created and erected in 1974. **MB**

SKELETON COAST

KUNENE, NAMIBIA

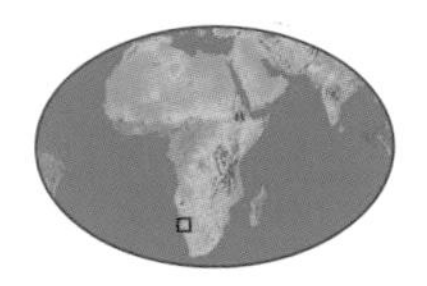

Length of Skeleton Coast: 300 mi (500 km)

Notable feature: shipwrecks

The Skeleton Coast is a long shoreline sandwiched between the Atlantic Ocean and the Namib Desert. It is bathed by the cold Benguela Current that sweeps icy water from the Antarctic northward along southern Africa's west coast. Offshore winds cause upwellings of nutrients from the deep sea, providing a wealth of food for unexpected numbers of animals. Fur seals and seabirds breed here, next to waters crammed with shoals of anchovies, sardines, and mullet. Brown hyenas visit the white sand beaches, and lions have been seen to scavenge on whale carcasses washed in on the tide. But it is the extraordinary number of wrecks for which the coast is famous. The coast is strewn with ocean liners, galleons, and clipper ships, victims of tumultuous seas, thick fogs, and jagged reefs. An anonymous "help" message on a piece of slate dated 1860, found beside 12 headless skeletons in 1943, is just one of many mysteries that haunt this shoreline. The popular name, however, arose when a Swiss pilot, on a flight from Cape Town to London, crashed here and a journalist suggested his bones would be somewhere on the "skeleton coast." The pilot was never found but the name stuck. **MB**

NAMIB DESERT

NAMIBIA

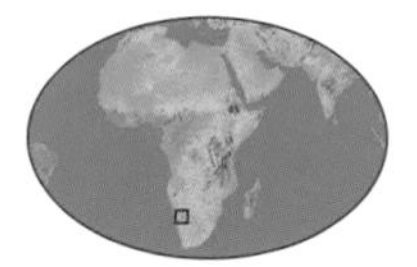

Coastal temperature: 50 to 60°F (10 to 16°C)

Inland temperature: 80°F (27°C)

Annual rainfall: 0.5 in (13 mm) at coast, 2 in (50 mm) at escarpment

Inland from Namibia's Skeleton Coast is the ancient Namib Desert. It stretches south from southern Angola to the Orange River in South Africa and extends east to the foot of the Great Escarpment of southern Africa. In the north, the bedrock has been carved by rivers into steep-sided gorges, while the south is covered with sand, yellow-gray at the coast and brick red inland. Parallel lines of sand dunes, running northwest to southeast, have individual dunes 20 miles (32 kilometers) long and up to 800 feet (244 meters) high. In the north, streams of sand reach the sea, but elsewhere they end in a *vlei*—a saltpan or mudflat—in the desert itself. For most of the year, water arrives on a southwest breeze in the form of thick fogs. However, dew is more important than rainfall for both plants and animals. The name Namib means "place where there is nothing" in the language of the Nama people, but there is actually plenty of wildlife. Plants include the *Welwitschia*, which produces just two enormously long leaves, and animals include gemsbok and springbok. The dune fields have few mammals but are home to a range of beetles, geckos, and snakes, all specialized to survive in these conditions. **MB**

BULL'S PARTY ROCKS

ERONGO, NAMIBIA

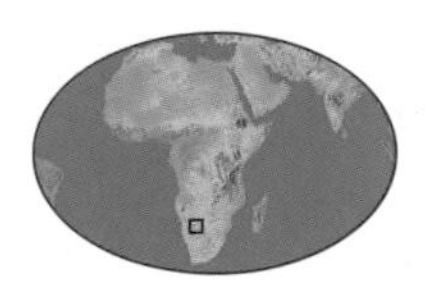

Height of Elephant's Head boulder: 52 ft (16 m)

Notable feature: San rock paintings

Bull's Party Rocks is a fascinating rock formation found on the Ameib Ranch on the southern edge of the Erongo Mountains. It is a collection of huge, rounded granite boulders that resemble a group of bulls facing each other. Nearby are several precariously balanced and mushroom-shaped boulders, and another unusually shaped rock outcrop that has been named the Elephant's Head.

As well as the rock formations, the Ameib Ranch is also well known for its San rock paintings. Phillip's Cave (an overhang) contains a superb painting of a white elephant, along with paintings of giraffe, ostrich, zebra, and humans. The paintings were described and first made famous in the book *Philipp Cave* by Abbe Henri Breuil.

The Erongo Mountains are approximately 25 miles (40 kilometers) north of Karibib and Usakos, and are the remains of a massive ancient volcano in central Namibia. Most of the mountain range is only accessible by four-wheel-drive vehicles, but it is only possible to reach Phillip's Cave on foot over a series of low hills. The Bull's Party Rocks are about three miles (five kilometers) from the main ranch. RC

MOON VALLEY

ERONGO, NAMIBIA

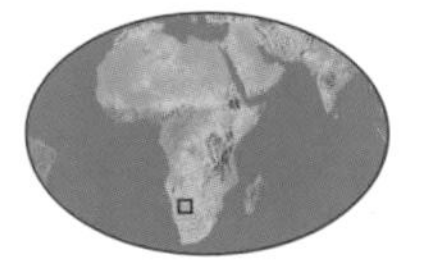

Age of Moon Valley: 450 million years

Age of Giant *Welwitschia* of the Husab: 1,500 years old

The Moon Valley (also known as the Ugab Rock Formations, Moon Landscape or Valley of the Moon), in the Swakop River Valley area of the Namib Desert, is an eerie landscape weathered by coastal fog and wind into bizarre rock formations. The Swakop River is one of the longest and broadest of Namibia's ephemeral watercourses. In the north, the eroded landscape consists of red lavas and yellow sandstones, whereas farther south the pale desert floor has contrasting dark ridges. These dark ridges (common in this part of the Namib) were formed by lava welling up through cracks in the rocks and then cooling into hard ridges that weathered proud of the surrounding granite.

The extraordinary *Welwitschia* plant is found in the nearby gravel plains and is unique to the Namib Desert. The plants have two large sprawling leathery gray leaves, usually torn into long ribbons by the wind, and a very deep taproot. It is thought that they may be able to live for up to 2,000 years. Moon Valley is located near to Swakopmund—a permit is required to visit the area and the *Welwitschias* (the same permit is valid for both sites). RC

BRANDBERG

ERONGO, NAMIBIA

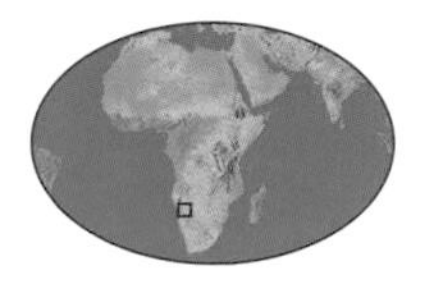

Namibia's highest mountain (Königstein): 8,858 ft (2,700 m)

Rock type: granite

Namibia's Brandberg is a great granite massif that rises to nearly 8,858 feet (2,700 meters) at the Königstein ("King's Stone" in German), the country's highest point. The mountain is nearly circular, with a radius of about 19 miles (30 kilometers), and consists of a wilderness of huge boulders and dramatic cliffs. Various gorges offer routes into this formidable world but there are no natural springs on the mountain. The midday heat can be intense even in the winter months, and in summer temperatures regualarly exceed 105°F (40°C). With little shade or water much of the mountain is barren. But life has found a way in even these harsh conditions. The mountain is home to several hardy species of desert plant, such as aloe. The rich colors of the landscape revealed at sunset are believed to have given rise to the mountain's name, which means "Fire Mountain" in Afrikaans.

Brandberg is home to a remarkable wealth of rock art, including one of southern Africa's best-known paintings, the *White Lady of the Brandberg*. Accurately dating rock paintings is notoriously difficult but many of those on the Brandberg are thought to be well over 2,000 years old. HL

SPITZKOPPE

OTJOZONDJUPA, NAMIBIA

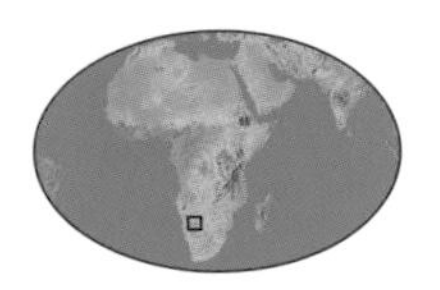

Age of Spitzkoppe: 150 million years

Rock type: granite inselberg

Notable feature: 1,804-ft (550-m) rock face on southwest

Namibia's Spitzkoppe is a remarkable granite inselberg or "island mountain," which appears to rise directly out of the gravel plains of southern Damaraland. It is one of the most photographed sites in the country. Sometimes known as Namibia's Matterhorn, the 5,577-foot (1,700-meter) peak reminded one early visitor of the tip of a giant elephant's tusk. From here one can experience some of the most beautiful sunsets and sunrises in Namibia, with the sun changing from a light orange to a dark yellow. The mountain itself is more than 700 million years old. Both Spitzkoppe and the Brandberg, which are 62 miles (100 kilometers) apart, were formed by the events leading to the separation of Gondwanaland into South America and Africa. Apparently the same vent, deep within Earth, caused these great upwellings of molten granite. It is strange to think that the Spitzkoppe was being formed underground as the surrounding landscape was gradually eroding away—the present peak rearing ever higher. Apart from the intrinsic beauty of the semi-desert surroundings, the Spitzkoppe region offers rock climbing, rock paintings, and precious and semi-precious stones. HL

NAUKLUFT NATURAL RESERVOIRS

HARDAP, NAMIBIA

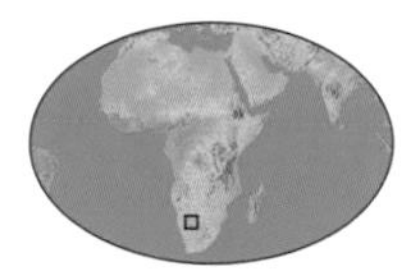

Age of Naukluft Natural Reservoirs: 2 to 4 million years

Depth of valley: up to 100 ft (30 m)

Length of valley: 1.5 mi (2.5 km)

As annual rains inundate the mountains along the northeastern corner of the Namib Desert in southwestern Namibia, the waters flow into Sesriem Canyon and renew the dried-out Tsauchab River, bringing life once again to one of the hottest and driest places on Earth. A lifeline into the Namib, this long fissure in a nub of sandstone connects the Naukluft Mountains to what is now called Namib–Naukluft National Park and features cup-like depressions and protected clefts which hold water deep enough to swim in. Lasting for weeks and even months after the distant mountain rains stop, these natural reservoirs provide desert denizens a relatively reliable source of water and sprout enough vegetation to support a surprising mix of animal life, from shovel-nosed lizards, sandgrouse, and red-necked falcons to ostriches, spear-horned oryxes, and mountain zebras.

An important source of water for early human inhabitants, the valley was frequented by Afrikaans-speaking settlers, who tied together six, or *ses*, lengths of rawhide rope, or *riem*, in order to lower buckets to collect drinking water, thereby giving the valley its name, Sesriem. DBB

SOSSUSVLEI & SESRIEM CANYON

HARDAP, NAMIBIA

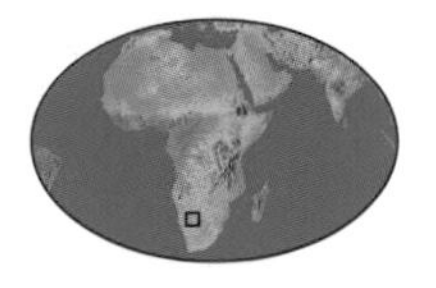

Area of Sossusvlei and Sesriem Canyon: 30 acres (12 ha)

Geology: Namib Desert clay pan

Notable feature: 1,000-ft- (300-m-) high shifting dunes

Along Namibia's coast, the cold waters of the Benguela Current skirt the Namib Desert. In the 62-mile- (100-kilometer-) wide region south of the port of Walvis Bay lie some of the highest mobile dunes in the world, towering over 1,000 feet (300 meters). With no permanent surface water, and practically no vegetation, this is an inhospitable region. The bed of the normally dry Tsauchab River forms one of the few viable entry points. Before entering the Namib desert, the Tsauchab forms the 1.2-mile- (2-kilometer-) long Sesriem Canyon. Farther west, the riverbed runs for about 30 miles (50 kilometers) deep into the heart of the Namib. It is flanked by spectacular dunes, best viewed as the rising sun emphasizes their flowing ocher curves. Sossusvlei is a clay pan, forming a cul-de-sac for the Tsauchab. Even after times of flood, the waters drain into the soil rather than evaporate, so that there is no excessive buildup of salt. The resulting sparse vegetation fringing the pans gives a home to the surprisingly rich fauna. **HL**

RIGHT *Sesriem Canyon in the Namib Desert, Namibia.*

THE FISH RIVER CANYON

HARDAP, NAMIBIA

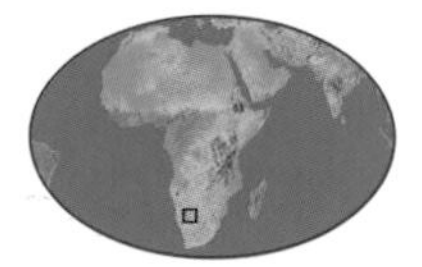

Length of Fish River Canyon: 40 mi (65 km)

Age: 500 million years

Namibia's Fish River Canyon is considered the world's second largest, after Arizona's Grand Canyon. There are, in fact, two canyons: The upper canyon was caused by geological faulting about 500 million years ago, while the lower was the result of erosion by the Fish River, which is still carving through the rock. The upper canyon is a wide north-south valley in which the Fish River flows in huge meanders. Viewed from above, the twists and turns remind one of a huge snake, and an early legend of the San people tells of a huge serpent slithering through the landscape creating the river valley in the process. The cliffs are at their most spectacular where the walls of the lower canyon cut in directly below those of the upper, giving a fall of some 2,000 feet (600 meters) from the plateau above to the 2.5-billion-year-old bedrocks below. The evening sun brings out colors ranging from a black band of limestone to a bewildering variety of tones and structures in the ancient basement rocks. Hot springs occur in the canyon bed, the best known being at *Ai-Ais* ("scalding"), where water emerges at 140°F (60°C). **HL**

KALAHARI DESERT

ANGOLA / BOTSWANA / NAMIBIA / SOUTH AFRICA / ZAMBIA / ZIMBABWE

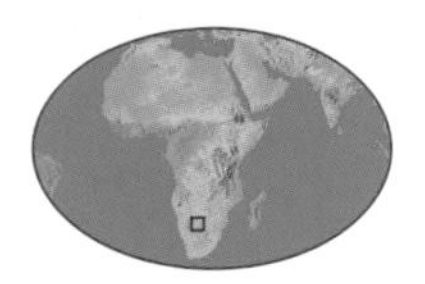

Summer temperature of the Kalahari: up to 122°F (50°C)

Annual rainfall: 16 to 18 in (406 to 458 mm) in the east, 12 to 14 in (305 to 356 mm) in the west

The greatest stretch of sand in the world is here in the Kalahari. It covers much of Botswana and reaches into Namibia, Angola, South Africa, Zambia, and Zimbabwe. Its underlying bedrock of molten lava was laid down about 65 million years ago. For the next 50 million years, it was eroded by wind and rain and buried in sand from the coast. Although the landscape is harsh and rugged, people came to live in the Kalahari 500,000 years ago. About 25,000 years ago, the nomadic San people started to live in the desert, and they are still here, the oldest human society in the world today. They share the desert with an abundance of wildlife, including meerkats, eagles, snakes, and gemsbok.

Huge numbers of springbok once migrated across the Kalahari; one herd is estimated to have been 130 miles (210 kilometers) long and 13 miles (21 kilometers) wide. Plants include the bushman's candle, a small succulent plant with a thick stem covered in spines and a cup-shaped flower. The bushman's candle gets its name because its stem is full of resin and burns with an aromatic smell. **MB**

BELOW *Sunset in the Kalahari Desert, Namibia.*

TSODILO HILLS

NGAMILAND, BOTSWANA

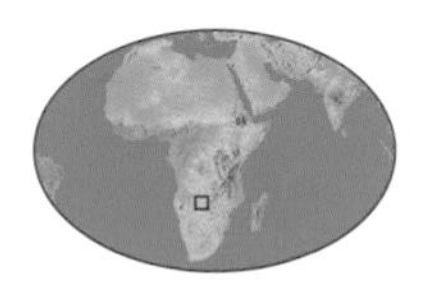

Height of southernmost hill: 4,577 ft (1,395 m)

Designated a World Heritage site: 2001

Situated in the northwestern corner of Botswana, the Tsodilo Hills became the country's first World Heritage site in 2001. When approached from the east, the hills rise sharply from the surrounding Kalahari bush, which provides a home to a wealth of African wildlife, including buffalo, springbok, and elephants. There are four separate hills, which are at their most dramatic when the sun is low in the sky in the morning or evening.

According to a legend of the local Kung people, the four hills once formed a family, and so are now known as Male, Female, Child, and Grandchild. The southernmost hill, Male, is also the highest point in Botswana at 4,577 feet (1,395 meters). Local communities in this harsh environment respect Tsodilo as a place of worship frequented by ancestral spirits.

The Tsodilo Hills boast one of the world's largest and most important collections of rock art, and are said to contain over 400 sites, with upward of 4,500 paintings, mostly dating to the period between 850 and 1100. A painting depicting what appears to be a whale is especially remarkable because the sea is 620 miles (1,000 kilometers) away. **HL**

OKAVANGO DELTA

NGAMILAND, BOTSWANA

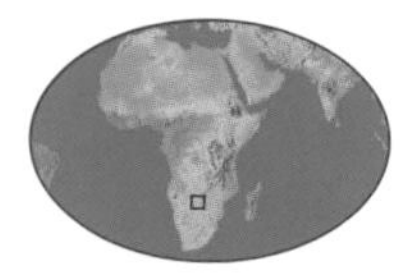

Area of Okavango Delta: 9,653 sq mi (25,000 sq km)

Vegetation: riverine papyrus, with woodland on islands

The Kavango River rises in the Angola Highlands and flows through Namibia before entering Botswana. Here it forms the Okavango Delta, the largest inland delta in the world. It is known as the "river that never finds the sea," because instead of flowing into the sea it flows slowly southward, spreading over extensive deposits of Kalahari sand. Eventually the water evaporates in Lake Ngami, the Boteti River, and parts of the Makgadikgadi Basin. When the river reaches the delta, it spreads outward through many channels and creates many shallow pans or depressions in the sand. The pattern of channels is ever-changing, as the river deposits silt and channels are blocked by vegetation; hippos move through the channels, keeping some open and re-opening others that have been closed.

The water and its nutrients enable plants such as papyrus to thrive; these in turn enable many animals and birds to live and breed in the delta in the middle of an otherwise dry habitat. Big game abounds and a great diversity of birds (over 400 species) provides a magnificent spectacle. However, the Okavango Delta is a fragile ecosystem that could easily be damaged or destroyed by dam-building upstream. PG

CHOBE RIVER

CHOBE, BOTSWANA / WESTERN STRIP, ZAMBIA

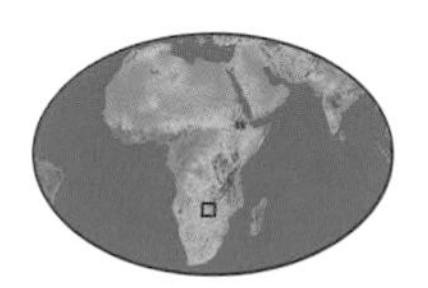

Chobe River vegetation: riverine woodlands, open grasslands, giant papyrus, reeds

Elephant population: 120,000 migrate to and from Chobe and other rivers

Southern Africa's largest river, the Zambezi, rises in Angola and flows southeast through Namibia's Caprivi region to join Botswana's Chobe River about 44 miles (70 kilometers) upstream from Victoria Falls. The summer rains falling in Angola usually reach the junction by April or May. As the Zambezi rises, it overwhelms the much smaller flow of the Chobe River, which is then pushed backward and appears to flow in reverse. The rivers are home to large herds of hippo and many giant Nile crocodiles. Female crocodiles may be seen protecting their nests on the banks of the Chobe. There are also large numbers of fish, of which the predatory tiger fish is the most voracious and probably the best known. When the waters recede, the relatively rare collared-pratincole is known to breed beside the rapids along the Chobe. The low-lying area between the rivers is mainly sediment deposited by the Zambezi—plants such as reeds have reclaimed these islands, but during major floods the rising waters inundate them. This forces the people and animals that live on the islands to move to ground outside the river valleys. The Chobe region has the highest concentration of elephants in Africa. PG

MAKGADIKGADI PANS

CENTRAL DISTRICT, BOTSWANA

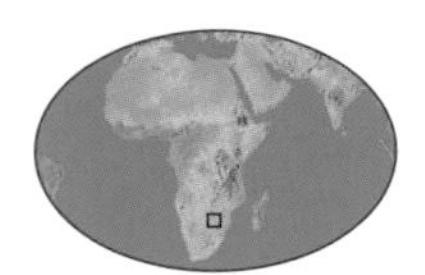

Area of Makgadikgadi Pans: 4,633 sq mi (12,000 sq km)

Area of Makgadikgadi Pans National Park: 1,892 sq mi (4,900 sq km)

Most of Botswana is covered by a mantle of wind-blown Kalahari sand that is over 330 feet (100 meters) thick in places. It forms the world's largest continuous sand surface covering 4,633 square miles (12,000 square kilometers). The sand was leveled by the wind, but as the climate became wetter, plants stabilized it. As a consequence, many areas have no slope, so water cannot run off. Where the Nata River (which rises in southwest Zimbabwe) runs into the Sua Pan, the river flows over soda-saline beds. When the water entering the area evaporates, this results in vast soda flats, such as the Makgadikgadi Pans between Maun and Nata. During the rainy season, the pans become a shallow lake, no more than three feet (0.9 meters) deep. As soon as they fill up with water, brine shrimp hatch by the million. This attracts an astonishing number of greater and lesser flamingoes (approximately 250,000) to the area to breed. When the birds are seen from a distance, they appear to turn the shoreline red in a spectacular sight. The surrounding grass cover, which is only three feet (0.9 meters) higher than the soda flats, attracts herds of Burchell's zebra and gemsbok. **PG**

DECEPTION VALLEY

GHANZI, BOTSWANA

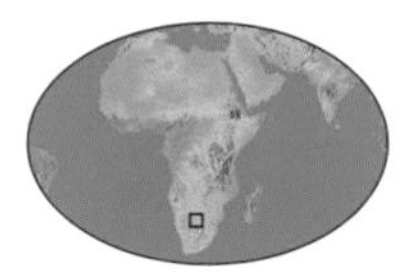

Length of Deception Valley: 50 mi (80 km)

Vegetation: short scrub and grass

Tree roots: 165 ft (50 m) below surface

Crossing the northern part of the Central Kalahari region from west to east is a series of deep flat-floored valleys. At first glance these appear to be river valleys in a semi-arid environment where there is not enough rain to cause water to run, let alone erode a huge valley. Deception Valley is, in fact, a "fossil" valley. It was formed by river erosion as a broad steep-sided river valley, before sand was blown in. The sand converted its V-shape into a much shallower valley with a flat floor, now covered with typical Kalahari scrub vegetation. This, coupled with occasional shallow pans that form at the time of the rains, attracts huge herds of antelope such as springbok and gemsbok—and the predators that prey on them. The rocks of the original valley are visible along its sides, and there are places where the Stone Age inhabitants gathered to make stone implements—today, many thousands, if not millions, of stone chips chipped off during the manufacture of stone cleavers can be found in the valley. Deception Valley was named by a De Beers mapping group in 1961: The team thought it was actually another valley farther south, as it did not appear on any of the maps at the time. **PG**

SIBEBE

SWAZILAND

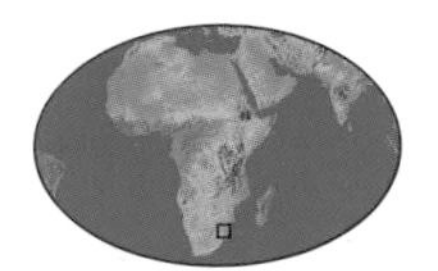

Sibebe rock type: granite pluton

Area of Mbuluzi River basin: 1,200 sq mi (3,100 sq km)

Elevation range of river: 410 to 5,000 ft (125 to 1,500 m)

Somewhere below Swaziland, following the cooling of a chamber of magma deep within Earth's crust, earth movements pushed a meteor-sized stone upward until it finally broke the surface millions of years later. Wind, rain, and a rushing river finished the job to expose Sibebe—also known as Bald Rock—in all its glory. Recognized as the world's largest exposed globe of granite and the second largest rock, Sibebe rests like an immense bowling ball within the greater Mbuluzi Mountain Range. It towers nearly 1,000 feet (300 meters) over the long and wide Mbuluzi River, a body of water which cuts clear across Swaziland into Mozambique.

Locals have named the guided three-hour hike up Sibebe's most precipitous façade "the steepest walk in the world." Once at the top, visitors are rewarded with clusters of massive boulders that form a network of caves. These natural shelters are decorated in prehistoric paintings created by the indigenous people who once called Sibebe home. Sibebe is located in Pine Valley about five miles (eight kilometers) from the capital city of Mbabane and is accessible by private transportation only. DBB

KRUGER—BAOBABWE

LIMPOPO / MPUMALANGA, SOUTH AFRICA

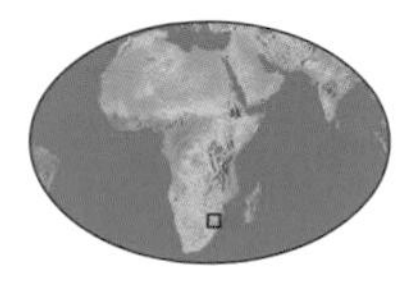

Area of Baobabwe: 31 sq mi (80 sq km)

Age of oldest baobab trees: at least 2,000 years

Vegetation: baobab trees

In the north of Kruger National Park (and also outside it), lies a region heavily dissected by streams and rivers. As the area is relatively dry, after heavy rains there is little groundcover to slow or stop erosion by flowing water. The most striking feature of this sandveld landscape is the incredible number of ancient baobab trees (so numerous that the area is often called Baobabwe), their sheer numbers appearing like a huge army traversing the landscape. These baobabs do not grow as large as those on the more fertile silts of river valleys like the Save, but their varied shapes are remarkable. The baobabs' massive squat trunks—which can grow up to an incredible 92 feet (28 meters) in girth—give rise to a crown of branches resembling a root system, which is why they are often referred to as the "upside-down trees." Their trunks have many holes that provide nest sites for barn owls and kestrels—some are totally hollow and provide a safe habitat for small animals and reptiles. When the baobabs blossom in early summer, their large, white, sweet-smelling flowers provide nectar for bats, birds, and insects. **PG**

BELOW *Ancient baobab trees at sunset.*

MODJADJI

LIMPOPO, SOUTH AFRICA

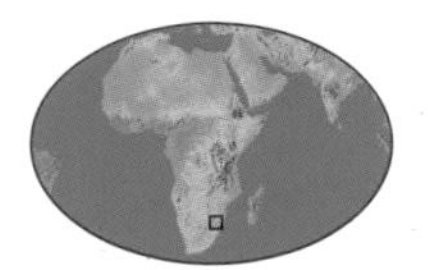

Area of Modjadji Nature Reserve: 1,310 acres (530 ha)

Height of Cycads: up to 40 ft (12 m)

Best time for seeds: December to February

Modjadji is home to the tallest cycads in southern Africa and the largest concentration of a single species of cycad, *Encephalartos transvenosus,* anywhere in the world. The cycad is an ancient plant, resembling a palm or tree fern, which first appeared over 200 million years ago and was very common during the age of the dinosaurs.

Today, Modjadji is a strange, primeval landscape of towering cycads, some of which are over 40 feet (12 meters) high. The female plants can produce giant seed cones that may weigh as much 75 pounds (34 kilograms).

According to local legend, the survival of the Modjadji cycads is due to the "rain queen." Over 400 years ago, a Shona lady called Dzugudini fell pregnant before marriage and had to flee her tribe. She fled south and settled near present-day Tzaneen, founding the Balobedu tribe who still live here. It was said that Dzugudini took with her the secret of making the rain fall, and so, although the tribe was very small, they were respected by the surrounding tribes and were never attacked. From the 1800s, the tribe has been ruled by a *Modjadji* or "rain queen," who is not allowed to marry, but may have children. **PG**

NYLSVLEY

LIMPOPO, SOUTH AFRICA

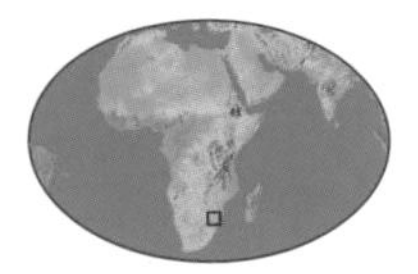

Area of Nylsvley Nature Reserve: 39,537 acres (16,000 ha)

Length of Nyl floodplain: 44 mi (70 km)

Notable feature: best-conserved floodplain in South Africa

When a group of early explorers moving northward through South Africa chanced upon a north-flowing river, they thought they had reached the headwaters of the Nile. Although this was not the case, they had discovered a feature rich in biological interest. The Nyl River rises in the hills of the Waterberg and flows through a relatively flat-floored valley between steep hills. Heavy summer storms flood the valley, causing the river's waters to spread out over a wide area, producing vast marshes. These form a 39,537 acre (16,000 hectare) grass floodplain, one of the largest in South Africa. Of this area, 7,660 acres (3,100 hectares) are now protected by the Nylsvley Nature Reserve, which is open to visitors. During the rainy season when the marshes are flooded, Nylsvley attracts an incredible variety and number of waterbirds. Over 100 species have been recorded on the river and 58 of them breed here. This is the greatest number of waterbird species breeding in one area in South Africa. It is also the only area in South Africa where some species, such as the rufous-bellied heron, breed. The Nyl River is unusual because it does not reach the sea but "dies out" near Potgietersrus. **PG**

KRUGER—THE RIVER LINES

LIMPOPO / MPUMALANGA, SOUTH AFRICA

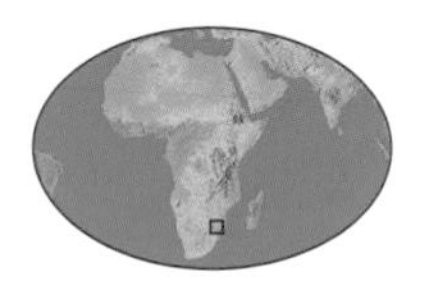

Area of Kruger National Park:
7,722 sq mi (20,000 sq km)

Established as a national park:
1898 (fenced partly 1961)

The Limpopo is the northernmost river of Kruger National Park, with the Crocodile River forming its southern boundary. They are in fact the same river. The Crocodile (or Krokodil) is the name for the upper Limpopo, which flows in a great arc, first north (forming the South Africa–Botswana border), and finally southeast through Mozambique to the Indian Ocean. In addition to these rivers, the Sabie, Letaba, Olifants, Luvuvhu, and Shingwidzi rivers provide the interior of the park with water. These rivers are essential sources of water for the thousands of animals living in Kruger—they are also important for irrigation and people living between the mountains and the park. The river lines are the most fertile, productive, and wettest parts of Kruger, and as such contain incredibly important habitats. Visitors can see a great diversity of wildlife: Riparian woodlands grow along the rivers, almost forest-like in places. Large trees, such as sycamore figs that can reach up to 70 feet (21 meters), provide nesting sites for birds, daytime retreats for leopards, and food for a huge variety of animals and insects. When in fruit, they attract monkeys, baboons, and fruit bats. **PG**

KRUGER—MOPANE VELD

LIMPOPO / MPUMALANGA, SOUTH AFRICA

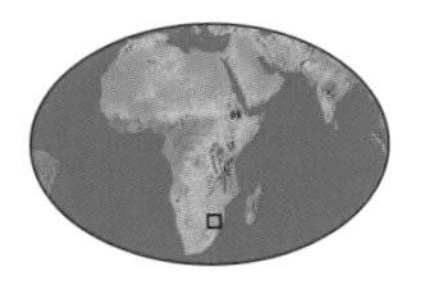

Height of mopane trees: up to 33 ft (10 m)

Other vegetation: sycamore fig, Natal mahogany, sausage tree, apple leaf, aloe, impala lily

The Kruger National Park is an enormous nature reserve stretching 218 miles (350 kilometers) along the Mozambique border. The northern half of the park (north of the Olifants River) is predominantly mopane veld. The mopane tree dominates large tracts of relatively flat country between the river valleys. Most areas have short trees—often no more than 5 feet (1.5 meters) high—but there are some places of true woodland where the tree crowns are 33 feet (10 meters) from the ground. These taller mopane trees have numerous holes, which are important breeding sites for birds, bats, and small mammals such as hares and rodents. The shorter trees were once dwarfed by bull elephants—it is among the mopane that many of the legendary huge bulls of Kruger were to be found. Although six of the original seven giant bulls have died, there are still many large bulls to see. They are growing and have started to become legends in their own right. They occur in small groups of three to seven individuals, usually dominated by one particularly large bull with enormous tusks. Fortunately there are still plenty of large bulls in the area, as well as breeding herds of elephants. **PG**

KRUGER—SOUTHERN HILLY COUNTRY

MPUMALANGA, SOUTH AFRICA

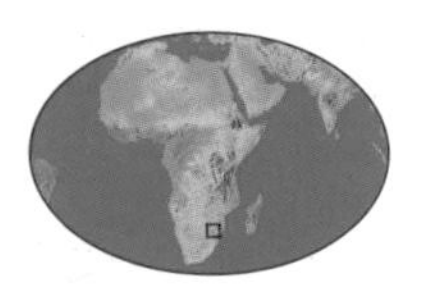

Highest point in Kruger (Khandzalive): 2,753 ft (839 m)

Kopjes rock type: granite

Trees in southern region of park: Cape chestnut, white pear, coral tree

The southern part of Kruger is quite unlike the other areas of South Africa in that it is very hilly. Intrusive granite is the main type of rock and it shows the effects of exfoliation—the splitting of rocks into a series of concentric shells—which has produced a landscape of hills and valleys covered by tall but sparse woodland. Some of the hills are small, bare rock domes, while many more have rocky slopes that are more or less covered with woodland. This part of Kruger is noted for its winding roads that offer dramatically changing landscapes quite different to the flat lands to the north. The alternation of hills and valleys has created a habitat in which game can thrive—particularly in the lower Sabie River region. Elephants, buffalos, lions, and leopards are here, as well as African painted dogs or African wild dogs, one of Africa's seriously endangered carnivores. These hills are also home to a unique antelope—the Klipspringer or rock jumper. This shy antelope is easier to spot here than anywhere else in South Africa. **PG**

KADISHI TUFA FALLS

MPUMALANGA, SOUTH AFRICA

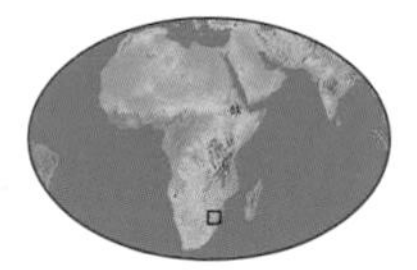

Height of tufa falls: about 660 ft (200 m)

Visiting bird species: fish eagle, brown-headed parrot, purple-crested loerie, chorister robin, olive bush shrike

Near the head of the Blyde Dam are the Kadishi Falls—at 660 feet (200 meters), the second highest tufa falls in the world. Tufa falls are a comparatively rare phenomenon; they form when water is saturated with calcium carbonate. Exposure to the air causes the water to evaporate, causing some of the dissolved calcium carbonate to be deposited as a white stalactite-like "fall" down the steep hillside. Few tufa falls are as easily viewed as these—visitors can approach the falls by boat, but because they are so high they are better viewed from a distance to take in the entire spectacle. More than 360 bird species have been recorded in the area. It is home to the world's third-largest Cape vulture colony, and has become home to several pairs of African finfoot—rare, elusive, and shy waterbirds. Like pterodactyls of the dinosaur age these unusual birds have a claw on the first digit of each wing, used to help them clamber through vegetation. A wide variety of aquatic species, including hippos, crocodiles, and otters, live in and around the rivers. In the surrounding forest visitors may see chacma baboons, vervet monkeys, rare samango monkeys, and maybe even the occasional leopard. PG

TSWAING CRATER—PRETORIA SALT PAN

GAUTENG, SOUTH AFRICA

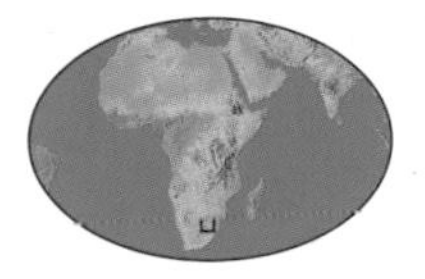

Width of Tswaing Crater: 0.7 mi (1.13 km)

Age of crater: 220,000 years

The Tswaing Crater was formerly known as the Pretoria Salt Pan. For many years it was considered to be a volcanic cone with a sunken floor containing a soda lake. The crater consists of a ring of rock rising some 200 feet (60 meters) above the surrounding land surface, and it was presumed that the hills were formed by volcanic rock pushed out of the central neck, with the subsequent collapse of the center into the magma chamber. However, the rocks forming this ring are granite, a deep-seated intrusive rock approximately three million years old, and not volcanic rock that has erupted at the surface. This crater is unique in South Africa—the central floor lies about 200 feet (60 meters) below the level of the surrounding land surface, and underlying the soda lake are a variety of fine-grained muds which contain fragments of rocks, some containing magnetite, an iron oxide. It is currently believed to be the impact crater of a meteorite that struck Earth approximately 220,000 years ago. PG

PILANESBERG

NORTHWEST PROVINCE, SOUTH AFRICA

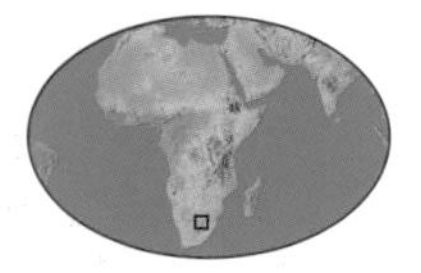

Area of Pilanesberg National Park: 135,900 acres (55,000 ha)

Established as a national park: 1979

Age of volcano: 1.2 billion years

This ancient feature is much more complex than it appears. It was originally a huge volcano (or series of volcanoes) that was either blown apart by a stupendous explosion or collapsed in on itself. It was then buried and subsequently exposed by erosion of the overlying rocks. The remaining volcanic cones are marked by radiating pattern of rivers: These rivers continue to erode downward into the older circular volcanic rocks to give an amazingly complex geological arrangement. The massive is 1.2 billion years old: the highest peak, the Pilanesberg, towers 2,000 feet (600 meters) above a cicular bowl, which is 12 miles (20 kilometers) in diameter and contains Mankwe Lake. It is here, in the heart of the ancient volcanic landscape, that the Pilanesberg National Park is found. In 1979 "Operation Genesis," one of the most ambitious game restocking projects ever undertaken, reintroduced large numbers of animals that had been wiped out by hunting. Today virtually all the animal species of southern Africa can be found, including lions, elephants, white and black hippos, buffaloes, and giraffes. Numerous Stone and Iron Age sites are scattered throughout the park. PG

BLYDE RIVER CANYON

KWAZULU-NATAL, SOUTH AFRICA

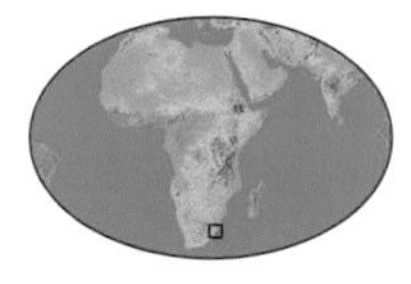

Length of Blyde River Canyon: 15 mi (24 km)

Depth of canyon: 2,640 ft (800 m)

Dominated by dramatic granite ridges, the 15-mile- (24-kilometer-) long Blyde River Canyon cuts through the northeastern part of the Great Escarpment and winds its way to the Blydepoort Dam at Swadini. At one time the waters of the Blyde River blasted through the rock, carving a 2,640-foot- (800-meter-) deep canyon—the third largest in the world—and one of the most beautiful landscapes in Africa.

Today the river threads through canyonsides cloaked in temperate rainforest and evergreen shrub-like plants known as fynbos. On one side stands the Three Rondavels (sometimes known as Three Sisters), three enormous spirals of dolomite rock that rise out of the canyon walls like giant space rockets. Their tops are iced with green vegetation and their sides stained with orange lichens. They get their local name from a similarity to the circular thatched huts common to indigenous peoples. Another feature, the Pinnacle, is a single quartzite column that rises out of the deep, wooded canyon. MB

BOURKE'S LUCK POTHOLES

KWAZULU-NATAL, SOUTH AFRICA

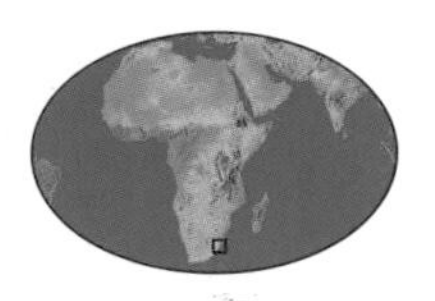

Maximum depth of potholes: 20 ft (6 m)

Rock type: dolomite

The Blyde River and the Treur River meet at Bourke's Luck Potholes. Here the Treur is channeled into a narrow cataract that turns almost 90 degrees into the Blyde. The abrupt change of direction causes the swirling water, and the boulders it carries from the hills upstream, to carve out a series of huge, bowl-shaped depressions or "potholes" up to 20 feet (6 meters) deep in the soft red and yellow-colored dolomite rock. At one time a farmer called Tom Bourke owned the site. He speculated that as prospectors upstream were successfully panning for gold, he should find gold nuggets in his potholes. He was right, and the feature became known as Bourke's Luck.

The names of the two rivers have their own story. In 1840, a party of Boer pioneers explored the area, looking for places to settle. The men struck out to the east, leaving their women and children encamped beside a river. When they failed to return on the arranged date the women thought their men had died and they called the river Treur, meaning "river of sorrow." Later, however, husbands and families were reunited at a second river, and they called it Blyde, meaning "river of joy." **MB**

KOSI BAY

KWAZULU-NATAL, SOUTH AFRICA

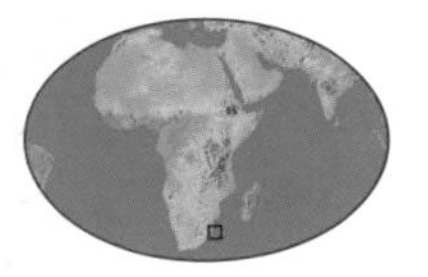

Trees in Kosi Bay: wild date, raffia palm, sycamore fig, mangrove

Marine/river animals: hippos, Nile crocodiles, leatherback turtles, humpback whales, bull sharks

In the farthest northeast corner of South Africa, near the border with Mozambique, lies a unique area of huge biological importance. Kosi Bay is a stunning mosaic of lakes and rivers, swamps and wet forests, that forms the most pristine river and lake system on the African coastline. Stretching 11 miles (18 kilometers), Kosi Bay consists of four lakes and a series of interconnecting channels that drain via a long sandy estuary into the Indian Ocean. The area is home to a great diversity of animals, birds, and plants. Near the estuary mouth, for example, the mangrove area features five different species of mangrove. In terms of fauna, Kosi Bay is home to the unusual two-armed mudskipper fish and the one-armed fiddler crab, as well as over 200 species of tropical fish; crocodiles and hippos can also be seen. During the winter, humpback whales pass the coast on their northern migratory route. The coastline is also one of the main breeding sites for sea turtles in South Africa—in December and January loggerhead and leatherback turtles come onto the beaches to lay their eggs. **PG**

RIGHT *South Africa's scenic Kosi Bay, near Mozambique.*

ISIMANGALISO WETLAND PARK

KWAZULU-NATAL, SOUTH AFRICA

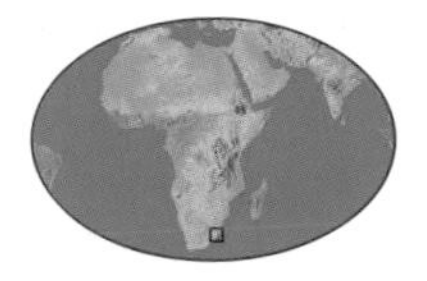

Depth of Lake St. Lucia: seldom more than 6 ft (1.8 m)

Area of wetland: 115 sq mi (300 sq km)

Vegetation: forest, bush and grasslands, with open water

Formerly know as St. Lucia Wetland Park, iSimangaliso Wetland Park on the eastern coast of KwaZulu-Natal stretches from Kosi Bay in the north to Cape St. Lucia in the south. The park—the first in South Africa to be declared a World Heritage site—comprises a huge lagoon-estuary running parallel to the coast. Massive forested sand dunes (the world's highest) prevent the Mkuze River from reaching the sea. The river is forced southward, creating the 37-mile (60-kilometer) long Lake St. Lucia. With its location between subtropical and tropical Africa, the park incorporates an astonishing variety of habitats, ranging from the Ubombo Mountains to floodplain grasslands, dune and coastal forests, saline swamps and salt marshes, mangroves, sandy beaches, and coral reefs. This has resulted in exceptional species diversity. The area is considered critical to the survival of a large number of species, including South Africa's largest populations of hippopotamus, crocodile, and the white-backed and pink-backed pelican. PG

GIANTS CASTLE

KWAZULU-NATAL, SOUTH AFRICA

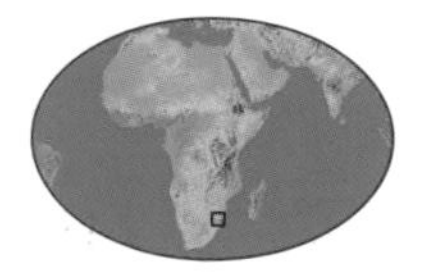

Giants Castle established as a protected area: 1903

Vegetation: montane bush and scrub

Highest point in South Africa (Injasuti Dome): 11,184 ft (3,409 m)

The Drakensberg Plateau owes its existence to extensive basaltic lavas laid down at the end of the Karoo period 190 million years ago. Erosion by rivers rising on this wet plateau has resulted in a massive line of cliffs capped by horizontal lava beds. This is best shown in the Giants Castle Reserve: The rolling foothills abruptly change to steep slopes that become massive cliffs rising over 10,000 feet (3,000 meters) into the sky. The San people believe that dragons lurk here and this gave the Drakensberg its name—visitors may also see the serrated peaks resembling the back of a dragon. The Zulu name for Giants Castle is *iNtabayikonjwa,* "the mountain at which one must not point." Pointing at the mountain is said to be disrespectful and will result in bad weather—indeed the Drakensberg has the wildest and loudest thunderstorms in southern Africa. Giants Castle is home to a variety of animals, such as the mountain eland and rhebok, but many people visit the reserve to see the majestic bearded vulture or lammergeier soaring overhead. This incredible bird is probably most famous for its habit of dropping large bones on a flat rock, breaking them open to feed on the marrow inside. PG

ORIBI GORGE

KWAZULU-NATAL, SOUTH AFRICA

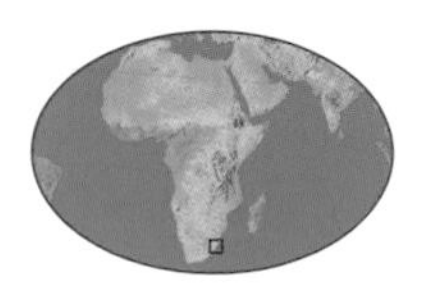

Length of Oribi Gorge: 15 mi (25 km)

Depth of gorge: 1,000 ft (300 m)

Rock type: sandstone above granite

One of South Africa's lesser known natural wonders, Oribi Gorge cuts deep into the sandstone of southern KwaZulu-Natal, about 12 miles (20 kilometers) inland from Port Shepstone. Here the Umzimkulwana River winds its way for nearly 16 miles (25 kilometers) past dramatic cliffs. There are beautiful views from spots at the top of these cliffs, but the gorge must be explored below to be fully appreciated; there are a number of trails of varying length throughout the gorge. The main river runs along a fairly level bed that has been eroded to a granite base that has been dated at over 1 billion years old. On either side of the gorge the 1,000-foot (300-meter) high cliffs provide sites for some impressive waterfalls. The ravine is largely filled with dense evergreen forest, but there are a variety of natural habitats. Due to this diversity and the relative inaccessibility of the gorge, it contains approximately 500 species of trees and a wealth of wildlife including the shy leopard and python, as well as the rare samango monkey. This is a good spot for bird lovers, who can hope to spot the elusive Narina trogon and a variety of raptors, including the martial and crowned eagles. HL

HLUHLUWE-UMFOLOZI

KWAZULU-NATAL, SOUTH AFRICA

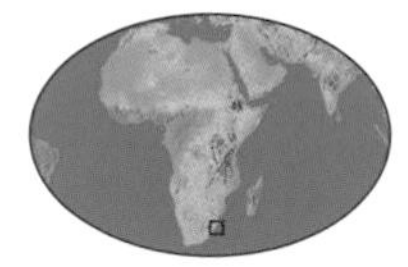

Area of Hluhluwe–Umfolozi Game Reserve: 237,206 acres (96,000 ha)

Numbers of rhino: black rhino: 350; white rhino: 1,800

The Hluhluwe–Umfolozi Game Reserve combines two world famous game parks—Hluhluwe in the north and Umfolozi in the south—that were originally Zulu royal hunting grounds. Umfolozi was the first game reserve to be established in Africa, and the first to introduce "wilderness trails" in which visitors could walk through the bush and camp under the stars. The main creature of interest is the extremely rare black rhino, and every effort is being made in the park to conserve the species. With rolling hills topped with grass separated by steep valleys lined with forest, the best savanna country in South Africa lies between the Black and White Umfolozi rivers, and it is here that both black and white rhino are flourishing. Game was devastated in the region at the beginning of the 19th century when the disease *nagana* spread by tsetse flies hit local livestock; farmers thought the only way to eliminate it was to kill the wildlife in the area. Gradually the parks have been restocked, but curiously the lions returned on their own. A male arrived unexpectedly in 1958 and a few years later a group of females mysteriously appeared. Today the pride controls the numbers of antelope in the park. MB

THE DRAKENSBERG

KWAZULU–NATAL, SOUTH AFRICA

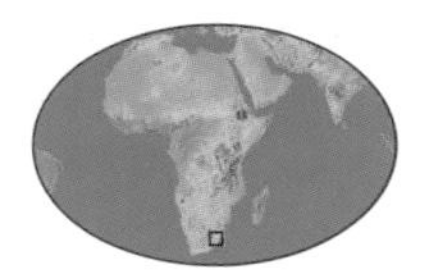

Length of the Drakensberg: 373 mi (600 km)

Highest point (Thaba Ntlenyana): 11,424 ft (3,482 m)

Geology: 5,000 ft (1,500 m) basalt above sandstone

Europeans have known South Africa's mightiest mountain range for two centuries as the Drakensberg—the Dragon's Mountain. To the Zulus the range is known as *uKhahlamba*, "the Barrier of Spears." Both names refer to the rocky basalt ridges and pinnacles that tower above the plains of KwaZulu–Natal. In 2000 the region was declared South Africa's fourth World Heritage site and given the name uKhahlamba/Drakensberg Park.

The Drakensberg acts as a vital watershed in the dry subcontinent with lush, green rolling grassland—home to vast grazing herds—that provides a welcome contrast to much of the surrounding parched, baked landscape. It stretches for around 620 miles (1,000 kilometers) from the Eastern Cape to the South Africa–Lesotho border and beyond. The underlying rock is a mixture of basalt and sandstone, the latter deposited around 220 million years ago, when the area was covered in a vast lake. Unusually

for a mountain range, the Drakensberg's jagged peaks and slopes were formed not by uplift or any other form of tectonic activity, but by eons of wind and water erosion. These processes have resulted in an area of exceptional natural beauty, with soaring buttresses, dramatic cutbacks, and golden sandstone ramparts. Its most famous feature is the awe-inspiring "amphitheater," a sheer, curved 400-foot- (120-meter-) long cliff shaped like a vast natural arena.

The range's highest summits all fall along the Lesotho border. Tallest of all, and the highest point in all southern Africa, is Thaba Ntlenyana, which climbs to 11,424 feet (3,482 meters). Around it rise various lesser peaks, many with evocative names, such as Giants Castle, Champagne Castle, Cathedral Peak, and The Old Woman Grinding Corn.

The area was inhabited for thousands of years by the San people, who left an incomparable collection of rock art in more than 500 caves scattered throughout the mountains. The oldest painting is 2,500 years old, while the most recent depicts a 19th-century scene of hunters on horseback. **HL**

BELOW *The lush green fields of the Drakensberg.*

MONT-AUX-SOURCES

KWAZULU-NATAL, SOUTH AFRICA

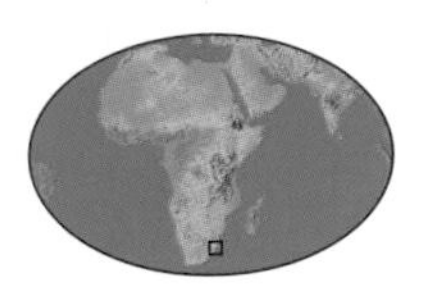

Mont-aux-Sources highest point: 10,882 ft (3,317 m)

Tugela Falls: 3,110 ft (948 m) total drop

In 1836, two French missionaries looking for the source of the Orange River arrived at a magnificent escarpment near a peak known to the Basotho as *Pofung*, "the Place of the Eland." Although the misunderstanding that the peak was the source of the Orange, Caledon, and Tugela rivers has now been modified, the name of Mont-aux-Sources has remained. Bounded by the Sentinel Peak to the northwest, the Western Buttress to the west, and the Eastern Buttress and the fearsome Devil's Tooth (which was not conquered until 1950) to the east, the "amphitheater" rears an awe-inspiring basalt face above the valley of the Tugela River. This giant rock face is 5 miles (8 kilometers) long and 2,640 feet (800 meters) high. The infant river, rising on the Mont-aux-Sources summit about 2 miles (3.2 kilometers) back from the escarpment, plunges dramatically over the edge in one of the world's highest falls, known as Tugela Falls. The great uplands—originally molten lava up to 4,620 feet (1,400 meters) thick—form an alpine ecosystem rare in Africa. Most of the vital rain falls in summer thunderstorms, but snowfall often deposits more than three feet (0.9 meters) of snow. HL

GOLDEN GATE

FREE STATE, SOUTH AFRICA

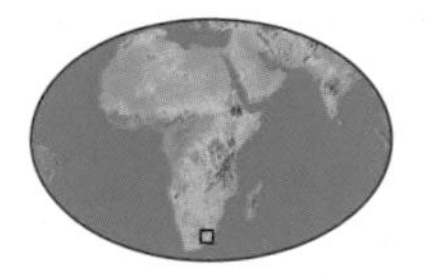

Area of Golden Gate National Park: 27,180 acres (11,600 ha)

Rock type: sandstone

Vegetation: indigenous grassland

The Golden Gate is a striking formation carved into the westward-facing ramparts of the Lesotho Mountains. It is formed from a fine wind-deposited sandstone, called "cave sandstone" due to the incredible number of caves formed. It is relatively easily eroded by water and the Little Caledon River has carved a deep valley through the area, leaving two sandstone bluffs guarding the entrance. The sandstone is naturally-colored, but in the setting sun the Golden Gate appears to be on fire, revealing a breathtaking tapestry of red, yellow, and purple hues. The cliffs themselves are over 330 feet (100 meters) high and the layering of the rock can be clearly seen. The more resistant layers actually appear to reach out from the cliffs. Also jutting out from the cliffs bordering the valley is the Brandwag buttress, a massive block of sandstone shaped like the prow of an ocean liner—it is said to be even more spectacular than the Golden Gate. The area boasts a variety of wildlife including antelope, eland, and Burchell's zebra, as well as the extremely elusive bearded vulture and bald ibis. A great number of rare and unusual plants grow in the highlands, such as the arum lily and fire lily. PG

THE RICHTERSVELD

NORTHERN CAPE, SOUTH AFRICA

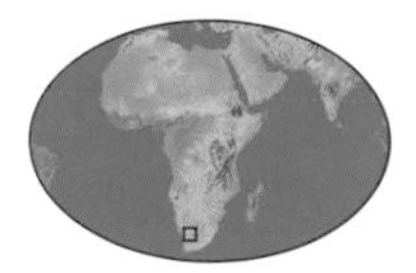

Highest peak in Richtersveld: 4,508 ft (1,374 m)

Annual rainfall: below 2 in (50 mm)

Vegetation: desert botany, notably succulents

In the northwest corner of South Africa, the Orange River follows a loop to the north for about 62 miles (100 kilometers) before it reaches the Atlantic at Oranjemund. On the south side of this loop lies the inhospitable, extremely rocky, and, until very recently, largely inaccessible Richtersveld. With the highest peaks reaching above 4,290 feet (1,300 meters), summer temperatures reaching 122°F (50°C), and an average annual rainfall of less than 2 inches (50 millimeters), the Richtersveld is, surprisingly, something of a botanist's paradise. While it is accurately described as a mountain desert, its rugged mountains and apparently barren valleys are home to approximately one third of South Africa's plant species. Among the more recognizable of these is the quiver tree, which takes its name from the fact that the San inhabitants of the region used its branches to make quivers for their arrows. Another distinctive plant is the *Pachypodium namaquanum*, known in Afrikaans as "halfmens" since legend has it that it is half plant and half human. On the border of the region, several rock-engraving sites on the black limestone rocks near the riverbed are thought to date back 2,000 years or more. HL

AUGRABIES FALLS

NORTHERN CAPE, SOUTH AFRICA

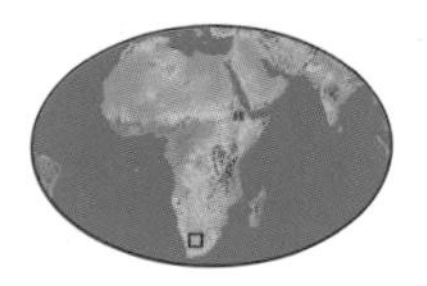

Height of first waterfall: 300 ft (90 m)

Height of second waterfall: 200 ft (60 m)

Rock type: granite

The Augrabies Falls, with their granite ravines and gorges, form an impressive feature on the Orange River in South Africa, roughly 50 miles (80 kilometers) upstream from the southeast corner of Namibia. Above the falls, the river flows through a relatively flat green valley, bounded by stark, rocky hills that stretch away into the distance. The formation of the landscape is such that the falls can only be approached readily from the south. From this direction visitors come upon the spectacular gorge with little warning. The water gathers speed as it runs down a series of rapids carved through the granite, giving an initial drop of about 300 feet (90 meters). After a final rush through a dogleg, the water hurtles over the last fall of nearly 200 feet (60 meters), into a deep surging pool below. The roar of the waters gave rise to its local Khoikhoi name "the place of great noise." The aura created by the chasm, which stretches 9.3 miles (15 kilometers) downstream, has given rise to the Khoikhoi legend of the great water snake. The snake is said to guard a fortune in diamonds, washed down from the mountains of Lesotho over the ages and now lying deep under the waters of the Orange River. **HL**

NAMAQUALAND

NORTHERN CAPE, SOUTH AFRICA

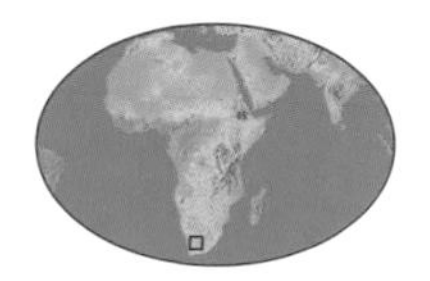

Highest point in Namaqualand: 5,597 ft (1,706 m)

Annual rainfall: 2 to 10 in (50 to 250 mm)

Notable feature: spectacular wildflower displays in spring

Namaqualand is the most northwesterly region of South Africa. Bounded by the Atlantic Ocean to the west and the Orange River to the north, it is a rather arid area. Home to the Nama people, from whom it takes its name, its succulent plants have provided food for their sheep for the last 2,000 years. The central region sweeps up to granite peaks but the rainfall rarely causes the rivers to run for any length of time. However, after good winter rains a seeming miracle occurs, as this somewhat uninviting landscape is transformed into an awe-inspiring floral wonderland. Visitors flock to the area to see the huge swathes of veld now splashed with color like a great Persian carpet. The spectacular orange and yellow Namaqualand daisies, which grow where the land has been overgrazed by the local sheep, are the most obvious of the new arrivals. However, the Namaqualand spring is much more than just a pretty vista. To the knowledgeable botanist, its true attraction lies in the untouched, more remote areas, where a huge variety of species and color can be found concealed among the drab bushes. **HL**

BELOW *Winter rains produce a stunning carpet of flowers.*

KGALAGADI TRANSFRONTIER PARK

SOUTH AFRICA / BOTSWANA

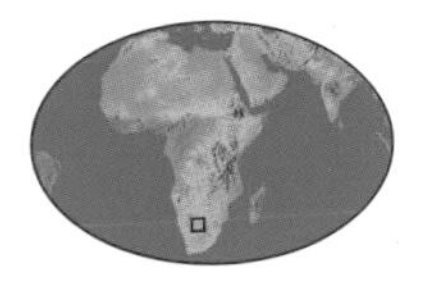

Area of Kgalagadi Transfrontier Park: 13,900 sq mi (36,000 sq km)

Vegetation: desert scrub with trees along river lines

Maximum temperature: 104°F (40°C)

Africa's first peace park—the Kgalagadi Transfrontier Park—was formed by the amalgamation of Kalahari Gemsbok in South Africa and Gemsbok National Park in Botswana. Spanning more than one country, the peace park unifies fragmented ecological habitats across national boundaries and promotes environmental and political stability. At the heart of the Kgalagadi Transfrontier Park are two rivers—the Auob and the Nossob—which flow southward to the Molopo River, carving valleys through the red sands of the Kalahari. Each year, the desert is transformed by the annual rains. Colorful flowers suddenly appear, and the plants along the riverbanks attract antelope, including red hartebeest, gemsbok, springbuck, and blue wildebeest. These, in turn, lure predators. The large black-maned lions are the most sought after by safari visitors, but cheetahs are often seen, as are spotted and brown hyenas and bat-eared foxes. Even leopards may be found resting in trees during the day. **PG**

WITSAND

NORTHERN CAPE, SOUTH AFRICA

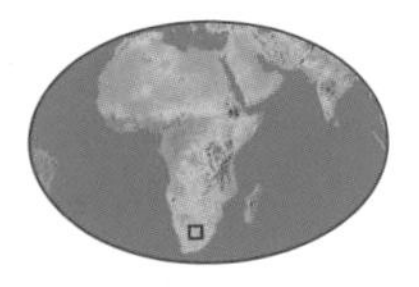

Area of Witsand Nature Reserve: 8,648 acres (3,500 ha)

Elevation of Witsand Nature Reserve: 4,000 ft (1,200 m)

Height of sand dunes: up to 200 ft (60 m)

The northwestern area of Northern Cape is Kalahari Desert country—large areas of semi-arid land, often covered with a layer of reddish, wind-deposited Kalahari sand. East of Upington, however, lies a series of large, white sand dunes that form a stark contrast to the red sands of the Kalahari Desert, and sit incongruously against the surrounding landscape. These dunes are called Witsand (White Sand) and are protected within their own nature reserve. It is not just their color that sets them apart. When the sand is dry and hot, any disturbance causes the dunes to "rumble," "roar," or "hum." The sand actually emits a very distinctive, low-pitched noise that has been likened to the roar of water from rapids or waterfalls heard from a distance. Because of this sound, the dunes have become known as the *Brulsand* or "roaring sands." Between the dunes is a series of valleys where, despite the dryness of the area, there are wet places knows as *vleis.* These are fed by water that seeps up from the underlying rock strata, creating a wholly different oasis environment which supports many plants and animals, and forms a startling contrast with the barren dunes on either side. **PG**

SOCIABLE WEAVER NEST

NORTHERN CAPE, SOUTH AFRICA

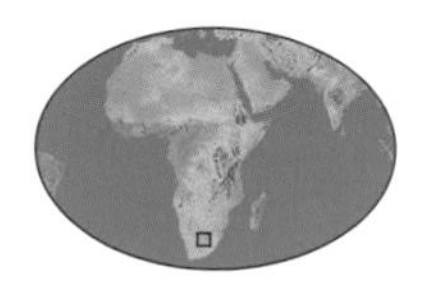

Length of sociable weaver:
5 in (12.5 cm)

Weight of sociable weaver:
1 oz (30 gm)

Diameter of nest chambers:
6 in (15 cm)

The nest of the sociable weaver is like no other bird nest. A small bird that lives in the dry northwest of South Africa, the sociable weaver lives in colonies of up to 300 birds, centred on a single huge communal nest. With incredible ingenuity, sociable weavers build nests that can be an astonishing 22 feet (7 meters) in length and weigh up to 2,200 pounds (1,000 kilograms)—sometimes nests can get so large that they break the trees in which they are built.

These ornithological architectural marvels are constructed largely from woven grass, with a rainproof roof made from coarse material, such as sticks supporting thick grass. Inside are up to 50 nest chambers—each about the size of a person's fist—lined with fine grasses, fur, and other soft material collected from the surrounding veld. Every nest has a downward-facing entrance tunnel, lined with grass straws pointing inward to prevent snakes and other predators from entering. Nest-building continues throughout the year, one straw at a time, as long as suitable grass straw is available. The nests are so well built, and the climate so dry, that some have been known to survive intact for over a hundred years. **PG**

LOWVELD FEVER TREES

SOUTH AFRICA / ZIMBABWE

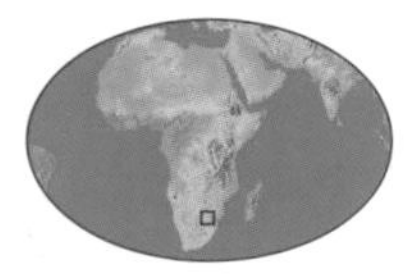

Area of Mkuze Game Reserve:
99,000 acres (40,000 ha)

Area of Gonarezhou National Park:
1,950 sq mi (5,050 sq km)

Immortalized by Rudyard Kipling in *The Elephant's Child*—"to the banks of the great, green, greasy Limpopo River all set about with fever trees"—the fever tree is a species of acacia that flourishes in swampy areas in the Lowveld areas of northwest South Africa and southeast Zimbabwe. The fever tree's bright yellow bark may account for its name, as malaria sufferers often turned yellow. However, it is more likely that the name was derived from the swampy areas where the trees are found—an ideal breeding ground for malarial mosquitoes during the summer. The bright yellow trunks create a surreal effect, especially during the winter when the trees are bare. During spring, they are covered with yellow ball-shaped flowers. One of the best places to see them is at the large forest in the Mkuze Game Reserve, South Africa. Here a boardwalk has been constructed to enable visitors to cross the wetter sections of this swampy habitat. There are also impressive fever tree forests along the lower Runde River in the Gonarezhou National Park in Zimbabwe. More than just an interesting spectacle, these forests are an important wildlife habitat, especially for birds such as the unusual green coucal. **PG**

BAVIAANSKLOOF

EASTERN CAPE, SOUTH AFRICA

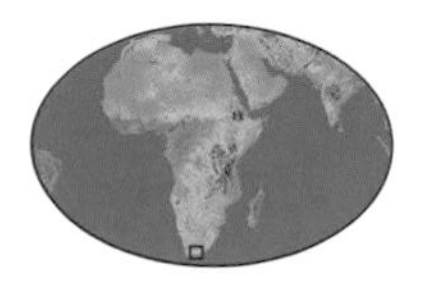

Length of Baviaanskloof: 62 mi (100 km)

Mountain height: 5,577 ft (1,700 m)

Rock type: folded Cape sandstone

Strangely, one sometimes finds wild and lonely regions in close proximity to well-populated centers. Baviaanskloof—the valley of baboons—is just such a place. With its eastern end only about 62 miles (100 kilometers) west of Port Elizabeth, it is bypassed by most people who head along the busy coastal route to Cape Town. Because few roads traverse the high ranges, there are remote valleys and uplands here that seem to stretch indefinitely. Baviaanskloof itself stretches east-to-west for over 62 miles (100 kilometers) between two mountain ranges, both with summits reaching above 5,577 feet (1,700 meters) at the eastern end of the Cape Fold mountain belt. The vegetation varies from the lovely proteas on the higher slopes to Afromontane ("African Mountain") forests overtopped by giant yellowwood trees and the strange cycads, which resemble spiky tree ferns. This remote area is home to a variety of animals including leopard, mountain zebra, lynx, kudu, and eland. The many overhangs and sandstone shelters also house a wealth of rock paintings and even now it is possible to come across an isolated shelter containing the remains of San hunting equipment. HL

VALLEY OF DESOLATION

EASTERN CAPE, SOUTH AFRICA

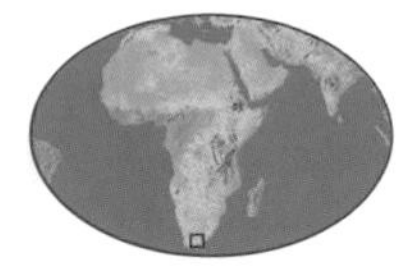

Depth of Valley of Desolation: 400 ft (120 m)

Rock type: dolerite

One of the more unusual aspects of the Valley of Desolation is that it is accessed from the top of a mountain. Situated in the Sneeuberg Range, which practically surrounds the town of Graaff-Reinet in South Africa's Eastern Cape, the valley was formed by the erosion of sedimentary rocks millions of years ago. Rather than suggesting "desolation," however, the valley offers both spectacular views and fascinating insights into the complex geological processes that created it.

As a result of erosion, the landscape consists of dramatic cliffs and jointed towers of hard dolerite rock which were left standing as the softer sediments crumbled and washed away. That not all the dolerite stood so firm is obvious from the confused jumble of boulders strewn across the valley floor below. For the visitor, the most impressive aspect of the scenery is formed by the startling rock pillars that tower some 400 feet (120 meters) from the floor; perhaps the "Towers of Silence" might have been a more apt name for the area. Beyond the pillars is Camdeboo Valley (from the Khoi for a "green hollow"), with the barren landscape of the Karoo stretching off into the distance. HL

THE COMPASSBERG

EASTERN CAPE, SOUTH AFRICA

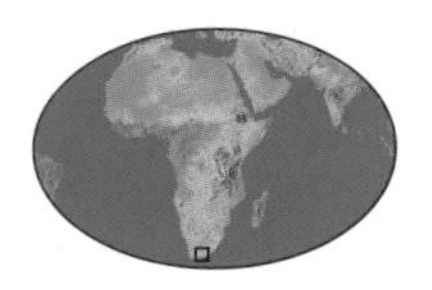

Height of the Compassberg: 8,215 ft (2,504 m)

Rock type: dolerite above sandstone

Age of fossils: 200 million years

First named during Governor van Plettenberg's historic inspection tour in the late 18th century, the Compassberg is a considerable mountain. It is, in fact, the highest summit in South Africa outside the great Drakensberg Range, and yet it remains one of South Africa's lesser-known peaks. Situated in the Sneeuberg Range north of Graaff-Reinet—near the point where the Western, Northern, and Eastern Cape provinces meet—the Compassberg forms part of the South African watershed. To the north and west, its slopes drain eventually to the Orange River and the Atlantic Ocean, while the other slopes drain to the Sundays River and the Indian Ocean. In the arid Karoo landscape, the sight of the streams running off toward all points of the compass is particularly impressive.

The mountain consists of sedimentary rocks—known to geologists as the Beaufort Group—that were laid down by the action of large rivers between the late Permian and the early Jurassic periods. The Compassberg and the surrounding Karoo landscape have been the source of important fossil finds dating back 200 million years, including early mammal-like reptiles. **HL**

HOGSBACK

EASTERN CAPE, SOUTH AFRICA

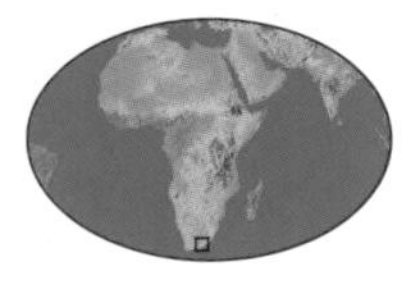

Height of Hogsback Mountain: 6,053 ft (1,845 m)

Rock type: shale and sandstone

The Amatola Mountains lie northwest of East London in South Africa's Eastern Cape province. Situated about 62 miles (100 kilometers) inland from the Indian Ocean, the range has considerable importance in the national history of the Xhosa people; it was a center for their resistance during the frontier wars of the 19th century. Situated on a spur of the Amatoles, the little village of Hogsback with its many waterfalls, yellowwood forests, high rainfall, and winter snows has an atmosphere that sets it apart from the hot, dry valleys below. Among the waterfalls, the Kettlespout Falls can be particularly impressive when the wind blows a plume of water high into the air, resembling the steam from a boiling kettle. The village is overlooked by four peaks with heights just below 6,600 feet (2,000 meters). One of these, Gaika's Kop, is named after the famous Xhosa chief of the late 18th to early 19th centuries, Ngqika. The other three are collectively known as "The Hogs" or the "Hogsback Mountain," and one in particular has high cliffs which resemble a hog. The Xhosa name is *Belekazana*, meaning "carrying on the back," as it is supposed to resemble a woman carrying her child on her back. **HL**

ADDO

EASTERN CAPE, SOUTH AFRICA

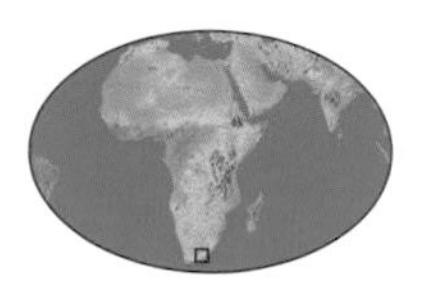

Area of Addo: 77 sq mi (200 sq km)
Vegetation: Addo scrub and thicket
Established as a national park: 1931

In 1931 the Addo region was made a protected area—the Addo Elephant National Park—to safeguard the last 11 elephants roaming the area. Today the park is home to approximately 350 African elephants—it has been so successful that it has actually become overcrowded. In response, surrounding land has been acquired—the new Greater Addo National Park will soon cover 1,875 square miles (4,856 square kilometers). As adult elephants deposit upward of 330 pounds (150 kilograms) of dung every day, dung beetles are extremely important to the ecology of the area—so much so that they are now a protected species. The flightless dung beetle is found almost exclusively in Addo. The unique vegetation of Addo is a dense, subtropical thicket. However, a peculiar succulent found in the park, called Spekboom, can grow up to 10 feet (3 meters) high. It is a quick-growing evergreen, with soft, fleshy leaves that are pleasant to squeeze and drink when thirsty. Recent research has shown that the Spekboom plant has remarkable properties in its ability to process carbon dioxide. It is thought to be one of the most efficient plants at removing the gas from the atmosphere. **PG**

HOLE IN THE WALL

EASTERN CAPE, SOUTH AFRICA

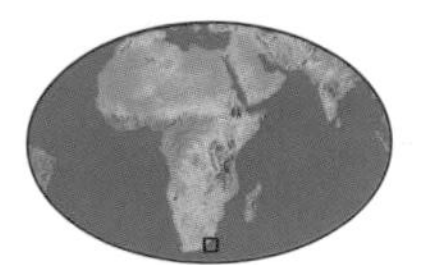

Height of Hole in the Wall: approx. 33 ft (10 m)

Width of arch: approx. 50 ft (15 m)

Rock type: dolerite

One of South Africa's most fascinating natural formations is to be found on the Eastern Cape's Wild Coast, about 31 miles (50 kilometers) southeast of Umtata. Here a massive dolerite island with sheer walls rises out of the surf, forming a rampart at the mouth of the Mpako River. Over the centuries, the combined action of the river and the surf has made a huge hole through the island, earning the formation the name "Hole in the Wall." To the Xhosa people of the region, the rock is known as *esiKhaleni*, "the place of thunder." At high tide during certain seasons, the waves clap through the opening in such a fashion that the concussion can be heard throughout the valley. The Xhosa tell the story of a beautiful girl who lived near the lagoon behind the rock rampart. One of the sea-people, creatures similar to mermaids, saw her and wanted her for himself. He brought a huge fish that battered its way through the rock, creating the Hole in the Wall. He took the girl, and she was never seen again. The rugged beauty of the coast in this region, with its grassy hills, steep cliffs falling into milkwood forests, and unspoiled beaches, forms the perfect backdrop to this awesome scene. **HL**

CAPE POINT

EASTERN CAPE, SOUTH AFRICA

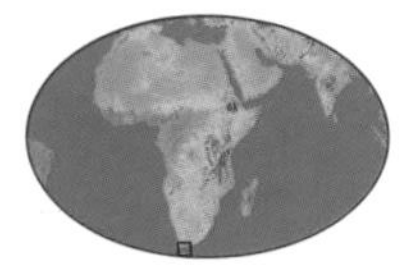

Height of Cape Point: approx. 820 ft (250 m)

Length of peninsula: 30 mi (50 km)

Rock type: Table Mountain sandstone

The Cape Peninsula runs south from Table Bay and Table Mountain, ending in an area of windswept beaches and soaring cliffs. The first man known to have seen it from the sea was Bartholomeu Dias in 1488. There are several names associated with the southern tip of the peninsula. Both the names Cape of Storms (*Cabo Tormentoso*) and Cape of Good Hope (*Cabo de Boa Esperança*) have been attributed to Dias. The region is shaped rather like a foot with the toe pointing east—the Cape of Good Hope now refers to the western end of the "heel," while the easternmost point of the "toe," with its great cliffs and magnificent views over False Bay, is known as Cape Point. Sir Francis Drake, rounding the Cape in 1580, referred to it as " … the fairest cape we saw in the whole circumference of the Earth." The legend of *The Flying Dutchman* originated here in the 17th century. The captain, Vanderdecken, had so much difficulty rounding the cape that he is said to have called on the devil to help him. As a result he and his ship were condemned to round the cape for "as long as time itself shall last." **HL**

RIGHT *Cape Point: beautiful but tricky to navigate.*

CAPE HANGKLIP

EASTERN CAPE, SOUTH AFRICA

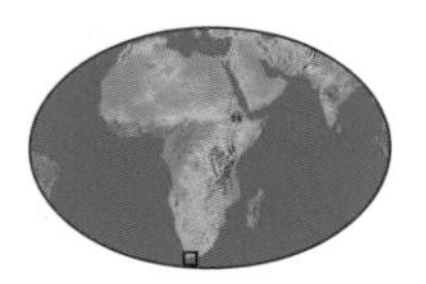

Height of Cape Hangklip: 1,500 ft (450 m)

Rock type: Table Mountain sandstone

Notable feature: often mistaken for Cape of Good Hope

Cape Hangklip forms the eastern end of the mouth of False Bay at Africa's southwestern tip. The name, "hanging rock," comes from the 1,500-foot- (450-meter-) high sandstone peak which towers nearby. From certain angles its cliffs appear to overhang the sea. The Portuguese—who were virtually the only mariners to sail these seas for the century following the discovery of the sea route to India in 1488—named the peak *Cabo Falso* or "False Cape." When sailing westwards, it was easy to mistake Hangklip for Cape Point and turn northwards into False Bay, rather than up the Atlantic coast, farther to the west. The area surrounding Hangklip has a number of attractive sandy coves, separated by rocky points. To the south stretches the ocean—often rough along this coast—while to the north there are magnificent views across False Bay to Table Mountain and the Cape Peninsula. Until the building of a road during World War II, this region was relatively isolated; escaping slaves used it as a sanctuary during the 18th century. It is a good bird-watching destination, and pathways along the shore provide a vantage point for whale-watching, especially between August and November. **HL**

CAPE AGULHAS

EASTERN CAPE, SOUTH AFRICA

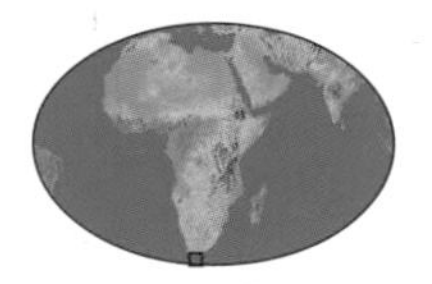

Length of Agulhas Bank: 125 mi (200 km)

Rock type: sandstone

Cape Agulhas does not have the awe-inspiring cliffs of Cape Point, but it is still of interest both as the southernmost tip of the African continent and the western end of the Agulhas Bank. The African continental shelf projects into the sea, south and east, for more than 125 miles (200 kilometers). At Agulhas, the sandstone, which forms such dramatic cliffs further up the coast, has become a low rock platform that extends long spurs into the surf. *Agulhas* is the Portuguese word for "needles,"—it is likely that the cape was named after these rocky outcrops. The vast Agulhas Bank, which stretches out unseen beneath the waves, has important consequences for the region. Partly because of it, the warm Agulhas Current is forced southward here and loses itself in a maze of swirls and eddies. The awkward currents, tides, and winds that result have led to there being more shipwrecks along the Agulhas coast than on any other open coastline in southern Africa. Early Portuguese navigators found this a strange and dangerous place. Cape Agulhas has one other important point of interest: Modern hydrographers have determined that this is where the Atlantic and Indian oceans meet. **HL**

TABLE MOUNTAIN

WESTERN CAPE, SOUTH AFRICA

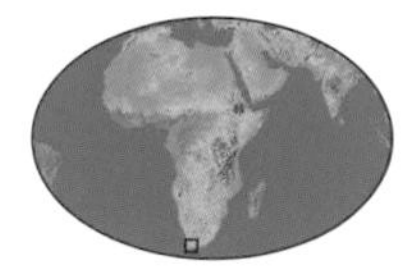

Length of Table Mountain plateau: 2 mi (3.2 km)

Age: 400 to 500 million years

Rock type: sandstone

Table Mountain can be seen and recognized from over 125 miles (200 kilometers) away, a reassuring beacon for sailors rounding the southern tip of Africa. In 1488, the Portuguese seafarer Bartholomeu Dias was the first European to see it, and today it is probably South Africa's most famous landmark. In geological terms, it is a huge block of sandstone that was laid down on the floor of a shallow sea 400 to 500 million years ago. Major earth movements lifted the rock to its present height of 3,560 feet (1,086 meters) above sea level. The "table" is about 2 miles (3.2 kilometers) long, with a distinct landform at each end—a conical hill known as Devil's Peak at one end and the distinctive Lion's Head at the other. In summer the top is sometimes veiled in a layer of cloud, giving the appearance of a tablecloth. Below the mountain, a network of paths crisscrosses green slopes scattered with wild flowers. The mountain is also home to a huge variety of fauna—some, like the Table Mountain Ghost Frog, are found nowhere else. A cable car takes less energetic visitors directly to the top. From here the city of Cape Town can be seen sprawling directly below, and, on a clear day, the Cape of Good Hope is just visible. **MB**

CEDARBERG RANGE

WESTERN CAPE, SOUTH AFRICA

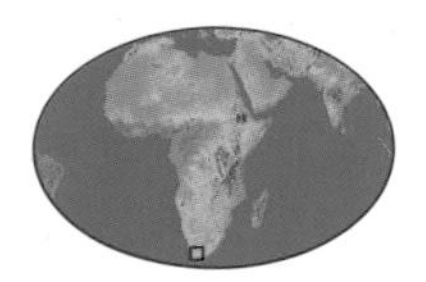

Length of Cedarberg Range: approx. 56 mi (90 km)

Width of Cedarberg Range: approx. 25 mi (40 km)

Rock type: Cape sandstone

Situated about 125 miles (200 kilometers) north of Cape Town, the Cedarberg Range rivals the Drakensberg as South Africa's best-known mountain area. With several peaks around 6,600 feet (2,000 meters) in height, the range is famous for its climbing and hiking opportunities, as well as a remarkable wealth of rock art—upward of 2,000 sites contain paintings which may be more than 5,000 years old, making it one of the richest areas of rock painting in the world.

The range takes its name from the Clanwilliam cedar, but fire and exploitation in earlier years have taken their toll and there are now few mature specimens. The highest peak in the range, Sneeuberg (Snow Mountain), towers over the landscape and can be seen from Table Mountain on a clear day. Its upper reaches are home to the rare snow protea, which only grows above the snowline on a few of the higher peaks. Among the many unusual rock formations in the area are the Maltese Cross (a dramatic 66-foot- [20-meter-] high pillar of weathered sandstone like a giant fist thrust into the sky), the Wolfberg Arch (a freestanding sandstone arch), and 100-foot- (30-meter-) high Wolfberg Cracks. HL

LANGEBAAN LAGOON

WESTERN CAPE, SOUTH AFRICA

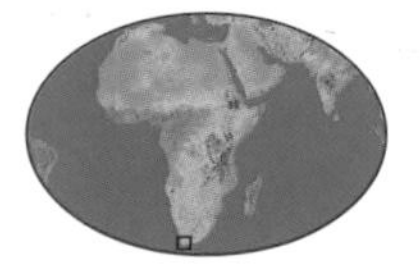

Length of Langebaan Lagoon: 10 mi (16 km)

Width of lagoon: 2.8 mi (4.5 km)

Situated on South Africa's Atlantic coast, just over 62 miles (100 kilometers) north of Cape Town, the Langebaan Lagoon is a 10-mile- (16-kilometer-) long southward extension of Saldanha Bay. Its shallow azure waters are home to a variety of fish species that, in their turn, attract great flocks of birds. The 18th-century naturalist le Vaillant described seeing an "impenetrable cloud of birds of every species and all colors," and even now, 200 years later, it is estimated that more than 100,000 birds can be seen in the spring and fall, when migrants from as far away as Siberia, Greenland, and Northern Europe arrive and leave. The salt marshes that line the lagoon, apart from attracting their own bird species, are also home to specialized succulent plants, and springtime is the season for spectacular displays of wildflowers, which draw many visitors. Birds and flowers are not the only living things to have inhabited the lagoon; people have frequented the area virtually since the birth of modern humans. Human footprints, preserved in slabs of sandstone near the water's edge, have been dated to 117,000 years ago, the period when modern humans are thought to have evolved in southern Africa. HL

THE HEX RIVER

WESTERN CAPE, SOUTH AFRICA

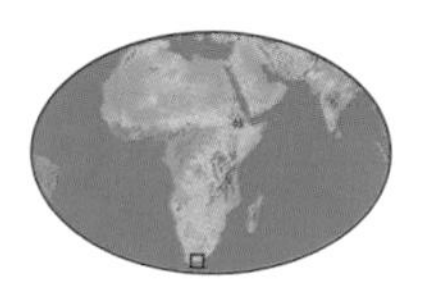

Highest peak in Hex River region (Matroosberg): 7,382 ft (2,250 m)

Notable feature: Cape mountain and valley

The name "Hex River" conjures up different images to different South Africans. To the traveler on the main highway linking Cape Town with Johannesburg or South Africa's famous Blue Train, it is a lush, beautiful valley and pass, linking the fertile farming areas of the Western Cape with the broad, arid expanses of the Great Karoo. To the farmer it refers to one of South Africa's foremost fruit and vineyard areas, famous for its export-quality produce, while to the mountaineer it calls to mind the magnificent peaks from which the river gets its water.

The Matroosberg, at over 7,382 feet (2,250 meters), is the highest point in the range and provides opportunities for winter skiing. Of Milner Peak it has been said that it is "an incredible surge of rock, thrust up like Atlas's shoulders to support the sky." The river that waters this valley is said to have originally taken its name from the letter X, referring to the number of times that the early track crossed through the water on its way up the valley. Later it was changed to Hex in memory of a young woman whose lover died in the mountains, where she had sent him to collect a rare flower. Her spirit is said to wander there still. HL

THE SWARTBERG

WESTERN CAPE, SOUTH AFRICA

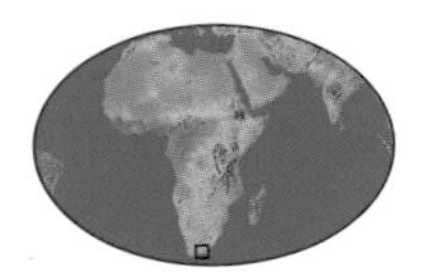

Width of Swartberg: about 12 mi (20 km)

Highest point: 7,628 ft (2,325 m)

Geology: dramatic sandstone folds

The Swartberg mountain range runs for over 125 miles (200 kilometers) west to east, forming a rampart between the Great Karoo and the Little Karoo. It is part of the Cape Fold mountain system and was initially formed by the movement of the African, South American, and Australian continental plates when they were all part of the Gondwanaland supercontinent 250 million years ago. The spectacular cliffs of folded layers of sandstone—of red, yellow, and ocher hues interspersed with green lichens—tower nearly 5,000 feet (1,500 meters) upward. The most accessible are at Meiringspoort. Farther to the west, Seven Weeks Poort, an area with even more dramatic rock-folding, is threaded by a primitive gravel road.

Near the eastern end of the range, Toorwaterpoort offers a route to the railway. Built a century ago, it was used to transport ostrich feathers from the Little Karoo to the coast. The name means "magical waters" and comes from nearby hot springs, where the mists forming over the water in the twilight appeared, to the early inhabitants, to be ghosts. To this day there is only one route for vehicles over the mountain—the Swartberg Pass—which remains a little-changed gravel road. **HL**

GAMKASKLOOF—THE HELL

WESTERN CAPE, SOUTH AFRICA

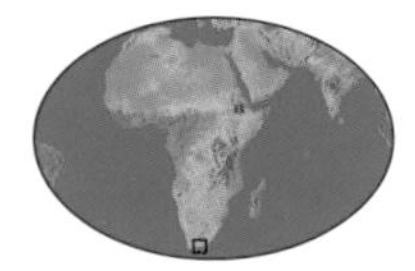

Width of Gamkaskloof: approx. 1.2 mi (2 km)

Depth of Gamkaskloof: approx. 2,000 ft (600 m)

Height of mountains: 5,610 ft (1,700 m)

Deep in the Swartberg Mountains of the Western Cape there is a 12-mile- (20-kilometer-) long valley which was successively the home of San, Khoikhoi, and Afrikaner farmers but had no road access until 1962. Officially known as Gamkaskloof, most people call it "The Hell" because of a remark by an early stock inspector who said that it was "a helluva place to get into or out of." *Gamka*, the Khoi word for "lion," is the name of the river that breaks through the 5,610-foot- (1,700-meter-) high mountains and brings life to the valley. Rock paintings from San days, together with tools left behind by earlier inhabitants during Stone Age times and pottery fragments from the Khoikhoi, are all to be found in the valley. In the early 19th century, Dutch farmers sought refuge here and stayed on with little change in lifestyle over the decades. During the Boer War, a group of Boers evading pursuit by the British took to the mountains and were astonished to come upon this isolated community who spoke an archaic form of Dutch. Sadly, the dramatic road, which runs for 30 miles (50 kilometers) from the top of the Swartberg Pass, was to prove the death knell for this way of life. **HL**

CANGO CAVES

WESTERN CAPE, SOUTH AFRICA

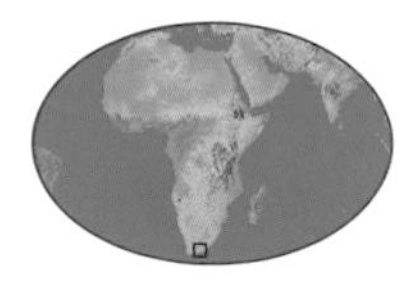

Length of Cango Caves: 3.3 mi (5.3 km)

Notable features: stalagmites, stalactites, hanging curtains

Deep below the foothills of South Africa's Swartberg Mountains is the labyrinth of caverns, tunnels, and underground lakes that forms the Cango Caves. They were discovered in 1780 by cattle herder Klaas Windvogel who, accompanied by his employer Mr. Van Zyl and local teacher Barend Oppel, was lowered into the first great chamber. By the light of their flickering torches they discovered a stalagmite 30 feet (9 meters) high, which was named Cleopatra's Needle. The first chamber, known as Van Zyl's Hall, is 330 feet (100 meters) long and 50 feet (15 meters) high, but in more recent years cave explorers have found more chambers, including one 1,000 feet (300 meters) long. With names like the Crystal Forest and the Throne Room, each chamber features stalagmites and stalactites made of calcite—a crystallized form of calcium carbonate or chalk—that grow into distinctive shapes. Bhota's Hall has hanging gothic curtains and a floor-to-ceiling column called the Leaning Tower of Pisa, while the Bridal Chamber has a structure similar to a four-poster bed. The formations are generally tinged red and pink by iron oxide, but some stalagmites lack the pigment and resemble white-hot pokers. **MB**

GREAT KAROO

WESTERN CAPE, SOUTH AFRICA

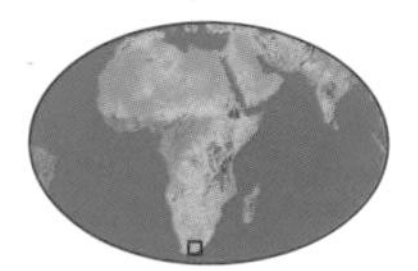

Area of Karoo National Park: 79,070 acres (32,000 ha)

Vegetation: open Karoo scrub with succulents

Age of fossils: up to 300 million years

Covering a huge swathe of southern South Africa—over 154,000 square miles (400,000 square kilometers)—is the semi-arid Great Karoo, part of which has been set aside for visitors in the Karoo National Park. Approximately 250 million years ago, the Great Karoo was a vast inland sea. As the climate changed, however, it evaporated, creating a swamp crawling with reptiles and amphibians. These swamps have long since disappeared, leaving a dry grassland, where, until the 19th century, large herds of zebra and antelope roamed, sharing their habitat with the Hottentot people, who at the time called the area "the place of great dryness." Over its long history, layer after layer of rock has been laid down on the plains of the Great Karoo, and now it is one of the richest sites for paleontologists in the world. The final geological act was one of cataclysmic volcanic activity, followed by a long period of erosion that slowly uncovered the succession of strange creatures that have called the Great Karoo home—from the bizarre *Pareiasaurus*, which looked like a cross between a hippopotamus and a crocodile, to the mammal-like reptiles and the early, rat-sized true mammals. **PG**

PLETTENBERG BAY

WESTERN CAPE, SOUTH AFRICA

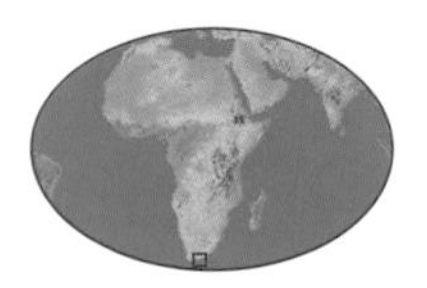

End of coastal whaling: 1916

Number of southern right whales in 1916: 40 females

Number of southern right whales today: 1,600 females

Plettenberg Bay is a striking horseshoe-shaped bay lying on the southern Cape coast, formed by the Robberg, a long promontory curving south and southeast. The early Portuguese explorers were so taken with the area that they named it *Bahia Formosa*, ("beautiful bay"), and, with its long beaches, dramatic peninsula, lagoons, and indigenous forests, it is still worthy of this title today. However, there is much more to Plettenberg than the view. This is where Atlantic and Indo-Pacific marine life meet and it has become well-known for its great diversity of whales and dolphins—perhaps the highest number of cetacean species anywhere in the world. The bay has long been recognized as a calving area for the southern right whale and for its huge schools of common dolphins—the largest estimated to contain up to 9,000 individuals. But recent research has revealed more—Bryde's and minke whales are residents, while humpback whales visit in June and July as they travel north and again from November to January on their return to the Antarctic. Killer whales are also seen regularly. It is no surprise, therefore, that the area has become a favorite with whale and dolphin watchers. PG

TSITSIKAMMA COAST

WESTERN CAPE, SOUTH AFRICA

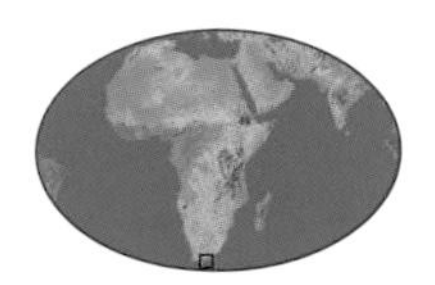

Length of Tsitsikamma National Park: 50 mi (80 km)

Walking trails: Otter Trail: 30 mi (50 km); Tsitsikamma Trail: 45 mi (72 km)

Vegetation: includes 800-year-old yellowwood trees

Lying along the southern coast of South Africa between Plettenberg Bay and Oyster Bay is the rocky Tsitsikamma coast, where cliffs drop vertically into the sea from a gently sloping plain about 660 feet (200 meters) above sea level. The plain, which stretches up to the base of the Tsitsikamma Range, was probably caused by the action of waves before the land rose and the sea level fell. Today, the rocky coast plunges below the waves to depths of over 100 feet (30 meters) in places. The action of water, both from the sea and rivers, has created a spectacular landscape. At the base of the cliffs the waves have eroded a new ledge, known as a wave-cut platform, while slicing into the coast are steep-sided ravines carved out by swift flowing rivers such as the Storms, Blaukraanz, and Groote. The sea has invaded some river mouths, while others, such as the Groote estuary, have been blocked by sand, leaving small lagoons and short stretches of beach. Where large pieces of cliff have collapsed, small islands have been formed. Near the mouth of the Storms River is the Schietklip, a rock that causes huge waves. PG

BELOW *Tsitsikamma is known for its beaches and lagoons.*

WILDERNESS LAKES

WESTERN CAPE, SOUTH AFRICA

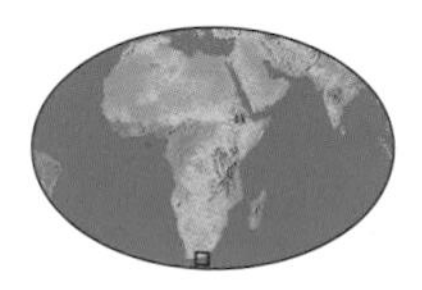

Area of Wilderness National Park: 6,454 acres (2,612 ha)—now part of Lakes Area National Park

Vegetation: fringing reeds, sedges

The Wilderness Lakes lie east of George, close to the southern Cape coast. In reality they are shallow lagoons—sometimes called estuarine lakes—rather than true lakes, and were formed by the infilling of the estuary by sediment, either deposited by the river or by wind-blown sand. They lie parallel to the coast, occupying an area 1 mile (1.6 kilometers) wide and 10 miles (16 kilometers) long. The Touw River formed the lagoons in the west of the area. These lakes are now separated from the sea by sand dunes that have been stabilized by the plants growing on them. When the rivers flood, however, they break through the coastal sands at the former river mouth and enter the sea. For a short time both the river and lakes are slightly tidal, as seawater flows into the lake at high tide. This allows fish into the lakes to spawn. Only one lake (Groenvlei) is totally isolated by the wind-blown sand, with neither a river flowing into it nor an outlet to the sea. Legend maintains that one of the lakes is home to a mermaid, and nearby San paintings depict a woman with a fish tail. Regardless of this reputed occupation, these lakes are a unique wetland habitat, now part of the Wilderness National Park. PG

LAKELAND—THE GARDEN ROUTE

EASTERN CAPE / WESTERN CAPE, SOUTH AFRICA

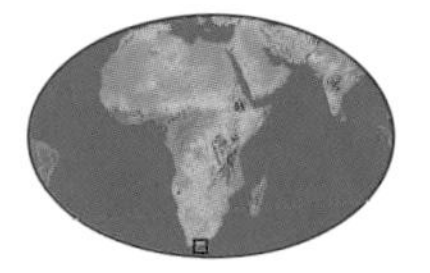

Skeletal remains of early man: 100,000 years old

Notable features: lakes, seashore, forests, mountains

Halfway along South Africa's southern coast—between the cities of Cape Town and Port Elizabeth—the sea coast and the mountain ranges narrow the coastal plain to a mere three miles (five kilometers) in places. The rapid 3,300-foot (1,000-meter) rise from the sea to the peaks, together with the prevailing winds off the warm currents of the Indian Ocean, has created a region of rare beauty. The combination of forest and lakes, of mountains and rivers, of beaches, cliffs, and lagoons is not one that occurs commonly on our planet's surface. Much of the land between the mountains and the sea is in the form of a plain about 660 feet (200 meters) above the sea. At the southern edge of the plain the land drops sharply, either over cliffs into the surf below or down wooded bluffs to the sandy soils of the coastal lowlands. Deeply incised valleys, carrying rivers of dark water, brown with dissolved tannin, dissect the upper plain.

At the town of Kansan, the river runs into the sea through two impressive cliffs, known as the Knysna Heads, but elsewhere sandbars often block the river mouths in the drier months. In the center of the region, heading east, lies the Garden Route, a chain of five lakes created in the valleys—from Wilderness to Knysna—between the sea and the Outeniqua Mountains. It is thought that these valleys were originally formed by the actions of wind and tide. There is increasing evidence, firstly from archeological digs and now from the genetic markers in our bloodlines, that it was in this area that modern humans, *Homo sapiens sapiens*, first lived. Coastal caves have yielded the remains of skeletons similar to our own, but dating to more than 100,000 years ago, while a 77,000-year-old pattern on a stone is claimed by some to be the world's earliest "art." More recently, in the mountains, the San people have produced remarkable paintings of creatures with human torsos and forked tails, reminiscent of mermaids in some cases, or of swifts or swallows in others.

The rapid rise from sea to peak, combined with the warm currents of the Indian Ocean, has created a region of rare beauty. At the southern edge of the plain the land drops sharply down wooded bluffs to the sandy soils of the coastal lowlands.

Today the region is home to a wealth of wildlife, including over 250 species of bird. Offshore, marine reserves provide a home to dolphins, seals, and southern right whales. **HL**

RIGHT *The rocky shores and wild seas of the Garden Route.*

ROBBERG

EASTERN CAPE / WESTERN CAPE, SOUTH AFRICA

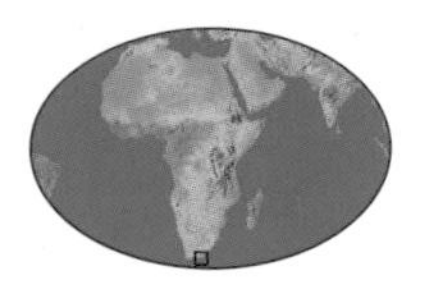

Width of Robberg: about 2,145 ft (650 m)

Rock type: sandstone

Notable features: superb coastal scenery, archeological sites

Peninsulas are comparatively rare on the southern African coast, but roughly 310 miles (500 kilometers) east of the cape toward Port Elizabeth, the beautiful headland known as Robberg (*Rob* being the Afrikaans word for "seal") thrusts 2.5 miles (4 kilometers) eastward into the Indian Ocean. The beauty of Robberg itself is complemented by magnificent vistas across Plettenberg Bay to the distant forests and mountains of the Tsitsikamma. It is no wonder that the early Portuguese mariners named this place *Bahia Formosa*—the "Beautiful Bay." The bay is created by Robberg, which runs parallel to the east-west line of the coast, but which gives the impression of jutting southward directly into the ocean. Here history links to prehistory. The underlying sandstone rocks link with the ancient southern continent of Gondwanaland. Scattered over the peninsula are Middle Stone Age artifacts, showing that early man was living here 100,000 years ago. More recently, the archeological record revealed in Nelson's Bay Cave gives a glimpse of receding shorelines followed by rising sea levels, as the last ice age firstly accumulated the waters in the polar ice caps, later releasing them to come flooding back. HL

THE MALDIVES

MALDIVE ATOLLS, INDIAN OCEAN

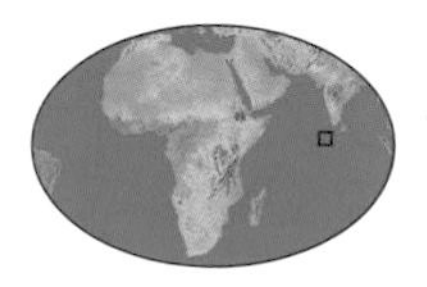

Area of the Maldives: 34,749 sq mi (90,000 sq km)—99 percent sea-covered

Length of archipelago: 510 miles (820 km)

Width of archipelago: 75 miles (120 km)

Lying in the Indian Ocean, southwest of Sri Lanka, this nation of 1,190 coral islands straddles the equator. Some 200 are inhabited, 87 as exclusive resort islands. Others are used only for drying fish and copra, while some remain totally uninhabited. The origins of the 270,000 native Maldivians are uncertain, but the islands have been populated for at least 7,000 years. Based on the geological feature known as the Laccadives–Chagos Ridge, the islands are divided among 26 atolls, each a ring-shaped coral reef enclosing a shallow lagoon (a form known locally as *faru*). Lagoons are often a beautiful blue or green, while the beaches are sparkling white sand. The average temperature hovers around 84 to 89°F (29 to 32°C). April is the hottest month, December the coolest. May to September is the wet (monsoon) season. However, strong storms are very rare events. The land-living biodiversity is poor. Plant specificity is low and bird diversity is also low, with 118 species, mostly seabirds. The glory of the Maldives is its coral reefs, rich in species of coral, fish, and strange marine invertebrates. **AB**

BELOW *The beautiful blue lagoons of the Maldives.*

ALDABRA ATOLL

ALDABRA GROUP, SEYCHELLES

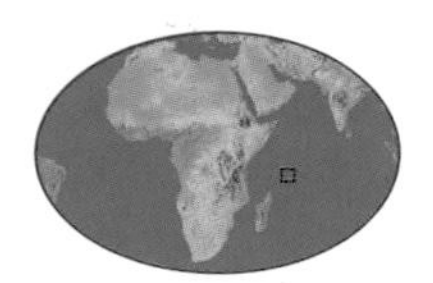

Age of Aldabra Atoll: approx. 125,000 years

Area of atoll: 59 sq mi (154 sq km)

Area of lagoon: 34,593 acres (14,000 ha)

The remote Aldabra Atoll is one of the world's largest coral atolls and home to the world's largest population of giant tortoises. The atoll comprises four main coral islands (Grand Terre, Malabar, Polymnie, and Picard) separated by narrow sea passages enclosing a large shallow lagoon. The surface of the ancient coral reef limestone is about 26 feet (8 meters) above sea level. It has been eroded by the weather, leaving razor-sharp rocks on which it is difficult to walk. The giant tortoises of Aldabra were almost wiped out by the end of the 19th century but have made a remarkable comeback, with more than 150,000 today. Endangered green and hawksbill turtles also come ashore to lay their eggs on Aldabra's beaches. The atoll is also an important breeding site for birds such as tropic birds, frigate birds, boobies, and terns. The Aldabran white-throated rail is the last flightless bird species to be found in the Indian Ocean. Robber crabs, which have disappeared from many other Seychelles islands, climb palm trees and scour the beaches for coconuts. The lagoon virtually empties when the tide is out, but the incoming tides fill the lagoon to a depth of 10 feet (3 meters). RC

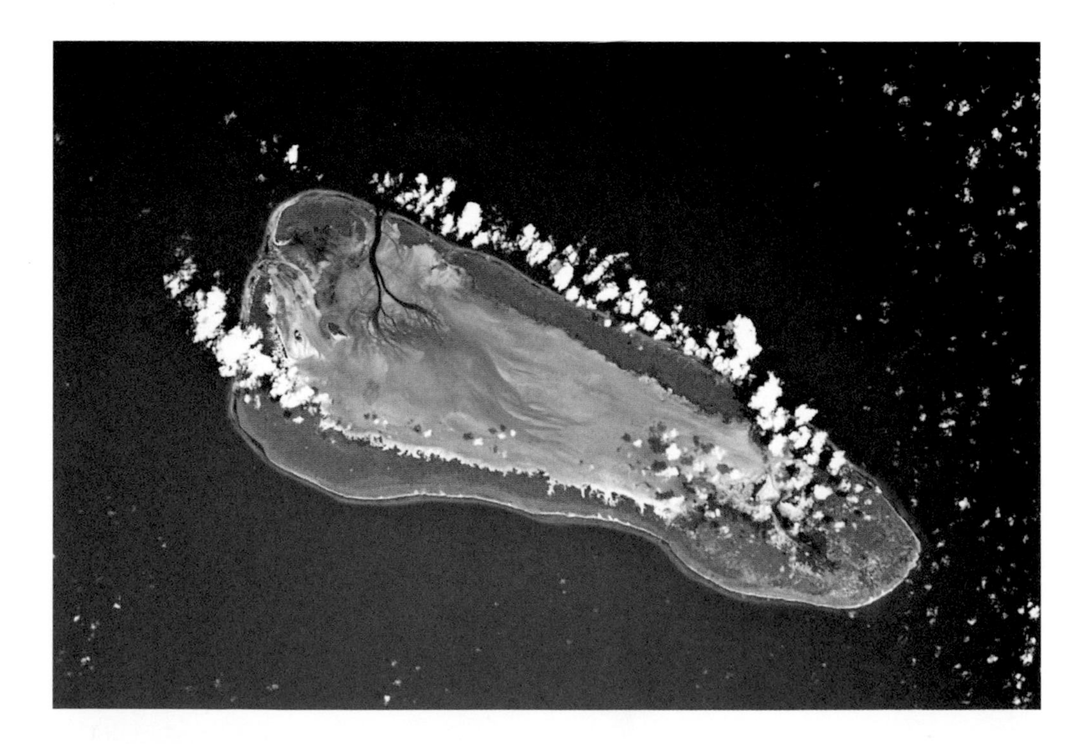

VALLÉE DE MAI

PRASLIN, SEYCHELLES

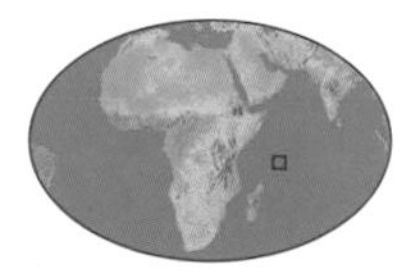

Area of Vallée de Mai: 48 acres (19.5 ha)

Area of Praslin: 16 sq mi (42 sq km)

Designated a World Heritage site: 1983

The Seychelles Islands have been isolated from any major landmass since the time of the dinosaurs and have many strange plants and animals. Granitic in origin, the second largest island is Praslin, and at its center is a mysterious valley full of palms—the Vallée de Mai. In places the canopy is so dense that no sunlight filters through. Freshwater crabs and giant crayfish inhabit the streams, and rare birds flit among rare trees. Many visitors have called it the Garden of Eden—an idea largely engendered by the extraordinary coco de mer palm. This tree produces a double nut, the largest in the world, shaped like a female pelvis. The spike-shaped male flower is also highly suggestive. At night, local people say the male trees lean over the female trees, and no human witnessing their behavior lives to tell the tale. A single nut can weigh 40 pounds (18 kilograms) and take 10 years to germinate. The palm also holds the record for the world's largest leaf—35 square feet (3.3 square meters) in size. Until the Vallée de Mai was found, people beyond the Seychelles thought the strange nuts came from the bottom of the sea—hence, the French name meaning "sea coconut." JD

TSINGY LANDS

MAHAJANGA, MADAGASCAR

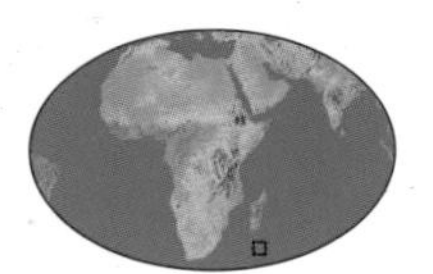

Bemaraba Cliff: 1,320 ft (400 m) above river valley

Maximum height of pinnacles: 100 ft (30 m)

In just two places—in the center of the Ankarana Plateau and in the Bemaraba Reserve—there is a type of landscape that is quite unique to the island of Madagascar. It is an unusual world where row after row of razor-sharp limestone rock pinnacles, some up to 100 feet (30 meters) high, have wafer-thin edges that could slice off an arm or a leg of a careless traveler. They are formed when heavy rain, averaging 70 inches (1,800 millimeters) a year, dissolves the soft upper limestone, leaving the harder lower parts as sharp pinnacles. They are known locally as *tsingy* because of the bell-like sound they make when struck, and they are so close together that the Malagasy say there is not one place between the pinnacles to set your foot safely. There are, however, creatures that walk and jump nonchalantly in the forest of sharp limestone needles and blades. They are lemurs, woolly primates unique to Madagascar, and many species are exceedingly rare. The inaccessible Tsingy Lands provide them and other rare animals, such as the chameleon *Brookesia peramata* and the gray-throated rail, with a haven. MB

RIGHT *The razor sharp pinnacles of the Tsingy Lands.*

ANKARANA PLATEAU

ANTSIRANANA, MADAGASCAR

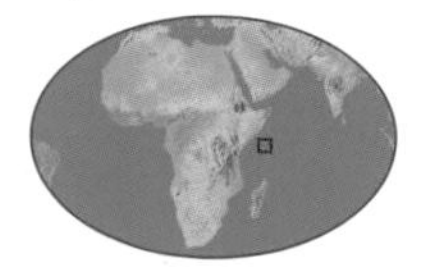

Area of Ankarana Plateau: 39 sq mi (100 sq km)

Rock type: limestone

Limestone thickness: 500 ft (150 m)

Madagascar is known as the "great red island" because of its soil color, and at its northern end is an African-style "lost world." The Ankarana Plateau is located 62 miles (100 kilometers) to the south of Diego Suarez. This is *karst* country, the main rock being limestone that has been eroded dramatically by water. Streams disappear into fissures and reappear deep underground in caverns and tunnels, such as the spectacular Grotte d'Andrafiabe—7 miles (11 kilometers) of spectacular galleries lined with impressive stalagmites and stalactites. The roofs of some caves have collapsed totally, leaving huge skylights hundreds of feet overhead. Here the sun pours in on the cavern floor encouraging isolated pockets of virgin forest, known as "sunken forests," to flourish. There are also thickly wooded, long, narrow canyons. They are home to lemurs—named after the Roman *lemures*, meaning "spirits of the dead"—as well as the aggressive cat-like fossa that hunts down lemurs, such as dwarf lemurs and sifakas. At certain times of the year, the underground waterways also hide even more dangerous animals, such as Nile crocodiles. MB

TROU AUX CERFS

PLAINES WILHEMS, MAURITIUS

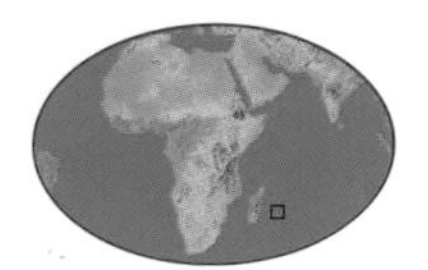

Elevation of Trou aux Cerfs: 2,145 ft (650 m)

Diameter of crater: 1,100 ft (335 m)

Depth of crater: 280 ft (85 m)

Trou aux Cerfs is a dormant volcano crater near the town of Curepipe that offers breathtaking panoramic views of the island of Mauritius. A road runs up the slope of the volcano to the crater, 2,100 feet (650 meters) above sea level. Near the rim of the volcano—which is 1,100 feet (335 meters) in diameter—there is a meteorological station that monitors cyclonic activity in the region. It is possible to climb down the inside of the deep, densely vegetated crater to the lake. The panoramic view looks out over the plateau towns and the mountains to the north and northwest. To the west are the three conical peaks of Trois Mamelles, and Montagne du Rempart, described by Mark Twain as a "dainty little vest-pocket Matterhorn." To the northwest is Mont St. Pierre and the Corps de Garde, while to the north are the Moka Mountains, the thumb-shaped peak of Le Pouce, and the extraordinary Pieter Both Mountain with a massive boulder balanced on its summit. Trou aux Cerfs is located to the west of Curepipe, a town known for its model shipbuilding and tea industries. Other attractions include a botanical garden and the nearby Tamarind Falls. RC

BLACK RIVER GORGES

BLACK RIVER, MAURITIUS

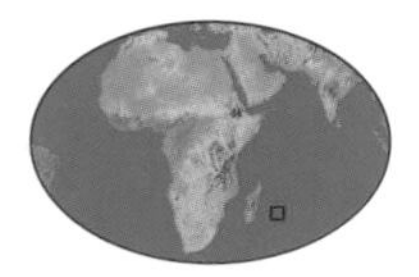

Area of Black River Gorges National Park: 16,240 acres (6,574 ha)

Area of Mauritius: 788 sq mi (2,040 sq km)

Highest point (Mt. Piton): 2,717 ft (828 m)

The Black River Gorges is a forest-clad area in the southwest of Mauritius. It was declared a national park in 1994 to protect much of the island's woodland—less than one percent of the island's original forest cover remains intact. The Black River Gorges National Park is the largest nature reserve on the island. There are nine species of bird and more than 150 species of plant that are found only on Mauritius. The national park has helped to ensure the survival of threatened native species such as the Mauritius kestrel and the pink pigeon. Visitors to the park, however, are more likely to see fruit bats and white-tailed tropic birds flying among the trees, or encounter monkeys, wild pigs, and deer. The gorges contain specimens of trees such as black ebony, tambalacoque (dodo tree), and the umbrella-shaped *bois de natte* tree. On a smaller scale there are many ferns and lichens, as well as flowering plants such as orchids and the national flower of Mauritius, known locally as *Boucle d'Oreille*. There are over 30 miles (50 kilometers) of trails in the Black River Gorges; these are easily accessible from Curepipe or Vacoas. The best time to visit is between September and January. RC

CHAMAREL COLORED EARTH & WATERFALL

BLACK RIVER, MAURITIUS

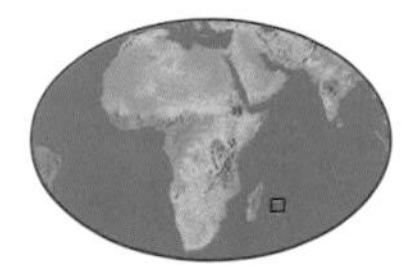

Height of Chamarel Waterfall: 272 ft (83m)

Chamarel earth colors: red, brown, violet, green, blue, purple, yellow

Best time to visit: sunrise

The Chamarel Colored Earth or *Terres de Couleurs de Chamarel* is a striking geological feature on the island of Mauritius. It is an area of undulating land with contrasting layers of colored earth. Known locally as "Seven Colored Earth," the seven hues are at their most striking at sunrise. This small area has no vegetation cover and is an exposed heap of volcanic ash, mineral oxides, and iron ore. The natural phenomenon is thought to be the result of weathering on unevenly cooled molten rocks, resulting in a lunar-type landscape. The ash is unusual because it is made up of elements that do not mix. Specimens of the colored earths in glass tubes are on sale at the site. Curiously, when the colored earths are mixed together in a tube, they separate out into distinct color bands again after a few days. There are several good vantage points to see the nearby *Cascade Chamarel* (Chamarel Waterfall) plunge down a high cliff face into the Riviere du Cap. It is the highest waterfall on Mauritius. The Seven Colored Earth and Chamarel Waterfall are popular tourist attractions about 2.5 miles (4 kilometers) south of the village of Chamarel in the southwest of the island. RC

LES CIRQUES

RÉUNION

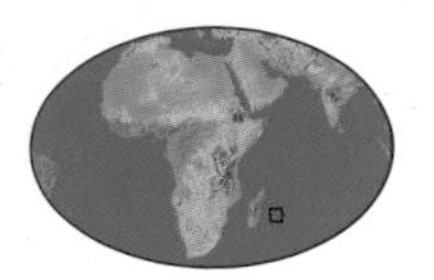

Area of Cirque de Cilaos: 21,600 acres (8,739 ha)

Area of Cirque de Salazie: 25,650 acres (10,382 ha)

Area of Cirque de Mafate: 24,710 acres (10,000 ha)

The cirques of Réunion are deep, bowl-shaped hollows that superficially resemble volcano craters but are actually caused by erosion. The three cirques—Cilaos, Salazie, and Mafate—surround the highest mountain on the island, Piton des Neiges. The wild and remote Cirque de Mafate is inaccessible by road. It is surrounded by the peaks of Gros Morne, Piton des Neiges, Grand Benare, and Roche Ecrite. The Cirque de Salazie, to the east of Mafate, has two main villages, Salazie and Hell-Bourg. The wettest of the three cirques, it has spectacular waterfalls such as the *Cascade du Voile de la Mariée* (Bridal Veil Falls). Cilaos is the most southerly of the cirques. The town of Cilaos is a spa resort whose *sources thermales* (thermal springs) are claimed to cure ailments such as rheumatism. The mountainous cirques region is a paradise for trekkers. Mafate cirque alone has more than 125 miles (200 kilometers) of marked trails. You can hike up Piton des Neiges from either Cilaos or Hell-Bourg, but the final ascent to the 10,069-foot (3,069-meter) high summit should be made very early in the morning. RC

PITON DE LA FOURNAISE VOLCANO

RÉUNION

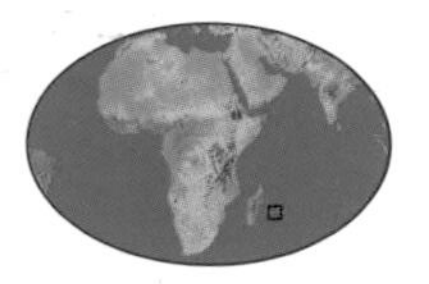

Height of Piton de la Fournaise Volcano: 8,632 ft (2,631 m)

Age: approx. 530,000 years

Area of Réunion: 972 sq mi (2,517 sq km)

Piton de la Fournaise or "The Furnace" is situated on the southeastern half of Réunion in the western Indian Ocean. Along with Kilauea in Hawaii, it is one of the world's most active volcanoes. It has erupted at least 153 times since 1640, most of which have been explosive and produced spectacular lava flows. An eruption occurs almost every year and volcanic activity is monitored by the Piton de la Fournaise Volcano Observatory.

The larger of the two main craters is known as Dolomieu or "Brulant." Only six eruptions (in 1708, 1774, 1776, 1800, 1977, and 1986) have originated from fissures on the outer slopes of the crater. The 1986 eruption added several acres of land to the southeast of the island as a result of lava flowing into the sea. Bory is the smaller of the two major craters. Several minor craters of the volcano have erupted recently, including the crater Zoe on the southeastern flank in 1992. RC

RIGHT *Molten lava spurts from a fissure on Réunion's Piton de la Fournaise Volcano.*

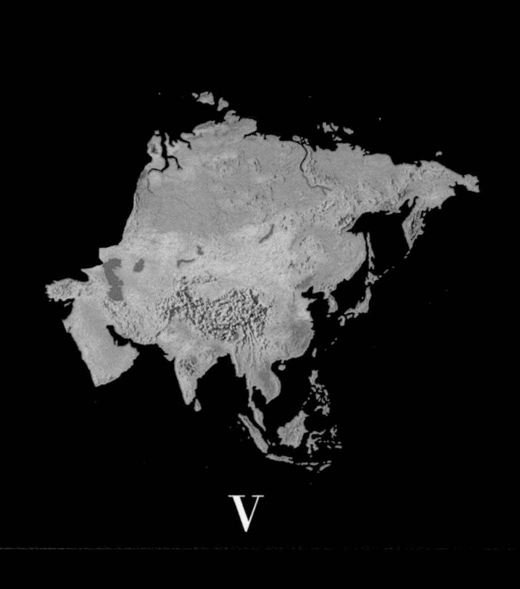

V

ASIA

The dual power of nature is demonstrated through the elements at play in Asia. Fire-spewing volcanoes raise new lands from their summits, yet cause damage with their violent births. Ocean worlds house water kingdoms and so much latent energy, while wind—the element that creates and reshapes landscapes—can also destroy in seconds. Finally, earth-bound and sky-stretching mountains—figures of inspiration, claimants of lives—remind us of both the compatibility and fragility of the human condition against such potentially terrifying forces.

LEFT *The titanic Mount Everest pierces through the Himalayan clouds.*

TAYMYR PENINSULA

TAYMYRSKIY (DOLGANO-NENETSKIY) AVTONOMNYY OKRUG, RUSSIA

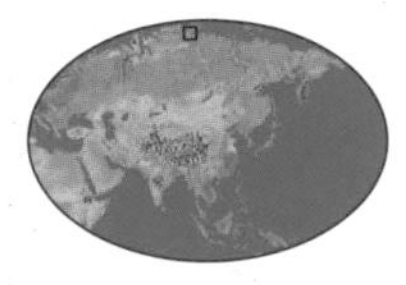

Area of Lake Taymyr: 2,700 sq mi (6,990 sq km)

Width of Siberian tundra: 2,000 mi (3,200 km)

Height of Byrranga Plateau: 5,000 ft (1,500 m)

The Taymyr Peninsula is the most northerly point on the Eurasian mainland and part of the vast Siberian tundra that borders the Arctic Ocean at the top of Russia. For three months in summer, the sun shines for 24 hours, yet the temperature barely rises above 40°F (5°C); in winter the sun may fail to appear at all, resulting in temperatures that plummet to -47°F (-44°C). During winter the soil is frozen, in some places down to a depth of 4,500 feet (1,370 meters), but in summer the top layer melts, creating one of the biggest bogs on Earth.

From the air, the Taymyr is seen to be divided naturally into a honeycomb-like landscape of banks and boggy pools, the result of constant freezing and thawing. On the ground, there is a chance of finding a mammoth tusk or even the frozen body of an entire prehistoric animal. A carpet of mosses and herbs and calf-high forests of dwarf willow cover the area.

The peninsula is dominated by the Byrranga Plateau, with Lake Taymyr—the Arctic's largest lake—at its southern edge. Though vast, it is shallow—no more than 10 feet (3 meters) deep. **MB**

NORTHERN STEPPE & SAIGA MIGRATION

RUSSIA / KAZAKHSTAN / CHINA

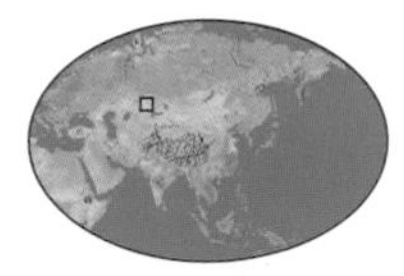

Height of adult saiga: 30 in (76 cm) at shoulders

Estimated world population: two million

Protected species since: 1923

Nature reserves protect large areas of the immense cold steppes and semi-deserts of central Asia, keeping them pristine. Steppe areas—some including wetlands—provide valuable refuge for steppe flora, a number of threatened bird species (such as the Siberian white crane and the Dalmatian pelican), and the critically endangered Saiga. This short-bodied, stout-legged goat antelope has large, sorrowful eyes and an enormously bulbous muzzle that keeps out dust in summer and warms cold air when breathing in winter.

In springtime, herds of Saiga follow the melting snows northward, constantly searching for freshly uncovered grasses, herbs, and sedges; indigenous flowers include tulips, irises, and anemones. The landscape in which they live is a stark, flat, dry, treeless grassland. It is, however, rich in herbs that grow on the black earth and, like the North American prairies, once supported huge numbers of plains animals such as tarpan and bison. Smaller species include rodents such as Bobak marmot, sousliks, and the vulnerable steppe pika, but moose, roe deer, and wild boar are also present. Although the Saiga survive, poaching is a major threat—their numbers have declined from 1 million in 1990 to less than 50,000. MB

SEA OF OKHOTSK

RUSSIA / JAPAN

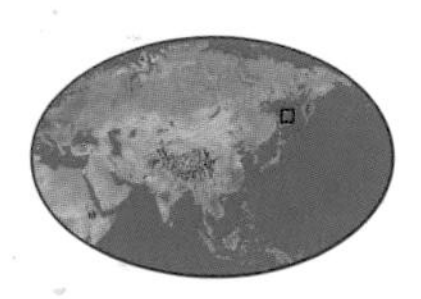

Deepest point of Sea of Okhotsk: 12,847 ft (3,916 m)

Average depth of sea: 2,923 ft (891 m)

The semi-enclosed Sea of Okhotsk is a large marine ecosystem at the margin of Russia and Northern Japan, bordered by Kamchatka to the east, Magadan to the north, Amur to the west, and Sakhalin and Hokkaido to the south. It is deepest near the Kuril Island chain in the southeast—enough to completely submerge Mount Fuji—and far shallower in the north.

The Sea of Okhotsk adjoins the Sea of Japan via the La Pérouse Strait, and Pacific waters enter through the northern Kuril Straits and return between the southern Kuril Islands. It is a highly productive ecosystem, and waters off Kamchatka and in the north and west Sea of Okhotsk are especially abundant in plankton. This extreme productivity has made the Sea of Okhotsk an important region for fish populations, which in turn attract seabirds and gray and bowhead whales. Commercial fisheries focus on walleye pollack, the most abundant species, with an estimated biomass of 10 to 15 million tons, but also include flounder, herring, salmon, halibut, sardine, saury, cod, capelin, crab, and shrimp. Today the sea is rich in oil and gas. **MBz**

RIGHT *The icy Sea of Okhotsk from Shiretoko Peninsula, Japan.*

TYULENII ISLAND

SAKHALINSKAYA OBLAST, RUSSIA

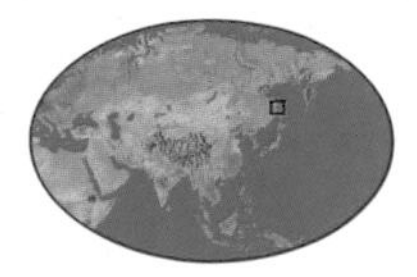

Length of Tyulenii Island: 0.6 mi (1 km)

Width of island: 0.3 mi (0.5 km)

Remote and isolated, Tyulenii Island is a low slab of rock rising 26 to 33 feet (8 to 10 meters) above sea level. Its beaches are crowded with northern fur seals who come in late summer to breed and give birth.

Toward the northern end, and along part of the east coast, huge bull Steller's sea lions battle over space, and females establish their harems and bully the fur seals. The fur seals extend their territory up the eastern slopes onto the flat rock table land of the island; here they share the open space with common murres.

So abundant are the murres that they seem to carpet the island in black and white, and toward the end of the summer thousands of discarded or lost eggs lie in heaps in gullies and crevices. No space is spared, no ledge left unused; even the roofs of the rangers' and researchers' huts are taken over as extensions of the nesting and resting grounds. Auklets nest beneath the floor and underneath the upturned boats that line the beach. It is hard to imagine a higher density of wildlife than exists on Tyulenii Island. **MBz**

VOLCANOES OF KAMCHATKA

KAMCHATSKAYA OBLAST, RUSSIA

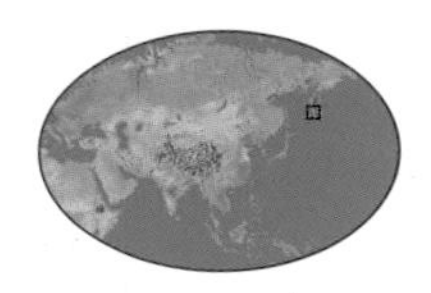

Number of volcanoes: over 300

Number of active volcanoes: 29

The Kamchatka Volcanoes are among the most outstanding in the world. More than 300 volcanoes are found on the Kamchatkan Peninsula, and 29 of them are currently active. They display a variety of types (Strombolian, Hawaiian, Pelean, Vesuvian, Plinian) as well as a diversity of related volcanic features such as geysers, mud pools, and mineralization. What sets this area apart from most other volcanic areas, however, is its biodiversity: 700 species of higher plant, a large marine component in the Bering Sea with many marine mammals and seabird colonies, and very high populations of wildlife species—including snow ram, northern deer, sable, wolverine, and such notables as the brown bear and Steller's sea eagle. In addition, Kamchatka's streams and lakes are some of the most productive salmon spawning areas in the Pacific. Part of the Pacific Rim of Fire, where most of the world's active volcanoes are found, these are among the best-studied volcanoes in the world. GD

BELOW *Kamchatka's powerful peaks puncture the landscape.*

KAMCHATKAN PENINSULA

KAMCHATSKAYA OBLAST, RUSSIA

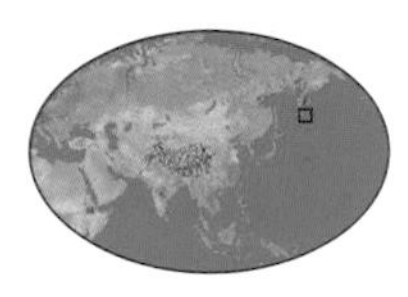

Human population: 400,000

Year first visited by outsiders: 1697

The Kamchatkan Peninsula hangs from the northeast Russian coast like a massive dagger pointing southward. Rugged and mountainous, it is pocked with active volcanoes that dominate a grand landscape with a multitude of thermal and mineral springs, geysers, and other phenomena of active volcanism further complemented by clear lakes, wild rivers, and spectacular coastline. Surrounded by the sea, the peninsula enjoys a moist climate leading to lush vegetation, and it is covered with Erman's birch, larch, poplar, and alder forest. The area is also home to a wealth of wildlife including brown bear, sable, and the magnificent Steller's sea eagle.

One of the last great wilderness areas on Earth, it was first visited by outsiders in 1697, and perhaps most famously by Vitus Bering and his naturalist Georg Wilhelm Steller on their voyage of discovery across the North Pacific. Today the peninsula is a tantalizing magnet for naturalists and fishermen; its remoteness and small population make it attractive but difficult to access. MBz

THE VALLEY OF THE GEYSERS

KAMCHATSKAYA OBLAST, RUSSIA

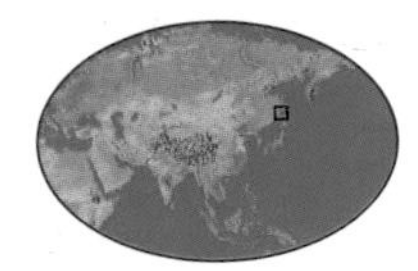

Length of Valley of the Geysers: 3.7 mi (6 km)

Area of geyser field: 1.5 sq mi (4 sq km)

Kamchatka's Shumnaya River ("Noisy River") rushes through rocky narrows and meanders across gravel shoals to the steamy world of the Valley of the Geysers. In April 1941, the Russian hydrologist Tatyana Ivanovna Ustinova stumbled across this steamy wilderness with her Itelmen (an inhabitant of the Kamchatka peninsula) guide, Anisfor Krupenin. Traveling along the riverbed of the Shumnaya, they discovered an intriguing side-stream where they found a bubbling landscape of sulfur springs, boiling mud, and active geysers. This side stream was later called the River Geysernaya.

The valley is a heady paradise—steaming waterfalls cascade down the valley walls, grassy banks breathe with life, geysers erupt jets of boiling water, and bubbling mudpots gurgle and pop. Multicolored clays and waterslides matted with algae mark the landscape, while wafting aromas bear witness to sulfur-belching springs.

The Valley of the Geysers is one of the most active geothermal regions on Earth. For approximately 3.7 miles (6 kilometers), the narrow winding Geysernaya steams, boils, erupts, and emits powerful smells. This single valley has more than 20 major geysers and dozens of smaller ones concentrated in just a 1.1-to-1.5-square-mile (3-to-4-square-kilometer) area. In the fall, colorful foliage adds natural beauty to this geological wonderland, but it is in winter that the landscape becomes truly magical, when all is pure snow white and the drifting steam from the valley coats the trees and shrubs with a delicate covering of hoarfrost crystals.

The heat generated by the geothermal activity of the area has an unusual effect on the surrounding landscape. Come springtime, trees and plants flower long before they do in any other region, and the river bank is home to warmth-loving plants such as water lilies and forget-me-nots.

The valley is a heady paradise–steaming waterfalls cascade down the valley walls, grassy banks breathe with life, geysers erupt jets of boiling water, and bubbling mudpots gurgle and pop. Multicolored clays and waterslides matted with algae mark the landscape.

In October 1981, the Valley of Geysers was hit by Typhoon Elsa. Torrential rains raised the water level in the River Geysernaya by several yards. The swell of water dragged ten-foot (three-meter) sized boulders along the river bed, destroying Geyser Pechya ("Big Oven") and seriously damaging Malakhitovi Grot ("Malachite Grotto"). Despite this, the area is still an amazing place to visit. MBz

RIGHT *One of the many steaming geysers in the Valley of the Geysers.*

LAKE BAIKAL

RESPUBLIKA BURYATIYA, RUSSIA

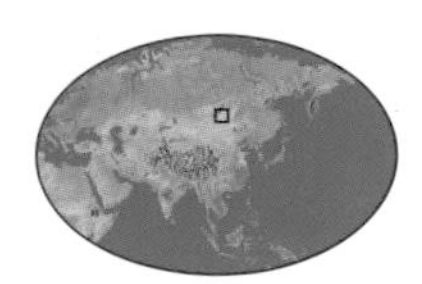

Length of Lake Baikal: 394 mi (635 km)

Width of lake: 30 mi (50 km)

Depth of lake 5,380 ft (1,640 m)

One fifth of the world's freshwater is contained in a single lake—Lake Baikal in Russia's southern Siberia. The lake is relatively small in surface area, ranking ninth in the world. It is 394 miles (635 kilometers) long and averages 30 miles (50 kilometers) wide, but it is extremely deep. It reaches a depth of 5,380 feet (1,640 meters) and contains about 5,500 cubic miles (23,000 cubic kilometers) of water; more than the entire combined contents of North America's Great Lakes.

The lake is also incredibly old. It was formed about 25 million years ago after a rift formed in Earth's crust. Hot springs on the lake bed indicate that the area is still geologically active. Annually, the region's seismic stations register up to 2,000 earthquake tremors.

In winter, the lake is frozen solid and local people drive out across the frozen landscape

to fish through holes they drill through the ice. In places where the ice was formed in calm conditions it is transparent and the fish can be seen swimming below. Although the ice is strong, daily temperature fluctuations cause intricate patterns of cracks that form gaping crevasses as much as three feet (0.9 meters) wide. In summer the ice splinters into tiny slithers, creating prisms of light that dance across the water. After the ice melts, the water can be so clear you can see down to a depth of 130 feet (40 meters) or more.

Many animals living here are unique to Baikal, such as the Baikal seal and the golomyanka fish, which gives birth to live young. The golomyanka fish can endure an amazing amount of pressure—at a depth of 3,300 to 4,620 feet (1,000 to 1,400 meters) it can move quite freely, whereas at such a depth even a cannon cannot shoot because of the enormous pressure. **MB**

BELOW *The shoreline of Lake Baikal silhouetted against the purple Russian sunset.*

YANKICHA—KURIL ISLANDS

SAKHALINSKAYA OBLAST, RUSSIA

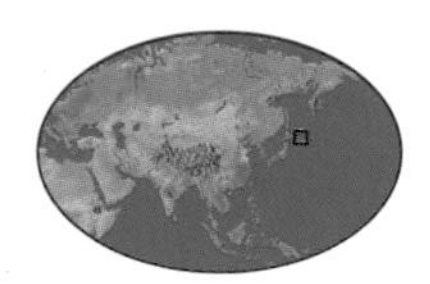

Diameter of Yankicha: 6,600 ft (2,000 m)

Height of Yankicha: 1,263 ft (388 m)

From beautiful Alaid Volcano on Atlasova Island in the north, to the peak within a peak of Tyatya Volcano on Kunashiri Island in the south, the Kuril chain is filled with so many natural geological and biological wonders that to single out one island is to overlook the hundreds of natural splendors on all the other islands. Yet amid so much natural beauty Yankicha Island stands supreme.

Tiny Yankicha, the emergent tip of an extinct volcano, is stunning. The steep-walled caldera is broken to the south and so is flooded by the sea. It forms a tranquil lagoon where harlequins and sea otters swim. The inner grassy slopes rise to the rocky rim where fulmars nest and the cliffs are covered with kittiwakes. Near the lip of the lagoon a scrape in the beach fills with thermal waters; no natural hot-spring bath can have a more spectacular setting. Migratory sperm whales frequent deep waters nearby, and to the north are the Srednego Rocks which attract northern fur seals and bellowing ranks of Steller's sea lions. Yankicha hosts a large colony of whiskered auklets. At dusk and dawn they sweep across the water like bees or smoke clouds as they return to their burrows. MBz

RIGHT *Hokkaido and the Kuril Islands.*

TAMGALY GORGE

ALMATY OBLAST, KAZAKHSTAN

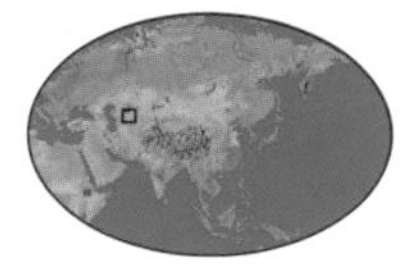

Notable feature: multilayered human settlements spanning twenty centuries

Oldest human habitation: Bronze Age

In the southeast of Kazakhstan, toward the western end of the Tienshan Mountains, the Chu-Ili mountain spur forms a canyon around the Tamgaly Gorge. An abundance of springs, rich vegetation, and shelter distinguishes the area from the arid mountains that fringe the border of Kazakhstan with Kyrgyzstan to the south and from the flat dry plains of central Kazakhstan to the north. Located just 100 miles (160 kilometers) northwest of Almaty, the low slate cliffs of the gorge and its surrounding rocky landscape, where glossy black stones rise up rhythmically in steps, have attracted pastoral communities since the Bronze Age and have come to be imbued with strong symbolic associations. "Tamgaly" in Kazakh means painted or marked place: This is an important archeological site with about 5,000 petroglyphs dating from the second half of the second millennium B.C.E. to the beginning of the 20th century. The central canyon contains the densest amount of carvings and what appear to be altars, suggesting that sacrificial offerings were made here. GD

SINGING SANDS

GOBI DESERT, MONGOLIA

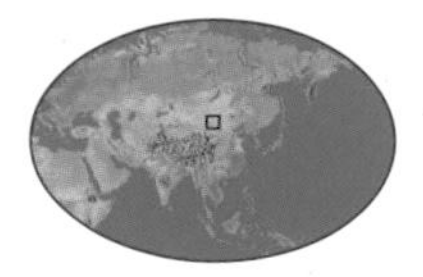

Length of Singing Sands: 120 mi (193 km)

Maximum height of dunes: 2,500 ft (800 m)

The name for the region in Mongolia is *Hongory Els*, meaning "Singing Sands." It refers to the noise made by sand grains as they pass over each other when wind moves them across the surface of the dunes. Unlike most sand particles, which are coarse and irregular, the particles of the Singing Sands are round and smooth. In dry weather conditions, these particles of sand rub against each other, creating an eerie musical sound.

The dunes extend for some 115 miles (185 kilometers) across the southern Gobi Desert between Mount Sevrei and Mount Zuulun (part of the Altati Range). One of at least 30 singing sand sites in the world, they are sensitive to pollution, which can micro-coat sand grains and kill the sonic effect.

The area is also famous for its oases and abundant wildlife, which includes wild sheep, ibex, and gazelles, as well as their predators, leopards and dhole (wild dogs), in addition to an abundance of birds. The most popular oasis is 150 miles (240 kilometers) from the fossil site at the famous Flaming Cliffs. **AB**

FLAMING CLIFFS

GOBI DESERT, MONGOLIA

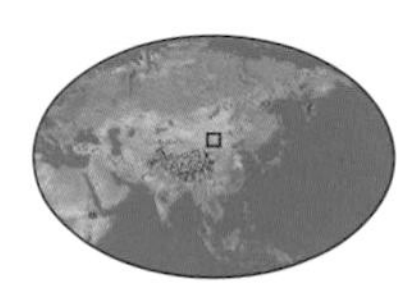

Rock type: fossil-rich sandstone

Age of fossils: 70 to 100 million years

Habitat: semi-desert

Situated in the southern Gobi Desert, this area's European name was given to it by the American paleontologist Roy Chapman Andrews, who was impressed by the glowing orange color of the rock when he was collecting dinosaur fossils there in the early 1920s. In Mongolian, it is known as *Bayanzag*, meaning "rich in saxauls," referring to a type of tree common in the region.

Overarched by a scorching sun, the arid desert and seasonal grassland is punctuated by the glowing red sandstones of the Djadokhta Formation. The area is a dinosaur hunters' paradise, and fragments of bone and dinosaur eggshells are often visible in the red rocks. In the 1920s, Chapman unearthed complete dinosaur skeletons, as well as the first known set of fossilized dinosaur eggs. The remains of early mammals are also found here. Fossil removal is illegal without appropriate permits.

Wildlife includes camels (domestic and feral), gazelles, wild ass, sakar falcons, desert warblers, and finches. For both logistical and bureaucratic reasons, it is exceedingly difficult to reach this area of the desert unless accompanied by a guide. AB

ALTAI MOUNTAINS

MONGOLIA / CHINA / RUSSIA / KAZAKHSTAN

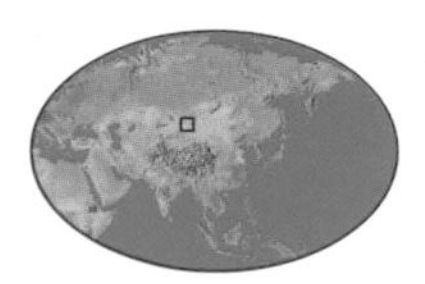

Highest point (twin-peaked Gora Belukha): 14,784 ft (4,506 m)

Climate: extremely cold, dry

Average temperatures: January: -11°F (-24°C); July: 54°F (12°C)

Rugged and dramatically picturesque with a wealth of habitats—coniferous and broadleaf forests, rich alpine pastures, barren ice plains, lakes, and thousands of glaciers—the Altai Mountains run diagonally northwest to southeast where China, Russia, Kazakhstan, and Mongolia meet. The highest peak here is Gora Belukha, which rises along the border of Russia and Kazakhstan. But it is not just the biodiversity that is rich: "Altai" means "gold" in both the Kazak and Mongolian languages because of the gold reserves in the mountains.

Known throughout history as the "cradle of the nomads," the mountains' pasturelands were home to the ancient nomads of China. The Huns (Xiongnu), Turks (Tujue), and Chengiskhan all lived here.

The Altai hit the headlines in the last century when archeologists discovered 2,500-year-old human mummies inside burial mounds. The fragments of the mummies' skin and tattoos, together with silk cloth and artifacts, had been preserved by permafrost. These days, people living in the mountains suffer much poverty, and many settlements and campsites have no electricity.

The region has abundant wildlife, including rare species such as the snow leopard, and many environmental and eco-tourist initiatives hope to integrate economic development projects with conservation. Tours to the mountains can be arranged from Barnaul, Russia, or Almaty, Kazakhstan. **RA**

TIEN SHAN MOUNTAINS

CHINA / KYRGYZSTAN / KAZAKHSTAN

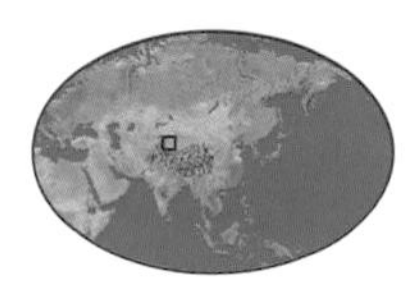

Length of Tien Shan Mountains: 1,800 mi (2,900 km)

Highest peak (Pik Pobedy): 24,406 ft (7,439 m)

Surface area of Lake Issyk-Kul: 64,585 sq ft (6,000 sq m)

Stretching 1,800 miles (2,900 kilometers) across central Asia, the Tien Shan ("Heavenly Mountains") rise from the desert and arid steppes to form a range with precipitous slopes, deep gorges, glaciers, and pure white snowfields. The highest peaks are Pik Pobedy, which rises to 24,406 feet (7,439 meters), and Hantengri Feng (22,949 feet, 6,995 meters), both near the border with Kazakhstan. The Russian Peter Semonyov first explored the mountains in 1865. Trekking from Alma-Ata in Kazakhstan, he first came to Lake Issyk-Kul, meaning "sacred lake," the world's largest mountain lake and one that never freezes. The following year he traveled on through the Santash Pass into the Tien Shan.

English traveler Charles Howard-Bury made a similar journey in 1913 and wrote how pansies of all colors grew so close together that every step taken by him and his party crushed them. He saw many plants familiar to an English country garden but growing wild, such as fruit trees, roses, and onions.

The slopes are also home to ibex, mountain sheep, wolves, wild boar, bear, and one of the world's rarest predators, the extremely shy snow leopard. **MB**

TAKLIMAKAN DESERT

XINJIANG UYGUR ZIZHIQU (SINKIANG), CHINA

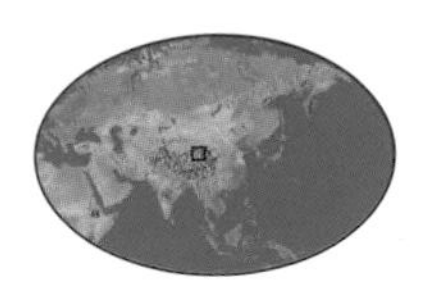

Area of Taklimakan Desert: approx. 96,000 sq mi (250,000 sq km)

Elevation of desert: 505 ft (154 m) below sea level

Height of dunes: 1,000 ft (300 m)

The Taklimakan Desert is a vast area of red, wind-blown sand. It covers an area greater than the size of the United Kingdom. The word *Taklimakan* means "if you go in, you won't come out," and travelers in camel caravans avoided it on the ancient Silk Route between the Mediterranean and the East. Confronted by great, pyramid-shaped sand dunes 1,000 feet (300 meters) high, and even greater piles of sand when hurricane-force winds pushed the sand still higher, the merchants skirted the area, relying on oases, such as Turpan and Kashi, on the eastern edge of the desert, for refreshment.

Turpan should be one of the world's most inhospitable places. It sits in the Turpan Depression, one of the lowest and hottest places on Earth. It is 505 feet (154 meters) below sea level and the daytime temperature soars regularly to 104°F (40°C). Curiously, despite these arid conditions, melons and grapes grow here in profusion. The Persians created an extraordinary system of wells and underground tunnels, called the *karez*, which channeled water from the heavenly Tien Shan Mountains. Today, many of the ancient towns along the Silk Route are no more than ruins. A sea route replaced the Silk Route in the 15th century. MB

YELLOW RIVER

CHINA

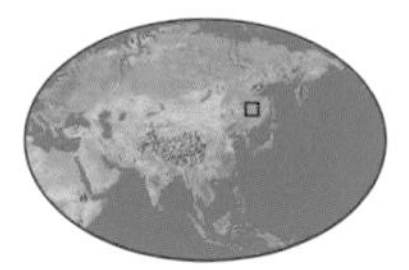

Length of Yellow River: 3,395 mi (5,464 km)

Extent of floods in 1931: 34,000 sq mi (88,000 sq km) flooded completely; 8,000 sq mi (21,000 sq km) flooded partially

The Yellow River, or Huang He, has its source in the springs and lakes of the Kunlun Mountains in Quigha Province and snakes its way across many Chinese provinces to form the second longest river in China (after the Yangtze). Its easterly journey starts off in a series of deep gorges before turning northeast at Lanzhou in Gansu Province. Here, spectacularly jagged and sheer cliffs and mountains border the lush river valley. It then flows through the Ordos Desert—an easterly extension of the Gobi Desert—before changing direction toward the south. It is here that the river cuts through loamy soils and picks up the yellow sediment from which it gets its name.

This area was thought to be the place where ancient Stone Age people settled—the cradle of Chinese civilization. But it is also "China's Sorrow," for the river sometimes causes severely disastrous floods. The worst in living memory was in 1931, when 80 million people lost their homes and one million drowned. The most easterly part of its course has changed several times, swapping between an outlet in the Yellow Sea and the present estuary in the Gulf of Bo Hai. MB

HUA SHAN MOUNTAIN

SHAANXI, CHINA

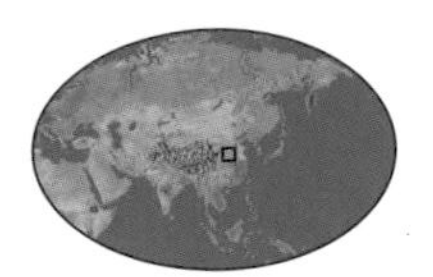

Height of Hua Shan Mountain: 7,220 ft (2,200 m)

Narrowest width of mountain path: 12 in (30 cm)

Terrain: rocky

China's Hua Shan Mountain is not among its highest, but this is difficult to believe when edging along the path at the face of Ear Touching Cliff. Its height is 7,220 feet (2,200 meters) but the feature that makes it one of "the five great mountains of China" is its precipitous nature. Some 75 miles (120 kilometers) east of Xi'an, the capital of Shaanxi province in central China, Hua Shan seems to rise straight off the plain and into the clouds. An ancient path, 7.5 miles (12 kilometers) long, winds from the bottom to the top by way of sheer cliffs and ledges with heart-stopping drops and breathtaking views. At the top are the five peaks that give the mountain its name—*Huashan* means "Five Flowers"—and along the way are Buddhist and Taoist temples and pavilions, as well as ruins of palaces. Various landmarks on Hua Shan include Ear Touching Cliff, Thousand Foot Precipice, Sky Leading Ladder, Sun and Moon Precipice, Facing Sun Peak, Axe Splitting Rock, and Fatal Cliff. But for anyone whose knees go weak at the thought of taking the path, there is also a cable car to the top. **DHel**

RIGHT *The sheer slopes of Hua Shan Mountain.*

WULINGYUAN

HUNAN, CHINA

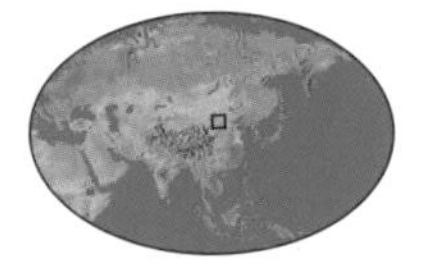

Number of sandstone pillars: over 3,000, many over 660 feet (200 meters) high

Notable feature: highest known natural bridge in the world

The poetic beauty of the Wulingyuan area lies in its quartzite sandstone pillars, ravines, pools, streams, and gorges, often shrouded in mist. While other natural areas boast similar pillars, the altitude here is lower and the climate subtropical, creating dense forests. There are also 40 caves, including the famous Yellow Dragon Cave with its 165-foot (50-meter) waterfall. Other attractions are the two very high natural bridges: The Bridge of the Immortals spans 85 feet (26 meters) and is roughly 5 feet (1.5 meters) wide at an elevation of 330 feet (100 meters), and the Bridge Across the Sky is even more impressive at 130 feet (40 meters) long and 33 feet (10 meters) wide, soaring 1,171 feet (357 meters) above the valley floor. This lush environment is home to several species threatened with extinction, such as the Chinese giant salamander, Asiatic black bear, and the clouded leopard. Of the 3,000 plants found in the site, 35 species are listed in the inventory of China's Rare Plants, and the area is also a "hotbed for fungi." Wulingyuan is an oasis of nature in the middle of a heavily populated agricultural area, but it has remained largely unspoiled thanks to its ruggedness and inaccessibility. **GD**

ZHOUKOUDIAN CAVES

BEIJING, CHINA

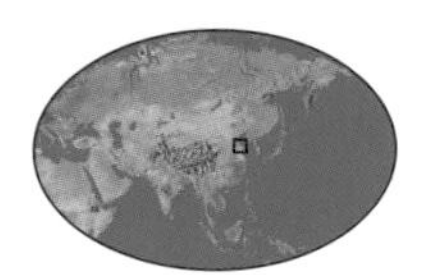

Maximum height of Zhoukoudian Caves: 130 ft (40 m)

Age of Homo Erectus remains: 200,000 to 500,000 years

Rock type: limestone

On the north face of Dragon Bone Hill are limestone caves and crevices that were home to some of our early ancestors. Their fossilized remains, together with stone and bone artifacts and signs of fire use, have been dated as 500,000 years old. The first discovery at Zhoukoudian was the skullcap and other bones of an apeman, known as "Peking Man," but which was labeled scientifically as *Homo erectus*, a direct ancestor of modern man. The bones were found here in 1929, but they mysteriously disappeared at the time of the Japanese invasion of China in World War II. Fortunately, casts were made and these can be seen in the American Museum of Natural History in New York.

Today at Zhoukoudian, there are four caves being explored and excavated, and the bones and tools discovered there can be viewed in a large exhibition center, along with hominid relics from other parts of China. Visitors can also enter the cave where the original fragments of Peking Man were found, together with the remains of 40 other individuals of different ages and sexes.

The caves are just 30 miles (50 kilometers) southwest of Beijing. MB

QINLING MOUNTAINS

SHAANXI, CHINA

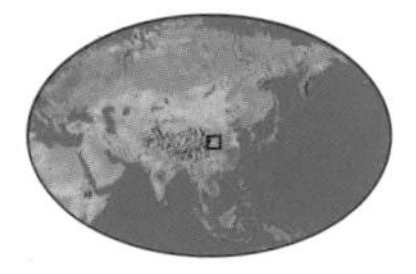

Area of Qinling Mountains: 29,537 sq mi (76,500 sq km)

Number of giant pandas: 200 to 300

Number of Sichuan golden monkeys: 4,000

Running from east to west, the Qinling Mountains not only separate the Sichuan Basin from the northern plains and Loess Plateau, but also form a watershed divide between China's mightiest rivers—the Yangtze and Yellow rivers. Peaks rise up to over 12,000 feet (3,600 meters), with lush subtropical forests to the south of the mountains and temperate vegetation on the northern slopes. At any given altitude, temperatures can be 55° F (13° C) cooler in the north. Warm rains in the south of the region encourage tree growth, including the Chinese mountain larch, Miaotai maple, Chinese yew, Qinling fir, and gingko—one of the oldest tree species in the world. Many animals here are rare, including an isolated population of giant panda (including also an individual with brown fur in Foping National Nature Reserve), as well as takin, giant salamander, crested ibis, and clouded leopard. Sharing its living space is the Sichuan golden monkey. This blue-faced monkey with golden hair lives in family groups that come together to form huge troops up to 500 strong. MB

RIGHT *Foping National Nature Reserve in the Qinling Mountains—home to the giant panda.*

JIUZHAIGOU

SICHUAN, CHINA

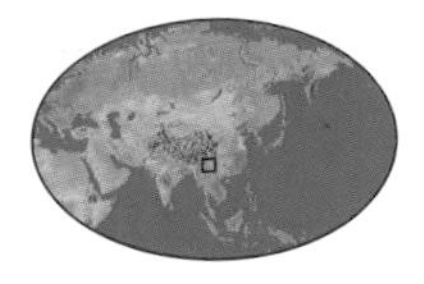

Area of Jiuzhaigou region: 280 sq mi (725 sq km)

Height of Xionguashai Falls: 256 ft (78 m)

Height of Zhengzhutan Falls: 92 ft (28 m)

Jiuzhaigou is a jagged region in the northern echelons of Sichuan Province. Its name means "nine-village valley," because it is said that there were once nine Tibetan villages along its length. Today, there are only six villages remaining, with a total population of about 800 people. Stretching across 280 square miles (725 square kilometers), Jiuzhaigou is comprised of superb landscapes of mountains, forests, spectacular limestone formations, lakes, and waterfalls. Some 140 bird species also live here, as well as a number of endangered mammal species, including the giant panda and golden snub-nosed monkey.

The best-known features in the area are the many lakes, which are famous for their high calcium content—fallen trees have been perfectly preserved in some of them for hundreds of years. Many are strings of classic ribbon lakes at the base of valleys that were once formed by glaciers and which have subsequently been dammed naturally by carbonate deposits. One such lake, Wolonghai, or Dragon Lake, has a calcareous dyke running through it that is clearly visible below the water surface, as the surrounding water is a darker color. In local folklore, this dyke is said to be a dragon sleeping on the bottom of the lake. RA

SHENNONGJIA

HUBEI, CHINA

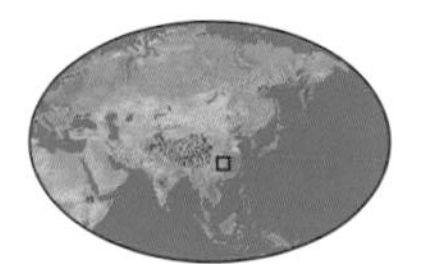

Area of Shennongjia National Natural Reserve: 27,207 sq mi (70,467 sq km)

Length of Hogpin Valley: 30 mi (50 km)

Length of Valley Lake: 9 mi (15 km)

Known as the "Roof of Central China," the Shennongjia National Natural Reserve has six peaks over 9,800 feet (2,987 meters), primitive forests, and the legend of the mysterious "wild man," a Yeti-like creature that is said to live in the forests of the Hubei Province. The climate is warm and humid, which promotes the growth of rare forest trees that, in turn, protect endangered animal species. Chinese dove trees and dawn redwood stand alongside 130-foot- (40-meter-) high firs, together with bamboo and cypress. The area is home to the South China tiger, musk deer, Himalayan black bear, and Reeve's pheasant.

There are curiosities—near the summit of Shennongjia's highest mountain, there have been reports of bears, deer, rats, snakes, and monkeys that are completely white, and in the remote forest, 16-inch (40-centimeter) long footprints, tufts of red-brown fur, and half-eaten corn cobs indicate the presence of a primitive ape-like creature—the *yeren*.

Landscape features include the Hongpin Valley, which is flanked by dangerously craggy peaks; Valley Lake, which is sandwiched between cliffs overgrown with forest; and Tianjing Cave, which has a natural skylight in its roof. The rest of the region features waterfalls, pools, streams, springs, cliffs, and large isolated boulders. MB

HUANGLONG NATURE RESERVE

SICHUAN, CHINA

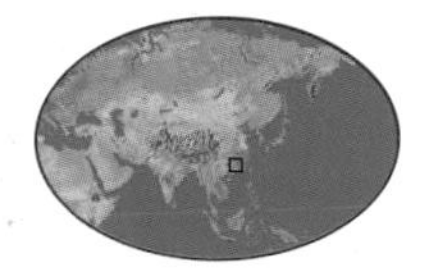

Area of Huanglong ponds: largest: 7,177 sq ft (667 sq m); smallest: 10.7 sq ft (1 sq m)

Depth of ponds: deepest: over 10 feet (3m); shallowest: 4 in (10 cm)

Huanglong Nature Reserve is a two-mile- (3.2-kilometer-) long valley, tucked away deep in the Sichuan Province among thick, primeval forests, that descends from 11,735 to 10,315 feet (3,578 to 3,145 meters) above sea level. The valley is covered with a thick, yellowish layer of carbonate of lime, which forms around differently shaped and sized ponds that join together like a terraced field. Huanglong comprises about 3,400 of these colored ponds. The algae and bacteria in the ponds cause the mineral-rich waters to glow in hues of creamy white, silvery gray, amber, pink, and blue—a process that is particularly striking when the weather is good.

Viewed from the air, the limestone valley resembles a massive yellow dragon, while the glistening pools look like dragon scales. *Huanglong* in Chinese means "yellow dragon." There are many other karst features to be found here that are worth seeing—the attractive caves and grottos coupled with the unusual pools themselves make the park a real geological paradise. In addition, the main tributaries of the Fujiang River flow through the Huanglong Nature Reserve, and there are a number of hot springs located here. Two of the most important springs are in the Mouri Gully—Kuang-quan and Feicui—and both are said to have healing properties due to their high mineral content.

Huanglong is situated at the intersection of four floral regions—the subtropical and tropical zones of the northern hemisphere, Eastern Asia, and the Himalayas—meaning that the plant life is diverse; the area supports over 1,500 species of plant. A large number of these are endangered: The region features 16 species of rhododendron alone that are under threat. Many endangered animals can also be found here, including the golden snub-nosed monkey, Asiatic black bear, Szechwan takin, common goral, and giant panda. Pandas have been seen to move in four or five distinct communities within the reserve. The valley is part of the Huanglong Scenic and Historic Interest Area World Heritage site. **RA**

Viewed from the air, the limestone valley resembles a massive yellow dragon, while the glistening pools look like dragon scales. The attractive caves and grottos coupled with the unusual pools themselves make the park a real geological paradise.

RIGHT *Warm fall sunlight reflects off the calcite deposits that build up on the pools at Huanglong.*

MOUNT LUSHAN

JIANGXI, CHINA

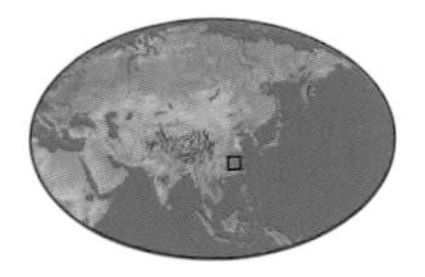

Area of Mt. Lushan:
135 sq mi (350 sq km)

Height of Hang Yang Peak:
4,836 ft (1,474 m)

Mount Lushan, located by Poyang Lakes, has been described as a magnificent mountain that "rises up out of lakes and rivers." It is considered to be a spiritual center of Chinese civilization, where both Buddhist and Taoist temples blend harmoniously with the landscape. It is a place of mighty peaks (the highest being Hang Yang Peak), roaring waterfalls, plunging gorges, and mysterious fogs that envelop and hang over the area for almost 200 days every year. All this, coupled with the mild climate, makes Mount Lushan one of the most popular tourist destinations in China. Lushan is known as "the Kingdom of Prose and Mountain of Poetry"—it has been the subject of many works of literature, and 4,000 pieces of verse are inscribed on the cliffs.

From the top of the 1,000-foot (300-meter) sheer face of Dragon's Cliff, visitors can look

out over the entire area and hear the roar of waterfalls in Stone Gate Ravine. This is a truly great experience, and recommended even for sufferers of vertigo.

Hidden in the eerie mists is the deeply incised Brocade Valley, where countless numbers of flowers are always in bloom. The line of five adjacent summits of the peaks of Five-Old-Man is also visible, the shape of which is thought to resemble five old men talking together. Five-Old-Man Cave is located near the top, and the peaks themselves are surrounded by the robust and unique-looking yingke pines. Behind this stunning vista is the Three-Cascades Waterfall that drops into Nine-Tier Gully.

Where Five-Old-Man meets the Hanyang Peaks is Hang Po Valley. There are splendid views of the sunrise from the highest peak of Five-Old-Man, and if foggy, it is said that people can hear "the sound of fog." **MB**

BELOW *The eerie mists hang in Brocade Valley, as seen from the heights of Mount Lushan.*

WOLONG NATURE RESERVE

SICHUAN, CHINA

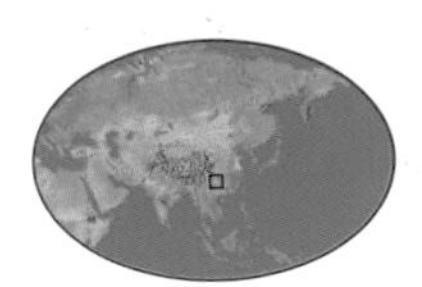

Area of Wolong Nature Reserve: 511,995 acres (207,210 ha)

Elevation of reserve: 4,000 to 20,500 ft (1,200 to 6,259 m)

Established as a nature reserve: 1963

Wolong is a mountainous area of temperate bamboo forest that is, for most of the year, enveloped in cloud, battered by torrential rain, or shrouded in thick fog. It was one of the first reserves to offer protection to and carry out research on the very rare giant panda, the internationally recognized icon of endangered animal campaigns. Today, it is home to a giant panda breeding center that is returning captive-bred animals to the wild. But the giant panda is not the only rare mammal living here. Its smaller cousin, the raccoon-like lesser or red panda, is present, along with 45 other large mammals, including the clouded leopard, takin, and white-lipped deer. Research is carried out on the ecology of bamboo—the giant panda's main food and the dominant plant of the area—but rare plants, such as dove trees and plum yews, also benefit from the panda's protection.

Wolong is overshadowed by Balang Shan, a mountain that rises about 15,092 feet (4,600 meters) above sea level. Here, trekkers can look down on soaring golden eagles and other magnificent birds of prey. MB

ZIGONG

SICHUAN, CHINA

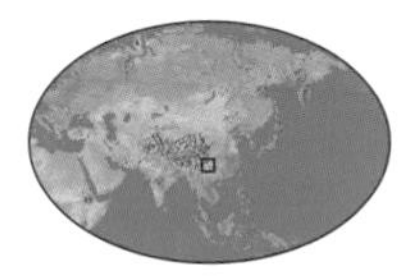

Age of fossils: 165 million years

Length of Shunosaurus dinosaur: 40 ft (12 m)

In 1979, paleontologists working for the British Museum were searching a low hill near Zigong (recently cleared for a vehicle parking lot) and were astonished to find the ground littered with fragments of dinosaur bones. They had chanced upon one of the world's great dinosaur graveyards, on a site near an ancient lake where aquatic grass was plentiful and trees grew tall. The site was especially remarkable because these fossils dated from the middle Jurassic, an age from which only a few dinosaurs were known. Subsequent excavations have yielded more than 6,000 dinosaur fossils from at least 100 individuals. Thousands of dinosaurs were buried here, in sand and mud laid down in a river delta. Most were plant eaters, including the sauropods, shedding light on how sauropods evolved to become the largest animals that ever walked the planet. One Zigong sauropod, *Shunosaurus*, grew to 40 feet (12 meters) long, and is the only sauropod known to have a club at the tip of its tail—presumably for defense against predators. Another, *Omeisaurus*, had a neck length of around half its 60-foot (18-meter) length. There were early stegosaurs, as well as an 11-foot (3.5-meter) long carnivore. **MW**

TIGER LEAPING GORGE

YUNNAN, CHINA

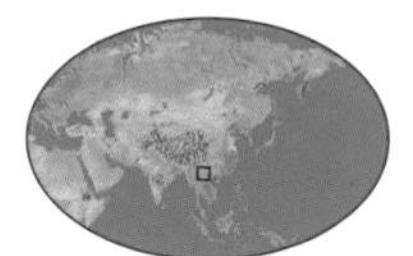

Length of Tiger Leaping Gorge: 10.5 mi (17 km)

Depth of gorge: 100 ft (30 m)

Vegetation: mountain meadows

At its narrowest, the world's deepest gorge is less than 100 feet (30 meters) from rim to rim. According to local legend, a tiger was once seen leaping across it, thus arose the name given to the deep gash that cuts between Dragon Snow and Jade Snow Mountains.

The Golden River, which carved the chasm between the two mountains over five million years, runs as white-water rapids through most of the gorge, dropping steeply in three sections, the third of which is among the roughest set of rapids in the world. But in spite of all the turbulence—or perhaps because of it—it is part of a tranquil and beautiful region of the eastern Himalayas. It was from the nearby city of Lijang that the Austrian–American scholar Dr. Joseph Rock wrote articles that are said to have inspired James Hilton to create Shangri-La in his novel *Lost Horizon.*

It is possible to walk the entire length of Leaping Tiger Gorge on a narrow path between a steep wall and a steep drop. One traveler has described the experience as "5,000 feet [1,500 meters] of vertical blackness above my head, 1,000 feet [300 meters] of vertical terror below me, and the foaming river roaring menacingly at the bottom." **DHel**

YANGTZE GORGES

CHONGQING, CHINA

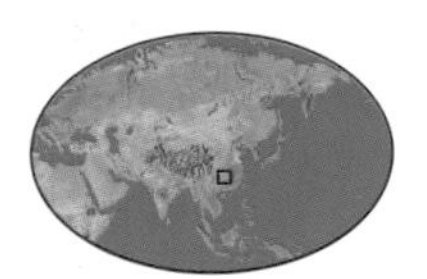

Length of Qutang Gorge: 5 mi (8 km)
Length of Wu Gorge: 25 mi (40 km)
Length of Xiling Gorge: 47 mi (75 km)

When spring floodwaters in the world's third longest river pour into the entrance of the Qutang Gorge, they can move at more than 20-miles (32-kilometers) per hour. Because the gorge is so narrow and steep—the cliffs on either side are twice the height of the Eiffel Tower and no more than 330 feet (100 meters) apart—this sudden influx can raise the river level by as much as 170 feet (50 meters).

Qutang is one of three gorges that stretch for 120 miles (190 kilometers) at the halfway stage in the Yangtze River's mighty 3,915-mile (6,300-kilometer) journey from the mountains to the sea. Here iron poles are embedded in the cliffs from which chains were once slung. The chains prevented raiders from going upriver and were also used to stop boats in order to collect a river toll.

The second gorge is the 25-mile- (40-kilometer-) long scenic Wu Gorge. It is lined by the Fairy Peaks, which, according to legend, were sent by the Queen of the Heavens to help cut the gorges. They are in fact large limestone columns that have been eroded into female-like shapes.

The third gorge is the Xiling, which is around 47 miles (75 kilometers) long. This was always thought to be the most dangerous of the three, with narrow passages, rapids, and whirlpools. Traveling along this stretch used to be an extremely hazardous affair. At one time, each junk had to be hauled through the rapids against the swiftly flowing current by 400 straining men on the riverbank. Stories of mishaps abound, including the tale of Père David, the French Christian missionary, who narrowly missed death when a junk racing at high speed almost collided with his boat, which was traveling upstream at the time. However, in the 1950s, all the rocks in the middle of the river were blown up, greatly calming the water's flow. Today, motorized ferries rather than sailing junks travel safely up and downstream.

The second gorge is the scenic Wu Gorge, lined by the Fairy Peaks which, according to legend, were sent by the Queen of the Heavens to help cut the gorges. They are in fact large limestone columns.

The Gezhouba Dam, just below the Xiling Gorge, has restricted the river, but the much higher Three Gorges Dam will, when it is finished in 2011, turn the Yangtze Gorges into an enormous reservoir. It will be the largest hydroelectric power station in the world.

In China the river is known as *Chang Jiang*, meaning, appropriately, "Long River." **MB**

RIGHT *The black ribbon of the Yangtze River slices its way through the steep Yangtze Gorges.*

GUILIN HILLS

GUANGXI ZHUANGZU ZIZHIQU, CHINA

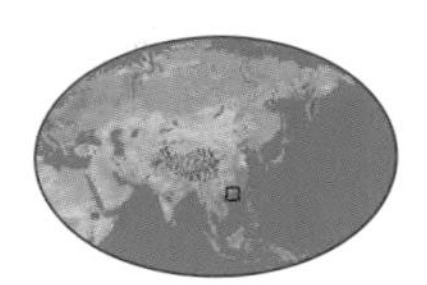

Length of Guilin Hills: 75 mi (120 km)

Highest peak (Piled Festoon Hill): 400 ft (120 m)

Age: 300 million years

Rows of steep-sided limestone hills rise from an otherwise flat landscape of paddy fields alongside a 75-mile (120-kilometer) stretch of the Li River in southern China. The limestone was formed at the bottom of a warm, shallow sea about 300 million years ago, but earth movements pushed up the strata, and the wind, waves, and rain sculpted them into the shapes we see today. Each hill has a descriptive name, such as "Five Tigers Catching a Goat," and one is called "Camel Hill" on one side because of its view and "Wine Jug Hill" from the other side. According to legend, "Elephant Trunk Hill" is said to be the elephant on which the King of Heaven toured the country. When it fell ill, a local farmer nursed it back to health, and in return, the elephant helped the farmer in his fields. The king was angry and turned the animal to stone. Its "trunk" dips into the Li River at the "Moon in the Water Arch." The highest hill is "Piled Festoon Hill," which is 400 feet (120 meters) tall. From August to October, when the cassia trees are in bloom, the area around the town of Guilin is bathed in the smell of cinnamon. MB

BELOW *The peaks of the Guilin Hills rise up behind the Li River.*

GUILIN CAVES

GUANGXI ZHUANGZU ZIZHIQU, CHINA

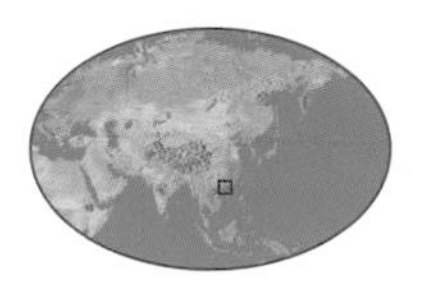

Length of Reed Flute Cave: 820 ft (250 m)

Width of Reed Flute Cave: 400 ft (120 m)

Age of rock: 350 million years

Underneath Guilin's conical hills is an extraordinary network of caves and tunnels that was cut into the limestone rock by underground rivers and streams. Some are enormous. The cathedral-like Gaoyan or "High Cave" and some of its neighbors have unusually large stalagmites and stalactites, some over 100 feet (30 meters) high. Even the caverns are vast. The Reed Flute Cave is 825 feet (250 meters) long and 400 feet (120 meters) wide and contains a feature known as the Old Scholar. According to legend, a poet sat down and tried to write about the splendor of the cave and its rock formations, but he was unable to find words that adequately described them. He pondered for so long that he eventually turned to stone.

The Reed Flute Cave gets its name from the reeds that once grew close to the entrance, which were cut and fashioned into flutes by the local people. Millions of visitors flock to Guilin's caves each year, but at one time they served an altogether more alarming and darker purpose. During World War II, the local population hid in the caves for protection when the towns and villages were bombed by Japanese aircraft. **MB**

HUANGGUOSHU FALLS

GUIZHOU, CHINA

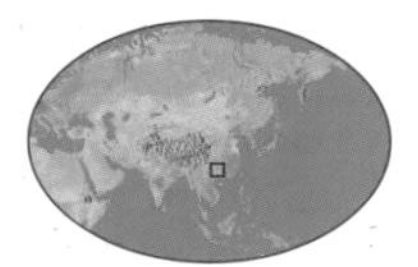

Height of Huangguoshu Falls: 224 ft (68 m)

Width of falls: 275 ft (84 m)

Length of Water Curtain Cave: 440 ft (134 m)

Huangguoshu Waterfall Scenic Area comprises more than 10 waterfalls, set both above- and belowground. At its heart lies the Huangguoshu Falls, the biggest waterfall in Asia, which measures some 224 feet (68 meters) tall and 275 feet (84 meters) wide. In the flood season, the falls plunge down with such force—the flow can reach a mighty 26,000 cubic feet (735 cubic meters) per second—that the steep cliffs tremble, while mist from the waterfall rises from the plunge pool, creating spectacular rainbows. At these times, the waters' thunderous roars can be heard from over 3 miles (5 kilometers) away. In contrast, during the dry season, water tumbles down in a much more sedate fashion, forming a series of smaller, separate streams from the overhanging cliffs.

Behind the waterfall, toward its base, is a long cave known as Water Curtain Cave. Visitors can reach it from a road on the mountainside. Inside, it is possible to hear, watch, and touch the waterfall.

Other interesting natural features in the area include the Tianxinqiao Stone Forest, whose "trees" are actually bizarrely shaped stalactites and stalagmites. RA

FOLDED BROCADE HILL

GUANGXI ZHUANG AUTONOMOUS REGION, CHINA

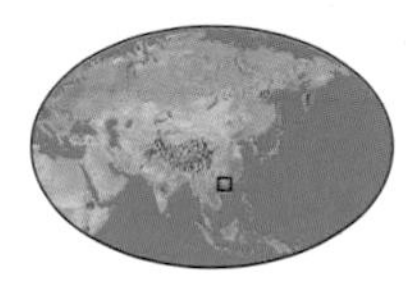

Area of Folded Brocade Hill: 494 acres (200 ha)

Height of Folded Brocade Hill: 240 ft (73 m)

Rock type: limestone, typical karst formation

Folded Brocade (or "Deicai") Hill, which is actually four hills—Yuyue Hill, Look-in-All-Direction Hill, Crane Peak, and Bright Moon Peak—is beside the Li River just north of the city of Guilin in China's Guangxi Zhuang Autonomous Region. The name comes from its layered rocks—the rock has been eroded in such a way as to resemble folds from a distance. To the Qin dynasty Chinese who founded the city 2,000 years ago, the hill looked like a stack of folded brocade.

The hill itself is an unusual and rather beautiful rock formation that has, over the centuries, been enhanced with carved-out chambers and archways. The vegetation clings onto the steep slopes and contrasts with the rock, contributing to the overall striking effect of the hill. Buddhist sculptures and pagodas are also featured, and tourists can visit a pavilion with a pinnacle that affords 360° views of other peaks, the city, and the countryside.

One of the more spectacular natural features here is Windy Cave, whose shape, with openings on two sides of the hill, causes a permanent wind to blow. In the cave are 90 images of the Buddha, from the Tang and Song dynasties. **DHel**

WAVE-SUBDUING HILL

GUANGXI ZHUANG AUTONOMOUS REGION, CHINA

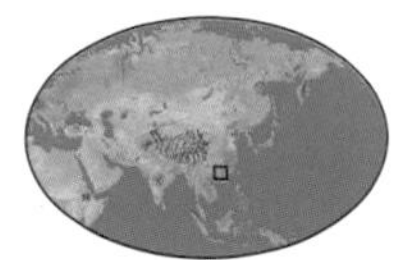

Height of Wave-subduing Hill: 699 ft (213 m)

Length of hill: 400 ft (120 m)

Notable feature: half-submerged hill

Among the several unusual landmarks along the Li River in the city of Guilin, in China's Guangxi Zhuang Autonomous Region, is Wave-subduing Hill. The hill is named thus because it slopes from the land down into the river itself and forms a barrier that the river's waves crash against. It is a block-shaped rock formation 400 feet (120 meters) long, 200 feet (60 meters) wide, and 699 feet (213 meters) high. It has many inscriptions and relics from the Tang, Song, Yuan, Ming, and Qing dynasties and, on its eastern side, a path that leads to the Listening-to-the-Waves Pavilion. On the hill's southern slope is a cave that, according to legend, was once illuminated by a giant pearl and inhabited by a dragon. One day, a fisherman stole the pearl, was overcome by shame, and brought it back—hence, the name, Pearl-returning Cave. Inside the cave is a boulder that hangs from the ceiling and does not quite touch the floor. This is called Sword-testing Rock, because in another legend it was once a pillar whose bottom was sliced away by an ancient general testing his sword. At the end of the cave is another smaller cave, containing 200 stone Buddha statues carved during the Tang Dynasty (618–907 C.E.). **DHel**

ELEPHANT TRUNK HILL

GUANGXI ZHUANG AUTONOMOUS REGION, CHINA

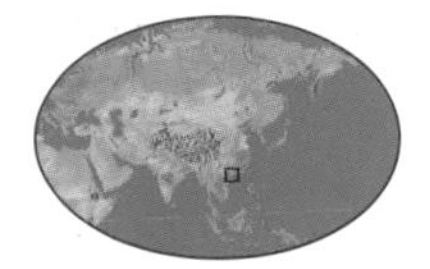

Overall height of Elephant Trunk Hill: 660 ft (200 m)

Height above river: 181 ft (55 m)

Age: 360 million years

If there is a region in the world that has more interesting rock formations than Guilin, in China's Guangxi Zhuang Autonomous Region, it could hardly have ones that are more imaginatively named. In addition to Folded Brocade Hill, Wave-subduing Hill, and others, there is Elephant Trunk Hill, which forms a huge arch on the city's Li River and in so doing resembles an enormous elephant drinking with its trunk. It is commonly regarded as the emblem of Guilin.

Measured from the bottom of the riverbed, the hill is 660 feet (200 meters) high, and from the surface, 181 feet (55 meters) high. Jutting out from the riverbank, it is 354 feet (108 meters) long and 330 feet (100 meters) wide. The arch itself—known as Moon-over-water Cave because the full moon, reflected on the water, seems to float inside—has more than 70 inscriptions from the Tang and Song Dynasties etched on its inner walls. On the land side is another cave with "windows," known as the elephant's eyes, which look out over the city. At the top of the hill is the Puxian Pagoda, built in the Ming Dynasty (1368–1644 C.E.) and shaped like the handle of a sword. DHel

DASHIWEI DOLINE

GUANGXI ZHUANG AUTONOMOUS REGION, CHINA

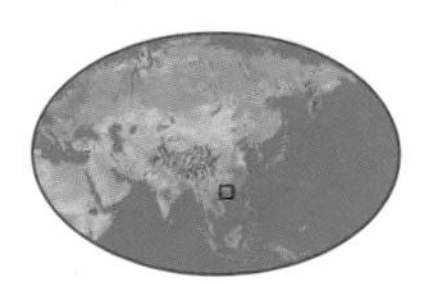

Depth of Dashiwei Doline: 2,011 ft (613 m)

Width of Dashiwei Doline: 1,378 ft (420 m)

Width of Xianozhai Doline: 2,165 ft (660 m)

Dashiwei Doline is one of the world's biggest dolines. Also known as a karst *tienkeng*, this giant doline is the remains of a large cave whose roof has crashed down, leaving a pit with almost vertical walls. At the bottom of the doline are scree slopes, a subterranean river whose waters often flood violently, and a large area of primeval forest. Plants native to the area include a unique species of tree fern, while animals living there include new species of blindfish, shrimp, crabs, spiders, and flying squirrels.

Dashiwei is one of 20 dolines in Leye County, in the Guangxi Zhuang Autonomous Region of southern China, the largest group of their kind in the world. Visitors can visit the dolines, although they are asked not to enter Dashiwei itself for fear of damaging the vegetation and disturbing the native birds. Similar gigantic dolines are to be found in Sichuan, including the world's biggest—Xiaozhai Doline—located on the upper reaches of the Yangtze River. MB

JADE DRAGON SNOW MOUNTAIN

YUNNAN, CHINA

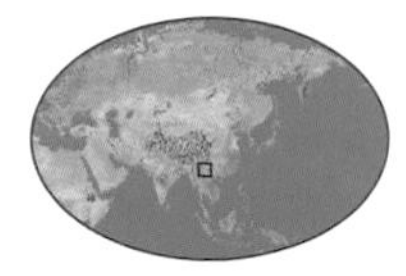

Highest peak of Jade Dragon Snow Mountain (Shanzidou): 18,360 ft (5,600 m)

Age: 230 million years

Vegetation: lush alpine

Jade Dragon Snow Mountain in China's Yunnan Province has 13 peaks, which from a distance resemble an undulating dragon's back. To complete the picture, in a certain light the mountain's snow appears to be green. The latter effect is probably caused by crystalized algae, and the former is the result of a flexure in Earth's crust around 230 million years ago. The shape the mountain takes today may have been influenced by events as recent as 12,000 years ago. The real distinction of the Jade Dragon is its biological wealth. Approximately 6,500 species of plant have been found on its slopes, including 50 species of azalea, 60 primroses, 50 rough gentian, and 20 lilies. There are so many different plants, with so many different flowering seasons, that for at least 10 months of the year the mountain below the snowline is carpeted in color. There are many kinds of rare animals, including red panda, musk deer, silver pheasant, and clouded leopard. Above the snow line, it is another world. The snow is permanent and the peaks are perpetually stormy. The highest is called Shanzidou, which is about two thirds the height of Everest; unlike Everest, however, Shanzidou has never been climbed. **DHel**

WONG LUNG WATERFALLS

HONG KONG, CHINA

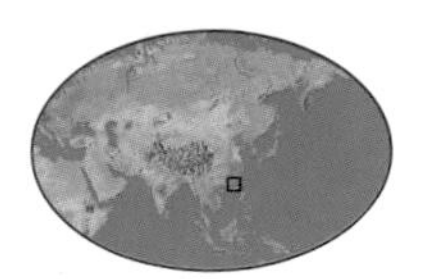

Height of Wong Lung Waterfalls: 300 ft (90 m)

Height of surrounding peaks: up to 2,850 ft (869 m)

In Hong Kong's Lantau North Country Park is a complex of valleys, gorges, and streams surrounded by mountains and known as Tung Chung Valley. The steepest mountain streams in the area are located here, all with names related to *Lung*, the Chinese word meaning "dragon." The main channel is the deep, densely wooded Wong Lung Valley, with its main stream—the Yellow Dragon. Its source is east of Sunset Peak, but much of its water comes from a number of tributaries known as the Five Dragons of Tung Chung. They feature deep gorges, sheer cliffs, deep clear pools, and towering waterfalls.

The Wong Tung Main Falls drop down a 66-foot (20-meter) rock slab, then two streams merge and tumble down into a pool of dark green water. One of the tributaries—Three Dragon Gorge—is hemmed in by 300-foot- (90-meter-) high cliffs, and has three spectacular waterfalls. Left Dragon Falls cascade down the cliff in two drops, while Right Dragon Falls is deflected over a convex rock face to create three drops. The last of the three is the 40-foot (12-meter) Dragon Tail Falls, which drop into a narrow, steep-sided gorge. **MB**

MOUNT MEILIXUESHAN

YUNNAN, CHINA

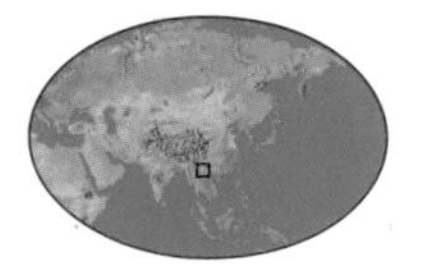

Mt. Meilixueshan's highest peak (Kawagebo): 22,113 ft (6,740 m)

Notable features: arid canyons, snow-capped peaks

Stand at the main vantage point to admire the dawn over Mount Meilixueshan—whose Tibetan name means "God of Snow Mountains"—and it is easy to see how this area helped inspire the fictional Shangri-La of James Hilton's novel *Lost Horizon.* In twilight, the long, serrated ridge glows cool white beneath the sky shot with stars. As the sun rises, the mountain's highest peak, Kawagebo, is suddenly tinged orange. The orange touches the other peaks, fades to whitish as the sun rises, and illuminates glaciers that snake down into the valleys through evergreen forests. In the gorge between the vantage point and the mountain, 13,200 feet (4,000 meters) below Kawagebo, the Mekong flows between arid hillsides that will later bake in the midday sun. Kawagebo and another lower peak are each—like the main summit in Shangri-La—an "almost perfect cone of snow." On a clear morning, Meilixueshan can similarly appear to be the loveliest mountain on Earth. To Tibetans, Meilixueshan is a sacred mountain.

The region around the mountain abounds in unique plants, and the forests are home to red pandas, Asiatic black bears, and musk deer. Snow leopards roam near the tree line. **MW**

MOUNT HUANGSHAN

CANTON, CHINA

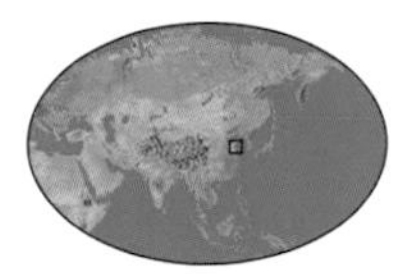

Height of Mt. Huangshan: three peaks over 6,000 ft (1,800 m)

Number of peaks: 72

Annual rainfall: 95 in (240 cm)

Pushing up through the mist to the south of the Yangtze River are 72 craggy peaks, known collectively as Mount Huangshan, or Yellow Mountain. They are made of granite that hardened from molten rock deep belowground. The overlying rocks have been eroded and washed away and the exposed granite weathered into ragged cliffs and untidy pinnacles. Pine trees cling to crevices in the rock face, some over 1,000 years old. Hot springs bubble from cracks, the temperature of the water a constant 108°F (42° C). Over 95 inches (240 centimeters) of rain falls each year, and the mountains are usually swathed in clouds and fog. Visitors are advised to wear wet weather gear in addition to warm clothing. The temperature in the mountains barely rises above 50° F (10° C) at best.

Many people come each year, for it is the ambition of China's huge population of more than 1.3 billion to visit these mountains at least once during their lives. Paths snake up among the peaks—the route to Tiandu Feng, meaning "Heavenly Capital Peak," involves a trek up 1,300 steps and a tricky crossing of a ridge just three feet (0.9 meters) wide, with only a chain to hang on to. MB

LUNAN STONE FOREST

YUNNAN, CHINA

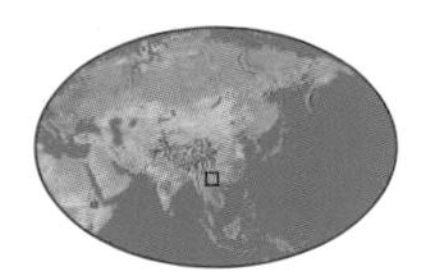

Area of Lunan Stone Forest: 2 sq mi (5 sq km)

Rock type: limestone

About 75 miles (120 kilometers) southeast of Kunming, capital of Yunnan Province, is a plateau with a strange "forest." It is made of stone. Like Madagascar's Tsingy Lands, the Lunan Stone Forest is made of blocks of limestone that have been weathered into hundreds of vertical-sided pillars and pinnacles with knife-like tops. Some are no higher than a person, while others are 100 feet (30 meters) high, and they can be found both clustered together and standing alone. A system of walkways has been constructed between the peaks, and visitors can relax in pavilions along the route. In time-honored fashion, the rocks are given names that perfectly describe their shapes, such as "Phoenix Preening Its Feathers" and "Layered Waterfall." Lichens and mosses cover them, and even creepers with red and pink flowers cling to the cracks and crevices. According to local legend, a pinnacle known as "Ashima Rock" is named after a young girl who was kidnapped by a wealthy aristocrat. Her lover tried to rescue her but she died and was turned to stone. Legend also tells how one of the Chinese Immortals created the forest. He passed young lovers courting in the open and thought they should have privacy. He therefore created a stone labyrinth in which lovers could hide. MB

KUNLUN MOUNTAINS

TIBET, CHINA

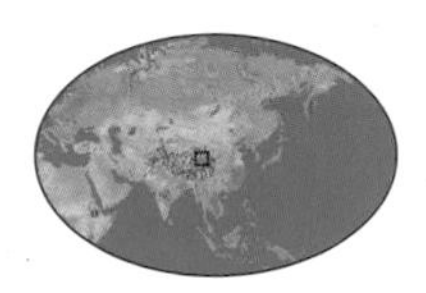

Height of Muztagata: 24,757 ft (7,546 m)

Height of Kongur Tagh: 25,325 ft (7,719 m)

Height of Dongbei: 25,017 ft (7,625 m)

The Kunlun Mountain Range is one of the longest mountain chains in Asia, extending 1,250 miles (2,000 kilometers) from the Pamir Mountains of Tajikistan eastward along the Tibet-Xinjiang to the Sino-Tibetan mountain ranges in the Qinghai province. The Kunlun Mountains divide the northern high Tibetan Plateau from the plains of Central Asia. The range has more than 200 peaks over 19,000 feet (5,791 meters) high. The highest peaks are Muztagata, Kongur Tagh, and Dongbei. The principal eastern section of the Kunlun Mountains is about 375 miles (600 kilometers) wide and comprises a complex of mountain chains and broad valleys. The smaller western section of the Kunluns is made up of three closely packed parallel mountain ranges and is only about 60 miles (95 kilometers) wide. Given its isolation from the Indian and Pacific monsoons, the region is very arid. There are large fluctuations in daily and seasonal temperatures and strong winds, particularly in the fall. Due to poor soils, lack of moisture, and cold temperatures, large sections of the Kunlun Mountains support very limited animal and plantlife. Much of the region is uninhabited and inaccessible. RC

NGARI

TIBET, CHINA

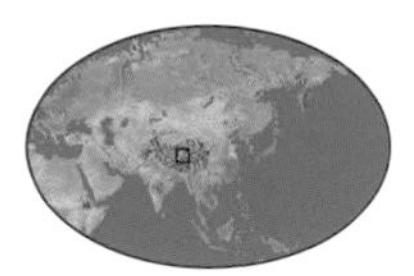

Average elevation of Tibetan Plateau: 15,000 ft (4,500 m)

Area of plateau: 131,000 sq mi (340,000 sq km)

The Tibetan Plateau is often described as the "roof of the world." The plateau rises towards the west in the Ngari region, a vast, sparsely populated area of mountain ranges, valleys, rivers, and lakes known as the "top of the roof of the world." As well as being a popular tourist destination for hikers, Ngari is also a pilgrimage destination for Tibetans and Hindus—the plateau is, for example, the birthplace of Bon, the indigenous pre-Buddhist religion of Tibet.

Aside from its spectacular scenery, the Ngari Prefecture has played an important role in the history of economic and cultural development in Tibet. In the western part of Ngari, Zhada County is famous for the ruins of the Guge Kingdom and the clay forests that surround them. Ngari is the largest prefecture in China in geographic area, but is the least densely populated. It is a haven for rare animals such as the wild yak, Tibetan wild ass, Tibetan antelope, and the Tibetan argali. One of the best-known sites for wildlife is Birds' Islet, located on Banggong Lake in the northern part of Ngari. The best time to see the huge numbers of migratory birds that gather here is from May to September. RC

YAMDROK YUMTSO LAKE

TIBET, CHINA

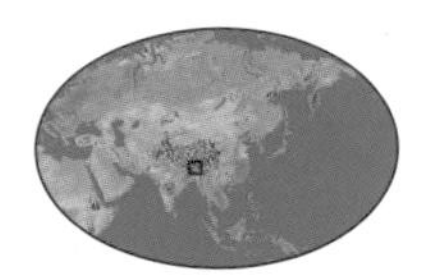

Area of Yamdrok Yumtso Lake: 246 sq mi (638 sq km)

Depth of lake: 100 to 130 ft (30 to 40 m)

Elevation of lake: 14,570 ft (4,441 m)

Yamdrok Yumtso (or Yamdrok-tso) is one of the three most sacred lakes in Tibet. According to legend, the beautiful turquoise lake was formed by the transformation of a goddess. The lake is broad in the south and more narrow to the north and is also known as "Coral Lake of the Highlands." To the west and north of the lake are snowcapped mountain ranges that are often shrouded in mist. There are dozens of juniper-cloaked small islands peppering the lake, on which flocks of many species of bird roost.

Local herdsmen ferry their livestock to the islands in hide boats at the beginning of summer and allow them to graze until the beginning of winter. The lake is also a pilgrimage site for Tibetans who trek here in summer to pray and receive blessings or to sit by the lake simply to contemplate. The waters of the lake are reputed to have healing powers: They are said to make the old feel younger, extend the life span of adults, and make children more intelligent.

To the south of the lake is the Sangding Monastery, which became famous as the residence of Dorje Phagmo, the only female high lama in Tibet. RC

BELOW *Yamdrok Yumtso snakes around the base of the hills.*

MOUNT KAILASH

TIBET, CHINA

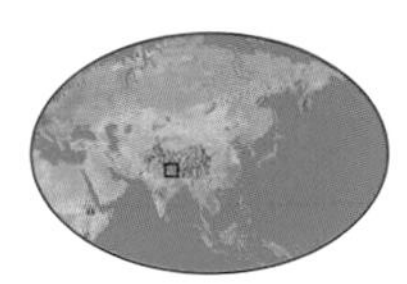

Height of Mt. Kailash:
21,778 ft (6,638 m)

Maximum altitude of ritual circuit:
18,400 ft (5,600 m)

Mount Kailash, located in the far west of Tibet, is one of the most sacred mountains in Asia. It is a site of worship for Buddhists, Hindus, Jains, and followers of Tibet's ancient Bon religion. Mount Kailash is the highest peak of the Gangdise Mountains and is known as *Gang Rinpoche*, meaning "Precious Jewel of Snow." Although not the highest mountain in the region, the diamond-shaped peak of Mount Kailash makes it stand out from the mountains that surround it. The 21,788-foot (6,638-meter) peak has more than 250 glaciers and is the source of four great rivers that drain the vast Tibetan Plateau: the Brahmaputra, the Indus, the Sutlej, and the Karnali (a tributary of the Ganges). Between the peaks of Mount Kailash and Mount Gurla Mandata lie Holy Lake (Lake Manasarova) and Ghost Lake (Lake Rakshastal). Although the lakes are connected via an underground tunnel, Lake Manasarova is a freshwater lake, whereas Lake Rakshastal is salty. For many centuries pilgrims have traveled to the region to make a ritual circuit of the mountain, said to erase the sins of a lifetime. The Kailash pilgrimage should also include a circuit of Lake Manasarova and a visit to the Tirthapuri Hot Springs. For many Tibetans, Mount Kailash is believed to be the hub of the universe. **RC**

TSANGPO CANYON

TIBET, CHINA

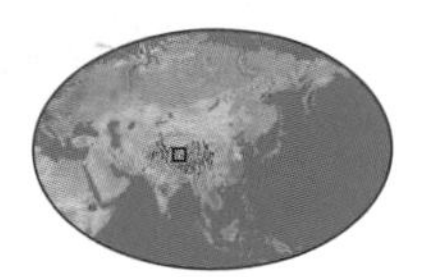

Alternative name: Ya-lu-tsang-pu Chiang

Total length of Tsangpo Canyon: 308 mi (496.3 km)

Depth of canyon: 17,395 ft (5,302 m)

With an average altitude of 10,000 feet (3,000 meters), the Tsangpo River (meaning the "Purifier") is the highest river in the world. Originating from the Chema-Yungdung Glacier in the northern Himalayas, it flows for 1,278 miles (2,057 kilometers) across the Tibetan Plateau before becoming the Brahmaputra in India. Toward the end of its journey across Tibet, the river turns sharply and is squeezed between the Namcha Barwa and Gyala Peri Mountains to form Tsangpo Canyon, the largest canyon in the world. In dramatic contrast, as Tsangpo arches around the easternmost part of the Himalayas, its narrowest point is no wider than New York's Fifth Avenue. In kayaking circles it has become known as the "Queen of Canyons" and is considered to be the world's most formidable: The river drops 9,000 feet (2,700 meters) in 1,500 miles (2,414 kilometers).

Very few people have ventured into the canyon, but an international team of canoeists led by Scott Lindgren completed its first navigation in November 2004, circumventing the 100-foot- (30-meter-) high Hidden Falls, a mysterious waterfall that nobody knew existed until as recently as 1998. **MB**

LAKE MANASAROVA

TIBET, CHINA

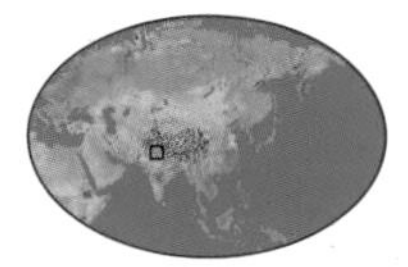

Elevation of Lake Manasarova: 15,049 ft (4,586 m)

Surface area of lake: 159 sq mi (412 sq km)

Depth of lake: 253 ft (77 m)

Lake Manasarova—one of Tibet's three sacred lakes—is the highest freshwater lake in the world at an altitude of 15,049 feet (4,586 meters). Its elevated location means that it freezes over completely every winter, only slowly thawing out in the spring. The surrounding area has a number of other water features, including another lake to the west, Lake Rakshastal, which is linked to Manasarova by a narrow channel. Several rivers, including the Indus, have their sources nearby.

Lake Manasarova is overlooked by holy Mount Kailash, whose snow clad peak is reflected in the pristine waters below. It is a vast and beautiful body of water, encompassing some 125 square miles (320 square kilometers) and sinking to a maximum depth of 300 feet (90 meters). The waters appear delicate blue near the shores, turning a rich emerald green in the center.

Manasarova has been a sacred site for over four thousand years and is said to have a cleansing and redeeming power. Mount Kailish and Lake Manasarova are places of pilgrimage for adherents of Hinduism, Tibetan Buddhism, and the indigenous religion, Bon. The lake is surrounded by eight monasteries. **AB**

DAISETSU

HOKKAIDO, JAPAN

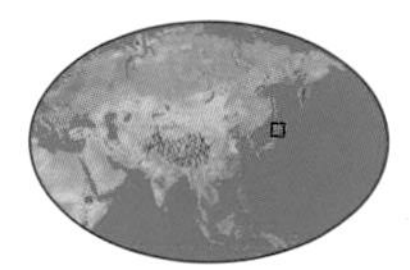

Area of Daisetsu: 892 sq mi (2,310 sq km)

Highest peak (Mt. Asahi): 7,513 ft (2,290 m)

Hokkaido is considered to be Japan's wild frontier, and at its center lies a truly untamed heart. The beauty, serenity, and sheer wildness of the varied peaks and volcanoes of Daisetsu were recognized in 1934 when it was designated Japan's largest national park.

In winter, this raised plateau, ringed with peaks, gorges, waterfalls, and steam vents, is a forbidding place, windswept and blanketed in snow. But its kinder summer face, when it is carpeted with wildflowers and ringing with birdsong, has led the indigenous Ainu people to regard it as the garden of the gods. The benevolent god-spirits are said to have roamed from towering Asahi-dake (Hokkaido's highest peak) to steaming Tokachi-dake and through the high alpine meadows and mixed boreal forests. Today a relict population of brown bears lives in one of its last remaining strongholds. Sika deer, red fox, Asiatic pika, and Siberian chipmunk are far more likely to be encountered here than bears, who shy away from visiting tourists.

From late June onward the tundra-like alpine floral zone presents scattered patches of colored brilliance like a fragmented quilt, but it is in fall that Daisetsu is at its most beautiful; then, the dense, dark green of the stone pines contrasts with the fiery reds of the rowans and the yellows of the mountain birches in a lavish palette of fall colors. **MBz**

KEGON FALLS & LAKE CHUZENJI

TOCHIGI, JAPAN

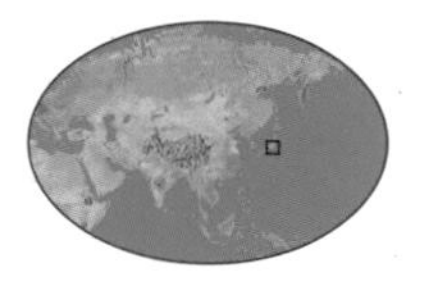

Height of Kegon Falls: 318 ft (97 m)

Width of falls: 23 ft (7 m)

Surface area of Lake Chuzenji: 5 sq mi (13 sq km)

Lake Chuzenji, a beautiful forested lake at the foot of the sacred Mount Nantai Volcano, was formed thousands of years ago by a lava flow that prevented water from draining from the mountain. Eventually the lava eroded and sprang a leak, creating the spectacular Kegon Falls—one of Japan's "big three."

Rainbows fill the valley at the base of the falls, and in winter the entire waterfall can freeze into a giant icicle. The main cascade plummets 320-feet (97-meters), and this is surrounded by 12 smaller cascades. Kegon Falls is one of the most powerful waterfalls in Japan—the water drops at a rate of three tons every second into a 16-foot (5-meter) deep pool.

Panoramic views of Kegon Falls and Lake Chuzenji are offered from the Akechidaira Plateau (accessible by ropeway or on foot). There is also a three-story viewing platform near the base of the falls, which is accessible via elevator. The lake and falls are located in the Nikko National Park. For about two weeks every year—during the koyo (fall) season—the mountainsides burst into an array of dazzling fall colors. This is an extremely popular time for visitors. RC

TEURI-JIMA

HOKKAIDO, JAPAN

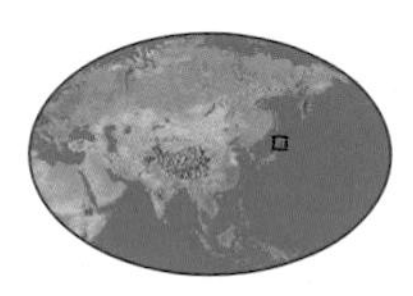

Circumference of Teuri-jima: 7.5 mi (12 km)

Height of Teuri-jima: 607 ft (185 m)

Teuri-jima is situated off the northwest coast of Hokkaido. It was once home to 2,500 people but now there are just 500 inhabitants—today the biggest population on the island is formed by one million seabirds. Along the east shore where the human fishing community clusters, the island is low and gently sloping, rising to precipitous cliffs along the wilder southern and western portions of the island where the seabirds dwell.

Since seabirds spend the majority of their lives at sea, they are clumsy land animals, and therefore select uninhabited—and predator free—terrain on which to raise their young. Teuri-jima's cliffs have pinnacles, such as Akaiwa, that project straight up from the ocean. There are also numerous ledges and narrow clefts, and clifftops carpeted with wildflowers. These rugged parts of the island thus form perfect homes to hordes of birds, foremost among which is the rhinoceros auklet.

This puffin-like seabird ranges only in the northern Pacific, and its main home is Teuri-jima. Living at sea for most of their lives, the auklets only visit land during their summer breeding season. To visit the southern end of the island at night is to witness a truly remarkable spectacle: The stocky seabirds, laden with food, come crashing ashore to visit their mates and young in their underground nesting burrows, flapping fast and furious. They are excellent divers but do not fly as well because they lack the precision control of their lighter relatives. This renders them at risk from predators, but by returning at night they can avoid such danger.

The rhinoceros auklet population has declined from an estimated 800,000 in 1963 to about 300,000 pairs in 2004, but even this reduced number still represents the largest colony of the species in the world. As such, there may be as many as 200 nesting holes in an area of 108 square feet (10 square meters). In addition to auklets, there are small numbers of guillemot and common murre. The common murre breeds uniquely on Teuri-jima Island. Once boasting numbers as high as 30,000 to 40,000, a 1999 survey registered just 12 birds living in the area around the island's Akaiwa Rock and a further seven living by the Byoubuiiwa Cliffs. Larger numbers of slaty-backed gull, Japanese cormorant, and as many as 10,000 pairs of black-tailed gulls make the island a true seabird paradise. A center was recently established on Teuri-jima to provide information for visitors and to help protect the region's threatened seabird species. **MBz**

Teuri-jima's cliffs have pinnacles that project straight up from the ocean. There are also numerous ledges and narrow clefts, making the island a perfect home to hordes of seabirds.

MOUNT FUJI

YAMANASHI / SHIZUOKA, JAPAN

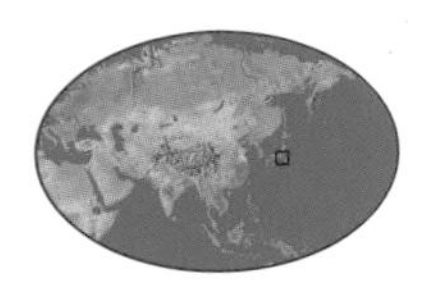

Height of Mt. Fuji: 12,388 ft (3,776 m)

Crater diameter: 2,295 ft (700 m)

Mount Fuji is known around the world as a potent and iconic symbol of Japanese beauty. Rising to 12,388 feet (3,776 meters) it is Japan's highest mountain—it stands alone, dominating the scenery of central Honshu. For centuries Mount Fuji's elegant silhouette has been revered by artists—it is said to suggest the mystery of the infinite—and today the mountain is recognized as a sacred space. The peak's simple, delicate lines, however, belie its violent past. As a dormant volcano, which last erupted in 1708, Mount Fuji's weathered surface consists largely of loose cinders. The peak is said to have reached its present shape 5,000 years ago.

The beauty of Mount Fuji lies in the distant views it affords, especially in winter when its

cap of snow extends down in a broad mantle. Above the tree line, Mount Fuji's alpine flora is impoverished; harsh weather and cinders give plants little chance to thrive, though its lower flanks are clothed in lush mixed forest. On summer evenings nightjars churr here, while early on summer mornings the same forests ring with the songs of flycatchers and buntings and the calls of four species of cuckoo. Red foxes or raccoon dogs also live in the area. From the summit of the mountain, the view looks down to the densely developed lowlands of central Japan. Mount Fuji itself is at its best when seen from the rocky coast to the south, with cherry blossoms in the foreground. In summer, before dawn, religious pilgrims and mountain hikers are known to surge from the huts that dot Mount Fuji's flanks to greet the sunrise at the summit—although it is said that only a fool climbs the summit twice. **MBz**

BELOW *Mount Fuji—the iconic symbol of Japan.*

YAKU-SHIMA

KAGOSHIMA, JAPAN

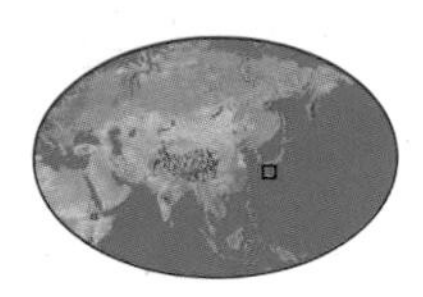

Circumference of Yaku-shima:
82 mi (132 km)

Highest point (Mt. Miyanoura):
6,348 ft (1,935 m)

The mountainous island of Yaku-shima rises from the East China Sea into the track of typhoons and forceful rains. In 1993, in recognition of its biodiversity and dramatic beauty, this small piece of coral-fringed paradise became one of Japan's first World Heritage sites. Yaku-shima has more than 40 granite peaks towering over 3,300 feet (1,000 meters) marking the highest point between the Japanese Alps of Honshu and the even higher peaks of Taiwan. The island presents a tremendous range of habitats; at higher altitudes, coniferous forest and sub-alpine species are found, while subtropical bougainvillea, bananas, and banyans thrive around the coastline. Yaku-shima's long-isolated peaks generate their own warm and rainy climate. The incredible rainfall provides mossy forests with sufficient moisture to keep rivers high year-round and waterfalls in spectacular flow. The island's damp, hushed forests support indigenous deer, macaques, and the immense Japanese cedars known as "sugi" that have made the island famous. Many are more than 1,000 years old, while the massive Jomon Sugi is reputedly the world's oldest tree—some say it is 7,200 years old. MBz

RIGHT *The gnarled trunk of Yaku-shima's giant camphor tree.*

THE RYUKYU ISLANDS

KAGOSHIMA / OKINAWA, JAPAN

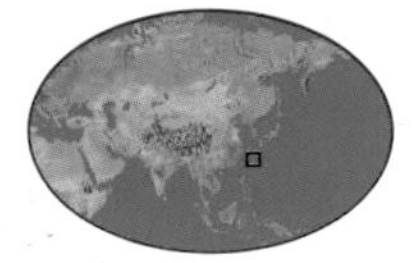

Land area of Ryukyu Islands:
2,389 sq km (922 sq mi)

Number of islands and islets:
more than 200

Rock type: volcanic

The Ryukyu Islands are a chain of more than 200 islands and islets spanning the warm waters between Japan's southern island of Kyushu and Taiwan. An archipelago within an archipelago, the islands are fascinatingly different from the rest of Japan.

The islands of Amami-o-shima and Okinawa support rich and varied ecosystems. Fringed with corals, they feature mangroves and subtropical forest with bizarre-looking cycad trees; their warm offshore waters attract whales, turtles, and sharks. The geological history of these islands has been a checkered one. At times they were connected to mainland Asia through basement rocks which rose to the surface of the ocean to create land bridges. When the land bridges collapsed, the islands were completely isolated. This process has occurred a number of times over a period of 250 million years. When connected, species reached here from Asia and moved between the islands. When separated, species diverged and evolved—unique forms such as the nocturnal black rabbit of Amami Island are the result of such a phenomenon. MBz

KUSHIRO MARSH

HOKKAIDO, JAPAN

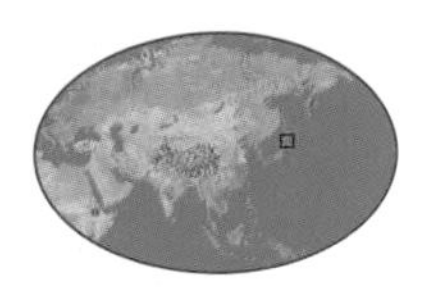

Area of Kushiro Shitsugen National Park: 104 sq mi (269 sq km)

Notable feature: red-crowned cranes

The largest remaining wetland in Japan, Kushiro Marsh is home to the graceful Ainu god of the marshes, "sarurun kamui," otherwise known as the crane. Japan's most symbolic creature, the rare elegant red-crowned crane, or *tancho,* is mentioned in folktales and fabled to live for 1,000 years. In the 1890s, the *tancho* was said to have been nearing extinction, until about a dozen near-starved cranes were discovered in the Kushiro Marsh in 1924. Due to the continuing contributions of local people, who help feed the cranes with corn and buckwheat each winter, the species has been saved from extinction—approximately 600 red-crowned cranes are now found in the Kushiro area.

The Kushiro Marsh is a long overgrown delta formed from the spoils of eastern Hokkaido's volcanoes; it is an enormous mire of peaty

pools and vast expanses of waving reed fronds that house fish, frogs, and dragonflies that are adapted to the cool summers and frigid winters of east Hokkaido.

Bordered by low wooded hills, with its back to the Pacific, the marsh faces north to the Akan volcanoes where the Kushiro River rises from beautiful Lake Kussharo. Above the vast swamp, a summer migrant from Australia, Latham's snipe (the "lightning bird"), buzzes and swoops, while Japanese cuckoos call boisterously. But the loudest call of all is the deep-noted duet of the *tancho*. In summer, the marsh comes alive with insects and birds; in winter, brittle reed stems rattle and keen in the wind, while frosted ponds crack and shatter. The *tancho* survives here year-round by hiding its nest in remote marshland reed beds. During winter, the birds gather at long-established sites where they forage and rekindle their pair bonds in a spectacular white-and-black feathered dance. MBz

BELOW *Home to the god of marshes, the red-crowned crane.*

IZUMI

JAPAN

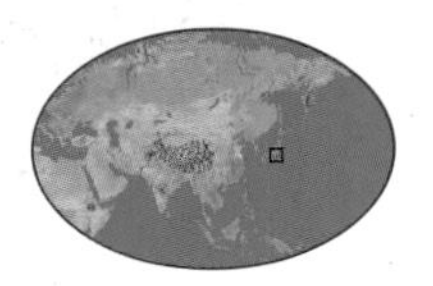

Number of crane species at Izumi: 2 to 5

Crane population: more than 11,000

It hardly seems believable that reclaimed land now used for farming is the setting for one of the world's most memorable wildlife experiences. Yet the tiny rice fields of the coastal plain of Arasaki, Japan, provide the site every year for the largest gathering of cranes in the whole of Asia. The cranes arrive from overseas, principally from northeast China and Russia, from where they make their long, laborious way down the continental coast and the Korean Peninsula before finally reaching Kyushu (Japan's southern main island) to spend the winter in Izumi.

With pink legs and a dark gray-and-white-striped neck, the white-necked cranes are possibly the most elegant of the crane species found here. Izumi's rice fields and agricultural lands provide perfect roosting and feeding conditions for them. However, Japan's dearth of flat, cultivable land has caused its problems. Over recent decades, thousands of acres of wetlands have been lost to agriculture, forcing the cranes to mass in ever smaller areas, and today they rely heavily on an artificial feeding station.

In this mass gathering of cranes, the white-necked versions are in fact greatly outnumbered by the more diminutive hooded cranes. More than 80 percent of the world's hooded crane population—approximately 8,000 birds—overwinter here. This species is well-known for its unison calling and mating dance in which the birds bow, jump, run, toss grass and sticks, and flap their wings.

The sights and sounds created by such huge numbers of cranes flying out from their roosts are astonishing. Ranks of birds march across the fields, while squadrons drain from massed flocks above. This is a crane gathering extraordinaire. The sheer volume of noise created by all these birds is nothing short of incredible.

It hardly seems believable that reclaimed land now used for farming is the setting for one of the world's most memorable wildlife experiences. Yet the tiny rice fields of the coastal plain of Arasaki are the site of the largest gathering of cranes in Asia.

While flocks of sandhill cranes in the United States or common cranes in Europe may outnumber the cranes found here, the gatherings in Izumi can offer both numbers and diversity. The great crane flock in Kyushu is like a magnet for wandering rarities, and it is not uncommon for East Asia's two commonest species to draw at least three other rarities. In 2004, common, sandhill, and Siberian cranes joined white-naped and hooded cranes to create a truly international gathering. MBz

RIGHT *Flocks of cranes on the fields of Izumi.*

9

MOUNT BAEKDUSAN / CHEONJI CALDERA LAKE

RYANGGANG, DEMOCRATIC PEOPLE'S REPUBLIC OF KOREA

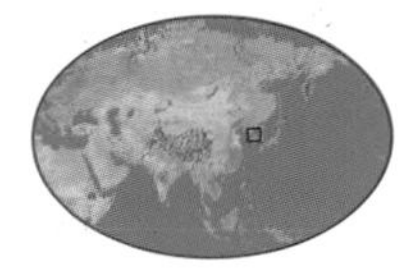

Average depth of Cheonji caldera lake: 699 ft (213 m)

Length of Yalu River: 490 mi (790 km)

Length of Tumen River: 324 mi (521 km)

On the country's border with China lies one of its most sacred mountains. Mount Baekdusan—known as the "Ever-white Mountain"—is an important symbol of Korean spirit. It is recorded as the mythical birthplace of the Korean people and even mentioned in the national anthem. As the highest peak in the country, it rises 9,003 feet (2,744 meters) above sea level. Although this dormant volcano is the source of several of the area's most important rivers, including the Yalu and Tumen, its main attraction is the caldera at it summit. This holds Cheonji, or the "Heavenly Lake," one of the largest and deepest caldera lakes in the world. It covers an area of 3.5 square miles (9 square kilometers) and has an estimated water storage capacity of 2 billion tons. Mount Baekdusan is surrounded by more than 20 densely forested peaks. The rugged lava ravines on the mountain slopes harbor bears, tigers, and leopards, as well as over 2,700 types of plant. Access is difficult due to a restricted infrastructure and border tensions, but special permits may be obtained via government tourist agencies. **AB**

YEONJUDAM POOLS

KANGWON, DEMOCRATIC PEOPLE'S REPUBLIC OF KOREA

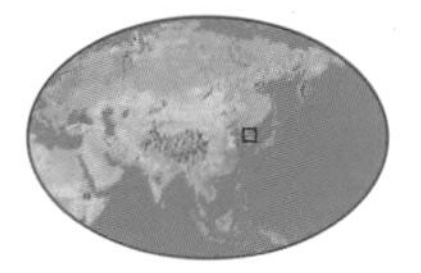

Highest peak in Geumgangsan Mountains: 5,373 ft (1,638 m)

Length of range: 25 mi (40 km)

Habitat: rocks, alpine, temperate woodland

Situated in the south of the country, the Yeonjudam Pools are a tranquil and ethereal sight. Delicate jade-colored pools contrast elegantly with the surrounding light-gray limestone rocks. In fall, the vivid reds and golds of the maples on the surrounding hills further enhance this spectacle. The creation myth of the Yeonjudam Pools tells of a heavenly fairy who dropped a necklace of emeralds; where the stones fell and the necklace broke, the pools were formed. The pools lie within the Geumgangsan Mountains, considered to be the most beautiful range in Korea; the serene peaks are often shrouded in a delicate mist. After a 50-year hiatus, the mountains are now open to foreign tourists. The range is divided into three areas: Neageumgang (Inner Diamond) Mountains, where the pools occur; Oegeumgang (Outer Diamond) Mountains, which includes the Manmulsang, the natural stone "Images of Ten Thousand Things"; and Haegeumgang (Sea Diamond) Mountains, where pine trees perch on stratified rock columns stretching out into the Sea of Japan. **AB**

GURYONG FALLS / GEUMGANGSAN MOUNTAINS

KANGWON, DEMOCRATIC PEOPLE'S REPUBLIC OF KOREA

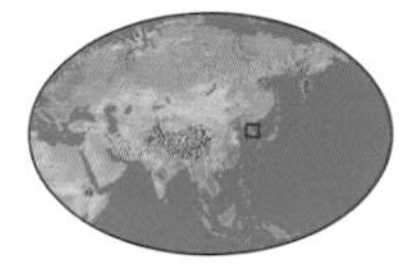

Height of Guryong Falls: 240 ft (74 m)

Height of engraved rock face: 60 ft (18 m)

Width of engraved rock face: 12 ft (3.6 m)

The Outer Diamond Mountains, or Oegeumgang region in the eastern part of the Geumgangsan Mountains, have several striking features including dramatic waterfalls and alpine pools. Guryong Falls, said to be the most breathtaking waterfall in the region, cascades like a long silk scarf over a huge granite cliff to the Guryongyeon Pool below. It has an unusual geological structure, for the cliff over which the water falls and the rock beneath the pool are made up of a single block of granite. From a lookout tower on a cliff next to the waterfall, the eight jewel-like Sangpaldam Pools can be seen, as well as the Guryongyeon Pool named after nine dragons said to guard the Geumgangsan Mountains. In 1919, the rock face of the granite cliff was engraved with the three characters "Mi-reuk-bul" (meaning "Buddha of the Future") by the calligrapher Kim Gyu-jin.

At the nearby Bibong Falls, water cascades over a striated rocky cliff and plunges down an awesome 450 feet (139 meters). The feathery white plumes of falling water are said to resemble the legendary phoenix in flight. **RC**

MANMULSANG

KANGWON, DEMOCRATIC PEOPLE'S REPUBLIC OF KOREA

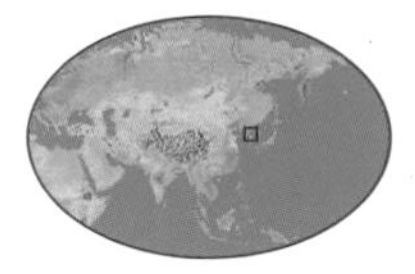

Area of Geumgangsan National Park: 9,600 acres (3,885 ha)

Rock type: limestone

The word *manmulsang* does not refer to a particular mountain, but to a part of the country's Obongsan region, north of Mount Geumgangsan in the Diamond Mountains. Manmulsang means "the world all in one place" and refers to the variety of shapes that can be imagined in the region's naturally sculpted limestone pinnacles.

Manmulsang is part of an extensive area of sedimentary rock that rotated so that the layers became exposed. Differential erosion of the multiple layers has resulted in this extraordinary natural gallery. Also known as "the place of 12,000 miracles," the area has many monasteries among the spectacular peaks. Surrounded by maple-rich deciduous forest, the mountains are especially beautiful in the fall when the gold-and-red foliage contrasts with the gray rock. The majority of the Manmulsang is included within the boundaries of Geumgangsan National Park. The area is very important to Korean culture, having inspired poets, artists, and mystics. The pinnacles are steep—many can only be ascended via specially placed ladders. **AB**

MANJANG-GUL & SEONGSAN ILCHULBAONG

JEJU-DO ISLAND, REPUBLIC OF KOREA

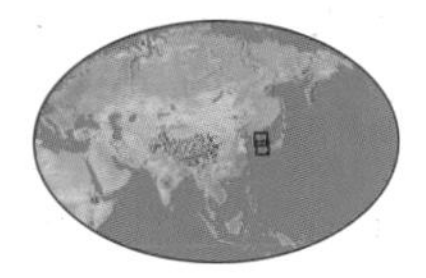

Height of Seongsan Ilchulbaong Peak: 300 ft (90 m)

Diameter of cone: 2,000 ft (600 m)

Notable feature: volcanic cone with sea cliffs, volcanic lava tube system

Situated on the eastern end of the volcanic island of Jeju-do, Seongsan Ilchulbaong Peak erupted from the sea 100,000 years ago. Today it is a dramatic crater 2,000 feet (600 meters) in diameter and 300 feet (90 meters) high, which can be reached via a trail from the nearby village. At the summit, visitors are rewarded with wonderful panoramic views of the island's eastern point and the waves splashing against the cliffs below. Another of Jeju-do's volcanic features, on the northeast coast of the island, is the Manjang-gul Cave system, considered to be one of the best caves in the world in terms of its length and structure. At 8.4 miles (13.4 kilometers), the Manjang-gul lava tube is the longest in the world, with passages varying in height and width from 10 to 66 feet (3 to 20 meters). Spectacular rippled surfaces and oddly-shaped congealed remnants indicate the original flow of lava. The cave system is home to highly specialized cave life. To protect this habitat, only 1.5 miles (2.5 kilometers) are open to visitors, but the caves are thoughtfully lit and have safe walkways. AB

BELOW *100,000-year-old Seongsan Ilchulbaong Peak.*

JUSANGJEOLLI COASTLINE

JEJU-DO ISLAND, REPUBLIC OF KOREA

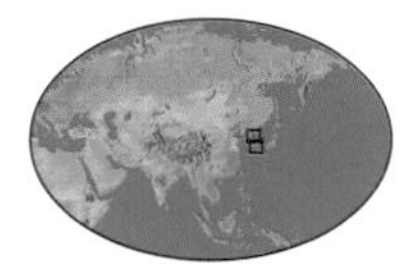

Length of Jusangjeolli Coast: 1.3 mi (2.1 km)

Height of cliffs: 66 ft (20 m)

On the southern coastline of the island of Jeju-do vertical stone pillars rise like giant hexagonal crystals from the sea, forming an extraordinary and striking cliff face. The Jusangjeolli are dark-gray hexagonal rock pillars, so regular in structure that it is hard to believe they are natural formations and were not carved by a stonemason. However, the pillars were formed from basaltic lava that came from Mount Hallasan. Subsequent erosion due to the constant battering of waves has resulted in the formation of natural staircases in some parts. When the tides are high the waves crash against the shore and spout up to 33 feet (10 meters) into the air. The district administration named the formation "Jisatgae Rocks"; however, the area's original name, Jisatgae Haean, means "Altar to the Gods." The Jusangjeolli extend along the coastline for approximately 1.2 miles (2 kilometers), between Jungmun and Daepo-dong in Seogwipo City. To reach the area visitors should travel southwest of Daepo-dong village through a pine forest to the cliffs. The area is a designated cultural monument and has become one of the most popular tourist sites in the country. RC

HWANSEON GUL CAVE

GANGWON-DO, REPUBLIC OF KOREA

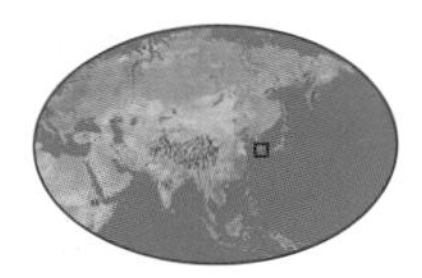

Length of Hwanseon Gul Cave: 3.8 mi (6.2 km)

Diameter of main chamber: 130 ft (40 m)

Average height of passages: 50 ft (15 m)

Lying on the east coast, the country's largest karst region (the Gangwon-do Province) contains 500 of the 1,000-plus caves that are found across the country. Within this mountain landscape, 2,700 feet (820 meters) above sea level, lies the largest limestone cave in Asia. Hwanseon Gul Cave is about four miles (6.5 kilometers) long, with passages approximately 50 feet (15 meters) high and 66 feet (20 meters) wide. However, its most impressive feature is the main chamber, a white sand oasis capable of holding thousands of people.

Hwanseon Gul Cave contains a wide variety of stalagmite formations, as well as a stalactite cascade known as the "Great Wall of China" and a structure in the main chamber known as the Okchwadae ("Royal Throne"). Although just 1 mile (1.6 kilometers) of the Hwanseon Gul Cave passage is open to the public, the cave tour enables visitors to see six stunning waterfalls and 10 clear cave pools. The temperature in the cave remains at about 52°F (11°C) throughout the year.

The karst cave region in Gangwon-do was designated a natural monument in 1966. Other nearby caves in the Taei-iri district of Samchok include Gwaneum Cave, renowned for its beautiful flowstone formations, Yangtumokse Cave, Dukbatse Cave, and Keunjaese Cave. **RC**

SUNRISE PEAK

CHOLLA-NAMDO / JEJU-DO ISLAND, REPUBLIC OF KOREA

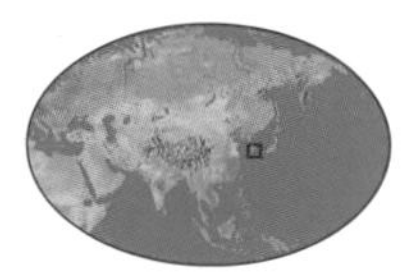

Elevation of Sunrise Peak: 597 ft (182 m)

Diameter of crater: 2,000 ft (600 m)

At the end of the Seongsan Peninsula, located off the eastern tip of Jeju-do Island, a volcanic crater known as Sunrise Peak rises from the sea. Its name was prompted by the breathtaking views offered from the volcano's summit at sunrise.

The peak emerged from beneath the ocean in a volcanic eruption about 100,000 years ago. Today the huge crater has 99 jagged rock pinnacles around the rim of the crater, which, from a distance, rise like a massive crown or castle from the sea. There are cliffs on the north and southeast sides of the peak, but a trail from Seongsan village leads up to the western edge of the crater. The views from the crater summit are spectacular, particularly when the sun appears to rise like a ball of fire from the sea. Sunrise Peak has been designated a natural monument and is open from sunrise to sunset. The setting is particularly beautiful in the spring, when the bright yellow rapeseed flowers blossom in the nearby fields. Each New Year's Eve crowds gather in the crater on Sunrise Peak to celebrate the rising sun and wish each other good luck for the year ahead. **RC**

MOUNT HALLASAN

JEJU-DO ISLAND, REPUBLIC OF KOREA

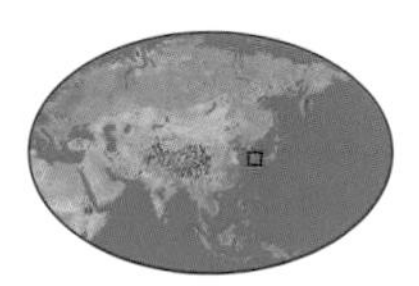

Alternative name: Che-judo

Elevation of Mt. Hallasan: 6,400 feet (1,950 m)

Type: sub-tropical volcanic island

Situated south of the coastal mainland, the lozenge-shaped Jeju-do Island is dominated by Mount Hallasan, the highest peak in the country. It can be seen from anywhere on the island, although its peak is often covered in clouds. A dormant volcano, Hallasan last erupted in 1007 C.E. As a result of its quiescence, its caldera now holds an exquisite crystal clear crater lake called *Baengnokdam* ("the water where the white deer play"). Legend tells of otherworldly men (said to be enlightened sages or mountain gods) that descended from heaven riding white roe deer to enjoy the beauty of the lake. Baengnokdam is surrounded by many grotesquely shaped rocks and cliffs that are said to protect this sacred area.

Mount Hallasan's rich volcanic soil has resulted in well-developed forests including subtropical broadleaved forest. At lower levels 70 tree species are found, including wild camellia, fig, forsythia, and orange; at higher elevations pine forests thrive; and near the top of the mountain, a wide variety of alpine plants appear. During spring, the mountain slopes are ablaze with deep pink royal azalea blossom, while in fall, tree foliage turns brilliant shades of red and gold. Black bear and deer can still be seen in the area. **AB**

KOPET DAG

IRAN / TURKMENISTAN

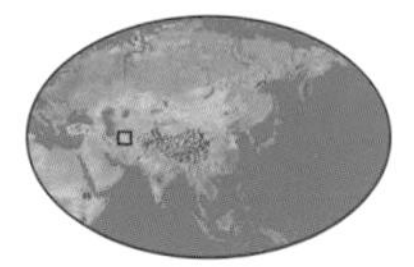

Elevation of Kuh-e Quchan: 10,466 ft (3,191 m)

Length of Bakharden Lake: 236 ft (72 m)

Width of Bakharden Lake: 100 ft (30 m)

On the border of Iran and Turkmenistan are the dry and barren mountains known locally as *Kopet Dag*, or alternately as the "Moon Mountains." Kopet Dag is a desert region where dust fills the air, plant life is almost entirely absent, and the sand dunes of the Kara Kum—the "black desert," one of the world's largest hot, sandy deserts—stretch away to the north. The mountains run for 400 miles (645 kilometers) from the Caspian Sea, along the Turkmenistan-Iran border, to the Harirud or Tejen River. The region is carved up by muddy-brown canyons and gorges with pitted and gouged walls.

The highest peak is Kuh-e Quchan at 10,466 feet (3,191 meters), but one treasure lies hidden below the slopes of the Kopet Dag. About 200 feet (60 meters) undergound in the Kov-ata Cave is a hot water mineral lake, known as Bakharden. The lake's water temperature is 97°F (36°C), and it is enveloped in sulfurous odors. *Kov-ata* means "father of caves," and this underground auditorium is certainly impressive. It lies 23 miles (37 kilometers) south of Ashgabat. **MB**

BAND-E AMIR LAKES

BAMIAN, AFGHANISTAN

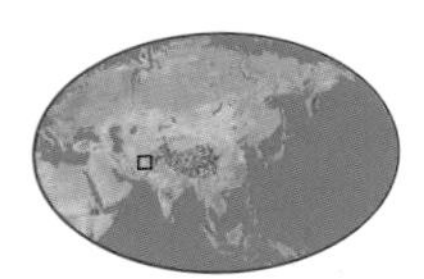

Elevation of Band-e Amir lakes: 10,000 ft (3,000 m)

Length of lake system: 7 mi (11 km)

About 10,000 feet (3,000 meters) high in the arid foothills of Afghanistan's Hindu Kush Mountains is a ribbon of lakes spread along a 7-mile (11-kilometer) stretch of the Band-e Amir River. Travertine—a coating of calcium carbonate on dead water plants that is deposited over thousands of years—forms natural dams in the Amir River, some of which are as high as a 20-foot (6-meter) house.

The dams break the river up into lakes that vary in color from blue to green to milky white, depending on the minerals and algae present. The lakes are fed by meltwater from the mountains, so even though the summer air temperature can reach 97°F (36°C), river and lake waters remain chilled. Some lakes are just 625 feet (190 meters) long, while others extend for up to 4 miles (6 kilometers). Surrounding them are cliffs of limestone and clay.

Folklore states that the lakes were formed when Ali, son-in-law of Muhammad, was held prisoner in the valley. In his fury, Ali triggered a landslide and blocked the river with a dam at Band-e Haibat. The lake area is so remote that it can only be reached by rough mountain track 50 miles (80 kilometers) from the town of Bamian. MB

INDUS RIVER

CHINA / INDIA / PAKISTAN

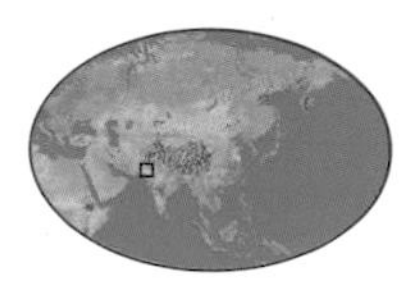

Length of Indus River:
1,740 mi (2,800 km)

Elevation of Lailas Range:
17,000 ft (5,200 m)

Height of Nanga Parbat:
26,660 ft (8,126 m)

The Indus River is known in Hindu legend as the Lion River because the river was fabled to pour from a lion's mouth. In the legend, the lion is representative of the Himalayas. The river's real source is in the Lailas Range, where, after passing through the Karakoram Range, it drops 12,000 feet (3,600 meters) through precipitous gorges that extend 350 miles (560 kilometers) and skirts the magnificent Nanga Parbat (or Naked Mountain) that towers above its surface. Here the Indus is hidden in exceptionally steep-sided gorges up to 15,000 feet (4,500 meters) deep, where the sun barely penetrates.

At the southern end of Attock Gorge—possibly a more appropriate lion's mouth—the river bursts onto the flatlands of the Punjab to form part of the largest irrigation system in the world. Seasonal monsoon rains and melting ice and snow combine to bring several months of serious flooding, and maybe even create a shallow, inland sea. At the end of its 1,740-mile (2,800-kilometer) journey, the Indus flows out into the Arabian Sea through a huge tropical mangrove-rich delta. MB

BELOW *The winding, mountainous course of the mighty Indus.*

K2

PAKISTAN / CHINA

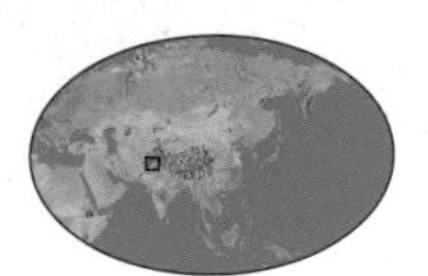

Height of K2: 28,251 ft (8,611 m)

Height of Karakorum Pass: 18,290 ft (5,575 m)

Height of Khunjerab Pass: 15,420 ft (4,700 m)

K2, or Mount Godwin-Austen, was first surveyed during the Survey of India in 1856 and named originally after the English topographer Henry Godwin-Austen. Locally it is known as Chogo Ri, which means "Great Mountain," and great it is—at 28,251-feet (8,611-meters) high, K2 is the second-highest mountain in the world after Mount Everest and is considered to be one of the most dangerous. Its rocky slopes tower up to 20,000 feet (6,000 meters) in height. Above this, deep snowfields extend to create carpeted white plains.

The name for the mountain most commonly used today is K2. This unusual name was derived from a survey that was conducted in the Karakorum Mountains on the Pakistan–China border. K2 was the second mountain that was surveyed.

The Karakorum Range stretches 300 miles (480 kilometers) between the Indus and Yarkut Rivers and is a southeast extension of the Hindu Kush mountains. The mountain's area contains disputed territory, contested by China in the north, Pakistan in the southwest, and India in the southeast. Its spectacular glaciers, including the Baltoro, one of the largest non-polar glaciers, feed many rivers, including the mighty Indus.

There are two main trade routes running through the mountains—Karakorum Pass, which is lodged 18,290 feet (5,575 meters) above sea level and Khunjerab Pass at 15,420 feet (4,700 meters). Both are above the perpetual snow line. K2 is the highest mountain in the Karakorum range.

The mountain's reputation for danger is derived in part from the numerous failed attempts that have been made to ascend its peak. In 1902, a first attempt was made by climbers who managed to scale 2,654 feet (809 meters). Seven years later, fuelled by the prospect of developing commercial alpine operations, the Duke of Abruzzi reached 24,606 feet (7,500 meters). The 1938 and 1939 American expeditions, however, ended in disaster. The peak was finally conquered by an Italian team led by Ardito Desio in 1954.

Even today, K2 has been scaled far fewer times than Everest and has a reputation among the rock-climbing community for being one of the most difficult mountains in the world to climb. **MB**

The name for the mountain most commonly used today is K2. This unusual name was derived from a survey that was conducted in the Karakorum Mountains on the Pakistan–China border. K2 was the second mountain that was surveyed.

RIGHT *K2's seemingly unreachable peak illuminated by the sun.*

HUNZA VALLEY

NORTHERN AREAS, PAKISTAN

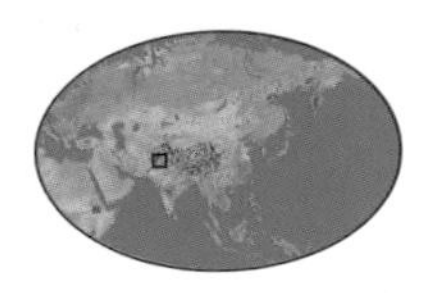

Elevation of Hunza Valley: 8,000 ft (2,438 m)

Elevation of Rakaposhi Peak: 25,550 ft (7,788 m)

Elevation of Ultar Peak: 24,240 ft (7,388 m)

The Hunza Valley is situated along the Karakoram Highway (KKH), and comprises part of the Pakistan–China Silk Route that has been legendary since the days of Marco Polo. When talking about the area, former viceroy Lord Curzon claimed that "the little state of Hunza contains more summits of over 20,000 feet (6,000 meters) than there are of over 10,000 feet (3,000 meters) in the entire Alps."

The Hunza River cuts straight through the Karakoram Range and is the largest tributary of the Gilgit River watershed. The KKH crosses the river via a swooping bridge, itself within walking distance from Haldikish ("Place of Rams"), which contains the Sacred Rock of the Hunza, a rock inscribed with the

scripts and carvings of many different eras. The Hunza Valley has a landscape of majestic snow-covered mountains, glaciers, plentiful orchards, and fields. The peaks of Rakaposhi and Ultar dominate the Hunza skyline, while the valley contains many glaciers, including the Passu and Batura.

The isolated and idyllic setting of the Hunza Valley was perhaps one inspiration for Shangri-La in the book *Lost Horizon* by James Hilton. The longevity of the people living here is supposedly attributable to the local diet, which includes large quantities of fruit and vegetables. Hunza water, which is also supposed to promote a long life, as well as vitality, is known locally as "mel."

The Hunza Valley is a superb area for both mountaineering and hiking. The main tourist season runs from May to October. The ruby mines are a popular tourist attraction. RC

BELOW *The rugged mountains of the Hunza Valley.*

KHYBER PASS

NORTH-WEST FRONTIER PROVINCE, PAKISTAN

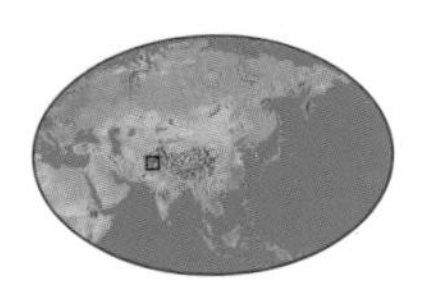

Length of Khyber Pass: 33 mi (53 km)

Width of Khyber Pass: 50 ft (15 m) to 460 ft (140 m)

Highest point: 3,500 ft (1,067 m)

One of the most famous mountain passes in the world, the historic Khyber Pass (meaning "across the river") runs through the Hindu Kush mountain range and connects Pakistan with Afghanistan. For centuries it has been a major trade route, and it is equally known as an invasion route from Central Asia to the Indian subcontinent. Alexander the Great and his army passed through the Khyber Pass in 326 B.C.E. to reach the plains of India. Persian, Mongol, and Tartar armies also invaded via the Khyber Pass in the 10th century, thus introducing Islam into India. The pass was an important area in the 19th-century Afghan Wars.

The pass is narrow and steep-sided, winding northwest through the Safed Koh mountains, and reaches its highest point at the border between Pakistan and Afghanistan. The mountains on either side can only be accessed in a few places. The pass is controlled by Pakistan and links the cities of Peshawar and Kabul. The British constructed the present road through the pass during the Afghan Wars, and there is also a traditional camel caravan route. For railway enthusiasts there is also the Khyber Railway, which passes through 34 tunnels and crosses 92 bridges to the head of the pass at the Afghan border. RC

NANGA PARBAT

NORTHERN AREAS, PAKISTAN

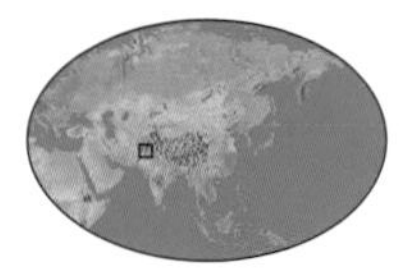

Height of Nanga Parbat: 26,660 ft (8,126 m)

Height of Rupal: more than 16,500 ft (5,000 m)

While most other Himalayan peaks are covered with snow, the 26,660-foot (8,126-meter) Nanga Parbat is the exception. Its steep walls and sharp ridges gather much less snow, which is why its name translates as the "Naked Mountain." It is the world's ninth highest mountain, and it stands remote and seemingly aloof at the western end of the Karakorum Range.

German explorers named the mountain "Murder Mountain." It was first drawn by the Schlagintweit brothers from Munich in 1854, but in 1857 one was murdered in Kashgar, thus initiating the legendary curse on Nanga Parbat. Since then, local Sherpas have dubbed it "Mountain of the Devil," for no other peak has claimed lives with such regularity. The mountain has three faces, each a challenge for mountaineers—Rakhiot, Diamir, and Rupal. At more than 16,500 feet (5,000 meters) high, Rupal is the most spectacular. The Austrian–Italian climber Reinhold Messner remarked that, "everyone who has ever stood at the foot of this face up above the Tap Alpe, studied it, or flown over it, could not help being amazed by its sheer size; it has become known as the highest rock and ice wall in the world." MB

KEOLADEO NATIONAL PARK

RAJASTHAN, INDIA

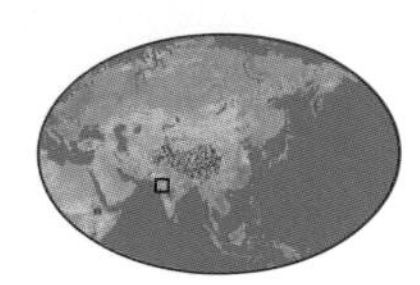

Area of Keoladeo National Park: 11 sq mi (29 sq km)

Number of bird species: approx. 400

Keoladeo National Park (formerly known as Bharatpur Bird Sanctuary) is one of the most important breeding and feeding grounds in the world for migratory birds. Although small, its shallow lakes and woodland support a vast population of birds, many of which migrate from as far away as Siberia and China.

The original marshland was modified during the 19th century by the Maharajah of Bharatpur, and subsequently became noted for the enormous numbers of duck that were shot (up to 4,000 in a single day). It was designated a Ramsar site (Wetland of International Importance) in 1981, a national park in 1982, and a World Heritage site in 1985.

The sanctuary attracts vast numbers of waterfowl such as ducks, geese, herons, storks, egrets, pelicans, cranes, and ibises. A major attraction is the arrival of the critically endangered Siberian crane, although only one or two pairs have visited in recent years. Other wildlife attractions include more than 30 bird of prey species and mammals such as nilgai, sambar, chita, and wild boar. RC

RIGHT *The famous avifauna sanctuary of Keoladeo National Park at sunset.*

SIACHEN GLACIER

JAMMU AND KASHMIR, INDIA

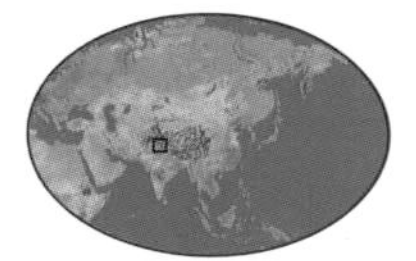

Length of Siachen Glacier: 45 mi (72 km)

Width of Siachen Glacier: 1.2 mi (2 km)

About 45 miles (72 kilometers) long and 1.2 miles (2 kilometers) wide, the Siachen Glacier is the largest glacier in the world outside the polar regions. It is located on the north-facing slopes of the Karakorum Range, near the India–Tibet border. The glacial meltwaters feed the Mutzgah and Shaksgam Rivers that run parallel to the mountains before plunging into Tibet.

Huge tributary glaciers, such as Shelkar Chorten and Mamostang, enter the main body from either side, forming a vast ice field. Icefalls mark many of the junctions. The sides of the glacier are strewn with rocks and boulders, but its center consists of an enormous snowfield. The sidewalls are steep and give rise to numerous avalanches. The Rimo Group of three glaciers is located east of the glacier. Known simply as North, Central, and South, these glaciers rise 20,000 to 23,000 feet (6,000 to 7,000 meters) above sea level. Between them, they consist of 270 square miles (700 square kilometers) of ice. Together with the Siachen Glacier, this concentration of glaciers covers about 772 square miles (2,000 square kilometers). The Siachen Glacier can be reached via Skardu in Ladakh. MB

RANTHAMBORE NATIONAL PARK

RAJASTHAN, INDIA

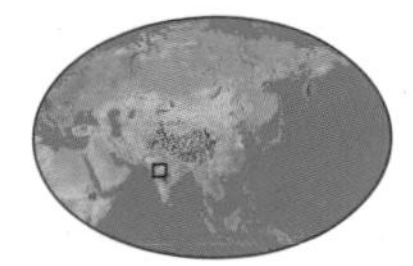

Area of Ranthambore National Park:
151 sq mi (392 sq km)

Number of recorded bird species:
272

Ranthambore National Park is one of the few places in the world where tigers can be observed at close quarters in the wild. The tigers have become accustomed to vehicles and can often be seen during the day. Situated where the Aravalli Hill Ranges and the Vindhya Plateau meet in the east of Rajasthan, Ranthambore was once a hunting preserve of the Maharajahs.

The landscape varies between dense, lush forest and open bushland, and is surrounded by steep slopes and massive rock formations. Ranthambore has high numbers of large mammals, including samba deer, chital, nilgai, leopard, and sloth bear, as well as a great variety of birds. The animals, especially deer, are easiest to spot around the lakes and pools. Ranthambore was declared a wildlife sanctuary in the 1950s and in the 1970s became part of the Project Tiger conservation program. It was awarded national park status in 1981. RC

BELOW *Ranthambore's lakes are the best place to spot animals.*

VALLEY OF FLOWERS

UTTARANCHAL, INDIA

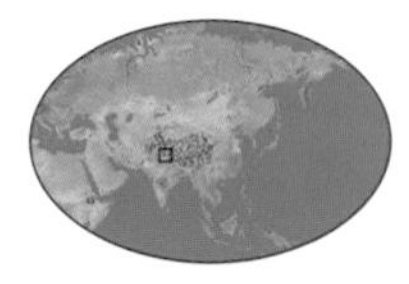

Area of Valley of Flowers: 34 sq mi (87.5 sq km)

Elevation of valley: 11,000 to 21,000 ft (3,500 to 6,500 m) above sea level

Established as a national park: 1982

The smallest national park in the Himalayas, the Valley of Flowers is an alpine valley surrounded by snow-covered peaks that in summer transforms into a brightly colored carpet of flowers and herbs: a true paradise for botanists. The valley was explored and named in the 1930s by Frank Smythe, whose book *The Valley of Flowers* made the area famous.

The valley is the catchment area of the Pushpawati River and has a unique microclimate. The north side is walled by steep cliffs, but slopes less steeply to the south, providing partial protection from the cold northern winds and the monsoon to the south. Rhododendron, rowan, and birch cover the northern slopes, while the southern slopes consist mainly of flower-rich alpine meadows. The rare Brahmal Kamal or "Lotus of the Gods" grows on the valley's upper slopes. The valley, near the village of Ghangaria, is only accessible during daylight hours in the summer, between June and October. RC

NANDA DEVI NATIONAL PARK

UTTARANCHAL, INDIA

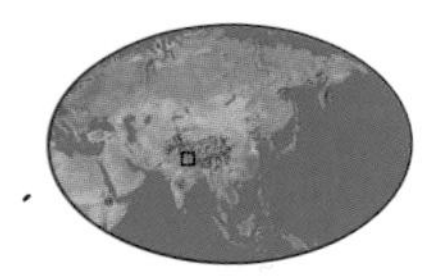

Elevation of Nanda Devi peak: 25,644 ft (7,816 m)

Area of Nanda Devi National Park: 243 sq mi (630 sq km)

Elevation of national park: 6,890 to 25,644 ft (2,100 to 7,816 m)

One of the most spectacular uninhabited wilderness areas in the Himalayas, Nanda Devi National Park is dominated by Nanda Devi, the second highest peak in India. The area is a vast glacial basin encircled by mountains and drained by the Rishi Ganga River. The first recorded attempt to enter the region was by W. W. Graham in 1883, but the inner sanctuary was not penetrated until 1934, by mountaineers Eric Shipton and Bill Tilman. In 1936, Tilman and Noel Odell made the first successful ascent of Nanda Devi. The area remained largely undisturbed until organized expeditions started in the 1950s. In 1983, the Indian government closed the sanctuary to protect the delicate ecology of the park.

Forest cover—mainly fir, rhododendron, birch, and juniper—is restricted to the Rishi Gorge. The vegetation in the drier inner sanctuary varies with altitude, from scrub and alpine meadows to glaciers with almost no vegetation. The park is renowned for its rare mammals, including snow leopard, Himalayan black bear, Himalayan musk deer, and bharal. The inner basin has a very distinctive microclimate—generally dry but with heavy rainfall during the monsoon season. RC

MILAM GLACIER

UTTARANCHAL, INDIA

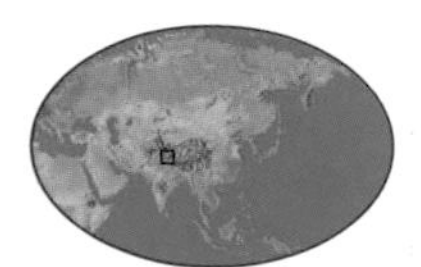

Length of Milam Glacier: 10 mi (16 km)

Area of glacier: 14 sq mi (37 sq km)

Elevation at snout of glacier: 12,400 ft (3,782 m)

Milam is one of the largest and best-known glaciers in the Kumaun region, and is spectacularly located on the eastern side of the Nanda Devi sanctuary on the south-facing slope of the main Himalayan range. The Milam Glacier originates from the Kohli slope and Trishul peaks, and is fed by subsidiary glaciers from a number of nearby peaks. The Goriganga River, one of the main arterial waterways of the Upper Kumaun Himalayas, arises in the valley of the Milam Glacier.

Treks to the glacier generally start from the village of Munsyari and follow the gorge of the Goriganga River from the forested lowland regions to high alpine meadows, with panoramic views of the mountain peaks. Milam Glacier is located about three miles (five kilometers) beyond Milam, which is one of the largest villages in Kumaun. The snout of the glacier, at an elevation of 12,400 feet (3,782 meters), is a 36-mile (58-kilometer) hike from Munsiyari and is the starting point for many treks and peak routes. The best times of year to hike the trail (which takes about 8 to 10 days) are from mid-April to June, before the monsoon, and from September to early November, after the rains. RC

MARBLE ROCKS

MADHYA PRADESH, INDIA

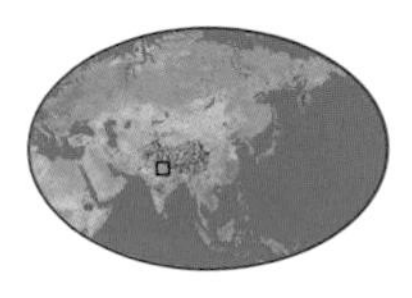

Length of Bhedaghat Gorge: 2 mi (3.2 km)

Height of cliffs: up to 100 ft (30 m)

The Narmada River passes through an impressive gorge of gleaming white, sheer marble cliffs at Bhedaghat, 14 miles (22 kilometers) to the west of Jabalpur. Nearby, the river narrows and plunges dramatically over the mighty Dhuandhar Falls, which are also known as the "Smoke Cascade." The perpendicular white magnesium limestone rocks, with darker volcanic seams of green and black, rise above the calm crystal-clear waters of the river. Captain J. Forsyth described the Marble Rocks in his classic book *Highlands of Central India*: "The eye never wearies of the effect produced by the broken and reflected sunlight, now glancing from a pinnacle of snow-white marble reared against the deep blue of the sky as from a point of silver; touching here and there with bright lights the prominences of the middle heights; and again losing itself in the soft bluish greys of their recesses." Interesting rock formations here include Elephant's Foot and the ledge known as Monkey's Leap. RC

KYLLANG ROCK & SYMPER ROCK

MEGHALAYA, INDIA

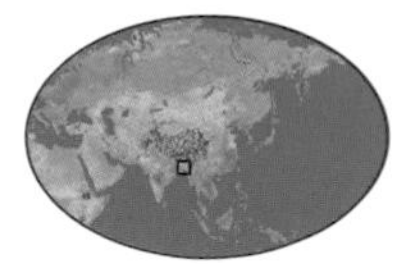

Height of Kyllang Rock: 720 ft (220 m)

Width of Kyllang Rock: 1,000 ft (300 m)

Elevation of Kyllang Rock: 5,400 ft (1,645 m)

Kyllang Rock is an imposing red granite dome that rises out of the rolling grass downs near the village of Mawni in Khadsawphra. The view from the top is impressive, especially toward the Himalayas to the north during the winter months. The rock is accessible from the north and east but is inaccessible from the nearly perpendicular southern flank. Sir Joseph Hooker (1817–1911) described Kyllang Rock's steep southern side as being "encumbered with enormous detached blocks," while the northern side is clothed with dense forests of rhododendrons and oaks.

Symper Rock is an almost flat-topped, loaf-shaped rocky dome (not unlike Kyllang Rock) that rises abruptly from the surrounding hillocks near Mawsynram. From the top of the hill there are superb views over the nearby hills, plains, and rivers of Bangladesh. Folktales claim that the gods U Kyllang and U Symper (who inhabited these rocks) fought a great battle. The victor was U Kyllang, so Kyllang Rock stands proud compared to the abode of the vanquished U Symper. The many holes at the base of Symper Rock are said to be evidence of the battle. RC

GIR NATIONAL PARK AND WILDLIFE SANCTUARY

GUJARAT, INDIA

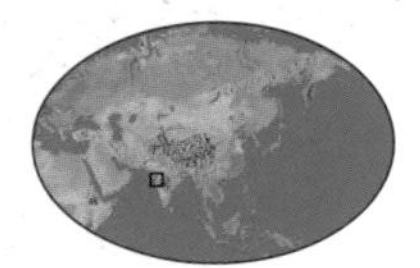

Area of Gir National Park: 545 sq mi (1,412 sq km)

Number of bird species: over 200

The Gir National Park, established in 1965 as a forest reserve to conserve the Asiatic lion, was declared a national park in 1974. The park is the only place in the world where Asiatic lions occur in the wild. Asiatic lions are slightly smaller than African lions and they have less prominent manes. They once ranged from Greece to central India, but by 1910 there were probably less than 30 left in the wild. The national park now has a population of about 300 lions. Gir also has one of the largest leopard populations of any park in India—they can be seen occasionally near lodges at night. The park and sanctuary are also home to four-horned antelope, deer, wild boar, jackals, hyenas, and marsh crocodiles.

The forest is mixed deciduous woodland (mainly teak) and the terrain is rugged and hilly. The best way to view wildlife is by four-wheel-drive vehicles, preferably around dawn or dusk when the lions are active. RC

WATERFALLS OF MEGHALAYA

MEGHALAYA, INDIA

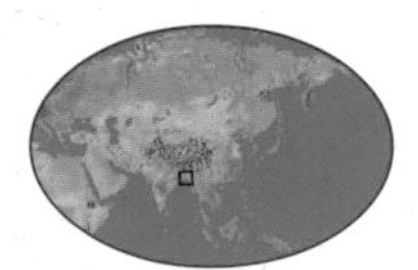

Average annual rainfall (Cherrapunji): 37.7 ft (11.5 m)

Average annual rainfall (Mawsynram): 39.0 ft (11.9 m)

The state of Meghalaya receives so much rain that it is unsurprising it has so many spectacular waterfalls. Two of the wettest places on Earth are located here: Cherrapunji (also known as Sohra) and Mawsynram. At Nohkalikai Falls near Cherrapunji—which is claimed to be the fourth-tallest waterfall in the world—a stream plunges down a rocky precipice into a deep gorge. Nearby are the Nohsngithian Falls and the Kshaid Dain Thlen Falls, where the mythical monster of Khasi legend known as Thlen was killed. Scars on the rocks are said to be axe marks where the monster was butchered. Crinoline Falls are located in the city of Shillong adjacent to the Lady Hydari Park. The spectacular two-tiered Elephant Falls—where a mountain stream cascades into fern-covered rocky dells—is situated 7 miles (12 kilometers) from Shillong. The beautiful Imilchang Dare Waterfall is located close to the Tura-Chokpot Road in West Garo Hills district. The stream here flows through a deep, narrow crevice that abruptly broadens into a cascade over a wide chasm. The large, deep pool at the base of the waterfall is a popular spot for picnics and swimming. RC

RIGHT *A waterfall cascading off of the Meghalaya plateau.*

WATERFALLS OF ORISSA

ORISSA, INDIA

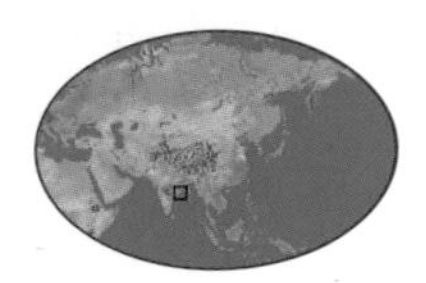

Height of Khandadhar Falls: 800 ft (244 m)

Height of Sanaghagra Falls: 100 ft (30 m)

Khandadhar Falls, Orissa's most famous waterfall, is located deep in the lush forest region of Sundargarh district, which is 35 miles (60 kilometers) from Keonjhar. Water from the Korapani Nala, a small perennial river, feeds the 800-foot- (244-meter-) high Khandadhar Falls. Khandadhar Falls is said to have acquired its name because it falls in the shape of a sword (a "khanda" is a type of double-edged sword). The waterfall is about 12 miles (20 kilometers) to the southeast of Bonaigarh via a fair-weather road, but the last mile (1.6 kilometers) is only negotiable on foot. The 200-foot- (60-meter-) high Badaghagra Falls are situated about 6 miles (10 kilometers) from Keonjhar and the area is one of the most popular picnic sites in the region. The 100-foot- (30-meter-) high Sanaghagra Falls are also located close to Keonjhar and are another popular tourist site. The spectacular Barehipani Falls is one of the tallest waterfalls in India, dropping 1,310 feet (399 meters) over a wide cliff in two tiers to a pool below. The waterfall is located in Simlipal National Park, along with the 500-foot- (150-meter-) high Joranda Falls. The Simlipal National Park, an important tiger reserve in Orissa, is open from November to June. RC

LONAR CRATER AND LAKE

MAHARASHTRA, INDIA

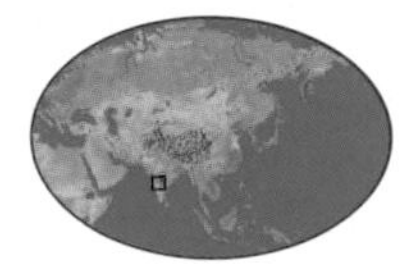

Diameter of Lonar Crater: 6,000 ft (1,800 m)

Diameter of Lonar Lake: 5,250 ft (1,600 m)

Diameter of Little Lonar (Amber Lake): 1,115 ft (340 m)

About 50,000 years ago a meteorite crashed into Earth near Lonar in the Buldhana district, forming a crater that is the third largest in the world. It is the oldest meteorite crater in the world and the only meteorite crater in basaltic rock (much like the craters on the moon). Lonar Crater contains a shallow blue-green lake that is extremely saline and highly alkaline. A freshwater stream feeds the lake throughout the year, but where the stream originates is unknown. There is no apparent outlet for the lake's water. The lake has two distinct regions that never mix—an outer neutral and an inner alkaline region. The mixed deciduous forest growing in the crater is home to a variety of animals including hundreds of peacocks and langur monkeys. Monitor lizards and geckos may also be seen, and there are many species of waterbird present, including flamingoes in the winter. The best time to visit is in winter when the weather is not too hot. Near the Lonar Crater is another smaller crater known as Little Lonar that contains Amber Lake. This crater is thought to have originated from the impact of a piece of the meteorite that split from the main body before hitting the ground. RC

PACHMARHI

MADHYA PRADESH, INDIA

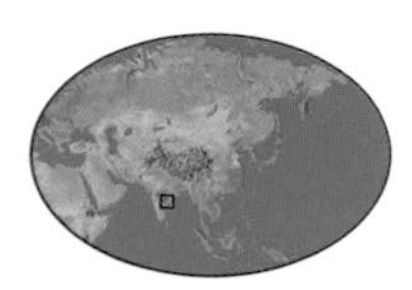

Area of Pachmarhi: 23 sq mi (59 sq km)

Elevation of Pachmarhi: 3,500 ft (1,067 m)

Height of Bee Falls: 100 ft (30 m)

Pachmarhi, an idyllic hill station in Madhya Pradesh, is a large plateau in the Satpura mountain range. This "verdant jewel of the hills" has a landscape of rugged hills, sal forests, spectacular waterfalls, beautiful pools, and deep ravines. Captain Forsyth, a Bengal Lancer, discovered Pachmarhi in 1857. Priyadarshini Point (originally Forsyth Point) is the point from which Forsyth initially set eyes on Pachmarhi and is a much-visited vantage point that provides superb views of the region. From Priyadarshini Point, there is a fantastic vista out over Handi Khoh, Pachmarhi's most impressive ravine, where Lord Shiva is believed to have imprisoned a serpent. The ravine is 330 feet (100 meters) deep with steep sides, and there are huge beehives under the shelter of its overhanging crags.

There are a number of spectacular waterfalls in Pachmari, of which Bee Falls (Rajat Prapat) is the most accessible. The source of this waterfall also supplies drinking water for Pachmarhi. The most beautiful of all the Pachmarhi plateau's waterfalls is Duchess Falls (Jalawataran) that threads down in three distinct cascades. Pachmarhi is situated halfway between Bhopal and Jabalpur. RC

CHILIKA LAKE

ORISSA, INDIA

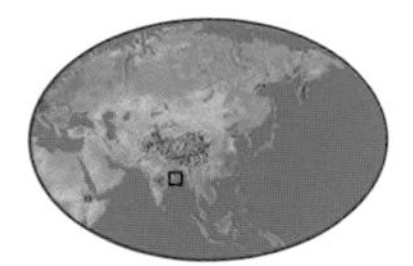

Depth of Chilika Lake: less than 20 in (50 cm), up to 12 ft (3.7 m)

Area of lake (monsoon): 450 sq mi (1,165 sq km)

Area of lake (summer): 350 sq mi (906 sq km)

Chilika Lake (situated along the east coast of India to the southwest of Puri) is the largest brackish water lagoon in Asia and the largest wintering ground for migratory waterfowl on the Indian subcontinent. More than 150,000 people depend on the lake's fisheries for their livelihood. The pear-shaped, shallow lagoon is connected to the Bay of Bengal to the northeast by a channel that runs parallel to the sea and is separated from it by a narrow sandbar.

The lagoon has a number of islands (including Honeymoon Island, Breakfast Island, Nalabana, Kalijai, and Birds Island) and a large area of marshland. The lagoon itself is a biodiversity hot spot and was designated a Ramsar site (Wetland of International Importance) in 1981. Nalaban Island has also been designated a bird sanctuary. A survey from 1985 to 1987 recorded over 800 animal species in and around the lagoon, including a number of globally threatened species. White-bellied sea eagles, grey-legged geese, peach-colored flamingoes, purple moorhens, and jacanas are just some of the birds to visit the lake. The lake provides the wintering ground for over one million migratory birds. RC

BELUM CAVES

ANDHRA PRADESH, INDIA

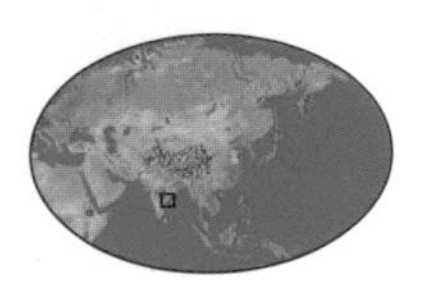

Length of Belum Caves: 10,580 ft (3,225 m)

Depth of caves: 30 to 95 ft (10 to 29 m)

The Belum Caves are the second-largest natural caves in the Indian subcontinent (after Meghalaya Caves), and constitute the longest cave system in the Indian plains region. The caves are located on flat agricultural land over limestone, in the Kurnool district. Enter the cave via the central of three well-like sinkholes; its steps lead down to the horizontal main passage, about 66 feet (20 meters) below ground level. The cave system extends up to 10,580 feet (3,225 meters), at a depth of 30 to 95 feet (10 to 29 meters). There are a number of large chambers, passages, alcoves, freshwater galleries, and siphons—and a recently discovered musical chamber, Saptasvarala Guha, in which the stalactites produce metallic sounds of different musical pitch when struck.

Robert Bruce Foote reported the existence of the caves in 1884, and they were subsequently explored by Daniel Gebauer in 1982 and 1983. It is thought that Jains and Buddhists occupied the caves centuries ago, and remains of vessels dating back to 4500 B.C.E. have been found in the caves. Belum Caves are located at Kolimigundla Mandal. Pathways and ventilation shafts have been created for the convenience of tourists. **RC**

WESTERN GHATS

MAHARASHTRA / KARNATAKA / GOA / KERALA / TAMIL NADU, INDIA

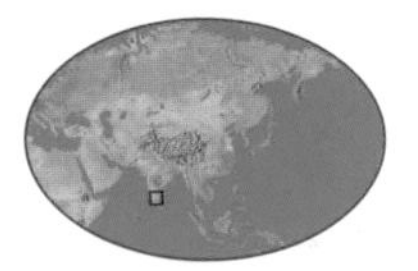

Area of Western Ghats: 61,776 sq mi (160,000 sq km)

Habitat: lowland evergreen monsoon forest, cloud forest, deciduous monsoon forest, dry-adapted scrublands

Situated on the westernmost tip of India, this 1,000-mile (1,600-kilometer) coastal range is famed both for its beauty and its biodiversity. There are 14 peaks above 6,600 feet (2,000 meters) and a rainfall pattern that varies from 118 inches (300 centimeters)—80 percent on the western slope—to 12 inches (30 centimeters) on the shadowed east. This combines to provide 11 different habitats that, in turn, support a great variety of animals and plants, many unique to these mountains. Of the 4,000 plant species, 35 percent are endemic, including 76 species of *Impatiens*, 308 of the 490 species of large tree, and nearly half the orchid species. Of the 125 mammal species, 23 are endemic. Nearly half of all India's unique mammals occur in this area, just five percent of the country's land area.

Key species include lion-tailed macaque, Nilgiri tahr, langur, marten, Travanacore flying squirrel, Malabar whistling thrush, parakeet, pigeon, hornbill, and the rufous-breasted laughing thrush. The region is protected by a logging ban which has been in place since the 1980s. **AB**

RIGHT *Pambadam shola forest in the Western Ghats.*

SILA THORANAM ARCH

ANDHRA PRADESH, INDIA

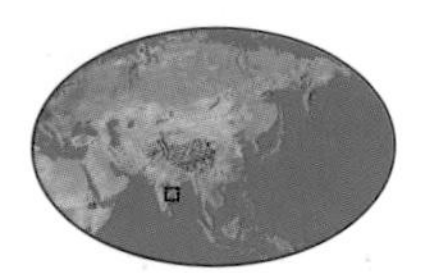

Length of Sila Thoranam Arch: 25 ft (7.5 m)

Height of arch: 10 ft (3 m)

Age of arch: 1.5 billion years

Sila Thoranam is an extraordinary natural rock formation situated on the holy hill of Tirumala in the southeast of Andhra Pradesh. This naturally formed arch of rock is the only one of its kind in Asia, and is a rare geological feature similar to the Rainbow Arch found in the United States in southern Utah. The Sila Thoranam Arch is believed to be 1.5 billion years old and to have been formed by weathering and wind erosion.

It is believed to be the place where Venkateshwara, an avatar of Vishnu, came down to Earth. Behind the arch are impressions in the rock resembling those of a foot and a wheel—the foot impression is said to be that of Vishnu. The arch is located about half a mile (one kilometer) to the north of the Sri Venkateshwara Temple, one of the most important pilgrimage centers in India. The temple claims to be the busiest in the world in sheer numbers of pilgrims (more than even Jerusalem, Mecca, or Rome). The nearby town of Tirupati is located at the base of the hill below Tirumala and is easily reached by air, road, or rail. RC

WATERFALLS OF KARNATAKA

KARNATAKA, INDIA

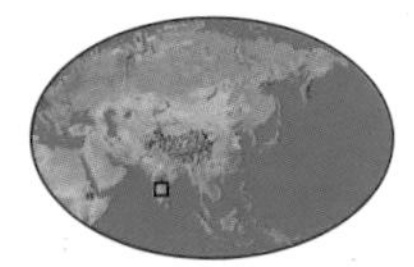

Height of Hebbe Falls: 250 ft (75 m)

Height of Unchalli Falls (Lushington Falls): 380 ft (116 m)

Height of Magod Falls: 660 ft (200 m)

The varied landscape of Karnataka State has many beautiful waterfalls, including the tallest waterfall in India (Jog Falls), but there are many others. The Shivasamudra Falls on the Cauvery River (near Mandya) plummets 350 feet (106 meters) into a rocky gorge with a deafening roar, forming the twin waterfalls of Barachukki and Gaganachukki. Hebbe Falls is a 250-foot- (75-meter-) high waterfall surrounded by coffee plantations near the Kemmannagundi hill station. The water gushes down in two stages, forming the Dodda Hebbe (Big Falls) and Chikka Hebbe (Small Falls). Unchalli Falls or Lushington Falls is situated in dense forest near Heggarne in Uttara Kannada district. The 380-foot (116-meter) falls, on the Aghanashini River, were discovered by J. D. Lushington, the Uttar Kanada district collector of the British government at the time. At the Gokak Falls, the Ghataprabha River plunges 170 feet (52 meters) over a horseshoe-shaped sandstone cliff into the gorge of the Gokak Valley. The Magod Falls are located in dense forest 50 miles (80 kilometers) from Karwar. Here, the Bedthi River plummets 250 feet (75 meters), then cascades down another 350 feet (106 meters) into a rocky ravine. All of Karnataka's waterfalls are impressive during the monsoons when the rivers are at full force. RC

JOG FALLS

KARNATAKA, INDIA

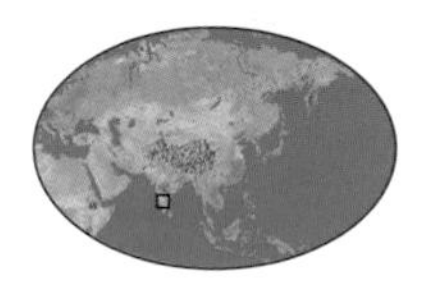

Average width of Jog Falls: 1,550 ft (472 m)

Depth of pool below falls: 130 ft (40 m)

Height of Raja Waterfall: 830 ft (253 m)

The Jog Falls, on the Sharavati River in Karnataka, are the highest waterfalls in India. The river plunges down in four separate cascades known as the Raja, the Rani, the Roarer, and the Rocket. The highest of the waterfalls is the Raja, which falls 830 feet (253 meters) into a 130-foot- (40-meter-) deep pool. The Roarer is situated next to the Raja, while the Rocket is a short distance to the south. The Rocket is so named because it sends great spurts of water out into the air away from the rock face. By contrast, the Rani falls gracefully over the rocks.

The Hirebhasgar Reservoir now controls the flow of the Sharavati River in order to generate hydroelectric power, and there is a vast difference between the wet and dry season flow of the waterfalls. During the dry season, it is possible to walk to the bottom of the falls and bathe in the pool below the falls, while in the wet season the falls may be shrouded in mist. The best time to visit is at the beginning of the cool season, just after the monsoon rains have finished (November to January) when the falls are at their most awesome. **RC**

HOGENAKKAL FALLS

TAMIL NADU, INDIA

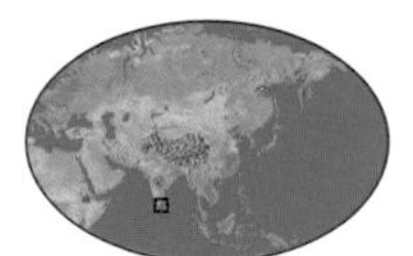

Height of Hogenakkal Falls: 66 ft (20 m)

Best time to visit: July to August

Hogenakkal Falls is a beautiful waterfall in the state of Tamil Nadu, situated on the Cauvery River, where it plunges down from the plateau edge to the plains. The river flows through a wooded valley, splitting and converging around small, forested islands and rocky outcrops before dropping vertically 66 feet (20 meters) to dash against the rocks below. The otherwise gentle river thunders down here, causing so much spray that it earned the name Hogenakkal, which means "smoking river" in the Dravidian language of Kannada. The waterfall is at its most impressive soon after the monsoon rains when the river is at full force in July and August.

At Hogenakkal, people cross the river in coracles, small circular basket-boats with bamboo frames covered in either buffalo hide or plastic. A trip to the base of the roaring falls in one of these flimsy-looking craft is an exhilarating experience. Hogenakkal also features a riverside spa, so you can calm your nerves after a boat ride by getting a massage from the local masseurs. Clients are massaged on slabs of rock and then sluiced down in bathing cubicles under cascades of water. The falls are located near Dharampuri on the Tamil Nadu-Karnataka border, about 80 miles (130 kilometers) from Bangalore. **RC**

KUDREMUKH NATIONAL PARK

KARNATAKA, INDIA

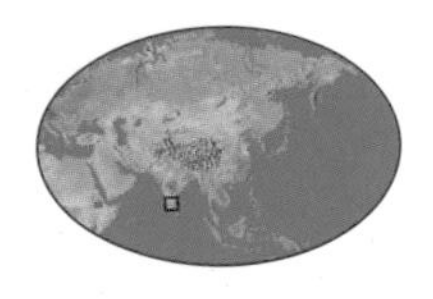

Area of Kudremukh National Park: 230 sq mi (600 sq km)

Average elevation of park: 3,300 ft (1,000 m)

Elevation of highest peak (Kudremukh): 6,214 ft (1,894 m)

The Kudremukh National Park is a relatively undiscovered trekker's paradise of hills and lush forests interspersed with rivers, waterfalls, and caves near the heart of the hill chains of the Western Ghats. The Kudremukh or "Horse Face" Hills overlook the Arabian Sea and acquired their name from the unusual shape of their highest peak. The wet climate and water-retentive soils give rise to thousands of streams that converge to form the three major rivers—the Tunga, the Bhadra, and the Nethravathi—that drain the region. Kudremukh National Park has been identified as a global biodiversity hot spot. The park has one of southern India's largest stretches of tropical wet evergreen forests interspersed with high-altitude grasslands and shola forests. Conservationists have recently won a campaign to prevent further mining of iron ore within the park. One of the park's major animal attractions is the endangered lion-tailed macaque, but the park is also home to tiger, leopard, sloth bear, giant squirrel, deer, porcupine, mongoose, snakes, tortoises, and about 195 species of bird. The national park is located in the districts of Dakshina Kannada, Udupi, and Chikmagalur, about 80 miles (130 kilometers) from the nearest airport and railroad at Mangalore. RA

ATHIRAPALLY & VAZHACHAL WATERFALLS

KERALA, INDIA

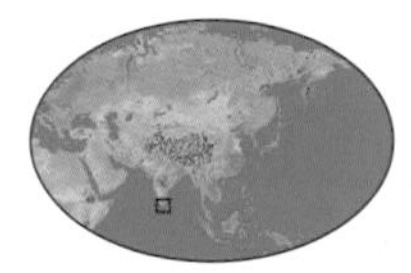

Height of Athirapally Waterfall:
83 ft (25 m)

Height of Vazhachal Waterfall:
100 ft (30 m)

The beautiful waterfalls at Athirapally and Vazhachal are the most famous in Kerala, although there are many other notable waterfalls in the state (particularly in the forests of the Western Ghats region). Athirapally Waterfall, east of Chalakudy in the Trichur district, is a breathtaking waterfall high up in the Sholayar Forest Range that fringes the rainforests of Kerala. Here, the Chalakudy River (the highest river in Kerala) plunges 83 feet (25 meters) into a picturesque gorge overhung with trees. The Athirapally Waterfall, which is a popular picnic spot, is 48 miles (78 kilometers) from Kochi (Cochin). The scenic Vazhachal Waterfall is just three miles (five kilometers) away from Athirapally, also on the Chalakudy River. The evergreen and semi-evergreen forests alongside the river have a high biodiversity, containing many unique and endangered species. The Vazhachal Waterfall is 55 miles (90 kilometers) from Kochi. A proposed hydroelectricity project on the Chalakudy River—which could affect both the Athirapally and Vazhachal waterfalls—has met with strong opposition from locals. RC

BELOW *Athirapally Waterfall, near Chalakudy.*

SIGIRIYA

NORTH-CENTRAL, SRI LANKA

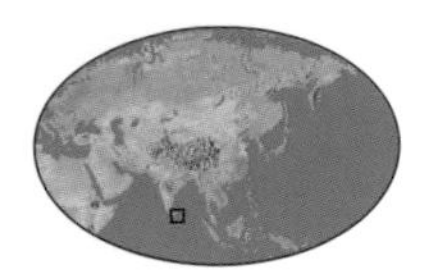

Height of Sigiriya: 660 ft (200 m)

Elevation of Sigiriya: 1,200 ft (370 m)

Designated a World Heritage site: 1982

Rising above the central jungles of Sri Lanka like a fortified, vertically-walled fortress, the block of granite known as Sigiriya is often referred to as the "eighth wonder of the world." Towering 1,200 feet (360 meters) over a landscape that is otherwise overwhelmingly flat, the tawny-colored rock veined in black not only dominates the dense forest at its foot, but is also a landmark visible from hundreds of miles away on the great Sri Lankan plains.

The giant rock once formed a magma plug in the center of a volcano. Over eons, however, the rest of the volcano eroded away, leaving only the plug standing.

Sigiriya's obvious defensive qualities led to it being picked as the site for a new fortress by King Kasyapa in the fifth century, following his usurpation of his father's throne. Fearing the wrath of his brother, Kasyapa believed that the rock offered him the best chance of escaping fraternal retaliation. The fortress was later used as a monastic complex before being abandoned in the 14th century, and was only rediscovered in 1907. **DBB**

RIGHT *The great rock of Sigiriya rises above the forest.*

DIYALUMA FALLS

UVA, SRI LANKA

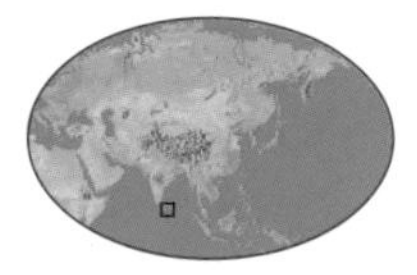

Height of Diyaluma Falls: 722 ft (220 m)

Rainy season: October to March

The Punagala Oya River weaves sleepily through the central highlands of Sri Lanka, past mixed conifer and pine forest peppered with centuries-old boulders. At the dramatic cliff known as Mahakanda Rock, the earth abruptly disappears from under the river, creating the shower of whitewater known as Diyaluma Falls. Although it is the second-highest waterfall in the country, Diyaluma is the most famous of Sri Lanka's more than 100 cascades, if not because it is often mistakenly claimed as the country's tallest falls, then due to its peaceful, inviting appearance. Aptly named "Watery Light," Diyaluma drops hundreds of feet against coal-black rock, the cascade's water droplets appearing to separate and float down from above like blowing white snow, contrasting against their backdrop.

Legend has it that, fleeing an enraged mob of citizens, a prince and his unsuitable peasant sweetheart scaled the cliff at Diyaluma Falls, only to have the young lady slip and fall to her death. The gods collected the prince's tears and continue to release them here to this day to weep forever in the wilderness, which is why some liken the falls here to a torrent of tears. **DBB**

BAMBARAKANDA FALLS

UVA, SRI LANKA

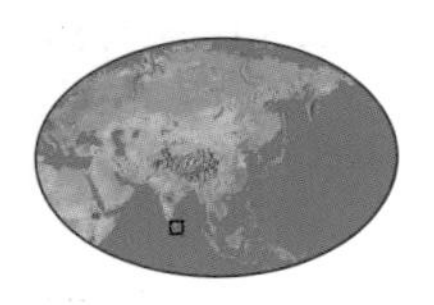

Height of Bambarakanda Falls: 863 ft (263 m)

Rainy season: October to March

The many faces of Bambarakanda Falls in south-central Sri Lanka are a spectacle worth seeking out. While at times they are an unbroken spout of water free-falling 863 feet (263 meters) over a vertical cliff, at other times the falls appear as a thin, veil-like curtain that hugs its host bluff and shifts from side to side like a slender belly dancer.

The tallest waterfall in Sri Lanka, this dynamic cascade changes dramatically with the seasons, transitioning between the wettest and driest periods of the year without losing an ounce of its charm. Between seasons, Bambarakanda wears a permanent rainbow necklace, because decreasing river water volumes cause the column to shift steadily closer to the cliff face until it crashes head-on into a rock outcropping about halfway down, shattering into a thick wall of mist.

While Bambarakanda may change gradually over the seasons, the ecology surrounding the falls changes abruptly from one end of its span to the next. Beginning its descent among a prickly shag of pines, the cascade eventually spills into a wide and narrow pool surrounded by broadleaved wild plantain, palm trees, and other tropical plants. **DBB**

VAVULPANE CAVES

UVA, SRI LANKA

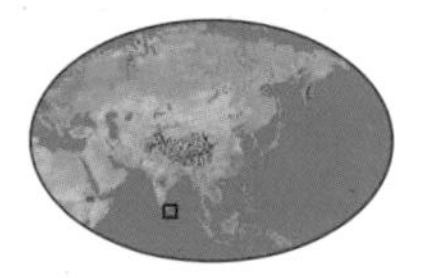

Elevation of Vavulpane Caves: 912 ft (278 m)

Length of Halwini Oya cave: 1,500 ft (450 m)

Notable feature: approx. 250,000 bats

The Vavulpane Caves lie on the eastern slopes of the Bulutota Range, in the gem-rich Ratnapura district of Sri Lanka. There are 12 caverns in total. The first and largest, Halwini Oya, is 1,500 feet (450 meters) long and has a domed roof from which white, cream, pink, and yellow stalactites that are mirrored in corresponding stalagmites hang.

The caves are fed by a natural hard-water spring that flows at 6 gallons (26 liters) per second. The concentration of calcium carbonate here is thought to be the greatest on the island. The water also contains iron oxide that stains the rocks a rusty orange, and which local folk believe to have medicinal properties. The stream disappears into the cave down a sinkhole that is just large enough for a person to crawl through.

Vavulpane, which means "cave of bats," is home to about 250,000 bats. There are six species represented, and their droppings provide food for cockroaches. Eel-like fish live in the stream and snakes are occasional visitors. White cobras have been seen catching bats. Not far away is a prehistoric forest with 4,000 giant tree ferns, one of the largest forests of its kind in the world. **MB**

SRI PADA

SABARAGAMUWA, SRI LANKA

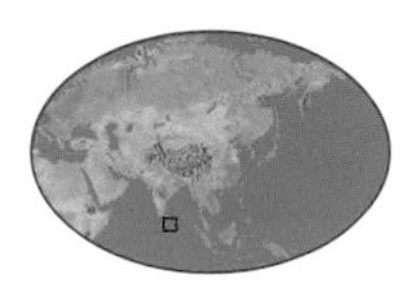

Alternative name: Adam's Peak

Area of Peak Wilderness Sanctuary: 55,950 acres (22,380 ha)

Shaped like a giant teardrop, Sri Pada is molded unlike any other peak in the world. Located in the heart of Sri Lanka's southern forests and indented at its summit with a curious depression in the shape of a giant footprint, this geologic formation has taken on a spiritual significance for all of Sri Lanka's religions. Also known as Adam's Peak, some Christians and Muslims claim this is where Adam first stepped down onto Earth. Hindus say it is the mark of Lord Shiva, and Buddhists believe it is derived from Siddhartha's third visit to the country. Religious conviction aside, the print has been filled in with concrete in order to preserve it in years to come.

Since 1940, Sri Prada has been protected as the Peak Wilderness Sanctuary and is just as much an ecological epicenter as a spiritual one. The convergence of montane rainforest, tropical lowland, and grassland here gives rise to three of the country's ten most important rivers. Home to two dozen endemic bird species as well as leopards, elephants, and rare frogs and insects, the sanctuary also attracts hundreds of thousands of pilgrims who trek to the top each year from Dalhousie to pray and meditate. **DBB**

LAKE BOLGODA

WESTERN, SRI LANKA

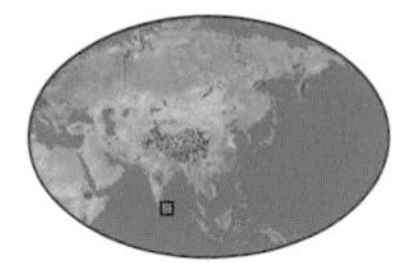

Surface area of Lake Bolgoda: 145 sq mi (374 sq km)

Area of wetlands: 350 acres (140 ha)

Biggest threats: sawdust dumping, fertilizer runoff, invasive species

Pinched between the Kalu and Kelaniya river basins and the Indian Ocean, Lake Bolgoda in southwestern Sri Lanka is a melting pot of water types, diverse species, and human interests. In the general shape of a dumbbell, the north and south bays are linked by a narrow channel. The lake's southern sector is connected to the sea by a small waterway, resulting in a spectrum of conditions ranging from seawater in the south to freshwater in the north. Vast enough to host eight islands, Bolgoda's southern shores remain almost untouched, while the north is a recreational hot spot complete with elegant hotels and restaurants. Bolgoda features a network of fringing wetlands that are recognized today as some the most important in Asia and are home to a long list of endangered creatures including Asiatic rock pythons, swamp crocodiles, and painted storks. By contrast, its banks are also peppered with timber mills and other factories, whose runoff waste product is beginning to overwhelm the ecosystem. Bolgoda's fate rests in the hands of a local environmental-cultural organization, which is currently pushing a comprehensive plan to benefit both industry and habitat. **DBB**

DUVILI ELLA

SABARAGAMUWA, SRI LANKA

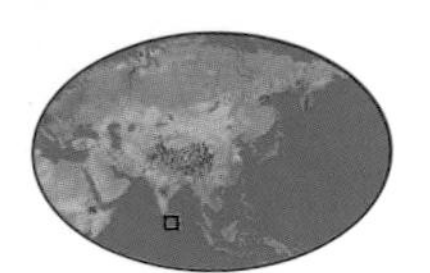

Alternative name: Dust Falls

Height of waterfall: 130 ft (40 m)

Width of waterfall: 80 ft (24 m)

Roaring down a chute of bare rock surrounded by dense jungle vegetation, Duvili Ella in southeastern Sri Lanka is the wildest accessible waterfall in the country. Wide and powerful, this raging jet of whitewater is a manifestation of nature that cannot help but make one feel small and powerless. Its great force is created when the steep banks of the rushing Walawa River abruptly narrow, resulting in a sudden increase in stream speed, just as the riverbed drops more than 100 feet (30 meters). The result is a torrent of water so strong that, upon crashing into a deep pool below, it sends a cloud of mist as high as the falls itself, thereby earning the name Duvili Ella, or "Dust Falls."

However, its intimidating presence does not end at the waterfall. After landing, the river is further squeezed by a narrow exit gorge, turning the basin below into a frothy tub of whirlpools. The only way to see Duvili is to hire a guide and hike a short, though arduous, distance. Keep your eyes and ears peeled for flying squirrels, plenty of birds and monkeys, and even forest elephants. Begin the hike in Kaltota about 17 miles (28 kilometers) from Balangoda. **DBB**

MANASLU

WESTERN NEPAL, NEPAL

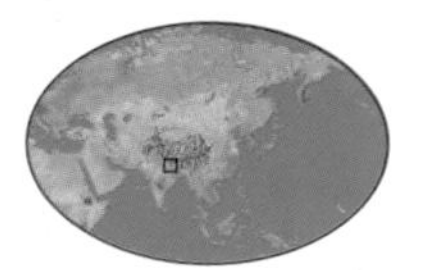

Highest peak (Manaslu): 26,758 ft (8,156 m)

Second highest peak (Manaslu East): 25,900 ft (7,894 m)

Alternative name: Kutang

Located about 40 miles (64 kilometers) east of Annapurna, Manaslu is the highest mountain in the Gurkha Massif and at 26,758 feet (8,156 meters) is the eighth highest mountain in the world. Towering over surrounding mountains, Manaslu's long ridges and valley glaciers culminate in an imposing peak. Its name is derived from the Sanskrit word *manasa*, meaning "Mountain of the Spirit." The mountain itself is not considered a technically difficult peak to climb, but the approach to the peak can be fraught with difficulties—the trek to base camp is arduous, and there is danger from inclement weather and avalanches.

In 1972, for example, all 16 members of a Korean expedition, including 10 Sherpas, were killed on the mountain when they were swept away by an avalanche at 22,800 feet (6,949 meters). Toshio Imanishi and Gyalzen Norbu made the first successful ascent with a Japanese expedition in 1956. In 1974, an all-female Japanese team reached the summit, making them the first women to conquer a 26,400-foot (8,000-meter) peak. Tragically, one of the party fell to her death on the descent between camps 4 and 5. **MB**

DHAULAGIRI

WESTERN NEPAL, NEPAL

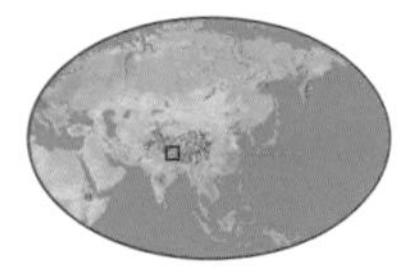

Height of Dhaulagiri: 26,906 ft (8,201 m)

Length of Dhaulagiri: 30 mi (50 km)

The seventh highest mountain in the world is close to the Nepal–Tibet border and is the highest mountain to be located entirely within Nepal. Dhaulagiri is one of the mountain's massifs that rises up on the western side of the Kali Gandack Gorge. Its name translates as "White Mountain." Upon its discovery by western surveyors in 1808, it replaced Ecuador's Chimborazo as the world's highest mountain—until Kanchenjunga and Mount Everest were discovered.

The crest of Dhaulagiri gives rise to several pyramid-shaped peaks, four of which are above 25,000 feet (7,620 meters). In 1950, an attempt was made on the highest peak by a French team led by Maurice Herzog, and the mountain was finally conquered on May 13, 1960, by a Swiss–Austrian expedition led by Max Eiselin. His was the first party to use an aircraft to reach the base of the mountain. Unfortunately, the aircraft crashed on approach and was later abandoned. The summit party included Austrian climber Kurt Diemberger, one of the few people to successfully ascend two mountains over 26,400 feet (8,000 meters), the other being Broad Peak in Pakistan in 1957. MB

ANNAPURNA

WESTERN NEPAL, NEPAL

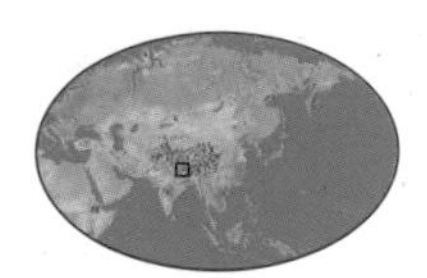

Highest peak (Annapurna I):
26,545 ft (8,091 m)

Second highest peak (Annapurna II):
26,040 ft (7,937 m)

Annapurna I, at 26,545 feet (8,091 meters), is the tenth highest mountain in the world. Located just north of Pokhara, it stands in central Nepal, and the glaciers on its western and northwestern sides drain into the Kali Gandaki Gorge.

The Annapurna Massif has many peaks, five of which contain "Annapurna" in their title. The highest, Annapurna I and II, stand like bookends at the western and eastern ends of the range. In 1950, Maurice Herzog's team ascended the North Face of Annapurna I, making it the first Himalayan peak over 26,400 feet (8,000 meters) to be climbed. Twenty years later Chris Bonington led a successful ascent of the South Face. Two American women, Irene Miller and Vira Komarkova, tackled the North Face successfully in 1978, making them the first Americans to reach the top, and in 1988 a large French team led by American climber Steve Boyer took the South Face. Trekking is best between April and October because routes become snowbound in winter. *Annapurna* is a Sanskrit name meaning "Provider" or "Goddess of the Harvest." MB

BELOW *Low-lying clouds ring Annapurna I.*

CHO OYU

TIBET, CHINA / NEPAL

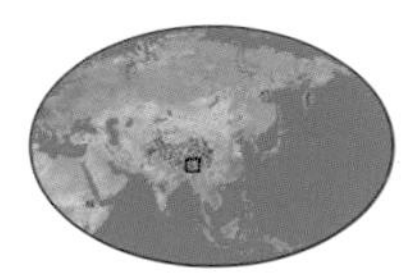

Height of Cho Oyu: 26,906 ft (8,201 m)
Height of Nagpa La: 19,000 ft (5,791 m)

On the border between eastern Nepal and Tibet stands the sixth highest mountain in the world—Cho Oyu, meaning "Turquoise Goddess." Located approximately 20 miles (32 kilometers) to the northwest of Mount Everest, with an elevation of 26,906 feet (8,201 meters), it stands out from smaller mountains that surround it and is a distinctive landmark for Everest climbers.

Just south of Cho Oyu peak is the Nagpa La Glacier saddle pass that, at 19,000 feet (5,791 meters), is the main trade route through this part of the Himalayas between Tibet and the valley of Khumbu. Consequently, it is considered by some mountaineers to be the "easiest" high peak to climb. Nevertheless, it has taken its toll—an international all-women expedition lost four members in an avalanche and two German climbers died of exhaustion at Camp 4, 25,000 feet (7,600 meters) above sea level. The first attempt at the peak was led by British climber Eric Shipton, who was beaten by an ice wall above 21,820 feet (6,650 meters) and forced to return. The summit was first reached on October 19, 1954, by an Austrian party, including Herbert Tichy, Sepp Jochler, and Sherpa Pasang Dawa Lama. MB

MOUNT EVEREST

TIBET, CHINA / NEPAL

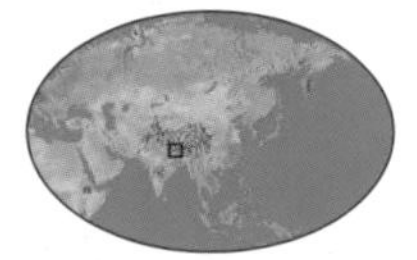

Elevation of Mt. Everest:
29,035 ft (8,850 m)

Local Tibetan name: *Chomolungma*
("mother goddess of the universe")

Local Nepalese name: *Sagarmatha*
("churning stick in the sea of existence")

The highest mountain in the world was once known as Peak XV—that is until military engineer Sir George Everest, Surveyor General of India, laid his eyes on it in 1865. He was almost thwarted by the Nepalese authorities in his attempts to survey the mountain, and so recruited *pundits*, or "learned experts," who secretly gathered information that enabled him to map the area accurately.

During the early part of the 20th century, westerners were allowed to climb Everest. They included the British mountain climber George Mallory who, when asked why he bothered, simply replied, "Because it's there." Unfortunately, he and many others who have attempted the ascent have died on the mountain, whose rapidly changing conditions can surprise even the most experienced of mountaineers.

The mountain was finally conquered in 1953, by a team led by Sir John Hunt who approached Mount Everest from the Tibetan side of the mountain. The victors were New Zealand beekeeper Sir Edmund Hillary and Nepalese Sherpa Tenzing Norgay, who set out from base camp early in the morning of May 29, 1953, and reached the summit about five hours later. MB

LHOTSE

TIBET, CHINA / NEPAL

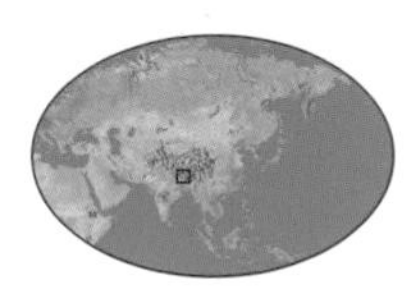

Elevation of Lhotse: 27,939 ft (8,516 m)

Height of Lhotse Shar peak: 27,503 ft (8,383 m)

Height of Nuptse peak: 25,849 ft (7,879 m)

Just to the south of Mount Everest, in the Khumbu Himal region of the Himalayas on the Nepal–Tibet border, is Lhotse, which means "South Peak" in Tibetan. At an elevation of 27,939 feet (8,516 meters), Lhotse is the fourth highest mountain in the Himalayas. It is connected to Mount Everest by the South Col, a vertical ridge running in an east–west direction between the two peaks. The ridge does not drop below 26,400 feet (8,000 meters), so Lhotse is often mistaken for a southern extension of Mount Everest. In fact, during the 1931 Survey of India it was given the survey symbol E1, denoting Everest 1. Lhotse was not climbed until some time after Everest itself had been conquered; even then, the initial ascent was an attempt to find a new route to the top of Mount Everest. On May 18, 1956, Fritz Luchsinger, Ernest Reiss, and two Swiss climbers finally made it to the summit.

The mountain has two subsidiary peaks—Lhotse Shar to the east of the main summit and Nuptse on the mountain's west ridge. The best trekking season for the area is April to May and late September to October. MB

MAKALU

TIBET, CHINA / NEPAL

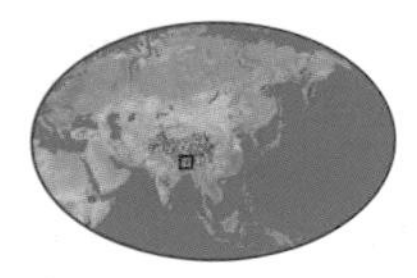

Height of Makalu peak: 27,765 ft (8,463 m)

Height of Chomo Lonzo peak: 25,650 ft (7,818 m)

Makalu is the world's fifth highest mountain peak, and is found about 14 miles (23 kilometers) southeast of Mount Everest, in the Khumbakarna Himal region of the Himalayas on the Nepal–Tibet border. It is a distinctly-shaped mountain that cannot be mistaken for any other. Pyramidal in shape, with glaciers and four sharp ridges, Makalu rises 27,765 feet (8,463 meters) above sea level.

A subsidiary peak—25,650-foot (7,818-meter) Chomo Lonzo—rises just north of the higher summit and is connected to Makalu by a saddle. Like most mountains in the area, Makalu had been previously admired by climbers, but was not attempted until after Everest had been climbed.

Makalu has proved itself to be a challenging mountain, with only five expeditions out of the first 16 attempts reaching the top. The first climb by an American expedition was in the spring of 1954, but a barrage of storms at 23,294 feet (7,100 meters) meant it had to turn back. It was not until May 15, 1955, that Jean Couzy and Lionel Terray, two members of a French party led by Jean Franco, reached the summit. Seven of their colleagues followed during the next couple of days. MB

BELOW *The peaks of Makalu rise above a cloudy haze.*

KANGCHENJUNGA

NEPAL / INDIA

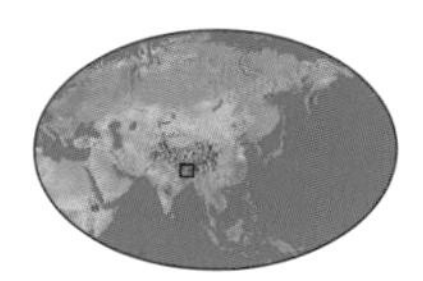

Elevation of Kangchenjunga: 28,169 ft (8,586 m)

Notable feature: third highest mountain in the world

Kangchenjunga is the third highest mountain in the world. It stands 28,169 feet (8,586 meters) above sea level on the Sikkim–Nepal border and comprises part of the Himalayan mountain range. In local dialect *Kangchenjunga* means "five treasures of the snow" for its five peaks. The mountain's most notable claim to fame is that it has never been conquered.

A British expedition led by Charles Evans came to within 5 feet (1.5 meters) of the main summit in 1925, but stopped in deference to the Sikkimese who consider the mountain sacred. Evans had undertaken not to disturb the area immediately around the summit. Most climbers, even today, consider that there is a cordon around the summit beyond which man dare not go. The people of Sikkim believe that the mountain deity looks down favorably on them. He has the power to destroy houses with floods and avalanches and obliterate crops with hailstorms. He is portrayed with a red, fiery face, wearing a crown with five skulls, and riding the mythical snow lion. An annual festival of dance dedicated to the mountains takes place in the early fall. Lamas wearing masks and brightly colored costumes dance and whirl against the backdrop of spectacular mountains. MB

KALI GANDAKI

WESTERN NEPAL, NEPAL

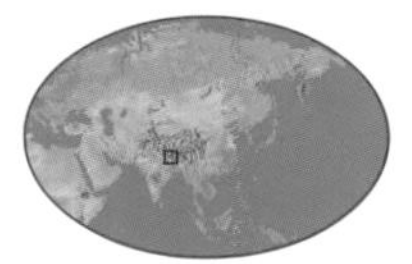

Age of Kali Gandaki valley: 50 million years

Habitat: from desert to semi-tropical forest

The Kali Gandaki is an old river. Once, it ran off the Tibetan Plateau and away to the sea, but then 50 million years ago major earth movements caused the continental plates to collide. The impact pushed up the magnificent Himalayas, but the river doggedly maintained its course. It cut down through the rocks and between the Annapurna and Dhaulagiri Mountains to create the deepest valley on our planet. The river, colored black with mud from deposits upstream, is edged about 14,520 feet (4,400 meters) below the highest peak. It created a valley that was not only a highway for traders, pilgrims, and soldiers between Tibet and Nepal, but was also a geographical enigma. At the north end of the valley the river passes through cold, barren desert while in the south it enters a landscape dominated by semi-tropical forest.

This is a place where people of diverse backgrounds have settled—including salt traders, stock rearers, and farmers. Today, the locals have capitalized on adventure tourism, yet they retain their social traditions and culture. Women take several husbands in order to ensure that families are safe and that children are always looked after in an environment where danger is always around the corner. MB

ROYAL CHITWAN NATIONAL PARK

WESTERN / CENTRAL NEPAL, NEPAL

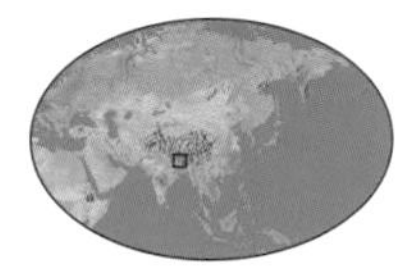

Area of Royal Chitwan National Park: 360 sq mi (932 sq km)

Number of bird species: more than 400

Number of amphibian and reptile species: 55

Nepal's best-known and most accessible national park, Royal Chitwan, lies in the lowlands to the southwest of Kathmandu and is an important refuge for tigers, Great Indian Rhinoceros, and gharial. Chitwan was a royal hunting reserve from 1846 to 1951, but habitat loss and rapidly declining animal numbers led to the southern part of the park becoming a rhinoceros sanctuary in 1963. In 1973, Royal Chitwan became Nepal's first national park, and in 1984 it was designated a World Heritage site.

The park has a diversity of habitats including "elephant grass" savannas, rivers, and sal forests. More than 50 species of mammals include such globally threatened animals as tiger, Great Indian Rhinoceros, leopard, wild dog, sloth bear, gaur, and Ganges River dolphin. More bird species have been recorded here than in any of Nepal's other protected areas. Reptiles found in the park include Indian python and two species of crocodile (mugger and gharial). Elephant safaris allow visitors to get close views of Indian rhinos and are a major attraction at Chitwan. The park can also be explored by four-wheel-drive vehicles, by guided jungle walks, or, alternately, by canoe along the Rapti River. RC

BLACK MOUNTAINS

CENTRAL BHUTAN

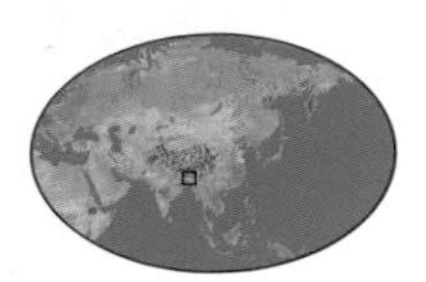

Number of black-necked cranes: about 350

Area of Bhutan: 18,150 sq mi (47,000 sq km)

Tucked into the eastern Himalayas, the country of Bhutan has some of the most rugged mountain terrain in the world. Landlocked between China and India and a bit smaller than Switzerland in terms of area, it is entirely mountainous except for a swath 8 to 10 miles (13 to 16 kilometers) wide along its southern border. Mountain peaks form a gigantic stairway from the south at an altitude of 1,000 feet (300 meters) to the highest summits in the north, towering over 23,000 feet (7,000 meters). In the center of the country at about 16,500 feet (5,000 meters), lie the slate-rich Black Mountains, forming a natural border between western and central Bhutan.

This area is a maze of forested hillsides and deep valleys. Turbulent rivers originating from the high Himalayas run south through this region, carving spectacular gorges before emptying into the southern plains. Many hillsides are too steep for farming and are covered in virgin conifer and broadleaf forest. There are lakes, alpine pastures, and stunning permanent ice peaks such as Dorshingla, which rises to 16,158 feet (4,925 meters). Pele La, at 11,483 feet (3,500 meters), is the most important pass through the Black Mountains, from Paro to Trashigang. Temperatures vary according to altitude, but there are five distinct seasons: summer, monsoon, autumn, winter, and spring. Here the climate is temperate with warm summers and cool winters.

The remote nature of the country, combined with its inaccessibility and a fundamental respect for life among its citizens, has made it a biodiversity hot spot. Fauna include red panda, wild boar, sambar, musk deer, Himalayan black bear, tiger, leopard, and the golden langur, endemic to Bhutan. There are some 449 species of bird in this region alone. One of the nation's most important wildlife preserves is in the Phobjikha Valley, a bowl-shaped glacial valley on the western slopes of the Black Mountains: It is the winter home of the endangered black-necked crane, which migrates from the colder Qinghai–Tibet plateau.

Early British visitors to Bhutan were struck by its stark beauty, writing of "dark and steep glens, and the high tops of mountains lost in the clouds ... altogether a scene of extraordinary magnificence and sublimity." **GD**

The Black Mountains are a maze of forested hillsides and deep valleys. Turbulent rivers originating from the high Himalayas run south through this region, carving spectacular gorges before emptying into the southern plains.

RIGHT *The forested hillsides of the Black Mountains near Lobding, Bhutan.*

PHOBJIKHA VALLEY & CRANES

WANGDUE PHODRANG, BHUTAN

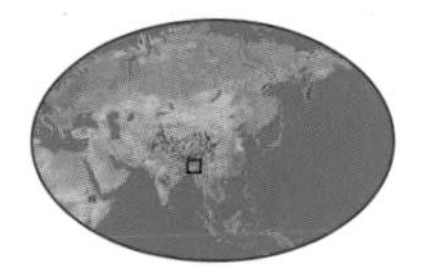

Adult crane height: approx. 5 ft (1.5 m)

Crane population in Phobjikha: approx. 440

Phobjikha is a high glacial valley in the eastern Himalayan kingdom of Bhutan. Located 9,500 feet (2,878 meters) above sea level, it is perhaps one of the most beautiful places on Earth. Steep hills covered in forest surround the valley, which is dominated by a golden-roofed 16th-century temple. On the valley floor is Bhutan's largest wetland—a dwarf bamboo fen with thick deposits of peat and a fast-flowing stream fed by Himalayan snows. In winter, large flocks of rare and beautiful black-necked cranes are found in the shallows just beneath the temple. They come to Phobjikha's wetlands to breed and to escape the harsh winters of the Tibetan Plateau. The birds arrive at their wintering grounds from mid-October and stay until mid-April.

The black-necked is the least known of the world's 15 species of cranes. It was first discovered in 1876 in Lake Koko-nor in the northeast corner of the Tibetan Plateau. The cranes can be found throughout much of the Himalayan region, although overall crane numbers have shrunk in recent decades because of the numerous threats facing them, including the cultivation of their breeding grounds. It is estimated that there are only around 6,000 cranes left in the world. However, they enjoy legal protection in India, China, and Bhutan. In Phobjikha, the birds figure prominently in the traditions of the local people and are therefore relatively safe from extinction.

In Bhutan the bird is known as *Thrung Thrung Karmo* and is revered as a heavenly bird (*lhab-bja*) in the Buddhist tradition of the local people. The *lhab-bja* appears in Bhutanese folklore, songs, dances, and history. Legend has it that the Bjakar Dzong ("White Bird") in nearby Bumthang was built when a white bird flew to a mound where the dzong ("monastery") stands today—that white bird is believed to have been the black-necked crane. They are considered so precious that anyone in Bhutan who injures a crane subsequently faces imprisonment for life. One of the most popular Bhutanese folk songs laments the time when the cranes depart in spring for Tibet. To celebrate this much loved and rare bird, the Bhutanese recently instituted a Crane Festival in Phobjikha on November 12. **JD**

Phobjikha is perhaps one of the most beautiful places on Earth. Steep hills covered in forest surround the valley, dominated by a golden-roofed 16th-century temple. On the valley floor is Bhutan's largest wetland.

MOUNT JHOMOLHARI

PARO, BHUTAN

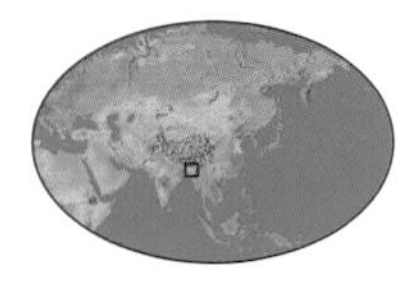

Height of Jhomolhari peak: 24,090 ft (7,300 m)

Age: 20 million years

Jhomolhari, Bhutan's deeply venerated guardian peak, rises on the Tibet–Bhutan border in the eastern Himalayas. Mountains have traditionally been revered as places of sacred power, and those belonging to this Himalayan mountain kingdom are no exception. To the Bhutanese, Jhomolhari is the physical embodiment of the goddess Jhomo, or Tsheringma, and is one of the country's two most respected and sacred peaks. The other is Gangkhar Puensum.

The deeply spiritual Buddhist Bhutanese believe that the gods live in these sacred mountains, so it is unusual for visitors to be permitted to explore and risk disturbing the peace of the gods. As a result, many of these peaks are still shrouded in mystery.

Mount Jhomolhari is one of the few exceptions. In 1939, F. Spencer Chapman successfully managed to scale it. In that year he wrote: "Jhomolhari gives a greater impression of sheer height and inaccessibility than any other mountain I know. It drops in a series of almost vertical rock precipices to the foothills beneath. It is thought by many to be the most beautiful mountain in the whole length of the Himalaya." At 24,740 feet (7,541 meters) Bhutan's loftiest peak, Gangkhar Puensum is now the world's highest unclimbed mountain—and will probably remain so for many years to come. JD

SUNDARBANS

BANGLADESH / INDIA

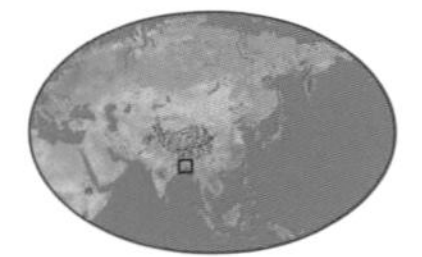

Area of Sundarbans: approx. 3,861 sq mi (10,000 sq km)

Area of delta: approx. 30,888 sq mi (80,000 sq km)

Area of Bangladesh Sundarbans: 2,297 sq mi (5,950 sq km)

The largest mangrove forest in the world lies in the delta of the Brahmaputra, Ganges, and Meghna rivers, and stretches across India and Bangladesh in the northern part of the Bay of Bengal. It consists of a changing complex of interconnecting tidal waterways, mudflats, islands with salt-tolerant mangroves, and the last stands of the virgin jungle that once covered the entire Ganges plain.

The Sundarbans is a wildlife sanctuary within the forest with spotted deer and wild boar and is renowned for its rhesus macaques, which fall prey to Royal Bengal tigers—some of which are known man-eaters. Saltwater crocodiles (the world's largest living reptiles), Indian pythons, and Ganges River dolphins also live here. In the meandering streams, creeks, and rivers, 260 species of bird have been documented, including migrants such as Siberian ducks. In a curious relationship between man and nature, smooth-coated otters living here are tamed by local fishermen to drive fish into their nets.

Rainfall can be heavy, humidity high, and temperatures in March can reach 109°F (43°C). There are no roads, so visitors must travel by boat. In Bangladesh, you can journey in a paddle steamer. MB

INLE LAKE

SHAN STATE, MYANMAR

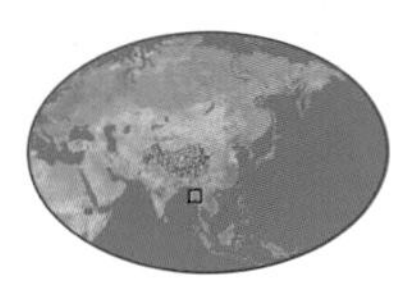

Elevation of Inle Lake: 3,000 ft (900 m)
Length of lake: 13.8 mi (22 km)
Width of lake: 6.9 mi (11 km)

On the Shan Plateau, in eastern Myanmar's Shan State, lies Inle Lake, or Nyaunagshwe, the country's second largest lake, encompassing an area of some 45 square miles (117 square kilometers). Its shallow waters support a great deal of fish life, including nine unique species. The lake also occupies a central role in the lives of the local Innsha people, who have developed a unique way of growing food on the lake's 5-foot- (1.5-meter-) deep waters. They tend "floating gardens" of aquatic vegetation anchored to the lakebed and separated by bamboo fences made of narrow strips of matted reeds. The garden-makers sell their wares in a novel way: They tow 60-foot (18-meter) lengths of garden behind their boats and cut off smaller sections for customers when they are required.

Some Innsha are not floating farmers but fishermen, who are equally famous for the way in which they propel their canoes. Standing in the stern, they paddle with one leg around the oar, leaving their hands free. They catch fish using a conical trap made of wood or bamboo. The trap is thrust into the water with the foot and when a fish strikes it, the net is freed with a pole. The Innsha's way of life is inextricably linked to the lake—their children are said to be able to swim long before they can walk. **MB**

GOLDEN ROCK

MON STATE, MYANMAR

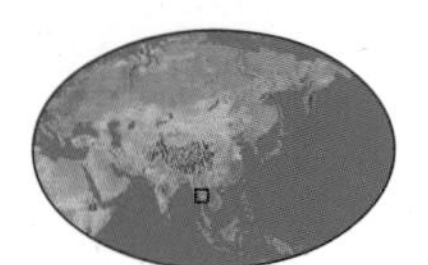

Elevation of Golden Rock: 4,000 ft (1,200 m)

Rock type: granite

Golden Rock is one of the world's unsolved natural mysteries—a shimmering, gold-colored granite boulder teetering on the edge of a cliff. The trek to this spectacle is a long though worthwhile hike—from the nearby town of Kyaik-to, a five-hour hike up a jungle path takes you from near sea level to a heady altitude of 4,000 feet (1,200 meters) and the settlement of Kin-pun, where the rock is located. However, the sight of this massive gilded boulder perched on the edge of a sheer cliff face is an astonishing reward—some even call it miraculous. Young boys push it to show how easily it can be moved, how precariously it is balanced.

How this rock stays perched on the edge is a complete mystery. The boulder does not even appear to touch the rock it is balanced upon. According to legend, the only reason it does not crash into the valley below is because of the 18-foot- (5.5-meter-) high Kaik-tiyo Pagoda shrine that sits atop it. The shrine is said to contain a single strand of the Buddha's hair, and it is this hair that maintains the rock's balance. The legend tells how King Tissa, who reigned in the 11th century, was given the hair by a hermit, but only on the rather specific conditions that the king find a huge rock resembling the hermit's head, move it to the edge of the cliff, build a shrine atop it, and place the hair inside it.

After a long search, the king finally found a suitable rock at the bottom of the sea. He raised the rock to the surface, carefully covered it in gold leaf, and transported it to the clifftop in a boat. The boat later turned to stone, and can be seen close to Kyaik-tiyo. As requested by the hermit, the king then built the shrine. Later, his queen, the beautiful Shwe-nan-kyin, was walking one day in the jungle just below the cliff when a tiger suddenly pounced out of the undergrowth at her feet. Rather than run, the queen instead fixed her eyes on the golden rock and committed herself to the hands of fate. Surprisingly, the tiger turned away as if threatened by the mysterious power of Golden Rock.

Golden Rock is a wonder to behold—a shimmering, gold-colored granite boulder teetering on the edge of a cliff. Young boys push it to show how easily it can be moved, how precariously it is balanced.

The full name for this shrine used to be "Kayaik-l-thi-ro," which literally translates as "pagoda carried by a hermit on the head." Over time, however, this has been abbreviated to "Kaik-tiyo." CM

RIGHT *The legendary Golden Rock, held in place solely by the Buddha's hair contained in its shrine.*

HA LONG BAY

QUANG NINH, VIETNAM

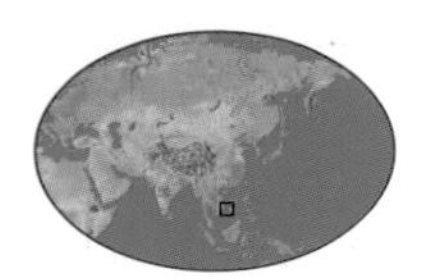

Extent of Ha Long Bay National Protection Area: 600 sq mi (1,553 sq km)

Number of islands: 1,969, of which 980 are named

Legend has it that the islands and islets peppering Ha Long Bay ("Bay of the Descending Dragon") were created by the dragons that helped defend Vietnam against attacking armies. Geologists tell a different tale: the area, once a limestone mountain range, was cut by rivers and collapsing caves to form the karstic landscape that was subsequently inundated by the sea. The islands are continuously being eroded into giant and fanciful statues that rise from the blue water—a pair of 40-foot- (12-meter-) high chickens, a 30-foot- (9-meter-) tall squatting toad, a great stone incense holder, and even an Egyptian pyramid. Caves, grottos, and sea arches abound; many of the caves are decorated with spectacular stalagmites and stalactites. Some 170 species of coral have been recorded here.

Plant life includes species found nowhere else on Earth, such as the Halong fan palm, which was first discovered in the late 1990s. One of the world's most endangered primates, the Cat Ba langur, known also as the golden-headed leaf monkey, inhabits the limestone forests of Cat Ba Island in the southwest of the bay. Despite conservation efforts, the langur is severely threatened by hunting. MW

BELOW *The sculpted islands of Ha Long Bay.*

HAI VAN PASS

DA NANG / THUA THIEN-HUE / QUANG NAM, VIETNAM

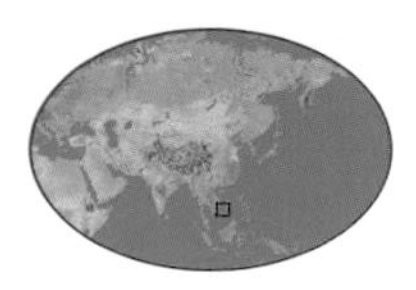

Highest point (Hai Van): 3,845 ft (1,172 m)

Area of Lang Co Lagoon: 3,706 acres (1,500 ha)

Area of Son Tra Island: 370 acres (150 ha)

Hai Van is the highest and longest pass in Vietnam. The main north to south national highway snakes through the pass for about 12 miles (20 kilometers), which also marks the border between the Thua Thien-Hoa, Quang Nam, and Da Nang Provinces, effectively separating the north and south parts of the country, both geographically and climatologically. The cold winds of the north are prevented from blowing farther south by the last section of the Thruong Son mountain range, which lunges down to the sea. Son Tra Island lies just offshore.

Hai Van translates as "wind and clouds," which is fitting given that its heights are usually shrouded in clouds that, according to the early 19th-century rebel poet Cao Ba Quat, seemed to "pour from heaven" and are buffeted by winds "like herds of galloping horses." On a clear day Son Tra Peninsula, Da Nang City, and endless long white beaches can be seen in the distance. The remains of fortified gateways delineate the highest point in the pass, and the gate facing Quang Nam is inscribed with the words "the most imposing gateway in the world." At the bottom of the northern face, streams feed the crystal-clear, turquoise waters of Lang Co lagoon. Local wildlife includes restricted-range birds, such as the Annam partridge and Edward's pheasant. MB

SON TRA PENINSULA

DA NANG, VIETNAM

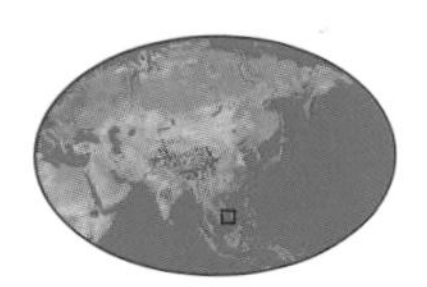

Length of Son Tra Peninsula: 10 mi (16 km)

Highest point (Son Tra): 2,274 ft (693 m)

Number of hard coral species: 129 species

Some say the shape of Son Tra Peninsula resembles that of a tortoise; others say it is like a mushroom—its cap Son Tra, its stalk the white sand beaches at its base. Son Tra, or Monkey Mountain, is regarded by locals as a special gift from God and was once known as "Tien Sa" because legend has it that fairies used to land here and amuse themselves with games of chess on the plateau of Chess Board Peak.

The peninsula was formed when silt deposits created a land bridge between the mainland and three islands—Nghe, Mo Dieu, and Co Ngua—forming a natural barrier that today protects Da Nang from storms and cyclones coming from the sea. Son Tra is a conservation area, with at least 12 square miles (30 square kilometers) of natural forest having survived the ravages of war. Rare monkeys, such as the red-shanked douc langur and a macaque which is said to be an intermediate stage between the crab-eating and rhesus macaques, live in the forests. The beaches are host to nesting olive ridley sea turtles. The fine corals and clear, subtropical waters, which are no more than 33 feet (10 meters) deep and lie 1 mile (1.6 kilometers) from the shore, attract divers from Da Nang. **MB**

MEKONG DELTA

VIETNAM / CAMBODIA

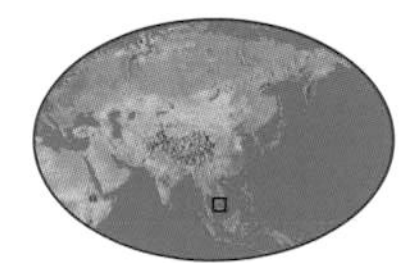

Area of Mekong Delta (Vietnam): 9.6 million acres (3.9 million ha)

Area of Mekong Delta (Cambodia): 4 million acres (1.6 million ha)

Percent of delta in semi-natural condition: 1.3 percent

One of the world's great rivers, the Mekong rises in northeastern Tibet, flows south through great gorges it has cut through mountains, then heads across the plains of Indochina before splitting into the nine channels that dominate the Mekong Delta and form the southern tip of Vietnam. The 12th longest of the world's rivers, the Mekong ranks third in terms of biodiversity, with many of its species found in the delta. The vast delta—which extends from Phnom Penh in Cambodia—was formerly dominated by seasonally flooded grasslands and forests, as well as swamps and mangroves. These natural habitats have been mostly destroyed by chemicals used during the Vietnam War and through conversion to farmland. But the areas that remain are still rich in wildlife, including endangered species. Tens of thousands of waterbirds such as egrets, herons, storks, cormorants, and ibises nest in colonies within patches of forest. During the dry season, around 500 sarus cranes migrate to the marshy grasslands. Other migratory waterbirds frequent the mangroves and coastal mudflats. Five species of dolphins occur, including the Irrawaddy dolphin. Threats include intensive land use, pollution, and current and planned development of dams. MW

PHONG NHA-KE BANG NATIONAL PARK

QUANG BINH, VIETNAM

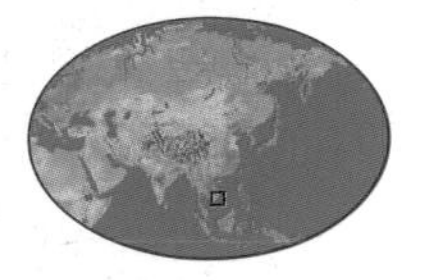

Area of Phong Nha-Ke Bang National Park core zone: 211,890 acres (85,754 ha)

Area of Phong Nha-Ke Bang National Park buffer zone: 466,667 acres (188,865 ha)

Phong Nha-Ke Bang National Park protects one of the world's great karstic landscapes and the oldest major karst area in Asia. Remnants of the tropical forests that once covered Indochina cling to the precipitous hills here and harbor some of the planet's rarest mammals. Beneath the tangle of peaks, plateaus, ridges, and valleys, over 44 miles (70 kilometers) of caves snake through the limestone. Ten species and sub-species of primate live in the park, including the vibrantly colored but endangered red-shanked douc langur and its close relative, the hatinh langur, unique to this limestone landscape. This is also a key refuge for the recently discovered saola, a curious mammal that resembles a small deer yet is more closely related to the cow. Endangered birds thrive here as well—the sooty babbler was first recognized in Laos in the 1920s but was not found again until researchers discovered it in Phong Nha-Ke Bang 70 years later. Despite legal protection, Phong Nha-Ke Bang suffers from deforestation and hunting. Tigers, Asian elephants, and wild cattle are almost extinct, and the saola are also under threat. MW

MARBLE MOUNTAINS

DA NANG, VIETNAM

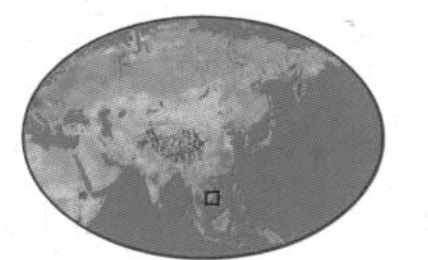

Average air temperature: 78°F (26°C)

Rock type: limestone

About 7.5 miles (12 kilometers) west of Da Nang City, five outcrops of limestone and marble tower over the surrounding flat countryside. Known as the Marble Mountains ("Ngu Hanh Son") or "Mountains of the Five Elements," each outcrop is named after a Vietnamese folklord—Thuy Son (water), Moc Son (wood), Kim Son (metal), Tho Son (earth), and Hoa Son (fire). The highest mountain, Thuy Son, is riddled with narrow, steep caves, which can be climbed all the way to the summit. From the top you can see China Beach and the slightly less idyllic charms of the Da Nang airbase.

Inside the caves are several features common to limestone landscapes, including numerous large stalactites. There is also plenty of evidence of human activity: shrines, statue guards, and Buddhas fashioned from the local green or white marble. It is also believed that the region's feudal lords once stored their gold and jewelry here, where they were guarded by monks from the local monastery. MB

RIGHT *One of the vast caverns of the Marble Mountain caves.*

PHU HIN BUN MOUNTAIN

BORIKHAMSAY, LAOS

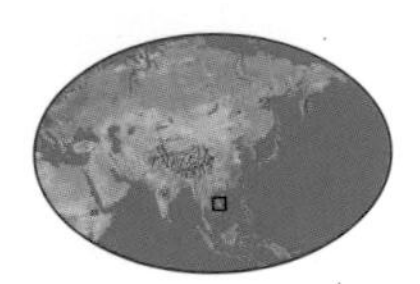

Length of Khong Lore Cave: 4.3 mi (7 km)

Extent of protected area: 610 sq mi (1,580 sq km)

Saddled in the neck of Laos is a huge wilderness of pure streams, deep forests, and striking karstic (limestone) scenery. Declared a National Biodiversity Conservation Area in 1993, Phu Hin Bun, meaning "limestone mountain," is one of the country's most accessible wild areas.

Notable throughout for its elemental beauty, this cave-riddled area has one particularly outstanding highlight. The Nam Hin Bun River meanders past many jagged outcrops before threading its way through the mountain itself. Its path has become Khong Lore Cave, a 4.3-mile (7-kilometer) natural tunnel up to 330 feet (100 meters) wide and in places just as high. The complete length of the river is navigable and takes at least an hour in a motorized canoe. It is also used as a transportation route by the local people.

Near the entrance, eerie formations likened to buffalo heads and elephant trunks are briefly illuminated, while stupa-shaped formations can be found in drier side tunnels and chambers. Filling the darkness with the sound of unseen rapids, the gravel-bed river finally exits into a new valley through a breathtaking stalactite-lined mouth. The vastness of Khlong Lore undoubtedly holds unexplored secrets. AH

PAK OU CAVES

LUANG PRABANG, LAOS

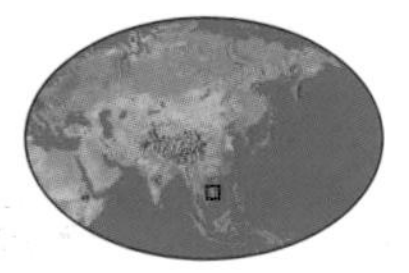

Highest point in Laos (Phou Bia): 9,242 ft (2,817 m)

Lowest point (Mekong): 230 ft (70 m) above sea level

In a huge overhanging limestone cliff at the confluence of the Ou River and the Mekong are two ragged caves. Tham Ting, the lower cave, is the smaller of the two and lies just 50 feet (15 meters) above the river; it is reached via a small bamboo jetty and a flight of carved stone steps. A path (and an extremely steep brick staircase) up the cliff face climbs to the higher cavern—Tham Phum—which is very deep and pitch dark. Inside the caves are more than 4,000 Buddha statues, some over 300 years old. The statues have been rendered in many different materials—wood, metal, clay, soapstone, limestone—and range in size from only a few inches to over 6 feet (1.8 meters). Most are standing or lying in the inky black recesses of Tham Phum and can be revealed only by torchlight, but those in the lower cave stare out eerily at the gray skies and brown Mekong River.

Pak Ou Caves is reachable either by boat—the 90-minute journey along the Mekong passing quiet rural villages and local fishermen in slender canoes—or in a jumbo (open taxi) from Luang Prabang. MB

NAM KHAN RIVER

LUANG PRABANG, LAOS

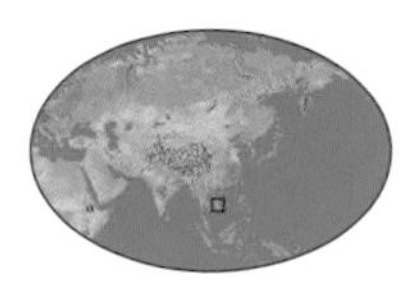

Depth of Nam Khan River: 5 ft (1.5 m)

Luang Prabang monastery established: 1353

Elevation of Luang Prabang promontory: 2,300 ft (700 m)

The Nam Khan, or "Creeping River," is best characterized by the sound of children playing in its shadows and the sight of terraced vegetable plots reaching down to the water's edge. A tributary of the Mekong, the Nam Khan meets the great river at Luang Prabang, the ancient royal capital of Laos and the second largest city in the country.

Luang Prabang sits on a promontory bordered on two sides by water and is home to Vat Xiengthong, one of the most beautiful monasteries in the region. The Nam Khan is relatively shallow and about 100 feet (30 meters) wide. It is bordered by mountains, as well as low, rounded, green-cloaked hills, and other hills showing the familiar towering limestone shapes typical of the karstic landscapes of southeast Asia. About 80 percent of the Nam Khan's blue waters flow slowly, but some stretches are grade-three rapids that are especially popular with canoeists and white-water rafters.

Just 19 miles (30 kilometers) south of the Nam Khan–Mekong confluence is a seemingly endless limestone staircase over which cascade the multistepped and mist-enshrouded Huang Si Falls. Picnic benches are not set up on the bank of the stream, but are actually in the water. MB

LUANG PRABANG WATERFALLS

LUANG PRABANG, LAOS

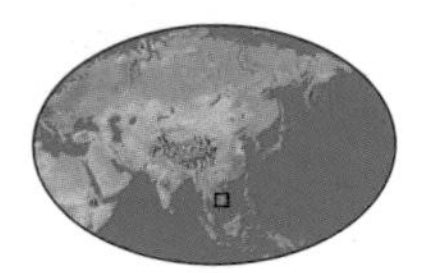

Height of Luang Prabang waterfalls: 200 ft (60 m)

Age of Luang Prabang kingdom: over 1,000 years

Luang Prabang is a tiny mountain kingdom more than 1,000 years old; it is a beautifully atmospheric Laotian "city" that combines the best of French colonial architecture and Buddhist heritage. It is nestled in a narrow valley where the Nam Khan and Mekong rivers join in an area with two fine waterfalls—Tat Kuang Si and Tat Sae. Tat Kuang Si is a wide, many-tiered waterfall tumbling over limestone rocks smoothed by calcite deposits and through a series of cool, blue pools. The base of the falls is neatly manicured as a public park, but a stream-side trail leads to the wilder second level, which has a cave behind the water curtain and is surrounded by forests with a meadow beyond. At Tat Sae, two streams converge to produce another serene, multi-level waterfall with shorter drops but more expansive and more numerous pools. Busy at weekends but quiet otherwise, the falls are wide and gently ease their way down the moderate slope. Downstream from Tat Sae lies the tomb of Henri Mouhot, the first European visitor to Angkor Temple in Cambodia. Appropriately, for many years the tomb was effectively consumed by jungle. AH

RIGHT *The veiled falls of Tat Kuang Si.*

CHAMPASAK WATERFALLS

CHAMPASAK, LAOS

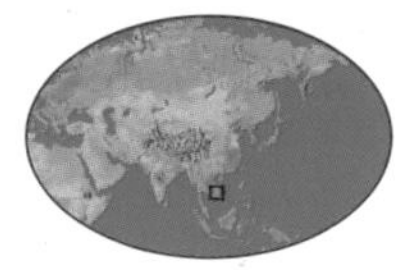

Height of Champasak Waterfalls: 400 ft (120 m)

Elevation of waterfalls: 4,000 ft (1,200 m)

The highest waterfall in Laos, Tat Fan, is found on the edge of the sandstone Bolaven Plateau in the southern province of Champasak. The area is most famous for growing coffee, plus the annual ceremony performed by Mon-Khmer ethnic groups in which water buffaloes are sacrificed.

The falls themselves drop over 400 feet (120 meters) in two powerful parallel streams. There are well-marked trails leading to their base but they are actually at their most spectacular when seen from a distance as a sudden vision of twin falls crashing into the gorge below. Further up the plateau, in a cooler climate, are the smaller falls of Tat Lo, which drop only 33 feet (10 meters) but which are wide, with a deep, luxuriant pool at the base. Dammed further upstream, the flow can suddenly double during the dry season when the authorities release water for electricity generation in the evenings. The falls are part of a 2,316-square mile (6,000-square kilometer) conservation area called Dong Hua Sao, which is still largely forested and wild beyond the coffee plantations. AH

THE MEKONG WATERDROP & SI PHAN DON DELTA

LAOS

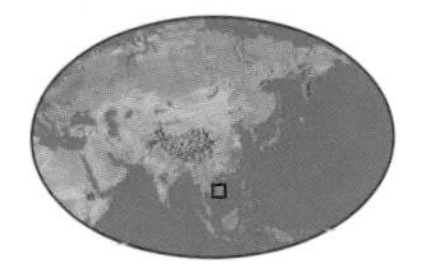

Alternative name: Khone Falls

Maximum width of Mekong River: 8.7 mi (14 km)

Maximum flow rate: 1,412,400 cu ft (40,000 cu m) per second

With a drop of a mere 33 feet (10 meters), the torrent of the Mekong Waterdrop is still so powerful that it actually hindered early travel. A short railway was constructed in an attempt to facilitate access, though it later proved an unsatisfactory bypass and was abandoned. The falls are also a major barrier to migrating fish, which get held up in the powerful torrents just below the drop. Local fishermen capitalize on this barrier, harvesting their catch by clambering on flimsy bamboo poles and ladders as whirlpools laden with fish swirl below.

This stretch of the Mekong River in southern Laos is an outstanding combination of diverse islands, channels, rapids, and waterfalls—it is no wonder that the livelihoods of so many people are tied to these natural splendors.

A pavilion overlooks the falls and provides a view of three of the many cascades that merge with a thunderous roar in a mist of foaming spray. This is also one of the best places to catch sight of one of the increasingly rare freshwater Irrawaddy dolphins that local fishermen believe are reincarnated people.

The Mekong Waterdrop and other nearby falls can also provide a considerable barrier to the waterflow itself, and during the height of the rainy season the broad section of the upstream river can stretch to a width of 8.7 miles (14 kilometers)—the very widest that the Mekong reaches on its 2,703-mile (4,350-kilometer) journey from Tibet to the sea. Permanent islands are formed along this stretch, but as the waters recede, many more islets and sandbars are exposed, giving the area the name of Si Phan Don, or "Four Thousand Islands."

The largest island, Don Khong, encompasses 8.5 square miles (22 square kilometers) and has around 55,000 residents, who live in two villages. The villages in this region are often named "head" or "tail" according to their position at the upstream or downstream end of the islands.

This intricate maze of islands and channels can only be fully appreciated from the air, but a boat journey also gives amazing views of the island landscape. This 30-mile (50-kilometer) stretch of the Mekong River in southern Laos is an outstanding combination of diverse islands, channels, rapids, and waterfalls—it is no wonder that the livelihoods of so many people are tied to these natural splendors. AH

RIGHT *The Mekong River en route to the Si Phan Don Delta.*

MAE SURIN WATERFALLS

MAE HONG SON, THAILAND

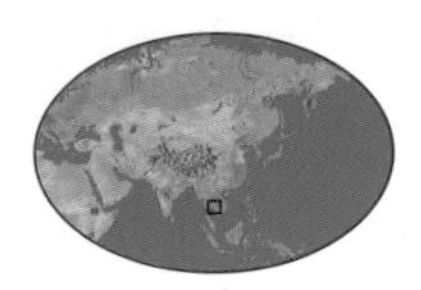

Height of Mae Surin Waterfalls: 262 ft (80 m)

Height of mountains: 5,747 ft (1,752 m)

Close to the Myanmar border in northwest Thailand are waterfalls of great beauty. The Mae Surin River plunges over a ledge in a single stream onto jumbled boulders 262 feet (80 meters) below, which makes it one of the country's highest single-tiered waterfalls. Dropping into a bowl-like depression lined with cliffs, the Mae Surin Waterfalls are equally impressive when viewed from across the valley or standing at the base where rainbows arc through the shimmering spray. The path to the falls passes natural pools and several villages of Hmong and Karen hill tribes. Many years ago, the area below the falls was deforested, leaving a large open area that now produces a golden display of wild sunflowers for two weeks each November. This spectacle is as dazzling as the waterfall itself—the flowers are actually an exotic weed originating from Mexico.

The area is surrounded by mountains up to 5,747 feet (1,752 meters) high, with cliffs and ravines covered in evergreen and pine forests. There is also the distinctive turtle-shaped limestone mountain of Doi Phu, which features a cave containing hot springs. **AH**

OB LUANG GORGE

CHIANG MAI, THAILAND

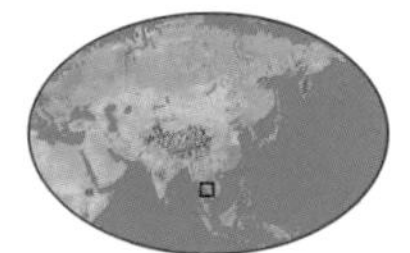

Length of Ob Luang Gorge: 1,000 ft (300 m)

Height of gorge: 130 ft (40 m)

Elevation of gorge: 6,494 ft (1,980 m)

Ob Luang Gorge near the northern city of Chiang Mai is an impressive sight. Rushing through a narrow granite ravine, barely 6.6 feet (2 meters) wide at its narrowest, the Mae Chaem River turns into a raging torrent with sheer 100-to-130-foot- (30-to-40-meter-) high cliffs on either side. Once known as Salak Hin, or "Sculptured Rocks River," the turbulent waters have created many strange rock formations along the 1,000-foot- (300-meter-) high gorge that rises above the river in a myriad of twisted shapes.

Near Ob Luang, meaning "big narrow," is a smaller gorge—Ob Noi, or "small narrow." The area has limestone caves used by Buddhist meditators, the 1,000-foot- (300-meter-) high Pa Chang, meaning "elephant cliff," and waterfalls such as the many-tiered Mae Chon, which drops 330 feet (100 meters), and the near-boiling Thep Phanom hot springs. Tied to these natural features is a rich archeological heritage. The narrow valley once had concentrated wildlife, and Stone Age hunter-gatherers left behind crude tools and rock paintings. Later came Bronze Age settlements, and in more modern times teak-loggers have used the river to transport their lumber to the lowlands. Well-signposted trails tell the natural and cultural history of the area. **AH**

DOI INTHANON MOUNTAIN

CHIANG MAI, THAILAND

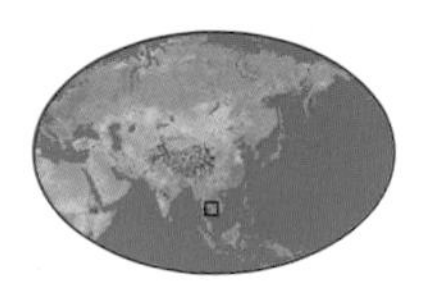

Elevation of Doi Inthanon Mountain: 9,108 ft (2,776 m)

Area of Doi Inthanon: 186 sq mi (482 sq km)

Height of Mae Ya waterfall: 820 ft (250 m)

Doi Inthanon is Thailand's highest mountain and its summit has the only true upper montane forest and sphagnum bog in the country. This granite massif with adjoining limestone outcrops is on the periphery of the mountaintop "islands" that stretch southeast from the Himalayas. Inthanon is a shortened name of the Chiang Mai Kingdom's last prince who gave importance to the mountain's watershed forests before he died in 1897.

Typically shrouded in mist, the "fog-drip" caused by condensation on trees helps feed numerous tributaries of the country's major river system. Vachirathan Waterfall has a single drop of 165 feet (50 meters), with a roar compared to stampeding elephants, while the even more impressive Mae Ya Waterfall tumbles 820 feet (250 meters) down hundreds of small steps. In contrast, the cavernous Borijinda Cave is dotted with natural skylights and serves as a peaceful meditation site for Buddhist monks.

The lower and middle elevations of Doi Inthanon have suffered through shifting agriculture—some 4,000 hill tribe villagers live on its expansive slopes. AH

KHLONG LAN WATERFALL & MOUNTAIN

KAMPHAENG PHET, THAILAND

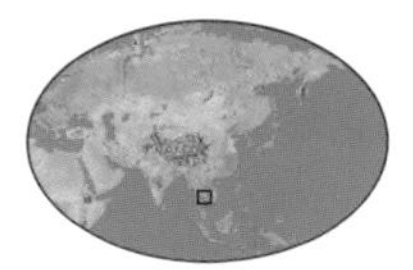

Height of Khlong Lan Waterfall: 312 ft (95 m)

Width of waterfall: 130 ft (40 m)

Height of Khun Khlong Lan Mountain: 4,721 ft (1,439 m)

Khlong Lan is a rugged and hilly area of northern Thailand full of waterfalls, rapids, and magnificent riverine scenery. Most famous is the 130-foot- (40-meter-) wide Khlong Lan Waterfall. The combined flow of five streams from Khun Khlong Lan Mountain drains into an upland lake, is then channeled through a narrow gorge for about 1.8 miles (3 kilometers), then leaps over a dramatic cliff face and drops 312 feet (95 meters) into a deep pool. Another picturesque area is Kaeng Kao Roi Rapids, which are studded with large boulders and thousands of small rocks and sandy beaches in an idyllic mountain setting.

Khun Khlong Lan Mountain used to have a varied mix of hill tribe villages on its slopes. However, the increasing number of residents threatened the integrity of this important watershed through clearing the forest for cultivation, especially for growing opium poppies. The whole population was relocated beyond the national park in 1986 to give the area a chance to recover. AH

MAE PING GORGE

CHIANG MAI, THAILAND

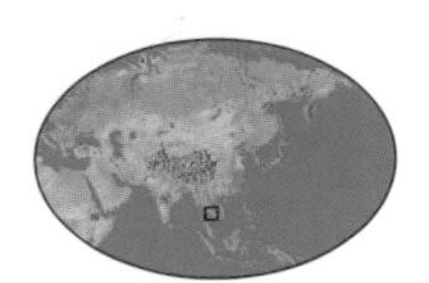

Elevation of Mae Ping Gorge: 4,060 ft (1,238 m)

Construction of Bhumidol Dam: 1964

Before the Bangkok–Chiang Mai railway was completed in 1921, travelers and traders faced an exhausting three-month river journey through the Mae Ping Gorge. The most daunting section was the hazardous Kaeng Song Rapids, located 75 miles (120 kilometers) south of Chiang Mai—boats had to be poled and pulled with ropes through the powerful torrent.

In 1964, long after these treacherous waters had ceased to be a significant commercial route, they were tamed by the construction of the Bhumibol Dam. However, even now in its gentler mood the gorge still retains its splendor as more tranquil waters glide between high cliffs marked with caves and extraordinary limestone formations before opening into the full expanses of a reservoir. Many of the cliffs are fringed with some of Thailand's finest deciduous forest, and around the seven steps of the beautiful Gor Luang Waterfall are stands of teak trees. One 19th-century traveler described the limestone mountains as "bold cliffs, crags and spires of spectacular beauty." A Buddhist place of worship stands as a monument to the many others that were lost beneath the reservoir. **AH**

ANG THONG ARCHIPELAGO

ANG THONG, THAILAND

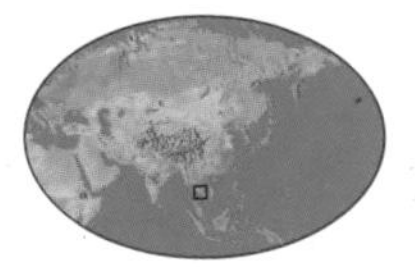

Number of islands in archipelago: 50

Total land area: 7 sq mi (18 sq km)

Nineteen miles (30 kilometers) from Samui Island lies a striking archipelago of craggy limestone islands with lagoons, beaches, caves, and coral reefs. Now a national park, most of the 50 islands are uninhabited, and for many years the area was protected from development because it was used by the Thai Navy.

The surrounding waters can be turbid but strong currents prevent sediment covering the corals. Overall, the closely grouped islands are surprisingly pristine on the gulf side of the peninsula. Ang Thong (which translates as "golden bowl") is a major breeding ground for Thailand's short-bodied mackerel, and its waters have been fished for more than 1,000 years. Two rocky paths climbing through forest—on the islands of Wua Talab and Sam Sao—lead to panoramic vistas of the curiously shaped, steep-sided islands, which have sculptured formations, including a rock arch off Sam Sao Island. On Koh Mae Koh Island, a steep climb over a high rim leads to a saltwater lake encircled by vertical cliffs. It is known as Thale Nai, or "inner sea," and contains waters that change from emerald to aquamarine. **AH**

RIGHT *The emerald-green waters of Ang Thong Archipelago.*

PHU KRADUNG MOUNTAIN

LOEI, THAILAND

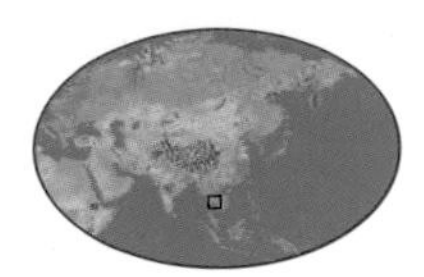

Elevation of Phu Kradung Mountain: 4,461 ft (1,360 m)

Height of waterfalls: 262 ft (80 m)

Rock age: 300 million years

If there is one mountain that Thai people aspire to climb simply for the satisfaction and pleasure, it is Phu Kradung in the northeast of the country. Every year, thousands make the arduous five-hour trek up the steep gradients of this sandstone table mountain to its 23-square-mile (60-square-kilometer) heart-shaped summit plateau. It is said that lovers who help each other reach the top will stay together but those who quarrel are likely to separate.

The views from the plateau of the rolling savanna and surrounding plains are reward enough. Hiking trails lead through open pinewoods, beautiful meadows, and lichen-covered rock gardens. Dense evergreen forest covers the mountain's wetter northern slopes, and numerous impressive waterfalls cascade from an overhanging ledge, dropping into the forest 262 feet (80 meters) below.

Phu Kradung means "bell mountain." Its distinctive bell shape is created by an erosion-resistant cap of orange-white sandstone from the Mesozoic Era. It was not until 1805 that the first person, a hunter on the trail of game, reached the top and brought back news of the plateau's isolated and unexpected secrets. **AH**

NAGA FIREBALLS

NONG KHAI, THAILAND

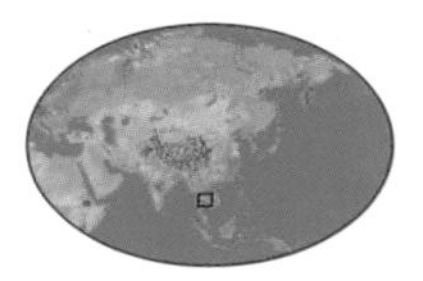

Distribution of Naga fireballs: 62 mi (100 km) along Mekong River

Age: at least 100 years

Every year on the night of the full moon of the eleventh lunar month, a phenomenon occurs along the Mekong River between northeastern Thailand and Laos that remains both mysterious and controversial. Small globes of fiery light, about the size of tennis balls, float up from the water's surface reaching heights of 330 feet (100 meters) or more before fading. Appearing sporadically along a 62-mile (100-kilometer) stretch of river, they create an eerie display that has begun to attract thousands of onlookers and almost as many explanations. Local mythology tells of great Naga serpents living in the depths of the Mekong who celebrate the Lord Buddha's return to Earth by spitting fire into the sky—the event coincides with the end of Buddhist Lent. More recent scientific explanations detail a complex combination of circumstances that lead to pockets of flammable gas being ignited. Skeptics, on the other hand, have publicized improbable theories of manmade rockets and tracer bullets that do not quite match the fireballs' description. Locally, there is a long history of sightings of Naga fireballs in the region, but investigations into the phenomenon have really only just begun. **AH**

KAENG SOPHA WATERFALL

PHITSANULOK, THAILAND

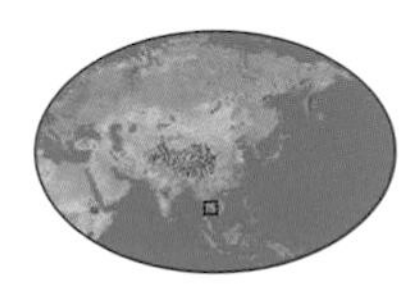

Height of Kaeng Sopha Waterfall: 130 ft (40 m)

Area of Thung Salaeng National Park: 469 sq mi (1,215 sq km)

Established as a national park: 1959

Kaeng Sopha is one of Thailand's most stunning waterfalls. Its three wide, gently curving steps have an aesthetically pleasing balance—whether channeling solid curtains of water during the rainy season or the more gentle flow of the dry months. Although deforestation in parts of the catchment has reduced the flow in modern times, its spray and thunder are still exhilarating.

Part of Thung Salaeng National Park, Kaeng Sopha lies on a limestone spine that guides the Khek River from the sandstone of the Petchabun Mountains to the dry plains below. The distinctive layering of limestone slabs that creates this staircase waterfall is evident when the river is not flowing at full force.

In contrast to Kaeng Sopha, the majority of the park is a landscape of pine-studded grasslands and dry expanses of barely vegetated rocky hardpan. Named after the Salaeng trees (snakewood) that produce strychnine-laced fruit, this area was probably the last stronghold of Schomburg's deer, which became extinct in the 1930s. Located in a relatively dry region, the accessibility and beauty of Kaeng Sopha Waterfall has made it a tourist magnet. AH

PHU RUA ROCK FORMATIONS

LOEI, THAILAND

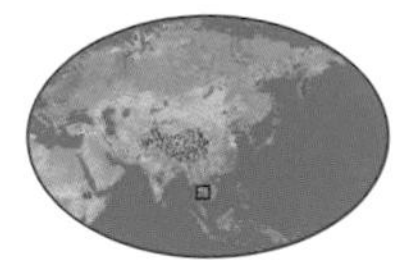

Elevation of Phu Rua formations: 4,477 ft (1,365 m)

Rock age: 50 to 150 million years

Area of Phu Rua National Park: 47 sq mi (121 sq km)

Phu Rua is a flat-topped sandstone mountain, or mesa, with some of the best examples of the strangely eroded rock pillars characteristic of northeastern Thailand. Its name means "boat mountain" and comes from a jutting cliff shaped like the bow of a Chinese junk. Impressive erect sandstone formations over 16.5 feet (5 meters) high are scattered throughout the area. Legend describes how a dispute between two cities over a royal marriage procession resulted in traditional dowry gifts being destroyed and turned into stone. Thus, among the many structures are shapes of a bowl, a cooking utensil, and a cow. Believing turtles to have low intelligence, a prince subsequently erected the Turtle Rock as a monument to the stupidity of war.

Cliffs are also a major feature and Pha Sap Thong, meaning "cliff that absorbs gold," is named after its covering of yellowish lichen. Crumbling cliff faces have created widespread natural rockeries smothered in March by snow-white rhododendron blooms that are shortly followed by swathes of orchids.

The trek to the top of Phu Rua is relatively easy, and on the summit is a Buddha statue overlooking nearby mountains and valleys. AH

KHAO YAI FORESTS & WATERFALLS

SARABURI / NAKHON NAYOK / NAKHON RATCHASIMA / PRACHINBURI, THAILAND

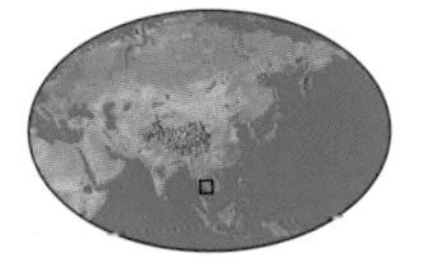

Area of Khao Yai National Park: 836 sq mi (2,168 sq km)

Elevation range of park: 200 ft to 4,431 ft (60 to 1,351m)

Vegetation: moist/dry evergreen and deciduous tropical forest, grassland

Thailand's showcase natural treasure is Khao Yai National Park, which is less than 124 miles (200 kilometers) northeast of Bangkok. Forming the western edge of the Dangrek Range that extends along the Cambodian border, Khao Yai, meaning "big mountain," has extensive forests with abundant and unusually visible wildlife. Rising abruptly from the central plains, volcanic rocks form the western part, with a rolling sandstone plateau to the east and some limestone outcrops to the north. These different rocks and the park's vast size of 836 square miles (2,168 square kilometers) have given Khao Yai's six main peaks a varied covering of evergreen and deciduous forests.

With a prodigious runoff from 118 inches (300 centimeters) of annual rainfall, Khao Yai plays an invaluable role in flood protection and water provision and also has many vigorous waterways and spectacular falls. Most impressive is Haew Narok or "Devil's Gorge," a mighty waterfall of two large steps that has swept elephants to their deaths. The falls are surrounded by steep, forested slopes and bare cliffs, with the conical peak of Khao Samorpoon in the background.

This whole vast area was uninhabited until 1902, when 30 families from the lowlands settled on the plateau. However, when the area began attracting bandits, the government forced them to move out. Later, the construction of a golf course in the middle of the forest sparked many protests and heated arguments that eventually led to its closure. This left open grassy areas that now make Khao Yai's wildlife so visible. A 31-mile (50-kilometer) network of marked forest trails, two watchtowers, and the fact that salt licks are located at the roadside offer good opportunities to see gibbons, elephants, and perhaps tigers.

Most impressive is Haew Narok or "Devil's Gorge," a mighty waterfall of two large steps that has swept elephants to their deaths. The falls are surrounded by steep, forested slopes and bare cliffs, with the conical peak of Khao Samorpoon in the background.

The many viewpoints, a bat-filled cave, and a total of 12 accessible waterfalls all add to the powerful attraction of Khao Yai. As the country's first national park, established in 1962, and located relatively close to Bangkok, it enjoys a special status in Thailand but one that it also earns through its outstanding forests. **AH**

RIGHT *Thailand's Khao Yai National Park abounds with waterfalls and lush forest.*

PHU HIN RONG KLA MOUNTAIN

LOEI / PHITSANULOK, THAILAND

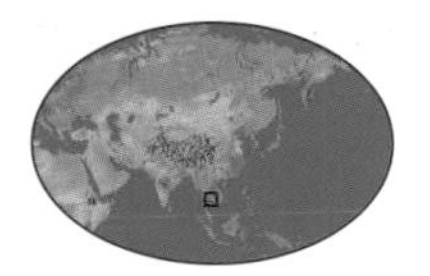

Maximum elevation of Phu Hin Rong Kla Mountain: 6,000 ft (1,800 m)

Rock age: 130 million years

Age of faults: 50 million years

The sandstone mountain of Phu Hin Rong Kla has some superb examples of the power of geological forces. Two contrasting rocky hardpans are the area's highlights. Lan Hin Daek, or "Broken Rock Field," has a series of deep parallel clefts, some narrow enough to step over but others too wide to jump, with smaller cracks at 90° effectively breaking the area into enormous rectangular blocks. These mini-faults run for 0.6 miles (1 kilometer) or more, and were caused by the stresses of tectonic movements cracking the surface crust. Open and sparsely vegetated, an aerial view shows a second series of long cleavages overlapping the main ones at a slight angle, indicating two planes of subterranean force. Nearby is Lan Hin Pum, or "Button Rock Field," which is an uncracked expanse of rock covered with rounded protrusions thought to have been molded by many centuries of sun, wind, and rain. Elsewhere, the appearance of Lan Hin Riap, "Smooth Rock Field," indicates how the other two areas may have looked many years ago, and the whole mountain is scattered with unusual rock formations. AH

PHA TAEM CLIFF

UBON RATCHATHANI, THAILAND

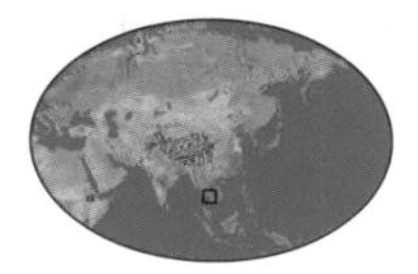

Height of Pha Taem Cliff: 330 ft (100 m)

Rock age: 50 to 150 million years

In one of its sedate phases along the Thailand–Laos border, the Mekong River eases through a wide valley defined by sandstone cliffs. The Thai side is known as Pha Taem and is famous for its sweeping views, prehistoric paintings, and natural rock sculptures. Above the cliffs is a rocky plain of dry forest, flower-rich meadows, and open rock gardens. Enormous mushroom-shaped blocks of eroded sandstone—dark gray above and yellow below—are common. Others are said to resemble a turtle, a camel's neck, and a group of pagodas. One whose shape resembles a "flying saucer," and is estimated to weigh 110,000 pounds (50,000 kilograms), is so finely balanced that it can be rocked with little effort. From several points along the clifftops are open vistas of the broad river and forested valley. Numerous streams tumble down gullies, with the sparkling Saeng Chan Waterfall the most attractive as it drops like a shaft of light through a hole in a rock slab. Along the base of the cliffs are 4,000-year-old paintings, including animals, people, and "tum" fishing baskets still used in the area. AH

MUKDAHAN ROCK FORMATIONS

MUKDAHAN, THAILAND

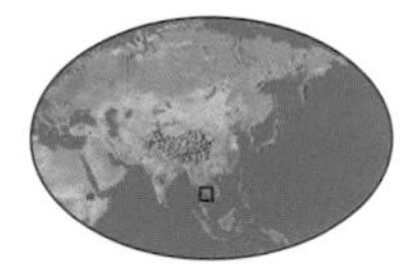

Area of Mukdahan National Park: 19 sq mi (49 sq km)

Rock age: 50 to 150 million years

In one of the least visited corners of Thailand lies a small national park, well known locally for its beautiful natural rock formations. Towering mushrooms and umbrellas of sandstone, with dark caps and yellow stalks, stand next to other shapes likened to airplanes, crowns, and a swan. The most stunning structures result from differential erosion, where lumps of relatively hard rock give the layers directly underneath some defense against weathering. However, this protection is only partial, and wind, water, and sun gradually eat away the softer rocks leaving a distinctive cap precariously balanced on a dwindling stalk that will eventually collapse. This area of Mukdahan Province consists of several hills with cliffs and loosely stacked rocks of great variety and striking shapes, plus waterfalls and small caves. A large outcrop called Phu Lang Se, surrounded by forest, is the best of several viewpoints in the area. The natural rock galleries provide small niches for plants that give a seasonally changing background to the solid rock outlines, as the different flowers bloom in sequence. AH

KAENG TANA RAPIDS

UBON RATCHATHANI, THAILAND

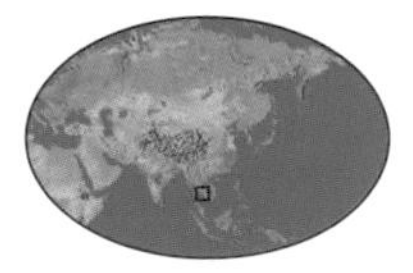

Length of Kaeng Tana Rapids: 0.6 mi (1 km)

Width of rapids: up to 1,000 ft (300 m)

Elevation of rapids: 660 ft (200 m)

The Moon River flows along the southern edge of northeastern Thailand's great sandstone plateau. Just before its confluence with the Mekong River are a series of rocky cataracts of which Kaeng Tana is the longest and most impressive. Edged by low cliffs and largely underwater at the height of the rainy season, for most of the year the water-sculpted bedrock lies exposed as irregular island platforms and protrusions. The natural drill of water eddies and pebbles has riddled the rock with holes, some large enough to serve as picnic hollows in the dry season. Before a controversial dam was built upstream, Kaeng Tana had a vital role in providing rich fishing grounds and helping to retain water during dry periods, but its main function now is as a fascinating natural playground. The Moon River was once an important transportation route, with this section known as Death Rapids. Just upstream lie sandy beaches and a suspension bridge that gives an overview of the area. Then shortly beyond the rapids lies the Two-colored River, where the gray-blue waters of the Moon River meet the Mekong. AH

THI LO SU WATERFALLS

TAK, THAILAND

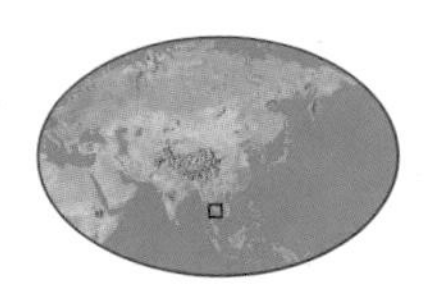

Height of Thi Lo Su Waterfalls: 500 ft (150 m)

Width of waterfalls: 1,000 ft (300 m)

A helicopter flying over a remote corner of Thailand's northern forests in 1986 stumbled on the amazing sight of Thi Lo Su Waterfalls. Only known previously to a few local villagers, the rumors of the immense and beautiful falls hidden in the jungle were confirmed. Once discovered, the gradual process of opening the area for more casual visitors began.

The falls are one of the most spectacular in Asia. White ribbons of water pour over high cliffs, surrounded by dense forest. Fed by the streams of "Strong-legged Mountain," several levels drop over calcium carbonate–encrusted limestone. Large trees perched along the falls' rim give the impression that the flow emanates from the forest itself.

The whole area is a wildlife sanctuary and contains waterfalls, rapids, caves, remote lakes, and forest-covered peaks. Initially, the falls were difficult to reach and few visitors made the two-day trek from the nearest town. Recent efforts to promote Thi Lo Su and improve access have been so successful that it has been necessary to reduce the number of visitors to 300 per day. The protection of this area in turn preserves the beauty of these waterfalls. AH

SAMUI ISLAND ROCKS

SURAT THANI, THAILAND

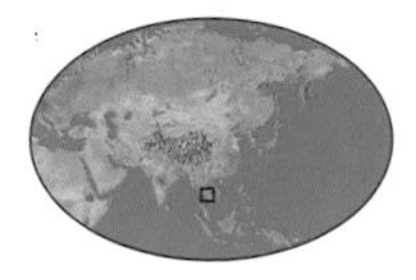

Area of Samui Island Rocks: 100 sq mi (264 sq km)

Rock age: 230 to 330 million years

The picture-book tropical island of Koh Samui, with its beautiful white beaches, was, until the 1970s, known simply as the biggest coconut plantation in the world. Two million coconuts were sent from Samui to Bangkok each month. Now the island's natural beauty has made it a major tourist destination. Rocky granite promontories separate the many varied beaches, providing a pleasingly aesthetic border to long stretches of white sand.

Beside the eastern beach of Lamai, nature has created Hin Ta and Hin Yai, "Grandfather" and "Grandmother" rocks in the shape of two extraordinarily accurate formations of the male and female genitalia. Hin Ta stands 13 feet (4 meters) tall and points skyward, while his companion is more modestly placed in a wave-splashed cleft. As natural novelties, these rocks must be among the most photographed in the world. Inland, the less crude overlap stone is a large block of granite precariously balanced on the edge of a cliff, which provides an excellent view of the surrounding area. AH

BELOW *The Samui Rocks silhouetted against the night sky.*

SAMUI ISLAND WATERFALLS

SURAT THANI, THAILAND

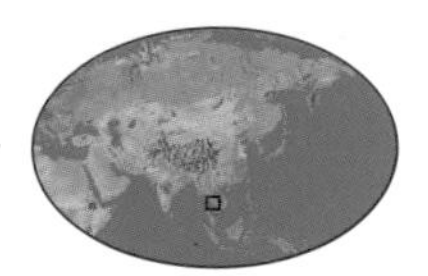

Height of Na Muang 1 Waterfall: 130 ft (40 m)

Area of Samui Island: 88 sq mi (229 sq km)

While the narrow, beach-lined fringe of Samui Island in southern Thailand is a well-populated playground, the interior is an impenetrable mix of coconut plantations and forest-covered hills. Relatively few people venture from the paradise coast, but further inland, the island has a series of attractive waterfalls. The most accessible of Samui's three main waterfalls is Na Muang 1 in the southwest corner. It cascades 130 feet (40 meters) down slabs of yellow limestone streaked with a mixture of green, orange, and brown that changes depending on the volume of water wetting the rock and the amount of algal growth. Tree roots and rocks form a natural staircase leading to a large pool at the base of the main drop, with some surprisingly sharp stones hidden beneath the foaming water. Located along a nearby rocky path is Na Muang 2, which is regarded as the most beautiful waterfall on the island. A longer hike through forest leads to the layered falls of Hin Lad, and further inland is Wang Saotong. **AH**

RIGHT *One of the many waterfalls on Samui Island.*

THUNG YAI NARESUAN & HUAY KHA KAENG FORESTS

KANCHANABURI, THAILAND

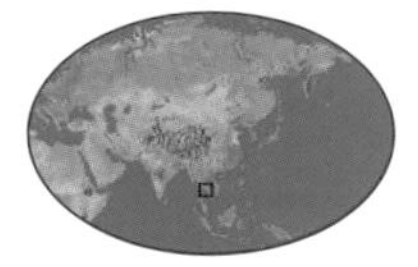

Elevation range of wildlife sanctuaries: 820 to 5,942 ft (250 to 1,811 m)

Vegetation: hill evergreen, mixed deciduous, savanna forests, bamboo groves, swamps, grassland

Forming an immense area of unspoiled forest in western Thailand, Thung Yai Naresuan and Huay Kha Kaeng wildlife sanctuaries are arguably the birthplace of Thai nature conservation. Covering 2,481 square miles (6,427 square kilometers) they represent the largest protected area in mainland southeast Asia, and their superb forests are a major stronghold for much of the region's wildlife. They also act as the heart of the western forest complex, a block of 17 closely linked parks and sanctuaries. Huay Kha Kaeng is unusual because it encompasses a complete catchment, with both the headwaters and the main channel of the sandy-bottomed Kha Kaeng River. Several other adjacent valleys and ridges then form Thung Yai Naresuan, with contrasting but equally pristine rivers. Much of the area is Paleozoic limestone, with the characteristic overhanging cliffs, sinkholes, and caverns typical of tropical karst landscapes. There are also granite intrusions that produce many mineral salt licks that help to sustain the abundant wildlife, which includes Thailand's last herd of wild water buffaloes. **AH**

KHAO CHONG PRAN BAT CAVE

RATCHABURI, THAILAND

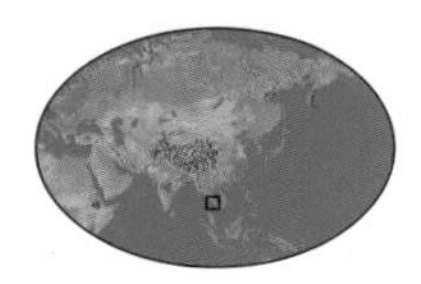

Area of Khao Chong Pran hill: 178 acres (72 ha)

Estimated bat population: 2,900,000

Maximum flyout rate: 2,100 bats per second

The evening exodus of wrinkle-lipped bats at the small limestone hill of Khao Chong Pran has become Thailand's most popular wildlife spectacle. From the comfort of a grassy field and open-air restaurants, over 120,000 visitors a year get to gaze at the sight of a long, continuous stream of bats emerging from the cave in the late afternoon. Twisting and bending as it snakes over the surrounding paddy fields, from a distance this dense cloud of bats is often mistaken for smoke. Conservative estimates suggest two to three million bats live here, but locally it is known as "the cave of a hundred million."

The first bats sometimes emerge a full two hours before sunset, and together with a cheer from the onlookers, these leaders are also met by an ambush of hungry hawks and falcons. The flow of bats leaving the cave then quickly increases and, at its peak, can reach over 2,000 per second. This mass flapping creates a pungent breeze and a sound something like a distant waterfall. By sunset the column will have dwindled considerably, although bats will continue to leave at a steady trickle for several more hours. Equally impressive, but witnessed by far fewer people, is their return around dawn, when a continuous and fast-flowing stream of bats pours into the cave from high above as if being sucked in by a subterranean force.

Khao Chong Pran translates as "Hunter's Hole Hill," and for generations these bats provided an unusual but valuable protein source for locals. However, far more valuable is the bats' guano, which makes one of the richest and most highly sought-after fertilizers because of its high nitrogen and phosphorus content. When improved transport, mist nets, and wide demand for bat meat threatened the very existence of the colony in the 1980s, it was given official protection. Carefully regulated guano-mining is still allowed, however—the villagers collect a staggering 800 large buckets per week. The profits from this by-product support the surrounding communities, which means that the local school has effectively been built from bat droppings.

The cave itself is lit through several holes in the roof that also provide a degree of ventilation. Visitors prefer to observe the bats from outside, avoiding the stench of the caves. Only the guano-miners seeking their "black gold" venture into the acrid atmosphere of the cave's depths. AH

Twisting and bending as it snakes over the surrounding paddy fields, from a distance this dense cloud of bats is often mistaken for smoke. Conservative estimates suggest two to three million bats live here, but locally it is known as "the cave of a hundred million."

THALE SAP LAKES

SONGKHLA, THAILAND

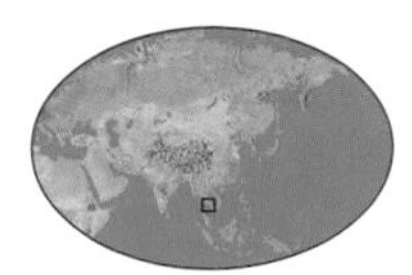

Area of Thale Sap Lakes: 580 sq mi (1,500 sq km)

Average depth of lakes: 6.6 ft (2 m)

Age: 5,000 years

On the Thai Peninsula there are four vast, interconnected lakes collectively known as Thale Sap or the "freshwater sea." With an average depth of barely 6.6 feet (2 meters), this enormous waterway stretches for 50 miles (80 kilometers). Its origins go back 5,000 years, but 150 years ago sediment from the Talung River engulfed the offshore island of Koh Yai. It left a very narrow channel, restricting flow and allowing the lakes to fill to their current size.

The waters are fresh in Thale Noi, salty in Thale Songkhla, and brackish in the intervening two lagoons, producing a sophisticated ecosystem where the balance is affected by the monsoon season. These graduated conditions have produced much natural diversity, most noticeably with the rich birdlife and gorgeous morning displays of flowering lotuses and lilies in Thale Noi. Thale Sap has become Thailand's main inland fishery, with an estimated 1.5 million people depending on the lakes. Unfortunately, this fragile ecosystem is under threat from pollution and development that threatens to reduce the water's natural flow. **AH**

KHAO KHITCHAKUT MOUNTAIN

CHANTHABURI, THAILAND

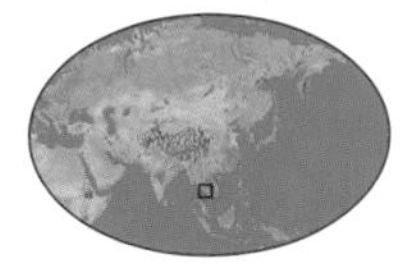

Area of Khao Khitchakut National Park: 22.7 sq mi (59 sq km)

Elevation of highest peak (Khao Phrabat): 3,559 ft (1,085 m)

Height of waterfall: 330 ft (100 m)

One of Thailand's smaller national parks, Khao Khitchakut is a steep, forest-covered granite lump on the edge of a range that includes the Cardamom Mountains in Cambodia. Its slopes bear numerous layered waterfalls. The most notable is Krathing Waterfall, which tumbles down 13 tiers of a boulder-strewn gully lined by gracefully arching bamboo. Khao Phrabat is the highest and most famous of Khao Khitchakut's several peaks. Its rounded crown is peppered with enormous spherical rocks and its religious significance attracts thousands of Buddhists every year. The pilgrims make the journey to the summit by foot each February and March to venerate what are said to be footprints of the Lord Buddha.

The broad, undulating summit of Khao Phrabat is strewn with smooth boulders that peek above the forest. These sculpted rocks resemble recognizable shapes such as an upside-down monk's alms bowl, a huge turtle, and an elephant. Many of these huge rocks provide alcoves and shelters for the mountain's plentiful shrines and Buddha images. **AH**

SIMILAN ISLANDS

PHANG-NGA, THAILAND

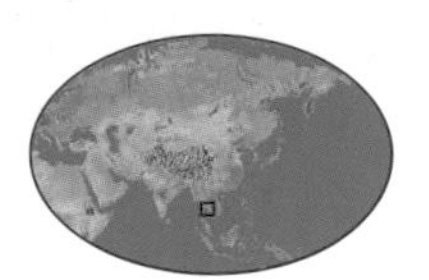

Number of Similan Islands: 9

Extent of protected area: 49.5 sq mi (128 sq km)

Rock age: 230 to 330 million years

The nine islands of the Similans are the jewels of the Andaman Sea. Although these granite outcrops are not large, they wear a cap of healthy forest patrolled by troops of monkeys. The islands were formed from extruded magma that cracked as it cooled. Over the years, erosion polished these large rocks into a jumble of smooth boulders.

However, it is beneath the waves that the real treasure lies, in the coral reefs that are rated in the world's top 10. The coral-encrusted boulders of the western reef tumble dramatically down to the seabed and are rich with sea life. Despite being exposed to the southwesterly monsoon storms, they are kept clear of sand by the currents providing clear waters for divers. The reefs also offer an exciting maze of passageways and cliffs.

In contrast, the calmer weather along the eastern shores has produced sandy beaches and gently sloping reefs, equally diverse but less challenging. Previously uninhabited, the establishment of a national park office in 1982 came just in time to stop the damaging practice of dynamite fishing and trawling. AH

RIGHT *The crystal waters of the Similan Islands.*

KHAO PHANOM BENCHA MOUNTAIN

KRABI, THAILAND

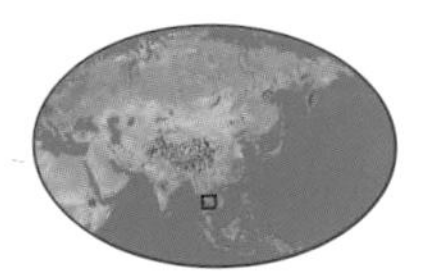

Elevation of Khao Phanom Bencha: 4,428 ft (1,350 m)

Vegetation: tropical rainforest

The grand granite mountain of Khao Phanom Bencha rises above the dazzling beaches and limestone karst of the coastal province of Krabi. Its name means "mountain praying on five points" and refers to the act of worship common to both Buddhists and Muslims who prostrate themselves on the ground. The five high peaks of Khao Phanom Bencha have the same alignment and point toward the area's most sacred site, Tiger Cave Temple. Khao Phanom Bencha dominates all inland views of the region. Its rich rainforest has huge, 130-foot (40-meter) trees supported by enormous buttress roots that form a dramatic contrast to the monotony of the surrounding rubber-tree plantations.

The clear pools, sandy beaches, and layered falls of Huay To have made it a popular film location. The falls of Huay Sakae and Yod Maphrao plunge down sheer cliffs, and are at their most impressive at the height of the rainy season when the force of the water sends up clouds of white spray. AH

KHAO SAM ROI YOT MOUNTAIN

PRACHUAP KHIRI KHAN, THAILAND

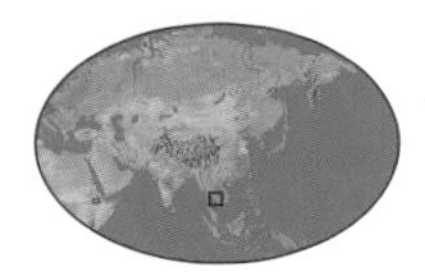

Area of Khao Sam Roi Yot: 38 sq mi (98 sq km)

Maximum elevation of Khao Sam Roi Yot: 2,000 ft (600 m)

Rock age: 225 to 280 million years

Surrounded by marshes and coastal plains, the jagged Khao Sam Roi mountain sits like a land-locked island on the upper "elbow" of the Thai Peninsula. Its name, *Sam Roi Yot*, has several translations. The first is "300 survivors" which refers to the survivors of a Chinese merchant shipwreck. The second translation is "300 shoots," which describes a local plant. But perhaps the most plausible translation is "mountain of 300 peaks." When viewed from the northern approach, numerous peaks form a rugged silhouette against the sky. The flat land that hugs the mountain's base makes its craggy cliffs and conical peaks look dramatic if slightly incongruous against its gentler marshy surroundings.

Evidence, such as raised beaches, undercut cliffs, and shell-rich soil, indicates that Khao Sam Roi Yot was once a formidable and inhospitable offshore island. Now enclosed by land and within easy reach, the diverse attractions of beaches, caves, bird-watching, and scenery has made this relatively small national park very popular. However, the bulk of the mountain remains inaccessible and merely forms an impressive backdrop to the activities around its fringe. AH

PHRA NANG PENINSULA

KRABI, THAILAND

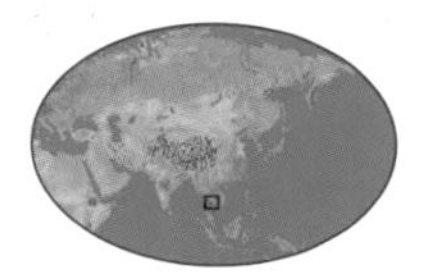

Rock age: 225 to 280 million years

Fossil age: 20 to 40 million years

Rated by many as one of the most beautiful beach settings in the world, the Phra Nang Peninsula earns its popularity through a combination of fine white sands and spectacular limestone outcrops. Beaches, coconut-lined at one end, hug awesome 820-foot (250-meter) cliffs at the other. The area's karst scenery stretches from dramatic landlocked hills to outcrops dipping their toes in the Andaman Sea to steep-sided islands with their own exquisite beaches.

Various local legends explain the peninsula's glorious landscape. One of the most fascinating is the tale of an angry hermit who turned a noisy wedding party and giant sea serpent into the limestone outcrops seen today. This crystalline limestone of Permian age was sculpted over millions of years, and in some areas the sea has isolated low-lying sea stacks. Both Buddhist and Muslim fisherman venerate the peninsula and make regular offerings to the shrine at Phra Nang Cave. A steep trail leads to an elevated viewpoint and the secluded, cliff-lined Princess Pool, which fills at high tide through an underground tunnel. For fossil enthuasiasts, "shell cemetery" is a treasure trove of fossil-filled sedimentary rock. AH

SRI PHANG-NGA WATERFALLS

PHANG-NGA, THAILAND

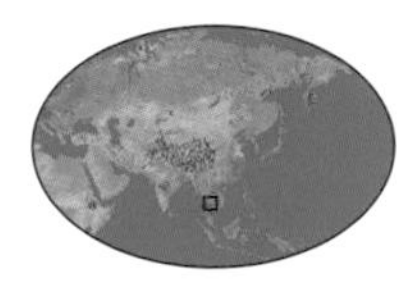

Height of Sri Phang-Nga Waterfalls: 207 ft (63 m)

Area of Sri Phang-Nga National Park: 95 sq mi (246 sq km)

The waterfalls of Sri Phang-Nga are often overshadowed by the many other areas of outstanding beauty along the Thai Peninsula. However, these falls are an overlooked treasure.

Sri Phang-Nga National Park was established in honor of a Thai king's 60th birthday and is a richly forested range of hills with many streams and cliffs, as well as three scenic waterfalls. Tam Nang is a wild torrent of water in the rainy season, which in this part of Thailand lasts for over half the year, from May to November. Its lower pools have schools of large fish that are regularly fed at the weekends by the local people. Ton Ton Sai cascades over a single large boulder, while the 148-foot- (45-meter-) high Ton Ton Toei plummets down a sheer cliff face.

The numerous waterfalls of Sri Phang-Nga are impressive enough, but the real charm of the area lies in the surrounding pristine moist evergreen forest, full of towering trees, dancing streams and winding creeks. The one main trail through the forest is a tough route filled with steep climbs, slithering descents, and stream crossings. However, the exhilarating natural surroundings are well worth the trek. **AH**

KHAO LUANG MOUNTAIN

NAKHON SI THAMMARAT, THAILAND

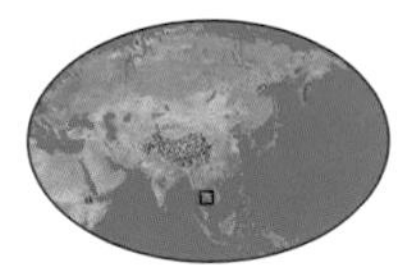

Height of Khao Luang Mountain: 6,019 ft (1,835 m)

Area of Khao Luang: 220 sq mi (570 sq km)

Known as the green roof of southern Thailand, Khao Luang Mountain consists of igneous rock with scattered limestone and is the highest mountain on the Thai Peninsula at 6,019 feet (1,835 meters). The mountain receives rains from both the eastern and western monsoons, and the rainy season can last for nine months, resulting in many vigorous streams and powerful waterfalls. Garom, Phrom Lok, and Tha Phae are all multi-leveled falls tumbling through tropical forest. The most famous is Narn Fon Sen Har, which for many years graced Thailand's highest denomination banknote. Khao Luang, which translates as "great mountain," has a significant place in modern Thai history. A combination of heavy rains and deforestation on some of the mountain's lower slopes caused catastrophic landslides in 1988 that killed over 300 villagers. As a consequence, a countrywide ban was placed on commercial logging.

A three-day round-trip to the summit follows the route of mountain streams running through superb evergreen forests. An indication of Khao Luang's biological riches is evident in the 300 species of orchid found on its fertile slopes. **AH**

PHANG-NGA BAY

PHANG-NGA, THAILAND

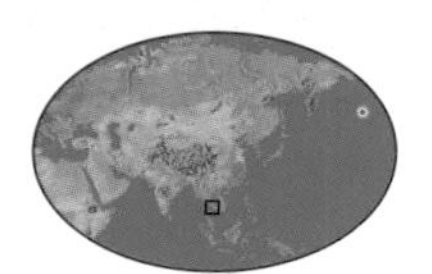

Area of Phang-Nga Bay: 135 sq mi (350 sq km)

Number of islands: 42

Rock age: 225 to 280 million years

In Phang-Nga Bay, on Thailand's Andaman Coast, spectacular limestone islands with sheer cliffs up to 1,320 feet (400 meters) in height emerge like scattered teeth from calm waters and extensive mangrove forest. Sheltered from the worst of the monsoon storms, the large, shallow bay contains around 42 rugged islands.

A highly fragmented spine of Permian limestone runs down western Thailand as part of the Tenasserim mountain range. Originally made from the bodies of billions of shellfish and corals in a sea stretching from China to Borneo, the limestone became fractured and exposed by the same upheaval that created the Himalayas. Waves and fluctuating sea levels, combined with mildly acidic humus and seeping water, gradually sculpted the rock into the breathtaking drowned karst seen in Phang-Nga Bay today.

With an average area of little more than 0.4 square mile (1 square kilometer), and typically capped with a sparse layer of forest scrub, many of these islands hide a stunning secret behind their pitted cliffs. Dark, sea-level tunnels lead through to hollow interiors. Here, open to the sky, are natural amphitheaters with central seawater pools and tiny, secluded beaches. Reachable by canoe, the tranquillity and seclusion of these chambers can be virtually guaranteed, because they are often only accessible at low tide.

The magnificent karst landscape of Phang-Nga Bay lies close to the tourist centers of Phuket and Krabi. Boat trips around the bay meander through winding channels of tangled mangrove forest, with the gigantic humps of protruding limestone punctuating the skyline. The dense mangrove eventually dwindles, leaving more open views of the island-studded seascape. Thanks to its eye-catching role in the James Bond film *The Man With the Golden Gun*, the most famous location is Koh Tapu—a single upright column of rock undercut by waves that looks almost ready to topple over.

Dark, sea-level tunnels lead through to hollow interiors. Here, open to the sky, are natural amphitheaters with central seawater pools and tiny, secluded beaches. Reachable by canoe, the tranquillity and seclusion of these chambers can be virtually guaranteed.

In the middle of the bay at Koh Pannyi, a Muslim village of 500 stilt-houses has been anchored to the island. Elsewhere are numerous caves; some contain 3,000-year-old rock paintings while others provide economic riches in the form of the highly coveted swiftlet nests, which are made into bird's-nest soup, a favorite local delicacy. **AH**

RIGHT *One of Phang-Nga Bay's spectacular pinnacles.*

KHAO LAMPI WATERFALLS

PHANG-NGA, THAILAND

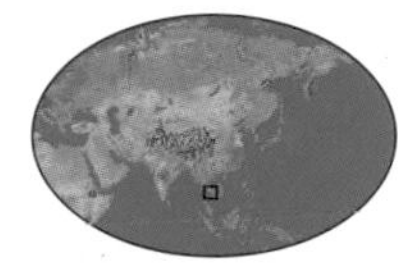

Height of Khao Lampi Waterfalls: 2,040 ft (622 m)

Age of rock: 60 to 140 million years

Sharing its national park designation with nearby Thai Muang Beach, the dome-shaped igneous granite mountain Khao Lampi sits inland in lush surroundings. Covered in tropical evergreen forest, this relatively small mountain has numerous streams with surprisingly powerful waterfalls running down its slopes. Khao Lampi's 60-to-140-million-year-old granite is rich in mica and quartz. The minerals are gradually eroded and transported seaward by streams and waterfalls toward Thai Muang Beach. The main Khao Lampi waterfall attracts many bathers to the pool at its base, but a difficult trail leads up to several quieter levels. Two more remote falls carry a smaller volume of water, but sit within pristine forest surroundings. Common species here include wild pig, common barking deer, and reticulated python, in addition to the endangered spiny terrapin and Siamese pit viper. Birds include the black-thighed falconet and the Oriental honey buzzard. "Flower watching" is also a popular pastime, given the exotic flora, such as banana flowers and orchids. AH

THAI MUANG BEACH

PHANG-NGA, THAILAND

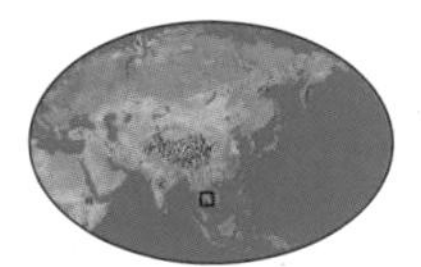

Length of Thai Muang Beach: 12.5 mi (20 km)

Turtle egg laying season: November to April

Thailand is famous for the white sands and calm, clear waters of its beaches, all located in the southern part of the country. A short distance north of the beaches on Phuket Island, running along the Andaman Coast of the mainland, lies Thai Muang beach. This impressive stretch of sand, lined with pine-like Casurina trees, has a gentle gradient and expansive width for half the year. However, during the monsoon season it is pummeled and reshaped by giant waves into a narrow, steeply sloped fringe, with strong offshore undertow currents. Flora here is based on mangrove forest on the eastern side, coconut trees, beach forest, and swamp forest, which thrives in the white sandy soil.

The beach is mainly protected as a breeding ground for two species of sea turtle: the leatherback and the olive ridley. When females come ashore to lay their eggs, park staff move the eggs to a nursery for safekeeping, as they are known as tasty culinary fare and will fetch a high price on the market, though illegal. The eggs incubate for about 60 days, and then the hatchlings are released into the sea. AH

KHAO LAK–LAM RU NATIONAL PARK

PHANG-NGA, THAILAND

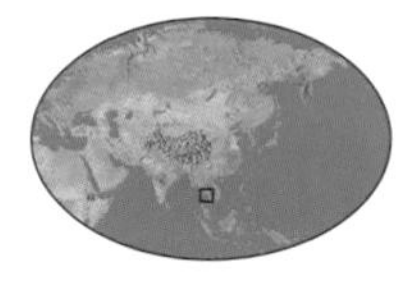

Elevation of Khao Lak-Lam Ru National Park: 3,533 ft (1,077 m)

Area of national park: 48 sq mi (125 sq km)

Habitat range: ocean to mountaintop forest

Khao Lak-Lam Ru National Park contains many stunning natural features. There are offshore islands and a boulder-strewn coast of secluded coves and long, sandy beaches, as well as the beautiful Laem Hin Chang ("Elephant Rock Peninsula").

Behind the coastal landscapes—which include mangrove forest, estuaries, and a freshwater lagoon—the land rises through plains, meadows, and rivers in forested valleys up to mountainous tropical forest. This broad range of habitats has caused some confusion as to whether to classify the park as terrestrial or marine. The area was established as a national park in 1989.

Once a rich tin-mining area, Khao Lak is now generally used as a pleasant stop-off on the way to more famous dive sites, with few people heading to the mountain itself. Inland, trails lead to various waterfalls: These include falls once used as a Buddhist retreat, the five-tiered Lam Ru, and the more distant Tansawan Falls. Tansawan's original name, "Bang-pisad," was changed to avoid inauspicious connections with monsters. AH

ERAWAN WATERFALLS

KANCHANABURI, THAILAND

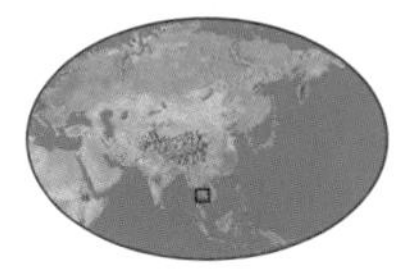

Height of Erawan Waterfalls: 500 ft (150 m)

Rock type: limestone

Dropping a total of 500 feet (150 meters), the seven tiers of Erawan Waterfalls form one of Thailand's most spectacular and popular attractions. They lie within the Erawan National Park and are reached via a 1.2 mile (2 kilometer) trek through tranquil jungle scenery. At the base of the falls, a giant ironwood tree grows, held sacred by pilgrims who draw water from the adjacent river for cures and blessings. Each subsequent level has deep, enticing pools, including a pool set within a natural amphitheater on the third tier. A lining of pale calcium carbonate—which has been dissolved and then redeposited from the surrounding limestone rock—gives the numerous pools a milky-blue hue. Mineral deposits also mirror the water flow over the many natural steps, giving the lower levels a smooth and softened outline. A trail reaches as far as the falls' sixth level; from here visitors must scramble up a cliff to reach the top. The seventh level of the falls resembles the head of the Erawan Elephant (the three-headed elephant from Hindu mythology), which gives the waterfall its name. AH

TA PROHM TEMPLE TREES

SIEM REAP, CAMBODIA

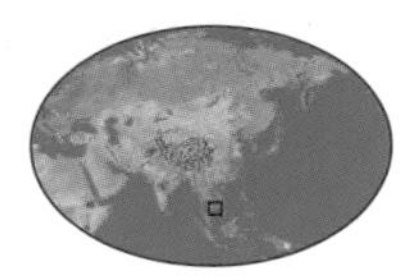

Area of Ta Prohm Temple: 173 acres (70 ha)

Vegetation: tropical lowland forest

Age of forest: 600 years

In 1860, the French explorer–naturalist Henri Mouhot stumbled upon the temple ruins of the old Khmer civilization hidden in the Cambodian jungle around Angkor. He was not the first European to visit these centuries-old remains, but his evocative accounts and drawings revealed to the outside world the true splendor of the magnificent architecture and stone carvings. In Mouhot's time, Angkor belonged to the forest. He wrote how "an exuberant vegetation has overgrown everything, galleries and towers, so that it is difficult to force a passage." A later visitor, Elie Lare, added that "with its millions of knotted limbs, the forest embraces the ruins with a violent love." Many others have since described how great trees enveloped the walls and statues with their powerful convoluted roots, slowly prying apart the massive stones and tearing down the walls in a colossal struggle measured in centuries. The process of restoring and protecting the cultural wonders of Angkor has meant relieving the forest of its grip, however, and most of the temples now stand surrounded but not dominated by trees.

Ta Prohm is a major temple complex that has deliberately been left in almost the same state as when Mouhot first struggled into its grounds. Built in the 12th century, this was once a great monastery spread out over a 173-acre (70 hectare) area, which, according to inscriptions, was served by 12,640 temple personnel. It was abandoned just 200 years later following the fall of the Khmer Empire, and the monsoon-fed forest quickly took over, as seeds of strangler figs, silk-cotton trees, and others rooted and matured on the very temple walls. Today, except for clearing a path for visitors and some structural strengthening to prevent further deterioration, Ta Prohm has been left in its "natural" state.

The great trees envelop the walls and statues with their powerful roots, slowly prying apart the massive stones and tearing down the walls in a colossal struggle measured in centuries. The combination of natural artistry and exquisite architecture is unparalleled.

Many tall, bulky trees balance on the crumbling but still solid temple walls, with their thick, entangled roots gradually forcing their way down through the cracks. The combination of natural artistry and exquisite architecture is unparalleled. When wandering through the confusingly jumbled grounds, with the massive stonework being both held together and torn apart by the sinuously beautiful trees, it is possible to share some of the wonder of those 19th-century explorers. AH

RIGHT *Trees dominate the Ta Prohm Temple.*

TONLE SAP LAKE

CAMBODIA

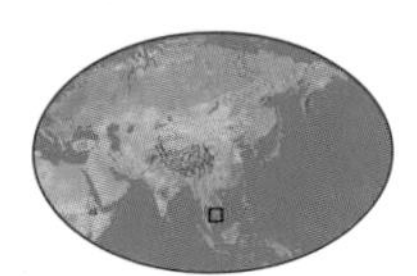

Area of Tonle Sap Lake: dry season: 1,004 sq mi (2,600 sq km); monsoon season: up to 5,018 sq mi (13,000 sq km)

Flow rate at Phnom Penh: 1,412,400 cu ft (40,000 cu m) per second

One of the most important inland fisheries in the world, Tonle Sap in Cambodia is the largest natural lake in southeast Asia. It came into existence less than 6,000 years ago during the most recent subsidence event of the Cambodian platform but now plays a vital role in buffering the Mekong River's floods. Every year, the monsoons swell the Mekong to a flow of 1,412,400 cubic feet (40,000 cubic meters) per second at Phnom Penh, which causes extensive flooding for up to seven months. The Tonle Sap River, a tributary of the Mekong, is a unique phenomenon in that it biannually reverses direction. As the waters rise, the 75-mile (120-kilometer) river that usually drains the great lake changes direction and begins filling it. The area of the lake quadruples in size, inundating surrounding forests and farmland. Eventually the rains subside and the Tonle Sap reverts to draining the lake. The influence of the seasonal ebb and flow of Tonle Sap stretches well beyond central Cambodia—this natural floodwater retention and release system reduces dry season saltwater intrusion in the Mekong Delta in Vietnam. AH

CARDAMOM AND ELEPHANT MOUNTAINS

CAMBODIA

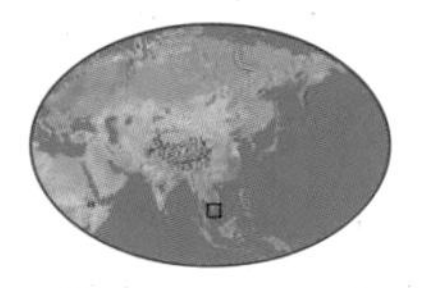

Area of Cardamom and Elephant Mountains: 3,860 sq mi (10,000 sq km)

Maximum height of Elephant Mountains: 5,810 ft (1,771 m)

Forest types: lowland and upland tropical forest

The mountains of southwestern Cambodia form a relatively unexplored, pristine wilderness rich in tigers, leopards, and rarer wildlife. The Elephant Mountains, reaching up to 5,810 feet (1,771 meters) at Mount Aural, are granite, while the adjoining Cardamoms are predominantly Mesozoic sandstone, with localized limestone and volcanic rock rich in gemstones. This variety of base rocks—and an annual rainfall that ranges from as much as 197 inches (500 centimeters) to just 79 inches (200 centimeters)—gives the area diversity in both landforms and forest types. Unusually for the region, the lowland forest between the two mountain ranges has remained reasonably intact. These isolated mountains were the last strongholds of the Khmer Rouge before their collapse in 1998. During their occupation, thousands of people were killed in the area but the troops living at the mountain's base were forbidden to shoot large animals. This vast area, originally preserved by Cambodia's relatively low population, has now gained official protection from the increasing threats of logging and hunting. AH

TAROKO GORGE

TAIWAN, CHINA

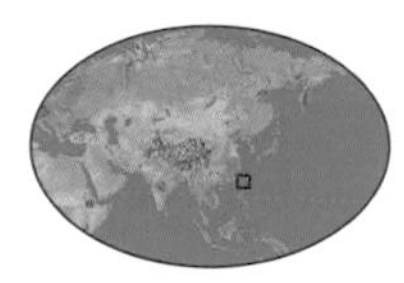

Age of Taroko Gorge: 70 million years

Highest peak: 12,140 ft (3,700 m)

The mountainous spine of the Formosan island of Taiwan is split by the spectacular, sheer-sided gorge of the Liwu River. A combination of uplifting and erosion over millions of years has formed dramatic cliffs. Taroko Gorge owes its existence to a concatenation of events beginning with the deposition of layer upon layer of calcareous sediments in the ocean and continuing with their lithification into limestone. Under intense pressure, subterranean heat, and immense time periods, the limestone deposits metamorphosed into marble before being forced up to form the mountains of the Taiwan ranges. Continued uplift of the hard marble strata—even as riverine erosion has taken its toll on the landscape—has carved out this incredible, ever-deepening, steep gorge, in places consisting of polished marble.

The area is protected as Taroko National Park, which supports many of Taiwan's indigenous wildlife species, including the Asiatic black bear, Formosan macaque, Formosan serow, and Formosan long-nosed tree squirrel, as well as various birds, including Mikado and Swinhoe's pheasants and the Formosan blue magpie. The park is blessed with stunning scenery and has many hiking trails, ranging from easy strolls to more difficult treks at higher altitudes, which take full advantage of these sublime vistas. **MBz**

PAGASANJAN FALLS

LUZON, PHILIPPINES

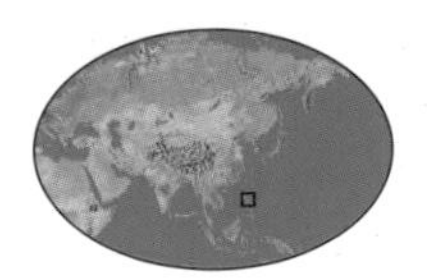

Depth of Pagasanjan Gorge: 300 ft (90 m)

Height of Pagasanjan Falls: 75 ft (23 m)

Rock type: volcanic

The dramatic Pagasanjan (also known as "Magdapio") Falls are southeast of Manila and reached after a journey through rice fields and coconut plantations. At Pagasanjan village, visitors transfer to either a traditional 20-foot- (6-meter-) long Filipino banca (dugout canoe made from the wood of the lawaan tree) or a fiberglass copy. At first, the river is slow and meandering, but eventually the valley becomes a steep-sided, 300-foot- (90-meter-) deep gorge, and the quiet waters become rock-strewn rapids. The journey upstream takes more than two hours, the bancas hauled manually by the local boatmen through the roaring water.

The cliffs on either side are made from ancient volcanic lahar, an aggregate of volcanic ash, mud, and boulders, and they are topped with lush forest. The gorge ends in a dead-end, box canyon where the Pagansanjan Falls tumble over the lip and drop into a deep, wide plunge pool. Bamboo rafts take the more courageous visitors right under the falls, where the full force of the drop can be felt. The return journey downstream through the series of 14 rapids is much quicker. Scenes from Francis Ford Copolla's *Apocalypse Now* were filmed here. **MB**

TAAL LAKE AND VOLCANO

LUZON, PHILIPPINES

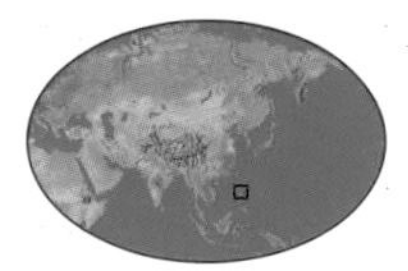

Area of Taal Lake: 94 sq mi (244 sq km)

Height of Volcano Island: 1,000 ft (300 m)

Height of Tagaytay Ridge: 2,000 ft (600 m)

Taal Lake (Lake Bonbon) is a lake within a volcano, and Taal Volcano is a volcano within a lake. The Taal Volcano rises out of an ancient water-filled caldera; one of the biggest volcanic depressions on Earth. The more recent volcano (Volcano Island) also has a crater lake and a smaller volcano with a lake (Yellow Lake) on its flanks. Most activity in historic times has come from Volcano Island, including at least 34 eruptions since 1572. The last major eruption was in 1977, although an event in 1965 was especially dangerous: A huge cloud of ash was ejected as high as 12 miles (20 kilometers) in the air and the blast destroyed palm trees 2.5 miles (4 kilometers) away. In 1754, lava buried four towns, and 1,344 lives were lost in an eruption in 1911. Today, the volcano is resting again, though since February 1999, 20-foot (6-meter) geysers of boiling water and mud have been spewing out close to a tourist trail.

Both crater and lake are best seen from a high vantage point, such as the north rim of the caldera, known as Tagaytay Ridge. The lake is home to a venomous sea snake that also breeds in freshwater and the world's only freshwater sardine. **MB**

CHOCOLATE HILLS

BOHOL SEA, PHILIPPINES

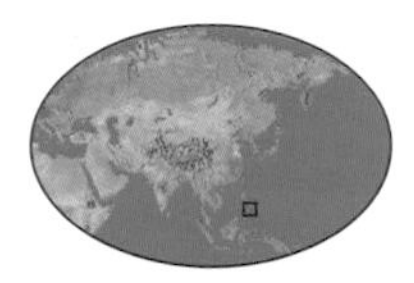

Height of Chocolate Hills: 100 to 300 ft (30 to 90 m)

Number of hills: 1,268

Rock type: limestone

On the Philippine island of Bohol there is an area of low, conical, and dome-shaped hills, known as the Chocolate Hills. They are covered with a carpet of rough grasses that are green when it rains but turn brown during the dry season, between February and May, hence the name. The mounds are packed together on a limestone plateau in central Bohol. Their origin is a mystery, for they have none of the features associated with limestone landscapes elsewhere in the world, such as caves or karstic formations. The accepted explanation at present is that they have been eroded by rain; however, local legends tell a different story. One tells how two angry giants hurled rocks at each other but eventually were reconciled, and the rocks were left where they fell. Another recounts the story of the giant Arago, who fell in love with and kidnapped a mortal named Aloya. Separated from her family, Aloya became desperately homesick and died. The Chocolate Hills have been eroded by the inconsolable Arago's torrents of tears. MB

PUERTO PRINCESA SUBTERRANEAN RIVER

PALAWAN, PHILIPPINES

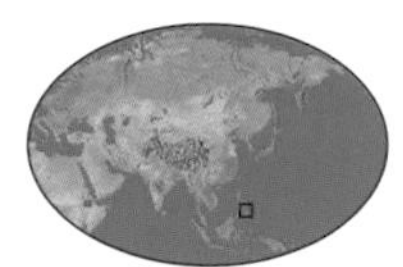

Length of Puerto Princesa subterranean river: 5 mi (8 km)

Height of cave chambers: 200 ft (60 m)

Area of national park: 78 sq mi (202 sq km)

A long and winding underground river meanders beneath the hills of Palawan in the southwest Philippines, emptying into the sea at a mangrove-fringed coast, with seagrass beds in shallow water and coral just offshore. A spectacular river cave has several large chambers, while aboveground the landscape is impressive, dominated by sharp karst ridges with limestone pinnacles.

The forest that covers swathes of hillside is among Asia's richest in tree diversity, and some of the Philippines' best remaining forest. It is home to species unique to this area, such as the Palawan porcupine, the Palawan peacock pheasant, and the Palawan tree shrew. Wildlife in the cave includes eight species of bat, while shrimps and fish inhabit the river. Although now within the Philippines archipelago, the area's fauna and flora is more closely allied to Borneo and southeast Asia, reflecting its geological history. The land that is now Palawan split from the Asian mainland around 32 million years ago, drifted toward the rest of the Philippines, and uplifted. When the sea level fell during the middle Pleistocene ice age, a landbridge formed between Borneo and Palawan, allowing many species to cross. MW

MOUNT KANLAON

NEGROS, PHILIPPINES

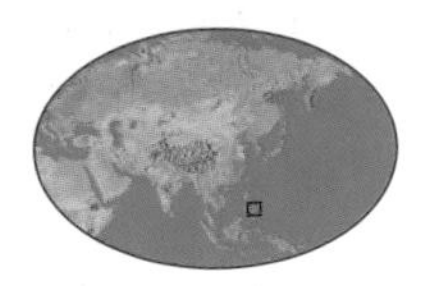

Height of Mt. Kanlaon: 8,040 ft (2,450 m)

Area of Mount Kanlaon National Park: 60,680 acres (24,557 ha)

Rock type: volcanic

Mount Kanlaon is one of the six most active volcanoes (and the tallest peak) in the central Philippines. Although there has not been a major eruption for 50 years, the mountain has become menacing again in recent years. In 2002, a cloud of steam rose 660 feet (200 meters) from the more active of its two craters, and the authorities are concerned for people's safety, for the volcano is prone to unpredictable pyroclastic flows—clouds of boiling hot gases that roll down the slopes at speeds of over 100 miles-per-hour (160 kilometers-per-hour). The slopes are deeply incised with rugged ravines and gorges and are cloaked in the last precious areas of virgin forest that are part of the Mount Kanlaon National Park.

The forests contain rare plants, such as pitcher plants, orchids, and the staghorn fern. Endangered yellow-backed sunbirds, blue-crowned racquet-tailed parrots, and Visayan tarictic hornbills occupy the canopy, and Visayan warty pigs and Negros spotted deer tramp the forest floor. Pythons and monitor lizards slither among the leaf litter, while the branches are home to dahoy pulay, a deadly venomous green tree snake. MB

CAGAYAN VALLEY CAVES

LUZON, PHILIPPINES

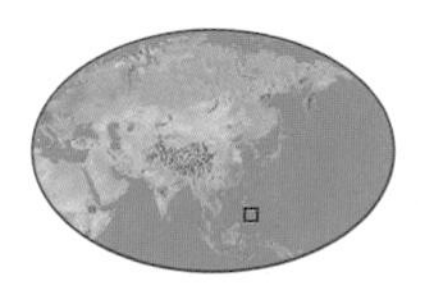

Depth of Jackpot Cave: 377 ft (115 m)

Length of Abbenditan Cave: 9.3 mi (15 km)

Area of Peñablanca Protected Landscape: 10,220 acres (4,136 ha)

The Cagayan River is the largest river in the Philippines and home to the rare and famous algae-eating "lurung" fish, its aromatic flesh a much sought-after gastronomic delight. But, more importantly, along its length are cave systems that are thought to have been home to the earliest humans in the region, between 100,000 and 400,000 years ago. The Aglipay Caves, located about 6 miles (10 kilometers) from Cabarroguis in Quirino Province, consist of 38 known chambers which contain six subterranean waterfalls, of which the source of water is still unknown. Callao Caves, located in the Peñablanca Protected Landscape in the Cagayan Province, has seven chambers, the first of which has been turned into a chapel. Jackpot Cave is the second-deepest cave in the Philippines, and the water-filled Abbenditan Cave is one of the longest. San Carlos Cave has the "Ice Cream Parlor" with a cluster of white stalagmites resembling scoops of vanilla ice cream, and Victoria Caves have seven adjacent cathedral-sized caverns. In all, it is estimated that there are over 350 cave systems in the area, of which only 75 have been documented. MB

BELOW *The wild forests that conceal Cagayan Valley Caves.*

TUBBATAHA REEFS

SULU SEA, PHILIPPINES

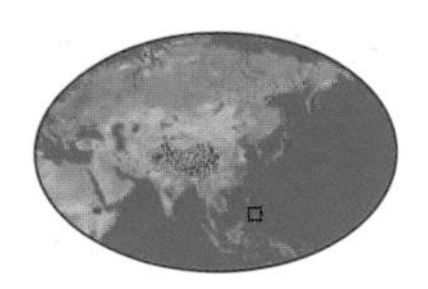

Area of Tubbataha National Marine Park: 128 sq mi (332 sq km)
Number of coral species: 300
Number of fish species: 379 species

The Tubbataha Reefs are formed of two atolls about 5 miles (8 kilometers) apart in the Sulu Sea. They are the largest atolls in the Philippines. Each is the eroded tip of a submerged, extinct volcano and has a classic atoll structure, with a shallow, sandy lagoon and sides that plunge steeply beneath the sea's surface to around 330 feet (100 meters).

Marine life abounds. There are butterflyfish, squirrelfish, sweetlips, and groupers, as well as blacktip and whitetip reef sharks, eagle and manta rays, and Napoleon wrasse. Dolphins feast on the fish shoals. Shellfish include giant clams. Turtles nest on the coral sand beaches, where there are also seabird colonies, with brown and red-footed boobies, brown noddies, and sooty and greater-crested terns.

Though a long way from land—lying 112 miles (181 kilometers) southeast of Puerto Princesa—the reefs were severely damaged by harmful fishing practices, including the use of dynamite and cyanide. They were proclaimed a national marine park in 1988; since then conservation efforts have reduced illegal and destructive fishing. MW

LAKE LANAO

MINDANAO, PHILIPPINES

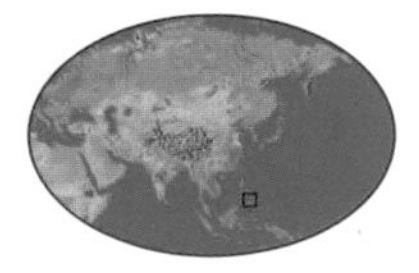

Area of Lake Lanao: 145 sq mi (375 sq km)
Maximum depth of lake: 367 ft (112 m)
Age: up to 20 million years

Lake Lanao is one of just 17 ancient lakes in the world estimated to be more than two million years old. It is also the second largest and deepest lake in the Philippines, sitting about 2,300 feet (700 meters) above sea level. Over the years, five small islands have erupted from its depths.

Local legend dictates that the lake was created by a group of "angels." Scientists however, claim that this ancient lake is the collapsed crater of an ancient volcano. Today, it is surrounded by farming villages with a backdrop of hills and mountains, most notable among which are the Signal and Arumpac Hills, Mount Mupo, and the appropriately named Sleeping Lady Mountain. The lake itself contains an unusual group of fish known as a species flock. This refers to 18 unique species of cyprinid fish living there today that have evolved from a single species—the spotted barb. The lake is also home to 41 species of freshwater crab that are unique to the region.

One of its outlets is the Agus River, one of the swiftest in the Philippines. Near the start of its 23-mile (37-kilometer) journey to the sea at Illana, its waters cascade over the Maria Cristina Falls, which were named after a Spanish queen and are now controlled by a hydroelectric system. MB

MOUNT APO

MINDANAO, PHILIPPINES

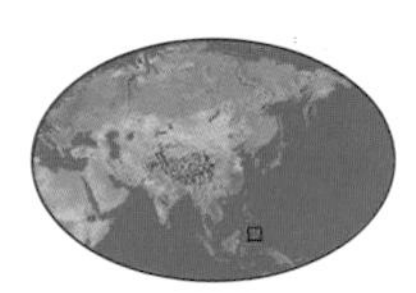

Height of Mt. Apo: 9,690 ft (2,954 m)

Area of Mount Apo National Park: 199,819 acres (80,864 ha)

At dawn on a clear morning, the summit of Mount Apo appears a symmetrical cone, high above Mindanao in the southern Philippines. But soon after dawn, mists swirl over the forests on Apo's slopes and rise to shroud the peak in cloud.

As its shape suggests, Apo is a volcano. Dormant for centuries, Apo still seeps heat, with sulfur-rich gases issuing from cracks around the summit to form bizarre yellow structures and hot springs creating streams that tumble down waterfalls. Apo is the highest mountain in the Philippines, and few plants other than bushes and grasses grow on the rugged and chilly peak area. Some 8,860 feet (2,700 meters) below, there is forest where moss covers the intertwining roots of small, tough trees. One species of moss grows up to 10 inches (25 centimeters) high; it is among the world's tallest mosses. The tree height and overall species diversity increase at lower altitudes, where there are impressive stands of tropical forest. The forest is the kingdom of one of the world's largest eagles, the endangered Philippine eagle. **MW**

KAMPUNG KUANTAN FIREFLIES

SELANGOR, MALAYSIA

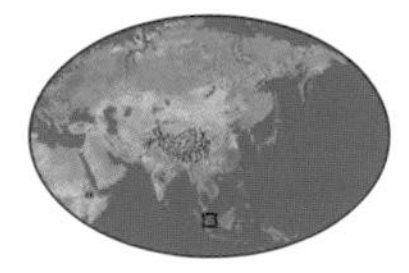

Vegetation: riverine forest dominated by mangroves

Distribution of fireflies: 0.6 mi (1 km) along Perak River

Along southern Malaysia's Perak River lies the small village of Kampung Kuantan. The riverine forests nearby host some of the world's largest concentrations of fireflies. Known locally as kelip-kelip, meaning "twinkle," the display is made by male beetles of the *Lampyridae* family, which are commonly known as fireflies and are just 0.25 inch (6 millimeters) long. The beetles settle on the riverside mangrove trees and flash a light organ on their abdomen to attract a mate. Each male flashes three times per minute. As thousands of males often congregate in the same tree and the display occurs over several thousand trees, the spectacle is extraordinary. At Tanjong Sari, 125 miles (200 kilometers) away, another congregation of male fireflies synchronize their flashes so that the forest appears to pulse with light. However, the Kampung Kuantan fireflies do not normally synchronize, resulting in a beautiful twinkling effect of green lights. The display is best viewed on a moonless night from a small wooden sampan on the river. The fireflies and their display trees are strictly protected by local custom and national law. **AB**

TAMAN NEGARA

PAHANG / KELANTAN / TERENGGANU, MALAYSIA

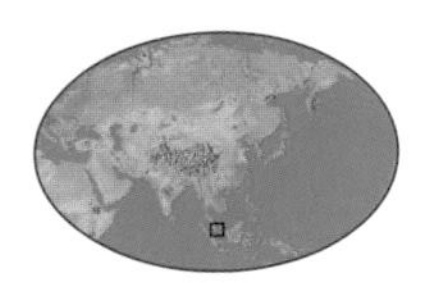

Area of Taman Negara: 1,677 sq mi (4,343 sq km)

Maximum height of dipterocarp tree: 250 ft (75 m)

Set in the heart of the Malaysian Peninsula, Taman Negara—which means simply "national park"—protects not only the world's oldest rainforest, but also Malaysia's oldest conservation area. But even though there has been land here for some 130 million years—much longer than the Amazon or Congo—and the area was not frozen during ice ages, the forest here is far from timeless. Its history stretches back to the dinosaur age, to the time when the first tiny ancestors of placental mammals were evolving and flowering plants were relative newcomers. The climate has cooled and warmed, and with the tropical heat and humidity predominating, evolution has run riot. The rainforest here ranks among the world's richest in its biodiversity. Dipterocarp forest dominates the skyline at lower elevations, and the "two-winged" seeds typical of this tree family carpet the forest floor. The tualung, southeast Asia's tallest tree, also grows here. Asian elephants, leopards, tapirs, Indochinese tigers, and sun bears roam the forest. Birds include the great argus—the male displaying fantastic wing and tail feathers. At higher elevations is cloud forest, with small trees, fan palms, and mounds of sphagnum moss. MW

BATU CAVES

SELANGOR, MALAYSIA

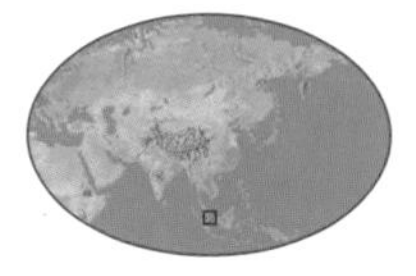

Alternative name: Gua Batu

Habitat: tropical limestone caves

Age: 60 to 100 million years

Though long known to local tribal people, this cave system was only made known to the Western world in 1878, when American explorer William Hornaby was taken to visit the caves. Only 8 miles (13 kilometers) from the national capital, Kuala Lumpur, Gua Batu consists of three major caves and several minor ones. All have been eroded from the local 400-million-year-old limestone. The largest cave is the 1,320-foot- (400-meter-) long "Temple" or "Cathedral," so named because the cave is considered to be sacred, and the ceiling is 330 feet (100 meters) high. Below this cave is Dark Cave, a 1.2-mile (2-kilometer) network of caves, grottos, and tunnels where the guano of five bat species fuels a complex invertebrate community of some 170 species. The caves are also home to a primitive trapdoor spider. Filled with splendid stalactites and stalagmites, the caves have major religious importance—they are the site of Thaipusam, a Hindu festival that takes place each January or February, and attracts over 800,000 people. The area has over 500 limestone hills and 700 caves, providing excellent climbing and caving opportunities. AB

KANCHING FALLS

SELANGOR, MALAYSIA

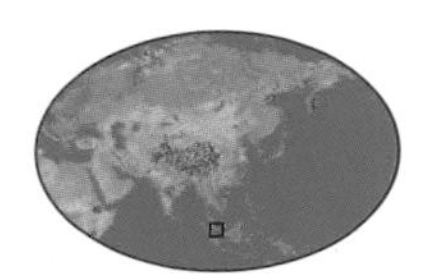

Area of Templar Park: 1,235 acres (500 ha)

Vegetation: lowland tropical rainforest

Rock type: limestone

Located just one hour by road from Malaysia's capital city, Kuala Lumpur, are Kanching Falls and Templer Park. Skirting the one-million-year-old limestone hill of Bukit Takun, the two parks offer picturesque scenery and a diversity of wildlife that is surprising given the proximity of Kuala Lumpur.

Kanching Park (also known as Hutan Lipur Kanching or Kanching Recreational Forest) has a series of seven waterfalls and a well-established trail system among the limestone hills and escarpments. The park is popular with weekend tourists, but the paths above the third waterfall are less busy. Above the final waterfall, Lata Bayas, is the forested plateau which is the source of the Kanching River, and where there is a well established (if much less frequented) trail system. The forest is a fine remnant of one that once covered Selanagor State, now one of Malaysia's most industrialized and populous states. There are jungle trails and several beautiful waterfalls cascading down the slopes and promontories of the naturally sculpted limestone. There are also a number of natural swimming holes. Among the wildlife that can be seen there are Kinloch's flying squirrel, bee-eaters, and buttonquails. **AB**

KINABALU

SABAH, MALAYSIAN BORNEO

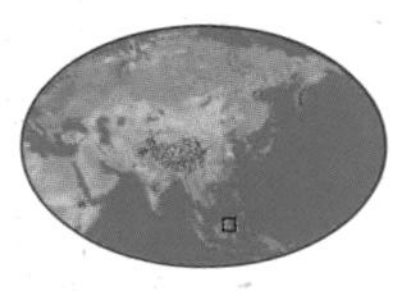

Height of Kinabalu: 13,455 ft (4,102 m)

Habitat: lowland tropical evergreen rainforest, cloud forest, alpine scrub, plant carpet

Area of national park: 1,677 sq mi (4,343 sq km)

Kinabalu, which is southeast Asia's highest mountain, is a flat-topped granite block that sits some 70 miles (113 kilometers) from Borneo's northern tip in the independent state of Sabah. Four altitudinally separate zones and a rich array of gullies, flatlands, and slopes provide such a diversity of habitats that Kinabalu has one of the richest and most varied floras in the world. Some 400 of the mountain's 4,000 plant species are unique to the area, including many of the 30 wild ginger, 750 orchid, 60 tree fern, and 15 pitcher plant species. The alpine zone has the world's largest carnivorous plant, the King Pitcher plant, which can hold 7 pints (3.3 liters) of liquid.

Lowland forests house the world's largest flower, the *Rafflesia.* Over 250 bird species have been recorded, including Kinabalu friendly warbler, chestnut-capped laughing thrush, Malaysian treepie, and, in the lower forests, the rhinoceros hornbill. Other animals include tree shrew, squirrel, clouded leopard, sun bear, slow loris, pangolin, the rare ferret-badger and lesser gymnure, and the Kinabalu angle-toed gecko. **AB**

RIGHT *The flat-topped granite peak of Kinabalu.*

GUA GOMANTONG

SABAH, MALAYSIAN BORNEO

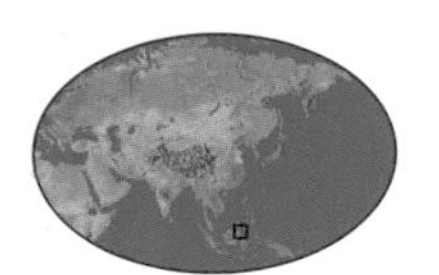

Age: 60 to 100 million years (limestone), 10 to 30 million years (caves)

Rock type: limestone

Vegetation: lowland forest

Gua Gomantong comprises two limestone cave complexes that are famed for their colonies of cave swiftlets. The birds nests are harvested biannually, in February to April and July to September, for the production of bird's nest soup.

The first cave, which is known as Simud Hitam and soars to 300 feet (90 meters), houses swift species that build nests mixed with feathers, the less valuable "black" nests. In the much less accessible second cave of Simud Putch, they produce the "white" nests. Made of almost pure cave swift saliva, the nests can fetch up to $680 per pound ($1,500 per kilogram). Once only accessible by river, the caves can now be reached by a seven-hour road trip from Sandakan. Permits are required to visit during the harvesting season and can be obtained from Sabah Forestry Department. Some two million bats spiral out of the caves at dusk. Bat hawks and cave-racer snakes feed on them as they pass, plucking them out of the air. Peregrine falcons do the same to the day-active swiftlets. The cave floors have a species-rich community of guano-loving invertebrates. To avoid damage to the fragile cave ecosystem, boardwalks have been constructed. **AB**

DANUM VALLEY

SABAH, MALAYSIAN BORNEO

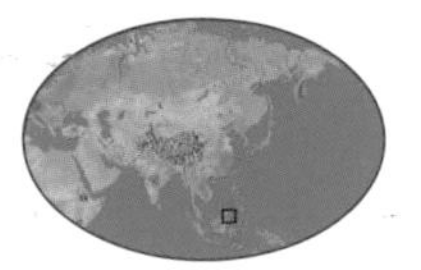

Area of Danum Valley: 170 sq mi (438 sq km)

Vegetation: lowland tropical rainforest

One of the last remaining areas of pristine rainforest in Borneo, and indeed southeast Asia, Danum Valley is renowned for its wildlife. Among the 124 species of mammal are the rare Sumatran rhino, bearded pig, mouse deer, Bornean bay cat, sun bear, clouded leopard, greater gymnure, pangolin, flying squirrel, and several species of primate, including tarsier, gibbon, and orangutan. More than 275 bird species can be found here, including Bulwer's pheasant, Bornean bristlehead, several pitas, barbet, broadbill, and hornbill. Ghost crayfish are common in the creeks.

The forest is dominated by several species of dipterocarp tree, but huge fig trees also occur. In the fruiting season these are a major attraction for local wildlife. Full moon hikes are spectacular, when many rare animals can be seen, such as the Malay weasel and sambar deer. The valley has a research center and forms part of the Danum Valley Conservation Area. There are over 30 miles (50 kilometers) of hiking trails, while a 90-foot- (27-meter-) high walkway provides an ideal vantage point from which to watch rainforest life. **AB**

RIGHT *The lush rainforest of Danum Valley.*

KINABATANGAN RIVER

SABAH, MALAYSIAN BORNEO

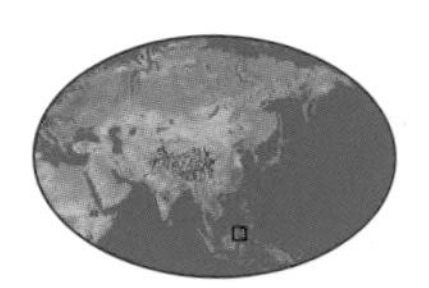

Length of Kinabatangan River channel: 350 mi (560 km)

Area of Kinabatangan River catchment: 6,500 sq mi (16,800 sq km)

The flood plain of the Kinabatangan River is among Borneo's richest areas for wildlife—and one of only two places in the world that is home to 10 species of primate. With up to 10 feet (3 meters) of rain a year, the river often floods, and stretches of rainforest give way first to swamp forest and then, toward the river mouth, to saltwater mangrove forests.

This is a major refuge for the proboscis monkey, a pot-bellied swamp denizen that leaps and wades across river channels and is named after the male's nose. The forests are also home to an important population of orangutans, as well as Bornean gibbons and maroon langurs, which, like the proboscis monkey, are only found in Borneo. The Borneo subspecies of the Asian elephant is also fairly common here. Threatened birds include Storm's stork. River life has been little studied, but includes freshwater sharks and rays. Estuarine (saltwater) crocodiles patrol the mangroves. There are caves holding millions of bats and swiftlets—these are Malaysia's main caves for harvesting birds' nests. Swathes of land have been deforested, mainly for oil palm plantations, but as floods can devastate plantations, owners have shown interest in forest conservation. Eco-tourism may encourage further conservation. MW

NIAH CAVES

SARAWAK, MALAYSIAN BORNEO

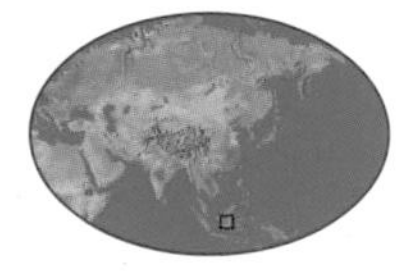

Height of Niah Caves: 250 ft (75 m)

Type of cave: karst

Niah Caves are quite simply enormous by any measure and are some of the most important caves in southeast Asia. Archeologists have discovered a long history of human habitation, including the skull of a young person dating back 40,000 years, and tools made from stone, bone, and iron, as well as numerous cave drawings. Today the caves are home to millions of bats as well as cave swiftlets.

The largest cave in the complex, the Great Cave, is an amphitheater with a floor area that could accommodate a total of three American ballparks. The caves were hewn by nature from a great block of limestone called Subis limestone, which covers an area of 6 square miles (16 square kilometers) and is located in northwestern Borneo. The caves are part of Sarawak's Niah National Park, which is covered by rich tropical forest and dominated by a great limestone peak called Gunung Subis. The park is accessible by longboat or vehicle from the town of Miri, and the caves are a two-mile (3.2-kilometer) trek along a planked walkway through the jungle. JK

CLEARWATER CAVE SYSTEM

SARAWAK, MALAYSIAN BORNEO

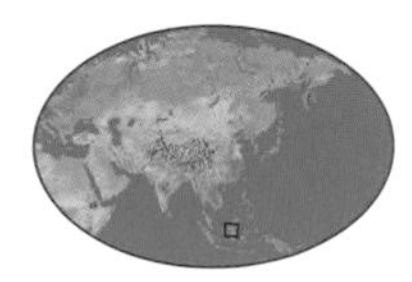

Length of Clearwater Cave System: 67 mi (108 km)

Height of Gunung Mulu: 7,800 ft (2,377 m)

Area of Gunung Mulu National Park: 130,622 acres (52,864 ha)

Of the vast cave and cavern honeycomb that underlies the limestone hills of Gunung Mulu National Park in northern Sarawak, over 183 miles (295 kilometers) of caves have been explored. A third of this total is located in the Clearwater Cave System, making it the world's eleventh longest cave system.

The Wind Cave is part of this system and is named for the notoriously cool breezes that blow through its narrower sections. The Wind Cave leads to the 32-mile- (51-kilometer-) long Clearwater Cave, the longest cave in southeast Asia. The lighting here illuminates such impressive stalactites and stalagmites that you certainly will not forget your visit.

Elsewhere, Mulu boasts superlatives. Deep in the hills is the world's largest natural chamber, the Sarawak Chamber. At 2,000 feet (600 meters) high and 50 feet (15 meters) wide, this chamber could house eight Boeing 747s lined up nose to tail. Another impressive cave, Deer Cave, is also in the area.

Mulu is also spectacular above ground. It is one of the world's richest areas for palms, with a total estimated at 108 species. Wildlife includes eight species of hornbill and a frog that breeds in the water reservoirs of pitcher plants. Four of the caves are accessible via guided tours. MW

MOUNT GUNUNG API & PINNACLES

SARAWAK, MALAYSIAN BORNEO

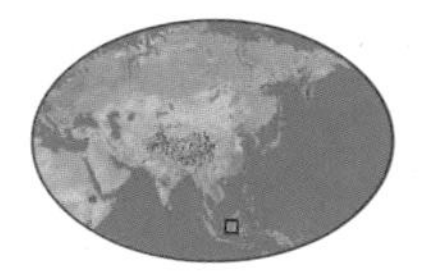

Height of pinnacles: 150 ft (45 m)

Length of jungle trek: 4.9 mi (7.8 km)

Length of steep climb: 3,300 ft (1,000 m)

The spectacular pinnacles at Mulu create a vast field of tall limestone needles that cling to the flanks of Gunung Api ("Fire Mountain"), a forest of silver blue stone encircled by dense green mountain rainforest. The pinnacles are formed by the virtually constant drenching of cloud forest rainfall that dissolves the joints within the limestone and erodes the rock into razor sharp pinnacles, with narrow crevices and intersecting ravines.

Gunung Api was first scaled only as recently as 1978, a testament to the treacherous, unforgiving terrain. Today, a steep, difficult trail ascends from the valley below up to the lowland rainforest into the eerie cloud forest. Here the path becomes a near vertical climb over sharp rocks covered in deep mats of moss. Orchids and carnivorous pitcher plants adorn the outcrops of limestone and stunted montane trees. At the pinnacles themselves, visitors often find themselves above the layer of clouds that mask the Melinau Gorge below. A visit to the pinnacles usually involves a return trip of three days, involving a boat ride, trek, and climb. DL

DEER CAVE / MULU

SARAWAK, MALAYSIAN BORNEO

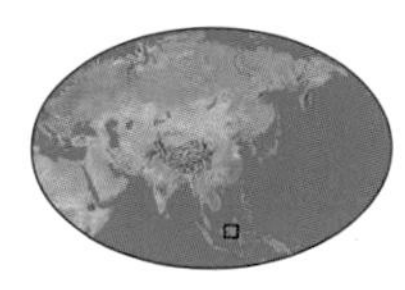

Length of Deer Cave: approx. 1.25 mi (2 km)

Width of cave: 574 ft (175 m)

Height of cave: 410 ft (125 m)

Cutting almost 1.25 miles (2 kilometers) through a rainforest-clad limestone peak, Borneo's Deer Cave is the largest cave passage in the world. It is so large, it could accommodate London's St. Paul's Cathedral five times over.

Deer Cave was named *Gua Payau* by the local Penan and Berawan people, who hunted sambur deer deep within the cavern for centuries. Although early visits were made by westerners in the 19th century, the cave was not formally mapped until 1961 by Dr. G. E. Wilford of the Borneo Geological Survey. Further exploration was later carried out by Britain's Royal Geographical Society expedition from 1977 to 1979.

This speleological wonder is the result of a unique combination of the Mulu area's very deep, easily eroded limestone and an extremely high annual tropical rainfall of 197 inches (5,000 millimeters). Fed by waterfalls that penetrate the roof high above, the small river that enters the cave from a surface sinkhole and meanders across the cave floor is gradually dissolving the bedrock and slowly increasing the spectacular dimensions of the passage. Not surprisingly, the cave is home to many different bat species. On most evenings, a vast cloud of bats—estimated to contain up to three million individual animals—pours out of the cave into the night sky in search of food from the rainforest. As the exodus begins, bat hawks leave their aeries high on the cliffs surrounding the entrance in order to soar and dive into the streams of winged mammals, certain of a meal.

Deer Cave is also home to hundreds of other species of cave creature. Spiders, millipedes, crickets, and cockroaches appear as a seething mass of life as they feast on the thousands of tons of bat guano. Accumulated over millennia, the decomposing guano gives rise to a visible mist of foul, stinking ammonia that is noticeable far from the cave entrance. High above, swiftlets build their saliva-cemented nests on precarious ledges. The cave racer snake, the only true cave reptile, has mastered the art of ambushing the bats, snatching them out of the air as they fly by in complete darkness and then constricting them in its tight coils.

The cave racer snake, the only true cave reptile, has mastered the art of ambushing bats, snatching them out of the air as they fly by in complete darkness and then constricting them in its tight coils.

Deer Cave is situated in the Gunung Mulu National Park, Sarawak's largest national park, on a jungle trail about 1.8 miles (3 kilometers) from the park headquarters. **DL**

SIPADAN ISLAND

MALAYSIAN BORNEO

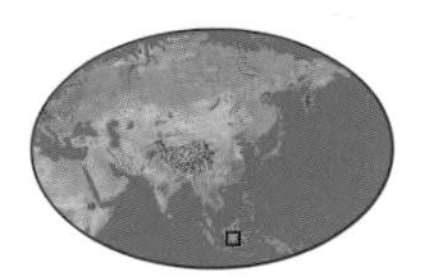

Type of island: volcanic atoll

Depth of sea: 2,000 ft (600 m)

Number of fish species: 3,000

The magical island of Sipadan, off the northern coast of Borneo, is situated in the heart of the world's richest seas and is one of the world's greatest diving sites. Sipadan is an oceanic island, formed by living coral growing on top of an extinct undersea volcano. The steep walls of the island plunge 2,000 feet (600 meters) into the deep blue sea and are a magnet for colorful sea life.

The "Drop Off" is a dive site on the northern side of the island—a few steps away from the sandy beach, the sea bottom appears to actually drop away. Within seconds of this drop, vast numbers of chevron barracuda, bigeye trevally, batfish, and a kaleidoscope selection of the 3,000 other species of dazzling fish that occur here surround divers.

The island is also a favorite nesting ground for green and hawksbill turtles. One of the most interesting dives here is the underwater limestone caverns, which have a labyrinth of tunnels and chambers, not to mention the skeletons of numerous turtles that have lost their way and drowned. JK

RIGHT *The clear, turquoise waters and lush, verdant forests of Sipadan Island.*

MOUNT MERAPI

JAWA TENGAH, INDONESIA

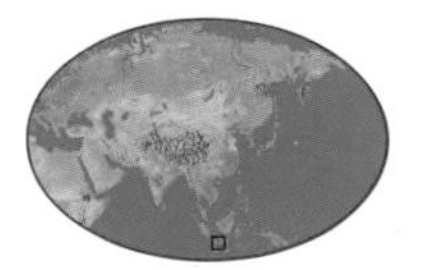

Height of Mt. Merapi: 9,551 ft (2,911 m)

Volcanic status: active

Most recent eruption: 1998

Mount Merapi, or "Mountain of Fire," is one of the most active and dangerous volcanoes in the world. Eruptions take place from Merapi every few years, loudly spewing ash and lava from inside the crater. Many eruptions are followed by pyroclastic flows (clouds of hot gas), which at 5,432°F (3,000°C) will melt or burn anything in their path. The local people call the heat cloud *wedus gembel*, meaning "curly sheep," because of its appearance.

In 1994, Merapi killed 66 people on the southwest slope. In spite of the significant dangers, 70,000 people live in the "forbidden zone" at the base of the volcano, to take advantage of the fertile farming land that the volcanic soils support. Each year on the Javanese New Year, villagers attempt to assuage the mountain by making Sedekah Gunung, a traditional offering.

The volcano can be climbed with the services of a guide. The strenuous ascent takes approximately six hours. In the chilly predawn, climbers can watch the lava and the stars, and at the summit can enjoy heavenly panoramic views taking in the peaks of other mountains and the blue ocean beyond. MM

GUNUNG RINJANI

LOMBOK ISLAND, INDONESIA

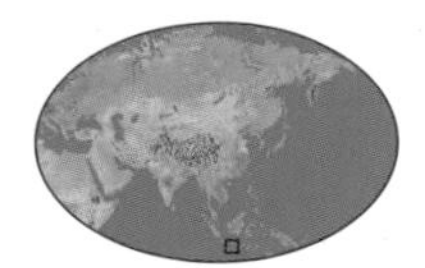

Height of Gunung Rinjani: 12,224 ft (3,726 m)

Caldera size: 5 mi (8km) from east to west; 3 mi (5km) north to south

Area of Segara Anak crater lake: 2,779 acres (1,125 ha)

Indonesia's second highest volcano, Gunung Rinjani, reigns over Wallacea—the world's foremost transition zone between biogeographical regions. Dominating the island of Lombok, it cools moist winds that blow from the northwest, forming rain clouds that nurture rainforest on the north of the volcano. A rain shadow leaves southeast Lombok parched. The interior of Gunung Rinjani is a great caldera, formed by the collapse of an even higher peak. Today, this holds a kidney shaped, 525-foot- (160-meter-) deep lake, Segara Anak, which is tinted blue-green by minerals.

A new cone, Batujai, erupted as recently as 1994, spewing lava into the lake, hurling boulders around the crater rim, and sprinkling ash over much of Lombok. Views from the crater rim are spectacular. Bali and other islands westward were linked to continental Asia during ice ages, so they share many plants and animals. But the strait between Bali and Lombok was too deep for a land bridge to form, and so east from Lombok the flora and fauna is increasingly Australasian. Lombok has relatively sparse animal life, with introduced species such as wild pig and Java deer. MW

LAKE TOBA

SUMATRA UTARA, INDONESIA

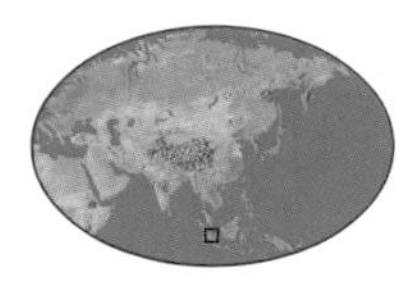

Dimensions of Lake Toba: 18 x 60 mi (30 x 100 km)

Depth of lake: 1,500 ft (450 m)

Ash produced 75,000 years ago: 672 cu mi (2,800 cu km)

The placid waters of Lake Toba—southeast Asia's largest lake—belie its location on the site of the greatest volcanic eruption of the past two million years. Around 75,000 years ago, a huge volcano exploded, with fissures expelling incandescent volcanic ash. This ash formed beds of tuff up to a third of a mile (0.5 kilometer) thick and has been found as far away as India.

The eruption was followed by a volcanic winter that lasted for six years, with global average temperatures lowered by up to 59°F (15°C). This event may have even changed the course of human evolution, with DNA evidence suggesting the entire human population was reduced to around 10,000 individuals in scattered groups. Locally, the eruption probably eradicated most life in a wide area. In its wake, the volcano collapsed to form the caldera that is today occupied by Lake Toba, one of the world's deepest crater lakes. Though there have been no eruptions in historic times, there are occasional earthquakes. The caldera dome is rising, forming an island larger than Singapore within the lake. MW

BELOW *Lake Toba seen from Samosir Island, Sumatra.*

KERINCI SEBLAT NATIONAL PARK

JAMBI / INDONESIA

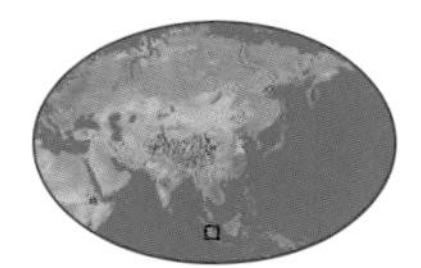

Area of Kerinci Seblat National Park: 3,397,488 acres (1,375,000 ha)

Height of Gunung Kerinci: 12,483 ft (3,805 m)

Lying south of the equator, and protecting habitats including lowland rainforest, cloud forest, and alpine meadows, Kerinci Seblat National Park is home to an astonishing diversity of plant and animal species. It is dominated by mountains, including the highest volcano in Indonesia, Gunung Kerinci. Though on an island (Sumatra), the park has many species that are similar to those in continental Asia, such as Asian elephants, tapirs, and clouded leopards, as well as Sumatran tigers and rhinoceroses. But many creatures have evolved to become distinct from their continental cousins, including the little known Sumatran rabbit.

The forest also nurtures gigantic flowers. *Rafflesia arnoldi* produces the world's largest single blooms—at three feet (0.9 meters) in diameter, they are the size of umbrellas. The world's tallest inflorescence, *Amorphophallus titanium,* is also here, reaching a height of 7 to 12 feet (2 to 3.7 meters) from the forest floor and weighing up to 170 pounds (77 kilograms). There are few tourist facilities, but if you visit, watch also for the orang pendek, an ape that walks upright and has often been reported, yet has so far eluded scientists. MW

ANAK KRAKATAU

BANTEN, INDONESIA

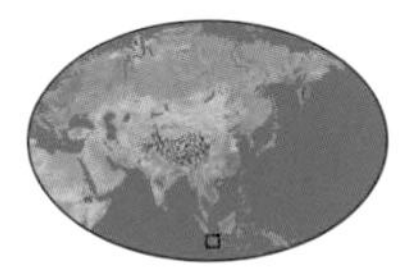

Most powerful explosion: volcanic explosion index (VEI) 6

Collapsed area: 8.9 sq mi (23 sq km)

Width of caldera: 3.7 mi (6 km)

Just as any child does, Anak Krakatau, "Child of Krakatau," is growing fast. The youngest of the four islands that comprise the Krakatau group, Anak Krakatau emerged above the Sunda Straits in the Indonesian seas in the early 1930s. Annual bouts of volcanic activity have raised the island to a considerable height, and it is now the second largest in the entire group.

The original Krakatau volcano famously blew apart in 1883, producing one of the most powerful explosions in recorded history and leaving behind fragments of the island. Much of the island collapsed underwater to depths of over 820 feet (250 meters). The immense tsunamis that followed the eruption and caldera collapse led to the deaths of at least 36,000 inhabitants on surrounding islands as far as 50 miles (80 kilometers) away.

Indonesia has more active volcanoes than any other country on Earth, the result of its location at the point where the Asian and Australian plates collide. Most of the islands lie along the arc traced by the two largest islands in the complex, Java and Sumatra. The Sunda Straits separate these two, marking a particularly active point in the fault line. It is only a matter of time before this child outgrows its parent. **NA**

GUNUNG GEDE–PANGRANGO

JAWA BARAT, INDONESIA

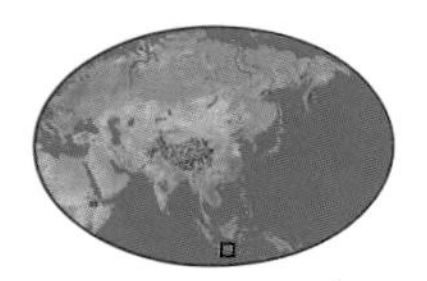

Area of Gunung Gede Pangrango National Park: 58 sq mi (152 sq km)

Height of Gunung Pangrango: 9,937 ft (3,029 m)

Height of Gunung Gede: 9,704 ft (2,958 m)

The mere sight of Java's twin peaks, Gede and Pangangro, suggests their violent origins, the lush diversity of the tropics, and even the island's former connections to north Asia and Europe. Both peaks are volcanoes. Pangrango is inactive, a stately cone covered in vegetation, with lava eroded into steep ravines that radiate from the summit. Gede, by contrast, is among the most active volcanoes on Java. There were major eruptions in 1747 and in 1840—when Gede hurled boulders that made craters over 13 feet (4 meters) deep. There have been 24 minor eruptions in the past 150 years.

Today, a 1,000-foot- (300-meter-) high, horseshoe-shaped cliff bounds the most recent crater, which hisses steam and sulfur. The lower reaches of the mountains hold remnants of the tropical rainforest that once covered much of Java. This is a key refuge for the Javan gibbon—also known as silvery gibbon for its soft gray fur—and Indonesia's national bird, the Javan hawk-eagle. Plants include over 200 species of orchid, tree ferns that grow to a height of 66 feet (20 meters), and, on the higher slopes, violets, primroses, buttercups, and Javan edelweiss. MW

GUNUNG AGUNG

BALI, INDONESIA

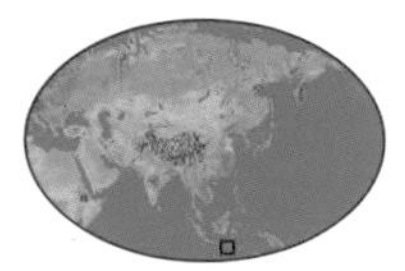

Height of Gunung Agung: 10,308 ft (3,142 m)

Last known eruption: 1963–64

The volcano Gunung Agung is the highest and most sacred mountain in Bali, and at 10,308 feet (3,142 meters), it towers over the eastern side of the island. At its base lies Bali's most important temple—Pura Besakih. Agung's phenomenal crater is seemingly bottomless and measures 1,650 feet (500 meters) across, occasionally venting smoke and steam. The last eruption took place in 1963, killing 2,000 people and making 100,000 homeless. Crops were destroyed all over the island, leading to famine. Agung is perhaps the most climbed high peak in Indonesia. The climb is non-technical, although is not suitable for inexperienced hikers, who are advised to employ a local guide to navigate the climb's early trails. The ascent takes four to six hours, beginning with a strenuous hike through very humid jungle, which eventually gives way to loose volcanic rock. To attain the summit, climbers must brave the precipitous summit ridge and biting winds. From the top, Mount Rinjiani on the neighboring island of Lombok is often visible, although clouds can obscure a view of the island below. MM

RIGHT *The sacred Gunung Agung towers over eastern Bali.*

KAWAH IJEN

JAWA TIMUR, INDONESIA

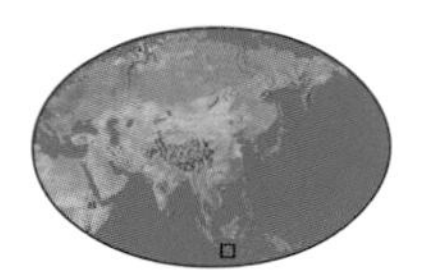

Elevation of crater lake: 7,710 ft (2,350 m)

Volume of crater lake: 1.3 billion cu ft (36 million cu m)

Area of Alas Purwo National Park: 107,290 acres (43,420 ha)

The stark landscape and swirling, steamy sulfurous gases around the rim of Kawah Ijen, the Lone Crater, are stark reminders of Java's position on the Pacific Ring of Fire. Ijen is one of a cluster of volcanoes that have emerged from a 12-mile- (20-kilometer-) wide caldera, itself the remnant of a giant, collapsed volcano, and is among 18 Javan volcanoes that have erupted since 1900. A lake of mineral-rich, turquoise water occupies the crater, which is around 1 mile (1.6 kilometers) across at its widest point. The water is warmed up to 108°F (42°C) by the hot rock beneath but occasionally becomes much hotter in steam-driven eruptions that have flung mud and sulfur to 2,300 feet (700 meters) above the crater rim.

Despite the toxic fumes and the threat of eruptions, men walk down into the crater to harvest newly deposited sulfur, carrying it in baskets slung over their shoulders. Kawah Ijen lies within the Alas Purwo National Park, which is home to animals including leopards, Asian wild dogs, and banteng, relatively slender wild oxen with white lower legs. The climb up the outer crater rim takes about an hour; the 20-minute descent into the crater is risky. MW

MOUNT BROMO & TENGGER HIGHLANDS

JAWA TIMUR, INDONESIA

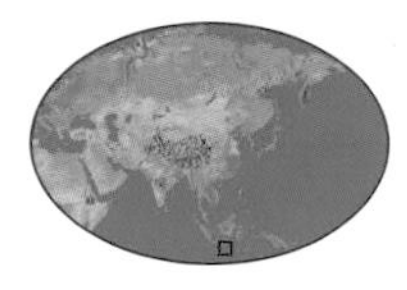

Area of Bromo Tengger Semeru National Park: 309 sq mi (800 sq km)

Height range of national park: 3,300 to 12,060 ft (1,000 to 3,676 m)

Diameter of Mt. Tengger: 6.2 mi (10 km)

Located in east Java, this region has five volcanoes, including 7,848-foot (2,392-meter) Mount Bromo, which seldom erupts, extinct Mount Batok, and Mount Semeru, a highly active volcano. Together they form the geological highlights of Bromo Tengger Semeru National Park.

The park includes a large area of volcanic sand, the Laut Pasir, caused by a past eruption of Mount Tengger. Four volcanoes, including Mount Batok and Mount Bromo, lie within Mount Tengger's caldera. Batok and Bromo are above the tree line, but the slopes of Semeru are clothed in forest. The area has a rich orchid flora of 157 species. There are over 400 other plant species in the park. Mammals include Timor and barking deer, silver leaf monkey, and wild pig. Birds include hornbill, crested serpent eagle, forest pittas, and a variety of waterfowl on the lakes in the extinct calderas of Ranu Pani and Ranu Regulo. AB

BELOW *The five great volcanoes of Bromo Tengger Semeru.*

KOMODO ISLAND

NUSA TENGGARA TIMUR, INDONESIA

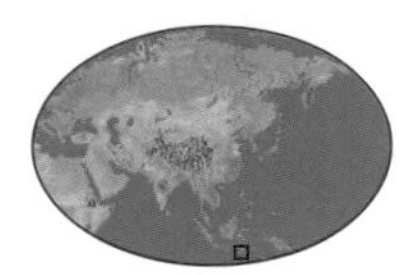

Area of Komodo Island: 108 sq mi (280 sq km)

Rock type: rhyolite porphyry

Komodo dragon population: approx. 2,500

Lying 300 miles (483 kilometers) east of Bali and forming part of the Republic of Indonesia, Komodo Island nestles between the large islands of Flores and Sumbawa. It was created in the great volcanic uplift that formed the major islands of Java, Bali, and Sumatra, along with the small islands of Padar and Rinca. Though sparsely populated, this hot, barren landscape supports an astonishing assortment of wildlife, including one of the greatest varieties of venomous snake in the world (which may go some way to explaining the lack of human inhabitants), masses of brightly colored birds—among them yellow-crested cockatoos and flightless megapode birds—as well as deer, wild boar, and the massive water buffalo.

It is most famous, however, for being the home of the fierce Komodo dragon, the largest lizard in the world. These reptilian giants measure 10 to 13 feet (3 to 4 meters) from nose to tail, can run as fast as a dog, and feed on large, live prey, such as monkeys and pigs. If a bite from their powerful jaws does not kill their quarry, the deadly bacteria in their saliva eventually will. While Komodo dragons are ferocious predators, there are no confirmed reports of one ever having killed a human being. In fact, they feed mostly on carrion.

The only way to reach Komodo Island is by boat. Nonetheless, every year thousands of visitors travel here to see the giant creatures. Some opt to observe the daily "feeding shows" when park rangers tempt the dragons with freshly killed meat, while others prefer to accompany the rangers into the hills on dragon-spotting hikes. All stay in simple guesthouses in the village of Komodo, the only center of human population on the island. The American explorer Douglas Burden, who visited the island in 1926, commented: "With its fantastic skyline, its sentinel palms, its volcanic chimneys bared to the stars, it was a fitting abode for the great saurians."

The island is also a popular diving destination, in part because of its unique climate, which boasts both tropical and temperate marine habitats. To the north, the waters are warm and home to many species of tropical fish, while the southern waters are cooled by icy ocean currents that travel up from Antarctica via the Indian Ocean. MM

Though sparsely populated, this hot, barren landscape supports an astonishing assortment of wildlife, including one of the greatest varieties of venomous snake in the world, masses of brightly colored birds, and the fierce Komodo dragon, the largest lizard in the world.

RIGHT *The rugged hills of Komodo, home to the world's largest lizard.*

VI

AUSTRALIA & OCEANIA

Australia is as plentiful and varied as the seas enclosing it—vibrant reef-worlds teem with marine life, and scorched outbacks boast legends as fluid as the mysteries, pinnacles, and rocks they tap. The Great Barrier Reef stretches to Papua New Guinea, while New Zealand's Tongariro volcano towers above waves of rugged hills. Waters spread as far as Hawaii's Mount Waialeale, all the while weaving through Oceania's full school of islands.

LEFT *Hardy and Hook reefs, part of Australia's giant Great Barrier Reef.*

WAIMEA CANYON

KAUAI, HAWAIIAN ISLANDS

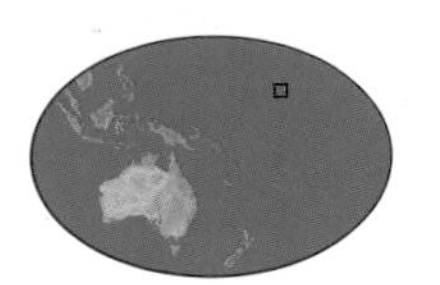

Length of Waimea Canyon: 10 mi (16 km)

Width of canyon: 1 mi (1.6 km)

Depth of canyon: 3,600 ft (1,097 m)

Carved over thousands of years from rivers and floods that flowed from the summit of Mount Waialeale, Waimea is the largest canyon in the Pacific. Mark Twain called it the "Grand Canyon of the Pacific" and although not as big, it is just as spectacular. The canyon is situated on Kauai Island and protected within the boundaries of Koke'e State Park. It was once part of an ancient volcano, but part of its flank collapsed thus enabling the Waimea River to cut through a weakness in the layers of volcanic rock. Over five million years the eroding waters have exposed different colored layers of lava, but the rocks have an overall tinge of red from the presence of iron. The red, green, blue, gray, and purple hues of the chasm highlight the canyon's dramatic crags, hills, and gorges. There are tracts of Kauai's rare upland forest of koa and red-blossoming ohia lehua trees, along with roses and siennas. Throughout the canyon, numerous lookout points provide excellent vantage points over the area; one of the best is the Kalalau lookout, with its panoramic view of the Kalalau Valley and Na Pali coast. For walking enthusiasts there are 45 miles (72 kilometers) of trails through the canyon and nearby Alakai Swamp. **MB**

HAWAIIAN WATERFALLS

HAWAIIAN ISLANDS

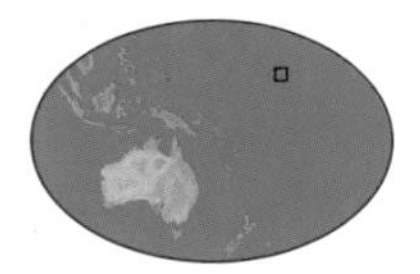

Number of waterfalls: over 24 major falls, over 200 smaller cascades

Longest cascade (Kahiwa Falls, Molokai): 1,750 ft (533 m)

Longest sheer drop (Akaka Falls, Hawaii): 442 ft (135 m)

The Hawaiian Islands stand squarely in the path of the wet, northeasterly trade winds that shed their moisture-laden burden on the land with unrivaled vigor—Mount Waialeale, on the island of Kauai, is the wettest place on Earth. This seasonal deluge, combined with a steep, porous volcanic landscape, bestows the islands with some of the most beautiful and dramatic waterfalls to be found anywhere. The flow of water perpetuates itself: As water carves out more of the land, waterfalls plunge ever farther. Waimea Falls are located in the beautiful 1,800-acre (728-hectare) Waimea Falls Park. Rainbow Falls in Wailuku River State Park on the island of Hawaii is one of the most impressive in the state. Most mornings dazzling rainbows flash from the rising spray—the falls are just 80 feet (24 meters) high, but their average daily flow is the greatest of any on the islands. Near Honolulu, Kapena Falls frequently fail to reach the bottom of the cliff as blustery trade winds catch the waters and hurl them upward. Some falls do not rely on seasonal rains. On Maui, Hanawi Falls, one of Hawaii's most beautiful, are fed by water from subterranean chambers and flow even in the driest weather. **DH**

MAUNA KEA

HAWAII, HAWAIIAN ISLANDS

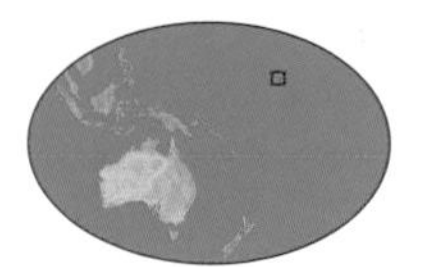

Area of Mauna Kea: 920 sq mi (2,383 sq km)

Age: about one million years

Eruptions: at least seven, between 6,000 and 4,500 years ago

Many people would find the notion of snow falling in Hawaii paradoxical, but every winter snow covers the summit of Mauna Kea, the tallest volcano on Hawaii, the main island. In fact, below the mountain's 13,795-foot (4,205-meter) summit, scientists have found glacial moraines from recent ice ages, even though they are absent from nearby Mauna Loa, just 114 feet (35 meters) lower. Mauna Kea began erupting on the seafloor some 800,000 years ago. Today, if you include the height of the volcano below the ocean surface, it is 5.6 miles (9 kilometers) high.

About 300,000 years ago, Mauna Kea produced the tall cinder cones and lava flows that cover most of its present-day surface, save for the summit, which is blanketed in glacial till. The volcano last erupted 4,500 years ago, but its quiescent periods between eruptions are long compared to those of its more active neighbors, Hualalai and Kilauea. For now it lies dormant, but researchers think it likely that it will erupt again, possibly as a consequence of a number of earthquakes. Mauna Kea's high, dry, pollution-free atmosphere also provides ideal viewing conditions for the world's largest astronomical observatory. DH

HALEAKALA CRATER

MAUI, HAWAIIAN ISLANDS

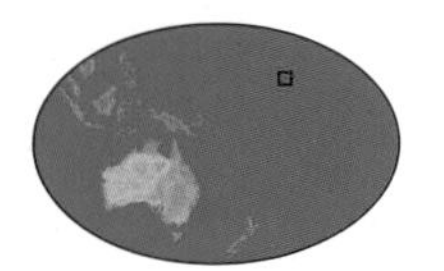

Area of national park: 46 sq mi (119 sq km)

Established as a national park: 1961

Designated a Biosphere Reserve: 1980

Deep beneath the Pacific Ocean, a part of the planet is grinding inexorably northwest. As it does, it is laying bare volcanic hot spots, which act as open wounds going straight to Earth's core. Magma bleeds from these lesions, welling upward, forming layer upon layer, until it eventually breaks the surface as a volcanic island. Along the Pacific Plate the process has spawned a long chain of islands extending from Hawaii toward Japan, one of the most notable of which is Maui.

In the Hawaiian group, Maui originally began as two separate volcanoes—Mauna Kahalawai and Haleakala; these gradually merged together over time. Haleakala, the larger of the two volcanoes, rises for some 30,000 feet (9,000 meters) from the ocean floor to 12,000 feet (3,600 meters) above sea level. Taking into account its entire height, it is one of the largest mountains in the world.

Maui's "sleeping" giant last erupted around 1790 C.E., when two small lava flows reached its southwest coast. Today the activity of the Pacific Plate has moved on, and Haleakala is now dormant and destined to become extinct, though tremors and earthquakes are still recorded in the area. The volcano is cool and studded with old cones traced with past flows of red, yellow, gray, and black lava, ash, and cinder courses. Eons of rains have created large amphitheaters near the summit and further cleft the flanks of the mountain with deep erosional scars. Those rains nourish lush forests on the windward slopes—the Kipahulu Valley is one of the most intact rainforest ecosystems in Hawaii.

Maui's "sleeping" giant last erupted around 1790 C.E. Today the activity of the Pacific Plate has moved on, and Haleakala is now dormant and destined to become extinct, though tremors and earthquakes are still recorded in the area.

At higher elevations, a vast native loa and 'oh'i'a rainforest still thrives. It is here that the endangered Maui nukupu'u, Maui parrotbill, and other rare native birds continue to be found. The rainforest oozes down the slope like lava for 35 miles (56 kilometers) until it reaches the sea. Wilderness trails wind their way alongside 400-foot (120-meter) waterfalls, tropical streams, and turquoise pools. On the leeward slopes, dry forest persists, despite browsing pests and fire, giving way at altitude to alpine shrubland, which is the realm of the rare Hawaiian goose, the nene. Only the hardiest of shrubs survive above this height, where the rain is absorbed by the dry, porous rock. In summer the rain persists every day, and in winter every night. DH

RIGHT *Hawaii's two sleeping volcanoes are slowly merging.*

MOUNT WAIALEALE

KAUAI, HAWAIIAN ISLANDS

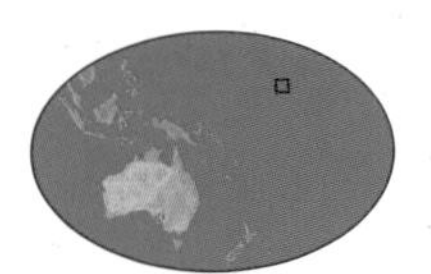

Height of Mt. Waialeale: 5,148 ft (1,569 m)

Average annual rainfall: 460 in (1,168 cm)

Kauai, the oldest of the Hawaiian Islands, was born eight million years ago as a single volcano that rose out of the ocean. The cone of Mount Waialeale is nestled into the island's central massif and bears mute testimony to the upheaval of its birth. Waialeale is one of the world's wettest mountains. An annual average of 460 inches (1,168 centimeters) of rain falls on its flanks. In 1982, a record rainfall of 666 inches (1,692 centimeters) fell at its peak, while 10 inches (25 centimeters) fell at the coast.

Over time, this incessant deluge has scoured some spectacular features, such as the gorge at Waimea, Hawaii's "Little Grand Canyon." Waialeale's catchment also feeds a labyrinth of streams, tumbling to the lowlands over numerous waterfalls, to swell the only navigable rivers in Hawaii: the Waimea, Wailua, Makaweli, and Hanapepe. Only well-adapted plants such as mosses, sedges, and grasses thrive on this high-altitude, sunlight-deprived, wet, and windy mountain. DH

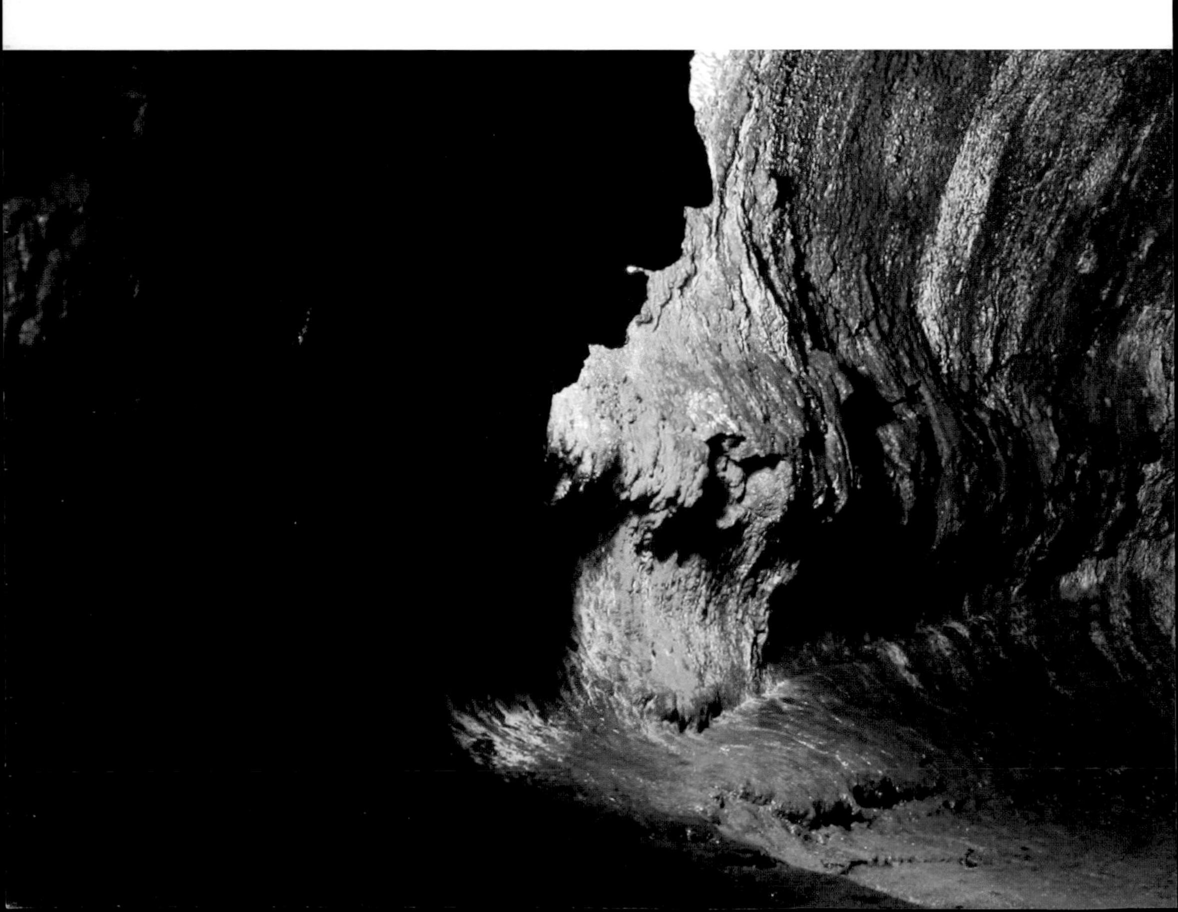

LAVA TUBES

HAWAII, HAWAIIAN ISLANDS

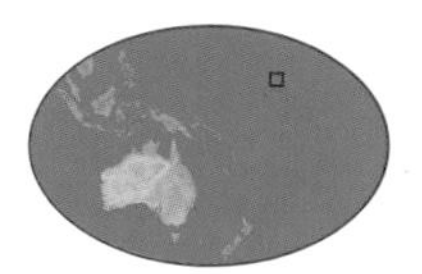

World's longest lava tube: Kazumura Cave

Length of Kazumura Cave: 36.9 mi (59.3 km)

Drop of Kazumura Cave: 3,605 ft (1,099 m)

Lava tubes form when molten lava flowing from volcanoes cools in different stages. While the surface layer solidifies, the hotter lava continues to flow, creating tubes beneath the surface.

Hawaii boasts the longest lava tubes in the world. When lava stops flowing from its source, the still molten contents of the tube will often drain downhill, exiting the tube and leaving an open cave at the end. Thurston Cave, in Hawaii Volcanoes National Park, is one such example, and the only navigable lava tube in the park. The lava tube was formed about 300 to 500 years ago, when a large vent called the Ai-laau Shield erupted on the east side of Kilauea's summit. Stalagmites and stalactites often form in these tubes, although some are made of solidified lava, and water collects to form underground pools. Most tubes are only a short distance beneath the surface, so tree roots often break through the roof. DH

BELOW *Thurston Cave in Hawaii Volcanoes National Park.*

MOUNT KILAUEA

HAWAII, HAWAIIAN ISLANDS

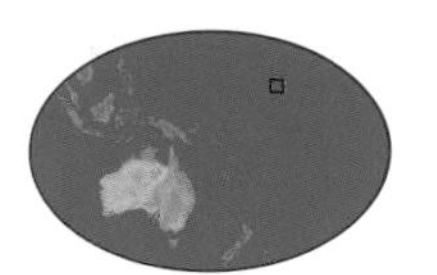

Lava output of Mt. Kilauea: 130,000 gal (492,104 l) per minute

Sulfur dioxide emissions: 2,500 tons per day

There are few better places to witness the powerful—and still active—forces that formed (and continue to form) our planet than Mount Kilauea, the world's most active volcano. In the longest-running eruption of its 200-year history, this fiery crown jewel of Hawaii Volcanoes National Park burst dramatically back into life in January 1983. It began spewing between 390,400 and 790,000 cubic yards (300,000 and 604,000 cubic meters) of lava each day from a fissure on its southeastern face called Pu'u O'o. In the decades since, Pu'u O'o's lava flows have already buried more than 39 square miles (101 square kilometers) of Kilauea's southern flank and added another 1 square mile (2.59 square kilometers) of land to the island. But Pu'u O'o has created and destroyed in equal measure. Tens of thousands of archeological features now lie buried, including temple sites, petroglyph fields, and old villages. On their relentless seaward journey, rivers of lava have consumed more than 180 homes, a church, a community center, and a power and phone network. They have also torched more than 25 square miles (65 square kilometers) of lowland rainforest, destroying the homes of rare hawks and honeycreepers, happyface spiders, and hoary bats. Despite its already considerable impact on the environment, Kilauea is considered a young volcano still in its shield-building stage. It currently stands 4,190 feet (1,277 meters) high, although most of its bulk is below sea level.

The present caldera (crater) formed around 1790 C.E. and cradles a pit crater called Halemaumau. Two rift zones extend east and southwest. However after many years of constant eruptions from the Kupaianaha vent, Kilauea is no longer flaunting its fire. Molten rivers of lava stopped flowing from the volcano to the sea in late 1991, and for now volcanic activity is limited mostly to inaccessible areas. According to native Hawaiians, volcanic eruptions are the outbursts of Pele, the tempestuous goddess of volcanoes. In her frequent moments of anger, Pele causes earthquakes by stamping her feet, and starts volcanic eruptions with her magic stick.

In 1980, UNESCO designated Hawaii Volcanoes National Park an international Biosphere Reserve and in 1982 it became a World Heritage site. DH

According to native Hawaiians, volcanic eruptions are the outbursts of Pele, the tempestuous goddess of volcanoes. In her frequent moments of anger, Pele causes earthquakes by stamping her feet, and starts volcanic eruptions with her magic stick.

RIGHT *Molten lava courses down Mount Kilauea's peak.*

MARIANA TRENCH

MICRONESIA, PACIFIC OCEAN

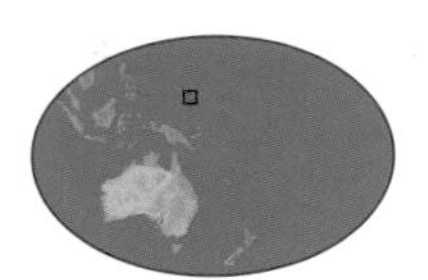

Length of Mariana Trench: 1,580 mi (2,550 km)

Width of trench: 43 mi (69 km)

Depth of Challenger Deep: 36,197 ft (11,033 m)

East of the Mariana Islands, near Japan, lies the world's deepest ocean trench. Formed as the Pacific tectonic plate was subducted under the Philippine plate, the deepest part of the depression—known as the Challenger Deep—is an estimated 36,197 feet (11,033 meters). Mount Everest could sit at the bottom of this trench with almost 1.5 miles (2.5 kilometers) of water to spare. Despite the intense cold, complete absence of sunlight, and crushing pressure, the Mariana Trench is home to a remarkable variety of life. Indeed, during the first descent of the Challenger Deep in 1960, researchers were surprised to discover a fish resembling a sole. Later studies revealed other fish species, including the bioluminescent anglerfish and crustaceans such as shrimp and crabs. Hydrothermal vents, places where superheated, mineral-rich seawater spews upward in columns of black smoke, are particular biological hot spots. They support a rich oasis of microorganisms that form the base of a complex but largely unknown food web. In the Mariana Trench, many species have a lifespan in excess of 100 years. NA

PALAU

MICRONESIA, PACIFIC OCEAN

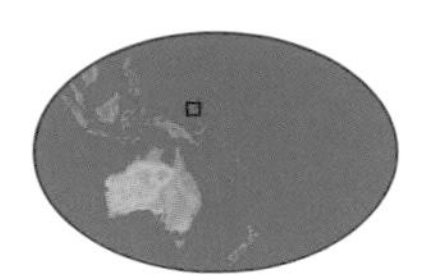

Land area of Palau: 177 sq mi (458 sq km)

Highest point (Mt. Ngerchelchuus): 794 ft (242 m)

Palau is a string of 343 islands in six groups and forms the westernmost archipelago in the Caroline chain, southeast of the Philippines. Thrust from the ocean more than 20 million years ago, ancient living reefs now form limestone islands, pockmarked with myriad fresh and saltwater lakes. Palau was named "the finest underwater wonder of the world" by an international organization of conservationists, divers, and marine scientists.

Its surrounding waters contain coral reefs, blue holes, hidden caves and tunnels, and over 60 vertical drop-offs. The salt lakes—protected by high bluffs and replenished only by seawater entering the narrow cracks and seeps—have become miniature marine ecosystems, each with its own unique physical, chemical, and biological processes. Some lakes support vast populations of jellyfish, which move daily from shore to shore, following the sun and phytoplankton. Landlocked and isolated from one another, these populations of jellyfish have evolved into separate species. DH

BELOW *Limestone islands of Palau dot surrounding salt lakes.*

NEW GUINEA

INDONESIA / PAPUA NEW GUINEA

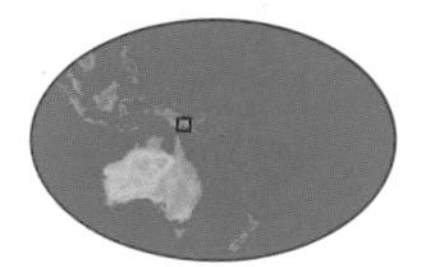

Area of New Guinea: 178,800 sq mi (463,000 sq km)

Terrain: mountains with coastal lowlands and rolling foothills

New Guinea, located north of Australia, is the world's second largest island. It encompasses the nation of Papua New Guinea on one side and the Indonesian provinces of Papua and West Irian Jaya on the other. The island is a unique combination of different environments, from high mountains to deep valleys and tropical forests to sandy shores. It also has far-reaching flat landscapes inundated with water. By the end of the rainy season in May, swollen and muddied river channels snake their way through savannas and dense forests.

The island has immense ecological value, with 11,000 plant species, nearly 600 unique bird species, over 400 amphibians, and 455 butterfly species—including the world's largest butterfly, the Queen Alexandra Birdwing. The real treat, however, are the birds, including giant flightless cassowaries, kokomos (hornbills), cockatoos, and the colorful birds of paradise. No large mammals are found on New Guinea, but around 250 species of small mammal have been recorded here, including the highly unusual tree kangaroo. Tree kangaroos are true kangaroos (they are members of the animal family *Macroodidae*). When in danger, they are capable of leaping 40 to 60 feet (12 to 18 meters) from the forest canopy to the ground with no ill effects. GM

RABAUL

EAST NEW BRITAIN, PAPUA NEW GUINEA

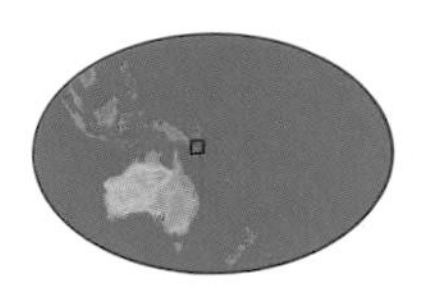

Height of Rabaul volcano: 2,257 ft (688 m)

Source of 1994 eruption: Mt. Tuvurvur

Rabaul is the name of an active volcano made up of a large central caldera surrounded by a range of peaks, the largest of which rises to 2,257 feet (688 meters). It is also the name of a town built between a picturesque harbor and the volcano. It proved to be a spectacularly ill-chosen site. In 1994, the volcano erupted, destroying the town.

Today, Rabaul is a strange wasteland buried in black volcanic ash. One writer described the broken structures of its buildings as "poking out of the mud like the wings of a dead bird," while he said the town itself resembled "a movie set for an apocalypse film ... with rubble and ruined buildings receding in every direction." From what was once a warm tropical lushness, Rabaul became a tangled mass of ruins covered in layers of muddy volcanic ash. But visitors still make their way here, despite the ever present danger. One of the volcano's vents, Tuvurvur, is constantly active, spewing a steady stream of ash into the air; it is out of bounds to visitors. All the other peaks, however, can be climbed. The hillsides contain numerous tunnels and caverns to explore; Japanese troops dug more than 310 miles (500 kilometers) of them during World War II. GM

FLY RIVER

WESTERN REGION, PAPUA NEW GUINEA

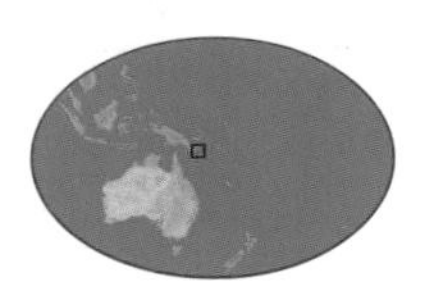

Length of Fly River with tributaries: more than 750 mi (1,200 km)

Species of bird: 387

The mighty Fly River is the longest river in the country, winding for almost 500 miles (800 kilometers) before it reaches the sea. The Fly River region contains areas of savanna and grassland, interspersed with woodland and monsoon forests, and touches the border between Papua New Guinea and the Indonesian province of Papua. The river floods regularly—annual rainfall reaches 33 feet (10 meters) in the uplands. The Fly rises in the western highlands where the nearby mountain peaks reach up to 13,200 feet (4,000 meters), and then flows southeast and down to the Gulf of Papua. As the freshwater from the Fly enters the sea it forms a mix of brackish water. These conditions have created the most extensive mangroves on Earth—more than 30 species of mangrove tree have been recorded in a single swamp. The trees provide habitat for many unusual species, including the saltwater crocodile and white-bellied mangrove snake. The river and its surroundings are home to some of Papua New Guinea's rarest plants, and 55 percent of the region's plant life is unique to the area. Of Papua New Guinea's 200 mammal species, 120 are found here, along with 387 species of bird. **GM**

HIGHLANDS

PAPUA NEW GUINEA

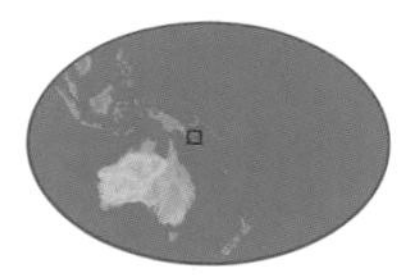

Area of Highlands: 70,000 sq mi (181,300 sq km)

Height of Mt. Wilhelm: 14,793 ft (4,509 m)

The Highlands of Papua New Guinea is a region of outstanding natural beauty, of steep valleys blanketed in lush vegetation and tall mountains—including Mount Wilhelm, the country's highest peak—crossed by fast-flowing rivers and tumbling waterfalls. The region was thought to be uninhabited until gold miners climbed the mountains in the 1930s. There they discovered 100,000 people living a subsistence existence, unaware of the outside world. These isolated highlanders were among the first agriculturalists in the world. Evidence from Kuk, a swampland in the upper Waghi Valley, revealed agricultural systems dating back 10,000 years, predating the cultivation of grain in the Middle East's "fertile crescent," traditionally regarded as the birthplace of agriculture. The southern highlands are spectacular, holding some of the most fascinating cultures of New Guinea.

The western highlands boast a large swathe of rainforest, enclosed within the Baiyer River Wildlife Sanctuary, whose leafy depths provide a home to the largest population of birds of paradise in the world. **GM**

RIGHT *Lush highland landscape of Papua New Guinea.*

OWEN STANLEY RANGE

PAPUA NEW GUINEA

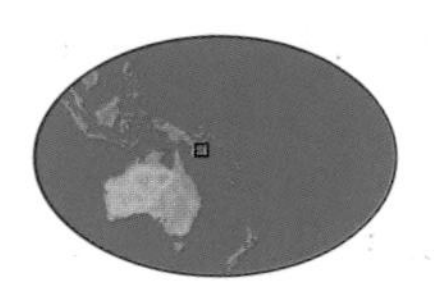

Vegetation in Owen Stanley Range: rainforest

Height of Mount Victoria: 13,360 ft (4,072 m)

The backbone of the southeastern tip of New Guinea is the rainforest-cloaked Owen Stanley Range, which forms part of a great mountain chain stretching across the center of the country.

With its rugged peaks and dense jungle, this is rough, challenging terrain, as the Japanese forces discovered in 1942 during their failed attempt to capture the Papua New Guinean capital, Port Moresby, through the backdoor. Today the Kokoda Trail, site of the famous battle between the Japanese troops and the Australian 7th Division, is a popular and spectacular five-day hike. It is fairly easy going on the eastern side of the coast to the village of Kokoda, located on a small plateau 1,320 feet (400 meters) above sea level, but it is flanked by mountains rising to more than 13,000 feet (3,900 meters). The highest peak is the 13,360-foot (4,072-meter) Mount Victoria. From the summit, it is possible to see all the way to the capital on a clear day. The trail then climbs over steep ridges adorned with mist-shrouded and stunted trees, through jungles of fern, orchids, and clean mountain streams, into steep valleys, and through dense rainforest until it runs down to the coastal plains. GM

SEPIK RIVER

EAST SEPIK / SANDAUN, PAPUA NEW GUINEA

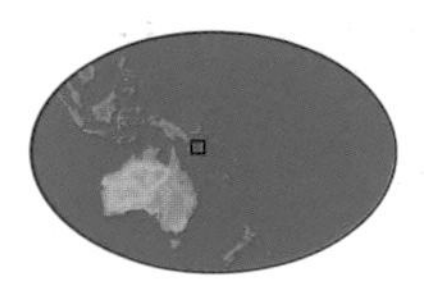

Length of Sepik River: 698 mi (1,123 km)

Altitude of river: from sea level to 11,480 ft (3,500 m)

The Sepik region is an immense grassland reserve dominated by one of the world's greatest rivers. The area derives its name from this mighty water source. The river—a meandering, oily-brown, vast flow of water—snakes its way for 698 miles (1,123 kilometers) from its headwaters in the highlands to the ocean, tearing great chunks of mud and vegetation out of its banks, which at times drift downstream like floating islands. As the Sepik has no actual river delta, it runs straight into the sea, tainting it brown for up to 30 miles (50 kilometers)—people living on islands off the coast are able to draw their freshwater straight from the ocean.

The Sepik floodplains contain around 1,500 lakes, which provide habitats for a number of unique species. The climate of the Sepik region is wet and tropical, although there is considerable variation both in the river's altitude and in local and regional climate. The Sepik River is navigable for almost its entire length, and the people who live near it depend on the river for their water, food, and transportation. Many experts believe the Sepik people create the best carvings in Papua New Guinea. GM

BOUGAINVILLE

NORTH SOLOMONS, PAPUA NEW GUINEA

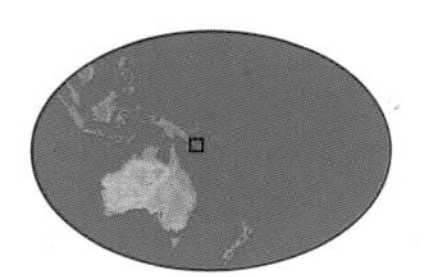

Area of Bougainville: 3,880 sq mi (10,050 sq km)

Human population: 200,000

Bougainville, a rugged and densely forested volcanic island, has white sandy beaches and striking coral reef systems. Further inland, pristine forests cover hills and vales soaring up to the mist-shrouded mountains, where several waterfalls can be seen cascading down deep mountain gorges. Bougainville has often been the scene of unpredictable natural violence. The island's volcanoes appear to lie dormant, but on a clear day smoke can be seen emanating from the two most well-known volcanoes, Mount Balbi in Wakunai and Mount Bagana in Torokina.

On the eastern part of the island, large stands of bamboo forest can be found as well as remnant stands of *Terminalia brassii* in swamp forests. Important wetland areas lie on the southern coast. Bougainville is home to many species that are native to the Solomon Islands to the south and southeast. Among the many interesting vertebrates is the little-known Bougainville honeyeater, unique to the island.

The tourist infrastructure of Bougainville is still in its infancy. War in the region has recently come to an end, but it is a good idea to check consular information before traveling here. GM

TROBRIAND ISLANDS

MILNE BAY, PAPUA NEW GUINEA

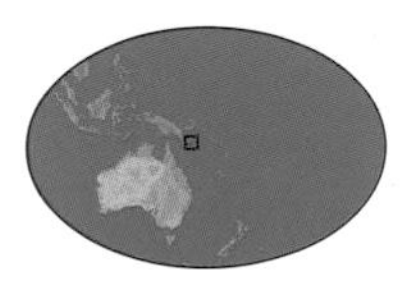

Largest Trobriand Island: Kiriwina

Notable feature: remote Pacific islands with unique species and habitats

The Trobriand Islands are a 170-square-mile (440-square-kilometer) coral archipelago off the eastern coast of Papua New Guinea. Once connected to the main island, the Trobriands separated off in the late Pleistocene period and have since then developed their own unique habitats. The islands' thick tropical rainforests support a good deal of flora and fauna that are found nowhere else, including several plant species, four species of mammal, and two species of bird of paradise. Unfortunately, the forests are coming under increasing threat, principally from encroaching agriculture and aggressive logging. This is of concern because the environments here have been little studied, certainly when compared with the native peoples, who have been drawing the attention of anthropologists for more than 100 years.

Despite the increasing number of visitors, the islanders have managed to preserve much of their culture, including their famous yam festival. Indeed, yams are so important to Trobriand life that the months of the year are named after each stage of this starchy root's growth. GM

NEW CALEDONIA

MELANESIA, PACIFIC OCEAN

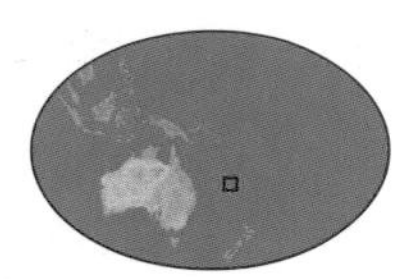

Area of main landmass: 6,180 sq mi (16,000 sq km)

Total area (including reefs and islands): 7,172 sq mi (18,576 sq km)

Maximum elevation (Mt. Panié): 5,518 ft (1,628 m)

A French Overseas Territory 932 miles (1,500 kilometers) east of Australia and 1,056 miles (1,700 kilometers) northeast of New Zealand, this highly isolated island group is not, like its neighbors Fiji and Vanuatu, of volcanic origin. Instead, it is a sliver of ancient Gondwanaland and, as such, is a living ark with a diversity and specialty of wildlife that often exceeds that of the much more famous Madagascar.

During its 56 to 80 million years of isolation, many unique forms evolved on the main island, Grand Terre, including 77 percent of the 3,322 vascular plants. Animals unique to the island include the world's largest gecko, a huge number of land snails, and a plethora of wonderful birds, such as the noutou pigeon (the largest living tree pigeon), the New Caledonian lorikeet, and the kagu (a bird so odd it has a family and an order all to itself). Of New Caledonia's 116 species of bird, 22 can be found only here.

There is also a 1,000-mile (1,600-kilometer) barrier reef (second only to the Great Barrier Reef), which provides a home to an enormous variety of marine life. There are dugongs in the lagoons, whales in the deeper waters, and nesting beaches for four species of sea turtle.

The island's land habitats are incredibly diverse, ranging from rainforest on the eastern side to dry forest on the western side of the central mountain chain, which has five peaks exceeding 5,000 feet (1,500 meters). There is also a lowland habitat rich in aromatic shrubs and mangrove vegetation. The rainforest is the richest habitat with 2,011 known plant species, compared to the dry forest's 379 known plant species. The forests on the limestone-covered, volcanically derived Loyalty Islands have a very different composition to those of Grand Terre. There are currently 25 reserves, including Mount Panié Special Botanical Reserve, which protects pristine rainforest and cloud forest, and the Rivière Bleue Reserve, which protects lowland rainforest.

New Caledonia is a sliver of ancient Gondwanaland and, as such, is a living ark with a diversity and specialty of wildlife that often exceeds the much more famous Madagascar. Of the island's 116 species of bird, 22 can be found only here.

In recognition of its biological importance, New Caledonia has received the highest conservation ranking in Oceania. It appears three times on WWF's Global 200 list of the world's most important places for biodiversity and was inscribed on the World Heritage List in 2008 in honor of the ecosystems and reef diversity of its lagoons. **AB**

RIGHT *The colorful underwater spectacle of New Caledonia.*

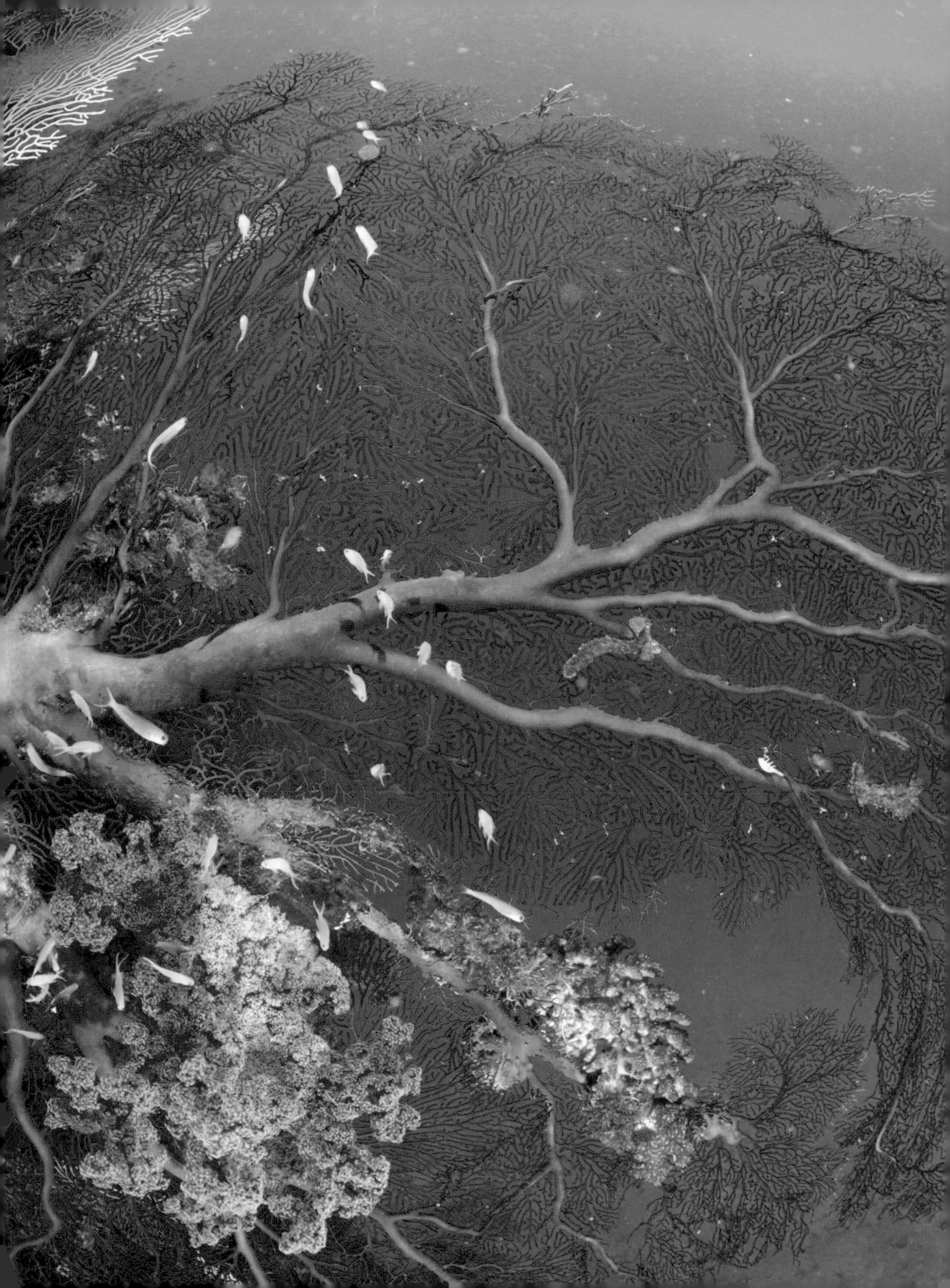

CASCADES DE FACHODA—TAHITI

FRENCH POLYNESIA

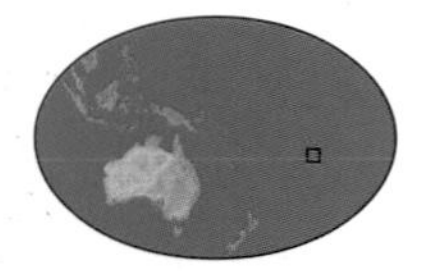

Highest point (Mt. Orohena): 7,352 ft (2,241 m)

Age of Tahiti Nui: 3 million years

Age of Tahiti Li: 500,000 years

One hundred days of rain each year ensures that Gauguin's and Loti's romantic island of Tahiti is, first and foremost, a place of waterfalls. The most striking waterfall is the Cascades de Fachoda on the Fautaua River. With a drop of approximately 1,000 feet (300 meters), it is one of the world's 25 highest waterfalls. However, it is a three-hour trek from Papeete. There is a choice of two tracks: The lower route follows the river to the foot of the falls, while the other reaches the lip. The more accessible but more modest 66-foot- (20-meter-) high Faarumai Falls are a short distance past the Arahoho Blowhole on the north coast. The second of the three cascades is a five-minute walk past the first, and the third is 30 minutes farther. The interior of Tahiti is a lush jungle with deep flower-filled valleys. Tahitian gardenia, hibiscus, and orchids abound and trees include coconut and pandanus palms, as well as Tahitian chestnut. Over 400 species of ferns grow here, including the "maire," which is an island symbol. **MB**

SAVAI'I ISLAND

SAMOA

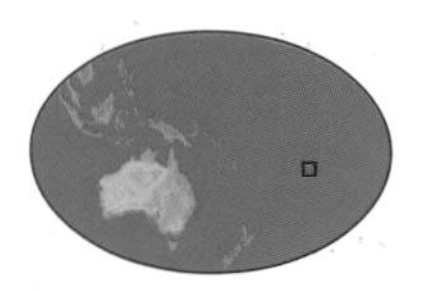

Area of Savai'i: 663 sq mi (1,717 sq km)

Highest point (Mauga Silisili): 6,096 ft (1,858 m)

Dimensions of Pulemeilei Mound: 200 ft (60 m) x 165 ft (50 m) at its base, 50 ft (15 m) high

Known as the "soul of Samoa," Savai'i is an unspoiled South Pacific island about 12 miles (20 kilometers) northwest of Upola. It is dominated by volcanic activity from the active but currently dormant Mount Matavanu. Moon-like lava fields from the 1905 to 1911 eruptions can be seen at Sale-aula, and elsewhere on the island are lava tubes and caves, such as Peapea Cave. On the coast are the Alofaaga Blowholes, where plumes of seawater shoot 100 feet (30 meters) into the air, and dangerous currents sweep black-sand beaches, such as Nu'u. Gataivai Falls cascade 16 feet (5 meters) directly into the sea.

Inland, the Afu Aau Waterfall tumbles into a freshwater lake surrounded by virgin forest. Due to its isolation, the island's more obvious wildlife is limited to bats, lizards, and 53 species of bird, including the rare tooth-billed pigeon. The island's main human-made feature is a gigantic pyramid, known as the Pulemeilei Mound, the largest archeological site in Polynesia. The pyramid is covered by forest, but its two-tier structure can be seen from a platform nearby. Savai'i is reached by air from Fagali'i Airport on Apia, or by sea from Multifanua Wharf on Upolu. **MB**

BORA BORA

FRENCH POLYNESIA

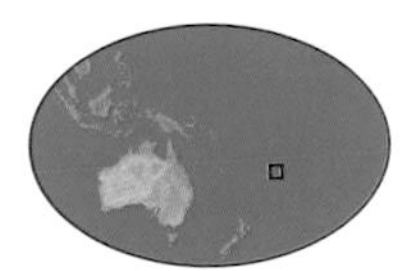

Length of Bora Bora: 6 mi (10 km)

Width of Bora Bora: 2.4 mi (4 km)

Highest peak (Mt. Otemanu): 2,385 ft (727 m)

A beautiful pearl on the string of Society Islands in French Polynesia, Bora Bora was immortalized by James A. Michener, who famously called it "the most beautiful island in the world."

Bora Bora rose from the sea as a volcano between three and four million years ago—just an infant by geological measure—but the craggy flanks of Mount Otemanu and the twin summits of Pahia and Hue are already heavily eroded. From their lofty black cliffs, tropical rainforest plunges west to the edge of an encircling lagoon, three times the area of the land. Beyond, a barrier reef and low motus, or coral spits, protect the island from Pacific swells. Within, the lagoon teems with tropical fish, lush corals, manta rays, and reef sharks.

The island sits squarely in the path of fierce tropical cyclones, which mostly develop as sea temperatures peak toward the end of the humid tropical summer. Such storms can pump around 2.2 million tons of air per second, driving wind speeds up to 187 miles (300 kilometers) per hour. Lightly constructed traditional homes suffer badly in such events, which have, unfortunately, struck Bora Bora many times since the late 1990s. DH

AITUTAKI ATOLL

COOK ISLANDS

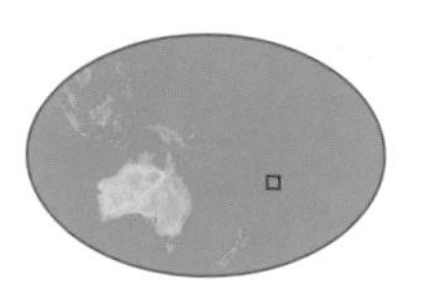

Area of Cook Islands: 772,200 sq mi (2 million sq km) of ocean

First settled: 800 to 900 C.E.

Aitutaki is the remnant of an old volcanic island, much of which has now subsided beneath the surface of the ocean, leaving behind one large island and a line of smaller ones, encircling a lagoon. Once the site of intense volcanic activity, this is now a place of low rolling hills—its highest hill, Maungapu, rises just 404 feet (123 meters)—blanketed by banana plantations and coconut groves. The beach of Tapuaetai, one of the smaller islands, is rated as one of the best in all Australasia. The roughly triangular lagoon encompasses an area of some 7 square miles (18 square kilometers) and boasts numerous coral spits, or motus. Beyond, the reef walls plunge straight down to the bed of the Pacific, 13,200 feet (4,000 meters) below.

Part of the far-flung Southern Group of the Cook Islands, Aitutaki was probably first settled around 800 to 900 C.E. The first European to sight the island was Captain Bligh. He arrived on the *Bounty* on April 11, 1789, just days before he was deposed during the famous mutiny. Bligh returned to Aitutaki in 1792, bringing the papaya (pawpaw) fruit with him. Today, papaya is a major Cook Islands export earner. **DH**

PALOLO SPAWNING

SAMOA

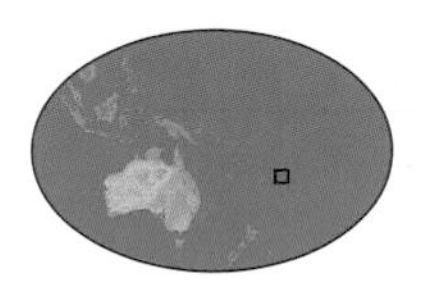

Length of Palolo worm: 12 in (30 cm)

Emergence times during October or November: Manu'a Islands: 10 pm; Tutuila: 1 am; Western Samoa: 4 to 5 am

On a single night, during the last quarter of the moon in October or November, millions of palolo worms spawn together in the Pacific Ocean. Although this remarkable event takes place close to many islands in the South Pacific, one of the most accessible island groups is Samoa.

The worms normally live on the seabed and burrow into the coral pavement. At an unknown signal, each worm divides into two parts. The hind end wriggles out of the burrow and floats to the surface. Each tail contains either eggs or sperm, and once afloat, sheds its cargo into the sea. The sperm fertilize the eggs, which then develop into larvae that, in turn, grow into new worms. The green bouillabaisse of millions upon millions of spent worms' tails attracts fish, sharks, birds, and people. The local Samoans consider the palolo feast a special delicacy, and they wade out into the shallow reef waters with hand nets, buckets, and tin cans to gather up the tails. They are eaten raw or fried with butter, onions, or eggs. The front ends, meanwhile, remain in their burrows and regenerate new tails, so that the reproductive extravaganza may repeat itself the following year. **MB**

LORD HOWE ISLAND GROUP

AUSTRALIA

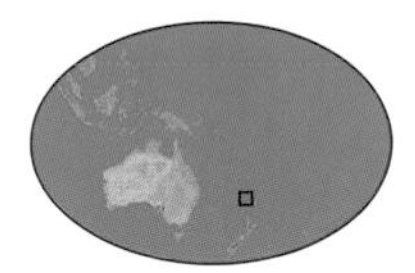

Length of Lord Howe Island: 6.8 mi (11 km)

Height of Mt. Lidgbird: 2,549 ft (777 m)

Height of Mt. Gower: 2,871 ft (875 m)

Lord Howe Island is an outstanding example of an oceanic island of volcanic origin. The island is crescent-shaped and one of a string of volcanic pinnacles that sit atop a submarine ridge running from the northern island of New Zealand. Other notable rocks or islands in the Lord Howe Island Group include Balls Pyramid, Gower Island, Mutton Bird and Sail Rock, Blackburn (Rabbit) Island, and the Admiralty Group. Lord Howe Island has two mountains—Mount Lidgbird and Mount Gower—that were pushed up almost seven million years ago when the geologic movement of an underwater plateau created a large shield volcano. The peaks are high enough for true cloud forest on the summits, including rare and unique palms and ferns. In the sea, a coral reef runs for four miles (6.4 kilometers) along the western side of Lord Howe and encloses a deep lagoon. Together with the Elizabeth and Middleton reefs, these are the southernmost coral reefs in the world. Their location in the Pacific Ocean is where tropical and temperate ocean currents meet. English mariners on HMS *Supply* discovered the island in 1788, on the way to the Norfolk Island penal colony. **GH**

RAPA NUI

DEPENDENCY OF CHILE

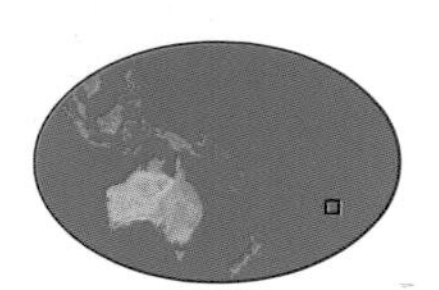

Area of Rapa Nui: 45 sq mi (117 sq km)

Distance to nearest inhabited land: 1,180 mi (1,900 km)

Human population: 2,000 (70 percent Polynesian)

One of the most isolated places on Earth, Rapa Nui (Easter Island) is a gaunt triangle of volcanic rock some 2,300 miles (3,700 kilometers) west of Chile, and 1,290 miles (2,075 kilometers) east of the Pitcairn Islands. A relatively young volcanic island, formed about 750,000 years ago, it is a place of rugged beauty, its center dominated by three extinct craters around which is a cracked, broken landscape of old lava flows and volcanic debris.

The island is best known for the giant stone monoliths, or moai, that line its barren coastal hills. The statues were carved from the island's soft tuff by the Rapa Nui, the original settlers who arrived around 1,200 years ago. The reason they were built, and in such huge numbers, remains unclear, but it is clear that their creation depleted the island's forests—the lumber was used as rollers to move the statues. As the population topped 4,000 people, resources to support them dwindled, and the Rapa Nui degenerated into warfare and cannibalism. When Captain James Cook arrived in 1775, he found just 630 people scratching out a marginal existence. In 1875, 100 years later, only 155 islanders remained. DH

RED CRAB SPAWNING

CHRISTMAS ISLAND, DEPENDENCY OF AUSTRALIA

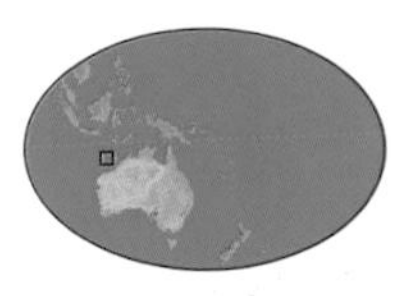

Area of Christmas Island: 52 sq mi (135 sq km)

Extent of protected area: 63 percent national park

Width of red land crabs: 5 in (12.5 cm)

This small island lodged into the eastern part of the Indian Ocean is host to an extraordinary procession each year. About 120 million red land crabs—the most conspicuous of the 14 species of land crab on the island—spend most of their lives in the forest, but come the wet season, during October to November, they begin to emerge from their dark hiding places and head for the coast.

The emergence is synchronized all over the island. A thick crab carpet spreads across the island, invading gardens, golf courses, roads, and railway tracks. Although they are land crabs, they need to return to the sea to breed, and males and females meet up on the shore to deposit and fertilize the eggs in the shallow waters. They do this at the last quarter of the moon, when there is the least amount of difference between high and low tides. Once they have spawned into the sea, they all head back to the forests again and simply disappear until the next year.

Meanwhile, the offspring develop in the sea as embryos and then emerge, barely formed as tiny, red crabs. They swarm over the rocks in their millions and make for the safety of the forest. MB

WET TROPICS OF QUEENSLAND

QUEENSLAND, AUSTRALIA

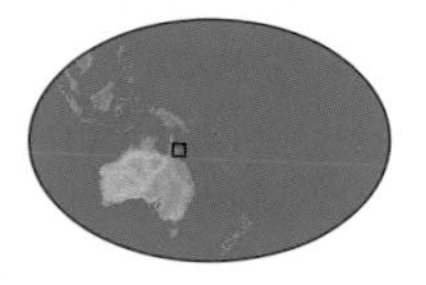

Extent of Wet Tropics of Queensland World Heritage site: 2,250,874 acres (910,900 ha)

Age: over 100 million years

Height of Wallaman Falls: 917 ft (280 m)

The Wet Tropics of Queensland World Heritage site, centered on the Daintree National Park and Cape Tribulation, contains one of the largest rainforest wilderness areas in Australia. It is a region of rugged mountains and mangrove forests, filled with deep gorges, fast-flowing rivers, and numerous waterfalls—the 1,115-foot (340-meter) Wallaman Falls has the highest sheer drop of any waterfall in Australia. The combination of fringing coral reefs and rainforest coastline at Cape Tribulation is unique in Australia; the site also borders on the Great Barrier Reef, another World Heritage site.

The Wet Tropics contain many examples of ongoing ecological processes and biological evolution, including exceptionally high levels of species diversity and uniqueness, reflecting long-isolated ancient habitats. They contain one of the most complete and diverse living records of the major stages in the evolution of land plants, as well as one of the most important living records of the history of marsupials and songbirds. The site also reflects eight of the major stages in the planet's evolutionary history, including: the ages of pteridophytes; conifers and cycads; angiosperms or flowering plants; the final break-up of Gondwanaland; the mixing of the wildlife and habitats on the Australian and Asian continental plates; and the impact of the many Pleistocene glacial periods, or ice ages, on tropical rainforest.

One fifth of Australia's bird species, one third of all its marsupial, reptile, and frog species, and two thirds of all its bat species are found here in an area that takes up less than one percent of the country's landmass.

The Wet Tropics of Queensland is a land of rugged mountains and mangrove forests, filled with deep gorges, fast-flowing rivers, and numerous waterfalls.

Aboriginal occupation is believed to date back approximately 50,000 years, and today the Wet Tropics continues to hold great significance for the oldest "rainforest people" on Earth. Where the rainforest meets the reef is also the site where Captain James Cook ran aground on the Great Barrier Reef in 1770. Captain Cook first sighted this stretch of coastline and named it Cape Tribulation "because here began all our troubles." In 1848 Edmund Kennedy became the first European to explore the Wet Tropics. GH

RIGHT *The winding Daintree River in the Wet Tropics of Queensland World Heritage site.*

LOW ISLETS

QUEENSLAND, AUSTRALIA

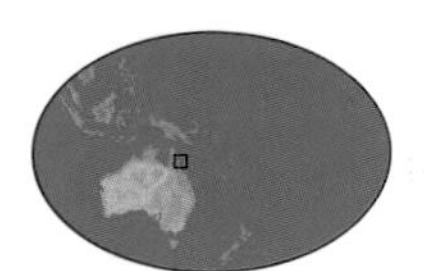

Distance covered by Low Islets: 1,437 mi (2,313 km)

Age: about 6,000 years

Vegetation: woodland, shrubbery

Low Islets is a group of two small sand cay islands 8 miles (13 kilometers) off Port Douglas, in the northern Great Barrier Reef. Coral cays are created when low-lying flat reefs collect enough sand and sediment to rise above sea level. The main islet's area is just 570 acres (231 hectares) and is sheltered from the powerful southeastern swell that breaks in violently on the outer ribbon reefs. Low Isle—which has a historic lighthouse dating from the 18th century—began life as a lagoonal reef in which the central depression has been filled with coral debris to become a low-tide platform. The corals of the islets are exposed by the tides. Abundant coral fish species are best viewed by snorkeling; however, the crown of thorns starfish, coral bleaching, and cyclone damage have caused declines in all families of hard corals.

The islets are among thousands of individual reefs that sweep north from the Tropic of Capricorn, some 1,437 miles (2,313 kilometers) to Torres Strait, where they merge with those along southern Papua New Guinea. All modern reefs have evolved in the last 6,000 years, since the sea returned to its present level. GH

MOUNT BARTLE FRERE

QUEENSLAND, AUSTRALIA

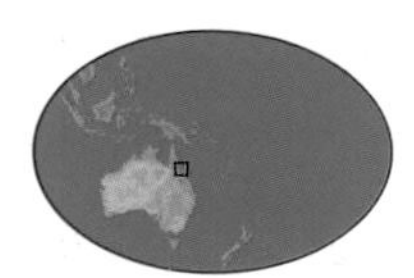

Summit of Mt. Bartle Frere: 5,325 ft (1,622 m)

Height of Bellenden Ker: 5,254 ft (1,592 m)

Area of national park: 196,437 acres (79,500 ha)

Northern Australia's highest mountain, Bartle Frere, is set amid the rugged, wet, and humid wilderness of the Bellenden Ker Range in Queensland's Wooroonooran National Park. Bartle Frere and Bellenden Ker dominate the landscape, and when not cloaked in mist and cloud the summit of Bartle Frere offers the chance to view the coastal lowlands and the Atherton Tableland. However, this wilderness area—and its associated Josephine Falls—are subject to extremes of cold, wind, rainfall, and leeches, and swimmers have died in the turbulent waters of the falls.

To the local Noongyanbudda Ngadjon people, the mountain is known as "Chooreechillum," their spiritual home. This jungle area was once so remote and impenetrable that it was only in 1886, and with Aboriginal aid, that a European first climbed the summit. Relict species, such as the rare skink, survive here. The rainforest is considered to be closest to the humid tropical lowland forests of southeast Asia. Vine fern forests and thickets occur on the slopes and summits of the high peaks. The canopy is low and dense and shows the streamlining effects of ongoing strong winds. GH

BARRON RIVER FALLS & GORGE

QUEENSLAND, AUSTRALIA

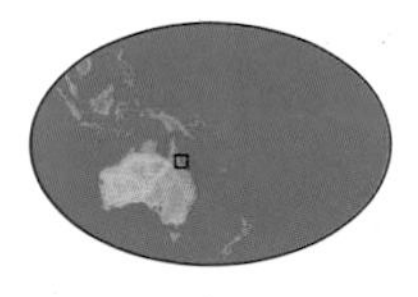

Area of Barron River Falls and Gorge: 6,870 acres (2,780 ha)

Height of falls and gorge: 858 ft (260 m)

Vegetation: tropical rainforest

The Barron Gorge is a rugged and hilly region lying approximately 19 miles (30 kilometers) northwest of Cairns. The wild rainforest valley contains diverse and unique ecosystems. It has a tall, closed canopy of vine forest with an open forest alliance of eucalyptus. The rainforest is easily accessible to walkers, with an increasing number of tracks and trails being opened. The historic Kuranda train and the Skyrail cableway also provide visitors with spectacular views of the gorge and the Barron River far below. At the top of the gorge, near Kuranda, lie the Barron Falls. Their once mighty flow is today diverted for hydroelectricity, so now they are only seen in full flow during the wet season from December through to March. During the dry season, the water over the falls reduces to a trickle. However, to allow Barron Gorge to maintain its status as a major tourist attraction, the floodgates of an upper dam are opened to allow the falls to tumble just as the Kuranda tourist train arrives. GH

HINCHINBROOK CHANNEL

QUEENSLAND, AUSTRALIA

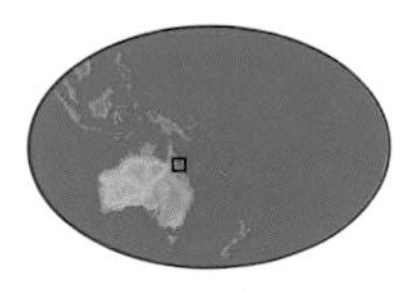

Area of Hinchinbrook Island: 32.5 mi (52 km) x 6.25 mi (10 km)

Height of Mt. Bowen: 3,530 ft (1,070 m)

Age: up to 260 million years

The Hinchinbrook Channel wetlands lie between Cardwell on the Queensland coast and Australia's largest island national park, Hinchinbrook Island. The highest peak on the island is Mount Bowen, the third highest peak in Queensland. Listed as a heritage site, the Hinchinbrook wetlands are an extensive and complex system of tall mangrove forests and swamp. The extensive flats and sinuous channels support vast seagrass beds, which are essential food for dugongs (sea cows), sea turtles, and an important habitat for the juveniles of a number of prawn species.

A number of dolphin varieties are found in the Hinchinbrook Channel including the Irrawaddy (river dolphin), the Indo-Pacific humpback dolphin, and the bottlenose dolphin. The area also has extremely high diversity of fish and crab species. At Scraggy Point, an extraordinary 2,000-year-old Aboriginal stone fishtrap complex can be seen. Scraggy Point is also an important habitat for the estuarine crocodile. Hinchinbrook Channel shares some of the rich bird diversity of the Wet Tropics but there is a special focus on some species, such as Torresian imperial pigeons and beach thick-knees. GH

MOSSMAN GORGE

QUEENSLAND, AUSTRALIA

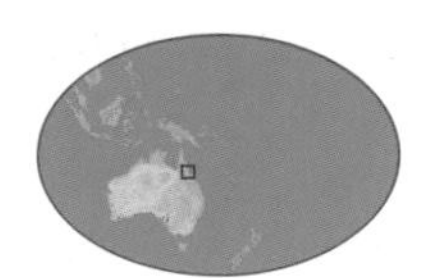

Length of Mossman River: 13 mi (20 km)

Area of Mossman Gorge: 218 sq mi (565 sq km)

Vegetation: lowland rainforest

Powered by its headquarters in the Main Coast Range, the Mossman River has created a deep, steep-sided gorge in its 13-mile (20-kilometer) flow to the sea. Located at the southern end of Daintree National Park, the gorge features a cool mountain stream fringed by primeval rainforest and strewn with giant granite boulders along its banks. Most of the gorge is inaccessible except to experienced hikers, but a two-mile (3.2-kilometer) track allows visitors into the rainforest.

Here, giant fig trees crowd each other and dense canopies of the rainforest block most light. Ferns and orchids live among the highest trees to seek out the sunlight. The gorge and rainforest are populated with one of the largest and most beautiful butterflies in Australia, the brilliant Blue Ulysses, which has a wingspan of more than 5 inches (12.5 centimeters). Also living here is the Hercules moth, the world's largest moth, with a wingspan of 10 inches (25 centimeters). Platypus and tortoise can be seen surfacing in the quieter sections of the Mossman. The gorge is the traditional home of the Kuku Yalanji Aboriginal people. GH

BELOW *Lush ferns and palms thrive in Mossman Gorge.*

LAWN HILL GORGE

QUEENSLAND, AUSTRALIA

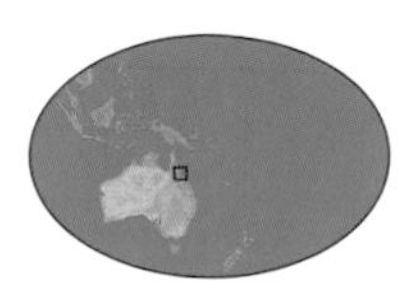

Depth of Lawn Hill Gorge: 230 ft (70 m)

Area of gorge: 43 sq mi (111 sq km)

Age of gorge: Precambrian

Lawn Hill Gorge is known for its spectacular, bright-red sandstone cliffs, lush riverine forests, and the emerald-colored waters of Lawn Hill Creek. Lawn Hill, or Boodjamulla, is a series of escarpments and gorges containing oasis springs, permanent waterholes, and continuous streams and rapids some 2.5 miles (4 kilometers) long. The gorges and creek have cut though truly ancient Precambrian sandstone cliffs, about 4.5 billion years old, flanked by melaleuca, pandanas, and cabbage palm vegetation. Aside from the main Lawn Hill Gorge, other areas of prized beauty include the limestone Colless Creek landscape areas and the "Grotto" area near Riversleigh. This ancient place, located in one of the most remote parts of the Australian outback, contains fossil sites formed in the Tertiary period, around 25 million years ago. The area is still a biodiversity hot spot, with high numbers of turtle species, amphibians, reptiles, wallabies, and kangaroos. It also provides a critical migratory bird habitat for the green pygmy goose. Ancient rock art sites are present, as is evidence of ancient Aboriginal occupation including fire-stick farming of vegetation. GH

CARNARVON NATIONAL PARK

QUEENSLAND, AUSTRALIA

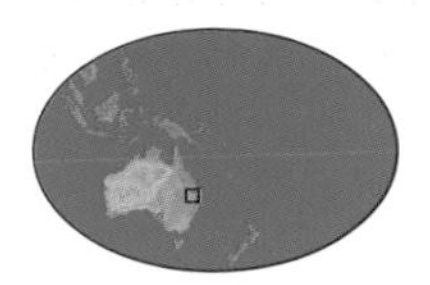

Area of Carnarvon National Park: 110 sq mi (285 sq km)

Vegetation: diverse forest

Queensland National Parks and Wildlife describes Carnarvon National Park as "a tangle of peaks, gorges, and sandstone cliffs—it is one of the wildest regions of the central western section of Queensland." The focus of this isolated park is the serpentine Carnarvon Gorge, with its precipitous 660-foot (200-meter) sandstone cliffs and eroded 18-mile- (29-kilometer-) long sandstone strata.

This striking gorge is home to gum trees, cabbage palms, cycads, and rare ferns. Near the waterfalls, elkhorns and lichens are more common. Weathering over millions of years has shaped and molded the landforms and habitat that have been home to generations of Aborigines for over 20,000 years. As such, Carnarvon's caves and cliff faces are inscribed with some of the country's best indigenous art—inscriptions of hands, axes, emu tracks, and boomerangs. The art is in the classic hand stencils of ocher and white, black, and yellow pigments. The park is 38 miles (61 kilometers) southwest of Rolleston. GH

BAYLISS CAVE

QUEENSLAND, AUSTRALIA

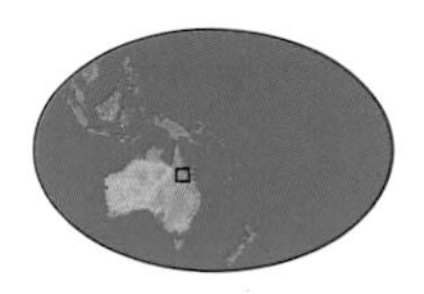

Elevation of Undara Crater: 400 ft (120 m) above sea level

Height of Bayliss Cave: 33 ft (10 m)

Length of Bayliss Cave: under 660 ft (200 m)

About 190,000 years ago, lava spewing from the Undara volcano spread over an area of 598 square miles (1,550 square kilometers). Caves and lava tubes were created when the rivers of lava were confined to valleys, forcing the lava to flow underground where it formed cylindrical tubes. Bayliss Cave is the longest single tube within the Undara lava tube system. The cave has a narrow shaft and a pungent, pitch-black chamber. Labeled by scientists a "bad air cave," its carbon dioxide levels have been measured at 5.9 percent—almost 200 times the normal concentration. Surprisingly, these extreme conditions support an abundance of life, with at least 52 resident animal species. The floor is covered with tiny creatures: silverfish-like isopods, white cockroaches, and scutigerids—strange cave-dwellers that move like millipedes and have no bodily pigment or eyesight. Inside the caves, many species of bat can be found; during the wet season they gather in huge nursery colonies to raise their young. While some caves in the Undara system are open to the public, access to the Bayliss Cave is unfortunately prohibited due to the heavy atmosphere of carbon dioxide and its significant biology. GH

GLASS HOUSE MOUNTAINS

QUEENSLAND, AUSTRALIA

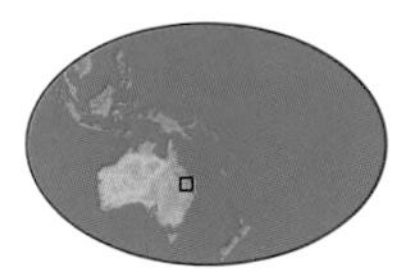

Area of Glass House Mountains: 4,673 acres (1,885 ha)

Age: 25 million years

Vegetation: rainforest, mallee

The Glass House Mountains are 16 ancient volcanic peaks that rise dramatically out of the surrounding coastal plains of the Tibrogaran, Ngungun, Coonowrin, and Beerwa National Parks. This set of steep-sided volcanic plugs of rhyolite and trachyte presides over the Sunshine Coast's tropical fruit farming area and ranges in height from 778 feet (237 meters) to 1,834 feet (556 meters). In 1770, Captain James Cook described the monoliths: "These hills lay but a little way inland and not far from each other; they are very remarkable on account of their singular form of elevation which very much resembles glass houses, which occasioned me giving them that name." But Archibald Meston, journalist and administrator, described them more aptly in 1895: "Each stands in gloomy isolation, silent and alone. One mass of rock (Tibrogaran) faces the railway line, cliff fronted, savage, defiant, towering majestically into the clear blue sky, the wild rough stone face scarred by the rains of ten thousand years." However, local Aboriginal belief considers the peaks as family members and the local streams as their tears of sorrow over an incident where a child showed cowardice. GH

NOOSA NATIONAL PARK

QUEENSLAND, AUSTRALIA

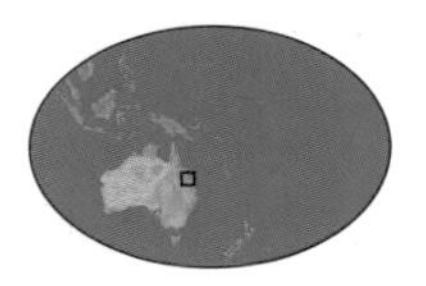

Area of Noosa National Park: 9 sq mi (23 sq km)

Age: 145 to 210 million years

Vegetation: coastal woodland, rainforest

The 660-foot- (200-meter-) high rocky headland and cliffs of Noosa National Park on Queensland's Sunshine Coast overlook fine ocean views, a string of sheltered beaches, varied grasslands, scrublands, open forest, and rainforest. This compact park is a natural oasis in contrast to the nearby holiday resort of Noosa Heads and is aimed at protecting what the Queensland Parks and Wildlife Service describes as "a small but important component of the flora and fauna of the Sunshine Coast."

This scenic area features a coastal walking track, palm grove circuit, and the Noosa Hill track. Some 13 plant communities have been mapped in the area including grassland, woodland, and open forest. Along the coast, wind-shearing salt spray and bushfires have created high dune heath. The high dunes are formed of sands that layer sediments with outcroppings of sandstone. Noosa Head contains examples of igneous intrusions, or quartz diorite, belonging to the Jurassic–Cretaceous age. The park's diversity provides a habitat for some 121 bird species, including the vulnerable red goshawk. GH

RIGHT *Early evening on the Sunshine Coast.*

CLARKE RANGE

QUEENSLAND, AUSTRALIA

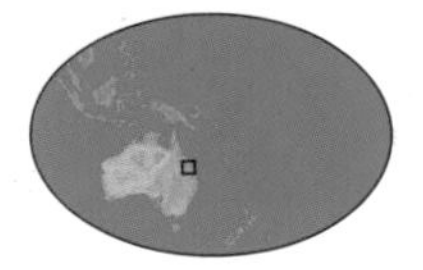

Area of Clarke Range: 567 sq mi (1,469 sq km)

Vegetation: tropical rainforest, eucalyptus woodland

The mist-shrouded mountains of the Clarke Range in Eungella National Park form the fourth-largest wilderness area in Queensland and the largest contiguous rainforest area in central Queensland. Eungella comes from an Aboriginal word meaning "Land of the Clouds." The forest ranges from 660 feet (200 meters) along the range's eastern side to 4,180 feet (1,274 meters) at the summit of Mount Dalrymple. Ancient geological violence has left a legacy of geological features as well as precipitous escarpments, including Broken River Gorge, Diamond Cliffs, and the Marling Spikes. Underlying these are ancient granitic rocks overlain with lava flows. The area has developed in isolation for thousands of years and is home to several forms that exist nowhere else: a bird, the Eungella honeyeater; a lizard, the orange-sided skink; a tree, the Mackay tulip oak; and three types of frog. Areas of exceptional fern growth are scattered throughout the park. Eungella features 12 miles (20 kilometers) of track, including the Palm Walk and Palm Grove, which meander through upland rainforest of red cedar and Mackay tulip oak, and groves of Piccabeen and Alexandra palms. GH

GREAT BARRIER REEF

QUEENSLAND, AUSTRALIA

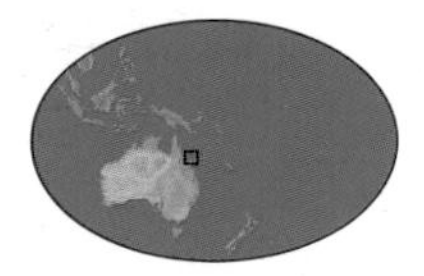

Length of Great Barrier Reef:
1,242 mi (2,000 km)

Area of Great Barrier Reef:
134,364 sq mi (348,000 sq km)

The Great Barrier Reef is listed as a World Heritage site of exceptional natural beauty, boasting some of the most spectacular scenery and marine wilderness on Earth. The world's largest such heritage site—with the most extensive coral reef system and one of the world's most biodiverse hot spots—the reef extends 1,242 miles (2,000 kilometers) and covers an area of 135,157 square miles (350,055 square kilometers) on the northeast continental shelf of Australia. The reef runs predominantly in a north–south direction, spanning a wide range of climates.

The Great Barrier Reef is listed as a World Heritage site of exceptional natural beauty, with some of the most spectacular scenery and marine wilderness on Earth, including the world's most extensive coral reef system.

Bigger than Italy, the reef system extends to Papua New Guinea and comprises a "broken maze" of vast turquoise lagoons and 3,400 individual reefs. These include 760 fringing reefs that vary in size from 2.5 acres (1 hectare) to over 24,710 acres (10,000 hectares) and in shape from flat platform reefs to elongated ribbon reefs. Woven among the reefs are about 618 continental islands, ranging from towering, rugged, and forested islands complete with freshwater streams, to 300 small coral and sand cays, and 44 low wooded islands and serpentine mangrove systems of exceptional beauty.

One third of the world's soft coral species exist here, as well as 1,500 species of reef fish. Six of the world's seven species of threatened marine turtle pass through on their way to their hunting or feeding grounds, and there are great areas of sea grass meadow that support one of the world's most important dugong (sea cow) populations.

The reef also has the largest green turtle breeding site in the world, more than 1,500 species of sponge, and over 4,000 species of mollusk. It is home to more than 30 species of mammal, including breeding humpback whales, and over 200 species of bird.

For all its myriad natural wonders, the site is also of great cultural importance, containing many middens, giant fish traps, and other archeological sites of Aboriginal or Torres Strait Islander origin. Noted examples are on Lizard and Hinchinbrook islands, and on the Stanley Cliff and the Clack Islands, which host spectacular rock art galleries. The reef is threatened by global warming and coral bleaching, which leading scientists claim are clearly damaging the reefs at an ever-increasing pace. GH

RIGHT *The watery maze of the Great Barrier Reef.*

CORAL SPAWNING

QUEENSLAND, AUSTRALIA

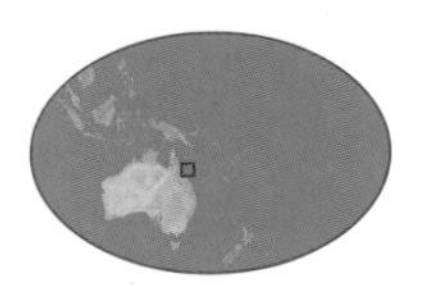

Occurrence: November to December after full moon

Location: along the entire length of the Great Barrier Reef

Every year for a few mysteriously chosen nights after a full moon in November and December, a spectacular natural event occurs in Australia's Great Barrier Reef in which millions upon millions of corals spawn en masse in a synchronized event. Described as being similar to an "upside-down snowstorm," many different coral species coordinate their sexual reproduction, releasing sperm and eggs into the ocean where they fertilize and float away to create new colonies of corals to reproduce the world's largest coral reef. When the ideal conditions are reached—warm water and a period of darkness of between four to six nights after a full moon—the vast numbers of coral polyps prepare to release their eggs and sperm. About 30 minutes before spawning, brightly colored pink or red bundles under the mouths of the polyps wait to escape. When the time is right, the bundles are squeezed out of the mouths to float to the surface where they drift in thick clouds colored red, pink, orange, and occasionally purple. Eventually they break apart and fertilize, the eggs dividing rapidly to form mobile larvae. The sudden abundance of food is a magnet for fish and other predators, but some larvae escape to found new reefs. GH

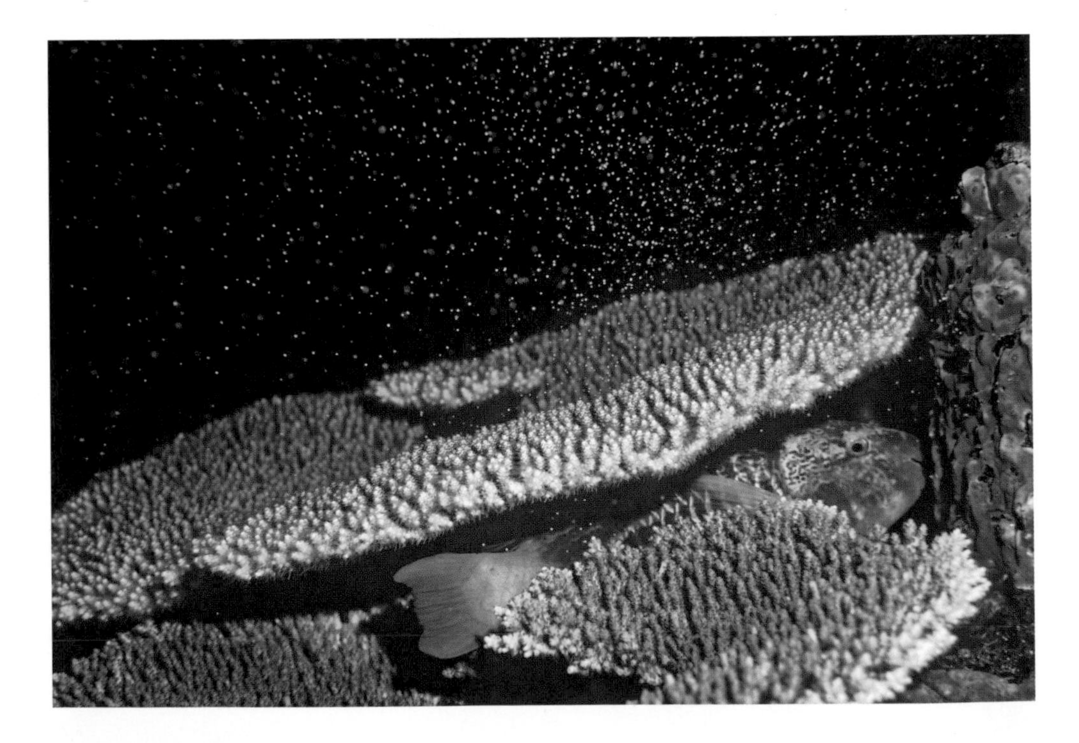

FRASER ISLAND

QUEENSLAND, AUSTRALIA

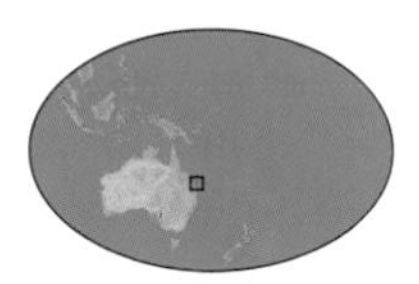

Dimensions of Fraser Island: 78 mi (125 km) long x 7.5 mi (12 km) wide

Area of island: 641 sq mi (1,660 sq km)

Location: off the coast from Hervey Bay, 190 mi (300 km) north of Brisbane

Fraser Island is the largest sand island and coastal dune system in the world. It formed during the last ice age when prevailing winds transported vast quantities of sand north from New South Wales and deposited it along the coast of Queensland. The fragile ecosystem consists of lush rainforest growing in pure sand—the only place in the world where this occurs. One plant species growing in the sand is the rare angiopter fern, one of the world's largest. The island is separated from the subtropical mainland by a narrow channel, and its huge tracts of sand dunes hold deep-blue freshwater lakes trapped 700 feet (213 meters) above sea level, forming half of all the world's perched freshwater dune lakes. Dingoes and wallabies live on the island, and 200 species of bird live within the eucalyptus forests, by the crystal clear streams and deserted white beaches. Whale-watching is possible from Hervey Bay. Fraser Island was named when Eliza Fraser was about to give birth aboard the brig *Sterling Castle* on May 13, 1836. The vessel struck the Great Barrier Reef, and although Eliza escaped drowning, she was later captured by Aborigines, but was eventually rescued. GM

HERON ISLAND

QUEENSLAND, AUSTRALIA

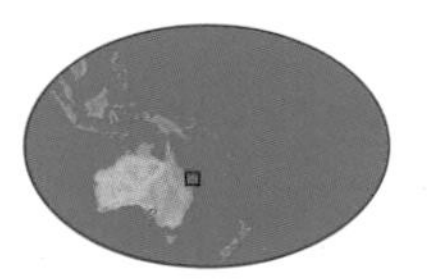

Area of Heron Island: 43 acres (17 ha)

Notable feature: coral cay surrounded by reefs that are exposed at low tide

Although two great English explorers—Captain James Cook in 1770 and Matthew Flinders in 1802—sailed past Heron Island on their journeys along the northern coast of Australia, neither located this small coral cay—possibly because they were avoiding the Great Barrier Reef. The island was not discovered by European settlers until January 1843, when HMS *Fly* anchored off its shores while attempting to find safe passage through the reef. The ship's naturalist named the island after what he believed were reef herons, but later turned out to be egrets.

Heron Island is located within the Great Barrier Reef Marine Park, and as well as the surrounding coral reef, the island has two main attractions: the presence from December to April each year of thousands of turtles that use the island as a breeding ground and, in the winter months from July to August, the large numbers of whales that pass through the channel between the cay and Wistari Reef. Bisected by the Tropic of Capricorn, the cay is surrounded by the sea, which provides a home to some 900 of the 1,500 species of fish and over 70 percent of the coral species found along the Great Barrier Reef. GM

WALLAMAN FALLS

QUEENSLAND, AUSTRALIA

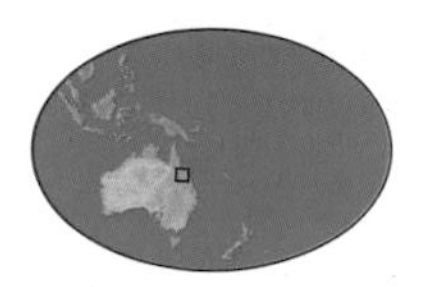

Height of Wallaman Falls: 1,000 ft (300 m)

Area of Girringun National Park: 315,000 acres (124,000 ha)

Tumbling through a rainbow-fringed cloud of mist to a large pool far below, the Wallaman Falls in northern Queensland is the highest year-round single-drop waterfall in Australia. Part of the Girringun National Park, they are the most accessible waterfalls on the coastal hinterland. Here the vegetation ranges from open eucalyptus bush to dense rainforest. Throughout Queensland's eastern highlands, the catchment areas of the many rivers are small, but their huge flows of rushing water from wet mountain ranges have gouged deep gorges and generated spectacular waterfalls.

Aborigines were the original inhabitants of these wet tropics, and more than 20 tribal groups have ongoing traditional ties to the site. The area therefore has enormous meaning and significance to its traditional occupants. The falls are 30 miles (50 kilometers) west of the town of Ingham, from where the road climbs steeply up the range. Walking tracks include a 1,000-foot (300-meter) climb to a lookout and a 1.2-mile (2-kilometer) hike to the bottom of the falls. The track back to the top is steep in places and requires strong lungs and legs. Rangers from the Girringun tribe maintain these tracks. GM

GONDWANA RAINFORESTS OF AUSTRALIA

QUEENSLAND / NEW SOUTH WALES, AUSTRALIA

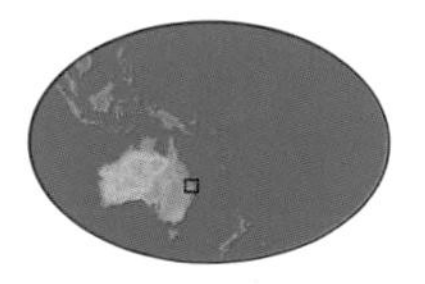

Rock age: up to 285 million years

Vegetation: subtropical, warm temperate, cool temperate rainforest

The Gondwana Rainforests of Australia World Heritage site is renowned for its dramatic mountain ranges, waterfalls, rivers, and important wildlife. The forests are remnant subtropical and temperate rainforest that run from Newcastle to Brisbane in discontinuous patches. While there are the remnants of exploded volcanoes up to 55 million years old, much of the underlying landforms were laid down up to 285 million years ago. Outstanding geological features can be seen around the shield volcanic craters.

This region comprises the largest areas of subtropical rainforest in the world. There are primitive plant families that are the direct descendants of flowering plants from over 100 million years ago, together with some of the world's oldest ferns and conifers. All this provides a home for more than 200 rare or threatened plant and animal species, including ancient lines of songbird, and the highest concentration of frog, snake, bird, and marsupial species in Australia. GH

SIMPSON DESERT

QUEENSLAND / NORTHERN TERRITORY / SOUTH AUSTRALIA, AUSTRALIA

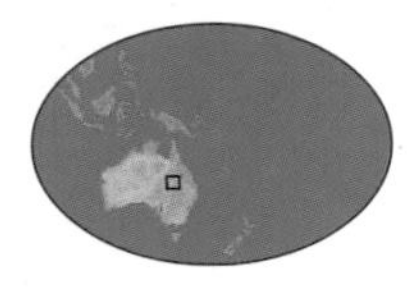

Area of Simpson Desert: 66,000 sq mi (170,000 sq km)

Age of desert: 40,000 years

From the air, the Simpson Desert is a striking sight, comprising some 66,000 square miles (170,000 square kilometers) of baked red earth. It is one of the world's best examples of dunal desert; its longitudinal dunes were formed about 40,000 years ago when the center of the continent became increasingly arid and surface sands blew across the landscape. The dunes are parallel and, with an average height of approximately 66 feet (20 meters), are among the biggest in the world. This earthy sea, with its waves of red sand ridges (the color comes from iron oxide), receives slightly more rainfall than the Sahara. Even so, precipitation is highly variable and unpredictable, and summer temperatures can exceed 122°F (50°C).

The desert stretches across the corners of three regions: Queensland, South Australia, and the Northern Territory. On its northern edge is a giant sandstone block, known as Chambers Pillar, which glows golden at sunrise. To the west is the Finke River, to the east the Georgina and Diamantina rivers, while to the south is Lake Eyre.

Seven Aboriginal tribes once occupied the desert, mainly around the watercourses on the desert boundaries. Many Aboriginal wells and stone arrangements throughout the central area, as well as the names of many of the desert's topographical features, indicate that Aborigines did traverse the entire region. Some of the wells are extensive and have 33-foot- (10-meter-) long tunnels dug at angles through sand to water-bearing layers.

More than 150 species of bird inhabit the Simpson Desert, including two rare species—the Eyrean grasswren, once thought to be extinct, and the Australian bustard. Wedge-tailed eagles, brown falcons, budgerigars, and zebra finches also live in the desert. Kites, crested pigeons, and galahs can be seen on the floodplains while waterbirds inhabit the playas, or inland basin lakes, when they fill. Many land animals are nocturnal and are rarely seen during the day. As well as a wide range of small marsupials such as the dunnart and mulgara, dingoes abound. In good seasons, kangaroos are known to inhabit the area. Among the feral animals are rabbits, foxes, camels, and donkeys. Most plants have short life cycles, growing, flowering, and setting seeds within a couple of months of rain. GM

The Simpson Desert is one of the world's best examples of dunal desert; its longitudinal dunes were formed about 40,000 years ago when the center of the continent became increasingly arid and surface sands blew across the landscape.

RIGHT *Chambers Pillar and the red sand of the Simpson Desert.*

ULURU

NORTHERN TERRITORY, AUSTRALIA

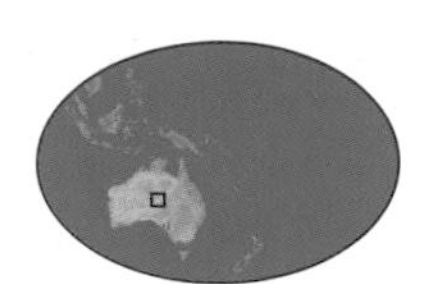

Area of Uluru-Kata Tjuta National Park: 510 sq mi (1,325 sq km)

Height of Uluru: 1,142 ft (348 m)

Age: 500 million years

Uluru, or "Ayres Rock" as it was known for much of the last two centuries, is a sacred place. For thousands of years it has been at the center of the life of its Aboriginal owners—Anangu—who still live there today and who share its management with Parks Australia. The rock itself is a sandstone monolith that rises above the flat dry plains of Australia's Northern Territory, almost at the center of the island continent. It is a formation known in geological terms as an *inselberg*, meaning "island mountain." It was pushed up by major earth movements about 500 million years ago, and most of it—like a vast terrestrial iceberg—remains hidden beneath the surrounding sea of sand.

Fine grooves cover the surface of the rock, while caves and crevices are worn into its flanks. Windblown sand causes the rock to erode, although on the rare occasions that it rains here, water cascades down the sides, leaving black veins on an otherwise uniformly red-colored surface. The rock's colors, however, are magnificent, and appear to change as the day goes on—glowing orange at sunrise, rusty

in early morning, amber at midday, and a dramatic deep crimson at sunset.

The surrounding countryside is dominated by many groves of mulga trees, blue-gray sandalwood, desert oaks, and bloodwood, a form of eucalyptus. Venomous snakes such as the king brown and western brown live among the dry-adapted vegetation, where they hunt marsupial moles, hopping mice, frogs, and lizards. A large pool on one side of the rock contains water for most of the year; the water snakes here are thought by A<u>n</u>angu to be guardians of the rock and pool.

The first European to see Ulu<u>r</u>u was William C. Gosse, who explored the area in the early 1870s. He named Ulu<u>r</u>u "Ayres Rock" after Sir Henry Ayres, who was Chief Secretary of South Australia at the time. Today, more than half a million people visit the rock each year, using the resort and tourist center at Yulara as a base. Aborigine guides lead short walking tours around the area. Circumnavigating the base can take four hours on foot, and climbing to the summit is discouraged because it is insensitive to local Aboriginal culture. Climbing is altogether banned when extreme heat, wind, or rain make it too dangerous. MB

BELOW *The sacred site of Ulu<u>r</u>u.*

KINGS CANYON

NORTHERN TERRITORY, AUSTRALIA

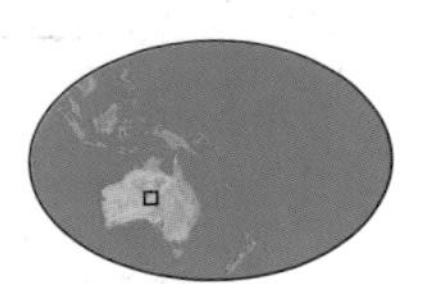

Depth of Kings Canyon: 330 ft (100 m)

Area of canyon: 521 sq mi (1,349 sq km)

Vegetation: desert oasis

Set within Watarrka National Park in Australia's Northern Territory, Kings Canyon is a giant natural amphitheater formed from the same terracotta-colored sandstone as the nearby Uluṟu and Kata Tjuṯa monoliths. It is the red rock area's deepest gorge, its ancient weather-eroded walls plunging down dramatically 330 feet (100 meters) from an escarpment on the surrounding plain to the west of the high plateau of the George Gill Range. Sunsets are absolutely spectacular.

Perched on a mesa next to Kings Canyon are some strange rock formations that look almost like beehives, their artificial appearance having earned them the nickname of the "Lost City." The surrounding Watarrka National Park, also known affectionately as "The Garden of Eden," is a scenic landscape of sand dunes, rugged ranges, rock holes, and moist gorges.

Kings Canyon is one of central Australia's richest botanical sites, providing a welcome refuge for animals from the surrounding desert. Three major biogeographical regions overlap here. There are 80 species of bird, 36 species of reptile, and 19 species of mammal, but it is the plants that command the most attention. The canyon sustains some 60 rare and relict plant species and has come to be regarded as a "living plant museum," principally because of its ancient stands of cycads and skeleton fork ferns. The latter have been found preserved among some of the area's 300-million-year-old rocks and have been termed "living fossils."

The exposed sandstone of the surrounding George Gill Range is a 350-million-year-old Mereenie formation—a rare type of sandstone that forms in dry, desert conditions. Beneath this is an underlying layer of 450-million-year-old Carmichael sandstone, which was formed by the more usual method of tidal or river deposition. The general scientific consensus is that the many deep cracks in the range occurred as the older sandstone was undercut, causing younger sandstone to break off and create today's sheer cliffs. Some of the older sandstone from ancient, wetter times has been preserved in the area's shady gullies and gorges. Watarrka has been the backyard of the local Luritja people for more than 20,000 years, so there are some well-preserved Aboriginal paintings and engravings in the area. The Kings Canyon Walk is a 6-mile (10-kilometer) trail with stunning views of the area. GH

Kings Canyon is one of central Australia's richest sites. There are 80 species of bird, 36 species of reptile, and 19 species of mammal, but it is the plants that command the most attention. The canyon has come to be regarded as a "living plant museum."

RIGHT *A rock escarpment at Kings Canyon.*

JIM JIM FALLS

NORTHERN TERRITORY, AUSTRALIA

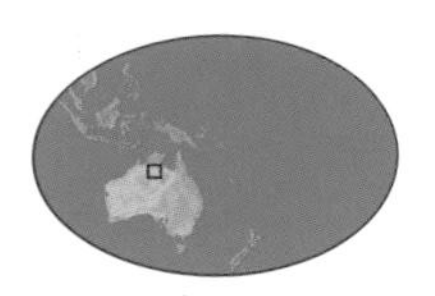

Area of national park: 7,336 sq mi (19,000 sq km)

Depth of falls: 660 ft (200 m)

Vegetation: monsoon forest

During the wet season, from October to May, Jim Jim Falls plunges 660 feet (200 meters) in a single drop from the Arnhem Land escarpment into a large, deep pool. Evening storms deluge the land and fall on the escarpment—once ancient sea cliffs—that rises 1,100 feet (330 meters) above the plains and extends over 310 miles (500 kilometers) along the eastern boundary of Kakadu National Park. The rains that feed Jim Jim Falls, one of the largest falls in the Northern Territory, create massive lakes that teem with birdlife. It is these birds that have won the park its international conservation status. However, the dominant sandstone wall of the Jim Jim face, along with its gorge walls, are spectacular even when the falls slow to a stream in the dry season, from June to September. The falls feed into a short, steep-sided gorge laden with huge rocks and covered in monsoon forest. They then transform into the broad Sandy Creek which feeds

open woodland and permanent billabongs decorated with water lilies and lotus lilies.

During the Mesozoic era, 140 million years ago, much of Kakadu was under a vast shallow sea. According to the Northern Territory Conservation Commission, the escarpment wall and the Arnhem Land plateau formed a flat land above the sea. Then, 100 million years ago, the seas receded. Rocks exposed from beneath the retreating Arnhem escarpment are volcanic in origin and are extremely old, about 2.5 billion years, or about half the age of Earth. Aboriginal artifacts found in the area are dated back 50,000 to 60,000 years, making these the oldest inhabited sites discovered in Australia. Kakadu is 162.5 miles (260 kilometers) east of Darwin on the sealed Arnhem Highway. To reach the falls there is a four-wheel-drive-only track and a difficult 0.6-mile (1-kilometer) walk. When the falls become a thunderous surge in the wet season, the track is closed to vehicles. During this period, the falls can only be accessed by air from nearby Jabiru and Cooinda. GH

BELOW *The scenic Jim Jim Falls during the dry season.*

FITZROY RIVER

WESTERN AUSTRALIA, AUSTRALIA

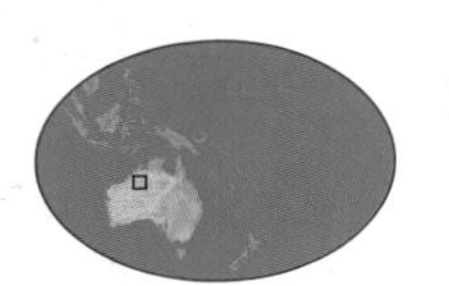

Length of Fitzroy River and its tributaries: 3,032 mi (4,880 km)

Age of reef complexes: up to 350 million years

Vegetation: open savanna and wetland

The Fitzroy River is among Australia's largest rivers and cuts through the 350-million-year-old Devonian Reef complexes to form the Kimberley's famous gorges. Together with the nearby Ord River, the mighty Fitzroy carries the largest volume of water of any river in Australia. When in flood, the Fitzroy is an awesome sight, with an annual average flow that would cover the whole of Australia in three feet (0.9 meters) of water. The Fitzroy's meandering and interconnecting channels are up to 40 feet (12 meters) deep; they sweep through the heart of the ancient Kimberley and encompass the internationally recognized Camballin Floodplain system of wetlands and swamps. Some 67 species of waterbird have been recorded on the extensive black soil floodplain adjoining the Fitzroy. Nineteen of them are listed on the JAMBA and CAMBA treaties that help protect migratory birds. The Fitzroy basin—including the Fitzroy, Mackenzie, Dawson, Connors, and Isaac rivers—is the only known habitat for the reclusive Fitzroy River turtle. A species famous for its method of respiration, it is capable of breathing through its bottom, leading to its nickname "the bum breathing turtle." **GH**

GEIKIE GORGE

WESTERN AUSTRALIA, AUSTRALIA

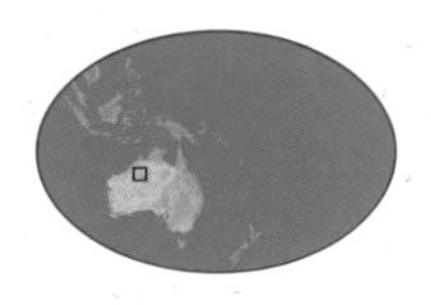

Age of Geikie Gorge: 350 million years

Height of gorge: 165 ft (50 m)

Length of gorge: 9 mi (15 km)

The exhumed reef of the Geikie Gorge is considered to be one of the most spectacular examples of a permanent, tropical, and gorge wetland ecosystem, and is also the longest gorge in the Kimberley region. Its sheer limestone walls, which rise up to 165 feet (50 meters) along its 9 miles (15 kilometers), are regarded as the best-preserved Paleozoic reef sequence known in the world. It is a heritage site for the "superb exposures" that show many of the important features, such as marine fossils, in the Devonian Reef complex of the West Kimberley District. The Fitzroy River carved through the ancient reefs, revealing fossils in the bare gorge walls. The permanent pools are home to at least 18 species of fish including Leichhardt's sawfish and the coach-whip stingray, both marine species, but which strangely occur 200 miles (320 kilometers) upstream. The pools are also a significant freshwater crocodile nursery area. Other animals include the rare purple-crowned fairy wren, gray falcon, Gouldian finch, peregrine falcon, and a specific orange horseshoe bat. The gorge was named after noted British geologist Sir Archibald Geikie by Edward Hardman, who explored the Kimberley region in 1883. **GH**

LITCHFIELD NATIONAL PARK

NORTHERN TERRITORY, AUSTRALIA

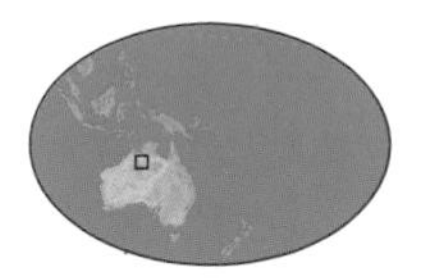

Area of Litchfield National Park: 570 sq mi (1,476 sq km)

Vegetation: eucalyptus woodlands, savanna

Litchfield National Park features numerous compact waterfalls that cascade from a sandstone plateau known as the Table Top Range. The park contains perennial spring-fed streams, permanent clear water holes, and falls up to 33 feet (10 meters) high, including the Buley Rockhole, Wangi, Tolmer, and Florence Falls. Litchfield is contained by steep slopes and encompassed by lowland plains of recent Quaternary sediment—within the weathered sandstone escarpments, plains, and undulating hills lie the weathered sandstone pillars of the so-called "Lost City." Another of the park's unique sights is the hundreds of termite mounds standing up to 6.6 feet (2 meters) high in a wide swathe of empty ground. They are called the magnetic mounds, as they are like enormous magnetic compasses with their thin edges pointing north-south and broad backs east-west—this minimizes their exposure to the sun, keeping them cool for the termites inside. Many of these features are of deep significance to the local Marununggu, Waray, Werat, and Koongurrukun peoples. Bandicoots, wallabies, quolls, honeyeaters, magpielarks, parrots, and rosellas live in the woodlands, and reptiles include skinks, geckos, and goannas. GH

KATHERINE GORGE

NORTHERN TERRITORY, AUSTRALIA

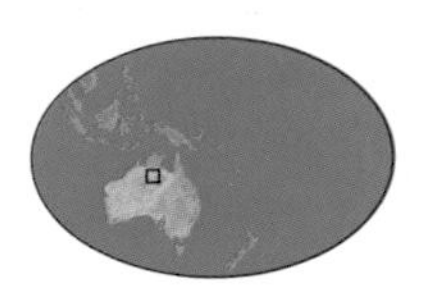

Height of Katherine Gorge: up to 200 ft (60 m)

Rock type: Proterozoic Eon sandstone

Area of national park: 1,127 sq mi (2,919 sq km)

The Nitmiluk (Katherine Gorge) National Park cradles the Katherine River, which has cut a series of 13 magnificent gorges through the surrounding desert landscape. These gorges, with their 200-foot- (60-meter-) high cliffs, rapids, sandy beaches, and long, calm pools have great ceremonial significance to the local Jawoyn people, who are the formal custodians of Nitmiluk and the nearby Edith Falls. The name *Nitmiluk* is a Jawoyn word meaning "the place of cicada dreaming." According to tradition, Bolung—the rainbow serpent—still inhabits the deep pools of the second gorge at Nitmiluk and care must be taken by visitors not to disturb him.

The name Nitmiluk *is a Jawoyn word meaning "the place of cicada dreaming." According to tradition, Bolung—the rainbow serpent—still inhabits the deep pools of the second gorge at Nitmiluk and care must be taken not to disturb him.*

The sheer cliff faces of the gorge show how the Katherine River has sliced down along ancient fault lines in sandstone laid down in the Proterozoic Eon, more than two billion years ago. The plateaus support low, open woodland of eucalyptus, swamps, and wetlands. Pockets of monsoon rainforest are present in the narrower gorges and are among the southernmost remnants of this type in the territory. The park provides a home to a number of rare or endangered plants, such as acacia and hibiscus. Rare or uncommon bird and mammal species include white-throated grass wren, hooded parrot, and rock ringtail possum. Katherine Gorge is also a prime breeding habitat for the freshwater crocodile. Local native mammals include kangaroos, euros, wallabies, bats, and dingoes.

More than 62 miles (100 kilometers) of tracks meander through the park; it is a five-day trek to Edith Falls on these tracks, but it is worth the effort. At the base of the falls, pandanus, melaleucas, and eucalypti fringe a large plunge pool. A number of bird species use the area as an oasis, including many waterfowl.

In 1862, John McDouall Stuart crossed the Katherine River and recorded in his diary: "Came upon another large creek, having a running stream to the south of west and coming from the north of east." Stuart—already on his sixth successful journey across the continent—named the river after one of the members of his sponsor's family. In the wet season, the water level can rise by 60 feet (18 meters) when the river becomes a torrent. GH

RIGHT *The black ribbon of Katherine River slices through Katherine Gorge.*

N'DHALA GORGES

NORTHERN TERRITORY, AUSTRALIA

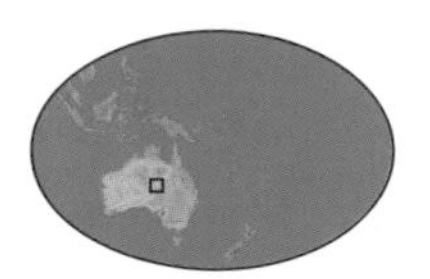

Vegetation of N'Dhala Gorges: rainforest refuge

Notable feature: Aboriginal petrographs

The N'Dhala Gorges are set in the sandstone of the Eastern MacDonnell Ranges that rise sharply out of the surrounding arid zone. N'Dhala Gorges Nature Park contains two eerily quiet gorges: a main gorge about 0.7 miles (1.1 kilometers) long and a side gorge about 2,640 feet (800 meters) long. They are famed for their rock art, or petroglyphs, created by the Eastern Arrernte Aboriginal people. There are over 5,900 prehistoric rock engravings, several art sites, shelter areas, painted areas, and hunting hides. Most are believed to have been created within the past 2,000 years, but some could be as old as 10,000 years. Indigenous custodians indicate that the designs are connected with a Caterpillar Aboriginal Dreaming story. Stone arrangements and other sites of cultural significance in the shady gorge are relics of the art and stories of the Eastern Arrernte people. The atmosphere of the area is characterized by its tumultuous history of bloody conflict between local Aborigines and intending white settlers during the 1880s. The area is also an important site for species of fire-sensitive plants including Hayes wattle and white cypress pine. GH

FINKE GORGE

NORTHERN TERRITORY, AUSTRALIA

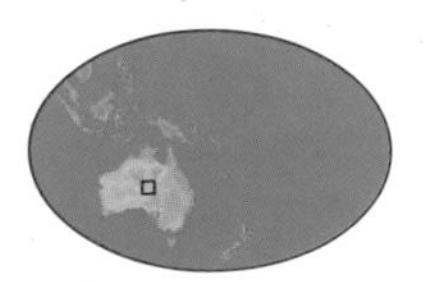

Age of Finke Gorge: approx. 285 million years

Area of national park: 177 sq mi (458 sq km)

Vegetation: rainforest relict

Finke Gorge National Park is the biological ark that carries the unique Palm Valley stand of "living fossils," which include ancient cycads, a relict-status water reed, and the distinctive red cabbage palm. The Finke River—officially claimed as the world's oldest river—cuts the Finke Gorge where it passes through the James Ranges. Wind erosion, the river, and its tributaries have cut features such as the Amphitheater and Initiation Rock in the ancient sandstone. A side stream known as Palm Creek enters the Finke, and its upper reaches provide the refuge known as Palm Valley. Its botanical relicts are from a previous age when the ancient center of Australia was much wetter. Some 3,000 adult red cabbage palms—which grow to a height of up to 82.5 feet (25 meters) and can be 300 years old—are restricted in distribution to the Finke and its tributaries. Known as "Pmolankinya" to the local Arrernte people, the palms and cycads are Dreamtime ancestors supposedly carried here by flames from a bushfire to the north. The suffering of the fire ancestors is represented by the palms' blackened trunks, while the palm leaves represent the long hair of the young men. GH

ORMISTON GORGE AND POUND

NORTHERN TERRITORY, AUSTRALIA

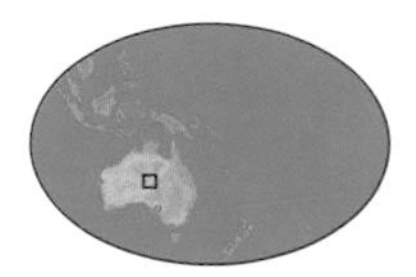

Height of Ormiston Gorge: 1,000 ft (300 m)

Area of gorge: 18 sq mi (47 sq km)

Age of gorge: approx. 500 million years

The beauty of Ormiston Gorge and Pound National Park has been immortalized in the watercolors of iconic Aboriginal artist Albert Namatjira. With its muted hues of gold, blue, and purple vegetation, river red gums, sacred waterholes, 1,000-foot- (300-meter-) high gorge, and folded rock formations, it is regarded by many as one of the most beautiful gorges in Central Australia. A single ghost gum tree stands sentinel above, clinging onto the towering crags of the gorge.

To local Aborigines, the gorge is part of the Emu Dreaming, and the waterhole is a registered sacred site. The creek is fed by what is said to be the world's oldest river, the Finke River, claimed to be 500 million years old. The creek also hosts species of fish threatened by the effects of global warming.

The first European to explore the area was Peter Egerton Warburton on his great trek (in 1873–1874) from Alice Springs to the Western Australian coast, across the Great Sandy Desert. Warburton named the Ormiston Gorge and Pound after his travels across this rugged landscape. The park itself is situated 82.5 miles (132 kilometers) west of Alice Springs. GH

WINDJANA GORGE

WESTERN AUSTRALIA, AUSTRALIA

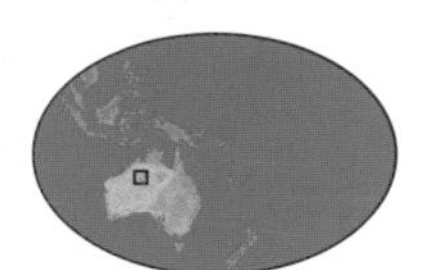

Age of Windjana Gorge:
350 million years

Height of gorge: 330 ft (100 m)

Length of gorge: 3 mi (5 km)

Windjana Gorge National Park, near the northern coast of Western Australia, takes its name from the Aboriginal Dreamtime Creator beings that dominate the local rock art. Geologically speaking, it is a remnant of ancient coral reefs that date back 350 million years to the Devonian period. The stunning Windjana, along with the area's other reef complexes of Geikie Gorge and Tunnel Creek National Park, was once part of a 620-mile (1,000-kilometer) barrier reef that lay submerged beneath an ocean that used to cover most of the Kimberley region.

With the ocean waters long since receded, the sheer black and red walls of Windjana now rise up suddenly to 330 feet (100 meters) from a wide alluvial floodplain and are dotted with distinctive boabs. The gorge was gouged out by the Lennard River, whose waters provide a refuge for wildlife and sustenance to the lush vegetation that grows along its three mile (five kilometer) extent. Primeval life-forms that lived in Devonian times can be seen preserved as fossils in Windjana's limestone walls. During the dry season, the Lennard's water levels drop, leaving behind a series of pools. Their lush collections of trees and shrubs have led to the gorge being described as "a little Shangri-La of shade trees."

The rich soils of the riverbank support native figs, the paper-barked cadjeputs, and the tall, broad-leaved Leichardt tree. Forest bordering the river contains gums, figs, and white cedar, while boabs appear here and there on the slopes. The gorge also provides shelter from the hot sun for many waterbirds, fruit bats, and noisy corellas. Windjana is also one of the best places to see freshwater crocodiles in their natural habitat.

During the dry season, the Lennard River's water levels drop, leaving behind a series of pools. Their lush collections of trees and shrubs have led to the gorge being described as "a little Shangri-La of shade trees."

The gorge's tranquillity masks its dark history as a site of a major battle fought in the 1890s between the police and a force of about 50 warriors led by an indigenous folk hero, Jandamarra. Jandamarra initially worked for the white colonists as a tracker for their police force, helping to round up sheep thieves. However, he turned on the police, attacking them and releasing his indigenous captives. A 10-year period of guerrilla warfare ensued, with Jandamarra using the rugged country of his traditional lands to elude the authorities. He was injured in clashes with police, but lived on. He was eventually shot near Tunnel Creek. GH

RIGHT *Windjana's limestone walls are rich in fossils.*

KATA TJUṮA

NORTHERN TERRITORY, AUSTRALIA

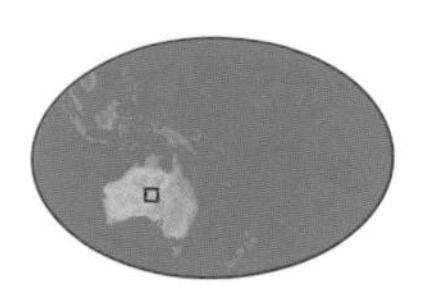

Area of Uluṟu-Kata Tjuṯa National Park: 510 sq mi (1,325 sq km)

Highest point: 1,791 ft (546 m)

The spectacular rock domes of Kata Tjuṯa—known as the Olgas to early European settlers—are named after a phrase in the local Aboriginal language meaning "many heads." Kata Tjuṯa is a group of 36 massive red rock outcrops, along with smaller domes and ridges outside the main grouping that total more than 60. The highest of these domes is 660 feet (200 meters) higher than the famed Uluṟu half an hour's drive away. Like Uluṟu, Kata Tjuṯa is part of the Uluṟu-Kata Tjuṯa National Park that is World Heritage listed for its cultural and natural significance. What are today giant rock mountains were originally laid down as sediments in a shallow inland sea. When, about 300 million years ago, the sea dried up and the area was covered by desert, a major upheaval forced the sedimentary rock up through the earth, and this rock has since been molded by the wind. Along with Uluru, Kata Tjuta is located in the traditional lands of the Pitjantjatjara and Yankunytjatjara Aboriginal people, known locally as Aṉangu. Aṉangu own the land and manage it jointly with Parks Australia. GM

BELOW *The sculpted red domes of Kata Tjuṯa.*

GOSSE BLUFF

NORTHERN TERRITORY, AUSTRALIA

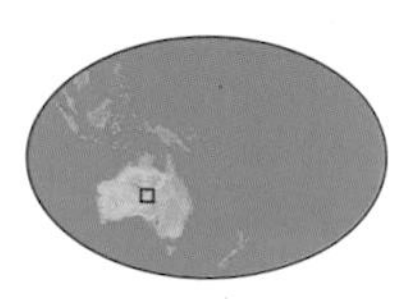

Height of Gosse Bluff crater: 500 ft (150 m)

Diameter of crater: 3 mi (5 km)

Age: 143 million years

Gosse Bluff is a registered sacred site for the Western Arrernte Aboriginal people. To them it is known as Tnorala, and, curiously, the legend of its origin shares similarities with the scientific explanation. Tnorala was created by a massive meteorite crashing to Earth about 143 million years ago. The impact blasted a crater of 8 square miles (20 square kilometers), creating one of the largest impact structures in the world. In Aboriginal folklore, the landscape was formed at the time of Creation. A group of women, of the Milky Way, danced across the sky. During the dance one woman placed her baby in a wooden baby carrier. As the dance progressed the carrier toppled over and crashed to Earth, its sides forming the steep walls of the crater. Since its dramatic beginnings, the crater has been eroding and is three miles (five kilometers) in diameter. It is located 109 miles (175 kilometers) west of Alice Springs in the southwest of the Northern Territory. The site is best appreciated from the air, but a good vista is still afforded from Tyler's Pass. The more pleasant times to visit the area are between April to October when the temperatures are cooler, although sometimes roads are impassable after heavy rain. **MB**

KAKADU NATIONAL PARK

NORTHERN TERRITORY, AUSTRALIA

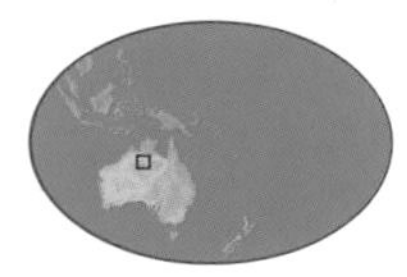

Area of Kakadu National Park: 7,646 sq mi (19,804 sq km)

Notable feature: rugged sandstone plateau 825 ft (250 m) high, extending over 375 mi (604 km)

The magnificent Kakadu National Park, located in Australia's Northern Territory, has been declared a World Heritage site because of its many important and diverse ecosystems and its living Aboriginal culture. Its most spectacular natural feature is a plateau—a huge, rugged sandstone formation that rises sharply to 825 feet (250 meters) and extends over 375 miles (604 kilometers). This area is marked by major waterfalls and deep gorges. Elsewhere in the park, tidal flats and floodplains shape the landscape. The park is at its most impressive in the monsoonal wet season, when violent electrical storms deluge the rugged landscape, creating vast floodplains that later teem with birdlife. However, the best times to visit Kakadu are just after the wet season when the waterfalls are in full flow, or at the end of the dry season, when animals can be seen around the diminished water holes. The area has been inhabited continuously for more than 40,000 years. Kakadu National Park is renowned for its ancient rock art paintings and carvings, and contains some of the earliest settlements in Australia. The name *Kakadu* comes from an Aboriginal Gagudju language. GM

WAVE ROCK

WESTERN AUSTRALIA, AUSTRALIA

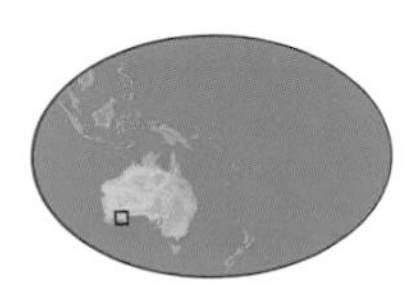

Height of Wave Rock: 50 ft (15 m)

Length of Wave Rock: 363 ft (110 m)

Surrounding vegetation: scrub

Wave Rock resembles a giant breaking wave frozen in time and has been described as "prehistoric surf"—an impression enhanced by the striking colored bands running vertically down the overhanging face. This geological wonder stands alongside some of Australia's other exposed, immovable tors, such as Uluṟu and Kata Tjuṯa, as a classic part of Australia's iconography.

Wave Rock is actually an overhanging wall on the northern side of Hayden Rock, a large granite remnant. It is an inselberg, an outcrop of rock that has been exposed by differential weathering—the surrounding rocks have been eroded by the elements over tens of millions of years. Dozens of these original granite outcrops have been exposed across the local wheat belt district, including Camel Peaks, The Humps, and King Rock.

Wave Rock is believed to be about 500 million years old. Springwater runs down the cliff face, shifting chemicals in the granite and leaving yellow, brown, red, and gray stains of carbonates and iron hydroxide, streaking the rock with lines of color and emphasizing the curvature and eerie heaviness of the overhang. GH

CAPE LE GRAND NATIONAL PARK

WESTERN AUSTRALIA, AUSTRALIA

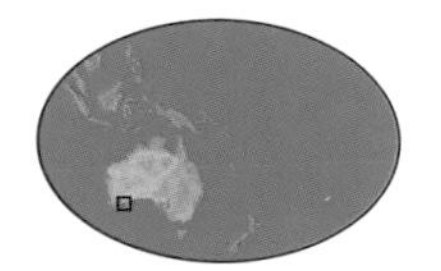

Area of Cape Le Grand National Park: 119 sq mi (308 sq km)

Age of peaks: 40 million years

Vegetation: coastal scrub

Cape Le Grand National Park is regarded as one of the most spectacular areas on Western Australia's coast. Its wild scenery comprises superb bays of white sand and brilliant blue water, framed by rugged granite peaks and sweeping heaths. Massive granite and gneiss outcrops form an impressive chain of peaks, including Mount Le Grand at 1,400 feet (345 meters), Frenchman Peak at 864 feet (262 meters), and Mississippi Hill at 600 feet (180 meters). The peaks are the result of movements in the planet's crust over the past 600 million years. These peaks were mostly submerged some 40 million years ago, when sea levels were over 1,000 feet (300 meters) above their present level. Today, the surrounding heathy sand plain, dotted with swamps and freshwater pools, provides a habitat for many small mammals, including the tiny honey possum. For walkers, the view from Frenchman Peak, across the bay and the park, is breathtaking, as are the views over Hellfire Bay, Thistle Cove, and Lucky Bay. GH

FITZGERALD RIVER NATIONAL PARK

WESTERN AUSTRALIA, AUSTRALIA

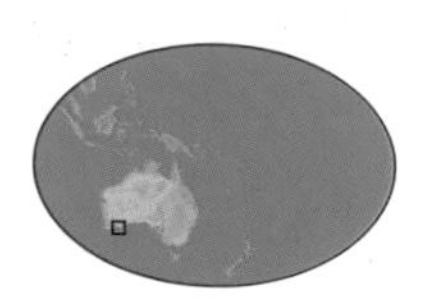

Area of Fitgerald River National Park: 1,274 sq mi (3,300 sq km)

Age of Fitzgerald River: 40 to 43 million years

Depth of Fitzgerald River: 1,500 ft (450 m)

The Fitzgerald River National Park is considered to be one of Australia's most valuable areas for rare and endangered species. It is a huge unspoiled wilderness, featuring four rivers with dramatic gorges and rugged cliffs, as well as wide marine plains fringed with pebbly beaches. A coastal mountain range known as the Barrens rises out of the ocean. Its multicolored cliffs, exposed along the Hamersley and Fitzgerald River valleys, were formed over 36 million years ago. The view across the plateau is regarded as one of the country's great wilderness scenes. To the north of the park, granite exposures distinguish the southerly edge of the ancient Yilgarn Block—a core of continental crust that underlies much of Western Australia. The park is so vast that the dibbler, a small marsupial with distinctive white eye rings, and a heath rat were only recently rediscovered. More than 1,800 species of flowering plant (best viewed from June to November), as well as a plethora of lichens, mosses, and fungi have been recorded. GH

WOLFE CREEK METEORITE CRATER

WESTERN AUSTRALIA, AUSTRALIA

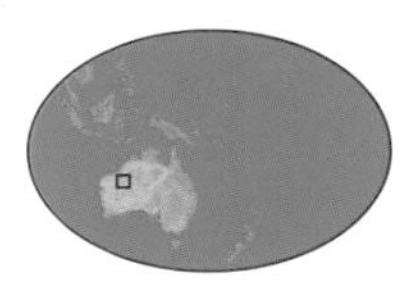

Age of Wolfe Creek Meteorite Crater: 300,000 years

Breadth of crater: 2,900 ft (880 m)

Depth of crater: 200 ft (60 m)

Long known to Aboriginal people as "Kandimalal," the Wolfe Creek crater was only discovered by Europeans in 1947 during an aerial survey. The Aboriginal Dreaming tells of two rainbow snakes that formed the nearby Sturt and Wolfe Creeks as they crossed the desert. The crater is said to be where one of the snakes emerged from the ground. Scientifically, this is the world's second largest crater from which fragments of meteorite have been collected. Datings of the crater rocks and the remains of the meteorite show it collided with Earth about 300,000 years ago. Weighing more than 55,000 tons, the impact of its huge mass is thought to have gouged a 400-foot (120-meter) hole in the desert plain, vaporized most of the iron meteorite, and caused a huge explosion that blasted meteorite fragments to a distance of about 2.5 miles (4 kilometers). The crater—which is only about 200 feet (60 meters) deep today—is home to a variety of wildlife, including brown ringtail dragons. GH

MURCHISON RIVER

WESTERN AUSTRALIA, AUSTRALIA

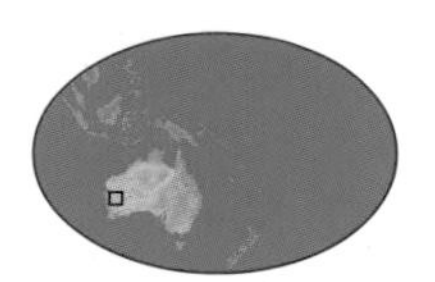

Depth of Murchison River: 429 ft (131 m)

Length of Murchison River: 50 mi (80 km)

Age of Murchison River: over 400 million years

With its origins approximately 300 miles (483 kilometers) inland, the Murchison River winds its way to the Indian Ocean by carving out a steep and twisted 50-mile (80-kilometer) gorge through ancient and jagged 400-million-year-old sandstone. The sandstone has contrasts of brownish red and purple sections against bands of white, and the layers are rich with trace fossils.

The thin-bedded, colored rock and rippled surfaces of the gorge at the foot of Red Bluff were caused by ancient waves moving over tidal flats. Geology students come from across the world to view the Lower Paleozoic (Ordovician) fluvial sandstones, which are thick with excellent trace fossil assemblages and trackways. Some beds, such as those at the Z-Bend and Loop, have a "can of worms" appearance caused by the fossilized burrows of worms. Other trails were left 400 million years ago by one of the first creatures to walk on land—a 6.6-foot- (2-meter-) long sea scorpion called a eurypterid.

Ospreys soar above the sea cliffs and wedge-tailed eagles patrol the gorges. Emus sip from the river's edge and dozens of black swans breed in the shallow pools. Whales frequent the river's entrance and dolphins play in the lower reaches.

The Murchison River is the main feature in Kalbarri National Park in Western Australia, 33 miles (53 kilometers) from Perth. The park contains numerous dramatic features, including giant river meanders such as the famed Loop, the acutely-twisting Z-Bend ravine, notoriously rugged sea stacks, and the Hawkes Head Gorge where the ancient river meets the ocean.

On the coast, some cliffs rise more than 330 feet (100 meters) above the ocean, while others, such as those at Rainbow Valley, frame rainbows in the sea mist or form precariously balanced rock formations. The cliffs to the north extend over 125 miles (200 kilometers) to the World Heritage site of Shark Bay.

The Murchison is home to nearly 200 species of birdlife, including eagles, songbirds, and wetland waders. Ospreys soar above the sea cliffs and wedge-tailed eagles patrol the gorges. Emus sip from the river's edge and dozens of black swans breed in the shallow pools. The Murchison's tidal waters abound with fish, including bream, whiting, and mulloway. Whales frequent the entrance and dolphins play in the lower reaches. Small but ferocious-looking thorny devil lizards thrive in Kalbarri. The park is also famous for its wildflowers, which bloom from late July through spring and into early summer. GH

RIGHT *The meandering clear waters of the Murchison River.*

MITCHELL RIVER AND FALLS

WESTERN AUSTRALIA, AUSTRALIA

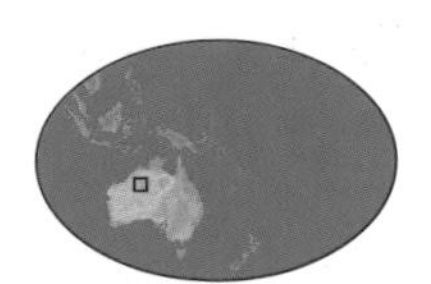

Age of Mitchell River: 1.8 billion years

Area of Mitchell River National Park: 445 sq mi (1,153 sq km)

Vegetation: rainforest patches

Mitchell River National Park contains one of the Kimberley region's most remote jewels, the Mitchell Falls. A series of waterfalls and pools—known as "Punamii-unpuu" by local Aborigines—plunges about 264 feet (80 meters) into a deep, dark freshwater pool. For the Wunambal people at Ngauwudu (Mitchell Plateau), the area is sacred. They believe that supernatural beings, Wungurr (or creator snakes), live in the deep pools below the Punamii-unpuu and Aunauya, so swimming is no longer allowed. To increase understanding of the traditional owners' customs and wishes, a cooperative management arrangement has been set up between the Aboriginal groups and the local authorities.

Punamii-unpuu is fed by the Mitchell River, which has carved gorges and waterfalls into the Mitchell Plateau, and eventually flows into Walmsley Bay and Admiralty Gulf. Small patches of temperate rainforest grow around the margins of the plateau and line some of the gorges. The palm fern grows profusely and giant kanooka trees grow out of the water amid a tangle of vines. The falls are accessed by a two-to-three-hour hike, although helicopter trips are also available. GH

BLUFF KNOLL

WESTERN AUSTRALIA, AUSTRALIA

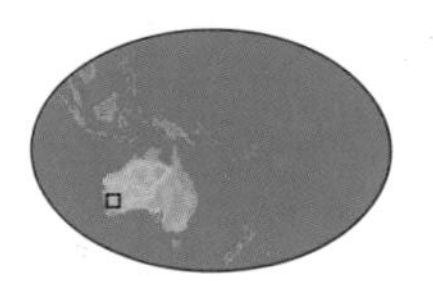

Age of Bluff Knoll: over 100 million years

Height of peak: 3,592 ft (1,095 m)

Length of peak: 40 mi (64 km)

Bluff Knoll in the Stirling Range National Park is the highest peak in the southwest of Western Australia and is known for its brooding beauty. The main face of the bluff forms one of the most impressive cliffs in Australia, with jagged peaks that stretch for 40 miles (64 kilometers). At 3,592 feet (1,095 meters) above sea level, the best views are from the top of Bluff Knoll, named "Pualaar Miial" or "great many-faced hill" by the local Aboriginal people in reference to its face-shaped rocks. The Qaaniyan and Koreng Aboriginal people once lived here. They wore long kangaroo skin cloaks and built conical huts in wet areas.

The peak is often curled in thick mists that float into the gullies and is believed to be an evil spirit called *Noatch*. In 1835, Surveyor-General John Septimus Roe glimpsed "some remarkable and elevated peaks" and named the range after Governor Sir James Stirling—its Aboriginal name is Koi Kyeunu-ruff. The rock here was once sand and silt from a river delta that flowed into a shallow sea. The park is one of the world's most important areas for plants, with 87 species found nowhere else on Earth. It is also famous for its mountain bells. GH

PORONGURUP NATIONAL PARK

WESTERN AUSTRALIA, AUSTRALIA

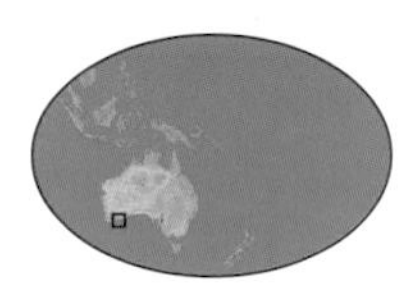

Age of Porongurups: 1.1 billion years

Height of hills: 2,640 ft (800 m)

Length of hills: 7.5 mi (12 km)

The Porongurups are regarded as the oldest hills in the world. This truly ancient and deeply eroded mountain belt lies in one of the oldest parts of Australia—formed by the collision of continents about 1.1 billion years ago. Renowned for their beauty, the Porongurups feature a dozen wooded peaks with bald summits and an elevation of over 2,000 feet (600 meters). The domes rise over the plain, 25 miles (40 kilometers) north of Albany. The highest point is the rugged Devil's Slide, which has been exposed by the slow weathering of the softer rocks surrounding the range. Its 7.5-mile- (12-kilometer-) long ridge catches any coastal moisture to support its "island" of remnant karri forest, leaving the loftier Stirling Range to the north dry and treeless. The karri forest is renowned for its majesty—the mighty trees grow up to 300 feet (90 meters) tall and are among the largest of all living things. In spring, the area is a blaze of color, with the purple flowers of tree hovea, the blues of the Australian bluebell, and the yellows of the pea-flowered narrow-leaved water bush. One of the best views in the Porongurups is from the summit of a tall granite outcrop called Castle Rock. GH

TWO PEOPLES BAY

WESTERN AUSTRALIA, AUSTRALIA

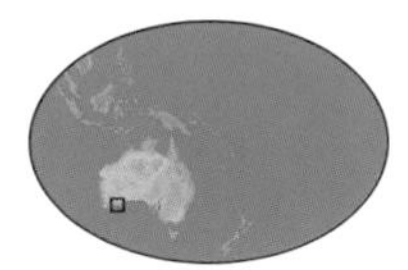

Area of Two Peoples Bay: 18 sq mi (47 sq km)

Age of bay: over 550 million years

World population of rat-kangaroo: approx. 40

Two Peoples Bay is a "mainland ark" for endangered species. It is nestled between the ancient granite massifs of Mount Gardner and Mount Many. The bay is protected from the heavy southern ocean swells by a headland of high rocky hills. A nature reserve covers the headland, islands, and isthmus, linking to a wetland system of lakes, streams, and swamps—remnants of a Pleistocene era estuary. The heath vegetation growing on Precambrian granitic gneiss, and the dense gullies and jarrah woodland, has led to the area's international recognition. Two presumed extinct species have been "rediscovered" here—a songbird and a rat-kangaroo. The reserve is one of Australia's most important regions for migratory birds, with 188 species recorded, including great-winged petrels and little penguins. Bird-watchers are also drawn to Two Peoples Bay by frequent sightings of rare noisy scrub-birds and western bristlebirds, especially at viewing spots such as Little Beach and the Heritage Trail. During the 1840s whalers sheltered in the bay to catch humpback and southern right whales—winter storms sometimes still expose whalebones along the shore. GH

KARIJINI NATIONAL PARK

WESTERN AUSTRALIA, AUSTRALIA

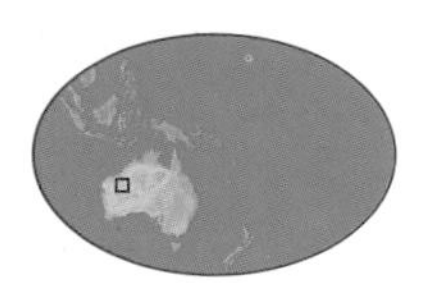

Age of Karijini National Park: 2.5 billion years

Area of national park: 2,420 sq mi (6,268 sq km)

Vegetation: semi-desert shrubland

Karijini National Park—in the heart of the famed Hamersley Range in Western Australia's Pilbara region—is among the country's most beautiful, wild, and dangerous parks, and is the second largest in Australia. Karijini is a tropical semi-desert, with summer thunderstorms and cyclones common over the stunning contours of the Hamersley Range. The park features eight spectacular red-rock gorges, filled with high waterfalls and fringed with lush eucalyptus forests and semi-desert shrubland.

In the north, small creeks—dry for most of the year—flow down abrupt chasms 330 feet (100 meters) deep. Downstream, the gorges expand and their sides transform from sheer cliffs to steep yet loose rocky slopes.

In Dales Gorge, streams, pools, waterfalls, and ferns contrast with the glowing terraced cliffs, weathered by centuries of exposure. The occasional snappy gum can be seen perched on rocky ledges. At Oxer Lookout, the junction of Weano, Red, Hancock, and Joffre gorges, tiers of banded rock tower over a pool at the bottom of the gorge. To explore these gorges you must be fit and prepared to submerge in near-freezing water, follow narrow paths, and cling to rock ledges. The rock of the gorges was a part of the sea floor 2.5 billion years ago, when bacteria and algae were the only forms of life on Earth. Karijini is also the traditional home of the Banyjima, Kurrama, and Innawonga peoples, with evidence of their occupation dating back more than 20,000 years. Termite mounds are a feature of the landscape, and rock piles created by the rare pebble mound mouse can also be seen. GH

BELOW *Red-rock gorges contrast with Karijini's shrubland.*

D'ENTRECASTEAUX NATIONAL PARK

WESTERN AUSTRALIA, AUSTRALIA

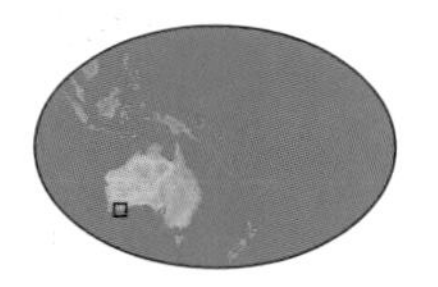

Area of Western Australia: 600 million acres (250 million ha)

Notable feature: coastal cliffs and massive basalt columns

Vegetation: karri forest to heath

Western Australia is vast, and contains some of the most ancient land on Earth. In the state's southwesterly corner lies the D'Entrecasteaux National Park, a region of dramatic coastal scenery—spectacular cliffs, wide beaches interspersed with low rocky headlands, and mobile sand dunes, one of which, the Yeagarup Dune, is an impressive 6.2 miles (10 kilometers) long. The dunes give way to a series of swamps and lakes, including Lake Jasper, the largest freshwater lake in the southern half of Western Australia. The cliffs are covered with pristine wildflower heaths; in more sheltered areas dense patches of karri forest can be found. Offshore, basalt columns stand west of Black Point. These striking landforms were formed from a volcanic lava flow 135 million years ago, an outpouring from the Earth that created a deep pool of molten rock. As it cooled, the lava cracked and shrank, forming perpendicular columns. The result was the close-packed hexagonal shapes now slowly being eroded by the ocean. GM

HOUTMAN ABROLHOS ISLANDS

WESTERN AUSTRALIA, AUSTRALIA

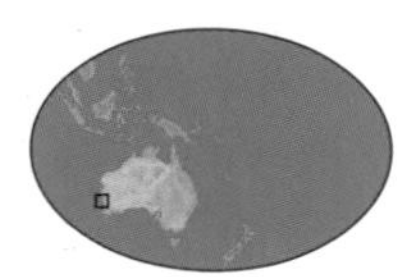

Water temperature during winter: 68 to 72°F (20 to 22°C)

Feature: coral reef surrounded by pellucid water

The Houtman Abrolhos Islands are part of an extensive coral reef system that stretches north to south across 62 miles (100 kilometers) of ocean. The *Abrolhos*—a Portuguese word meaning "keep your eyes open"—form one of the richest marine areas in Australia, as well as a significant site for breeding colonies of swallows, oyster catchers, and fairy terns. As the 120 coral islands lie in the stream of Western Australia's warm, southward-flowing Leeuwin Current, the waters surrounding the Abrolhos have become a meeting place for tropical and temperate sealife. During the winter, water temperatures stay between 68 to 72°F (20 to 22°C), enabling corals and tropical species of fish and invertebrates to thrive in latitudes where they would not normally survive. Only 11 of the Abrolhos Islands are inhabited, and this is just for four months of the year when fishermen pull Western Rock lobsters from the ocean. The remainder of the year the islands are almost deserted. The reefs were frequently the first part of Australia that European sailors chanced upon. Bound for the Dutch East Indies, they were driven off course by the Roaring Forties, and often met their deaths on the reefs. GM

KARRI FOREST

WESTERN AUSTRALIA, AUSTRALIA

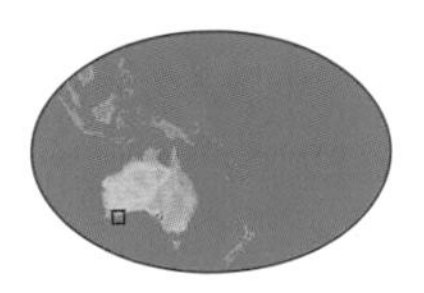

Maximum height of Karri tree: 300 ft (90 m)

Notable feature: karri trees (third tallest trees in the world)

Walking through a lush karri forest in the wilds of Western Australia provides an unforgettable experience of solitude and peace as Australia's tallest hardwood trees soar into the heavens. Well described by its Latin name of *Eucalyptus diversicolor*, the karri may grow to a height of almost 300 feet (90 meters), making it the third tallest tree on Earth. The tree is native to the wetter areas of southwestern Australia and occurs in patches that are often mixed with other tree species to make up a mosaic of vegetation, interspersed with sedgelands and heathlands.

The karri forest understory is soft and lush, especially in areas that have high rainfalls of 43 inches (110 centimeters) or more a year, with a wide variety of colorful plants. Blue wisteria vines and red coral vines grow among the trees. The Boranup Forest near the south coast of Western Australia is the farthest west that karri grows. It is isolated from the main body of the karri belt, more than 62 miles (100 kilometers) to the east, by gray infertile sands caused by lower rainfall. In the southwest, karri grows almost exclusively on deep red clay loams, whereas at Boranup it survives on limestone-based soils. GM

KENNEDY RANGE

WESTERN AUSTRALIA, AUSTRALIA

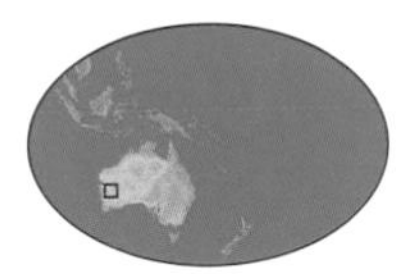

Notable feature of Kennedy Range: huge mesa dominating the surrounding plains

Vegetation: spinifex, wattle, and other shrubs on the mesa; grasses on the plains

In the harsh interior of Western Australia's far northwest, the extraordinary sandstone battlements of the Kennedy Range rise up from the surrounding plains. Some 250 million years ago the region was a shallow ocean basin off the edge of the ancient Australian continent. It was subsequently raised above sea level, where erosion has stripped away much of the rock. Today, marine fossils found in the range's sandstone strata reflect the park's early geological history. This huge mesa extends north for almost 125 miles (200 kilometers) from the Gascoyne Junction region. The national park contains gorges with precipitous sides and a vast plateau topped with ancient dune fields. The apparently endless rows of waterless red sand dunes were formed from underlying sandstone approximately 15,000 years ago. Stabilized by the roots of spinifex with scattered shrubs, the dunes rose up to 60 feet (18 meters) above the troughs between them. Freshwater springs on the western side of the range support abundant wildlife and were the main source of food and water for the Aboriginal people who lived on the range. In the months after winter rains, the dusty red landscape is carpeted with wildflowers. GM

PINNACLES DESERT

WESTERN AUSTRALIA, AUSTRALIA

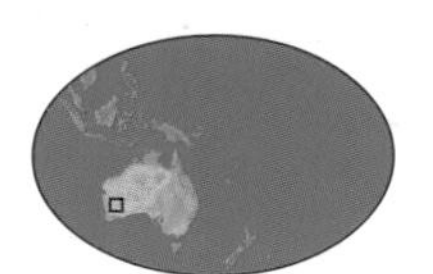

Notable feature: eroded limestone pillars rising up from desert

Vegetation: from trees to heath flowering plants

Among Australia's best-known landscapes is the extraordinary Pinnacles Desert, near the town of Cervantes in Western Australia. Here thousands of huge limestone pillars rise up to 12 feet (3.5 meters) above shifting yellow sands. Many of these strange columns are jagged with sharp edges, while others look like surreal tombstones. The pillars of limestone were formed over tens of thousands of years and are the eroded remnants of formerly thick beds of rock. Seashells, broken down into lime-rich sands, were brought ashore by waves and then carried inland by the wind to form high mobile dunes. These dried out and combined with the sand in the lower levels of the dunes, eventually producing a hard limestone rock. Curiously, while the process of erosion took millennia, the pinnacles are thought to have been exposed only in relatively recent times.

Scientists believe the pinnacles were first exposed about 6,000 years ago but were then covered over by shifting sands before being exposed again in the last few centuries. This process is still occurring, driven by the predominantly southerly winds in the northern part of the desert, which are blowing a steady covering of sand over the pinnacles in the south. Over time, all the limestone spires will probably be swallowed again and the cycle repeated, creating and then covering up more strange and wonderful shapes.

Located in the Nambung National Park, the Pinnacles Desert is a striking contrast to other areas of the park, which include beautiful beaches, coastal dunes, shady groves of tuart trees, and low heath rich in flowering plants. The heath plants burst into flower from August to October, creating a striking spectacle for thousands of visitors. Aboriginal artifacts at least 6,000 years old have been found in the desert, although Aboriginal people have probably not lived in the area for hundreds of years. *Nambung* is an Aboriginal word that means "crooked" or "winding" and it was from the Nambung River that the park was named. Yet the pinnacles remained relatively unknown to Australians until the late 1960s, when the Western Australia Department of Lands and Surveys added the area to an established national park. Today, 150,000 people visit the park each year. GM

Among Australia's best-known landscapes is the extraordinary Pinnacles Desert, where thousands of huge limestone pillars rise above shifting yellow sands. Many are jagged with sharp edges, while others look like surreal tombstones.

RIGHT *The jagged pillars of Pinnacles Desert rise from a yellow sea of sand.*

MARGARET RIVER CAVES

WESTERN AUSTRALIA, AUSTRALIA

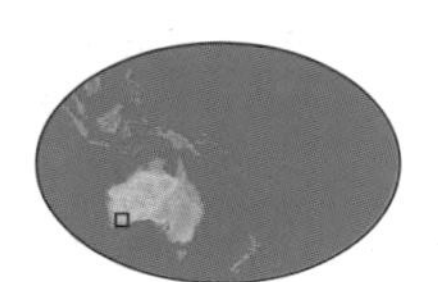

Average temperature of caves: 63°F (17°C)

Number of limestone caves: 350

A band of limestone passing through the Margaret River region has created some 350 caves, all of which are located within the Leeuwin-Naturaliste National Park. Four of these spectacular caves are open to the public. The caves include long, steep ascents and are extremely humid, with temperatures around 63°F (17°C).

Mammoth Cave reveals prehistoric fauna from over 35,000 years ago—it provides easy viewing and a chance to see fossil remains of the extinct *Zygomaturus trilobus*, a wombat-like creature. Lake Cave, with its aura of peace and mystery, is a pristine chamber deep beneath the earth. Within the cave lies a tranquil lake reflecting delicate limestone formations. Probably the most startling cave, however, is Jewel Cave. Its lofty chambers hold intricate decorations that glow with a golden light. It has one of the longest straw stalactites—20 feet (6 meters) long—found in any tourist cave. Fossil remains have also been found here, making it an important site for research. Moondyne is the most recently developed cave, where visitors can experience "adventure caving," wearing overalls, helmets, and miners' lights to explore the hidden depths. GM

MOUNT AUGUSTUS

WESTERN AUSTRALIA, AUSTRALIA

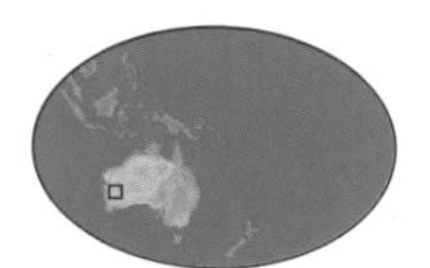

Height of Mt. Augustus: 2,366 ft (717 m)

Notable feature: world's biggest rock—largest and most solitary peak on Earth

One of the most solitary peaks on Earth is Mount Augustus, located 530 miles (853 kilometers) north of Perth. The mountain is actually a huge rock—the biggest in the world—and rises 2,366 feet (717 meters) above a stony, red sand plain of arid shrubland. Standing out on the plain plateau, it is clearly visible from the air for more than 100 miles (160 kilometers). It is in fact twice the size of Australia's famed Uluṟu, five miles (eight kilometers) long and covering an area of 19 square miles (49 square kilometers). Mount Augustus and its surrounding terrain were formed from sediment deposited on an ancient sea floor 100 million years ago. The deposits formed sandstone and conglomerate strata that eventually, with movement in the planet's crust, folded and uplifted. The underlying granite is 1.65 billion years old. Not only is the mountain twice the size of Uluṟu, it is considerably older. Surrounding the rock are white-barked river gums, while species of wattle spread across the red sand plain where honeyeaters and babblers forage for food. Nearby, emus seek fruits, while bustards snatch insects and small reptiles from the ground. GM

MUNDARING WEIR

WESTERN AUSTRALIA, AUSTRALIA

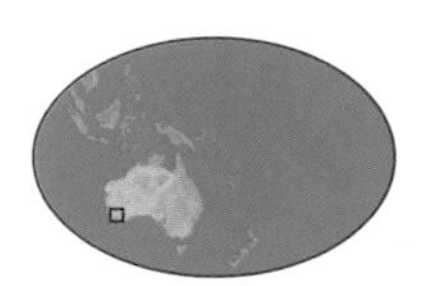

Area of Mundaring Weir reservoir: 27.2 million cu yards (21 million cu m)

Notable feature: system that pumps water 438 mi (705 km) inland

Situated in an attractive bush setting not far from the center of Perth is one of the world's greatest engineering projects, which now provides a showcase for some stunning natural wonders. Completed in 1903, the Mundaring Weir forms the start of a pipeline that carries water 438 miles (705 kilometers) inland both to agricultural areas and to the Coolgardie and Kalgoorlie goldfields. In the 1890s, thousands of prospectors were drawn to the infant goldfields, but in the searing heat, water was in desperately short supply. Given the task of getting freshwater to the miners, the Engineer-in-Chief of Public Works decided to build a storage reservoir in the hills near Perth and pump the water inland. The project involved constructing a 27.2-million-cubic-yard (21-million-cubic-meter) storage reservoir at Mundaring, as well as eight large steam-driven pumping stations to pump the water through a steel pipeline. Today the 404-mile (650-kilometer) Golden Pipeline Heritage Trail is a feature of the area, passing through areas of forest that provide a home to an abundance of nature, such as kangaroos, emus, and jarrah trees—including a 600-year-old specimen known as the King Jarrah. GM

RUDALL RIVER NATIONAL PARK

WESTERN AUSTRALIA, AUSTRALIA

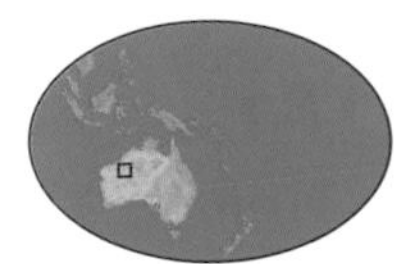

Notable feature of national park: huge desert and massive sand dunes

Vegetation: a mosaic of tree and shrub steppe covering sand dunes and rocky hills

The Rudall River National Park is one of the biggest and most remote national parks in the world and is located on the boundary between the Great Sandy and Little Sandy Deserts. The predominantly flat region is subject to cyclones and intense rainstorms—the mottled ground bears testament to frequent fires started when lightning strikes. The park follows the course of the Rudall River, which rises in rugged hills, then flows northeast through sand-dune country into Lake Dora on the edge of the Great Sandy Desert.

The region is characterized by patchy areas of grass and desert vegetation. It is an undeveloped national park, set aside for conservation and research rather than tourism—the area is accessible only to those with a permit. Along the watercourses of the Rudall River lie a number of permanent water holes, which are rare in the region. They serve as an oasis for a rich and diverse range of flora and fauna, including birds, mammals, and reptiles. GM

SERPENTINE NATIONAL PARK

WESTERN AUSTRALIA, AUSTRALIA

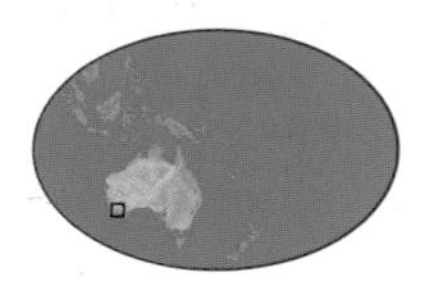

Area of Serpentine National Park: 10,300 acres (4,300 ha)

Notable features: river and dramatic waterfall

The Serpentine National Park is located on a geological feature known as the Darling Scarp, at the western edge of a huge 2.5-billion-year-old plateau that forms the base of much of southwestern Australia. The park starts at the foot of the scarp and rises up the steep slopes of the Serpentine River valley, past a granite wall polished smooth by the rushing waters of the Serpentine Falls. In winter, the white waters of the Serpentine River cascade into a swirling, rock-rimmed pool.

Originally, the Serpentine River was a series of lakes that joined in the rainy season to snake down the Peel-Harvey estuary. When Europeans settled here, they dug long, deep channels between the lakes, effectively "straightening" the natural watercourse. Today, environmental programs are underway to "put the bends back" using logs and strategic planting, so that the river mimics natural-flowing streams.

A 0.3-mile (0.5-kilometer) trail runs along the Serpentine River to the falls. The park contains stunning jarrah forests, and hillsides become a blaze of color as wildflowers bloom from July to November. The park lies 32 miles (52 kilometers) southeast of Perth. GM

TORNDIRRUP PENINSULA

WESTERN AUSTRALIA, AUSTRALIA

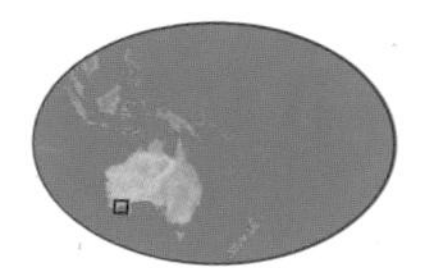

Notable features: oldest rocks on earth and blowholes

Native mammals: pygmy possum, kangaroo

The Torndirrup Peninsula is composed of three major rock types, the oldest of which assumed its current form as a result of enormous pressures and temperatures between 1.3 and 1.6 billion years ago. This means that the rocks predate almost all life on Earth. Yet despite their extreme age, these gneisses were formed during the second half of Earth's history.

The southern ocean has sculpted a natural bridge in the coastal granites on the peninsula. The Torndirrup Gap is a fearsome place where the waves rush in and out with tremendous ferocity. Here at the blowholes, air blows through a crack in the granite, creating a noise loud enough to shock the eardrums. Along this dramatic terrain, windswept coastal heath gives way to huge granite outcrops, sheer cliffs, and steep sandy slopes and dunes. In spring the heath puts on colorful displays of wildflowers. A rare flowering plant, the Albany woollybus, has also been found in the park. The vegetation on the peninsula provides habitats for a number of native mammals, such as pygmy possums and kangaroos. Offshore, whales can be seen from the cliffs during winter, and seals sometimes visit the coast. GM

PURNULULU

WESTERN AUSTRALIA, AUSTRALIA

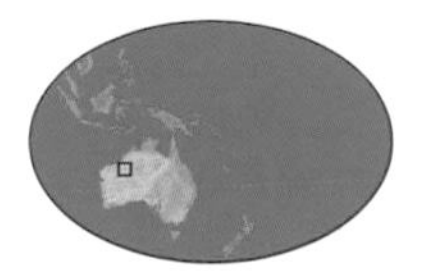

Elevation of Purnululu: up to 1,896 ft (578 m)

Age: 360 million years (Devonian)

Vegetation: tropical savanna

Purnululu (known to some as the "Bungle Bungle Mountains") is a complex of gigantic and ancient beehive-shaped towers, with striking horizontal bands of layered black lichens and orange silica. The vast massif rises to a height of about 1,000 feet (300 meters) above the neighboring woodland and grass-covered plain, with steep cliffs on its western face. These rock towers have been World Heritage listed, as they are regarded as one of the most extensive examples of sandstone towers in the world. The domes were created through a complex process of sedimentation, compaction, and uplift caused by the collision of Gondwanaland and Laurasia about 300 million years ago, the convergence of the Indo-Australian Plate and the Pacific Plates 20 million years ago, and millions of years of weathering. The rocks sit amid a dramatic plateau and are cut by 660-foot- (200-meter-) deep sheer-sided gorges, such as Echidna Chasm, Cathedral Gorge, and Piccaninny Gorge. These are dotted with numerous waterfalls and springs and vast caves. Rainforest plants such as palms and ferns are found in the deeper valleys. Aboriginal tribes have lived here for at least 40,000 years. GH

NINGALOO REEF

WESTERN AUSTRALIA, AUSTRALIA

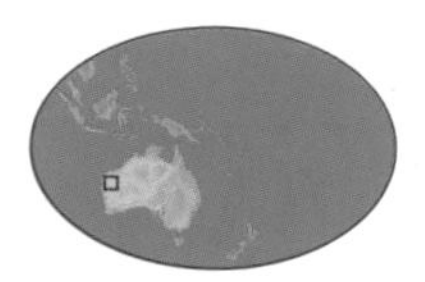

Length of Ningaloo Reef: over 161 mi (260 km)

Width of reef: up to 660 ft (200 m)

Length of whale shark: 40 ft (12 m)

Ningaloo Reef is a virtually untouched fringing reef protecting a shallow, brilliant white sandy lagoon of clear tropical waters. Every March and April, just days after the mass spawning of corals, whale sharks arrive to feed on a profusion of marine life that itself feeds on great clouds of coral spawn. These gentle ocean giants are about the length of a large bus but are the least feared of all sharks because they feed only on zooplankton, squid, and small fish. Ningaloo Reef is one of the few places in the world that the world's largest fish congregates regularly in significant numbers.

Visitors can witness the event on controlled sightseeing tours offshore. Boats observe an 820-foot (250-meter) protected zone around the sharks, with only one vessel at a time in the contact zone, and none for longer than 90 minutes. People swimming with whale sharks must not touch or ride on the animals. They must remain 3.3 feet (1 meter) clear of the fish's head or body and 13 feet (4 meters) away from its powerful tail flukes. During the winter months the waters beyond the reef are the migratory route for dolphins, dugongs, and humpback whales moving north to calve in warmer waters. GH

SHARK BAY & STROMATOLITES

WESTERN AUSTRALIA, AUSTRALIA

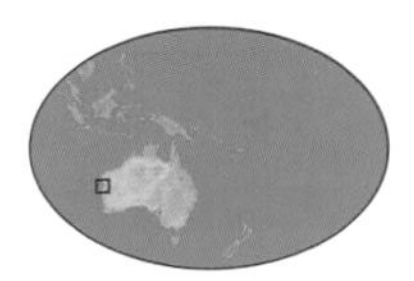

Area of Hamelin Pool Nature Reserve: 510 sq mi (1,320 sq km)

Height of stromatolite columns: up to 5 ft (1.5 m)

Living stromatolites discovered: 1956

On the northwest coast of Australia is the aptly named Shark Bay, famous for seagrass and sharks. At Hamelin Pool, one corner of this vast bay, there lies something even more interesting—living stromatolites. Here, a sandbar covered by seagrass impedes the movement of the tide, so evaporation by wind and sun contributes to make the waters highly salty. Grazing mollusks that would normally keep organisms such as blue-green algae under control are absent, so the blue-greens thrive unchecked. Like corals, they secrete lime, forming calcium carbonate cushions in the shallows and tall columns in deeper water. These cushions have been found as fossils in rocks, and have been shown to be over 2 billion years old. The scene at Hamelin Pool at low tide, therefore, is a window on a time when the simplest organisms we see today dominated the planet. A boardwalk enables visitors to view the living stromatolites without damaging them. The best time to visit Shark Bay is from June through October, when winds are light and temperatures pleasant. **MB**

ALLIGATOR GORGE

SOUTH AUSTRALIA, AUSTRALIA

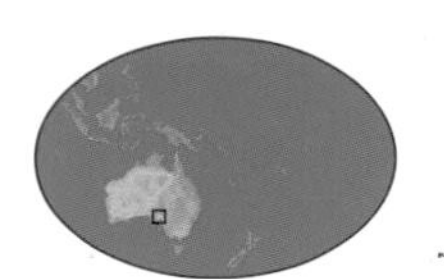

Length of Alligator Gorge: 3 mi (5 km)

Area of Mount Remarkable National Park: 43,000 acres (17,500 ha)

Located in the north of Mount Remarkable National Park, Alligator Gorge offers some of the most beautiful and spectacular views in the Flinders Ranges. The sheer red-brown quartzite cliffs tower 100 feet (30 meters), and steps leading down into the rugged gorge mark the start of two walking trails. One track follows the creek upstream to "The Terraces." Here the gorge floor features rocks patterned with ripple lines. These were created 500 to 600 million years ago when the Flinders Ranges formed part of an ancient seabed. A second trail leads to "The Narrows," where the walls of the gorge are only 10 feet (3 meters) apart and it is possible for just two or three people to pass at a time. This section of the gorge can become flooded, but there are large stepping stones to prevent visitors from getting their feet wet. Ferns and eucalyptus trees fight for light in the shady crevices, but in more sunlit areas dramatic displays of wildflowers can be seen—throughout the gorge it may be possible to spot kangaroos and emus. Until the mid-1960s this area was used for sheep grazing and timber production. GM

BLUE LAKE

SOUTH AUSTRALIA, AUSTRALIA

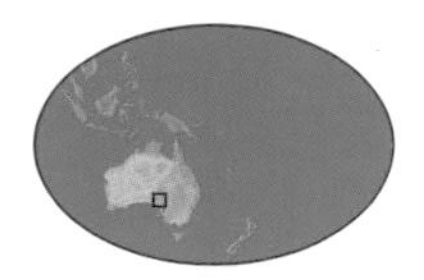

Depth of Blue Lake: 250 ft (75 m)

Age of lake: 5,000 years

During the winter, Blue Lake is a dull shade of gray, but incredibly each September the lake takes on a spectacular transformation, turning a deep azure. This vibrant color remains all summer until it gradually fades the following March. It is not known why the lake changes color. Some scientists believe that microscopic blue organisms rise to the surface in the warmer months, giving rise to its bold color. Others claim that as the temperature rises at the surface, dissolved calcium carbonate salts precipitate as extremely fine particles that scatter light at the blue end of the spectrum. Blue Lake—thought to be at least 250 feet (75 meters) deep—is one of three crater lakes on Mount Gambier, a volcano that last erupted about 5,000 years ago. Although the crater rim measures 66 feet (20 meters), due to the lake's great depth, the bottom of the lake is 100 feet (30 meters) below the level of the main street of the nearby town of Mount Gambier. Each year the town holds a festival when the lake turns blue. GM

BELOW *Blue Lake's stunning color is best viewed in summer.*

CANUNDA NATIONAL PARK

SOUTH AUSTRALIA, AUSTRALIA

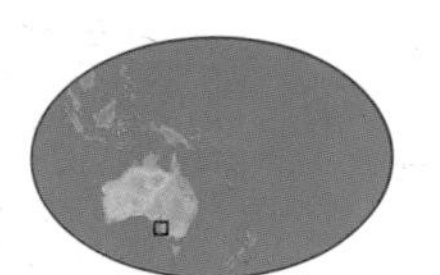

Area of Canunda National Park: 43 sq mi (110 sq km)

Rare bird species: orange-bellied parrot, hooded plover

Dominated by vast sand dunes and a stunning shoreline, the Canunda National Park runs along South Australia's southeast coast for 25 miles (40 kilometers). It is the largest coastal park in the state, covering 43 square miles (110 square kilometers), and lies between the sea and Lake Bonney. The park's northern area is characterized by low limestone cliffs facing the turbulent southern ocean, where sea stacks, offshore islands, and reefs dot the seascape. In the south, the landscape is dominated by huge mobile sand dunes and long stretches of dangerous surf. On the sandy shores, one of the world's rarest bird species, the orange-bellied parrot, spends the winter months feeding on the abundant crop of sea rocket. Between August and January, another rarity, the hooded plover, nests just above the high-water mark. Vegetation varies greatly from stunted and twisted coastal plants along the cliff tops, to reeds and tea trees on the inland swamps. Large Aboriginal middens that point to human habitation in the area for tens of thousands of years have been identified and protected. Huge dune drifts that change the topography from year to year sometimes reveal the remains of Aboriginal campsites. GM

THE COORONG

SOUTH AUSTRALIA, AUSTRALIA

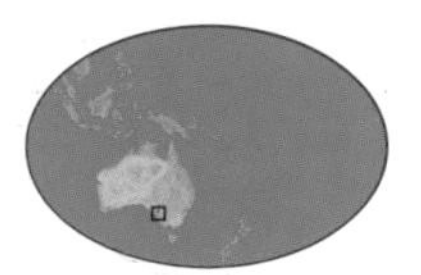

Area of the Coorong: 120,000 acres (50,000 ha)

Length of the Coorong: 62 mi (100 km)

Notable feature: refuge for waterbirds

Adjacent to the southern ocean and near the mouth of the Murray River in South Australia is a range of parallel sand dunes. In the depression between the dunes there is a saltwater lagoon formed tens of thousands of years ago. This is the Coorong, a 120,000 acre (50,000 hectare) national park, significant as the temporary home for migratory wading birds attracted to its coastal and salt-influenced terrestrial and aquatic habitats. A popular destination for visitors is the Younghusband Peninsula. The Coorong was created in 1966 to conserve the landscape of dunes, lagoons, wetlands, and coastal vegetation, as well as the huge variety of birds, animals, and fish that live in or visit the area. The park's immense ecological value was recognized when it was put on the list of "Wetlands of International Importance especially as Waterfowl Habitat," maintained by the International Union for the Conservation of Nature and Natural Resources in 1975. Six years later Australia, Japan, and China signed an agreement to protect migratory birds, birds in danger of extinction, and their environments. GM

RIGHT *The Coorong is an important habitat for waterfowl.*

GAWLER RANGES

SOUTH AUSTRALIA, AUSTRALIA

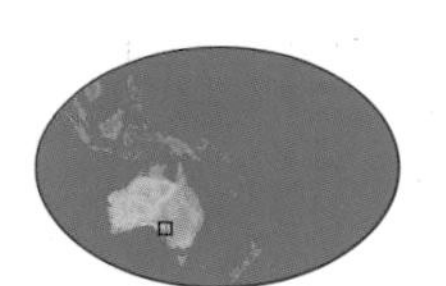

Area of the Gawler Ranges: 6,600 sq mi (17,000 sq km)

Vegetation: low-lying saltbush scrub, heavily wooded eucalyptus country

On the northern side of South Australia's Eyre Peninsula are the Gawler Ranges, a dry region with gorges, weathered rocky outcrops, and seasonal waterfalls. This is a wilderness area where vast domes of volcanic rock present a vivid array of colors in striking contrast to the pure white of the many saltwater lakes that dot the region. The ranges are renowned for their display of wildflowers in the spring; the first recorded sighting of South Australia's floral emblem—Sturt's Desert Pea, a brilliant flower with bulging black "eyes"—was made here in 1839 by explorer Edward John Eyre. Some 140 species of bird have been recorded, from the long-legged emu to the soaring wedge-tailed eagle, from the screeching Major Mitchell cockatoo to tiny flycatchers and rainbow bee eaters. It is an excellent place for viewing wildlife, being one of the few areas where three of Australia's five kangaroo species can be seen in one location (the red, western gray, and euro), as well as other marsupials such as southern hairy-nosed wombats, hopping mice, and pygmy possums. GM

GREAT AUSTRALIAN BIGHT

SOUTH AUSTRALIA, AUSTRALIA

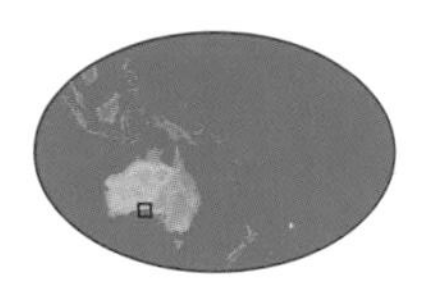

Length of Great Australian Bight: 725 mi (1,160 km)

Notable feature: world's longest line of sea cliffs

The largest island in the world also features the longest line of sea cliffs in the world, known as the Great Australian Bight. White-colored rock near the base of these cliffs was formed on the seabed some 40 million years ago. To Edmund Delisser, a 19th-century surveyor, it looked as if the body of an immense whale was embedded here. The bight forms a huge open bay on the southern edge of the continent, adjacent to the Nullarbor Plain. Aborigines lived on its shores for millennia, but it was not until the 19th century that the first European sailed into the bight. A narrow strip of ocean—measuring 20 miles (32 kilometers) wide and 200 miles (320 kilometers) long—was declared a protected marine park in 1998. Many large sea animals live in or visit the Great Australian Bight's waters, the most famous being the southern right whale, which breeds and calves here, as well as the rare Australian sea lion and the notorious great white shark. Southern bluefin tuna also migrate through the bight. GM

BELOW *The long stretch of cliffs of the Great Australian Bight.*

KANGAROO ISLAND

SOUTH AUSTRALIA, AUSTRALIA

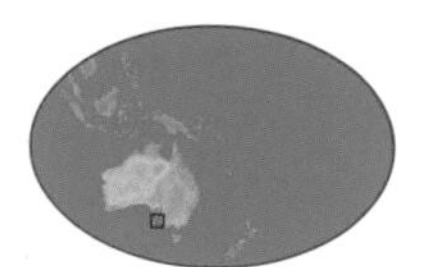

Length of Kangaroo Island: 100 mi (160 km)

Width of island: 34 mi (55 km)

Vegetation: thick eucalyptus scrub, herbaceous plants

Located off the South Australian coast, across the narrow Backstairs Passage from the Fleurieu Peninsula, Kangaroo Island is Australia's third largest island. More than half the land has never been cleared of vegetation and a third has been conserved in national and conservation parks, including five significant wilderness protection areas. Free of foxes and rabbits, wildlife is plentiful, and tourists flock to the island to see the many animals and birds that thrive there.

Aborigines were the first to settle on Kangaroo Island. However, the history of their occupation is scant and archeologists believe they left 3,000 years ago for reasons that may never be determined. Isolation, poor soils, a scrubby vegetation, and an average annual rainfall of less than 29 inches (75 centimeters) mean the island has remained largely underdeveloped. Matthew Flinders was the first European to land on Kangaroo Island. During his circumnavigation of Australia in 1802, he explored, charted, and then named the island after he and his crew had killed 31 strange, hopping animals for their meat and to make soup from their massive tails. Kangaroos still inhabit the island today. GM

LIMESTONE COAST

SOUTH AUSTRALIA, AUSTRALIA

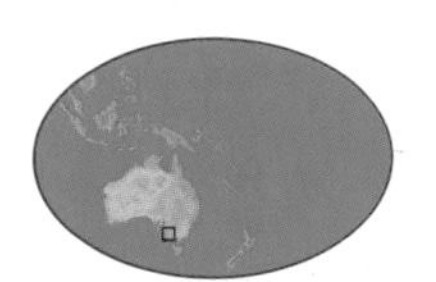

Area of Limestone Coast: 8,000 sq mi (21,000 sq km)

Notable features: vast network of caves and wetlands

Halfway between Adelaide and Melbourne is the Limestone Coast. Named after the bedrock that provides a natural filter for the many vineyards that thrive in the area, the limestone gives the region the unique terra-rossa soil of the Coonawarra, renowned for producing some of the world's finest red wines. The presence of limestone has also led to the formation of many of the area's caves and wetlands, which are rated of international importance. Beautiful pastures conceal a remarkable honeycomb of caves that contain layers of prehistoric animal fossils. Aboveground, pine forests and extinct volcanoes dot the landscape.

The Limestone Coast has a plentiful supply of underground water complemented by good rainfall. This, combined with the fertile soil, temperate climate, and flat land, makes it highly suitable for industry and agriculture. The region produces 10 percent of Australia's national wine grape crush but accounts for 20 percent of the country's premium wine production. It is also famous for the quality of its crayfish, specifically rock lobster, harvested from the cold waters of the southern ocean. GM

NARACOORTE CAVES NATIONAL PARK

SOUTH AUSTRALIA, AUSTRALIA

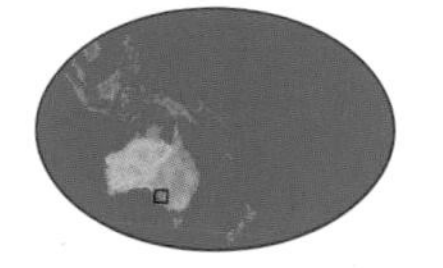

Area of Naracoorte Caves National Park: 1,440 acres (600 ha)

Notable features: fossil specimens of species from very small frogs to buffalo-sized marsupials; Australian ice age megafauna

The origins of the Naracoorte Caves date back to when Australia broke away from the ancient supercontinent of Gondwanaland 50 million years ago. At this time, the Southern Ocean reached 62 miles (100 kilometers) further inland than its current position and, over millions of years, a thick layer of limestone was formed on the seabed. As the sea retreated, groundwater slowly dissolved the limestone to form a wide network of underground caves. Today, many of these caverns have magnificent growths of stalactites and stalagmites. One of the most spectacular is Alexandra Cave, which was only discovered in 1908.

For more than 500,000 years the caves have acted as pitfall traps and predator dens, preserving an exceptional fossil record of the thousands of animals that lived in and around the caves. Many of these ancient animals have been recreated at the Wonambi Fossil Center. The fossil chamber in the Victoria Fossil Cave, which is the largest and most extensively studied cavern, has given scientists a unique insight into the region's prehistoric climate, vegetation, and environment. The studies have also provided a better understanding of the evolution of Australian fauna, particularly the megafauna, such as the giant kangaroo and marsupial lion, which lived here until around 47,000 years ago.

Many caves still provide important habitats for bats and other cave-dwelling creatures. Tens of thousands of bentwing bats live within the depths of the Bat Cave. Infrared cameras offer a unique opportunity to observe bats in their natural environment without disturbing them. The females arrive in spring to give birth to their young, and the cameras placed throughout the cave allow visitors to see what is happening. This is the only place in the world where such technology has been used to view bats in their natural habitat. The bats can be seen drinking water from tiny straw stalactites in the cave, grooming themselves, and suckling their young, called pups. There are also sightings of albino bats, and there has been one sighting of a 6.6-foot (two-meter) brown snake in the cave. GM

For more than 500,000 years, the caves have acted as pitfall traps and predator dens, preserving a rich fossil record of thousands of animals. Many caves still provide important habitats for bats and other cave-dwelling creatures.

RIGHT *Delicate stalactites in Victoria Fossil Cave.*

LAKE EYRE BASIN

NORTHERN TERRITORY / QUEENSLAND / SOUTH AUSTRALIA, AUSTRALIA

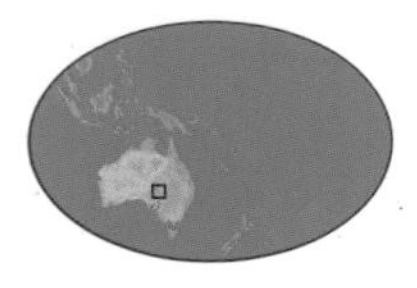

Area of Lake Eyre: 3,900 sq mi (10,000 sq km)

Area of river system: 400,000 sq mi (1,036,000 sq km)

Annual rainfall: 0.5 in (1.25 cm)

Covering an extraordinary one sixth of the Australian continent, the Lake Eyre Basin contains some of the rarest, least exploited ecosystems on Earth. The basin has been described as one of the world's last unregulated, wild river systems, spreading out over more than 400,000 square miles (1,036,000 square kilometers) of the country's arid and semi-arid heart. Unlike other rivers, however, its flows are highly varied and unpredictable. At 50 feet (15 meters) below sea level, Lake Eyre is Australia's lowest point and, while it usually contains little or no water, it is also the fifth-largest terminal lake on Earth. The rivers and creeks that flow into the lake run for only short periods after drenching rain, and most of the time they are dry. When the water evaporates, large salt pans remain. There are few trees in the region and little moisture for most of the year—temperatures regularly exceed 122°F (50°C) and have been known

to reach 140°F (60°C); the average annual rainfall is only 0.5 inches (1.25 centimeters). Even when the water does move, its flow steadily decreases because of a vast network of braided channels, floodplains, water holes, and wetlands. The large permanent water holes provide important habitats for wildlife and are indispensable to the region's towns, smaller communities, and cattle ranches. The highest recorded water level in Lake Eyre was in 1974, but it would take the average flow of Australia's largest river, the Murray, to maintain that level. The Eyre Basin is part of Australia's arid zone and the ecosystems it supports are varied and often unique, making it an important conservation area, containing wetlands, grasslands, and deserts. A number of endangered species also call this area home. Rare marsupials, the greater bilby and the kowari are both found in this area, as is the *Acacia peuce* (waddywood), one of Australia's rarest and most striking trees. The area is culturally significant and contains a wealth of Aboriginal and non-Aboriginal history. GM

BELOW *The Lake Eyre Basin is characterized by salt pans.*

NULLARBOR PLAIN

SOUTH AUSTRALIA, AUSTRALIA

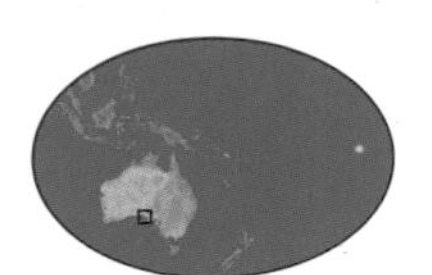

Area of Nullarbor Plain: 105,000 sq mi (272,000 sq km)

Notable features: world's largest single piece of limestone with underground caves and lakes

Australia's Nullarbor Plain is the world's largest single piece of limestone: a vast, flat, treeless landscape that divides the eastern and western regions of Australia. It rises up to 660 feet (200 meters) above sea level and extends for 1,250 miles (2,000 kilometers) across the southern parts of South and Western Australia. The plain runs south of the Great Victoria Desert and ends abruptly with the cliffs of the Great Australian Bight. The driest place in Australia is found in this region—the Farina settlement receives an annual rainfall of just 5.6 inches (142 millimeters). The plain's name means, appropriately enough, "no trees." Instead, the most abundant form of vegetation found in this arid and semi-arid landscape is small scrub.

Twenty-five million years ago, this was the bed of an ancient sea but, over time, it has been uplifted by earth movements and scoured flat by wind and rain. It is Australia's largest karst or weathered limestone terrain. In places the limestone layer is up to 200 feet (60 meters) deep. Continuing water erosion has created many collapsed sinkholes that lead down to deep cave systems. One is the Cocklebiddy Cave, which consists of a straight tunnel over 3.7 miles (6 kilometers) long, of which 90 percent is flooded by an aquifer that lies 300 feet (90 meters) below the surface.

This amazing natural wonder has prompted the creation of a number of human ones. The Transcontinental Railway runs across the Nullarbor from Port Augusta in South Australia to Perth in Western Australia and boasts the longest section of straight track in the world—297 miles (478 kilometers). The Eyre Highway also passes through the plain's southernmost area and has a 94-mile (150-kilometer) stretch that is said to be the longest straight section of tarred road on Earth. Along the highway are five of the most spectacular lookouts on the Australian coastline, where towering, vertical limestone cliffs are assaulted by the Southern Ocean. The waves flow into caves and some have become blowholes. Subterranean rivers course through them, forming large underground lakes. The waters abound with small crustaceans, spiders, and beetles, many of which have adapted to the constant darkness and are blind. Important fossils have also been found in these caves, including the skeleton of a giant lion. GM

Nullarbor Plain is the world's largest single piece of limestone: a vast, flat, treeless landscape that divides the eastern and western regions of Australia. It rises up to 660 feet (200 meters) above sea level and extends for 1,250 miles (2,000 kilometers).

RIGHT *The Nullarbor Plain ends abruptly at towering cliffs.*

WILPENA POUND

SOUTH AUSTRALIA, AUSTRALIA

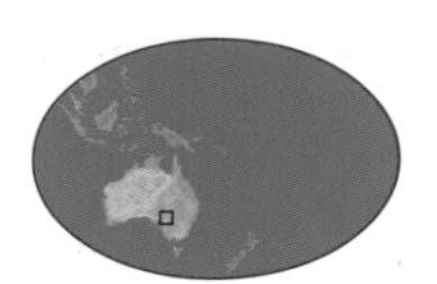

Area of Wilpena Pound: 39.5 sq mi (100 sq km)

Location: Flinders Ranges

Height of St. Mary's Peak: 3,842 ft (1,170 m)

At the heart of the Flinders Range lies Wilpena Pound, a huge, crater-like formation rising from the flat plains. From the sky it is easier to see the traces of the vast geological forces that formed it. A huge dome of rock was pushed up from the ocean floor 650 million years ago, creating mountains of Himalayan proportions. Over the millennia, the processes of erosion have left the comparative stumps that now make up the rim of the valley. Today the rock basin covers 39.5 square miles (100 square kilometers) and reaches a height of 1,650 feet (500 meters). The walls are made of quartzite, a rock highly resistant to weathering. First discovered by Europeans in 1802, it was another 50 years before Wilpena was settled and used for early sheep runs—the local farmers named it Wilpena Pound due to its resemblance to a sheep pen. In 1972, it became part of the Flinders Ranges National Park. A favorite with hikers and bird watchers, the pound is home to kangaroos and euros, at least 97 species of bird, and many beautiful plants. One of the most striking trees in Wilpena is the native or white cypress pine, used extensively for buildings and fences due to its natural resistance to termites. GM

LAKE MUNGO

NEW SOUTH WALES, AUSTRALIA

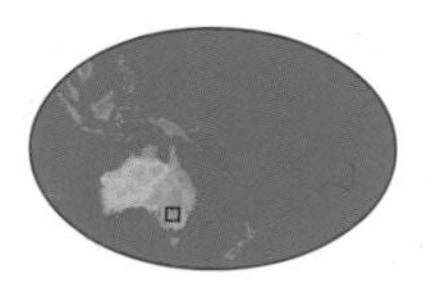

Age of Lake Mungo: about 2 million years

Vegetation surrounding lake: mallee scrub

Height of Walls of China: 100 ft (30 m)

Today, Mungo National Park—613 miles (987 kilometers) west of Sydney—resembles a vast moonscape. It is an eerie landscape of desert sands, fluted dunes and ridges, and sparse but hardy vegetation. About 45,000 years ago, however, Lake Mungo was a massive freshwater lake. It dried up about 19,000 years ago, leaving an extraordinarily rich deposit of fossils, including a 40,000-year-old burial site, with a body dusted in red ocher and a nearby grave containing the earliest known human cremation. Carbon dating shows that Aboriginal tribes first inhabited the area between 45,000 and 60,000 years ago, making it the site of the oldest known human occupation in Australia.

The remains of extinct creatures—giant kangaroos, Tasmanian tigers, and a little-understood ox-sized creature—have also been found. Walls extend around the old lake and rise above the surrounding plain. They are known as the "Walls of China," and are the most famous example of a lunette in the world—and one of the largest. The lunette, or crescent-shaped sand dune, was formed by quartz sands and palletized clay, blown from the lake by continental winds. GH

WILLANDRA LAKES

NEW SOUTH WALES, AUSTRALIA

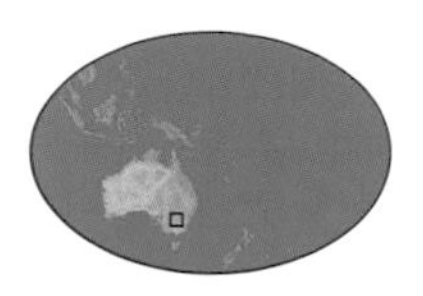

Area of Willandra Lakes: 926 sq mi (2,400 sq km)

Age: up to 2 million years

Surrounding vegetation: mallee, spinifex

When last full of water, over 19,000 years ago, the Willandra Lakes region covered over 386 square miles (1,000 square kilometers) and was an abundant source of freshwater and food for early man. This system of Pleistocene lakes was created during the last 2 million years and most are today fringed on the seaward shore by a dune—or lunette—created by winds sweeping across the continent from the interior. The lakes region is now a mosaic of semi-arid landscape approximately 926 square miles (2,400 square kilometers) in area, comprising dried saline lakebeds with saltbush vegetation, and fringing sand dunes and woodlands. Located in the Murray–Darling Basin area, the region was recognized internationally for representing the major stages in the planet's evolutionary history, significant ongoing geological processes, and for "bearing an exceptional testimony to a past civilization." Willandra provides some of the earliest substantial evidence of fully modern humans and their adaptation to the local environment. Aborigines lived on the shores of the Willandra Lakes from 45,000 to 60,000 years ago, and are believed to have exploited now extinct super-kangaroos, among other species. GH

CUNNINGHAM'S GAP

NEW SOUTH WALES / QUEENSLAND, AUSTRALIA

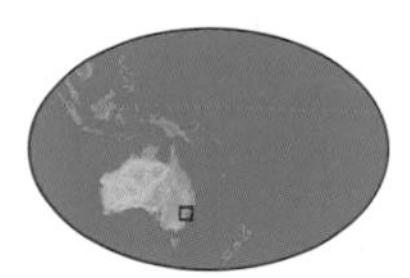

Age of Cunningham's Gap: up to 33 million years

Vegetation: tall open rainforest, woodland, scrubland

Main Range National Park is a system of volcanic ridges that rises steeply on one side, with impressive gaps, cliffs, peaks, and rock faces often covered in rainforest. The park—part of Australia's Great Dividing Range—is a section of a giant arc of mountains near Brisbane and extends to the Queensland-New South Wales border. Cunningham's Gap, named after the botanist and explorer Allan Cunningham, who discovered the passage through the Main Range in 1827, became a major trade route that linked the early colonies of Sydney and Melbourne to Brisbane. It has since become a passage for a major interstate highway, but the remaining area is rich in pioneering history, centered especially on the iconic cattle country of the nearby Darling Downs. The extensive rainforests of the Great Dividing Range and Cunningham's Gap were seen as a critical timber resource and by the early 1900s most of the area's red cedar had been felled. Despite being logged until the early 1990s, and still being an important recreation and landscape resource for local people, the region remains of immense biogeographic importance, providing habitat for numerous rare species of plants and animals. GH

MYALL LAKES

NEW SOUTH WALES, AUSTRALIA

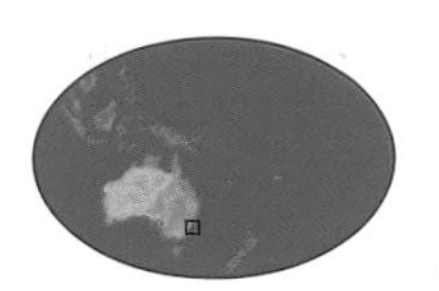

Area of Myall Lakes National Park: 77,991 acres (31,562 ha)

Area of coastal lagoons: 24,710 acres (10,000 ha)

Vegetation: swampland, forest

The Myall Lakes National Park is one of the largest and most complex lake systems in Australia. An area of outstanding natural beauty on the north coast of New South Wales, it has four main lakes or lagoons, together with swamps, high grassy dunes, grasslands, woodlands, open forest, rainforests, and coastline. Narrow straits that create a continuous waterway to Port Stephens from the lower Myall River join the lakes. Along the coast are 25 miles (40 kilometers) of almost unbroken beaches. *Myall* is actually an Aboriginal word meaning "wild." The 13-mile (21-kilometer) Mungo Track takes visitors to Mungo Brush, a popular camping and picnic spot. The area was the land of the Worimi and Birpai peoples and many ancient middens remain. The Mayall Lakes play host to a vast array of wildlife, including kangaroos, koalas, sugar gliders, ring-tailed possums, echidnae, lace monitors, and carpet pythons, in addition to numerous bird species. Bird watchers can spot honeyeaters, kookaburras, and ground parrots; the tawny frogmouth is most prolific in the Broadwater area. The area has retained much of its natural condition except for some sand mining. GH

BELMORE FALLS

NEW SOUTH WALES, AUSTRALIA

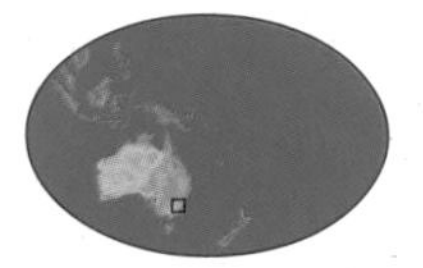

Height of Belmore Falls: 330 ft (100 m)

Depth of Shoalhaven Gorge: 1,837 ft (560 m)

Depth of Kangaroo Valley: 1,000 ft (300 m)

Belmore Falls drops 330 feet (100 meters) to feed two major creeks that have cut down through classic New South Wales highland escarpment—part of Australia's mountainous spine, the Great Dividing Range. Located south of the equally breathtaking Hindmarsh Lookout, Belmore Falls has panoramic views over Morton National Park and Kangaroo Valley. The falls drop into the Barrengarry Creek Valley that joins the Kangaroo Creek and the upper reaches of the Shoalhaven River.

Morton National Park is a series of high plateaus and deep gorges and features the Shoalhaven River. The river has sliced down 1,837 feet (560 meters) through the sandstone plateau to form the Shoalhaven Gorge. The complex geology and altitudinal differences of the area's landforms mean that dry sclerophyll forest occupies the plateau, wet sclerophyll forest covers the valley sides, and rainforest sits on the valley floors.

Kangaroo Valley is framed by sandstone cliffs. The northern cliff face features natural amphitheaters, with streams that descend from the surrounding mesas in waterfalls. The name of one local town, Bundanoon, comes from the Aboriginal for "place of deep gullies." GH

GREATER BLUE MOUNTAINS

NEW SOUTH WALES, AUSTRALIA

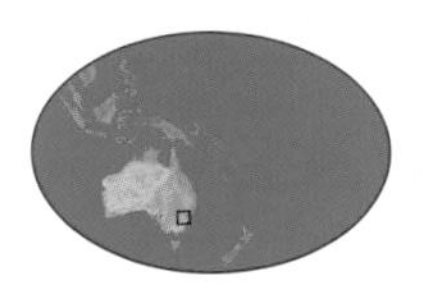

Age of Greater Blue Mountains: about 440 million years

Maximum height of peaks: approx. 4,000 ft (1,200 m)

Vegetation: forest

The Greater Blue Mountains—which form the dramatic backdrop for Sydney's western horizon—are regarded as the largest, most unspoiled, and most scenic wilderness areas in New South Wales.

This vast area of breathtaking viewpoints over rough tablelands, sheer cliffs, giant rocky monoliths, gorges, impassable valleys, forests, and life-abundant swamps is World Heritage listed as a biodiversity hot spot. More specifically, it is recognized for recording the evolution of the country's eucalypt vegetation and wildlife and provides a vital living space for "living fossils" dating to a time when Australia was part of the supercontinent of Gondwanaland. The mountains, which are deeply cut sandstone plateaus, include the iconic Three Sisters, Katoomba Falls, and Leura Cascades.

More than 400 different animal types are present in the mountains, including rare reptiles such as the green and golden bell frog. Aboriginal association with the land dates back at least 14,000 (and maybe as much as 22,000) years. There are nearly 700 sites indicating rock shelters, artwork, axe grinding grooves, stone implements, and tool production. GH

FITZROY FALLS

NEW SOUTH WALES, AUSTRALIA

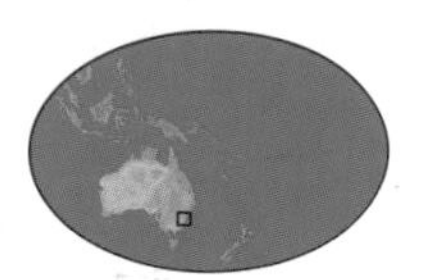

Height of Fitzroy Falls: 269 ft (82 m)

Vegetation: varied forest

Fitzroy Falls are located within Morton National Park. The park, one of the largest in New South Wales, features rugged sandstone cliffs, deep protected valleys with rainforest canopies, and four rivers: the Clyde, Endrick, Shoalhaven, and Kangaroo. Fitzroy Falls drops dramatically down a steep escarpment and plunges 260 feet (80 meters) into the creek below, set in the bottom of a ravine. On good days, the falls can be spectacular, while on others the huge gorge is filled with somber mist. The West Rim Walking Track offers the best view of the falls. They are surrounded by old-growth forest—woodland, stringybark, and old man banksia trees—which drops to rainforest level through gullies covered in wildflowers.

Originally called Throsby's Waterfall, after one of the Southern Highlands' explorers, Charles Throsby, it was renamed in 1850 after a visit by Sir Charles Fitzroy, Governor of New South Wales and Governor-General of the Colonies. Wildlife is abundant: Hawks and eagles soar on thermals, while parrots and lorikeets flit about the treetops. Visitors to the area may be able to spot kangaroos, echidnae, and dingoes, as well as snakes and lizards. GH

KANANGRA WALLS

NEW SOUTH WALES, AUSTRALIA

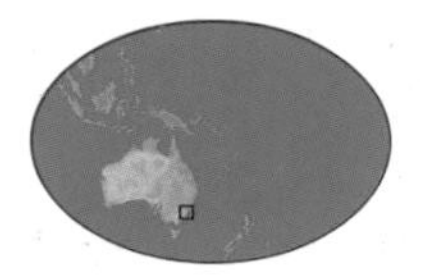

Age of Kanangra Walls: at least 285 million years

Vegetation: mountain gum woodland

The Kanangra Walls are a majestic set of sandstone cliffs and ramparts that tower 430 feet (130 meters) over the vast gorges, rugged plateaus, and wild scenic rivers of the Kanangra-Boyd National Park. Here hikers can experience waterfalls plunging 330 feet (100 meters) down sheer cliffs set in great walls of sandstone.

The views from the Kanangra Walls Lookout over to the Grand Gorge and Mount High and Mighty, Mount Cloudmaker, and Mount Stormbreaker are breathtaking and justifiably described as the best in the Greater Blue Mountains. Peppermint, mountain gum, silvertop ash, and stringybark cover much of the area. From the Kanangra Walls car park there are two lookouts: One takes in the Kanangra Creek Gorge, with Mount Cloudmaker and the main ridge of the Greater Blue Mountains in the distance, while the other overlooks the Kanangra Falls and ravines at the top of the gorge. A highlight of this area is the Kowmung River, regarded as one of the last wild untouched rivers in New South Wales. The Kanangra Walls are 122 miles (197 kilometers) west of Sydney, and a 45-minute drive from the Jenolan Caves. GH

BEN BOYD NATIONAL PARK

NEW SOUTH WALES, AUSTRALIA

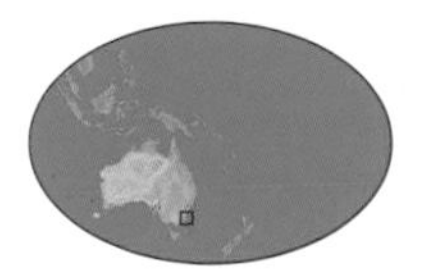

Age of gorge and outcrops: 345 to 410 million years

Vegetation: coastal woodlands, rainforest

Ben Boyd National Park, on New South Wales's most southerly coast, contains some of the region's most dramatic and rugged scenery, with striking white cliffs, a coastal gorge, and rocky outcrops that reach into the sea. The rocks were laid down in the Devonian period, when sediments in estuaries were transformed into arches and curves after being compressed, heated, folded, and twisted. Above these strata lie much younger Tertiary sand, gravel, clay, ironstone, and quartzite. Much of the cliff area is now covered with heath and banksia shrub. A popular local feature, "The Pinnacles," is an unusual formation of soft white sand topped with a thick cap of vivid red clay that dates back 65 million years—this is best viewed from the opposing cliff top. Between October and December, Boyd's Tower provides a sweeping vantage point to watch whales migrating along the coast. During the 19th century, a whaling station was established at Twofold Bay. A killer whale, known as Old Tom, used to herd migrating whales onto the harpoons of local whalers. Old Tom and his fellow orcas were rewarded with the lips and tongues of the stricken whales. Today there is a killer whale museum in nearby Eden. GH

MOUNT KOSCIUSZKO

NEW SOUTH WALES, AUSTRALIA

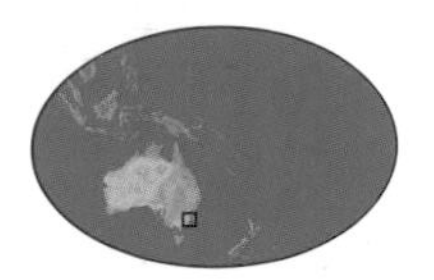

Height of Mt. Kosciuszko: 7,309 ft (2,228 m)

Height of Mt. Twynam: 7,204 ft (2,196 m)

Mount Kosciuszko is regarded as one of the world's great national parks. It covers six wilderness areas—Byadbo, Pilot, Jagungal, Bogong Peaks, Goobarragandra, and Bimberi—as well as encompassing the highest mountains in Australia and the iconic Snowy River. This region is known as "the roof of Australia": Granite boulders or tors dominate the landscape, and Mount Kosciuszko and Mount Twynam have snowcapped peaks all year round. In other areas of the park, limestone gorges and spectacular caves are interspersed with stepped ponds and glacial alpine lakes such as Blue Lake, Lake Albina, and Hedley Tarn. The grasslands and woodlands provide habitats for the rare mountain pygmy possum and corroboree frog, as well as many rare plant species.

The best views of the beautiful Snowy Mountains are from the Alpine Way on the western side of the range, where there are two lookouts—Olsen's Lookout and Scammell's Lookout. The Tumut area to the northwest is also dramatic, as are the Buddong Falls and Yarrangobilly Caves. The summit of Mount Kosciuszko can be accessed on foot via a number of walks of varying difficulty. GH

WARRUMBUNGLE NATIONAL PARK

NEW SOUTH WALES, AUSTRALIA

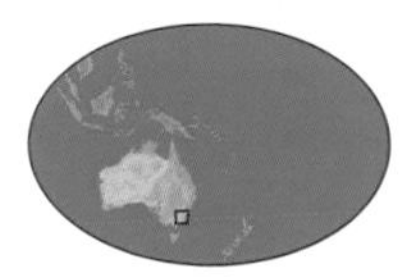

Age of Warrumbungle National Park: up to 17 million years

Height of Warrumbungles: over 3,300 ft (1,000 m)

Height of The Breadknife: 300 ft (90 m)

The Warrumbungles—meaning "crooked mountains"—are a series of volcanic plugs, dikes, spires, domes, and cliffs that rise sharply from the surrounding tableland and plains. About 17 million years ago, hot magma exploded through large shallow lakes to form a massive shield-shaped volcano. The strange rock formations seen today are the cores of extinct volcanoes that have been weathered by the wind and rain. A particularly striking volcanic feature is known as The Breadknife—a vertical sheet of igneous rock forced up through the surrounding rocks and gradually exposed as the surrounding soft rock was eroded.

Many of the peaks in the Warrumbungles tower more than 3,300 feet (1,000 meters) above the rugged landscape. The rich diversity of plant life in the park is a reflection of the varied landscape and soil patterns—it is a place where the flora of the dry western plains overlaps with that of the moist east coast. The area is vegetated with white gums and narrow-leafed ironbark of the eucalyptus family, as well as black cypress pine. The Warrumbungles contain almost one third of Australia's parrot and cockatoo species. GH

SYDNEY HARBOR

NEW SOUTH WALES, AUSTRALIA

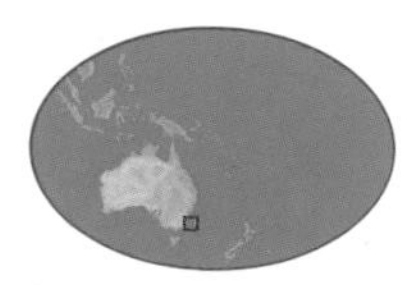

Area of Sydney Harbor: 21 sq mi (55 sq km)

Depth of Sydney Harbor: 30 to 155 ft (9 to 47 m)

Sydney Harbor—one of the world's most famous harbors—is in fact a "drowned valley" that extends 13 miles (20 kilometers) inland, where it meets the Parramatta River. The harbor's depth ranges from 30 to 155 feet (9 to 47 meters) at low water, and its irregular foreshores cover more than 150 miles (241 kilometers). The entrance to the harbor is dominated by two rugged sandstone headlands that open out to the Pacific Ocean. In 1788, a report of Captain Arthur Phillip's arrival with the First Fleet of convict ships to sail into Sydney declared: "It must have been like entering paradise ... when the sea-worn convoy passed through the dun and barren headlands into the untouched harbor—the water brilliantly blue, the shores high and wooded without being precipitous, a scattering of islands, sandy beaches, the trees shimmering under the sun." Today Sydney Harbor National Park offers hiking with spectacular views of the area's innumerable islands and secluded sandy bays framed by heath-covered cliffs and remnants of subtropical rainforest. Fauna includes seabirds nesting on cliff edges, as well as endangered little penguins—the harbor shelters a colony. GM

BARMAH–MILLEWA FORESTS & WETLANDS

VICTORIA, AUSTRALIA

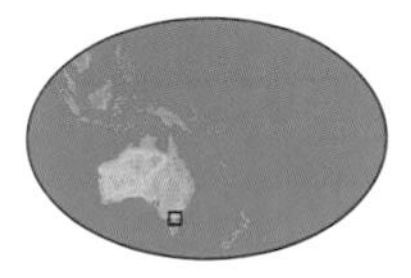

Area of Barmah-Millewa Forests and Wetlands: 170,000 acres (70,000 ha)

Notable feature: part of largest red gum eucalyptus forest in Australia

The Barmah–Millewa forests and wetlands are among the most important in the world, forming part of the largest periodically flooded red gum forest in Australia. The region is a very popular place for bird-watching, as huge numbers of ibis and other nesting waterbirds such as cormorants and spoonbills use the wetlands for breeding; the tranquil lakes and billabongs, seasonally flooded grassland, and dense forest are home to half of all the threatened species found in the region. In 1936, the ecology of the wetlands came under threat after the building of a huge dam interrupted the natural water cycle. The flow ran high in summer for irrigation and low in winter when the dam refilled. This reversed natural flow patterns, leaving the red gums with wet roots in summer and dry in winter. Over the years this caused dramatic tree dieback and choking overgrowth. In 1999, a new strategy reintroduced winter flooding. This has resulted in the most successful bird-breeding in the area since the 1970s. GM

CROAJINGOLONG NATIONAL PARK

VICTORIA, AUSTRALIA

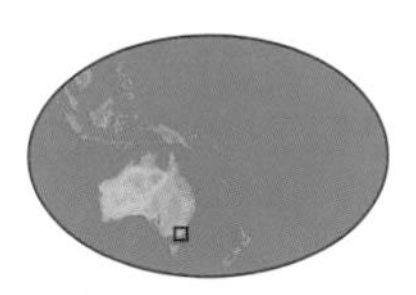

Area of Croajingolong National Park: 2.1 million acres (850,000 ha)

Notable feature: one of Victoria's three Biosphere Reserves

The Croajingolong National Park—one of Victoria's three Biosphere Reserves—extends for 62 miles (100 kilometers) along the wilderness coast of East Gippsland. Aborigines have lived in the area for at least 40,000 years. The local tribe, the Krauatungalung, gave its name to the park, and descendants of the original people still live in the area. In the 1900s, two national parks were set aside and these were greatly enlarged and combined to form Croajingolong in 1979. The park contains remote beaches fringed with rainforest, estuaries surrounded by heathland, and dramatic granite peaks. There are a number of popular short walking tracks along the coastline. For visitors looking out for wildlife, 52 mammal species and 26 reptiles have been recorded in the area; with 306 species of bird—a third of Australia's total species—it is a bird-watcher's paradise. The wetlands attract over 40 species of migratory seabirds and waders, while in the forests six owl species have been recorded. GM

GIPPSLAND LAKES

VICTORIA, AUSTRALIA

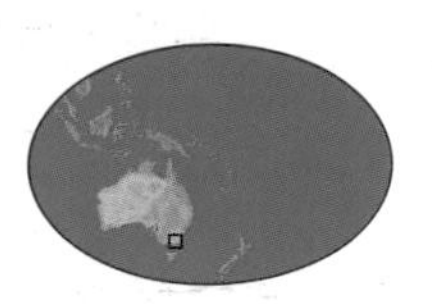

Area of Gippsland Lakes system: 72,000 acres (30,000 ha)

Length of system: 32 mi (51 km)

The Gippsland Lakes system is a vast 230-square-mile (600-square-kilometer) network of rivers, lakes, lagoons, and islands in a popular resort area of Victoria in southeast Australia. It was once part of a massive bay, but over many thousands of years, wave-deposited sands gradually formed a set of massive coastal barriers up to 125 feet (38 meters) high. These became the Sperm Whale Head Peninsula and Rotamah and Little Rotamah islands. The outer barrier eventually enclosed and formed the Ninety Mile Beach.

Fed by no fewer than six separate rivers, the Lakes National Park contains Lake Wellington, the park's largest body of water, which is connected via McLennons Straight to Lake King and to the man-made Lakes Entrance, which adjoins the seawaters of Bass Strait. The Gippsland Lakes area is peaceful and attracts large numbers of visitors for its abundant fishing and bird-watching opportunities.

Rotamah Island, fringed by the waters of Lake Victoria and Lake Reeve, is scattered with bird hides, giving visitors the opportunity to spot some of the 190 species found on the island. Lake Reeve is visited each spring and summer by thousands of migratory waders, some of which arrive from as far afield as Siberia, and is internationally recognized as an important wildlife habitat. Both lakes are host to breeding colonies of the vulnerable fairy terns and little terns, while Ninety Mile Beach is a good place to observe shore and ocean birds such as shearwaters and gulls.

The lakes boast a number of unusual geographical features. The most notable include the Mitchell River silt jetties, which run far into the lakes and are among the longest in the world. They were formed by silt washed down by rivers more than a million years ago. The five-mile- (eight-kilometer-) long jetties are technically a form of delta known as a "finger-delta" and are among the finest examples of their kind in the world.

The park contains plenty of gentle walking tracks for visitors who want to enjoy the peace and quiet. Wildlife includes swamp wallabies, bush-tailed possums, and wombats. The first human inhabitants lived in a paradise with both fresh and saltwater, abundant fish, and wildlife. The area sustained the Gunai Kurnai nation for at least 18,000 years and Aborigines still live in or near the park. GM

The Gippsland Lakes system is a vast network of rivers, lakes, lagoons, and islands. It was once part of a massive bay, but over thousands of years, wave-deposited sands formed coastal barriers. It is internationally recognized as an important wildlife habitat.

RIGHT *The Gippsland Lakes alongside Ninety Mile Beach.*

THE GRAMPIANS

VICTORIA, AUSTRALIA

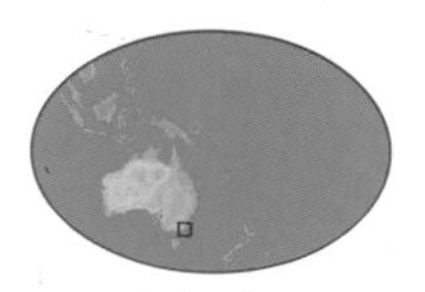

Age of the Grampian Ranges: 400 million years

Area of the Grampian National Park: 645 sq mi (1,670 sq km)

Height of Mt. William: 3,830 ft (1,170 m)

Rising from the flat farmlands of Victoria's western district, the Grampians comprise four hard, red sandstone ridges, laid down some 400 million years ago. These forest-covered peaks, dotted with waterfalls, streams, and swathes of wildflowers, are protected by the 645-square-mile (1,670-square-kilometer) Grampians National Park.

Mount William is the highest peak in the Grampians at 3,830 feet (1,170 meters), but Mount Arapiles near Horsham is regarded as one of the premier rock climbing destinations in Australia. Other striking rock formations include a Grand Canyon and heavily eroded outcrops known as the Balconies (Jaws of Death) and the Giant Stairway.

Visitors can canoe along the Wimmera River or follow a trail to Mackenzie Falls, one of the largest waterfalls in the state, immersing themselves in the dramatic landscape. With over 100 miles (160 kilometers) of trails throughout the park, and countless scenic viewpoints such as Reid's and Lakeview Lookouts, bushwalkers are spoiled for choice. The ranges also contain exceptional Aboriginal rock art—the region contains over 80 percent of sites found in Victoria. GM

LAKE EILDON NATIONAL PARK

VICTORIA, AUSTRALIA

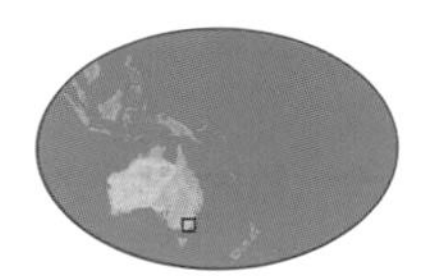

Area of Lake Eildon National Park: 107 sq mi (275 sq km)

Vegetation: open woodland to dense forest

Lake Eildon National Park lies nestled in the northern foothills of Victoria's central highlands and protects an area of 107 square miles (277 square kilometers). The park was established when Lake Eildon, Victoria's largest human-made lake, was constructed for irrigation and hydroelectricty in the 1950s. The lake was formed by the damming of four rivers, creating a water storage area with a capacity six times that of Sydney Harbor. Today, Lake Eildon is a hugely popular destination for hiking, fishing, and camping; numerous trails of varying lengths are offered along the lake's 320-mile (515-kilometer) shoreline. People seeking panoramic views of the lake can attempt the Blowhard Spur Trail or visit Foggs Lookout on the way to Mount Pittinger. The sense of peace and isolation in the park is emphasized by the majestic rugged hills and steep-sided valleys. As visitors explore the open scrub bush on the north side of the lake or the dense forest on the eastern and southern sides, they can see an abundance of wildlife. The park is well known for its eastern gray kangaroos, koalas, and wombats, and it is also home to a number of rare bird species, including wedge-tailed eagles and king parrots. GM

MURRAY RIVER

NEW SOUTH WALES / SOUTH AUSTRALIA / VICTORIA, AUSTRALIA

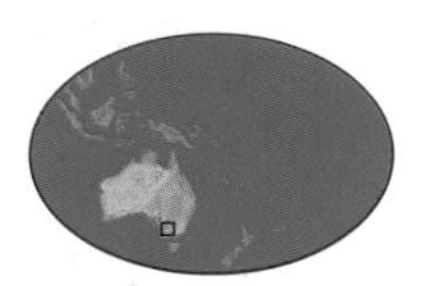

Length of Murray River: 1,625 mi (2,600 km)

Age: 20 million years

Area of Murray–Darling Basin: 390,000 sq mi (1 million sq km)

For 20 million years, Australia's Murray River has been flowing from the high reaches of the mountains to the wide expanse of the ocean. From its watershed in the Snowy Mountains, more than 6,600 feet (2,000 meters) above sea level, the Murray runs down to the Southern Ocean 1,625 miles (2,615 kilometers) away, making it Australia's longest waterway and the seventh longest river in the world.

Australia's longest waterway and the seventh longest river in the world, the Murray River and its various tributaries drain an enormous area of land: The Murray–Darling Basin covers almost one seventh of the total area of Australia.

The Murray River and its various tributaries, which include the 864-mile- (1,390-kilometer-) long Darling River, drain an enormous area of land: The Murray–Darling Basin covers almost one seventh of the total area of Australia. It takes in half of Victoria, three quarters of New South Wales (it forms the boundary between the two provinces), part of South Australia, and an area of Queensland greater than the whole of Victoria. At its source, the Murray River falls 5,000 feet (1,500 meters) in 125 miles (200 kilometers), slowing as it snakes its way toward the ocean. It then flows through floodplains cut by numerous distributaries, anabranches, and billabongs lined with large river red gums (the archetypal tree of the system), while sandy beaches are found in the nooks and hollows along its path. Finally, the river enters shallow Lake Alexandrina before arriving at its narrow mouth in Encounter Bay, near Adelaide. The whole area is home to a great wealth of wildlife, including numerous species of fish, frog, and snake, as well as nearly 400 species of bird.

The Murray carries only a small fraction of the water of comparably sized rivers in other parts of the world, and with great annual variability of its flow it has even been known to dry up completely in periods of drought. Today, the flow of the Murray and its tributaries has been curtailed even further by numerous reservoirs, dams, and irrigation schemes that supply Australia's richest agricultural area. But salinity and pollution through agricultural runoff has become a national problem, and a campaign is underway to plant 10 million trees to help combat these threats.

The Murray River has a tremendous cultural importance for the local Aborigines. According to the indigenous peoples of Lake Alexandrina, the Murray was created by the tracks of the Great Ancestor, Ngurunderi, as he pursued Ponde, a legendary Murray cod. GM

RIGHT *Trees reflected in the slow-moving waters of the Murray River.*

OTWAY RANGES

VICTORIA, AUSTRALIA

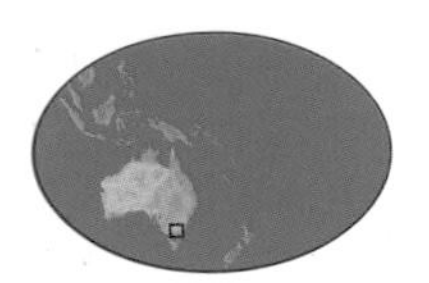

Area of Otway Ranges: 19.5 sq mi (50 sq km)

Vegetation: temperate rainforest

The Otway Ranges run along the southern coast of Victoria. This lush region of cool temperate rainforest originally dates back 140 million years to Gondwanaland, when dinosaurs roamed the planet. The high rainfall in the area encourages dense rainforest growth, in particular myrtle beech and tree ferns, with an understory of low ferns and mosses. Visitors hiking through the ranges can see some of the tallest trees in the world and numerous beautiful waterfalls—more than half of all the falls in Victoria are found in the Otway Ranges. In the late 1800s and early 1900s, the forests were extensively logged; today many trails through the valleys follow the remains of the old timber tramways.

However, the local conservation movement has called for more protection of the falls and the ranges. A small national park was recently extended after an area of forest was almost wiped out by extensive logging. Growing public concern caused the state government to declare a large portion of the range a national park in 2004, extending it from a small area around Cape Otway to take in the major sections of rainforest and the main waterfalls. GM

PHILLIP ISLAND

VICTORIA, AUSTRALIA

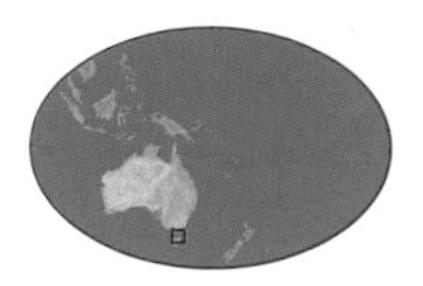

Area of Phillip Island: 101 sq mi (260 sq km)

Location: 87 mi (140 km) from Melbourne

Notable feature: seasonal penguin parade at dusk

The rugged southern coastline of Phillip Island provides some of the best surf beaches in Victoria. Located 87 miles (140 kilometers) southeast of Melbourne, its southern shores face the tumultuous Bass Strait. The Phillip Island Nature Park is a reserve designed primarily to protect and promote awareness of the fairy penguins. The island is home to one of Australia's best-known and most popular wildlife attractions—the "penguin parade." Each year, thousands of people arrive to watch a colony of little penguins returning ashore from the Bass Strait waters each night. Depending on the season, 300 to 750 penguins arrive on the southwestern edge of the island at sunset. They swim up to 32 miles (51 kilometers) a day to hunt and, to protect themselves from predators such as sharks and seals, they move in packs known as "rafts" that can include up to 300 birds. Between August and March, as the sun fades in the sky, the penguins tumble ashore, waddling up beach tracks to their burrows in the sand dunes in order to breed—hopefully raising two clutches in a good season. GM

BELOW *Seal rocks off the coast of The Nobbies on Phillip Island.*

PORT PHILLIP BAY

VICTORIA, AUSTRALIA

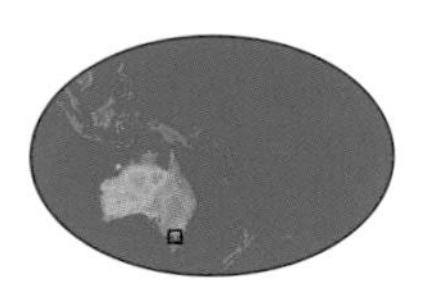

Age of Port Phillip Bay: 10,000 years

Average depth of bay: 43 ft (13 m)

In 1802, British explorer Matthew Flinders, in an account of his circumnavigation of Australia, described how he entered a large expanse of water in what was to become the southern Australian state of Victoria, and the location of Melbourne: "I congratulated myself on having made a new and useful discovery; but here I was in error. This place, as I afterward learned at Port Jackson, had been discovered 10 weeks before by Lieutenant John Murray ... of the *Lady Nelson*. He had given it the name of Port Phillip."

Port Phillip Bay extends nearly 38 miles (61 kilometers) north to south, and 42 miles (68 kilometers) east to west, encompassing a total area of around 745 square miles (1,930 square kilometers). The city of Melbourne lies on its northern shore at the mouth of the Yarra River. Known in geographical terms as a graben—an area of depressed land set between two parallel faults—the bay was formed at the end of the last glacial period, nearly 10,000 years ago, when a huge area of land subsided along a fault line on one side of what is now the Mornington Peninsula, causing the ocean to enter the bay. A second graben formed when the land sank and the ocean flowed in to create Westernport Bay.

The bay is clear and shallow—the average depth is only 43 feet (13 meters)—so light penetrates through to much of the seafloor. A narrow isthmus of land separates the bay from Bass Strait, which is ranked as one of the world's most dangerous stretches of water. It is known as "The Rip," and nearly 100 ships have foundered here over the past 160 years. To help ships navigate its narrow, reef-fringed opening and unpredictable waves, pilots use two lighthouses to align vessels with the center of the Great Ship Channel as they enter and leave the bay. In the 19th century, many huge sailing ships were wrecked trying to make their way into the bay, and today taking a pilot on board all large vessels is mandatory. The rough seas of this stretch of water have left their indelible mark on the coastline, sculpting sandy beaches and creating rugged, rocky basalt headlands.

The Bass Strait is one of the world's most dangerous stretches of water. The rough seas here have left their indelible mark on the coastline, sculpting sandy beaches and creating rugged, rocky basalt headlands.

Point Nepean is a highlight of the Mornington Peninsula National Park and features outstanding natural coastal scenery and panoramic views over the turbulent entrance of Port Phillip Bay. GM

RIGHT *The dangerous waters of Port Phillip Bay.*

TWELVE APOSTLES

VICTORIA, AUSTRALIA

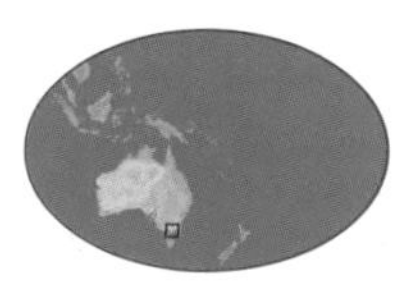

Height of Twelve Apostles: rock stacks up to 150 ft (45 m)

Type of rock: limestone

The mighty Twelve Apostles are iconic rock stacks along Victoria's Great Ocean Road. These giant stacks rise up from the southern ocean resembling huge skyscrapers and are the most dramatic feature of the spectacular Port Campbell National Park. The tallest rock stack is 150 feet (45 meters) high, and the limestone cliffs that form the backdrop to the apostles tower up to 230 feet (70 meters).

The limestone rock was formed from the skeletons of marine creatures that fell to the seafloor. As the ocean retreated during the last ice age, the limestone was exposed. Over the next 20 million years, the wind and seawater eroded the limestone into the stacks on view today. The relentless ocean and blasting winds gradually eroded the softer limestone, forming caves that eventually became arches. When the arches collapsed, tall rock islands of hard sandstone and limestone were left isolated from the shore. This created the Apostles, as well as many other stunning natural features, along a remarkable stretch of coastline. The cliffs are still being eroded at a rate of about 1 inch (2.5 centimeters) per year. This may eventually form more stacks from other rocky headlands lining the coastline. GM

TOWER HILL

VICTORIA, AUSTRALIA

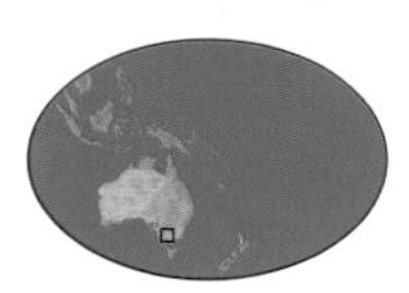

Age of Tower Hill: 25,000 to 30,000 years

Notable feature: extinct volcano

Many of Australia's most recent volcanic eruptions took place across the wide plains of southwestern Victoria and South Australia. Tower Hill, near the western Victorian coast, was formed as a result of hot magma rising from beneath the surface and exploding when it met water-saturated rocks. The violent explosion created a funnel-shaped crater, later filled by a lake. The molten lava was less viscous than other volcanic eruptions and flowed over a large distance. Called a nested caldera or maar volcano, Tower Hill consists of an outer rim with a series of small cones rising from a water-filled depression, created by the collapse of the major cone. Artifacts found in the ash layers around the volcano revealed that Aborigines were living in the area at the time. The region offered a rich source of foods for the Koroitgundidj people, whose descendants retain links with the land. Tower Hill was declared Victoria's first national park in 1892 in an attempt to halt the decline of the once majestic volcano. Vegetation was originally very diverse but early settlers removed much of it. In recent times, volunteers have planted more than 300,000 trees, and the plan is to re-introduce indigenous ferns and grasses. GM

WESTERNPORT BAY

VICTORIA, AUSTRALIA

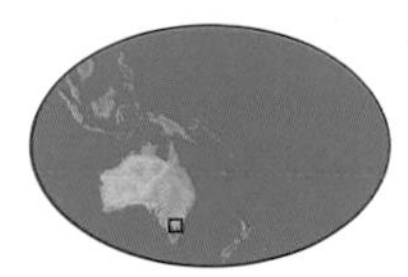

Length of Westernport Bay: 28 mi (45 km)

Age of bay: 10,000 years

This large inlet on the south Victorian coast was formed as a graben—a block of rock thrown down between two parallel faults. Located on one side of the Mornington Peninsula, it extends 28 miles (45 kilometers) inland, with a width varying from 10 to 22 miles (16 to 35 kilometers). George Bass, friend and fellow explorer of Matthew Flinders, the first man to circumnavigate Australia, is thought to have been the first European to sail into Westernport Bay. He was on one of his voyages of discovery at the end of 1797 when he entered the bay in a whaleboat that he had sailed down the coast from Sydney.

Today, because Westernport Bay is a short distance from Melbourne, it serves not only as a city-escape but also as a commercial fishing port and a valuable site for industry. Excessive fishing and pollution from industrial waste have killed other bays in Australia, and Westernport Bay is also at risk. However, considerable efforts are being made to preserve its mangrove-lined shores and protect its waters from pollution. Marine scientists are particularly interested in the bay's populations of crustaceans known as "Balmain bugs," as well as gummy sharks, sea pens, and ghost shrimps. GM

WILSONS PROMONTORY

VICTORIA, AUSTRALIA

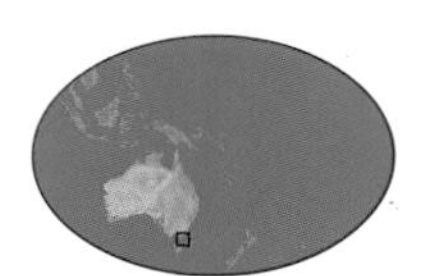

Area of Wilsons Promontory: 37,300 acres (15,550 ha)

Established as a national park: 1898

Designated a Biosphere Reserve: 1982

On the southernmost point of the Australian mainland lies Wilsons Promontory, the largest coastal wilderness in Victoria. Declared a national park in 1898, the dramatic peninsula—also known as "the Prom"—covers a vast stretch of bays, coves, inlets, and untouched beaches along its 81-mile (130-kilometer) coastline. Inland there are mountainous areas dotted with forests and fern gullies. The history of Aboriginal occupation dates back at least 6,500 years, and the land still has spiritual significance for local Aboriginal groups. The promontory's wildlife and plants were a valuable food source, particularly in summer, and it may have served as part of a landbridge to reach Tasmania during past ice ages. The explorers George Bass and Matthew Flinders were probably the first Europeans to see "the Prom" on their 1798 voyage from Sydney. They recognized its commercial value, and exploitation of seals, whales, timber, and, later, cattle grazing, were to continue for 100 years. The park's range of vegetation includes large areas of temperate rainforest and heath; in the swamp regions, the most southerly mangroves in the world can be found. GM

WYPERFELD NATIONAL PARK

VICTORIA, AUSTRALIA

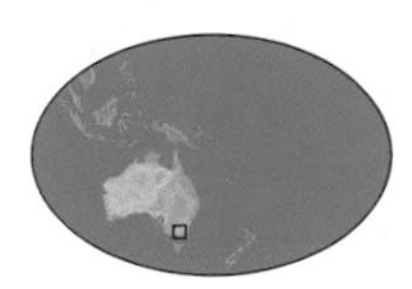

Area of Wyperfeld National Park: 1,338 sq mi (3,465 sq km)

Number of plant species: approx. 520

Some 25 million years ago, northwest Victoria was submerged beneath a shallow sea. As the ocean slowly retreated, westerly winds blew sand over the exposed inland areas, building a complex of rolling dunes; those that can be seen today were formed between 15,000 and 40,000 years ago. Spread out over this flat, semi-arid region is Wyperfeld National Park. A vast landscape, the park's major feature is a chain of lakebeds connected by Outlet Creek, the northern extension of the Wimmera River. The lakes fill when the Wimmera River floods and then, when it rains, the semi-arid countryside is transformed by the blooming of tiny desert plants that sprout from long-dormant seeds, carpeting the ground with clusters of flowers. Some 520 plant species are found in the park, and they occur in distinct communities. A dominant species is the mallee—shrubby eucalyptus with numerous stems arising from underground roots that store food and send up new stems if those above ground die. Aborigines occupied the area for at least 6,000 years, and they regularly moved north along Outlet Creek in search of food. Due to the unreliable water supply, they rarely stayed in one place for long. GM

AUSTRALIAN ALPS WALKING TRACK

AUSTRALIAN CAPITAL TERRITORY / NEW SOUTH WALES / VICTORIA, AUSTRALIA

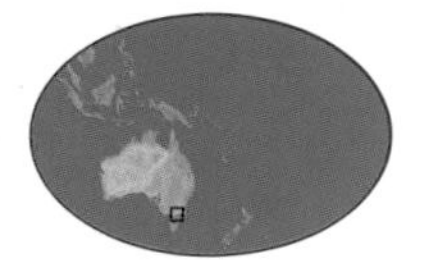

Length of Australian Alps Walking Track: 400 mi (644 km)

Length of time to walk entire track: 8 to 10 weeks, one way

Best time: November to May

The best walking and bush country in Australia is crossed by the 400-mile (644-kilometer) Australian Alps Walking Track, the nation's longest high-country trail. The track starts in Walhalla in Victoria's south and passes some of Australia's highest peaks, as well as crossing rivers and creeks on its way north to Tharwa, just south of Canberra. The track climbs up and crosses exposed regions such as the Bogong High Plains and the Jagungal Wilderness. It passes through magnificent tall forests and stunted snow gum woodlands. Remarkably for such a remote trail, it has been partly signposted. The track is distant from towns or other settlements, but hikers can join it at many places between Walhalla and Canberra, as it links with popular walking tracks in the Baw Baw, Alpine, Kosciuszko, and Namadgi national parks. Although it can be covered in eight weeks, many people walk shorter sections of the trail, such as the Baw Baw Plateau, the Bogong High Plains, and the Jagungal Wilderness Area. GM

ALPINE NATIONAL PARK

VICTORIA, AUSTRALIA

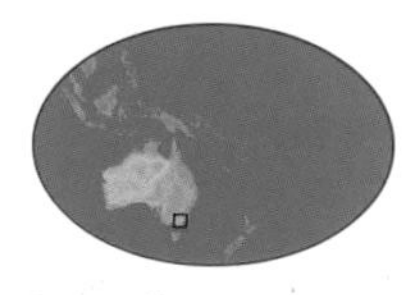

Area of Alpine National Park: 155,000 acres (646,000 ha)

Location: 138 mi (220 km) from Melbourne

Highest peak (Mt. Bogong): 6,516 ft (1,986 m)

Victoria's Alpine National Park is the state's largest and protects its highest mountains and varied alpine environments. The adjoining national parks in New South Wales and the Australian Capital Territory form a protected area covering most of Australia's high country. Because of the mountainous terrain, the park contains extensive snowfields, but the warmer months bring striking displays of wildflowers. There are plenty of opportunities to explore the area via hiking or four-wheel-drive vehicle.

In a continent as dry as Australia, the Australian Alps are vitally important as a source of water. It is here that most of the major rivers of southeastern Australia begin their journeys to the ocean. Victoria's Alpine National Park is one of eight such areas across the nation's high country that are managed cooperatively between the states to ensure the mainland alpine and subalpine environments are protected consistently and that policies and guidelines across borders are compatible. Aboriginal people have occupied the alpine region for tens of thousands of years. Today they visit the Alps in summer to hold ceremonies and gather to cook the nutritious bogong moths that shelter there. GM

CRADLE MOUNTAIN & LAKE ST. CLAIR

TASMANIA, AUSTRALIA

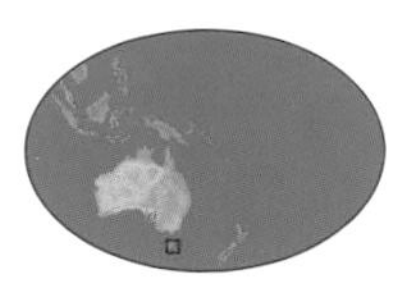

Area of national park: 386,000 acres (161,000 ha)

Depth of Lake St. Clair: 660 ft (200 m)

Vegetation: temperate rainforest, alpine heathlands, buttongrass, deciduous beech

Cradle Mountain is a spectacular peak, its jagged appearance formed by the glaciers that gouged and sculpted it into the shape of a baby's cradle some 10,000 years ago. It rises to 5,100 feet (1,554 meters) above sea level and dominates the north of the Cradle Mountain–Lake St. Clair National Park. At the park's southern end lies Lake St. Clair, Australia's deepest natural freshwater lake.

The Overland Track that runs from Cradle Mountain to Lake St. Clair is one of Australia's best-known bush trails and attracts walkers from all over the world. The track is 50 miles (80 kilometers) long and takes about five days to complete. It winds through breathtaking scenery—rugged mountains, ancient pine forests, and alpine plains, passing icy streams and still, glacial lakes on the way. There are also plenty of shorter bush walks around Lake St. Clair that take just a few hours. A popular option involves taking a ferry cruise on Lake St. Clair, followed by a two-to-five-hour walk through the rainforest beside the lake. GM

BEN LOMOND NATIONAL PARK

TASMANIA, AUSTRALIA

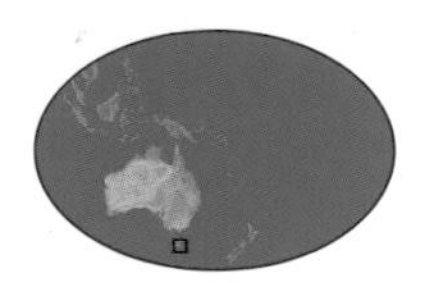

Area of Ben Lomond National Park: 45,000 acres (18,000 ha)

Height above sea level: 5,160 ft (1,573 m)

Notable feature: Tasmania's largest alpine area

The Ben Lomond National Park is set on a large dolerite plateau. It is Tasmania's largest alpine area, with a huge variety of glacial and periglacial features that are considered to be of national significance. During the ice age, massive glaciers thrust their way across the land, forcing the soft rock into spectacular mountains and valleys, and creating lakes and rivers that are said to be among the purest on Earth. Rising to 5,160 feet (1,573 meters), Ben Lomond is a tough, challenging mountain—its sheer escarpments, deep valleys, and shocking weather are said to offer the most unforgiving hiking in Australia. Eucalyptus is dominant on its lower slopes, with ancient forest and rainforest remaining in fire-protected areas. In spring and summer the region produces a dazzling display of alpine wildflowers, and the park is home to an abundant array of plant, animal, and birdlife, including the rare Ben Lomond leek orchid and the wedge-tailed eagle. If visitors can brave driving along the infamous Jacobs Ladder—a series of hairpin bends—the lookout at the top provides spectacular panoramic views of Flinders Island and Strickland Gorge. GM

FLINDERS ISLAND

TASMANIA, AUSTRALIA

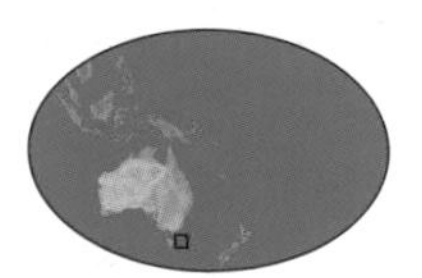

Area of Flinders Island: 522 sq mi (1,376 sq km)

Notable features: mountains and coastal plains

Flinders Island is the largest of the Furneaux Islands, a group of 52 islands that stretch across Bass Strait between Tasmania and Australia. The Furneaux Islands began as mountains on a landbridge that joined Tasmania to mainland Australia. With flooding at the end of the last ice age, the mountain group became 52 mountains in the sea. Located 12 miles (20 kilometers) off the northeast tip of Tasmania, Flinders Island is 40 miles (64 kilometers) long and 18 miles (30 kilometers) wide. Although it is mountainous, with Mount Strzelecki being the highest peak, about half the island is coastal sand dunes. The breathtaking geography, marine climate, sparkling beaches, and spellbinding array of wildflowers give the region its name as the Mediterranean of the Pacific. The abundant wildlife includes over 200 species of bird. The Cape Barren goose, regarded as one of the world's rarest species of goose, is prolific here. One of the geological treasures is Killiecrankie Diamond, which is actually a hard form of topaz. Belying its beauty, Flinders has a horrific history of atrocities against Aborigines by white settlers in Tasmania in the early 1800s. GM

EAGLEHAWK NECK

TASMANIA, AUSTRALIA

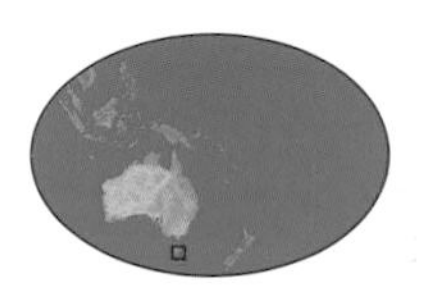

Width of Eaglehawk Neck: 300 ft (90 m)

Notable feature: four spectacular rock formations

Eaglehawk Neck is a narrow isthmus less than 300 feet (90 meters) wide that connects the Tasman Peninsula and the Forestier Peninsula. It was originally formed from wave-deposited sand from Pirate's Bay to the east and Norfolk Bay to the west. In the early 1800s, Eaglehawk Neck was manned by prison guards. The guards ran a chain from one side of the neck to the other, to which they tethered savage dogs to stop convicts escaping from the nearby penal settlement. Today, visitors are drawn to the region's dramatic landscape, in particular four spectacular natural formations.

The Tasman Arch is a huge natural stone arch that has been shaped by wave erosion. At the Devil's Kitchen, waves roar as they crash onto rocks 200 feet (60 meters) below, while at the Blowhole, the sea rushes in under the rocks and is shot high into the air. The Tessellated Pavement is a highly unusual natural feature—it looks like the work of a bricklayer, but is in fact created by the ocean. The rocks were fractured by three sets of earth movements, which produced the tiled appearance. Their flatness is due to the action of waves carrying sand and gravel and erosion by seawater. GM

FREYCINET PENINSULA

TASMANIA, AUSTRALIA

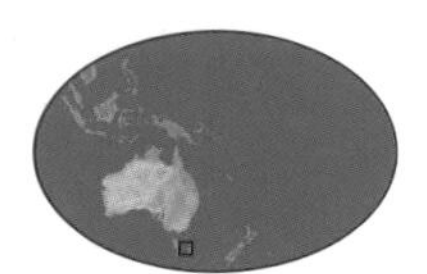

Area of Freycinet Peninsula: 25 sq mi (65 sq km)

Height of Mt. Freycinet: 2,011 ft (613 m)

Notable feature: scenic coastal area

Jutting out into the ocean on Tasmania's mild east coast is the rugged and beautiful Freycinet Peninsula, part of the Freycinet National Park. At the entrance is Coles Bay and nearby are the 1,000-foot- (300-meter-) high pink granite outcrops called the "Hazards." The peninsula consists of granite mountains that run down to blue inlets, such as the exquisitely shaped Wineglass Bay, fringed by white sandy beaches and curious, orange-colored rocks—an effect produced by a specific lichen. As one of Tasmania's most attractive coastal areas, with a typically mild, maritime climate, the region and its scenic beauty attract many visitors each year.

The park offers some magnificent walks, the longest being the peninsula circuit, which is a 17-mile (27-kilometer) trail. Many species of bird live on the peninsula, or stop there, including white-bellied sea eagles and large Australasian gannets. Black swans and wild ducks inhabit the Moulting Lagoon Game Reserve. Offshore, whales and dolphins can be spotted in the clear blue waters. The peninsula was named after Louis de Freycinet, a French cartographer who produced the first detailed maps of Australia's uncharted coastline. GM

FRANKLIN–GORDON WILD RIVERS NATIONAL PARK

TASMANIA, AUSTRALIA

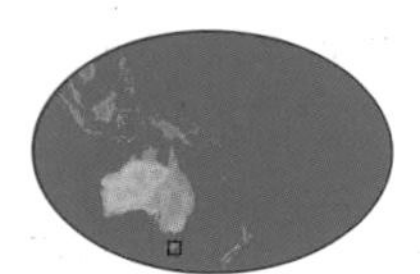

Area of national park: 105,800 acres (441,000 ha)

Features: wilderness region of high peaks, rainforests, gorges

Vegetation: temperate rainforest, pines, deciduous beech

White water rapids, icy streams cascading down steep mountainsides, and still glacial lakes are typical of the Franklin–Gordon Wild Rivers National Park. The 1,702-square-mile (4,408-square-kilometer) park is a World Heritage site and comprises a dramatic region of quartz peaks, remnant stands of Huon pine, and the Kutilina Caves, in which Aboriginal stone tools dating back more than 5,000 years were found. With its magnificent white quartzite dome, Frenchman's Cap is the most prominent peak in the region, rising 4,734 feet (1,443 meters). It is a difficult climb, however, and only very experienced hikers should attempt it. The Franklin River is one of the great wilderness rivers, and is the only major wild river system in Tasmania that has not been dammed. It flows for 75 miles (121 kilometers) down to the majestic Gordon River, from an altitude of 4,620 feet (1,408 meters) almost to sea level. The river becomes a raging torrent as it passes through ancient heaths, deep gorges, and rainforests. GM

QUEENSTOWN

TASMANIA, AUSTRALIA

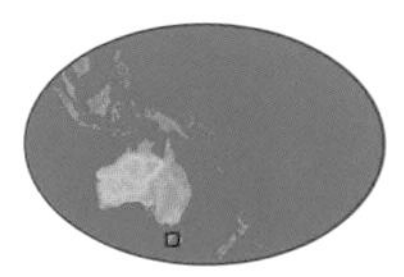

Location of Queenstown: 160 mi (256 km) west of Hobart

Notable feature: startling hills denuded from pollution

It can be difficult to decide whether Queenstown, west of Hobart, is a terrible place or a beautiful one. Perhaps it is best described as both. It is an old copper and gold mining town, first opened in the late 1800s, set in a surreal, barren lunar landscape. The surrounding hills were long ago stripped of their dense rainforests to fire the belching smelters and then permanently denuded by poisonous sulfurous fumes. Heavy rainfall washed away the topsoil, leaving eroded channels and exposing the striking purple and gold rocks beneath. Yet a short walk from the nightmarish landscape leads into temperate rainforest with waterfalls (including Tasmania's highest, the 338-foot [103-meter] Montezuma Falls) and spectacular views over the Franklin River valley, which gives some idea of what Queenstown must have looked like before the mine.

Despite the environmental cost of the now abandoned mine, such is the surrounding landscape's appeal to visitors that some local residents actively oppose the reforestation of the bare hills, fearing that the return of vegetation would "spoil" the moonscape and end its existence as a tourist attraction. GM

WALLS OF JERUSALEM

TASMANIA, AUSTRALIA

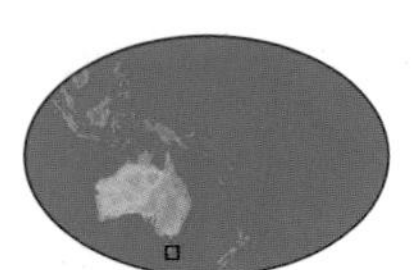

Area of Walls of Jerusalem: 197 sq mi (510 sq km)

Vegetation: range of subalpine and alpine plants

Regarded by many as the jewel in the crown of Tasmania's national parks, the remote Walls of Jerusalem National Park has a ring of mountains that forms a natural amphitheater in Tasmania's central plateau. The alpine region is dominated by dolerite peaks and high-mountain vegetation, and is exposed to the harsh extremes of Tasmania's changeable weather. The park contains the largest stand of the specific pencil pines, some 1,000 years old. These trees have been saved from the bushfires that have destroyed the species elsewhere on the plateau. In line with its biblical title, the park has many appropriately named geological sites: Herod's Gate, the entrance to the center of the park; tiny lakes known as Solomon's Jewels; and King David's Peak, which dominates the area. The park is spread out over 197 square miles (510 square kilometers) and access is possible only on foot. Hikers need to know how to read a map and navigate, particularly around the central plateau conservation area, as whiteouts occur, either from cloud or snow. From the end of the ice age, Aborigines occupied the area for more than 11,000 years, but by 1831, the Big River tribe was reduced to a mere 26 individuals. GM

BALLS PYRAMID

TASMAN SEA, AUSTRALIA

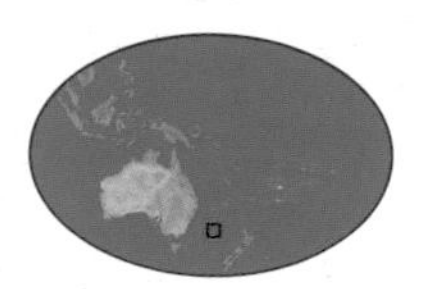

Height of Balls Pyramid: 1,811 ft (552 m)

Age: 80 to 60 million years

First climbed: 1965

Balls Pyramid is the remote cap of an undersea volcano, or seamount, 14 miles (23 kilometers) southeast of Lord Howe Island, 440 miles (708 kilometers) northwest off the coast of Sydney. It rises more than 5,900 feet (1,798 meters) from the floor of the Pacific Ocean before towering a near-vertical 1,811 feet (552 meters) above the ocean surface. While the pyramid is the highest sea stack in the world, it tapers to a summit just 13 feet (4 meters) across. First climbed as recently as 1965, it is a remnant of a volcanic caldera that formed about seven million years ago and has been eroding ever since.

Balls Pyramid is home to tens of thousands of seabirds, including gray ternlets, wedge-tailed shearwaters, and black-winged petrels. It also boasts venomous centipedes and an animal that was believed to be extinct—the Lord Howe Island stick insect. In 2001, an expedition discovered a colony of the insects under a single bush on Balls Pyramid. This giant insect is a true heavyweight of the arthropod world. Its body is the color and length of a large cigar, and it grows to a length of 5 inches (12.5 centimeters). Efforts are now underway to ensure its protection. DH

MOUNT TARANAKI

NORTH ISLAND, NEW ZEALAND

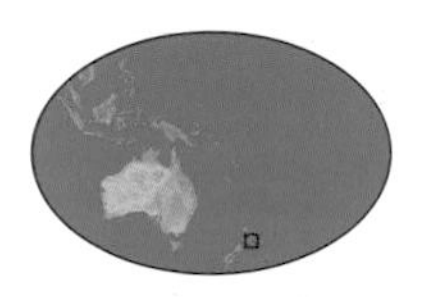

Age of Mt. Taranaki: 120,000 years

Extent of ejected material: 77 sq mi (200 sq km)

Average annual rainfall: 118 in (300 cm)

The elegant volcanic cone of Mount Taranaki—formerly Mount Egmont—stands alone amid the near-circular remnant of native forest in Egmont National Park. Taranaki is dormant for now, but its 120,000-year life has been violent and changeful. Built from frequent eruptions of lava and tephra, Taranaki once stood more than 9,000 feet (2,700 meters) high, but the mountain has a habit of blowing itself apart. It has erupted eight times in the past 500 years—the last occurred 250 years ago—and volcano experts say it will certainly do so again. These self-destructive episodes, coupled with the erosive, drenching west coast rains, have lowered Taranaki to 8,261 feet (2,518 meters). Maori tales say Taranaki once stood among the other volcanoes in the heart of the North Island. All the volcanoes were gods and warriors and lusted for Pihanga, who stood just beyond their reach. So the mountains fought for her, searing the land in their battles. Tongariro triumphed, and the vanquished volcanoes left the lovers to their solitude. In the night they departed. Taranaki, after gouging the chasm that is now the Whanganui River, stopped at the sea, to gaze back upon Pihanga from afar. DH

TONGARIRO NATIONAL PARK

NORTH ISLAND, NEW ZEALAND

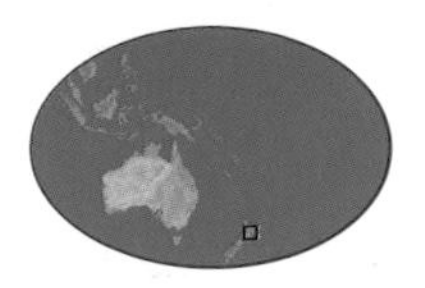

Height of Tongariro volcano: 6,456 ft (1,968 m)

Height of Ngauruhoe volcano: 7,516 ft (2,291 m)

Height of Ruapehu volcano: 9,176 ft (2,797 m)

Tongariro, in the center of New Zealand's North Island, was established as the country's first national park in 1887, when Maori Paramount Chief Horonuku te Heuheu Tukino IV bequeathed the sacred peaks of the Ngati Tuwharetoa people to the British Crown. Today, Tongariro is a mixed cultural and national World Heritage site, a status that recognizes both the park's important Maori cultural and spiritual values as well as its remarkable volcanic features.

A place of herb fields and forests, lakes and deserts, it is also home to some of New Zealand's rarest native fauna, such as short- and long-tailed bats. Kiwi, the national animal of New Zealand, still live here, as do kaka, the forest parrot, and the native falcon. But the verdant tranquillity aboveground belies the molten chaos more than 62 miles (100 kilometers) below the surface.

The park is best known for its trio of andesitic volcanoes—Tongariro, Ngauruhoe, and Ruapehu—the latter two ranking among the most active composite volcanoes in the world. Ruapehu, at 9,176 feet (2,797 meters), is the tallest mountain on North Island, a massive, complex stratovolcano that cradles an acidic crater lake in an active vent near its summit. Built up by successive lava and tephra eruptions over the last 200,000 years, it still carries the remnants of receding glaciers. In 1995, and again in 1996, Ruapehu erupted spectacularly, sending towering clouds of ash and steam skyward and covering the surrounding area with a thick film of ash.

Tongariro, at 6,456 feet (1,968 meters), is a large andesitic volcanic massif comprised of more than a dozen cones. An alpine trek taking in the Red Crater, Blue Lake, and North Crater is one of the world's best and most popular day walks. At Ketetahi, on Tongariro's northern flank, more than 40 fumaroles (vents that emit steam and gas) generate nearly the same energy as the nearby Wairakei geothermal power station—in the region of 130 megawatts.

A place of herb fields, forests, and lakes, the verdant tranquillity aboveground belies the molten chaos more than 62 miles (100 kilometers) below the surface. The park is best known for its trio of andesitic volcanoes–Tongariro, Ngauruhoe, and Ruapehu.

The youngest volcano in the park, Ngauruhoe, has grown to 7,516 feet (2,291 meters) since its birth about 2,500 years ago and still retains a perfect, almost archetypically volcanic, conical shape. It produced spectacular lava flows in 1949 and 1954, followed by ash eruptions in the mid-1970s. DH

RIGHT *Snowy trails meander down the gray barren slopes in Tongariro National Park.*

CAPE KIDNAPPERS

NORTH ISLAND, NEW ZEALAND

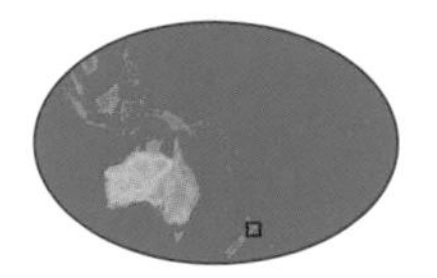

Total number of gannets: 5,200 nesting pairs

Australasian gannet: wing span 6.6 ft (2 m), average weight 4.4 lb (2 kg)

At the southern end of Hawke's Bay, on the east coast of North Island, is the world's biggest mainland colony of nesting gannets. The gannet is more usually an island breeder, but at Cape Kidnappers it congregates in huge numbers on the mainland.

The headland gets its name from Captain Cook, who, in 1769, almost lost his Tahitian interpreter to the local Maoris who kidnapped him. To the Maoris, the Australasian gannet is known as Takapu, and before the 19th century the bird populations were low in number. The gannets established themselves on the saddle of the promontory in the 1850s and today 2,200 pairs nest there. Spillover sites on a nearby plateau and at Black Reef have increased the total by over 3,000 pairs. Offshore, the birds can be seen fishing. They dive into the sea from a height of 100 feet (30 meters) at speeds of 90 miles per hour (145 kilometers per hour), like white bullets from the sky, in pursuit of shoals of fish.

The cape is near Clifton Domain—it can be approached along the beach at low tide; a track takes visitors up onto the headland and close to the plateau colony. The saddle and Black Reef colonies are closed at all times. **MB**

ROTORUA GEOTHERMAL REGION

NORTH ISLAND, NEW ZEALAND

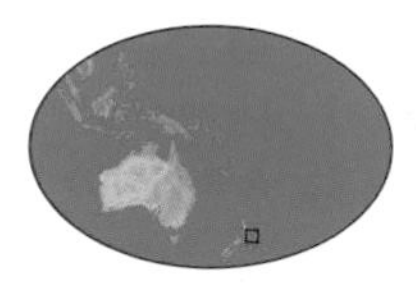

Last eruption at Rotorua: 1,800 years ago

Eruption column: 45,778 cu yd (35,000 cu m) of ash and gas

Height of eruption column: 30 mi (50 km)

Deep beneath New Zealand, two giant tectonic plates, the Pacific Plate and the Indo-Australian Plate, are engaged in a colossal struggle. As the Pacific Plate grinds 62 miles (100 kilometers) below the surface, it creates enough friction and heat to melt itself, turning into 1,832°F (1,000°C) magma. At that temperature, magma starts rising through cracks in the plate, meeting cold ground water on the way. Around Rotorua, that turmoil finds expression as more than 1,200 geothermal features—geysers, hot springs, mud pools, fumaroles, silica terraces, and salt deposits. This extreme environment has attracted its own unique life-forms—multihued lichens, mosses, and specially adapted heat-tolerant plants. The algae that thrive beside the boiling waters have changed little since the beginning of life on Earth. For more than a century, Rotorua's geothermal wonders have drawn visitors from around the world. However, of more than 200 geysers active in the 1950s, just over 40 remain. DH

OPARARA ARCH

SOUTH ISLAND, NEW ZEALAND

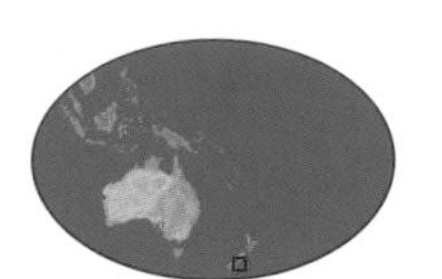

Span of Oparara Arch: 165 ft (50 m)

Height of arch: 141 ft (43 m)

Rock type: limestone

Nestled between the Tasman Sea and the densely forested mountains of Kahurangi National Park, the Oparara Basin has a number of features unique to this region of New Zealand's South Island. The Oparara River meanders through limestone outcrops, carving out narrow gorges and ravines. Where small caves once carried the volume of the river through steep cliffs, the water has dissolved the rock and enlarged the cavities into a series of three massive arches. The largest and most spectacular of these is the Oparara Arch. Covered in lush temperate-rainforest foliage, the arch is the largest natural archway in the Southern Hemisphere. Below, large black eels bask in dark pools in the boulder-strewn waters. High above, in the dark recesses of the cavernous tunnel, shimmering points of light indicate a colony of glowworms, their sticky threads ready to snare unwary flying insects. Long-legged, cave-dwelling spiders scuttle across the dank walls, and alien-like wetas (giant indigenous crickets) are watchful for predators and prey. The Oparara Arch can be reached from a rough four-wheel-drive track from Karamea, followed by a kayak journey along the Oparara River. DL

MARLBOROUGH SOUNDS

SOUTH ISLAND, NEW ZEALAND

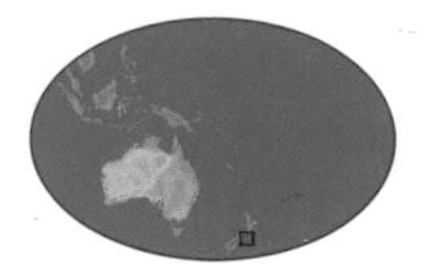

Age of Marlborough Sounds: 15 to 20 million years

Rate of movement: 0.25 in (0.6 cm) per year

The Marlborough Sounds, lying between fingers of land that point from the southern tip of New Zealand's South Island into the turbulent Cook Strait, are a classic "drowned valley" landscape. During a mid-Miocene mountain-building episode 15 to 20 million years ago, the whole area was tilted and the sea poured in, flooding the valleys and turning mountain peaks into islands. Then, as the last ice age came to an end, meltwater swelled the new lakes and sounds. The two biggest valley systems, carved along an outrider of the Alpine Fault, became the Pelorus and Queen Charlotte Sounds. At places such as French Pass, during an outgoing tide the land funnels the racing waters into currents of up to seven knots. Today, the sounds are unique in New Zealand for being the only area where the land is sinking into the sea. But their journey is not only downward; riding on the boundary of the Pacific and Indo-Australian Plates, they have traveled some 32 miles (52 kilometers) north at a rate of 0.25 inch (0.6 centimeter) per year since the Pliocene period, around seven million years ago. DH

RIGHT *The drowned valleys of Marlborough Sounds.*

FIORDLAND

SOUTH ISLAND, NEW ZEALAND

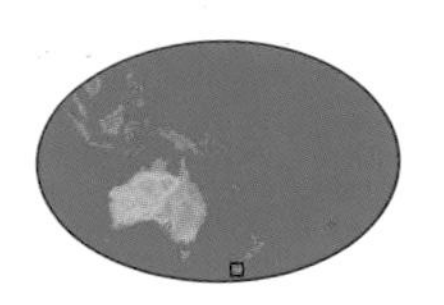

Length of Fiordland: 143 mi (230 km) from northeast to southwest

Width of Fiordland: 50 mi (80 km)

Established as a national park: 1952

Southwest New Zealand is one of the world's greatest wilderness regions, epitomized by the mountains, glaciers, forests, and fjords of Fiordland National Park. Declared part of a World Heritage site in 1990, Fiordland is a place of extravagant beauty, carved by wind, ice, rain, and sea. Granite mountains, the oldest in New Zealand, plunge straight down from the snowline to sea level and are usually beset by thunderous rain clouds. Long fingers of sea reach deep into the land between vertical walls festooned with waterfalls. Fiordland's best-known attraction is Milford Sound, described by Rudyard Kipling as the "eighth wonder of the world." Carved by glaciers over successive ice ages, it plunges to a depth of 869 feet (265 meters) beneath the stark tower of Mitre Peak. Even the pioneers could not break this land, and the foothills still carry virgin rainforest. The rich, peat soils they nourish flow under torrential rains to the sea, where they turn the waters dark and brown. This gloomy freshwater

layer darkens the depths, coaxing normally deepwater species to much shallower depths than elsewhere. In the fjords, black corals, sea pens, and other rare marine organisms can be found at depths of just 16 feet (5 meters).

Above the surface, a rich variety of undisturbed habitats support creatures and plants that were once found on the ancient land of Gondwana; the takahe, a large flightless rail once thought extinct, still persists high in the Murchison Mountains, and more than 700 unique plants are found here. Fiordland has been wrenched, folded, and tipped in the collision between two of Earth's crustal plates. Buried under ocean sediment for millions of years, it was then released, thrust to the surface and exposed to the ravages of the ice ages. Cracked by faults, rocked by earthquakes, and ground under thick icecaps, the hard granite has endured, although few summits reach more than 6,600 feet (2,000 meters). Behind the mountains to the east, the dark, still lakes of Te Anau and Manapouri run deep—over 1,320 feet (400 meters)—and well below sea level. DH

BELOW *Low-level clouds hover over the glassy waters of Fiordland.*

WEST COAST GLACIERS

SOUTH ISLAND, NEW ZEALAND

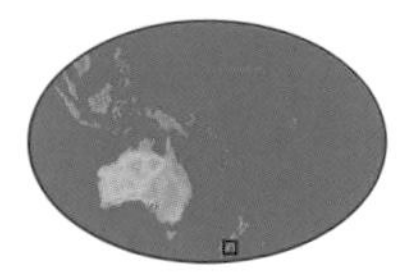

Length of Franz Josef Glacier: 6.8 mi (11 km)

Length of Fox Glacier: 8 mi (13 km)

Average advance rate: Franz Josef Glacier, 6.5 to 10 ft (2 to 3 m) per day

On the wild West Coast of New Zealand, shed from the Southern Alps, the Fox and Franz Josef Glaciers plunge more than 8,200 feet (2,500 meters) through dramatic glacial valleys to flow into rainforest. These unique relics of the last ice age bend and shatter into spectacular icefalls of crevasses and pinnacles. The glaciers move fast—Franz Josef averages 6.5 to 10 feet (2 to 3 meters) a day, 10 times the typical valley glacier speed. The West Coast's great rains fuel this rapid travel—some 118 inches (300 centimeters) a year fall as snow above the glaciers. Both glaciers respond to exceptionally wet seasons with bursts of advance, some very rapid and with potentially disastrous consequences—Franz Josef grinds to a halt just 820 feet (250 meters) above visitor facilities and 12 miles (20 kilometers) from the sea. But, like most of the world's glaciers, they are in overall retreat as the planet warms. Early Maori people called Franz Josef *Ka Roimata o Hinehukatere*, meaning "Tears Of The Avalanche Girl." According to Maori legend, Hinehukatere loved climbing in the mountains and persuaded her lover, Tawe, to climb with her. Tawe fell to his death and Hinehukatere's many tears froze to form the glacier. DH

SOUTHERN ALPS

SOUTH ISLAND, NEW ZEALAND

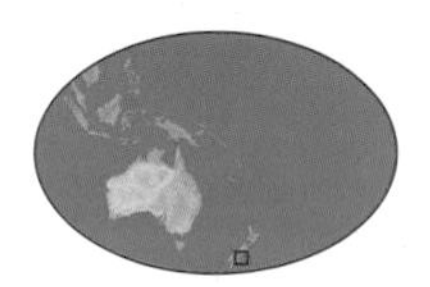

Total length of Southern Alps: 403 mi (649 km)

Highest point (Aoraki/Mt. Cook): 12,316 ft (3,754 m)

Rate of uplift: 0.4 in (1 cm) per year

Running like vertebrae down the back of New Zealand's South Island, the Southern Alps betray a subterranean turmoil deep below. This spectacular range, stretching 403 unbroken miles (649 kilometers) from Milford Sound in the south to Blenheim in the north, is the child of an unholy union between two of Earth's great tectonic plates: the Indo-Pacific and the Australian. As the plates collide with Herculean force, the land above is buckled and rent as mountains are pushed up by as much as 0.4 inch (1 centimeter) per year. By that reckoning, tectonic forces have lifted the Alps by nearly 16 miles (25 kilometers) in the last five million years. But New Zealand is a wet and windy place, and erosion by the elements—as well as frequent earthquakes triggered by the tectonic plates—have kept the mountains to a high point of 12,316 feet (3,754 meters) at Aoraki (or Mount Cook), the highest point in New Zealand. Aoraki itself was lowered 33 feet (10 meters) in 1991, when a 4.5-mile- (7-kilometer-) long avalanche of rock and ice fell from its summit. Twenty-six other peaks along the range tower to over 10,000 feet (3,000 meters), with hundreds more falling just short. DH

SUTHERLAND FALLS

SOUTH ISLAND, NEW ZEALAND

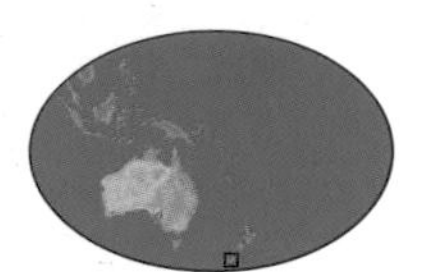

Total cascade of Sutherland Falls: 1,904 ft (580 m)

Location: 14 mi (23 km) southeast of Milford Sound

Sutherland Falls—which is actually a series of three spectacular cataracts—leaps from the headwaters of the Arthur River, southeast of Milford Sound in southwestern New Zealand. It is the second highest waterfall in the Southern Hemisphere, and the fifth highest in the world, with a total drop of 1,904 feet (580 meters). Normally, it makes three bounds of 815 feet (248 meters), 751 feet (229 meters), and 338 feet (103 meters), but in flood (a not uncommon state in rainy Fiordland) it melds into a single giant plume. At times like this, the falling cataract whips the air into gusting winds, and the impact of water on rock is a booming roar, heard from afar. The falls are fed by snowmelt from Lake Quill, named after William Quill, who managed to climb Sutherland Falls in 1890. The falls themselves bear the name of one Donald Sutherland, a prospector who came this way in 1880. On his death, he was buried according to his wish beneath the falls, where his wife Elizabeth later joined him. Not long after, however, a deluge unearthed them both and delivered them to the deep waters of the Sound. The Maori call Sutherland Falls *Te Tautea*, meaning "The White Thread." **DH**

POOR KNIGHTS ISLANDS

OFFSHORE ISLANDS, NEW ZEALAND

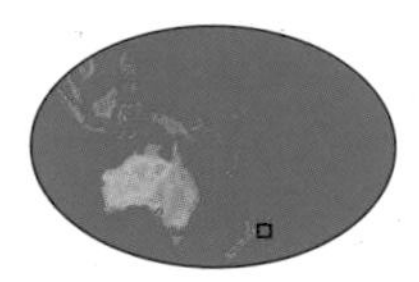

Awarded marine reserve status: 1998

Length of Rikoriko Cave: 165 ft (50 m)

The Poor Knights Islands, vestiges of a chain of ancient volcanoes, rise from the edge of New Zealand's northeastern continental shelf. Swept by the warm East Auckland Current, they support a rich mix of subtropical and temperate marine fauna found nowhere else in the country. They are widely considered to be one of the world's top 10 dive sites. During the last ice age, surf pounded the rocky beaches, now submerged, scouring sea caves (Rikoriko is one of the greatest in the world), tunnels, and arches in the softest rock. Today, plankton-rich currents sweep through them, nourishing sponge gardens, ranks of soft corals, anemones, zooanthids, and gorgonian fields. Lush kelp forests carpet the steep undersea cliffs, offering shelter to over 150 pelagic and reef fish species, including subtropical rarities. Black coral reaches up from the dark depths and schools of stingrays hang in the swirling currents. Above the surface, hordes of seabirds, including 2.5 million Bullers shearwaters, nest in the steep cliffs, which they may share with the ancient tuatara lizard. Venomous centipedes and giant weta, a specific bush cricket, share the scrubby cliff tops with flax snails and geckos. **DH**

WHITE ISLAND

OFFSHORE ISLANDS, NEW ZEALAND

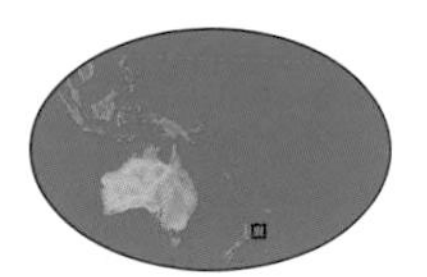

Summit height of White Island:
1,053 ft (321 m)

Age: 100,000 to 200,000 years

One of New Zealand's most active volcanoes, White Island rises 1,053 feet (321 meters) from the Pacific Ocean, 29 miles (47 kilometers) from the mainland Bay of Plenty. It is the country's only live marine volcano, and scientists and volcanologists come from around the world to study its unique features. It is thought to be between 100,000 and 200,000 years old, but the small portion of the island above sea level is estimated to be only 16,000 years old in its present form. White Island is in fact the summit of two overlapping stratovolcanoes. The island has erupted more than 35 times since 1826, most spectacularly on July 27, 2000, when a new crater 500 feet (150 meters) across was created. The eruption covered the eastern half of the island with up to 11 inches (28 centimeters) of ash and pyroclastics, including large blocks of semi-molten pumice. The interior of White Island is a harsh, poisonous wasteland, all but devoid of life. Instead, yellow and white sulfur crystals grow in lush beds beside steaming, bubbling fumaroles. The ruins of an old sulfur works, a rusted testament to many failed mining ventures, are slowly succumbing to the corrosive air. DH

VII

POLAR REGIONS

An icy wilderness where the frozen fingertips of polar icecaps feel their way across stretches of land and water, the Antarctic and Arctic form the polar regions of our globe. Giant icebergs calved from the frosty cliffs of huge ice sheets turn the seas into icy soups; emperor penguins huddle against the bitter Antarctic winds; and polar bears negotiate the floating ice floes in search of food.

LEFT *A blue iceberg in the Weddell Sea, Antarctica.*

GREENLAND ICECAP

GREENLAND, ARCTIC

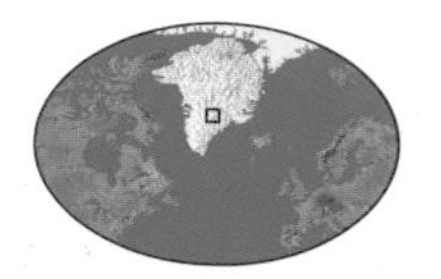

Surface area of Greenland Icecap: 708,086 sq mi (1,833,900 sq km)

Length of icecap: 1,570 mi (2,350 km)

Average thickness of ice: 5,000 ft (1,500 m)

The Greenland Icecap began forming about three million years ago and has today accumulated to a depth of 2 miles (3.2 kilometers) in places. This enormous ice sheet covers 85 percent of Greenland and is the world's second largest mass of ice. Only the Antarctic ice sheet is larger. The Greenland Icecap gives the impression of being a gigantic, static mass, frozen in time, but that is far from the truth. This is a dynamic, constantly changing geological formation. The sheer weight of the ice at higher elevations pushes down on the rest and forces the ice to flow outward from the interior to the sea. In some places near its outer edge the ice is moving 66 to 100 feet (20 to 30 meters) per day.

Where the flowing ice meets the sea, spectacular icebergs calve into the water. One estimate indicates that more than 1 billion tons of ice are discharged into the sea each year and that if all of Greenland's ice were to melt it would raise world sea levels by 23 feet (7 meters). JK

BELOW *The icy fingers of the Greenland Icecap.*

SØNDRE STRØMFJORD

GREENLAND, ARCTIC

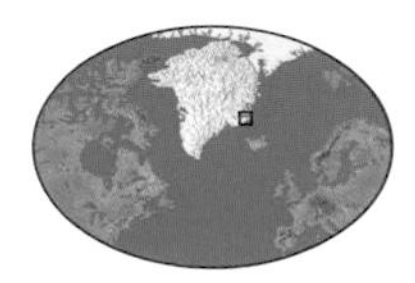

Alternative name: Kangerlussuaq

Length of Søndre Strømfjord: 100 mi (160 km)

Width of fjord: 3 mi (5 km)

The Inuit name for Søndre Strømfjord is Kangerlussuaq, which translates as "the long fjord." Located on Greenland's southwest coast, this 100-mile (160-kilometer) fjord cuts a straight path into the interior of Greenland. It is one of the longest fjords in the world, and lies framed by dramatic, glacial-carved mountains, 40 miles (64 kilometers) north of the Arctic Circle. The dry, low Arctic area has diverse habitats such as heaths, salt lakes, and mountain tundra, which are home to a rich mix of wildlife, including caribou, musk oxen, and arctic fox. In the icy green waters of Kangerlussuaq Fjord, narwhals can sometimes be spotted, while polar bears patrol the shore and ivory gulls fly overhead. In some places along the fjord, glaciers have reached right down to the water, creating a fabulous landscape of ice cliffs and icebergs. The town of Kangerlussuaq offers easy access to the inland icecap, with the only road in Greenland that leads directly to it. Greenland is a world of extreme beauty and extreme contrasts. Winter temperatures can fall to -58°F (-50°C); in summer they can rise to 82°F (28°C). **JK**

IKKA FJORD

GREENLAND, ARCTIC

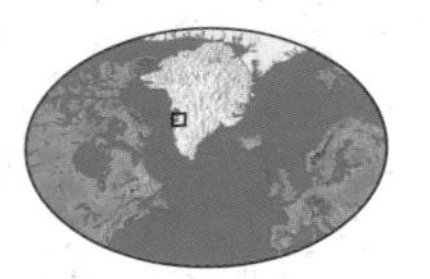

Average water temperature of Ikka Fjord: 37°F (3°C)

Maximum depth of fjord: 100 ft (30 m)

Number of mineral columns: approx. 700

The underwater world of Ikka Fjord in southwestern Greenland is like no other, for it contains a forest of remarkable mineral columns growing on the seabed. Although first described several decades ago, research begun in 1995 has determined that the towers are made of a form of calcium carbonate called ikaite. This rare mineral only occurs under special conditions, when bicarbonates in freshwater from seabed springs mix with calcium in seawater. Due to the extremely low temperatures, precipitation is inhibited and ikaite is formed. Ikaite is a very fragile mineral that crumbles when exposed to air, but underwater it can grow into an amazing variety of shapes. The tips of the columns are particularly beautiful, with crests and spires that take on different designs depending on the temperature and salinity of the water. More than 700 columns have been counted over a 1.25-mile (2-kilometer) stretch. Many are over 66 feet (20 meters) high, and their tops are visible at low tide. The growth of the columns is phenomenal—about 1.6 feet (0.5 meter) per year. It is possible to dive in the fjord if you bring your own equipment. JK

MACQUARIE ISLAND

SOUTHERN OCEAN

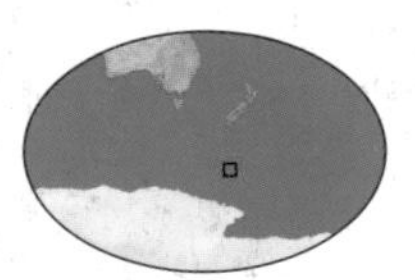

Highest point of Macquarie Island (Mt. Hamilton): 1,420 ft (433 m)

Age of island: 600,000 years

Vegetation: tussock, mire, feldmark

Remote, wind-blasted Macquarie Island is on the way to Antarctica in the "Furious Fifties" at about 55° south. It is believed that the island, volcanic in origin, began as a spreading ridge under the ocean between 11 and 30 million years ago. As the spreading halted and the crust began to contract, rock was squeezed upward until it breached the surface of the ocean about 600,000 years ago. Since then it has been sculpted by fierce and unrelenting wave action, in contrast to other sub-Antarctic islands that have been worked by glaciers. A wave-formed platform with deep peat beds forms a coastal terrace around the island, with ancient sea stacks offshore. Behind this terrace, sharp escarpments rise 660 feet (200 meters) to a central plateau dominated by Mount Hamilton. Countless lakes, tarns, and pools sit on the edge of the plateau overlooking a vast and featureless sea. But there is life here, and plenty of it. A huge rookery of 850,000 royal penguins represents most of the world's population. There are also albatross and elephant seals; offshore, leopard seals patrol. Macquarie is about 930 miles (1,500 kilometers) south-southeast of Australia's southernmost state, Tasmania. GH

RIGHT *The isthmus of Macquarie Island.*

HEARD & McDONALD ISLANDS

SOUTHERN OCEAN

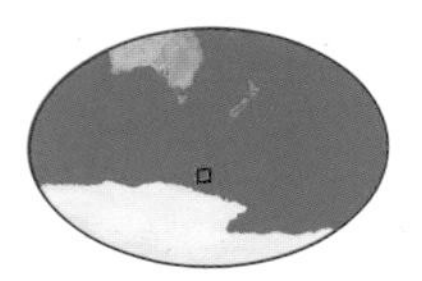

Height of Big Ben: 9,006 ft (2,745 m)

Thickness of Big Ben icecap: 500 ft (150 m)

McDonald Islands' highest point: 755 ft (230 m)

Heard Island and the McDonald Islands lie on the Kerguelen–Heard submarine plateau and rise from the Southern Ocean just south of the border between icy southern and warmer northern waters—the Antarctic Convergence. The continuously glaciated Big Ben dominates Heard Island. Its summit is Mawson Peak, an active, towering volcano covered in snow and glacial ice that contrasts with the black volcanic rock. It has been climbed only three times due to its height, remoteness, and the ferocity of its conditions. The glaciers are said to be the most dynamic in the world, and the ice cliffs overhang the sea. The McDonald Islands are 27 miles (44 kilometers) to the west of Heard and are also volcanic in origin; all provide undisturbed habitat for sub-Antarctic plants and animals. In the mid-19th century, sealers discovered the seals of Heard Island and within 30 years wiped out almost the entire fur seal population and most of the elephant seals. These colonies are only beginning to reestablish themselves 150 years later. GH

ZAVODOVSKI ISLAND

SOUTH SANDWICH ISLANDS, SOUTHERN OCEAN

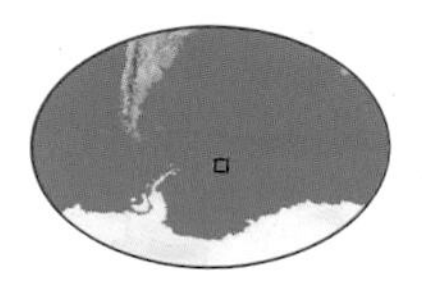

Area of South Sandwich Islands: 120 sq mi (310 sq km)

Height of Mt. Asphyxia: 6,000 ft (1,800 m)

About 1,000 miles (1,600 kilometers) from the tip of the Antarctic Peninsula is an active volcanic islet only 4 miles (6.4 kilometers) across, yet each summer it is the temporary home of up to 21 million penguins, one of the largest gatherings of its kind in the world. Most of the birds are chinstrap penguins, recognized by the black band of feathers under their beaks, and these are joined by yellow-plumed macaroni penguins. The birds come to this remote spot in the southern Atlantic Ocean to nest on volcanic ash from the aptly named Mount Asphyxia. Each nest is no more than 31 inches (80 centimeters) from its neighbor, so from a distance the island resembles a dense black-and-white carpet. The volcano on which the birds roost erupts gently almost daily, generating great swirls of smoke and steam. The warmth, however, keeps the snow away for most of the year, extending the penguins' breeding period. MB

BELOW *The active volcanic islet of Zavodovski Island.*

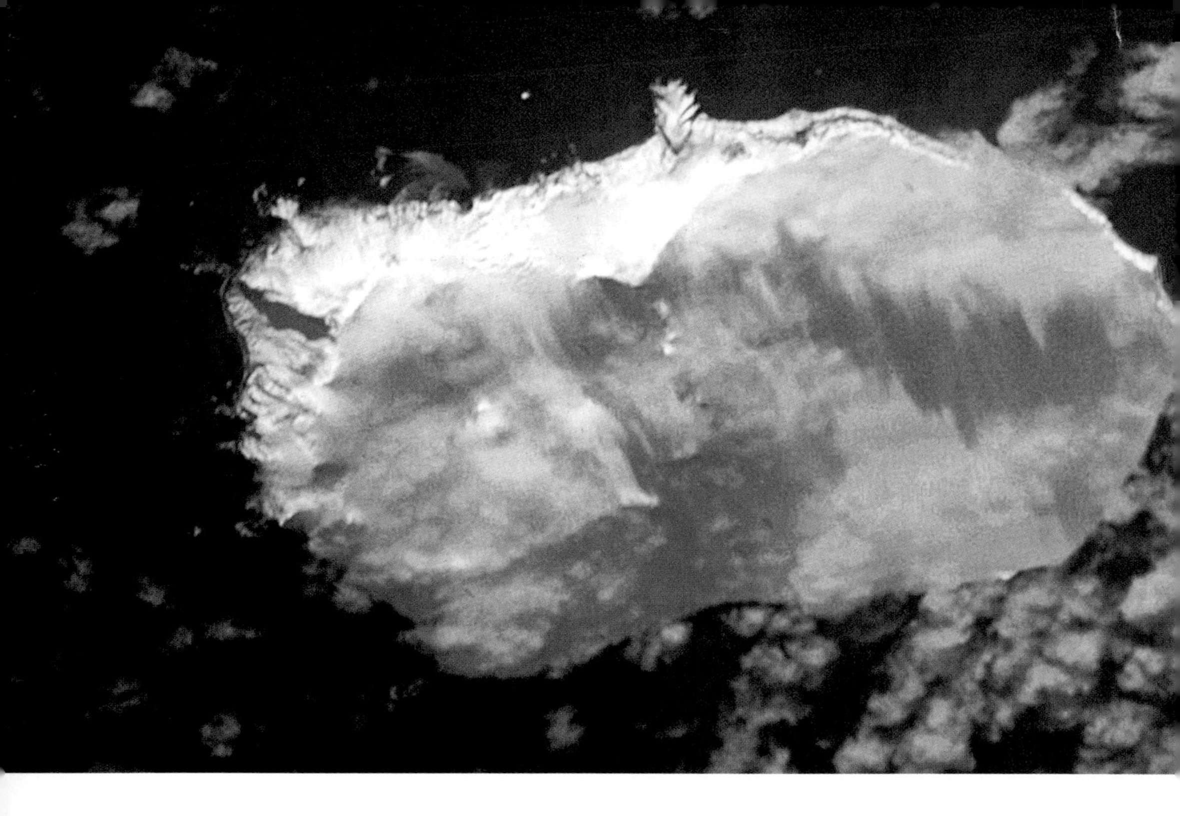

BOUVET ISLAND

SOUTHERN OCEAN

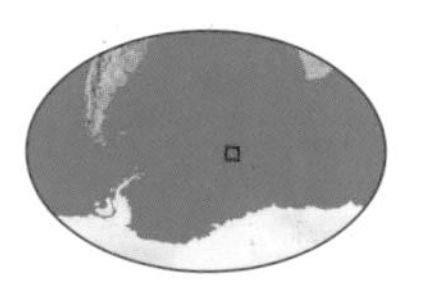

Area of Bouvet Island: 23 sq mi (60 sq km)

Highest point: 3,068 ft (935 m)

Bouvet Island is the southernmost island of the Mid-Atlantic Ridge. It lies 1,370 miles (2,205 kilometers) southwest of Cape Agulhas at the tip of South Africa and 1,020 miles (1,642 kilometers) southeast of Gough Island in the South Atlantic, making it the most isolated piece of land on Earth. It is, according to sailors who frequent these waters, one of the most fearsome places on the planet. Surrounded by vertical ice cliffs, outcrops of sheer volcanic rocks, skerries, and reefs, it is difficult to land and equally difficult to leave, and rock and ice falls occur continuously. Set in the "Furious Fifties," the island is pounded by storms, yet fulmars, Cape pigeons, prions, and blue petrels fly here, and humpback whales and fur seals feed here. In 1739, French navigator Bouvet de Lozier, who was unable to land, discovered Bouvet. The remote island was not sighted again until 1808 when a whaling vessel, the *Swan*, charted its exact position, but again the crew was unable to land. The first humans to put ashore were from another whaler, the *Sprightly*, in 1825, but these men were marooned here for a week. In 1927, the Norwegians landed and later annexed the island. MB

SUB-ANTARCTIC ISLANDS

ANTARCTICA

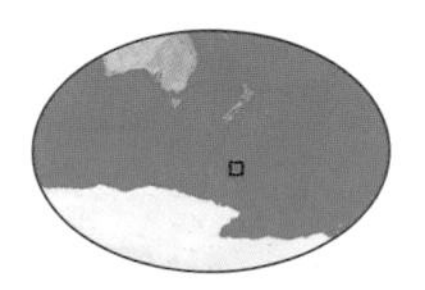

Total land area of sub-Antarctic Islands: 295 sq mi (764 sq km)

Terrestrial flora: 35 unique species

Fauna: 126 bird species, including 5 that breed nowhere else in the world

New Zealand's five sub-Antarctic island groups are tiny oases of land and life in the vast Southern Ocean. The Auckland, Bounty, Snares, Antipodes, and Campbell groups endure horrendous weather, with storm conditions prevailing for most of the year, yet they are home to an abundance of life. The islands contain some of the southernmost forests in the world, species at the limit of their physical endurance. They provide essential habitats for vast seabird colonies. The world's largest breeding populations of wandering albatross and shy albatross nest on the Auckland Islands, while Campbell Island is home to the world's largest breeding colony of royal albatross. The Auckland Islands are also the main breeding rookery of one of the world's rarest seals, the New Zealand (Hooker's) sea lion. The islands lie along the migratory route of a number of whale species. A breeding population of at least 100 southern right whales gathers in Port Ross in the Auckland Islands between June and September each year. In 1986 the islands were declared national nature reserves, and in 1998 they received international recognition as a World Heritage site. DH

DRY VALLEYS

ANTARCTICA

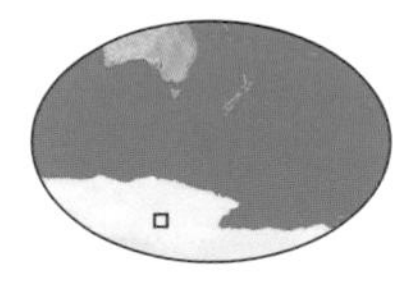

Total area of the Dry Valleys: 1,853 sq mi (4,800 sq km)

Mean annual air temperature: 1.4 to -4°F (-17 to -20°C)

Depth of permafrost: 787 to 3,182 ft (240 to 970 m)

The Dry Valleys—the largest ice-free area in Antarctica—are located within 1,853 square miles (4,800 square kilometers) of frozen lakes, ephemeral streams, arid rocky soils, and permafrost. This surreal landscape is so punishing that NASA chose to test its Viking Mars probe here.

Glaciers retreated through here around four million years ago, shaping the harsh landscape. The major Dry Valleys share certain physical characteristics: They are on average 3 to 6 miles (5 to 10 kilometers) wide and 9 to 31 miles (15 to 50 kilometers) long. The valleys are parched because the Transantarctic Mountains block the flowing ice of the Polar Plateau, and they receive little precipitation—no rain has fallen here for at least two million years, and the 4 inches (10 centimeters) or so of snow a year that does fall usually sublimates (changes from a solid to a gas) immediately.

In this area only mosses can survive in the stony soils; there are no vascular plants or vertebrates and very few insects. The region is so dry that 3,000-year-old mummified seal carcasses have been found, preserved by the desiccating air. DH

CAPE ADARE

ANTARCTICA

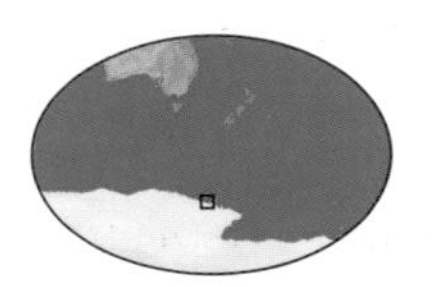

Admiralty Mountains' highest peak (Mt. Minto): 13,668 ft (4,166 m)

Coordinates of Favreau Pillar: 71.57S, 171.07E

At the tip of Victoria Land and backed by the Admiralty Range—the highest mountain range in the Antarctic—Cape Adare is at the edge of the Ross Sea and is the nearest part of Antarctica to New Zealand. Its large, flat spit of black pebbles is approached from Robertson Bay and is home to between one-half and one million adelie penguins, the largest adelie rookery in the Antarctic. These residents show no fear of human visitors, and tourists can sit among the birds, watching courtship behavior, chick feeding, and territorial disputes without disrupting the natural order. The cape is also the site of the oldest habitation in the Antarctic: the hut built and occupied by Carsten Borchgrevink, a Norwegian who led the first expedition to overwinter on the continent in 1899. Sadly, also here is the oldest grave, that of Nicolai Hansen, a member of Borchgrevink's "Southern Cross" expedition. He died on October 14, 1899, and was buried 1,000 feet (300 meters) above Ridley Beach. Not far away, to the east of Foyn Island (one of the Possession Islands), is the curious Favreau Pillar, a sheer, rock monolith rising vertically from the icy sea. MB

TRANSANTARCTIC MOUNTAINS

ANTARCTICA

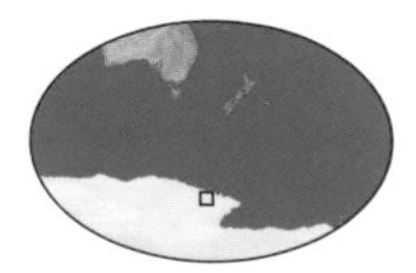

Area of Transantarctic Mountains: 225,461 sq mi (583,943 sq km)

Highest peak (Mt. Markham): 14,275 ft (4,351 m)

The Transantarctic Mountains stretch 3,000 miles (4,800 kilometers) across the continent from Victoria Land on the Ross Sea to the Coats Land on the Weddell Sea, splitting it into two distinct geographical and geological regions. The larger, eastern subcontinent rests on ancient Precambrian rock mostly above sea level. The smaller western region lies mostly below sea level. Basement rocks similar to those found in Australia, South Africa, and South America confirm its birthplace—Gondwanaland. The range is the longest in Antarctica, and one of the longest in the world, although in many places it lies buried beneath a deep ice sheet, with just the summits of many mountains left exposed. Such peaks are known as nunataks. The mountains are geologically complex, with layers of Jurassic dolerite sandwiched between much older bands of 200-to-400-million-year-old sandstone. Born during the upheaval of the Cenozoic Era about 65 million years ago, the mountains were further wrenched by more recent rifting and tilting and have yielded fossils that say much about Antarctica's history. Today these rocks are exposed once more in the steep eastern faces of the Royal Society Range. DH

THE POLAR PLATEAU

ANTARCTICA

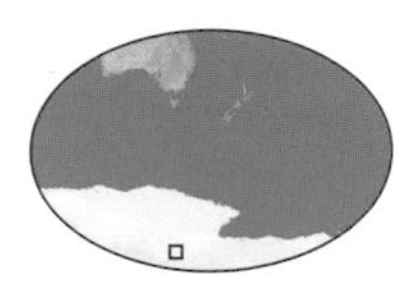

Total area of Antarctic ice sheet: 5.1 million sq mi (13.2 million sq km)

Average depth of ice: 1.5 mi (2.5 km)

Maximum depth of ice: 3 mi (5 km) on Wilkes Land

Nearly 1 mile (1.6 kilometers) above sea level, the Polar Plateau, in the center of the East Antarctic Ice Sheet, is one of the coldest and driest places on Earth. In the perpetual polar winter, temperatures at the Russian Federation Vostok research station routinely dip below -58°F (-50°C)—on July 21, 1983, a world record was set with a low of -129°F (-89.4°C). Antarctica is the world's highest continent, averaging 7,546 feet (2,300 meters), and it carries 9.2 million square miles (24 million square kilometers) of ice (70 percent of the world's freshwater). The bitterly cold air on the plateau contains virtually no water vapor, making the Antarctic interior the world's biggest desert.

Ice forms at a rate of 2 to 35 inches (5 to 89 centimeters) a year, which gives a clue to the age of the ice sheet—the ice of the Polar Plateau probably formed during the Miocene Epoch and is at least 15 million years old. The sheet contains over 5 million cubic miles (20 million cubic kilometers) of ice, and weighs so much that in many areas it pushes the land below sea level. Without this burden, Antarctica would eventually rise another 1,500 feet (450 meters) above sea level. DH

MOUNT EREBUS

ANTARCTICA

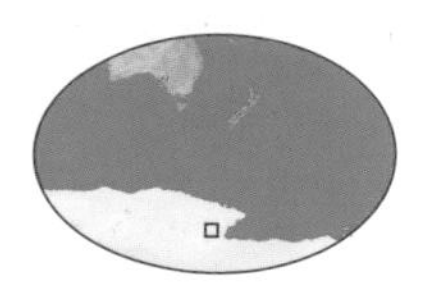

Height of Mt. Erebus: 12,447 ft (3,794 m)

Width of outer crater: 2,132 ft (650 m)

Mount Erebus is the world's southernmost volcano, and the largest, most active volcano on the Antarctic continent. A cloud of vapor perpetually wafts from its 12,447-foot (3,794-meter) summit crater, the high point of Ross Island. Erebus has erupted at least eight times in the last 100 years; the most recent activity began in 1972 and continues today. Volcanic bombs as large as 26 feet (8 meters) across have been blasted from the summit. Within its 330-foot- (100-meter-) deep outer crater lies a similarly deep inner crater, about 820 feet (250 meters) wide, cradling a lake of molten lava.

The first to witness the wrath of Erebus were members of Shackleton's 1908 expedition, who recorded a "vast abyss" emitting a steam column 1,000 feet (300 meters) high. During a brief clearing, they observed "lumps of lava, large feldspar crystals, and fragments of pumice." Erebus supports a broad glacier system. From its flanks, a number of ice sheets grind inexorably to the edge of Ross Island, where they either form abrupt cliffs that periodically collapse into the Ross Sea on the north and west or merge gently with the Ross Ice Shelf along the eastern coast. DH

ANTARCTIC SEA ICE

ANTARCTICA

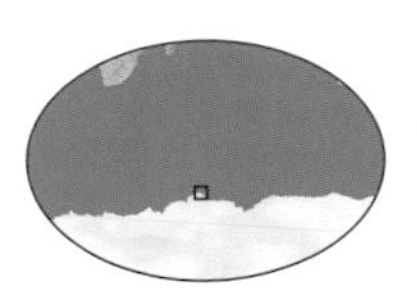

Total area of ice in winter:
7.7 million sq mi (20 million sq km)

Total area of ice in summer:
1.5 million sq mi (4 million sq km)

Every austral winter, the Antarctic ice sheet swells to form a 7.7-million-square-mile (20-million-square-kilometer) frozen fringe, eventually covering an area larger than the continent itself. This freeze is Earth's biggest natural event, and the most significant for the world's climate. An enormous reflector, it repels up to 80 percent of solar radiation and restricts the heat transferred between ocean and atmosphere. Each day, the ice advances about 3 miles (5 kilometers), creating an extra 3,861 square miles (10,000 square kilometers) of ice. In still water, hexagonal crystals form on the surface, creating an oily sheen known as "grease-ice." "Frazil ice" follows as the slurry thickens; this can then form into plates of "pancake ice." As snow falls from above and the sea freezes beneath, the ice gradually thickens, smothering the sea under a solid mass. But out on the edge of the pack, ocean swells and wind break the ice into large pieces—called pack ice—that float before the wind and currents. By the end of summer, the entire mass has shrunk again to just 1.5 million square miles (4 million square kilometers). DH

BELOW *The towering cliffs of the Antarctic sea ice.*

LEMAIRE CHANNEL

ANTARCTICA

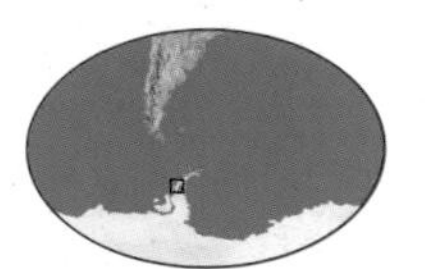

Best time of day to visit: sunset

Size of Antarctica: 5.5 million sq mi (14 million sq km)

Along the vast white expanse of Antarctica, Lemaire Channel is a spectacular narrow strait between the continental mainland and Booth Island. Just 5,250 feet (1,600 meters) wide and seven miles (11 kilometers) long, its waters are protected by sheer cliffs rising 3,000 feet (900 meters) from the sea, and it is almost impossible to see it before entering. It is highly popular with cruise ships, attracted by the calm waters and stunning scenery that have earned it the name "Kodak Gap." It runs from the northeast to the southwest, and the southern end culminates in an archipelago of icy islands; at the northern end at Cape Renard are two tall, rounded peaks, often topped with snow. The passage is sometimes blocked by ice, forcing navigators to backtrack around Booth Island. Orca and humpback whales are commonly seen, as well as penguins, elephant seals, and seabirds such as snow petrels, south polar skuas, and blue-eyed shag. The channel was first discovered by the Belgian explorer Adrien de Gerlache in 1898, who oddly named it after the Belgian Charles Lemaire, known for exploring parts of the Congo. GD

ANTARCTIC PENINSULA

ANTARCTICA

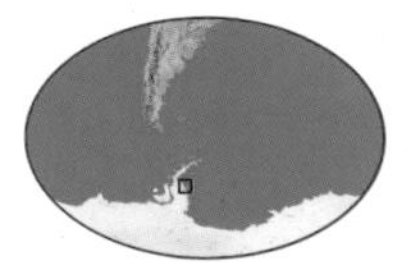

Length of the Antarctic Peninsula: 800 mi (1,287 km)

Age of peninsula: 225 million years

Highest peak (Mt. Français on Anvers Island): 9,258 ft (2,822 m)

The Antarctic Peninsula sweeps in an 800-mile (1,287-kilometer) arc north from the Antarctic Continent, fused by a spine of 8,200-to-10,000-foot (2,500-to-3,000-meter) peaks—the second longest mountain chain in Antarctica. An extension of the Andes, the peninsula joins South America via the partially submerged Scotia Ridge—a 1,988-mile (3,200-kilometer) chain that breaks the surface as the South Orkney, South Sandwich, and South Georgia islands. The peninsula's climate is slightly kinder than that of the continent, allowing for a modest flora of mosses, lichens, and green algae, which grow best on the west coast islands. Flanked by sea ice, deep fjords, ice falls, steep glaciers, and myriad offshore islands, the peninsula is a rich breeding ground for seabirds, seals, and penguins. During the summer, orca, humpback, and sperm whales come to feast on krill and the animals it attracts. In the 19th century, this bounty also attracted sealers and whalers who were responsible for much of the discovery and mapping of the region. Britain, Argentina, and Chile all claim a territorial interest in the peninsula. On January 7, 1978, the first "native" Antarctican was born here. DH

RIGHT *The Antarctic Peninsula and its icy sea.*

Contributors

Rachel Ashton has worked for twelve years with *BBC Wildlife Magazine*, but with her strong interest in natural history, especially marine biology, and the environment, she now runs a marine-wildlife travel company, *Ocean Wanderers*, which enables people to visit the world's marine hot spots to swim with whale sharks, manatees, and see whales, dolphins, and sharks.

Nick Atkinson trained as a biologist, studying zoology at Nottingham University and gaining a doctorate from the University of Edinburgh. His PhD thesis was based on the hybrid zone between two European species of fire-bellied toad, and required extensive periods of field work in Central Europe, from Poland in the north, through Slovakia, Hungary and Romania, though concentrating mainly on the Balkan states. Currently based at the University of Newcastle, Nick is a contributor to several popular science publications, including *BBC Wildlife Magazine*, *Natural History*, and *The Scientist*.

Adrian Barnett is a tropical biologist and journalist who has worked and traveled in 23 different countries. Having worked on remote area biological surveys in West Africa and South and Central America, and tour-guided in Asia, he is now working on a primate conservation research program in Amazonian black water swamp forests. He lives in California.

Mark Brazil, author, columnist, and nature guide, has worked in many countries, but currently lives in Japan. He is probably best known for his books *The Birds of Japan* and *The Whooper Swan*. Fascinated by island biology, he is a leading authority on the natural history of Japan, especially Hokkaido, where he has worked as a professor of biodiversity and conservation at Rakuno Gakuen University near Sapporo. He has also worked on projects for TRAFFIC (Japan), the Wild Bird Society of Japan, and WWF Japan. He has contributed regularly to *BBC Wildlife Magazine* and *The Japan Times*.

Michael Bright was an award-winning producer with the BBC's Natural History Unit for 30 years. He and his colleagues are based in Bristol, England, but they scour the globe for stories and images. Bright is the author of over 80 natural history books, including *Andes to Amazon: A Guide to South America*, the producer of numerous television and radio programs, including the award-winning television series *Natural World* and *Wildlife on One*, and has been editorial consultant on reference works such as *Discovering the Wonders of Our World* and *The Wildlife Year*.

Dave Brian Butvill steers clear of short, guided, all-you-can-see "whiplash tours," preferring to linger in remote areas to really get a feel for them. In this way, the Wisconsin native first scoured the American West, from Arizona's cactus forests through Yellowstone and Glacier national parks to the Alaskan tundra, before moving to California and adopting Yosemite and the High Sierra as a second home. Butvill recently moved to Central America. Having tasted Nicaragua, Panama, and Guatemala, he now lives in, and writes from, the central mountains of Costa Rica.

Chris Cole has been traveling the far-flung corners of the world ever since he left home—from the jungles of Madagascar to the Arizona desert, the coral reefs of the Maldives to the rhododendron forests of the Himalayas. Much of this travel has been while working for the BBC's Natural History Unit, most recently exploring closer to home for a series about the wildlife and countryside of the British Isles. As well as making wildlife films, Chris has contributed articles and pictures to many BBC publications, but has not yet managed to persuade anyone to fund his lifelong ambition to see the Northern Lights.

Rob Collis is a research librarian specializing in natural history at the BBC. He is a graduate in ecology and information science and provides information to documentary program-makers at the BBC Natural History Unit on all aspects of wildlife and the natural world. His travels are confined mostly to the U.K. and Europe, including the Republic of Ireland, France, Spain, Greece, Belgium, Italy, and Denmark.

Tamsin Constable is a freelance writer with a particular interest in natural history. She grew up in Tanzania, Cameroon, and Mali and has traveled extensively in Africa. She studied zoology and psychology before training as a journalist at London's City University. After six years with *BBC Wildlife Magazine*, she went freelance. Her work includes documentary scripts, a book on chimpanzees, travel features for the national press, copy-writing, and editing. She lives and works in Leeds, West Yorkshire.

Andrew Cooper is a broadcaster and award-winning television producer working with the BBC Natural History Unit. He has traveled worldwide. In the last fifteen years alone he has visited over thirty different countries. His programs regularly feature in the top ten most watched BBC wildlife documentaries and have been widely sold overseas. He is also the author of six books, some of which, published by the BBC, have ranked among the top ten best-selling hardback books in Britain.

Jenny Devitt has traveled all her life. Born in North Wales, she was soon transported to the adventures of a childhood in the wild bush country of Southern Africa. Since then she has visited, lived, and worked in many countries: from the isolated islands of Cape Verde to Mexico, Australia, Bosnia, the Seychelles, and Himalayan Nepal and Bhutan. She has written and broadcast internationally, and is currently living in the southern French Pyrenees, restoring an old house in a remote village.

Gina Doubleday is a freelance writer, editor, and translator living in Paris. She has contributed to many titles, including InSight Guides, Libération, ELLE, and Vogue. She also works extensively for the UNESCO World Heritage Center, writing magazine articles and news releases about World Heritage sites for various media.

Teresa Farino is a British writer specializing in environmental issues who has been living in the Picos de Europa (northern Spain) since 1986. She is author of many wildlife and travel books, including *Sharks—the Ultimate Predators* (1990), *The Photographic Encyclopedia of Wildflowers* (1991), *Landscapes of Northern Spain: Picos de Europa* (1996), *Landscapes of Barcelona and the Costa Brava* (2003), and *Traveler's Nature Guide: Spain* (2004). Teresa also leads wildlife holidays in Spain and Portugal on a regular basis.

Peter Ginn recently relocated from Zimbabwe where he taught at Peterhouse School for 28 years, and is now running safaris in Zimbabwe, Zambia, Botswana, and Madagascar. Although initially these were birding safaris, today most cover general natural history with birds forming one of a number of major subjects. Peter has written seven bird books and was general editor of *The Complete Book of Southern African Birds.* He has over 500 bird species on slides. Peter is a teacher by vocation but uses his bird knowledge and photographs to get people interested in our natural environment.

Dave Hansford is an award-winning photographer, writer, and cameraman based in Wellington, New Zealand. He specializes in natural history, Earth sciences, adventure travel, and the environment. After 14 years as a press photographer, he formed a small multimedia partnership—Origin Natural History Media—with his partner Marieke. His work has appeared in *BBC Wildlife Magazine, Australia Nature, Action Asia, NZ Geographic, Destinations, NZ Business, Wilderness, Forest & Bird,* and *Seafood,* as well as in newspapers throughout New Zealand and Australia. He has filmed for Natural History NZ Ltd. and worked on the BBC production *Life of Birds.* His hunger for adventure has taken him to Antarctica, Africa, Australia, China, and Pakistan.

Guy Healy is a freelance correspondent for *BBC Wildlife Magazine* based in Australia. He has traveled through the jungles of Australia's Cape York, explored the opal mines of South Australia's outback, and managed to survive a four-wheel-drive trip up the famed Gibb River track in Western Australia's Kimberley region. He also managed to live through a sailing adventure from Hobart in Tasmania to Sydney and looks forward—with a sense of trepidation—to many more years of exploring his native land, Australia.

David Helton is an American who has lived in Japan, Mexico, Saudi Arabia, Greece, Ireland, Italy, Spain, and now Britain. He is a freelance science writer for newspapers, magazines, and television. He has worked for *The Times* and *BBC Wildlife Magazine,* has had a novel published, and has written or edited some 180 documentary scripts for television.

Adrian Hillman trained as an ecologist and worked in nature conservation in England for several years before joining Voluntary Service Overseas and moving to Thailand. He originally intended to stay for two years, but the culture, food, climate, and wildlife (especially the bats) persuaded him to hang around for longer.

Joe Kennedy is a writer and television documentary producer of natural history, science, and adventure films. He has traveled extensively around the world, in Africa, Asia, Europe, and North America. The nature of his work means that he spends several months a year in some of the most remote —but beautiful—locations on Earth. When he is not filming he takes still photography of landscapes. And being Canadian he also has a special passion for the Arctic and the Inuit people who live there.

David Lazenby is a photographer, writer, artist, and designer. He was born a Yorkshireman, grew up in South Africa, and has lived in Denmark, Australia, and the United States. His passion for the world of caves, rainforests, and archeology has taken him on numerous expeditions to far-flung, wild places around the globe. From searching for World War II remains in the caves of South Pacific islands to exploring the mysterious ritual caverns of the ancient Maya, his adventures are often the subject matter for the museum and natural science exhibits he designs. David's photography and articles regularly appear in international magazines, books, and travel guides.

Hugo Leggatt was born in Exeter, Devon, in 1940. Shortly after World War II he moved to South Africa. He graduated in Physics at Cape Town University and has spent most of his working life teaching in South Africa. He married Helen, of Russian-Greek ancestry, in Kampala, Uganda in 1964. They have three sons and four grandchildren. Travel has always been a significant interest and he has traveled widely in South and East Africa and Europe, as well as Russia, Turkey, Israel, Jordan, and India. Now retired, he is spending a lot of time on the rock paintings in the southern area of South Africa.

Mairi Macleod has traveled the globe in pursuit of the study of animal behavior. She has studied the behavior of langur monkeys in India, humpback whales off the east coast of Australia, and rehabilitated monkeys and other animals in an Amazonian forest reserve. She earned her doctorate after spending two years following samango monkeys around the dune forest of Kwa-Zulu Natal in South Africa. She has also traveled in East Africa, Indonesia, and the South Pacific. Mairi currently works as a freelance journalist.

Geoffrey Maslen is a Melbourne-based, freelance journalist who writes for newspapers and magazines in various countries. He has traveled to all states and territories in Australia and has visited most of the continents on Earth.

Chris Mosey is an English writer and photographer living in France. He previously spent 11 years in Scandinavia where he was a correspondent for the English newspapers *The Observer, The Times,* and the *Daily Mail.* He has traveled extensively in Europe and southeast Asia, where in 1987 he reported from Thailand and Myanmar for the Observer Foreign News Service.

Charlotte Scott has traveled around the world several times. She has worked in Australia, Borneo, North America, and Kenya as both a marine biologist and a television producer, and traveled to Madagascar, Ecuador, Botswana, and Oman as a keen photographer and explorer. She currently produces wildlife films for the BBC and contributed to the book of the hit TV series *British Isles: A Natural History.*

Penny Turner Educated at the University of Aberdeen, Penny now lives in Northern Greece, where she is a courier for natural history tours in Greece and the Balkans. She has worked for all the major environmental organizations in Greece and traveled widely as a consultant for International League for the Protection of Horses. She opened and ran a riding school in Greece and traveled by horse over the Greek mountains. She recently completed a 1,000-mile trek across Greece with her horse George. She shared first prize in *BBC Wildlife Magazine*'s Nature Writer of the Year competition.

Martin Williams is a writer and photographer with a strong interest in wildlife and conservation. During the 1980s, he led bird migration studies at Beidaihe, east China, and settled in Hong Kong—which he finds a fascinating place, and a great base for traveling in east Asia. On journeys to see birds, write articles, conduct biodiversity surveys, and enjoy the wilder side of life, he has explored places ranging from Indonesian volcanoes to the steppes of Inner Mongolia, from Malaysian rainforests to the eastern Himalayas.

Glossary

Aboriginal Dreaming: The Dreaming is the framework of Aboriginal creation stories, which give meaning to human life. It has different meanings for different Aboriginal groups. The Dreaming establishes strong links between people, animals, and the land—the same forces that created the world and everything in it are thought to be responsible for the birth of a child.

Acid: a descriptive term applied to igneous rocks with more than 60 percent silica.

Acacia: any of various spiny trees or shrubs of the genus Acacia.

Active Volcano: a volcano that is erupting. Also, a volcano that is not presently erupting, but that has erupted within historical time and is considered likely to do so in the future.

Alkaline: relating to or containing an alkali; having a pH greater than 7.

Alluvial Fan: the alluvial deposit of a stream where it issues from a gorge upon a plain or of a tributary stream at its junction with the main stream.

Ammonite: one of the coiled chambered fossil shells of extinct mollusks.

Andesite: a dark gray volcanic rock.

Anhydrate: a mineral of a white or slightly bluish color, usually massive. It is anhydrous sulfate of lime, and differs from gypsum in not containing water.

Anomalure: large scaly-tailed flying squirrels.

Aquifer: a body of rock that contains significant quantities of water that can be tapped by wells or springs.

Archipelago: a group of many islands in a large body of water.

Arête: a sharp-crested ridge in rugged mountains.

Arribada: the arrival of a mass nesting of turtles.

Arribe: a river gorge.

Atoll: a coral island consisting of a reef surrounding a lagoon.

Auklet: any of several small Auks (black and white diving birds) of northern Pacific coasts.

Basalt: volcanic rock (or lava) that characteristically is dark in color, contains 45 to 54 percent silica, and generally is rich in iron and magnesium.

Batholith: a great mass of intruded igneous rock that for the most part stopped its rise a considerable distance below the surface.

Block: angular chunk of solid rock ejected during an eruption.

Blue Hole: water-filled caves and sinkholes in which water is usually azure.

Boab: a large spreading tree with branches that radiate from its swollen barrel-like trunk.

Bomb: fragment of molten or semi-molten rock, 2.5 inches (6.3 centimeters) to many feet (several meters) in diameter, which is blown out during an eruption. Because of their plastic condition, bombs are often modified in shape during flight or upon impact.

Bongo: large forest antelope of central Africa having a reddish-brown coat with white stripes and spiral horns.

Bore: a rush of water that advances upstream with a wave-like front, caused by the progress of an incoming tide from a wide-mouthed bay into its narrower portion.

Bromine: a non-metallic, largely pentavalent, heavy, volatile, corrosive, dark brown liquid element belonging to the halogens; found in seawater.

Bronze Age: a period between the Stone and Iron ages, characterized by the manufacture and use of bronze tools and weapons.

Bushbuck: antelope with white markings like a harness and twisted horns.

Caldera: a large circular volcanic depression, often originating due to collapse; a crater.

Calving: the loss of glacier mass when ice breaks off into a large water body such as an ocean or lake.

Carbonatite: a carbonate rock of apparent magmatic origin, generally associated with kimberlites and alkalic rocks. Carbonatites have been variously explained as derived from magmatic melt, solid flow, hydrothermal solution, and gaseous transfer.

Carboniferous age: the age immediately following the Devonian, or Age of fishes, and characterized by the vegetation which formed the coal beds. This age embraces three periods: the Subcarboniferous, the Carboniferous, and Permian.

Cay: a low island or reef of sand or coral.

Cenotes: vertical sinkholes made of limestone and filled with water.

Cerrado: a Brazilian type of parkland composed of scattered trees in dense grass.

Cetacean: any member of the group of marine mammals that includes whales, dolphins, and porpoises.

Chert: a variety of silica containing microcrystalline quartz.

Chlorophyll: any of a group of green pigments found in photosynthetic organisms.

Chough: a European corvine bird of small or medium size with red legs and glossy black plumage.

Cichlid: any of a family of freshwater fish of tropical America and Africa and Asia similar to American sunfish. Some are food fish; many small ones are popular in aquariums.

Cirque: deep, steep-walled basin on a mountain, usually forming the blunt end of a valley.

Composite Volcano: a volcano composed of interbedded lava and pyroclastic material commonly with steep slopes.

Continental Crust: solid, outer layers of the Earth, including the rocks of the continents.

Coucal: Old World ground-living cuckoo having a long, dagger-like hind claw.

Crater: a steep-sided, usually circular depression formed by either explosion or collapse at a volcanic vent.

Curlew: large migratory shorebirds of the sandpiper family; closely related to woodcocks but having a down-curved bill.

Cycads: an ancient group of seed plants with a crown of large compound leaves and a stout trunk. They are a minor component of the flora in tropical and subtropical regions today, but during the Jurassic Period they were a common sight in many parts of the world.

Devonian: a period in the Paleozoic Era that covered the time span between 400 and 345 million years ago.

Diorite: a granular crystalline intrusive rock.

Dipterocarp: evergreen tree of the family Dipterocarpaceae (chiefly tropical Asian trees with two-winged fruits, which yield valuable woods and aromatic oils and resins).

Dolerite: a form of volcanic rock rather similar to basalt, but with finer rock crystals.

Doline: a special geological landscape found in karst regions and formed by repeated cave-ins of the tops of underground caves.

Dolmen: a prehistoric megalith typically having two upright stones and a capstone.

Dome: a steep-sided mass of viscous (doughy) lava extruded from a volcanic vent (often circular in plane view) and spiny, rounded, or flat on top. Its surface is often rough and blocky as a result of fragmentation of the cooler, outer crust during growth of the dome.

Dormant Volcano: literally, "sleeping." The term is used to describe a volcano that is presently inactive, but which may erupt again. Most of the major Cascade volcanoes are believed to be dormant rather than extinct.

Dugong: an aquatic herbivorous mammal of a monotypic genus (*Dugong*) that has a bilobed tail and, in the male, upper incisors altered into short tusks. It is related to the manatee and inhabits warm coastal regions—also called *sea cow*.

Duiker: any of about 19 small to medium-sized antelope species native to sub-Saharan Africa.

Dunlin: small common sandpiper that breeds in northern or arctic regions and winters in the southern United States or Mediterranean regions.

Echinae: an oviparous, spiny-coated, toothless, burrowing, nocturnal monotreme mammal (*Tachyglossus aculeatus*) of Australia, Tasmania, and New Guinea that has a long extensile tongue, long heavy claws, and feeds chiefly on ants.

Echinoderm: marine invertebrates with tube feet and calcite-covered, five-part radially symmetrical bodies.

Eland: either of two large African antelopes of the genus Taurotragus having short, spirally twisted horns in both sexes.

Eocene: from 58 million to 40 million years ago; presence of modern mammals.

Episode: a volcanic event that is distinguished by its duration or style.

Erratic: rock fragment carried by glacial ice or by floating ice, deposited at some distance from the outcrop from which it was derived and generally, though not necessarily, resting on bedrock of different lithology.

Eruption: the process by which solid, liquid, and gaseous materials are ejected into the Earth's atmosphere and onto the Earth's surface by volcanic activity. Eruptions range from the quiet overflow of liquid rock to the tremendously violent expulsion of pyroclastics.

Euros: Adnyamadhanha (Australian aboriginal language) *yuru*; marsupials belonging to the Macropodidae family.

Extinct Volcano: a volcano that is not presently erupting and is not likely to do so for a very long time in the future.

Extrusion: the emission of magmatic material at the Earth's surface. Also, the structure or form produced by the process (e.g., a lava flow, volcanic dome, or certain pyroclastic rocks).

Feldspar: any of a group of hard crystalline minerals that consist of aluminum silicates of potassium, sodium, calcium, or barium.

Fissures: cracks or divides.

Fjord: a long narrow inlet of the sea between steep cliffs.

Floodplain: level that is submerged by floodwater, or a plain built up by steam deposition.

Flowstone: a general term for a type of cave decoration or speleothem that encrusts floors or walls of caves.

Formation: a body of rock identified by lithic characteristics and stratigraphic position that is mappable at the Earth's surface or traceable in the subsurface.

Francolin: a spurred partridge of the genus Francolinus and allied genera of Asia and Africa. The common species (*F. vulgaris*) was formerly widespread in southern Europe, but is now nearly restricted to Asia.

Fumarole: a hole or spot in a volcanic or other region, from which fumes issue.

Garganey: a scarce and very secretive breeding duck in the United Kingdom, smaller than a mallard and slightly bigger than a teal.

Gemsbok: large South African oryx with a broad black band along its flanks.

Genet: agile Old World viverrid having a spotted coat and long, ringed tail.

Geode: a hollow rock or nodule with the cavity usually lined with crystals.

Geothermal Energy: energy derived from the internal heat of the Earth.

Geyser: a spring that ejects intermittent jets of heated water and steam.

Gneiss: a laminated metamorphic rock similar to granite.

Gondwanaland: also called Gondwana; hypothetical former supercontinent in the Southern Hemisphere, which included South America, Africa, peninsular India, Australia, and Antarctica.

Goshawk: a large hawk of Eurasia and North America used in falconry.

Graben: a depressed segment of the crust of the Earth or a celestial body (as the moon) bounded on at least two sides by faults.

Grike: a surface landform comprising a solutional trench cut into the limestone along a joint.

Groundsel: Eurasian weed with heads of small yellow flowers.

Guenon: small, slender African monkey having long hind limbs and tail and long hair around the face.

Gypsum: widely distributed mineral consisting of hydrous calcium sulfate, used as soil amendment and plaster of paris; drywall.

Hanging Valley: a secondary valley that enters a main valley at an elevation well above the main valley's floor. These features are the result of past erosion caused by alpine glaciers.

Head (of glacier): top of a glacier valley.

Hominid: any of the family of erect bipedal primate mammals comprising recent humans together with extinct ancestral or related forms.

Hot Spot (volcanic): a place in the upper mantle of the Earth at which hot magma from the lower mantle upwells to melt through the crust, usually in the interior of a tectonic plate, to form a volcanic feature; a place in the crust overlying a hot spot.

Hoodoo: a spire.

Ice Age: any period of time during which glaciers covered a large part of the Earth's surface.

Ignimbrite: the rock formed by the widespread deposition and consolidation of ash flows and Nuees Ardentes. The term was originally applied only to densely welded deposits but now includes non-welded deposits.

Intrusion: the process of emplacement of magma in pre-existing rock. Also refers to igneous rock mass so formed within the surrounding rock.

Ionosphere: the outer region of the Earth's atmosphere; contains a high concentration of free electrons.

Iron Age: the period following the Bronze Age; characterized by the rapid creation and spread of iron tools and weapons.

Jurassic Period: from 135 million to 190 million years ago; characterized by dinosaurs and conifers.

Karren: a general term used to describe the total complex of superficial micro-solutional features of limestone pavement.

Karst: a type of topography formed by dissolution of rocks like limestone and gypsum that is characterized by sinkholes, caves, and subterranean passages. It is a term originating from a limestone region in the former Yugoslavia, and is derived from the Slovenian word *kras*, meaning a bleak, waterless place.

Kudu: either of two spiral-horned antelopes of the African bush.

Lahar: moving fluid composed of volcanic debris and water.

Lava: molten rock that issues from a volcano or from a fissure in the surface of a planet.

Lava Dome: mass of lava, created by many individual flows, that has built a dome-shaped pile of lava.

Lava Flow: an outpouring of lava onto the land surface from a vent or fissure. Also, a solidified tongue-like or sheet-like body formed by outpouring lava.

Lava Tube: a tunnel formed when the surface of a lava flow cools and solidifies while the still-molten interior flows through and drains away.

Lechwe: tawny-colored African antelope inhabiting wet grassy plains; a threatened species.

Lee: facing in the direction of motion of an overriding glacier—used especially of a hillside.

Liana: any of various woody vines, especially of tropical rainforests that root in the ground.

Liverwort: any of numerous small green nonvascular plants of the class Hepaticopsida growing in wet places.

Llano: an open grassy plain in Latin America.

Loma: a small hill.

Mafic: of, relating to, or being a group of usually dark-colored minerals rich in magnesium and iron.

Magma: molten rock beneath the surface of the Earth.

Magma Chamber: the subterranean cavity containing the gas-rich liquid magma that feeds a volcano.

Mantle: the zone of the Earth below the crust and above the core.

Maquis Vegetation: low, sclerophyllous, evergreen, heath-like formation, largely restricted to ultrabasic substrates at various altitudes.

Massif: a principal mountain mass; a block of the Earth's crust bounded by faults or flexures and displaced as a unit without internal change.

Menhir: a tall upright megalith; found mainly in northern France and England.

Mesa: isolated, relatively flat-topped natural elevation.

Mesozoic Era: from 230 million to 63 million years ago.

Midden: a small pile (as of seeds, bones, or leaves) gathered by a rodent.

Miocene: an epoch in Earth's history from about 24 to 5 million years ago. Also refers to rocks formed in that epoch.

Mire: wet spongy earth (as of a bog or marsh).

Moraine: accumulated earth and stones deposited by a glacier.

Monitor lizard: any of various large, tropical, carnivorous lizards of Africa, Asia, and Australia; fabled to warn of crocodiles.

Mudflow: a flowage of water-saturated earth material possessing a high degree of fluidity during movement. A less-saturated flowing mass is often called a debris flow. A mudflow originating on the flank of a volcano is properly called a lahar.

Mudpot: a hot spring with limited water supply. The water in a mudpot is highly acidic and it dissolves nearby rock into small pieces of clay. This clay then mixes with the hot water to create mud. Hot steam rising from below causes the mud to bubble and pop as the steam is released into the air.

Nagana: an unscientific but convenient name for trypanosomiasis (the animal version of sleeping sickness) transmitted by tsetse flies (Glossina species) in Africa. The symptoms include anemia, intermittent fever, and a slow, progressive emaciation.

Narwhal: an arctic cetacean (*Monodon monoceros*) about 20 feet (6 meters) long, with the male having a long twisted ivory tusk.

Neolithic: of or relating to the most recent period of the Stone Age (following the mesolithic).

Okapi: similar to the giraffe but smaller, with much shorter neck and stripe on the legs.

Oligocene Age: from 40 million to 25 million years ago; appearance of sabertoothed cats.

Ordovician Age: from 500 million to 425 million years ago; conodonts, ostracods, algae, and seaweeds.

Outlier: something (as a geological feature) that is situated away from or classed differently from a main or related body.

Oxbow Lake: a crescent-shaped lake (often temporary) that is formed when a meander of a river is cut off from the main channel.

Paleolithic Age: of or relating to the second period of the Stone Age (following the eolithic).

Pangolin: toothless mammal of southern Africa and Asia having a body covered with horny scales and a long snout for feeding on ants and termites.

Páramo: a treeless alpine plateau of the Andes and tropical South America.

Permian Period: 235 to 230 million years ago.

Phreatomagmatic: an explosive volcanic eruption that results from the interaction of surface or subsurface water and magma.

Phytoplankton: photosynthetic or plant constituent of plankton; mainly unicellular algae.

Pipit: a small songbird resembling a lark.

Plankton: the passively floating or weakly swimming (usually minute) animal and plant life of a body of water.

Plasma: the gaseous state of hot ionized material consisting of ions and electrons, present in the stars and fusion reactors; sometimes regarded as a fourth state of matter distinct from normal gases.

Pleistocene: an epoch in Earth history from about 1.6 million years to 10,000 years ago. Also refers to the rocks and sediment deposited in that epoch.

Precambrian: all geologic time from the beginning of Earth history to 570 million years ago. Also refers to the rocks that formed in that epoch.

Porphyry: any igneous rock with crystals embedded in a finer groundmass of minerals.

Potash: a potassium compound often used in agriculture and industry.

Prion: a protein particle that lacks nucleic acid and is believed to be the cause of various infectious diseases of the nervous system

Protea: a South African shrub whose flowers when open are cup- or goblet-shaped, resembling globe artichokes.

Proterozoic: of, relating to, or being the eon of geologic time or the corresponding system of rocks that includes the interval between the Archean and Phanerozoic eons; perhaps exceeds in length all of subsequent geological time, and is marked by rocks that contain fossils indicating the first appearance of eukaryotic organisms (as algae).

Ptarmigan: large arctic and subarctic grouse with feathered feet and usually white winter plumage.

Pumice: light-colored, frothy volcanic rock, usually of dacite or rhyolite composition, formed by the expansion of gas in erupting lava. Commonly seen as lumps or fragments of pea-size and larger, but can also occur abundantly as ash-sized particles.

Pyrite: a common mineral (iron disulfide) that has a pale yellow color.

Pyroclastic: formed by or involving fragmentation as a result of volcanic or igneous action.

Pyroclastic Flow: lateral flowage of a turbulent mixture of hot gases and unsorted pyroclastic material (volcanic fragments, crystals, ash, pumice, and glass shards) that can move at high speed (50 to 100 miles per hour [80 to 160 kilometers per hour]). The term also can refer to the deposit so formed.

Quaternary: the period of Earth's history from about 2 million years ago to the present; also, the rocks and deposits of that age.

Quorum: a select group.

Reedbuck: a South African antelope (Cervicapra arundinacea); so called from its frequenting dry places covered with high grass or reeds. Its color is yellowish brown. Called also inghalla and rietbok.

Relict: a surviving species of an otherwise extinct group of organisms; a relief feature or rock remaining after other parts have disappeared.

Ring of Fire: the regions of mountain-building earthquakes and volcanoes that surround the Pacific Ocean.

Roaring Forties: a tract of ocean between roughly 40 and 50 degrees latitude south characterized by strong westerly winds and rough seas.

Ruff: a medium-sized wading bird with a long neck, small head, short, slightly droopy bill, and medium-long orange or reddish leg.

Rhyolite: a very acid volcanic rock that is the lava form of granite.

Salt pan: an undrained natural depression in which water gathers, leaving a deposit of salt upon evaporation.

Salt plug: the salt core of a salt dome.

Schist: any metamorphic rock that can be split into thin layers.

Scoria: a bomb-size pyroclast that is irregular in form and generally very vesicular. It is usually heavier, darker, and more crystalline than pumice.

Seamount: an underwater mountain rising above the ocean floor.

Shoebill: a large African wading bird allied to storks and herons, and remarkable for its enormous broad swollen bill. It inhabits the valley of the White Nile.

Sienna: an earthy substance containing oxides of iron and usually of manganese that is brownish yellow when raw and orange red or reddish brown when burnt.

Sierra: a range of mountains, especially with a serrated or irregular outline.

Silic: of, relating to, or derived from silica or silicon.

Silica: a chemical combination of silicon and oxygen.

Sill: a tabular body of intrusive igneous rock, parallel to the layering of the rocks into which it intrudes.

Sinkhole: a hollow in a limestone region that communicates with a cavern or passage.

Sitatunga: a spectacular aquatic antelope.

Solar wind: a stream of protons moving radially from the sun.

Spelunker: a person who studies or explores caves.

Stalactite: a cone-shaped deposit of minerals hanging from the roof of a cavern.

Stalagmite: a deposit of calcium carbonate like an inverted stalactite formed on the floor of a cave by the drip of calcareous water.

Steppe: an extensive plain without trees (associated with eastern Russia and Siberia).

Stratovolcano: a volcano composed of explosively erupted cinders and ash with occasional lava flows.

Strike-slip Fault: a type of fault whose surface is typically vertical or nearly so.

Stromatolites: carbonate-rich layered structures built up by bacteria.

Stone Age: the earliest known period of human culture, characterized by the use of stone implements.

Talus: a sloping mass of rocks at the base of a cliff

Tarn: small, steep-banked mountain lake or pool.

Tephra: solid material ejected into the air during a volcanic eruption.

Tepui: sandstone plateau (mesa).

Toadflax: common European perennial having showy yellow and orange flowers; a naturalized weed in North America.

Toe (of glacier): the terminus of a glacier.

Tor: a high craggy hill.

Trachyte: a usually light-colored volcanic rock consisting chiefly of potash feldspar.

Tramontana: a cold dry wind that blows south out of the mountains into Italy and the western Mediterranean.

Travertine: mineral consisting of a massive, usually layered calcium carbonate formed by deposits from spring waters or hot springs.

Triassic: from 230 million to 190 million years ago; characterized by dinosaurs, marine reptiles, and volcanic activity.

Tsetse: blood-sucking African fly; transmits sleeping sickness.

Tufa: a soft, porous rock consisting of calcium carbonate deposited from springs rich in lime.

Tuff Cone: rock composed of finer types of volcanic detritus fused together by heat.

Unconformity: a substantial break or gap in the geologic record where a rock unit is overlain by another that is not next in stratigraphic sucession, such as an interruption in continuity of a depositional sequence of sedimentary rocks or a break between eroded igneous rocks and younger sedimentary strata. It results from a change that caused deposition to cease for a considerable time, and it normally implies uplift and erosion with loss of the previous formed record.

U-shaped valley: formed by the process of glaciation. It has a characteristic U-shape, with steep, straight sides, and a flat bottom.

Veld: southern African term for natural vegetation, usually grassland or wooded grassland.

Vent: the opening at the Earth's surface through which volcanic materials issue forth.

Vicuñas: humpless camels, found chiefly in the Andes from Peru to Argentina.

Vleis: an area of marshland.

Vulcanian: a type of eruption consisting of the explosive ejection of incandescent fragments of new viscous lava, usually in the form of blocks.

Waterbuck: any of several large African antelopes of the genus Kobus having curved ridged horns and frequenting swamps and rivers.

Xeric woodland: woodland characterized by, relating to, or requiring only a small amount of moisture.

General Index

Q

R

S

Index of UNESCO World Heritage sites

The entries on UNESCO World Heritage sites in this book are identified by symbols next to their titles. Listed below are the official UNESCO World Heritage names with the dates of the sites' inscription, each followed by the relevant entries in this book and their page references.

Picture Credits

2 Gavin Hellier/naturepl.com 18 Ron Watts/Photolibrary 20 David Noton/naturepl.com 21 Staffan Widstrand/naturepl.com 23 Andre Gallant/Getty 24 Hemera Technologies/Jupiter 27 David Noton/naturepl.com 29 Grant Faint/Getty 30 David Noton/naturepl.com 33 Radius Images/Photolibrary 34 Andre Gallant/Getty 35 Sue Flood/naturepl.com 36 Thomas Lazar/naturepl.com 38 Justine Evans/naturepl.com 39 blickwinkel/Alamy 40 Ulli Steer/Getty 41 Michael Melford/Getty 43 Harvey Lloyd/Getty 45 Nancy Simmerman/Getty 46 Lynn M. Stone/naturepl.com 49 Aflo/naturepl.com 51 Harold Sund/Getty 52 Jack Dykinga/Getty 54 Barrie Britton/naturepl.com 55 Richard H. Smith/Getty 57 Alan Kearney/Getty 59 Walter Bibikow/Getty 60 Gary Randall/Getty 61 Dave Schiefelbein/Getty 62 Michael Hanson/Oxford Scientific/Photolibrary 65 Gary Randall/Getty 66 Jeff Foott/naturepl.com 67 Jeff Foott/naturepl.com 69 Torsten Brehm/naturepl.com 70 Jeff Foott/naturepl.com 73 James Balog/Getty 74 David Noton/naturepl.com 75 David Hanson/Getty 77 Marc Muench/Getty 78 Jack Dykinga/Getty 79 Art Wolfe/Getty 81 James Randklev/Getty 82 Doug Wechsler/naturepl.com 83 Afl/naturepl.com 85 Ingo Arndt/naturepl.com 86 Gavin Hellier/naturepl.com 87 William Smithey Jr/Getty 89 Jeff Foott/naturepl.com 91 Gavin Hellier/naturepl.com 93 Gavin Hellier/naturepl.com 94 Tom Mackie/Getty 95 Aflo/naturepl.com 96 Tim Barnett/Getty 97 Ruth Tomlinson/Getty 99 Aflo/naturepl.com 100 Gavin Hellier/naturepl.com 101 Gavin Hellier/naturepl.com 102 Jeff Foott/naturepl.com 103 David Meunch/Corbis 105 Jeff Foott/naturepl.com 109 Marc Muench/Getty 110 David Noton/naturepl.com 113 Harvey Lloyd/Getty 114 Tom Bean/Getty 115 Mike Hill/Getty 117 Gavin Hellier/naturepl.com 118 Rob Atkins/Getty 121 Harvey Lloyd/Getty 122 Doug Wechsler/naturepl.com 125 Aflo/naturepl.com 127 Mark Newman/Lonely Planet 128 Grant Faint/Getty 133 Laurance B. Aiuppy/Getty 136 Jack Dykinga/Getty 139 Jeff Foott/naturepl.com 140 Hanne & Jens Eriksen/naturepl.com 142 Robert Freck/Getty 147 George Lepp/Getty 148 Jurgen Freund/naturepl.com 149 Suzanne Murphy/Getty 151 Frans Lemmens/Getty 155 Simeone Huber/Getty 159 Tony Waltham/Getty 160 Jerry Driendl/Getty 163 Jeff Rotman/naturepl.com 164 Doug Perrine/naturepl.com 165 Kevin Schafer/Getty 167 Don Herbert/Getty 171 Georgette Douwma/naturepl.com 173 Gavin Hellier/Getty 174 Richard Elliott/Getty 175 Tom Bean/Corbis 178 Darrell Jones 179 Bill Hickey/Getty 181 Pete Turner/Getty 183 Brooke Slezak/Getty 185 The Travel Library Limited/Photolibrary 186 Steve Vidler/Photolibrary 188 Pete Turner/Getty 191 eStock Photo/Alamy 192 Thomas Schmitt/Getty 193 Lee Dalton/Alamy 194 Hermann Brehm/naturepl.com 197 David Welling/naturepl.com 198 Juan Silva/Getty 200 SA Team/Foto Natura/Getty 201 Pete Oxford/naturepl.com 202 Pete Oxford/naturepl.com 204 Kazuyoshi Nonmachi/Corbis 206 Galen Rowell/Corbis 209 Solvin Zankl/natur epl.com 210 Jim Clare/naturepl.com 213 Staffan Widstrand/naturepl.com 214 Peter Oxford/naturepl.com 217 Silvestre Machado/Getty 218 Macduff Everton/Getty 221 Peter Oxford/naturepl.com 222 Luis Veiga/Getty 225 Russell Kaye/Getty 227 Doug Allan/naturepl.com 229 Frans Lanting/Corbis 231 Micheal Simpson/Getty 233 Alejandro Balaguer/Getty 234 Michael Dunning/Getty 237 Staffan Widstrand/naturepl.com 238 George Steinmetz/Corbis 239 Frans Lanting/Corbis 240 Hermann Brehm/naturepl.com 242 Hubert Stadler/Corbis 245 Doug Allan/naturepl.com 246 Art Wolfe/Getty 247 Rob Mcleod/Getty 248 Rhonda Klevansky/naturepl.com 250 William J Hebert/Getty 251 Chris Gomersall/Getty 253 Tony Arruza/Getty 254 David Noton/Getty 256 Hanne & Jens Eriksen/naturepl.com 259 Daniel Gomez/naturepl.com 260 Aflo/naturepl.com 262 Genevieve Vallee/Alamy 263 WorldFoto/Alamy 265 Pete Oxford/naturepl.com 266 Gabriel Rojo/naturepl.com 268 Joel Damase/Photononstop/Photolibrary 270 Siqui Sanchez/Getty 272 George Kavanagh/Getty 273 Pal Hermansen/Getty 275 Neil Lucas/naturepl.com 276 Ernst Haas/Getty 277 Ernst Haas/Getty 278 Doug Allan /naturepl.com 281 Asgeir Helgestad/naturepl.com 282 Andreas Stirnberg/Getty 283 Terje Rakke/Getty 284 Florian Graner/naturepl.com 285 Gavin Hellier/naturepl.com 287 Florian Graner/naturepl.com 290 Michael Lander/Getty 291 Chad Ehlers/Getty 293 Hans Strand/Getty 294 Panoramic Images/Getty 297 Jorma Luhta/naturepl.com 298 David Tipling/naturepl.com 300 Colin Palmer Photography/Alamy 301 Rick Price/naturepl.com 302 David Noton/naturepl.com 303 Juan Manuel Borrero/naturepl.com 305 Bernard Castelein/naturepl.com 306 Richard Ashworth/Getty 309 Geoff Dore/naturepl.com 310 Geoff Simpson/naturepl.com 312 Bernard Castelein/naturepl.com 313 Bernard Castelein/naturepl.com 314 David Cottridge/naturepl.com 316 Hans Christoph Kappel/naturepl.com 319 Nick Turne/naturepl.com 323 Neale Clarke/Getty 324 The Wrekin from Willstone Hill, by John Bentley 327 Roy Rainford/Getty 329 Wooky Hole Caves 330 Jon Arnold/Getty 331 Charles Bowman/Getty 332 David Noton/naturepl.com 333 Colin Varndell/naturepl.com 335 Guy Edwardes/Getty 336 David Noton/naturepl.com 337 David Hunter/Robert Harding Travel/Photolibrary 338 Ross Hoddinott /naturepl.com 339 David Noton/Getty 340 The Photolibrary Wales/Alamy 341 Derek P Redfearn/Getty 343 Guy Edwardes/Getty 344 Gavin Hellier /naturepl.com 348 Derek P Redfearn/Getty 349 Tim Edwards/naturepl.com 350 Bernd Mellmann/Alamy 351 Ronald Wittek/Photolibrary 353 Walter Bibikow/Getty 354 Christoph Becker/naturepl.com 355 Stephen Studd/Getty 357 Mike Read/naturepl.com 361 Martin Dohrn /naturepl.com 362 Jean E Roche/naturepl.com 363 Doc White/naturepl.com 364 ImagesEurope/Alamy 367 Scott Markewitz/Getty 369 Jean E Roche/naturepl.com 370 Jess Stock/Getty 373 David Hughes/Getty 374 Robert Harding Picture Library Ltd/Alamy 375 Michael Busselle/Getty 376 Jean E. Roche/naturepl.com 377 Nicolas Thibaut/Photononstop/Photolibrary 379 John Miller/Getty 380 John Miller/Getty 383 Yannick Le Gal/Getty 385 Jean E Roche/naturepl.com 389 Liba Taylor/Corbis 393 Paul Trummer/Getty 395 Paul Trummer/Getty 398 Mike Potts/naturepl.com 399 Christoph Becker/naturepl.com 401 Ingo Arndt/naturepl.com 402 Jon Arnold Images Ltd/Alamy 403 Aflo/naturepl.com 405 Dan Santillo/Alamy 407 Tim Edwards/naturepl.com 408 Francesco Ruggeri/Getty 411 Ingo Arndt/naturepl.com 412 Martin Gabriel/naturepl.com 415 Hemis/Corbis 417 Gavin Hellier/naturepl.com 418 Cristian Baitg Schreiweis/Alamy 420 Jose B. Ruiz/naturepl.com 421 Jose B. Ruiz/naturepl.com 425 Jose Luis Gomez de Francisco/naturepl.com 426 Jose B. Ruiz/naturepl.com 429 Jose B. Ruiz/naturepl.com 430 Jose B. Ruiz/naturepl.com 432 Alan Dawson Photography/Alamy 434 Jose B. Ruiz/naturepl.com 435 Jose B. Ruiz /naturepl.com 437 Iain Lowson/Alamy 439 Jose B. Ruiz/naturepl.com 441 Teresa Farino 442 Javier Muñoz Gutiérrez/Alamy 443 Nigel Bean/naturepl.com 444 f1 online/Alamy 447 Teresa Farino 451 Walter Bibikow/Getty 454 Andrea Pistolesi/Getty 456 Marco Simoni/Getty 459 Bernard Castelein/naturepl.com 461 Anne & Jens Eriksen/naturepl.com 463 Nigel Marven/naturepl.com 465 Carolyn Brow/Getty 467 Fred Friberg/Getty 468 SEUX Paule/HEMIS/Photolibrary 470 Harvey Lloyd/Getty 473 Hanne & Jens Eriksen/naturepl.com 474 Hanne & Jens Eriksen/naturepl.com 475 Hanne & Jens Eriksen/naturepl.com 477 Michele Falzone/JAI/Corbis 478 Bildagentur Rm/Photolibrary 481 Jurgen Freund/naturepl.com 482 Bruce Davidson/naturepl.com 484 Jose B. Ruiz/naturepl.com 486 imagebroker/Alamy 487 Jose B. Ruiz/naturepl.com 488 Jose B. Ruiz/naturepl.com 491 Nick Barwick/naturepl.com 498 Doug Allan/naturepl.com 500 MJ Photography/Alamy 503 George Chan/naturepl.com 505 Jon Hicks/Corbis 508 Mitch Reardon/Lonely Planet 511 Justine Evans/naturepl.com 512 Jose B. Ruiz /naturepl.com 514 Anup Shah/naturepl.com 517 Bruce Davidson/naturepl.com 519 Bruce Davidson/naturepl.com 520 Mark Deeble & Victoria Stone/Photolibrary 523 Bruce Davidson/naturepl.com 524 Daniel J. Cox/Getty 525 NASA-ISS/digital version by Science Faction/Getty 527 Giles Bracher/naturepl.com 529 Jose B. Ruiz/naturepl.com 531 Mitsuaki Iwago/Minden Pictures/Getty 535 Aflo/naturepl.com 536 Pete Oxford/naturepl.com 538 Peter Ginn 539 Peter Ginn 540 Peter Ginn 541 Peter Ginn 543 Peter Ginn 544 Peter Ginn 545 Peter Ginn 546 Vincent Munier/naturepl.com 547 Vincent Munier/naturepl.com 548 Richard du Toit/naturepl.com 549 David Noton/naturepl.com 551 Vincent Munier/naturepl.com 552 David Noton/naturepl.com 553 Bildagentur/Tips Italia/Photolibrary 555 Pichugin Dmitry/Shutterstock 556 Pichugin Dmitry/Shutterstock 558 Richard du Toit/naturepl.com 559 Richard Du Toit/naturepl.com 560 Frans Lanting/Corbis 561 Peter Ginn 562 Peter Ginn 565 Peter Ginn 567 Peter Ginn 569 Andreas Stirnberg/Getty 571 Stephanie Lamberti/

ABPL/Photolibrary **574** John Lamb/Getty **577** Neil Nightingale/naturepl.com **578** Pete Oxford/naturepl.com **581** Tony Heald/naturepl.com **582** Pete Oxford/naturepl.com **585** Walter Bibikow/Getty **587** Walter Bibikow/Getty **589** Steve Bloom/Getty **591** Laurence Hughes/Getty **593** Frans Lemmens/Getty **594** Fraser Hall/Getty **595** Fraser Hall/Getty **596** Ed Collacott/Getty **599** Fraser Hall/Getty **600** Caroline Caroline/MauritiusImages/Photolibrary **602** NASA/Corbis **603** Berndt Fischer/Oxford Scientific/Photolibrary **605** Pete Oxford/naturepl.com **607** Fraser Hall/Getty **609** Sylvain Grandadam/Getty **610** Ethel Davies/Photolibrary **612** Nigel Marven/naturepl.com **613** N. A. Callow/Robert Harding World Imagery/Corbis **615** National Geographic/Getty **616** Nigel Marven/naturepl.com **619** Nigel Marven/naturepl.com **620** Konstantin Mikhailov/naturepl.com **623** Corbis **624** Gertrud & Helmut Denzau/naturepl.com **625** Art Wolfe/Getty **626** Konstantin Mikhailov/naturepl.com **627** John Sparks/naturepl.com **628** China Tourism Press/Getty **629** China Tourism Press/Getty **631** China Tourism Press/Getty **633** Gavin Maxwell/naturepl.com **634** China Tourism Press/Getty **635** Sylvia Cordaiy Photo Library Ltd/Alamy **637** China Tourism Press/Getty **638** China Tourism Press/Getty **640** DLILLC/Corbis **643** Xi Zhi Nong/naturepl.com **644** David Noton/naturepl.com **646** Yann Layma/Getty **649** Pete Oxford/naturepl.com **651** China Tourism Press/Getty **652** Peter Oxford/naturepl.com **654** Alexander Walter/Getty **657** Aflo/naturepl.com **658** David Pike/naturepl.com **660** David Pike/naturepl.com **663** Aflo/naturepl.com **664** Aflo/naturepl.com **667** Japan Travel Bureau/Photolibrary **670** Core Agency/Getty **674** Mahaux Photography/Getty **677** Getty **678** Christopher Klettermayer/Jupiter **680** Christina Gascoigne/Getty **681** Paula Bronstein/Getty **683** Hitendra Sinkar Photography/Alamy **684** Toby Sinclair/naturepl.com **689** Nick Haslam/Alamy **693** Ian Lockwood/naturepl.com **696** Hornbil Images/Alamy **699** Martin Puddy/Jupiter **703** Colin Monteath/Photolibrary **704** David Paterson/Getty **706** Tony Waltham/Getty **707** Chris Noble/Getty **708** Roger Mear/Getty **710** Bernard Castelein/naturepl.com **711** Maximilian Weinzierl/Alamy **713** travelib asia/Alamy **715** David Shaw/Alamy **716** Frédéric Soltan/Corbis **717** Arco Images GmbH/Alamy **719** Jerry Alexander/Lonely Planet **720** Geoffrey Clifford/Getty **722** Steve Raymer/Getty **723** Nevada Wier/Getty **725** LOOK Die Bildagentur der Fotografen GmbH/Alamy **727** Jerry Alexander/Getty **729** Jerry Alexander/Getty **731** Nevada Wier/Getty **735** Michele Falzone/Jon Arnold Travel/Photolibrary **739** WoodyStock/Alamy **742** Neil Emmerson/Getty **745** Justin Pumfrey/Getty **749** Stephen Frink/Getty **753** Pete Turner/Getty **757** Gavin Hellier/naturepl.com **759** Robin Smith/Photolibrary **762** Stuart Dee/Getty **766** Ingo Arndt/naturepl.com **767** Japan Travel Bureau/Photolibrary **769** Christer Fredriksson/Lonely Planet **771** Daniel J. Cox/Getty **773** Robbie Shone/Alamy **774** David Poole/Getty **777** Ed Robinson/Pacific Stock/Photolibrary **778** Paul Nevin/Photolibrary **780** Wayne Lawler; Ecoscene/Corbis **781** Art Wolfe/Getty **783** Hugh Sitton/Getty **784** Michael Pitts/naturepl.com **787** Wolfgang Kaehler/Corbis **788** Peter Harrison/Photolibrary **791** G. Brad Lewis/Getty **793** James Randklev/Getty **794** Dave Jepson/Alamy **797** G. Brad Lewis/Getty **798** Stuart & Michele Westmorland/Getty **800** Gary Bell/zefa/Corbis **801** AFP/Getty **803** Martin Dohrn/naturepl.com **807** Lionel Isy-Schwart/Getty **809** Lionel Isy-Schwart/Getty **810** Peter Hendrie/Getty **812** Peter Hendrie/Getty **813** Jurgen Freund/naturepl.com **815** David Wall/Alamy **818** Panoramic Images/Getty **820** Jason Edwards/Bio-Images **823** Jason Edwards/Bio-Images **825** Travel Pix/Getty **826** Clive Bromhall/Oxford Scientific/Photolibrary **827** Panoramic Images/Getty **828** Gerry Ellis/Getty **831** Ted Mead/Getty **832** Navaswan/Getty **835** Stefan Mokrzecki/Photolibrary **836** David Curl/naturepl.com **839** David Noton/naturepl.com **841** Jason Edwards **843** Slick Shoots/Alamy **845** William Osborn/naturepl.com **846** David Noton/Getty **848** Hanne & Jens Eriksen/naturepl.com **849** Thomas Schmitt/Getty **851** William Osborn/naturepl.com **853** Martin Gabriel/naturepl.com **856** Ted Mead/Photolibrary **861** Gavin Hellier/Robert Harding/Getty **865** John William Banagan/Getty **866** Doug Pearson/JAI/Corbis **867** Steven David Miller/naturepl.com **868** Steven David Miller/naturepl.com **871** Jason Edwards/Bio-Images **872** Brian Connett/fotolibra **874** Robin Smith/Getty **875** Tim Edwards/naturepl.com **877** Diego Lezama Orezzoli/Corbis **878** Chris Sattlberger/Getty **881** Jason Edwards/Bio-Images **885** Michael Townsend/Getty **887** Jason Edwards/Bio-Images **888** William Osborn/naturepl.com **889** Cephas Picture Library/Alamy **890** Peter M. Wilson/Alamy **893** Bill Bachman/Alamy **894** Robert Francis/Getty **895** Bill Bachman/Alamy **897** Jason Edwards/Bio-Images **898** Tomek Sikora/Getty **901** Shutterstock **902** Andreas Stirnberg/Getty **904** Peter Hendrie/Getty **905** Jen & Des Bartlett/Oxford Scientific/Photolibrary **907** Geoffrey Clifford/Getty **909** Bill Bachman/Alamy **911** Paul A. Souders/Corbis **913** Kirk Anderson/Getty **915** Hideo Kurihara/Alamy **916** Holger Leue/Lonely Planet **917** Travel Pix/Getty **919** James L. Amos/Corbis **920** David Noton/Getty **922** Jeremy Walker/Getty **923** Pete Turner/Getty **925** Kim Westerskov/Getty **926** Winfried Wisniewski/Photolibrary **928** R H Productions/Getty **931** Grant Dixon/Lonely Planet **932** Doug Allan /naturepl.com **934** NASA Johnson Space Center—Earth Sciences and Image Analysis **937** Frans Lanting/Corbis **938** Kevin Schafer/Getty **941** Geoff Renner/Robert Harding/Getty

Acknowledgments

Quintessence would like to thank the UNESCO World Heritage Center for its invaluable guidance and support during the creation of this book.

If you have any inquiries you would like to address to the organization, please contact:

World Heritage Center
UNESCO
7, Place de Fontenoy
75007 Paris
France

Tel:+33 1 45 68 15 71
Fax:+33 1 45 68 55 70
Email: wh-info@unesco.org
http://whc.unesco.org

Quintessence would also like to thank the following individuals for their assistance in producing the book:

2009 Edition:
Joe Fullman, Becky Gee, Ann Marangos, Frank Ritter

Original Edition:
Laura Goodchild, Lewis Miller, Manuel Navarrete, Teresa Riley, Darryl Walles, Jodie Wallis